Savannas, Barrens, and Rock Outcrop Plant Communities of North America

Savannas and barrens were major components of the historic North American landscape before it was extensively altered by agricultural and urban development during the past century. Rock outcrop plant communities and serpentine barrens are of interest because they are refugia for endemic species adapted to extreme environmental conditions. Many of these communities are currently reduced to less than one percent of their original area and are imperiled ecosystems.

This book provides a coherent, readable summary of the technical information available on savannas, barrens, and rock outcrop plant communities. It is organized by region, into four parts: eastern/southeastern region, central/midwestern region, western/southwestern region, and northern region. Written by internationally recognized regional specialists, each chapter includes a description of the climate, geology, and soils associated with the community and information about its historic and current vegetation.

This book will be a useful text for graduate and advanced undergraduate students studying vegetation ecology, as well as a valuable reference for professional and amateur naturalists interested in the conservation, restoration, and management of these communities.

DR. ROGER C. ANDERSON, Distinguished Professor of Biology at Illinois State University, is on the editorial board of Restoration Ecology and is a Fellow of the Illinois Academy of Science.

DR. JAMES S. FRALISH, Professor emeritus, Southern Illinois University, Carbondale, serves as a research consultant with the Center for Field Biology, Austin Peay State University, Clarksville, Tennessee, and is certified by the Ecological Society of America as a Senior Ecologist.

DR. JERRY M. BASKIN is Professor of Biological Sciences at the University of Kentucky and serves on the editorial board of Seed Science Research.

T0171982

Savannas, Barrens, and Rock Outcrop Plant Communities of North America

Edited by

ROGER C. ANDERSON
Illinois State University, Normal

JAMES S. FRALISH
Southern Illinois University–Carbondale

JERRY M. BASKIN
University of Kentucky, Lexington

CAMBRIDGE
UNIVERSITY PRESS

CAMBRIDGE UNIVERSITY PRESS
Cambridge, New York, Melbourne, Madrid, Cape Town, Singapore, São Paulo

Cambridge University Press
The Edinburgh Building, Cambridge CB2 8RU, UK

Published in the United States of America by Cambridge University Press, New York

www.cambridge.org
Information on this title: www.cambridge.org/9780521573221

First published 1999
This digitally printed first paperback version (with corrections) 2007

A catalogue record for this publication is available from the British Library

Library of Congress Cataloguing in Publication data

Savanna, barrens, and rock outcrops plant communities of North America
   / edited by Roger C. Anderson, James S. Fralish, Jerry M. Baskin.
      p. cm.
   Includes bibliographical references
   ISBN 0-521-57322-X (hardbound)
   1. Savanna plants – United States. 2. Rock plants – United States.
3. Plant communities – United States. 4. Plant ecology – United
States. 5. Savanna plants – Canada. 6. Rock plants – Canada.
7. Plant communities – Canada. 8. Plant ecology – Canada.
I. Anderson, Roger C, (Roger Clark) II. Fralish, James S. (James
Steven) III. Baskin, Jerry M. (Jerry Mack)
QK115.S38 1999
581.7'0973 – dc21
                                                    98–25688
                                                       CIP

ISBN 978-0-521-57322-1 hardback
ISBN 978-0-521-03581-1 paperback

Contents

List of Contributors   *page* vii

   Introduction
   ROGER C. ANDERSON, JAMES S. FRALISH, AND JERRY M. BASKIN   1

   EASTERN/SOUTHEASTERN REGION

1 Ecology and Conservation of Florida Scrub
   ERIC S. MENGES   7

2 Southeastern Pine Savannas
   WILLIAM J. PLATT   23

3 New Jersey Pine Plains: The "True Barrens" of the
   New Jersey Pine Barrens
   DAVID J. GIBSON, ROBERT A. ZAMPELLA,
   AND ANDREW G. WINDISCH   52

4 Vegetation, Flora, and Plant Physiological Ecology
   of Serpentine Barrens of Eastern North America
   R. WAYNE TYNDALL AND JAMES C. HULL   67

5 The Mid-Appalachian Shale Barrens
   SUZANNE H. BRAUNSCHWEIG, ERIK T. NILSEN,
   AND THOMAS F. WIEBOLDT   83

6 Granite Outcrops of the Southeastern United States
   DONALD J. SHURE   99

7 High-Elevation Outcrops and Barrens of the
   Southern Appalachian Mountains
   SUSAN K. WISER AND PETER S. WHITE   119

   CENTRAL/MIDWEST REGION

8 Dry Soil Oak Savanna in the Great Lakes Region
   SUSAN WILL-WOLF AND FOREST STEARNS   135

9 Deep-Soil Savannas and Barrens of the
   Midwestern United States
   ROGER C. ANDERSON AND MARLIN L. BOWLES   155

10 Open Woodland Communities of
   Southern Illinois, Western Kentucky,
   and Middle Tennessee
   JAMES S. FRALISH, SCOTT B. FRANKLIN, AND DAVID D. CLOSE   171

11 The Big Barrens Region of Kentucky and Tennessee
   JERRY M. BASKIN, CAROL C. BASKIN,
   AND EDWARD W. CHESTER   190

12   Cedar Glades of the Southeastern United States
     JERRY M. BASKIN AND CAROL C. BASKIN   206

13   Savanna, Barrens, and Glade Communities
     of the Ozark Plateaus Province
     ALICE LONG HEIKENS   220

14   The Cross Timbers
     B. W. HOAGLAND, I. H. BUTLER, F. L. JOHNSON,
     AND S. GLENN   231

     WESTERN/SOUTHWESTERN REGION

15   Ponderosa and Limber Pine Woodlands
     DENNIS H. KNIGHT   249

16   The Sand Shinnery Oak (*Quercus havardii*)
     Communities of the Llano Estacado:
     History, Structure, Ecology, and Restoration
     SHIVCHARN S. DHILLION AND MICHELE H. MILLS   262

17   Oak Savanna in the American Southwest
     MITCHEL P. McCLARAN AND GUY R. McPHERSON   275

18   Juniper–Piñon Savannas and
     Woodlands of Western North America
     NEIL E. WEST   288

19   Serpentine Barrens of Western North America
     A. R. KRUCKEBERG   309

20   California Oak Savanna
     BARBARA ALLEN-DIAZ, JAMES W. BARTOLOME,
     AND MITCHEL P. McCLARAN   322

     NORTHERN REGION

21   Jack Pine Barrens of the Northern
     Great Lakes Region
     KURT S. PREGITZER AND SARI C. SAUNDERS   343

22   The Cliff Ecosystem of the Niagara Escarpment
     D. W. LARSON, U. MATTHES-SEARS, AND P. E. KELLY   362

23   Alvars of the Great Lakes Region
     PAUL M. CATLING AND VIVIAN R. BROWNELL   375

24   The Flora and Ecology of
     Southern Ontario Granite Barrens
     PAUL M. CATLING AND VIVIAN R. BROWNELL   392

25   The Aspen Parkland of Canada
     O. W. ARCHIBOLD   406

26   Subarctic Lichen Woodlands
     E. A. JOHNSON AND K. MIYANISHI   421

Index of Plants   439

Index of Animals   458

Topic Index   463

List of Contributors

Dr. Barbara Allen-Diaz
Department of Environmental
   Sciences and Policy Management
151 Hilgard Hall #3110
University of California
Berkeley, CA 94720-3110

Dr. Roger C. Anderson
Department of Biological Sciences
Campus Box 4120
Illinois State University
Normal, IL 61790-4120

Dr. O. W. Archibold
Department of Geography
University of Saskatoon
Saskatoon, Saskatchewan,
Canada S7N 5A5

Dr. James W. Bartolome
Department of Environmental
   Sciences and Policy Management
151 Hilgard Hall #3110
University of California
Berkeley, CA 94720-3110

Dr. Carol C. Baskin
Department of Biological Sciences
University of Kentucky
Lexington, KY 40506-0225

Dr. Jerry M. Baskin
Department of Biological Sciences
University of Kentucky
Lexington, KY 40506-0225

Mr. Marlin L. Bowles
The Morton Arboretum
Rt. 53
Lisle, IL 60532

Dr. Suzanne Hill Braunschweig
Department of Biology
Virginia Polytechnic Institute
Blacksburg, VA 24061-0406

Ms. Vivian R. Brownell
Agriculture Canada
Center for Land & Biological Resources
The William Saunder Building
Central Experimental Farm
Ottawa, Ontario, Canada K1A 0C6

Mr. I. H. Butler
Oklahoma Natural Heritage Inventory
University of Oklahoma
111 E. Chesapeake Street
Norman, OK 73019

Dr. Paul M. Catling
Agriculture Canada
Center for Land & Biological Resources
The William Saunder Building
Central Experimental Farm
Ottawa, Ontario, Canada K1A 0C6

Dr. Edward W. Chester
Department of Biology
Austin Peay State University
Clarksville, TN 37044

Mr. David D. Close
Department of Forestry
Southern Illinois University
Carbondale, IL 62901-4411

Dr. Shivcharn S. Dhillion
Department of Biology & Nature Conservation
Agricultural University of Norway (NLH-Aas)
pb 5014
Aas, N-1432, Norway

Dr. James S. Fralish
Departments of Forestry and Plant Biology
Southern Illinois University
Carbondale, IL 62901-4411

Dr. Scott B. Franklin
Department of Biology
University of Memphis
Memphis, TN 38152

Dr. David J. Gibson
Department of Plant Biology 6509
Southern Illinois University
Carbondale, IL 62901-6509

Dr. Susan Glenn
Department of Forest Sciences
Faculty of Forestry
University of British Columbia
Vancouver, BC, Canada V6T 1Z4

Dr. Alice Long Heikens
Department of Biology
Franklin College
Franklin, IN 46131-2598

Dr. Bruce W. Hoagland
Oklahoma Natural Heritage Inventory
111 E. Chesapeake Street
Department of Geography
University of Oklahoma
Norman, OK 73019

Dr. James C. Hull
Department of Biological Science
Towson State University
Baltimore, MD 21204

Dr. Edward A. Johnson
Department of Biological Sciences
University of Calgary
Calgary, Alberta, Canada T2N 1N4

Dr. F. L. Johnson
Oklahoma Biological Survey
625 Elm Street, #302
University of Oklahoma
Norman, OK 73019

Mr. Peter Edwin Kelly
Cliff Ecology Research
University of Guelph
Guelph, Ontario, Canada N1G 2W1

Dr. Dennis H. Knight
Department of Botany
University of Wyoming
Laramie, WY 82071-3165

Dr. A. R. Kruckeberg
Department of Botany
University of Washington
Seattle, WA 98195

Dr. Douglas W. Larson
Department of Botany
University of Guelph
Guelph, Ontario, Canada N1G 2W1

Dr. Uta Ursula Matthes-Sears
Cliff Ecology Research Group
University of Guelph
Guelph, Ontario, Canada N1G 2W1

Dr. Mitchel P. McClaran
Division of Range Management
301 Biological Sciences East
University of Arizona
Tucson, AZ 85721

Dr. Guy R. McPherson
Forest Watershed Program
School of Natural Resources
University of Arizona
Tucson, AZ 85721

Dr. Eric S. Menges
Archbold Biological Station
PO Box 2057
Lake Placid, FL 33852

Ms. Michele H. Mills
Division of Biological Sciences
Section of Plant Biology
University of California
Davis, CA 95616

Dr. K. Miyanishi
Department of Geography
University of Guelph
Guelph, Ontario, Canada N1G 2W1

Dr. Erik T. Nilsen
Department of Biology
Virginia Polytechnic Institute
Blacksburg, VA 24061-0406

Dr. William J. Platt
Department of Botany
Louisiana State University
Baton Rouge, LA 70803

Dr. Kurt S. Pregitzer
School of Forestry and Wood Products
Michigan Technological University
Houghton, MI 49931-1295

Dr. Sari C. Saunders
School of Forestry and Wood Products
Michigan Technological University
Houghton, MI 49931-1295

Dr. Donald J. Shure
Department of Biology
Emory University
Atlanta, GA 30322

Dr. Forest Stearns
Emeritus Professor of Biological Sciences
University of Wisconsin–Milwaukee
Forestry Sciences Center
North Central Forest Experiment Station
USDA Forest Service
5985 County Highway K
Rhinelander, WI 54501

Dr. R. Wayne Tyndall
Maryland Department of Natural Resources
909 Wye Mills Road (P.O. Box 68)
Wye Mills, MD 21679

Dr. Neil E. West
Department of Rangeland Resources
Utah State University
Logan, UT 84322-5230

Dr. Peter S. White
Department of Biology, CB# 3280
University of North Carolina
Chapel Hill, NC 27599-3280

Mr. Thomas F. Wieboldt
Department of Biology
Virginia Polytechnic Institute
Blacksburg, VA 24061-0406

Dr. Susan Will-Wolf
Department of Botany
University of Wisconsin-Madison
430 Lincoln Drive
Madison, WI 53706-1381

Dr. Andrew G. Windisch
The Nature Conservancy
New Jersey Natural Heritage Program
Department of Environmental Protection
CN 404
Trenton, NJ 08064

Dr. Susan K. Wiser
Landcare Research
P. O. Box 69
Lincoln 8152
Christchurch, New Zealand

Dr. Robert A. Zampella
State of New Jersey
Pine Lands, Commission
PO Box 7
New Lisbon, NJ 08064

Acknowledgments

We thank the following individuals for assistance in the preparation of this book: M. Rebecca Anderson, Dale Birkenholz, Jennifer Biser, Lauren Brown, Angelo Capparella, Erica Corbett, Gail Corbett, David Doss, Nancy Doss, Christopher Dunn, Christina Kirk, James Krupa, Edward Mockford, Debra Nelson, Mark Omi, Richard Swigart, Charles Thompson, Douglas Whitman, and Cheryl Winchester. Financial support for this project was generated from contributors to and participants in the North American Conference on Savannas and Barrens that was held at Illinois State University on October 15–16, 1994.

Introduction

R. C. ANDERSON, J. S. FRALISH,
AND J. M. BASKIN

Savannas, barrens, and rock outcrop plant communities are the topic of numerous research and technical articles. Rock outcrop plant communities and serpentine barrens are of interest because they are refugia for endemic species adapted to extreme environmental conditions. Savannas and barrens were major components of the historic landscape before it was extensively altered by agricultural and urban development during the past century. Many of these communities were reduced to less than one percent of their original area and are imperiled ecosystems. There has been relatively little synthesis of information about these ecosystems from papers published in scientific journals, conference proceedings, or technical reports from state and federal governmental agencies and private organizations. Our book synthesizes this technical knowledge and will increase awareness of these vegetation types and communities and aid in their conservation and restoration.

The savannas covered in this volume occur in diverse and geographically distant regions of the continent. They include pine savannas of the southeastern Gulf Coastal Plain; aspen parklands of the Canadian provinces of Alberta, Manitoba, and Saskatchewan; California oak savannas; juniper/piñon savannas; subarctic lichen woodland of northwestern Canada; and others. Some of the savanna types cover broad geographical areas, such as the cross timbers that extended from Kansas into Texas and the southern Gulf Coastal pine savannas that occurred from North Carolina to Texas along the coastal plain. All of these communities have unique ecological features. However, they share a common feature in having an environment that restricts tree growth and prevents development of closed-canopy forests.

Fire is an important element that retards tree establishment in most savannas and barrens, but shallow soils over bedrock, and extremes of climate or microclimate, also are important factors. In portions of North America, tree development is limited by precipitation and by climatic factors influenced by elevation. In arid regions, high-elevation mountain forests have a transition through savanna that grade into shrublands or grasslands at lower elevations. Savanna-like vegetation also can occur at the upper elevational or latitudinal limit of tree distribution. For example, limber pine (*Pinus flexilis*) can form transitional vegetation between alpine tundra and coniferous forest in portions of the Rocky Mountains. Similarly, open-canopied conifers with a ground cover of lichens dominate the subarctic woodland lichen vegetation that occurs between the closed conifer forest and tundra across much of Canada.

The terms *barren* and *savanna* have no precise definition, and their meanings vary regionally (Heikens and Robertson 1994). The term *barren* appears to have been used historically to describe areas with restricted tree growth. The word *barren* also creates some difficulty in that it has been used as a singular or plural noun. The context of the

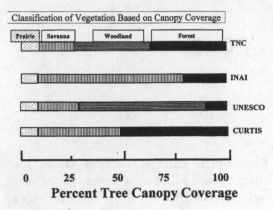

Percent Tree Canopy Coverage

Figure I.1. Classification of prairie, savanna, woodland, and forest based on canopy coverage of trees, modified from Taft (1997). (Sources of information: INAR = Illinois Natural Areas Inventory; TNC = The Nature Conservancy; UNESCO = United Nations Scientific and Cultural Organization; Curtis 1959.)

sentence allows the reader to determine the number forms of the word.

Barrens developed because of repeated fires on sites that would support forest in the absence of fire, or in locations where edaphic factors were the primary cause for absence of closed forests. Sites with edaphic factors limiting tree growth include the serpentine barrens of the east and west coasts of North America. However, on eastern serpentine barrens, fire was essential for their maintenance. Shale barrens of the eastern United States occur on unstable, eroding slopes, with large rock fragments in the soil. Open tree canopies on these sites result in an excessively high level of irradiance on the soil surface. These conditions generate high temperatures that apparently are detrimental to many plant species. Excessive drainage, coupled with low nutrient availability, and frequent fires are important factors in maintaining the New Jersey pine barrens and the jack pine–Hill's oak barrens in the Great Lakes states of Michigan and Wisconsin. Similarly, the shinnery oak savannas of Texas, Oklahoma, and New Mexico occur on sandy soils with low water-holding capacity.

The term *savanna* has been used almost interchangeably with *barrens* to describe the same type of vegetation in different parts of North America. According to Kline (1997), the term *savanna* comes from the Spanish *sabana* (earlier *zabana*), a word 16th-century Spanish colonists adopted from natives in the Caribbean islands. The word was used to describe flat, treeless areas that were dominated by grasses. Some of the earliest European visitors to Illinois applied the term *savanna* to grasslands with or without trees (White 1994). Midwestern U.S. savannas recently have been described as grasslands with trees by some authors (Packard and Mutel 1997), but the term is applied to a range of vegetation types. Generally, savannas appear to be part of the vegetation continuum between forests and grasslands. The portion of the continuum occupied by savanna varies depending upon how different authors have defined forest and grassland. Frequently, prairie, forest, and savanna are delineated based on the cover of tree canopies or the density of trees. However, there is no agreed upon criteria for tree canopy coverages or densities to separate the three vegetation types. For example, most authors set the lower limit of canopy coverage of savannas at 5%–10%, with areas with lower tree canopy cover classified as grasslands. However, the defined upper limit of tree canopy cover of savannas can range from 25% to 80%, Figure I.1 (Taft 1997).

Exclusion of fire from communities during the past 50–100 years has resulted in conversion of savannas and barrens to woodlands. Associated with this conversion has been a concomitant loss of the savanna herbaceous component and its replacement by herbs of closed forests. In juniper–piñon savanna of the southwestern United States, loss of the understory has occurred without replacement. In many instances, trees dominating closed woodlands are less fire tolerant than those of the savanna. In the southwestern United States, fire suppression during this century has encouraged development of closed woodlands of juniper and piñon pine, resulting in conditions highly susceptible to catastrophic fires. These dense woodlands readily support crown fires and are accurately described as "timebombs" waiting to explode. Thus, fire suppression followed by fuel buildup is responsible for many of the "fire storms" that have occurred in the southwestern United States in recent years. Often, reintroduction of fire is possible and can be used to rehabilitate the savanna/barrens vegetation. However, in some instances management with prescribed burning is not possible without prior reduction in fuel load. For example, in ponderosa pine savannas, douglas fir (*Pseudotsuga menziesii*) and white fir (*Abies concolor*) can form dense stands of trees that contribute to high fuel loading. In other localities, conversion to woodlands is complete. Where barrens historically occurred on two kinds of sites – those with deep soil, and also those on thin and rocky soil – fire exclusion has left barrens only on the edaphically restricted sites that retarded woodland development. Invading trees have reduced even the remaining areas of barrens, and barrens persist only where edaphic factors are most restrictive to tree growth.

Savannas are economically important in California and the southwestern United States, where they provide watershed protection and forage for grazing, and serve as habitat for many species of wildlife. In other portions of North America, savannas have been essentially obliterated by agriculture or urban development coupled with fire exclusion.

Rock outcrop plant communities are ecologically related to barrens where soils are shallow and bedrock is exposed on the surface. The rock outcrop communities have more exposed bedrock than barrens, and lichens may form a dominant component of the vegetation. Fires often are not important for maintenance of rock outcrop plant communities. Somewhat distinctive vegetation develops on rock outcrop as a function of

microhabitat differences between rock outcrop sites and the surrounding habitats with deeper soil. The rock outcrop communities considered in this volume include the cliff communities of the Niagara Escarpment of Ontario, Canada. These cliffs support an ancient forest dominated by eastern white cedar trees (*Thuja occidentalis*) that are more than 1,600 years old, as well as a consistent assemblage of herbaceous vascular plants, ferns, bryophytes, lichens, and endolithic organisms. These cliff ecosystems and granite rock outcrops of eastern Canada support no endemic species, presumably as a result of recent Pleistocene glaciation that may have disrupted these habitats. In contrast, cedar glades and granite outcrop communities of the eastern United States, which are also considered in this book, support a substantial number of endemic species. These outcrop species, as well as the nonendemic plants, are adapted to the extreme environmental conditions associated with exposed bedrock. Soil formation from bedrock is extremely slow, and soil depth varies depending on depression in the undulating bedrock surface or exfoliation bedrock that forms shallow soil traps. Soils are often no more than a few centimeters deep but can extend deeply in narrow rock crevices in the bedrock. Trees and shrubs often are restricted to fissures in the rock crevices or to the deepest depressions where sufficient soil has accumulated. Some species are common to rock outcrop plant communities and savannas and barrens, and they may have ecological similarities with some types of barrens. For example, the alvars community of Canada, which occurs on limestone bedrock, has a remarkable floristic similarity to some of the savannas and barrens of the Great Lakes and Midwest, strengthening the links between these vegetation types. Cedar glades included in this volume are limited to areas south of the maximum advance of Pleistocene glaciation in the eastern United States.

The information on savannas, barrens, and

rock outcrop plant communities of North America in this book is in a form that will be used by the scientific community and by university classes. However, it will also be of interest to persons involved in restoration and preservation of these communities and to professional and amateur naturalists.

REFERENCES

Curtis, J. T. 1959. *The Vegetation of Wisconsin*. Madison, Wis.: The University of Wisconsin Press.

Heikens, A. L., and Robertson, P. A. 1994. Barrens of the Midwest: A review of the literature. *Castanea* 59:184–194.

Kline, V. 1997. Orchards of oaks and a sea of grass. In *The Tallgrass Restoration Handbook*, ed. S. Packard and Cornelia F. Mutel, pp. 3–21. Washington, D.C.: Island Press.

Packard, S., and Mutel, C. F., eds. 1997. *The Tallgrass Restoration Handbook*. Washington, D.C.: Island Press.

Taft, J. B. 1997. Savanna and open-woodland communities. In *Conservation in Highly Fragmented Landscapes*, ed. M. Schwartz, pp. 24–54. Chicago, Ill.: Chapman and Hall.

White, J. 1994. How the terms savanna, barrens and oak openings were used in early Illinois. In *Proceedings of the North American Conference on Savannas and Barrens*, ed. J. S. Fralish, R. C. Anderson, J. E. Ebinger, and R. Szafoni, pp. 25–63. Chicago, Ill.: U.S. Environmental Protection Agency, Great Lakes National Program Office.

EASTERN/SOUTHEASTERN REGION

1 Ecology and Conservation of Florida Scrub

ERIC S. MENGES

Introduction

Florida scrub exhibits a multitude of contradictions. It is a shrubland dominated by xeromorphic plants in a region of subtropical temperatures, abundant rainfall, and luxuriant primary productivity. Florida scrub is resilient to fire, but sensitive to short fire return intervals and fire suppression. Although scrub has low species diversity, its level of endemism is among the highest of any North American plant community. To many, Florida scrub appears harsh and unappealing, but it inspires rhapsody among its defenders. Finally, although scrub plants and animals have persisted for millennia of drought, fire, and infertility, the unique biota of Florida scrub is vulnerable to continuing human development.

Scientific knowledge of Florida scrub is as patchy as the remaining distribution of this endangered ecosystem. Many studies on community responses to fire have been completed or are ongoing, but data suggesting the normal range of fire return intervals are sketchy. Information on nutrient cycling, herbivory, and belowground competition is scarce. The distributions and habitat requirements of several endemic plants are well known, although the reasons for high endemism remain controversial. New species of invertebrates, and even new trophic interactions, are being described continually. A sense of urgency pervades the basic scientific study of Florida scrub because (1) habitat destruction is proceeding rapidly, and (2) it is

clear that unanswered land management questions will be crucial to conserving the biodiversity and ecological integrity of the remnant scrub.

Florida Scrub: Vegetation Description and Distribution

Physiognomic Description

Florida scrub is an ecosystem dominated by large xeromorphic shrubs, notably oak (*Quercus* spp.; nomenclature follows Wunderlin [1982] unless otherwise indicated), palmetto (*Serenoa repens, Sabal etonia*), and ericaceous shrubs (e.g., *Vaccinium* spp., *Lyonia* spp.). Large trees (mainly *Pinus clausa*) and herbs are secondary components (Stout and Marion 1993; Menges and Hawkes 1998). Florida scrub occurs on excessively well drained soils throughout Florida and in parts of Georgia and Alabama. It is also characterized by moderate- to low-frequency, high-intensity fires that top-kill most of the vegetation. Although several common life history responses to fire are represented, common species typically recover by resprouting (Menges and Kohfeldt 1995). Consequently, dominant shrubs may be considerably older and more massive belowground than is apparent from their aboveground size (Abrahamson 1995).

Geology, Climate, and Soils

Florida scrub occurs on relict beach ridges and bars (Stout and Marion 1993) in several regions of Florida (Figure 1.1). The Lake

Roger C. Anderson, James S. Fralish, and Jerry M. Baskin, eds. *Savannas, Barrens, and Rock Outcrop Plant Communities of North America*. Copyright © 1999 Cambridge University Press. All rights reserved.

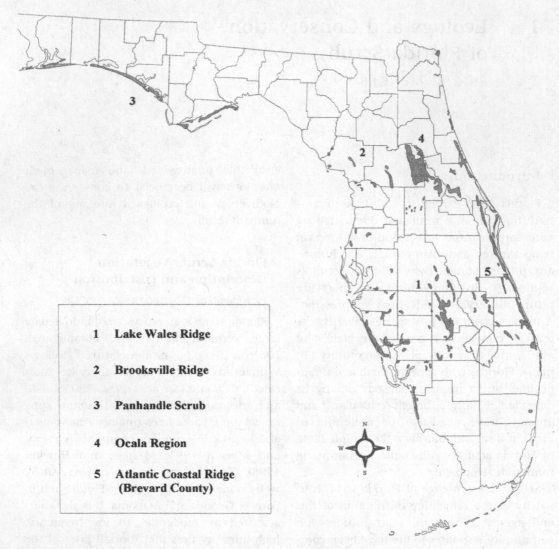

1   Lake Wales Ridge

2   Brooksville Ridge

3   Panhandle Scrub

4   Ocala Region

5   Atlantic Coastal Ridge
     (Brevard County)

Figure 1.1. Generalized map of scrub in Florida. Compiled from several data sources by Nancy Deyrup of Archbold Biological Station. Some smaller scrub patches are not shown. Most major upland ridges and other locations mentioned in the text are labeled.

Wales Ridge (a 160-km-long narrow ridge in south-central Florida) probably dates to the late Pliocene, several million years ago, when sand dunes and beaches were deposited along a shoreline 25 m above mean sea level (Watts and Hansen 1994). Other scrub ridges are considerably younger. For example, in Brevard County on the Atlantic Coastal Ridge, Merritt Island is 240,000 years old, and the barrier island at Cape Canaveral is 7,000 years old (Schmalzer and Hinkle 1990).

Florida scrub is an ancient plant community, having persisted for at least 50,000 years (Watts and Hansen 1994), several times longer than communities in glaciated areas. Abundances of scrub species have fluctuated with glacial events, with xeric elements being considerably more abundant during much of the Pleistocene (13,700 years before present and earlier) than in more recent times (Watts and Hansen 1994).

Today, the climate of peninsular Florida is

humid and subtropical. Most of the peninsula has 125 days or more with maximum temperatures exceeding 31 °C and subfreezing days are uncommon. Annual rainfall, which exceeds 1,200 mm, supports forests on mesic and wet sites. Winters are mild and fairly dry, with the majority of rain falling in June through September due to convective thunderstorms and tropical storm systems. Precipitation exceeds potential evaporation statewide (Fernald and Purdum 1992).

Lightning was the primary ignition source for presettlement fires, with 70–90 thunderstorm days annually (Fernald and Purdum 1992). Lightning strikes occur mainly in summer, but lightning during late spring droughts creates the peak natural fire season (May in central Florida). The spatial and temporal distribution of lightning strikes may create heterogeneity in fire ignitions that can shape the vegetation mosaic.

Florida scrub and related xeric upland communities such as sandhill and xeric hammocks are found on moderately to excessively drained acid quartzipsamment soil (Entisols derived from quartz sands; Abrahamson et al. 1984; Kalisz and Stone 1984). Mesic communities such as flatwoods are distinguished by subtle changes in soil drainage or depth to the water table (Abrahamson et al. 1984). Drought stress in Florida scrub is most important during winter and early spring (Menges and Gallo 1991), and frequent fog during these seasons may relieve drought stress in scrub plants (Menges 1994). The xeromorphic adaptations of the dominant plants (e.g., small, curled, waxy, hairy, evergreen leaves, high biomass allocation to belowground parts in oaks; Abrahamson et al. 1984; Johnson, Abrahamson and McCrea 1986) may be adaptations to periodic drought, low soil fertility, or both (Huck 1987; Menges and Gallo 1991). Scrub soil is quite infertile (Kalisz and Stone 1984; Huck 1987), although coastal soil often retains shell fragments, which increase the pH and soil nutrient levels (Schmalzer and Hinkle 1990). Tissue nutri-

ents and standing crop are similar to other shrublands and vary little with time-since-fire (Schmalzer and Hinkle 1996), although concentration of tissue nutrients may increase temporarily postfire (Anderson and Menges 1997). Shifts in species composition occur in response to soil type, particularly differences between xeric yellow and white sands despite their chemical similarity (Kalisz and Stone 1984). Litter fall rates are high and litter turnover rates are low (Lugo and Zucca 1983), thus litter tends to build up for several years following a fire (Schmalzer and Hinkle 1996).

Types of Florida Scrub Vegetation

Several types of Florida scrub can develop on xeric uplands (Table 1.1). One of the common types is sand pine scrub, an infrequently burned community with a sand pine canopy occurring on xeric sites with white or yellow sands. Sand pine often has serotinous cones, although this characteristic varies regionally (Myers and White 1987; Stout and Marion 1993; Huck et al. 1996). Rosemary scrub (rosemary phase of sand pine scrub; Abrahamson et al. 1984) is dominated by the shrub Florida rosemary (*Ceratiola ericoides*) on xeric white sand (e.g., St. Lucie, Archbold) and forms the most open structured scrub. Sand pine scrub and rosemary scrub are found in interior and coastal areas (Johnson and Muller 1993). On the Lake Wales Ridge, rosemary scrub includes many endemic plants (Hawkes and Menges 1996). If fires occur frequently, Florida rosemary may be eliminated because seed production in this shrub does not begin until about 10 years postfire (Johnson 1982) and individuals are killed by fire (Menges and Kohfeldt 1995).

Scrubs dominated by evergreen shrub oaks (*Quercus myrtifolia, Q. chapmanii, Q. geminata, Q. inopina*), with or without scattered sand pines or south Florida slash pines (*P. elliottii var. densa*), are known as scrubby flatwoods or oak scrub. These occur throughout Florida

Table 1.1. *Major vegetation types found in Florida's xeric uplands.*

Vegetation type	Soils	Tree layer	Shrub dominants	Herb layer	FRI[1] (yrs)
Sand pine scrub	Xeric, yellow or white	Sand pine (*Pinus clausa*)	Dense oaks (*Quercus* spp.), fetterbush (*Lyonia* spp.), palmettos (*Serenoa repens, Sabal etonia*)	Sparse	15–100
Rosemary scrub	Xeric, white	Sand pine lacking	Florida rosemary (*Ceratiola ericoides*)	Forb dominated	15–100
Scrubby flatwoods (oak scrub)	Xeric or dry–mesic, white or gray	Pines or lacking	Dense oaks, palmettos, fetterbush	Sparse	5–20
Oak–palmetto scrub	Dry–mesic, gray	Slash pine (*P. elliottii*)	Dense oaks, saw palmetto	Sparse	5–20
Sandhill (high pine)	Xeric, yellow	Longleaf pine (*P. palustris*)	Scattered turkey oak (*Q. laevis*)	Grass/forb dominated	1–10
Oak–hickory scrub	Xeric, yellow	Slash pine	Scrub hickory (*Carya floridana*), myrtle oak (*Q. myrtifolia*)	Sparse	*
Xeric hammock	Xeric, yellow or white	Oaks (*Quercus* spp.), others	Oaks, others	Sparse	**

[1] Fire return interval.
* Not known. May have developed from infrequently burned sandhill communities, but once established it may be stable over a range of fire return intervals.
** Develops gradually from unburned or infrequently burned xeric communities.

on excessively or moderately well drained soil (Table 1.1). Scrubby flatwoods are intermediate between flatwoods and rosemary scrub in vegetation and soil characteristics (Abrahamson and Hartnett 1990; Stout and Marion 1993). Scrubby flatwood communities burn every 5–20 years, support few endangered plant species, and are the primary habitat for the threatened Florida scrub-jay (*Aphelocoma coerulescens*).

In Georgia and Alabama, a scrubby flatwoods-like community known as evergreen scrub forest is dominated by shrubby oak with a ground cover of lichens. Dwarf oak-evergreen shrub forests in Georgia resembling rosemary scrub are dominated by turkey oak (*Q. laevis*), Florida rosemary, Ashe's savory (*Calamintha ashei*), sand spikemoss (*Selaginella arenicola*), and other plants (Wharton 1978).

Several scrub types have been described only from coastal areas. Oak–palmetto scrub (Table 1.1), an intermediate community between mesic flatwoods and xeric scrub, recovers rapidly from fire and requires frequent burning to maintain physiognomy for the Florida scrub-jay (Schmalzer and Hinkle 1992). Woody goldenrod (*Chrysoma pauciflosculosa* (Michx.) Greene) scrubs are found only in the coastal Florida panhandle (Johnson and Muller 1993). Barrier island scrub dynamics may be influenced by hurricane disturbance and sand movements to a greater extent, and fire to a lesser extent, than inland scrubs (Myers 1990; Gibson and Menges 1994).

In Florida, scrub also intergrades with other xeric communities such as sandhill (high pine), which is usually dominated by longleaf pine (*Pinus palustris*), turkey oak, wiregrass (*Aristida beyrichiana* Trinius and Ruprecht), and herbs. Sandhill vegetation burns more frequently than scrub (Table 1.1). In the absence of fire, it may be invaded by scrub plants such as sand pine or oak (Myers and White 1987). Sandhill and sand pine scrub may alternate over time on xeric yellow soil depending on fire history (Kalisz and Stone 1984; Myers 1985). Also, some vegetation types appear to be intermediate between sandhill and scrub. On the southern Lake Wales Ridge, the hickory phase of southern Ridge sandhill (Abrahamson et al. 1984) has some floristic elements of sandhill, but because of species dominance (scrub hickory [*Carya floridana*] and myrtle oak [*Quercus myrtifolia*]), vegetation structure, postfire recovery, and presence of endemic scrub plants not found in sandhill, for example *Dicerandra frutescens* (Menges 1992), it is best termed oak–hickory scrub.

In the absence of fire, scrub may be invaded by mesic hammock (forest) species in northern Florida (Myers and White 1987), but such sites farther south develop into xeric hammocks with little species turnover (Menges et al. 1993; Table 1.1). This difference may be related to lack of local seed sources for mesic hammock species in south-central Florida.

Fire Ecology

As with most shrublands, Florida scrub is profoundly affected by fire. Fire intensity varies depending on fuel, weather, and ignition pattern. Many scrub species depend on fire and some are most abundant shortly after fire (Johnson and Abrahamson 1990; Menges and Kohfeldt 1995). For example, narrowly endemic herbs such as *Eryngium cuneifolium* (Menges and Kimmich 1996), *Bonamia grandiflora* (Hartnett and Richardson 1989), and *Warea carteri* (Menges and Gordon 1996) have peaks in demographic parameters such as survival, recruitment, and densities shortly after fire.

Resprouting and clonal spread of dominant shrubs (Menges and Kohfeldt 1995) returns postfire cover rapidly to prefire levels (Abrahamson 1984a; Abrahamson 1984b; Schmalzer and Hinkle 1992). Low-growing blueberries (*Vaccinium darrowii* and *V. myrsinites*) have aggressive clonal growth (Menges and Kohfeldt 1995) that allows them to peak in cover and fruiting within a few years postfire (Abrahamson 1984b). Flowering peaked within 7 years after fire among five resprouting plants of scrubby flatwoods (*Befaria racemosa, Hypericum reductum, Palafoxia feayi, Quercus inopina, Vaccinium myrsinites*) (Ostertag and Menges 1994). Oaks are slower to recover, but after 5 years or so they suppress blueberries and other small shrubs.

Seedling recruitment also occurs following fire, and for obligate seeders this is the only means of postfire population recovery. Obligate seeders tend to be xeric-site specialists, open microsite specialists, herbs, and endemics to a greater degree than species that can resprout (Menges and Kohfeldt 1995). Perhaps resprouters displace seeders to sites with less postfire competition. Seeders often recover via germination from a dormant seed bank. The cues for germination appear unrelated to specific effects of fire (heat or smoke, for example), although these mechanisms have not been well studied. Seed banks vary by site and season, but densities are very low (Carrington 1996).

A few species (e.g., *Polygonella basiramia* (Small) Nesom and Bates) appear to recover from fire by dispersing seeds into burned sites from unburned patches; their seeds do not occur in seed banks, and population recovery is delayed on burned sites.

Fires often are patchy in Florida scrub, with patchiness varying depending on the scrub type. Heterogeneous fires tend to occur most often in xeric scrub types with patchy fuels (especially rosemary scrub) and under moderate weather conditions. Fire patchiness influences subsequent community recovery. Unburned patches (or domes) of scrub oak may grow tall enough to become fire resistant (Guerin 1993). These patches provide cover, nest sites, and acorns for Florida scrub-jays during the first few years postfire; thus, patchily burned territories are not abandoned (Fitzpatrick, Woolfenden and Kopeny 1991) as are completely scorched areas.

Without fire, slow structural and compositional changes occur in Florida scrub vegetation. The trend is toward local loss of diversity, with smaller shrubs and herbs dropping out and a few shrubs increasing in dominance, notably myrtle oak and sand live oak (Menges et al. 1993).

Vertebrates are also affected by structural changes between fires. For the Florida scrub-jay, optimal habitat includes oaks less than 2 m tall, few pine trees, and patches of open sand for acorn caching (Fitzpatrick, Woolfenden and Kopeny 1991; Breininger et al. 1995). As shrubs grow taller, as pines recruit and grow, and as open sand patches decrease in size (Hawkes and Menges 1996), the demographic performance of Florida scrub-jays decreases, and they eventually abandon the area (Fitzpatrick, Woolfenden and Kopeny 1991). Abandonment takes about 20 years in interior scrub, but it may occur more rapidly in coastal areas (Breininger and Schmalzer 1990). Resulting population viability in fire-suppressed landscapes is low (Root 1996). Reptile communities, particularly endemic reptiles, are more abundant following fire and other disturbances than in unburned, mature sand pine scrub. Many appear to specialize in open-sand microsites (Greenberg, Neary and Harris 1994).

Microsites and Patch Dynamics

Gap Specialization

Gaps among dominant shrubs of rosemary scrub provide an important microhabitat for herbaceous plants and obligate seeders (Menges and Hawkes 1998). Gaps persist between fires due to slow biomass recovery of Florida rosemary, an obligate seeder, compared to resprouting oaks that dominate scrubby flatwoods (Johnson, Abrahamson and McCrea 1986). Allelopathy by Florida rosemary (Fischer et al. 1994) may also play a role in maintaining the openness of rosemary scrub.

Several herbaceous species are specialized for gaps. Individual survival for *Eryngium cuneifolium* is twice as high for plants 2 m or more distant from shrubs as for plants closer than 50 cm. Survival also increases in large gaps and in the absence of Florida rosemary and Ashe's savory (Menges and Kimmich 1996). Demographic statistics for *Hypericum cumulicola* are explained by time-since-fire and microhabitat within gaps (Quintana-Ascencio and Morales-Hernandez 1997). Finally, *Polygonella basiramia* density and fecundity increase with gap area (Hawkes and Menges 1995). *Eryngium*, *Hypericum*, and *Polygonella* seem to form a sequence from greater to lesser (1) sensitivity to shrub canopies, (2) edaphic specialization, (3) metapopulation dynamics (Quintana-Ascencio and Menges 1996), and (4) geographic range (Christman and Judd 1990). The narrower geographic ranges of *Eryngium* and, to some extent, *Hypericum* may be caused by these microsite and metapopulation differences. Gap specialist species also occur among small-statured plants in other types of Florida scrub (e.g., *Dicerandra frutescens*, Menges 1992; *Chrysopsis floridana*, Lambert and Menges 1996).

In scrub microsites following typical spring fires, conditions are more severe than those in typical fall postfire microsites in California chaparral. In chaparral, a greater proportion of species specialize in postfire seedling establishment compared to scrub (Carrington 1996). Most scrub species establish seedlings during winter, when soil temperatures are favorable for survival. In scrub, the main effects of fire on seeds and seedlings are probably due to alterations of microsites. Fire kills seeds on the soil's surface, but buried seeds that survive benefit from reduced litter depth and greater canopy openness (Carrington 1996).

Microsites, Lichens, and Soil Crusts

Lichens and soil crusts may affect microsites for vascular plant seedling recruitment. Most lichens increase in abundance during fire-free intervals (Menges and Kohfeldt 1995; Hawkes and Menges 1996). The mechanisms for lichen increases are poorly known, and colonization by vegetative fragments from unburned patches could be important. Soil crusts are formed from various mixtures of algae, fungi, cyanobacteria, lichens, bacteria, and mosses (Johansen 1993; Eldridge and Greene 1994). In deserts, soil crusts increase soil organic matter (Johansen 1993) nitrogen (Evans and Ehleringer 1993; Belnap, Harper and Warren 1994), and they affect the soil moisture regime (Johansen 1993). Soil crusts in deserts are destroyed by natural (fire, animal movements) and human-caused (trampling, vehicle use) disturbances and probably recover slowly. Such effects are yet to be determined for Florida scrub. In Florida scrub, algae in soil crusts are grazed by burrowing, flightless, endemic pygmy mole crickets (Deyrup and Eisner 1996).

Metapopulation and Landscape Dynamics

Florida scrub is patchy due to its interdigitation with sandhill and other vegetation, the mixture of different types of scrub, fire history, and a mosaic of fire intensity, including unburned patches. Landscape-level dynamics affect the functioning of individual patches and overall abundance patterns of species in the scrub landscape (Menges and Hawkes 1998). Landscape-level spatial patterns have been studied in Florida scrub-jays and rosemary scrub plants.

Florida Scrub-Jay Spatial Dynamics

Habitat loss and fragmentation are the leading causes of an ongoing rangewide reduction in the Florida scrub-jay (Fitzpatrick, Woolfenden and Kopeny 1991; Stith et al. 1996). Compared to several other vertebrate species endemic to Florida scrub (Florida mouse, *Podomys floridanus*; sand skink, *Neoseps reynoldsi*; scrub lizard, *Sceloporus woodi*; bluetail mole skink, *Eumeces egregius lividus*), the jay seems most sensitive to these landscape alterations. Scrub-jays require considerable expanses of habitat and disappear as scrub habitat becomes overgrown or fragmented.

Florida scrub-jays live as family groups in large (ca. 10 ha) territories. Groups include prebreeding offspring that assist their parents in rearing new young. Dispersal is extremely limited, and many offspring become breeders in a portion of their natal territory. Florida scrub-jays prey on a variety of arthropod species, but acorns are the only plant food critical to their existence. Several oak species that mast-fruit asynchronously provide a bountiful annual crop, some of which is cached for use during winter and early spring (DeGange et al. 1989).

Data from long-term studies provocatively demonstrate the importance of landscape characteristics to Florida scrub-jay population viability (Fitzpatrick, Woolfenden and Kopeny 1991; Stith et al. 1996). Under optimal conditions, an individual scrub patch must be at least about 300 ha to maintain an independent viable population. Smaller patches are sinks and must be within a few miles of immigrant sources. Although various intervening habitats affect dispersal, virtually no dispersal >5 km occurs across any habitat besides oak scrub (Stith et al. 1996). Jays are also very sensitive to recent fire history. Areas burned within the last few years provide inadequate cover and insufficient acorns, but habitats unburned for two or more decades represent population sinks (Woolfenden and Fitzpatrick 1984; Fitzpatrick, Woolfenden and Kopeny 1991) and will not support viable populations (Root 1996). Because fire patchiness can be at a finer scale than jay territories, jay families may occupy territories that represent a useful mosaic of postfire ages.

Current studies of Florida scrub-jays surviving in suburban habitat reveal important differences from more natural situations (Bowman 1996). Compared to "rural" jays, suburban jays live at higher densities, invest more in reproduction, and lay larger and more clutches. However, they produce comparable numbers of fledglings with higher post-fledging mortality, compared to jays in more natural landscapes. For Florida scrub-jays, suburban habitats are population sinks.

Detailed data on dispersal and demography and related metapopulation models suggest that the Florida scrub-jay is distributed as 42 separate metapopulations, only half of which are viable (Stith et al. 1996). Individual populations include viable "mainland" populations, isolated "island" populations, and intermediates, grouped in various arrangements. Scrub-jay population viability can be improved by restoration of current habitat (Root 1996).

Metapopulation Dynamics of Plants

Although dispersal information on plants is difficult to obtain, their spatial distributions

can be analyzed to infer metapopulation dynamics. Incidence of 62 species of plants in rosemary scrub patches was analyzed in relation to patch size, isolation, and time-since-fire (Quintana-Ascencio and Menges 1996). Those species (e.g., *Eryngium cuneifolium*, *Hypericum cumulicola*) most sensitive to the spatial structure of patches, and presumed to have the greatest potential for metapopulation dynamics, tended to be specialists for rosemary scrub (Quintana-Ascencio and Menges 1996). This potential for metapopulation dynamics has strong implications for conservation of plants, since the spatial arrangement of preserves and of habitat patches within preserves can affect the viability of populations and their distribution.

The Landscape and Fire

Fires in the Florida scrub landscape (which may include several habitats, such as sandhills, flatwoods, and wetlands) have complex dynamics. Fires can burn frequently through grassy fuels (e.g., seasonal ponds dominated by cutthroat, *Panicum abscissum*, can re-burn within three weeks). However, scrub vegetation does not accumulate sufficient fuels to burn until about 5 years postfire (longer in the most xeric sites). Therefore, the juxtaposition of habitat types in the landscape affects fire in scrub.

Conservation Biology of Florida Scrub
Biogeography and Rarity of Scrub Species

The particulars of the spatial distribution of Florida scrub and its species must be considered in conservation efforts. Many plants and invertebrates are endemic to subsets of the Florida scrub, such as particular ridges and parts of ridges. Highlands County, on the southern Lake Wales Ridge, was recently identified in a GAP analysis as 1 of 11 counties that were hot spots for two or more groups of endangered species, in this case plants and herptiles (Dobson et al. 1997). The

Lake Wales Ridge and neighboring small ridges in central Florida are the most endemic rich, containing 17 endemic plants and 40 plants found only in Florida (Christman and Judd 1990). These include the recently rediscovered *Ziziphus celata* Judd and D. Hall (DeLaney, Wunderlin and Hansen 1989), the recently described *Dicerandra christmanii* Huck and Judd (Huck et al. 1989), *Crotalaria avonensis* DeLaney and Wunderlin (DeLaney and Wunderlin 1989), the chemically distinctive *Dicerandra frutescens* (McCormick et al. 1993), and the first lichen listed as endangered in the United States, *Cladonia perforata* (Buckley and Hendrickson 1988), now known from beyond the central Florida ridges. Other ridges in Florida support fewer endemic plants. Florida scrub endemics are among the rarest plants in the southeastern United States, and their conservation may require intensive management efforts.

Many amphibians and reptiles are associated with Florida scrub, including the gopher tortoise (*Gopherus polyphemus*), which acts as a keystone species by supporting numerous burrow commensals (Stout and Marion 1993). Three reptiles – the Lake Wales Ridge endemic, bluetail mole skink (*Eumeces egregius*); the sand skink (*Neoseps reynoldsi*); and the Florida scrub lizard (*Sclenoporus woodii*) – are endemic to Florida sandhill and scrub (Mohler 1992). Types of rarity differ between scrub herptiles and scrub plants (McCoy and Mushinsky 1992). Plants tend to have narrower geographic ranges than herptiles, and herptiles are more habitat-restricted with smaller populations. These results suggest that scrub reserve designs aimed at particular groups may not be appropriate for protecting other taxa.

Invertebrates may include the most narrowly endemic scrub organisms. As with plants, the Lake Wales Ridge harbors the greatest endemism for scrub invertebrates (Deyrup 1990). However, narrow endemics can be found on a number of ridges. For example, among flightless, endemic grasshop-

pers in the family Acrididae, different narrowly endemic species are found on each of the Lake Wales, Atlantic Coast, Ocala, Trail, and St. John River Ridges (Deyrup and Franz 1994). Most of these ridges do not support narrowly endemic plants. More new taxa await discovery because many arthropods of scrub and sandhill habitats are cryptic and little studied (Deyrup 1990).

Genetics of Scrub Plants

Little is known about the genetics of scrub organisms, and thus the implications of genetics to conservation remain to be determined. Scrub is naturally fragmented, and many species have fragmented distributions within scrub habitat. Consequently one might expect isolated populations to be genetically differentiated and small populations to be genetically depauperate. However, the recent history of habitat loss may not yet be reflected in genetic patterns. Based on analysis of only a few populations of three endemic taxa, McDonald and Hamrick (1996) found considerable genetic variation for the scrub endemic *Eryngium cuneifolium*. However, population-level genetic variation is lower than expected. Differentiation among populations of six scrub species is also generally low, with overall genetic variation typical of narrow endemics (Menges et al. 1998). Two endemic *Polygonella* species had more variation than their widespread congeners, a pattern attributable to postglacial migration and repeated founder effects in the widespread species (Lewis and Crawford 1995). Sand pine, common throughout Florida scrub, had relatively low genetic variation for a pine, with significant genetic differences between two geographically separated varieties (Parker and Hamrick 1996).

Human Effects on Florida Scrub

Humans have had three major effects on Florida scrub: habitat loss, habitat fragmentation, and fire suppression. These effects have the potential to alter critical components of scrub ecosystems (Table 1.2). On the Lake Wales Ridge, about 85% of the original area of scrub has been lost (Christman and Judd 1990), and even less remains in some coastal areas. Extant scrub patches now number only in the hundreds, and most are quite small, isolated, and unprotected, particularly in coastal areas (Johnson and Muller 1993) and near Orlando. Although there are many problems with small scrub parcels, edge effects do not seem to be a major concern for plants (Blanchard 1992; Halpern and Menges unpublished data). No scrub vertebrates were excluded from all small parcels (McCoy and Mushinsky 1994), although most individual fragments may have small, declining, or non-viable populations.

Management Issues

Fire at appropriate intervals is necessary for scrub species persistence; too frequent fire and (especially) fire suppression can alter Florida scrub communities. Even if fires re-occur in areas long unburned, it may cause different vegetation responses than in frequently-burned areas (Abrahamson and Abrahamson 1996). Small, privately held parcels are particularly overgrown, with open-space specialists often reduced to small populations on edges or openings. Although the largest areas of scrub are in Ocala National Forest, this public land is managed for timber production, with little prescribed burning of unharvested areas. Managers of newly purchased lands often face the daunting prospect of reintroducing fire to a fire-suppressed landscape where scrub fires can burn extremely intensely and be difficult to control (Myers 1990). Predictions of flame length made for Florida scrub using a fuel model parameterized for California chaparral consistently underestimate fire intensity in Florida scrub. An oft-mentioned challenge to prescribed burners of Florida scrub is to identify the narrow window between conditions where the scrub will not burn at all and conditions where control of an ignited fire is impossible.

Mechanical pretreatments such as roller-chopping can increase the range of conditions for subsequent prescribed burning. Mechanical treatments also mimic some effects of fire (Greenberg, Neary and Harris 1994; Greenberg et al. 1995), but further study is critical to assess the consequences of them on Florida scrub. Mechanical treatments cause more soil disturbance than fire and may leave scrub areas vulnerable to invasion by exotic and weedy species. These treatments alone, without subsequent fire, will leave abundant litter that will likely inhibit the recruitment of many endangered plants that often increase following fire. However, small-scale mechanical treatments, which create open space and do not leave litter, may be appropriate as intensive management aimed at recovering individual populations of gap-requiring endangered plants.

Other scrub management issues have received little consideration. Firelane width and methods of creating firelanes (plowing, disking, mowing, wet lines) vary widely among management agencies. Permanent firelanes remove scrub habitat, but also allow effective control of fires and act as refugia for gap specialists. Trampling may be an unrecognized concern for lichens and soil crusts (Table 1.2). Exotic plants seldom invade undisturbed scrub habitats (Halpern and Menges unpublished data), but invasion can occur following fires or soil disturbance. For example, natal grass (*Rhynchelytrum repens*) can invade many soil-disturbed scrub areas; Brazilian pepper (*Schinus terebinthifolius*) has invaded some south Florida scrub preserves; and cogongrass (*Imperata cylindrica*) has the potential to affect scrub sites statewide. Exotic plants are most likely to cause serious problems in the small urban preserves along Florida's coasts and in the Orlando area. Feral hogs cause excessive soil disturbance.

Severely disturbed scrub areas do not

Table 1.2. *Potentially critical components and processes in Florida scrub ecosystems, their key roles, and major threats associated with them.*

Component	Key roles	Associated major threats
Xeric soils	Severe conditions limit productivity, diversity	Disturbance encourages exotic plant invasion
Oaks	Major biomass, wildlife food, less flammable	Land clearing
Palmettos	Major biomass, wildlife food, moderately flammable	Land clearing
Grasses	Fine fuel increases fire frequency, rate of spread	Fire suppression reduces grasses, alters fire regime
Gopher tortoise	Supports burrow commensals	Human predation, disease
Gaps among shrubs	Microhabitat for endemic plants and invertebrates	Altered fire regime
Lichens, soil crusts	Affect seedlings, water, nutrients	Trampling, soil disturbance
Fire	Controls vegetation structure, spatial heterogeneity, species abundances	Fire suppression, burns out of natural fire season
Landscape pattern	Affects metapopulation dynamics, fire spread, gene flow	Habitat fragmentation threatens metapopulations, alters fire spread, disconnects gene flow
Isolated scrub ridges	Have promoted evolution of unique taxa	Locally intensive loss before protection

recover without active restoration, probably due to limited recolonization abilities of many species (Abrahamson 1995). Restoration of abandoned phosphate mines and citrus groves is being attempted (Poppleton, Clewell and Shuey 1983), but it is too soon to know if any reasonable success can be anticipated. Given the lack of knowledge of species interactions, soil biota, invertebrate fauna, and many other aspects of scrub ecology, current scrub restoration is experimental at best.

Conservation Strategies

Land acquisition can directly alleviate the threats of habitat loss and fragmentation and make fire management more practical.

Several ambitious land acquisition programs target Florida scrub, which has been considered a priority for land acquisition in Florida (Cox et al. 1994).

On the endemic-rich Lake Wales Ridge, several key parcels have been identified for protection. Although central Florida has one of the smallest percentages of conservation lands of any region in the state (Cox et al. 1994), land acquisition is proceeding. Many tracts are being purchased under the state's Conservation and Recreational Lands (CARL) Program, through a bond initiative known as P-2000. Plans are to spend $300 million per year over 10 years to buy conservation lands (Cerulean 1991). A project to protect an archipelago of scrub sites on the Lake Wales

Ridge ranks first among CARL projects. The Lake Wales Ridge National Wildlife Refuge also seeks to protect an archipelago of significant scrub sites (U.S. Fish and Wildlife Service 1993). Private institutions such as The Nature Conservancy and Archbold Biological Station have also been active in purchasing land on the Lake Wales Ridge.

In coastal uplands, a detailed inventory has been used to prioritize sites (Johnson and Muller 1993). County land acquisition programs are most active in coastal counties (Cerulean 1991), where most remaining scrub parcels are small and isolated. Particularly noteworthy are efforts in Brevard County, on the Atlantic Coastal Ridge. Considerable scientific effort, combined with political activity, has led to significant scrub land acquisition efforts (Swain, Hinkle and Schmalzer 1993). A team of scientists developed a regional habitat conservation plan for scrub that focused on the Florida scrub-jay as a key indicator species (Swain et al. 1995). Analyses of existing scrub habitat, scrub-jay spatial distribution, landscape spatial structure, and various reserve designs and their effects on scrub-jay viability led to four alternative reserve designs.

Protection of multiple sites is often recommended, but is extremely crucial for Florida scrub for a number of reasons. Many endemic species occur only at scattered sites, and therefore protection of only a few scrub sites will doom some species to extinction. It also appears that protection of sites from throughout species' ranges will be necessary to protect genetic diversity (Menges et al. 1998). Multiple sites will be necessary to protect species that function as metapopulations, especially if many individual habitat patches cannot support viable populations by themselves (Quintana-Ascencio and Menges 1996; Stith et al. 1996). Finally, fire managers of small scrub patches may find it difficult to maintain patchiness in the face of fire sup-

pression or, on the other hand, escaped or accidental fires that completely burn individual patches. To provide spatial heterogeneity that will provide habitat for many species with a range of life histories, multiple scrub patches within species dispersal distances may need to be managed in a coordinated fashion.

Finally, Florida scrub, like many shrublands, barrens, and savannas, does not immediately capture the hearts of the general populace in the manner of an ancient forest. Instilling knowledge about the intricacies of scrub ecology is key to nurturing positive attitudes for Florida scrub among the public, and is essential to its persistence. Education is a necessary prerequisite for widespread scrub appreciation and stewardship.

Summary

Florida scrub is a xeric shrubland that has survived for over 50,000 years on dry ridges in Florida. This ecosystem has many narrow endemics on the most ancient ridges such as the Lake Wales Ridge. Vegetation varies with soil drainage, soil type, and geographic location, but oaks, ericads, and palmettos are the most common shrub dominants. Pine canopies vary from none to closed. Florida scrub is characterized by infrequent, high-intensity fire ignited by lightning, and most of its dominant species resprout following fire. Gaps among shrubs are important microhabitats for subshrubs and herbs, many of which are narrowly endemic. Such gaps are most common after fire and in the most xeric vegetation dominated by the obligate seeder Florida rosemary. Lichens and soil crusts are common, but their effects on other biota are poorly understood. The Florida scrub-jay and rosemary scrub plants form metapopulations, with landscape spatial structure and habitat fragmentation likely affecting their persistence. Many endemic plants, birds, reptiles, and invertebrates are specialized for xeric Florida habitats, but their distributions are often discontinuous, idiosyncratic, or poorly known. These species are threatened by continuing habitat loss, habitat fragmentation, fire suppression, and invasion by exotic organisms. Appropriate land management, purchase of critical habitat, landscape-level conservation planning, and public education are increasing, and provide real hope for Florida scrub conservation.

Acknowledgments

This chapter was improved by the comments of Warren Abrahamson, Roger Anderson, Jerry Baskin, Reed Bowman, Margaret Evans, James Fralish, Doria Gordon, Christine Hawkes, Vickie Larson, Pedro Quintana-Ascencio, Kurt Reinhart, Hilary Swain, Glen Woolfenden, and Rebecca Yahr. Nancy Deyrup and Christina Casado put together the scrub map in Figure 1.1.

REFERENCES

Abrahamson, W. G. 1984a. Postfire recovery of Florida Lake Wales Ridge vegetation. *American Journal of Botany* 71:9–21.

Abrahamson, W. G. 1984b. Species responses to fire on the Florida Lake Wales Ridge. *American Journal of Botany* 71:35–43.

Abrahamson, W. G. 1995. Habitat distribution and competitive neighborhoods of two Florida palmettos. *Bulletin of the Torrey Botanical Club* 122:1–14.

Abrahamson, W. G., and Abrahamson, J. R. 1996. Effects of a low-intensity winter fire on long-unburned Florida sand pine scrub. *Natural Areas Journal* 16:171–183.

Abrahamson, W. G., and Hartnett, D. C. 1990. Pine flatwoods and dry prairies. In *Ecosystems of Florida*, ed. R. L. Myers and J. J. Ewel, pp. 103–149. Orlando,Fla.: University of Central Florida Press.

Abrahamson, W. G., Johnson, A. F., Layne, J. N., and Peroni, P. A. 1984. Vegetation of the Archbold Biological Station, Florida: an example of the southern Lake Wales Ridge. *Florida Scientist* 47:209–250.

Anderson, R. C., and Menges, E. S. 1997. Effects of fire on sandhill herbs: nutrients, mycor-

rhizae, and biomass allocation. *American Journal of Botany*. 84:938–948.

Belnap, J., Harper, K. T., and Warren, S. D. 1994. Surface disturbance of cryptobiotic soil crusts and nitrogenase activity, chlorophyll content, and chlorophyll degradation. *Arid Soil Research and Rehabilitation* 8:1–8.

Blanchard, J. D. 1992. Light, vegetation structure, and fruit production on edges of clearcut sand pine scrub in Ocala National Forest, Florida. M.S. thesis, University of Florida, Gainesville, Fla..

Bowman, R. 1996. *Contributions of Suburban Jay Populations to the Metapopulation Dynamics of Florida Scrub-Jays*. Tallahassee, Fla.: Interim Report to Florida Game and Fresh Water Fish Commission.

Breininger, D. R., Larson, V. L., Duncan, B. W., Smith, R. B., Oddy, D. M., and Goodchild, M. F. 1995. Landscape patterns of Florida scrub jay habitat use and demographic success. *Conservation Biology* 9:1442–1453.

Breininger, D. R., and Schmalzer, P. A. 1990. Effects of fire and disturbance on plants and birds in a Florida oak/palmetto scrub community. *The American Midland Naturalist* 123:64–74.

Buckley, A., and Hendrickson, T. 1988. The distribution of *Cladonia perforata* Evans on the southern Lake Wales Ridge in Highlands County, Florida. *The Bryologist* 91:354–356.

Carrington, M. E. 1996. Postfire recruitment in Florida sand pine scrub in comparison with California chaparral. Ph.D. dissertation, University of Florida, Gainesville.

Cerulean, S. I. 1991. *The Preservation 2000 Report. Florida's Natural Areas, What have We Got to Lose?* Winter Park, Fla.: The Nature Conservancy.

Christman, S. P., and Judd, W. S. 1990. Notes on plants endemic to Florida scrub. *Florida Scientist* 53:52–73.

Cox, J., Kautz, R., MacLaughlin, M., and Gilbert, T. 1994. *Closing the Gaps in Florida's Wildlife Habitat Conservation System*. Tallahassee,Fla.: Florida Game and Fresh Water Fish Commission.

DeGange, A. R., Fitzpatrick, J. W., Layne, J. N., and Woolfenden, G. E. 1989. Acorn harvesting by Florida Scrub-Jays. *Ecology* 70:348–356.

DeLaney, K. R., and Wunderlin, R. P. 1989. A new species of *Crotalaria* (Fabaceae) from the Florida central ridge. *Sida* 13:315–324.

DeLaney, K. R., Wunderlin, R. P., and Hansen, B. F. 1989. Rediscovery of *Ziziphus celata* (Rhamnaceae). *Sida* 13:325–330.

Deyrup, M. 1990. Arthropod footprints in the sands of time. *Florida Entomologist* 73:529–538.

Deyrup, M., and Eisner, T. 1996. Photosynthesis beneath the sand in the land of the pygmy mole cricket. *Pacific Discovery* 49:44–45.

Deyrup, M., and Franz, R. 1994. *Rare and Endangered Biota of Florida.* Vol. IV, *Invertebrates.* Gainesville,Fla.: University Presses of Florida.

Dobson, A. P., Rodriguez, J. P., Roberts., W. M., and Wilcove, D. S. 1997. Geographic distribution of endangered species in the United States. *Science* 275:550–553.

Eldridge, D. J., and Greene, R. S. B. 1994. Microbiotic crusts: a review of their roles in soil and ecological processes in the rangelands of Australia. *Australian Journal of Soil Resources* 32:389–415.

Evans, R. D., and Ehleringer, J. R. 1993. A break in the nitrogen cycle in aridlands? Evidence from ^{15}N of soils. *Oecologia* 94:314–317.

Fernald, E. A., and Purdum, E. D. 1992. *Atlas of Florida.* Gainesville: University Press of Florida.

Fischer, N. H., Williamson, G. B., Weidenhamer, J. D., and Richardson, D. R. 1994. In search of allelopathy in the Florida scrub: the role of terpenoids. *Journal of Chemical Ecology* 20:1355–1380.

Fitzpatrick, J. W., Woolfenden, G. E., and Kopeny, M. T. 1991. *Ecology and Development Related Habitat Guidelines of the Florida Scrub-Jay (Aphelocoma coerulescens coerulescens).* Tallahassee,Fla.: Florida Nongame Wildlife Program, Technical Report No.8.

Gibson, D. J., and Menges, E. S. 1994. Population structure and spatial pattern in the dioecious shrub *Ceratiola ericoides. Journal of Vegetation Science* 5:337–346.

Greenberg, C. H., Neary, D. G., and Harris, L. D. 1994. Effect of high-intensity wildfire and silvicultural treatments on reptile communities in sand-pine scrub. *Conservation Biology* 8:1047–1057.

Greenberg, C. H., Neary, D. G., Harris, L. D., and Linda, S. P. 1995. Vegetation recovery following high-intensity wildfire and silvicultural treatments in sand pine scrub. *The American Midland Naturalist* 133:149–163.

Guerin, D. N. 1993. Oak dome clonal structure and fire ecology in a Florida longleaf pine dominated community. *Bulletin of the Torrey Botanical Club* 120:107–114.

Hartnett, D. C., and Richardson, D. R. 1989. Population biology of *Bonamia grandiflora* (Convolvulaceae): effects of fire on plant and seed bank dynamics. *American Journal of Botany* 76:361–369.

Hawkes, C. V., and Menges, E. S. 1995. Density and seed production of a Florida endemic, *Polygonella basiramia,* in relation to time since fire and open sand. *The American Midland Naturalist* 133:138–148.

Hawkes, C. V., and Menges, E. S. 1996. The relationship between open space and fire for species in a xeric Florida shrubland. *Bulletin of the Torrey Botanical Club* 123:81–92.

Huck, R. B. 1987. Plant communities along an edaphic continuum in a central Florida watershed. *Florida Scientist* 50:112–128.

Huck, R. B., Johnson, A. F., Parker, A. J., Parker, K. C., Platt, W. J., and Ward, D. B. 1996. Management of natural communities of Choctawhatchee sand pine (*Pinus clausa* (Engelm.) Sarg. var *immuginata* Ward) in the Florida panhandle. *Resource Management Notes* 8:89–91.

Huck, R.B., Judd, W. S., Whitten, W. M., Skean, J. D., Jr., Wunderlin, R. P., and DeLaney, K. R. 1989. A new *Dicerandra* (Labiatae) from the Lake Wales Ridge of Florida, with a cladistic analysis and discussion of endemism. *Systematic Botany* 14:197–213.

Johansen, J. K. 1993. Cryptogamic crusts of semi-arid and arid lands of North America. *Journal of Phycology* 29:140–147.

Johnson, A. F. 1982. Some demographic characteristics of the Florida rosemary *Ceratiola ericoides* Michx. *The American Midland Naturalist* 108:170–174.

Johnson, A. F., and Abrahamson, W. G. 1990. A note on the fire responses of species in rosemary scrubs on the southern Lake Wales Ridge. *Florida Scientist* 53:138–143.

Johnson, A. F., Abrahamson, W. G., and McCrea, K. D. 1986. Comparison of biomass recovery after fire of a seeder (*Ceratiola ericoides*) and a sprouter (*Quercus inopina*) species from south-central Florida. *The American Midland Naturalist* 116:423–428.

Johnson, A. F., and Muller, J. W. 1993. An assessment of Florida's remaining coastal upland natural communities. Florida Natural Areas Inventory, Tallahassee.

Kalisz, P. J., and Stone, E. L. 1984. The longleaf pine islands of the Ocala National Forest, Florida: a soil study. *Ecology* 65:1743–1754.

Lambert, B. B., and Menges, E. S. 1996. The

effects of light, soil disturbance and presence of organic litter on the field germination and survival of the Florida goldenaster, *Chrysopsis floridana* Small. *Florida Scientist* 59:121–137.

Lewis, P. O., and Crawford, D. J. 1995. Pleistocene refugium endemics exhibit greater allozymic diversity than widespread congeners in the genus *Polygonella* (Polygonaceae). *American Journal of Botany* 82:141–149.

Lugo, A. E., and Zucca, C. P. 1983. Comparison of litter fall and turnover in two Florida ecosystems. *Florida Scientist* 46:101–110.

McCormick, K. D., Deyrup, M. A., Menges, E. S., Wallace, S. R., Meinwald, J., and Eisner, T. 1993. Relevance of chemistry to conservation of isolated populations: the case of volatile leaf components of *Dicerandra* mints. *Proceedings of the National Academy of Sciences, USA* 90:7701–7705.

McCoy, E. D., and Mushinsky, H. R. 1992. Rarity of organisms in the sand pine scrub habitat of Florida. *Conservation Biology* 6:537–548.

McCoy, E. D., and Mushinsky, H. R. 1994. Effects of fragmentation on the richness of vertebrates in the Florida scrub habitat. *Ecology* 75:446–457.

McDonald, D. B., and Hamrick, J. L. 1996. Genetic variation in some plants in Florida scrub. *American Journal of Botany* 83:21–27.

Menges, E. S. 1992. Habitat preferences and response to disturbance for *Dicerandra frutescens*, a Lake Wales Ridge (Florida) endemic plant. *Bulletin of the Torrey Botanical Club* 119:308–313.

Menges, E. S. 1994. Fog temporarily increases water potential in Florida scrub oaks. *Florida Scientist* 57:65–74.

Menges, E. S., Abrahamson, W. G., Givens, K. T., Gallo, N. P., and Layne, J. N. 1993. Twenty years of vegetation change in five long-unburned Florida plant communities. *Journal of Vegetation Science* 4:375–386.

Menges, E. S., Dolan, R. W., Gordan, D. R., Evans, M. E. K., and Yahr, R. 1998. *Demography, Ecology, and Preserve Design for Endemic Plants of the Lake Wales Ridge, Florida.* Final report to The Nature Conservancy's Ecosystem Research Program.

Menges, E. S., and Gallo, N. P. 1991. Water relations of scrub oaks on the Lake Wales Ridge, Florida. *Florida Scientist* 54:69–79.

Menges, E. S., and Gordon, D. R. 1996. Three levels of monitoring intensity for rare plant species. *Natural Areas Journal* 16:227–237.

Menges, E. S., and Hawkes, C. V. 1998. Interactive effects of fire and microhabitat on plants of Florida scrub. *Ecological Applications.*

Menges, E. S., and J. Kimmich. 1996. Microhabitat and time-since-fire: effects on demography of *Eryngium cuneifolium* (Apiaceae), a Florida scrub endemic plant. *American Journal of Botany* 83:185–191.

Menges, E. S., and Kohfeldt, N. 1995. Life history strategies of Florida scrub plants in relation to fire. *Bulletin of the Torrey Botanical Club* 122:282–297.

Mohler, P. E. 1992. *Rare and Endangered Biota of Florida.* Vol. III, *Amphibians and Reptiles.* Gainesville,Fla.: University Presses of Florida.

Myers, R. L. 1985. Fire and the dynamic relationship between Florida sandhill and sand pine scrub vegetation. *Bulletin of the Torrey Botanical Club* 112:241–252.

Myers, R. L. 1990. Scrub and high pine. In *Ecosystems of Florida*, ed. R. L. Myers and J. J. Ewel, pp. 150–193. Orlando: University of Central Florida Press.

Myers, R. L., and White, D. L. 1987. Landscape history and changes in sandhill vegetation in north-central and south-central Florida. *Bulletin of the Torrey Botanical Club* 114:21–32.

Ostertag, R., and Menges, E. S. 1994. Patterns of reproductive effort with time since fire in Florida scrub plants. *Journal of Vegetation Science* 5:303–310.

Parker, K. C., and Hamrick, J. L. 1996. Genetic variation in sand pine (*Pinus clausa*). *Canadian Journal of Forest Research* 26:244–254.

Poppleton, J., Clewell, A., and Shuey, A. 1983. Sand pine scrub restoration at a reclaimed phosphate mine in Florida. In *Symposium on Surface Mining, Hydrology, Sedimentology and Reclamation*, ed. D. Graves, pp. 395–398. Lexington, Ky.: OES Publications, College of Engineering, University of Kentucky.

Quintana-Ascencio, P. F., and Menges, E. S. 1996. Inferring metapopulation dynamics from patch-level incidence of Florida scrub plants. *Conservation Biology* 10:1210–1219.

Quintana-Ascencio, P., and Morales-Hernandez, M. 1997. Fire-mediated effects of shrubs, lichens, and herbs on the demography of *Hypericum cumulicola* in patchy Florida scrub. *Oecologia* 112:263–271 .

Root, K. V. 1996. Population viability analysis for the Florida Scrub-Jay (*Aphelocoma coerulescens coerulescens*) in Brevard County, Florida. Ph.D.

dissertation, Florida Institute of Technology, Melbourne, Fla.

Schmalzer, P. A., and Hinkle, C. R. 1990. *Geology, Geohydrology and Soils of Kennedy Space Center: A Review.* Kennedy Space Center, Fla., NASA Technical Memorandum 103813.

Schmalzer, P. A., and Hinkle, C. R. 1992. Recovery of oak–saw palmetto scrub after fire. *Castanea* 57:158–173.

Schmalzer, P. A., and Hinkle, C. R. 1996. Biomass and nutrients in aboveground vegetation and soils of Florida oak-saw palmetto scrub. *Castanea* 61:168–193.

Stith, B. M., Fitzpatrick, J. W., Woolfenden, G. E., and Pranty, B. 1996. Classification and conservation of metapopulations: a case study of the Florida Scrub-Jay. In *Metapopulations and Wildlife Conservation*, ed. D. McCollough, pp. 187–215. Washington D.C.: Island Press.

Stout, I. J., and Marion, W. R. 1993. Pine flatwoods and xeric pine forests of the southern (lower) coastal plain. In *Biodiversity of the Southeastern United States. Lowland Terrestrial Communities*, ed. W. H. Martin, S. G. Boyce, and A. C. Echternacht, pp. 373–446. New York: John Wiley and Sons.

Swain, H., Hinkle, C. R., and Schmalzer, P. A. 1993. Stewardship at the local level: a case study from Brevard County, Florida. In *Partners in Stewardship: Proceedings of the 7th Conference on Research and Resource Management in Parks and on Public Lands*, ed. W. E. Brown and S. D. Veirs, Jr., pp. 452–462. Hancock, Mich.: George Wright Society.

Swain, H. M., Schmalzer, P. A., Breininger, D. R., Root, K. V., Bergen, S. A., Boyle, S. R., and. MacCaffree, S. 1995. Scrub conservation and development plan, Brevard County, Appendix B, Biological Consultant's Report. Florida Institute of Technology, Melbourne.

U.S. Fish and Wildlife Service. 1993. *Final Land Protection Plan. Proposed Establishment of Lake Wales Ridge National Wildlife Refuge.* Southeast Regional Office, Atlanta, Ga.

Watts, W. A., and Hansen, B. C. S. 1994. Pre-Holocene and Holocene pollen records of vegetation history from the Florida peninsula and their climatic implications. *Palaeogeography, Palaeoclimatology, and Palaeoecology* 109:163–176.

Wharton, C. H. 1978. *The Natural Environments of Georgia.* Atlanta: Georgia Department of Natural Resources.

Woolfenden, G. E., and Fitzpatrick, J. W. 1984. The Florida Scrub-Jay: demography of a cooperative-breeding bird. *Monographs in Population Biology* No. 20. Princeton, N.J.: Princeton University Press.

Wunderlin, R. 1982. *Guide to the Vascular Plants of Central Florida.* Tampa, Fla.: University Presses of Florida.

2 Southeastern Pine Savannas

WILLIAM J. PLATT

Introduction

When Ponce de Leon first landed in Florida, overstory pines and a ground cover containing grasses, forbs, and shrubs were widespread over southeastern North America. The earliest descriptions were strikingly consistent. Journals of early explorers (e.g., Alvar Nunez Cabeza de Vaca and Hernando de Soto in the 1500s), travelers (e.g., John Latrobe, Mark Catesby, John Williams), and naturalists (e.g., John and William Bartram, Samuel Lockett, Andre Michaux, Thomas Nuttall in the 1700s and early 1800s) all depicted landscapes as open, with a low herbaceous ground cover and visibility for more than a kilometer through pines that most often did not form a complete overstory (Williams 1827, 1837; Small 1921a, b; Harper 1948; Tebo 1985; Frost 1993; Harcombe et al. 1993; Means 1996). More recent descriptions by field biologists reinforced the concept of open landscapes dominated by pines and herbaceous ground-cover plants as the prominent upland landscape in the southeastern United States (e.g., Nash 1895; Schwarz 1907; Harper 1911, 1914, 1927, 1943; Harshberger 1914; Wells 1928; Wells and Shunk 1931). These descriptions still apply to the few old-growth stands present today (e.g., Platt, Evans and Rathbun 1988; Doren, Platt and Whiteaker 1993; Gilliam, Yurish and Goodwin 1993; Noel, Platt and Moser 1998).

These fragments of natural history all suggest that southeastern landscapes were predominately savannas. The savanna physiognomy, clumps of trees or sparsely distributed trees not forming continuous canopies, and ground covers dominated by warm-season grasses, is maintained by seasonal climates, herbivory and frequent fire (Werner 1991; McPherson 1997; Scholes and Archer 1997). Southeastern savannas (hereafter collectively called pine savannas, although cypress, palms, and oaks were sometimes present) occurred within a >2,000,000-km^2 region, from 25° to 37° N latitude and 75° to 95° W longitude, extending 3,000 km westward from southeastern Virginia to eastern Texas and 1,000 km southward from northern Georgia and Alabama to the Florida keys (Figure 2.1; Frost 1993; Peet and Allard 1993; Stout and Marion 1993). Thus, pine savannas occurred from mountain slopes in Alabama and Georgia and dissected hilltops of Louisiana and Texas to seasonally flooded flatwoods along Atlantic and Gulf coasts and limestone outcroppings in the everglades and keys of south Florida (Stout and Marion 1993; Ware, Frost and Doerr 1993).

Southeastern pine savannas began to change in character and disappear from the landscape following European and African settlement of the region (Croker 1987). These communities were utilized for open-range grazing and timber products (i.e., naval stores) to varying degrees prior to the advent of railroads (Frost 1993). Over a little more than a half-century following the Civil War, virtually all large pine timber was harvested from the Coastal Plain

Roger C. Anderson, James S. Fralish, and Jerry M. Baskin, eds. *Savannas, Barrens, and Rock Outcrop Plant Communities of North America.* Copyright © 1999 Cambridge University Press. All rights reserved.

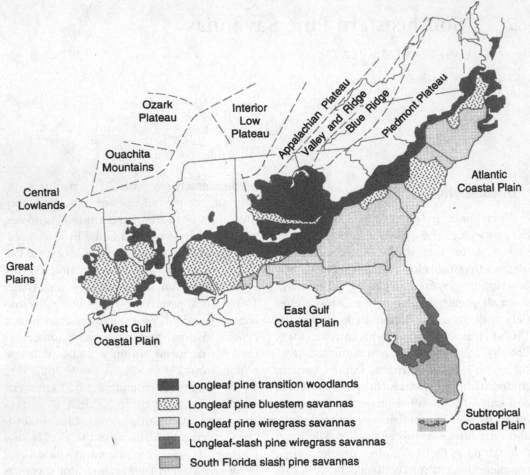

Figure 2.1. The range and distribution of southeastern pine savannas in the southeastern United States. Pine savannas constitute the dominant upland vegetation in most of the Atlantic, East and West Gulf, and Subtropical Regions of the Coastal Plain Physiographic Province and extend northward into the Appalachian, Valley and Ridge, and Piedmont Plateau Physiographic Provinces. The five general divisions of southeastern pine savannas include longleaf pine transition woodlands, primarily along northern and western boundaries; longleaf pine bluestem savannas ·in disjunct regions of the eastern Coastal Plain and through-out the western Coastal Plain; longleaf pine wiregrass savannas in disjunct regions along the Atlantic Coast and in part of the east Gulf Coastal Plain, extending south into central Florida; longleaf–slash pine wiregrass savannas in central Florida; and south Florida slash pine savannas in subtropical southern Florida and the southernmost Florida keys. These ranges, modified from Frost (1993) and Craighead (1971) are intended not to reflect local distributions of pine savannas, but to indicate the types of pine savannas that might be expected in a general region of the southeastern United States.

(Croker 1987; Frost 1993; Ware, Frost and Doerr 1993); rates of deforestation were more rapid than those currently occurring in the tropics (Noss 1989; Simberloff 1993). Fire suppression also became widespread early in the 20th century, resulting in further degradation of remaining pine savannas (Vogl 1973; Komarek 1974; Peet and Allard 1993).

At present, virtually all southeastern savannas have disappeared (Ware, Frost and Doerr 1993; Noss, LaRoe and Scott 1995; Means 1996). About 4% of the original area containing savannas was present in 1985 (Kelly and Bechtold 1990); a decade later, 1,200,000 ha (3%) remained. Of that area, about half was in public lands and, at most, 3,000 ha were old-growth pine savannas (Outcalt and Outcalt 1994; Means 1996). Second-growth pine stands, which do not resemble stands in old-growth pine savannas, have come to be regarded as forests, not savannas (Noel, Platt and Moser 1998). Most are fragmented and highly degraded, often little more than pine plantations with a depauperate ground cover and none of the characteristic fauna (Means and Grow 1985; Noss 1989; Means 1996).

Disappearance of old-growth savannas and extensive alterations of second-growth savannas following fire suppression have resulted in ecological concepts of these communities as successional to hardwood forests. Results of studies conducted over the past two decades in remaining intact old- and second-growth pine savannas do not support this concept. Instead, they reinforce and extend concepts first expressed by Chapman (1932a, b) of pine savannas as persistent communities. Frequent fires ignited by lightning and spread by pyrogenic vegetation, especially during annual transitions from dry to wet seasons, coupled with low-nutrient soils, appear to be responsible for the persistence of southeastern savannas. Recent studies also indicate that (1) unique southeastern savanna environments have a long evolutionary history, resulting in many endemic and specialized species of plants and animals, and (2) pine savannas are species-rich habitats varying regionally and locally in plant species composition in ways that rival tropical forests (on very different spatial scales).

Physical Environment

Climate of the Southeastern Coastal Plain

Temperatures. The southeastern United States is characterized by long growing seasons (W. H. Martin and Boyce 1993). The transition from subtropical to warm temperate climates along the latitudinal gradient in the southeast involves decreases in mean annual temperatures and lengths of growing seasons and increases in seasonalities of temperatures (Greller 1980, 1989; Chen and Gerber 1990; Winsberg 1990). Although mean July temperatures are 25–27 °C, mean January temperatures decrease from 16 °C in southern Florida to 10–12 °C in northern Florida and about 5 °C in North Carolina (Stout and Marion 1993). Freezing temperatures occur more often northward along the latitudinal gradient; average frost-free days decrease from 365 in southern Florida to about 300 along the Gulf coast and around 250 in North Carolina (Chen and Gerber 1990; Stout and Marion 1993).

Precipitation. Annual precipitation in the southeast is distinctive in three ways: total annual amount, seasonality, and number of rain-free days. Annual precipitation is the highest in eastern North America; 1,000–1,600 mm rainfall typically occurs annually throughout most of the southeast. This annual precipitation declines northward from the Gulf coast and westward through Louisiana and Texas (Chen and Gerber 1990; Winsberg 1990; W. H. Martin and Boyce 1993).

Rainfall is distinctively seasonal. Throughout the southeastern coastal plain, peak rainfall, typically between 150 and 300 mm/month, occurs during the middle of the growing season (June–September). In contrast, minimum rainfall, usually less than 150 mm and often only 30–60 mm/month, occurs during the early part of the dormant season (November–January). A bimodal peak in rainfall occurs along the Gulf coast, with a lesser rainfall peak in January–March. Southward along the latitudinal gradient in Florida, there is a transition from bimodal rainfall to single wet (June–October) and dry (December–April) seasons, with >80% of the annual rainfall during the wet season (Chen and Gerber 1990). Annual droughts thus are a characteristic of the climatic regime, especially in central and southern Florida. Longer-term droughts of one to two decades of below average rainfall have also have been documented (e.g., Stahle, Cleaveland and Hehr 1988).

The southeast also is characterized by the largest number of rain-free days, often 70–110/year, in eastern North America (Jordan 1984; Stout and Marion 1993; Ware, Frost and Doerr 1993). The intervals between successive rainfalls change seasonally (Olson and Platt 1995). Along the Gulf coast, minimal intervals between successive rainfalls are less than a week during winter and midsummer, but increase to about two weeks during late spring and fall. In south Florida, intervals between successive rainfalls are short during the summer wet season, but often extend several weeks during the dry season (Snyder 1986, 1991; Snyder, Herndon and Robertson 1990; Gunderson and Snyder 1994).

Large-scale disturbances. Violent windstorms (hurricanes and tornadoes) occur frequently in the southeast (Simpson and Lawrence 1971; Grazulis 1984; Neumann et al. 1993). Return times of tornadoes, which affect small areas, are on the magnitude of many centuries to millennia (Glitzenstein and Harcombe 1988). In contrast, hurricanes, which affect large areas, have return times varying from one to two decades in Florida and along the Gulf Coastal Plain (Armentano et al. 1995; Batista and Platt 1997). The large amounts of rainfall that accompany hurricanes often affect local hydrology. High winds also cause mortality of trees, enhancing the open nature of pine savannas in coastal regions (Platt, Evans and Rathbun 1988; Platt and Rathbun 1993; Armentano et al. 1995). Still, southeastern savanna conifers appear more wind resistant than hardwoods or forest pines (Bromley 1939; Gresham, Williams and Lipscomb 1991; Platt and Rathbun 1993; Armentano et al. 1995; Noel et al. 1995).

Geology, Soils, and Hydrology

Geological history. Almost all southeastern pine savannas occur within the Coastal Plain (Figure 2.1), one of 22 broad physiographic regions of the continental United States (Brouillet and Whetstone 1993). The Coastal Plain is underlain by limestone dating from the late Cretaceous to early Cenozoic (Thornbury 1965; W. H. Martin and Boyce 1993). Early Tertiary orogeny, while the North American plate moved from tropical to temperate latitudes, produced uplifting that resulted in deposition of clastic sediments over the limestone base of the Atlantic and Gulf coastal shelves (Delcourt et al. 1993). Thus, most pine savannas occur on soils comprised almost entirely of sands, loams, or clays of marine origin, although sometimes overlain by eolian or fluvial deposits (Stout and Marion 1993; Ware, Frost and Doerr 1993). The major exceptions occur in the everglades and lower keys regions of south Florida, where pine savannas occur on limestone outcroppings, and in northern Alabama and Georgia, where pine savannas occur on rocky outcrops and mountain slopes.

Landscapes. Gentle topographic gradients (generally <10%) are the most prominent local feature of Coastal Plain landscapes. Local topographic differences may be only a few centimeters, are rarely more than a few meters, and at most are 15–35 m (Brown, Stone and Carlisle 1990; Stout and Marion 1993; Ware, Frost and Doerr 1993). Still, subtle physiognomic differences occur between inner and outer Coastal Plain landscapes. Inner Coastal Plain sediments along Atlantic and Gulf coasts have remained above sea level since the Miocene (H. J. Walker and Coleman 1987); the rolling topography has resulted from weathering and gradual erosion of marine and fluvial deposits (H. J. Walker and Coleman 1987; Harcombe et al. 1993; Peet and Allard 1993). Greater drainage in the inner Coastal Plain results in nonriverine hydric sites being uncommon; where present, they are typically associated with local, impermeable, often indurated soil horizons (Bridges and Orzell 1989; Harcombe et al. 1993; Peet and Allard 1993).

The shoreline of the outer Coastal Plain has varied markedly since the Miocene. Along most coastlines, sediments have been periodically exposed and submerged, resulting in extensive reworking of soils (Walker and Coleman 1987; Stout and Marion 1993). Thus, outer Coastal Plain soils are often unconsolidated sands; subtle substrate differences in a flat landscape result from Pleistocene-aged dunes and bars of coarse sands interwoven with old embayments containing finer sands and clays (Brown, Stone and Carlisle 1990; W. H. Martin and Boyce 1993).

Soils. Four principal soil orders (based on classifications in *Soil Taxonomy*, Soil Survey Staff 1975) are associated with southeastern pine savannas (Brown, Stone and Carlisle 1990; W. H. Martin and Boyce 1993). All four have very narrow A horizons of mineral topsoil containing little organic material; three have eluviated subsurface horizons altered by downward leaching and movement of minerals and soil particles over many years of high rainfall. *Ultisols* and *Alfisols* are characterized by argillic horizons, subsurface zones formed by downward translocation of clay-size particles from surface horizons. Ultisols, highly leached soils, are predominant in the inner southeastern Coastal Plain; these soils have <35% base saturation (the proportion of chemical exchange sites on soil particles occupied by basic cations rather than acidic cations). Alfisols, most often present in the western portion of the eastern Gulf Coastal Plain and in the west Gulf Coastal Plain, have a base saturation >35% and may have fragipans, cemented silt layers resulting in perched water tables and slowly drained soils. *Spodosols*, which occur primarily in Florida, are characterized by an illuvial sub-

surface spodic horizon containing mixtures of metals (iron, aluminum) and organic matter derived from leaching of acidic compounds in decomposing vegetation when water levels in soil are high. Frequently, an argillic horizon occurs beneath the spodic horizon. *Entisols*, which are characterized by a thin A horizon over unconsolidated sand or bedrock (usually limestone) characterize recent soils where there has been insufficient time for weathering to occur (i.e., along Gulf and Atlantic coasts).

Hydrology. The hydrology of southeastern pine savannas is determined by interactions between landforms, soils, and climate. Upland soils often are very porous and drain quickly following rains, producing xeric habitats. Where argillic horizons are well developed or where multiple layers of varied texture occur in the soil profile (such as coarse sands above finer soils), subsurface soil layers may retain moisture, producing mesic habitats. If sands overlie some impermeable layer, then water may flow laterally until it emerges at the surface along the slope in a hydric seepage zone (Means and Mohler 1978; Means 1996). Where argillic horizons or fragipans are near the surface, surface soils may become saturated during wet seasons, especially in flat landscapes (Brown, Stone and Carlisle 1990). Such saturation commonly occurs along east and west Gulf coasts where a wet season coincides with the dormant season, when transpiration is reduced. Because these same soils commonly experience water deficits during seasonally dry springs and falls, plants in these habitats are adapted for both annual inundation and drought. High sodium levels in surface soils add an additional stress in some areas of the West Gulf Coastal Plain.

Nutrient availability. Weathering of Coastal Plain soils has occurred over many years of seasonally high rainfall in porous soils. High rates of leaching and cation exchange have resulted in acidic soils and low supplies of most mineral nutrients. Thus, neither water nor nutrients may be readily available for plants, especially in deep sands (J. Walker and Peet 1983; Brown, Stone and Carlisle 1990). Reduced availability of nutrients (particularly nitrogen and phosphorous) influences growth rates and physiognomy of the vegetation (J. Walker and Peet 1983; Brown, Stone and Carlisle 1990; Christensen 1993). Nitrogen-fixing plants (e.g., legumes, *Myrica cerifera*) and ericaceous shrubs (e.g., *Vaccinium, Gaylussacia*) are common in the ground cover.

The most notable low-nutrient, acidic habitats in both east and west Gulf Coastal Plains are seepage savannas or bogs located where clay layers, argillic horizons, fragipans, or rocks are close to the surface (Plummer 1963; Eleuterius 1968; Means and Mohler 1978). These extremely open savanna habitats, which contain few pines or woody shrubs (most likely a result of frequent growing season fires and standing water; Christensen 1981, 1985; Olson and Platt 1995), are characterized by hundreds of herbaceous species, including several species of different genera of carnivorous plants, such as pitcher plants (*Sarracenia*), sundews (*Drosera*), bladderworts (*Utricularia*), and butterworts (*Pinguicula*) (Folkerts 1982).

Where limestone is close to the surface and no argillic horizons are present, soils are less acidic and more fertile (Ware, Frost and Doerr 1993). Palms (especially *Sabal palmetto*, the cabbage palm), often indicate calcium- and phosphate-rich soils, such as in central Florida dry prairies and savannas (Brown, Stone and Carlisle 1990). In general, more hardwood species occur on calcium-rich than noncalcareous soils (Myers 1990; Platt and Schwartz 1990). High-diversity pine savannas also occur on limestone outcroppings in Everglades National Park (Snyder 1986; Snyder, Herndon and Robertson 1990; W. J. Platt, unpublished data).

Biota

Geological History of Savanna Species

Pine savannas and associated species have been present in the southeast for some time. Palynological records, fossil records of vertebrate species, and the biogeography of several taxonomic groups are consistent with the hypothesis that southeastern pine savannas have been present, at least in Florida, since the late Tertiary or early Quaternary. Thus, pine-dominated savannas predated humans in the southeastern Coastal Plain.

Pollen records from lake sediments indicate that pine- and oak-dominated savannas were present in the southeast during the Wisconsin glaciation (Delcourt and Delcourt 1987). Pines and oaks have alternated in dominance over the past 20,000 years at several Coastal Plain sites (Platt and Schwartz 1990) and over the past 40,000 years at Camel Lake in northern Florida (Watts 1992); the most recent ascendancy by pines to dominance began about 8,000 years ago. However, species of pines and oaks can not be determined from pollen.

The presence of savannas in the southeast can be inferred from the ages of fossils of some associated vertebrate species. Several sinkhole fossil assemblages dating to the early Pleistocene are consistent with the presence of xeric pine savannas. Species of lizards from an Irvingtonian site (Inglis IA) in Citrus County, Florida, dating about 1.6–2.0 million years ago, suggest xeric pine savanna interspersed with xeric forest, assuming similar habitat preferences of the species (Meylan 1982). Fossils of gopher tortoises and rattlesnakes from a later Irvingtonian site (Leisey Shell Pit, Hillsborough County, Florida) also suggest xeric pine savanna (Meylan 1995). In addition, fossils of birds characteristic of pine savannas (red-cockaded woodpecker, brown-headed nuthatch) occur in Florida Pleistocene deposits that date from >10,000 years ago (Woolfenden 1959; Means 1996). Even older savanna habitats that may date to 20 million years ago are suggested by faunal assemblages from the karst region of central Florida (Webb 1990).

Biogeographic evidence for the antiquity of southeastern pine savannas comes from association of many plant and animal species with taxa whose center of distribution is the southwestern United States and northern Mexico. Such affinities occur among many plant species, including dominant pines (Mirov 1967) and grasses, amphibians and reptiles from xeric and mesic habitats (Guyer and Bailey 1993), and xeric pine savanna arthropods (arachnids and a variety of insects) (Folkerts, Deyrup and Sisson 1993). These patterns suggest migrations from the southwest into the southeastern Coastal Plain, perhaps during the late Tertiary or early Quaternary (P. S. Martin and Harrell 1957; Webb and Wilkins 1984; Webb 1990). The considerable endemism among plants and animals further suggests that savannas have occurred somewhere in the southeastern Coastal Plain for even longer periods of time (Snyder, Herndon and Robertson 1990; Webb 1990; Folkerts, Deyrup and Sisson 1993; Guyer and Bailey 1993). Moreover, many endemic southeastern species could have a long evolutionary history associated with a general savanna type of habitat, but not necessarily the specific communities currently present in the landscape (Platt and Schwartz 1990; Folkerts, Deyrup and Sisson 1993).

Regional Distribution and Composition of Vegetation

In presettlement times, pine savannas were widespread over two-thirds to more than three-fourths of the land surface area in the southeastern Coastal Plain (Myers 1990; Frost 1993; Schwartz 1994). Longleaf pine (*Pinus palustris*) was the dominant tree (Figure 2.1) over most of the Coastal Plain, often forming single species overstories (Wahlenberg 1946,

Myers 1990; Ware, Frost and Doerr 1993; Frost 1993; Stout and Marion 1993). One of the best remaining examples is the Wade Tract in southern Georgia (Figure 2.2, upper). The extent to which longleaf pine or a mixture of species dominated sites is uncertain for many areas. Along the northern and western edges of the range of longleaf pine (Figure 2.1), shortleaf (*Pinus echinata*) and loblolly pines (*Pinus taeda*) were important on ridges and slopes (Lockett 1874; Mohr 1897; Brown 1944; Marks and Harcombe 1981; D. L. Martin and Smith 1991; Frost 1993; Harcombe et al. 1993). North Florida slash pine (*Pinus elliottii* var. *elliottii*) and pond pine (*Pinus serotina*) were important in low, wet savannas of the outer eastern Coastal Plain (Peet and Allard 1993; Stout and Marion 1993). In the subtropical Coastal Plain, pine savannas (Figure 2.1) often contained a mixture of longleaf and south Florida slash pine (*P. elliottii* var. *densa*) (Stout and Marion 1993; Snyder, Herndon and Robertson 1990). From Lake Okeechobee southward, south Florida slash pine–dominated savannas (Figure 2.1) were intermixed with prairies or seasonally inundated savannas dominated by palms (*Sabal palmetto*) or pond cypress (*Taxodium ascendens*) (Robertson 1953; Craighead 1971; Abrahamson and Hartnett 1990; Snyder, Herndon and Robertson 1990; Doren, Platt and Whiteaker 1993). One of the best remaining examples is Lostman's Pines in Big Cypress National Preserve (Figure 2.2, lower).

Hardwood species occurred in stands dominated by savanna pines. Despite some regional differences in species composition, large numbers of woody species occurred in southeastern pine savannas. Peet and Allard (1993) recorded 61 woody plant species present in at least 5% or more of the 0.1-ha plots they studied in 216 stands in the east Gulf Coastal Plain (Table 2.1). Moreover, more than one-fourth of these species were present in more than 50% of the plots, and more than one-half were present in at least 30% of

Figure 2.2. Upper: Old-growth longleaf pine wiregrass savanna (Wade Tract) in Thomas County, Georgia. Crowns of overstory pines of variable age and size are above 30% of the ground cover, which is dominated by *Aristida beyrichiana, Schizachyrium scoparium, Andropogon gerardi,* and *Sorghastrum secundum.* Photograph was taken by W. J. Platt 5 months after a prescribed fire in May, 1991. Lower: Old-growth south Florida slash pine savanna (Lostman's Pines) in Big Cypress National Preserve, Monroe County, Florida. Crowns of overstory pines of variable age and size are above 15% of the ground cover, which is dominated by *Sabal palmetto, Cladium jamaicense, Muhlenbergia capillaris,* and *Schizachyrium rhizomatum.* Photo was taken by W. J. Platt in January, 1994, about 18 months after the eyewalls of Hurricane Andrew crossed the site; no fires had occurred within 3 years.

the plots. Other than pines, the most abundant woody species were "scrub oaks," often small in stature and with many epicormic sprouts as a result of fire damage (Table 2.1). Turkey (*Quercus laevis*), sand-post (*Q. margaretta*), and bluejack (*Q. incana*) oaks were prominent on east Gulf Coastal Plain sandy soils (Rebertus, Williamson and Moser 1989a, b; Peet and Allard 1993); blackjack (*Q. marilandica*), post (*Q. stellata*), southern red (*Q. falcata*), and water (*Q. nigra*) oaks occurred on more mesic sites with more silt or clay in the soil (Gilliam, Yurish and Goodwin 1993; Peet and Allard 1993). Other tree species (e.g., *Carya alba, Diospyros virginiana, Nyssa sylvatica,* and *Prunus serotina*) were more abundant on mesic than on either hydric or xeric/subxeric sites (Table 2.1). Also, woody shrubs differed in relative abundances at different elevations (Table 2.1). Similar patterns of distribution of woody species occurred in the west Gulf Coastal Plain (Marks and Harcombe 1981; Harcombe et al. 1993).

Subtropical pan-Caribbean hardwoods occurred in south Florida slash pine savannas on limestone outcroppings (Phillips 1940; Alexander 1967; Tomlinson 1980). Over 60 woody species were recorded from pine savannas of Everglades National Park (Table 2.2). Most of these species were tropical; only 5 of these species also were abundant in longleaf pine savannas (Table 2.1). Some of these species (e.g., *Trema micrantha, Lysiloma latisiliquum, Cassia deeringiana*) have dormant seeds that germinate following disturbances (Tomlinson 1980). A few subtropical hardwoods are resistant to fires (e.g., *Bursera simaruba*, which has layered bark, and palms, such as *Sabal palmetto* and *Serenoa repens*). Most subtropical hardwoods in south Florida resemble temperate hardwoods in that they resprout following fires (Robertson 1953; Alexander 1967).

The ground cover of southeastern pine savannas, based on recent quantitative descriptions that are consistent with older general observations, contained a diverse mixture of grasses, forbs, and shrubs (Christensen 1988; Bridges and Orzell 1989; Harcombe et al. 1993; Peet and Allard 1993). Comparisons of floras in longleaf pine savannas indicate extensive regional variation, both between east and west Gulf Coastal Plains, within these regions, and locally. Extensive species lists and relative frequencies of ground-cover and overstory species in different community types are provided for east and west Gulf Coastal Plains by Peet and Allard (1993) and Bridges and Orzell (1989), respectively. In all sites, warm season grasses (including *Andropogon, Aristida, Ctenium, Schizachyrium, Sorghastrum, Sporobolus*) dominated the ground cover. Two species of wiregrass (*Aristida stricta, A. beyrichiana*; Peet 1993) were dominant in large disjunct areas of the eastern part of the Coastal Plain (Figure 2.1). *Aristida stricta* was dominant along the upper Atlantic Coastal Plain in North Carolina and northern South Carolina; *A. beyrichiana* was dominant in large areas extending from southern South Carolina westward to coastal Alabama and Mississippi and southward to southern Florida (R. K. Peet, unpublished data). An example of an old-growth longleaf pine savanna with wiregrass dominant in the ground cover is the Wade Tract (Figure 2.2, upper). Little bluestem (*Schizachyrium scoparium*) was most commonly the dominant upland grass in more western pine savannas and in some large disjunct areas in the eastern pine savannas (Figure 2.1) (Harcombe et al. 1993; Peet and Allard 1993). In south Florida, wiregrass was the dominant grass on sandy soils, and firegrass (*Andropogon cabanisii*) was dominant on limestone outcroppings (Craighead 1971; Snyder, Herndon and Robertson 1990). Other warm-season grasses (e.g., *Digitaria, Muhlenbergia, Paspalum, Schizachyrium, Sorghastrum*) and sedges (e.g., *Cladium jamaicense*) also were prominent on rockridge limestone outcroppings (W. J. Platt and J. K. DeCoster, unpublished data). An example of

Table 2. 1. *Average frequencies of occurrence of 60 species of trees and woody shrubs in xeric, mesic, and hydric longleaf pine savannas of the eastern Coastal Plain, based on data in Peet and Allard (1993). Frequencies, expressed as a percent, were obtained as means of values recorded within 0.1-ha plots in xeric and subxeric (n = 9), mesic (n = 3), and hydric (n = 5) types of communities. Data were collected from 216 sample sites in 44 counties from North Carolina to eastern Louisiana and from Northern Alabama to central Florida (Peet and Allard 1993).*

Species	Xeric & subxeric sites	Mesic sites	Hydric & seepage bog sites
Acer rubrum	3. 3	20. 6	37. 3
Arundinaria gigantea	3. 8	2. 4	30. 7
Callicarpa americana	10. 0	22. 2	0. 0
Carya alba	6. 7	45. 2	0. 0
Carya pallida	8. 1	16. 7	1. 3
Clethra alnifolia	7. 2	26. 2	28. 9
Cornus florida	22. 1	31. 8	1. 3
Crataegus uniflora	10. 7	7. 1	0. 0
Cyrilla racemiflora	0. 0	2. 4	14. 2
Diospyros virginiana	64. 4	97. 6	16. 1
Gaylussacia dumosa	54. 4	88. 6	41. 3
Gaylussacia frondosa	20. 5	45. 2	50. 3
Ilex coriacea	0. 0	15. 9	11. 3
Ilex glabra	21. 0	23. 0	81. 6
Ilex opaca	17. 1	13. 5	6. 5
Ilex vomitoria	28. 6	22. 2	0. 0
Kalmia latifolia	2. 8	33. 3	0. 0
Licania michauxii	19. 5	0. 0	0. 0
Liquidambar styraciflua	5. 0	18. 2	12. 8
Lyonia ligustrina	0. 0	7. 1	19. 6
Lyonia mariana	16. 7	52. 4	31. 0
Magnolia virginiana	8. 9	9. 5	27. 5
Myrica cerifera	37. 6	50. 0	44. 3
Myrica heterophylla	0. 0	4. 8	38. 4
Nyssa biflora	1. 1	0. 0	13. 0
Nyssa sylvatica	10. 6	83. 3	11. 5
Osmanthus americanus	16. 1	11. 1	0. 0
Oxydendrum arboreum	3. 9	19. 0	2. 5
Persea borbonia	12. 8	13. 5	25. 4
Pinus echinata	6. 1	11. 1	1. 3
Pinus palustris	100. 0	100. 0	100. 0
Pinus serotina	1. 1	7. 1	36. 7
Pinus taeda	27. 4	20. 6	9. 7
Prunus serotina	22. 1	64. 3	3. 8
Quercus falcata	33. 3	35. 7	1. 1
Quercus geminata	18. 3	0. 0	0. 6
Quercus hemisphaerica	34. 3	13. 5	2. 4
Quercus incana	76. 9	41. 3	6. 6
Quercus laevis	80. 6	56. 3	3. 8
Quercus margaretta	57. 0	56. 3	3. 8
Quercus marilandica	20. 0	100. 0	17. 3
Quercus nigra	4. 4	38. 1	5. 3

Species	Xeric & subxeric sites	Mesic sites	Hydric & seepage bog sites
Quercus pumila	8. 9	0. 0	1. 7
Quercus stellata	9. 4	31. 8	1. 7
Rhododendron atlanticum	0. 6	4. 8	6. 8
Rhus copallinum	39. 5	81. 0	23. 6
Sassafras albidum	44. 1	66. 7	26. 2
Serenoa repens	18. 1	0. 0	0. 0
Symplocos tinctoria	15. 0	2. 4	8. 2
Toxicodendron pubescens	31. 4	61. 9	6. 3
Toxicodendron vernix	5. 6	0. 0	9. 5
Vaccinium arboreum	47. 8	40. 5	0. 0
Vaccinium crassifolium	10. 6	42. 9	43. 3
Vaccinium darrowii	2. 2	22. 3	0. 0
Vaccinium elliottii	26. 2	35. 7	0. 0
Vaccinium formosum	2. 8	7. 1	20. 4
Vaccinium fuscatum	31. 0	4. 8	17. 8
Vaccinium myrsinites	6. 1	0. 0	0. 0
Vaccinium stamineum	42. 7	33. 3	0. 6
Vaccinium tenellum	27. 7	66. 7	6. 1

an old-growth south Florida slash pine savanna with *Muhlenbergia capillaris, Schizachyrium rhizomatum,* and *C. jamaicense* in the ground cover is Lostman's Pines (Figure 2.2, lower).

Shrubs were prominent in the ground cover of pine savannas. Common temperate shrubs in longleaf pine savannas (Table 2.1) included sprouts of trees, especially oaks, as well as runner oaks (*Q. minima, Q. pumila*), sumac (*Rhus*), ericaceous shrubs (*Vaccinium, Gaylussacia*), palms (*Serenoa, Sabal*), wax myrtle (*Myrica*), and hollies (*Ilex*), among others (Peet and Allard 1993; Olson and Platt 1995). In south Florida (Table 2.2), pan-Caribbean subtropical palms (e.g., *Serenoa, Sabal, Thrinax, Coccothrinax*), woody shrubs (*Guettarda, Dodonaea, Morinda, Byrsonima, Myrsine, Ardisia, Acacia*) and suffructescent shrubs (*Lantana, Croton, Solanum*) were common (Snyder 1986; Snyder, Herndon and Robertson 1990).

Hundreds of forb species occurred in pine savannas (Christensen 1988; Platt, Evans and Davis 1988, Hardin and White 1989, J. Walker 1993). Dominant families included Asteraceae, Lamiaceae, Fabaceae, Liliaceae, Scrophulariaceae, Ericaceae, Orchidaceae, and Xyridaceae (Christensen 1988; J. Walker 1993). Sizable proportions of the flora were endemic to one of four regions identified by Walker (1993) as distinctive (with different endemics and rare species) within the overall range of longleaf pine: south Atlantic Coastal Plain, peninsular Florida, and the east and west Gulf Coastal Plains. About 200 of these species were noted as rare (J. Walker 1993). Approximately 15% of the south Florida pine savanna vascular plant species were endemic to that region; about 50 species were rare (Snyder, Herndon and Robertson 1990).

Local Distribution and Composition of Vegetation

Pine savanna plant species appear sensitive to small changes in elevation. The controlling physical factors have been the subject of debate, as subtle effects of combinations of soil characteristics, hydrology, nutrient availability, and fires appear over small changes in elevation, often as little as 10 cm (Marks and Harcombe 1981; Platt and Schwartz 1990;

Table 2.2. *Average frequencies of 74 species of suffructescent and woody shrubs and trees in high and low south Florida slash pine savannas on limestone outcroppings in Everglades National Park, based on unpublished data (W. J. Platt and J. K. DeCoster). Frequencies, expressed as percent, are means of values recorded within 0.1-ha plots at high (n = 15) and low (n = 9) elevations within different regions of Long Pine Key, Everglades National Park, Dade County, Florida.*

Species	High elevation (2–4 m above msl	Low elevation (1–2 m above msl)
Acacia pinetorum	13.3	77.8
Angadenia sagraei	100.0	88.9
Annona glabra	13.3	11.1
Ardisia escallonioides	100.0	100.0
Argythamnia blodgettii	13.3	0.0
Ayenia euphrasifolia	40.0	22.2
Baccharis halmifolia	80.0	100.0
Bumelia reclinata	53.3	100.0
Bumelia salicifolia	100.0	100.0
Bursera simaruba	0.0	11.1
Byrsonima lucida	100.0	88.9
Callicarpa americana	40.0	22.2
Cassia deeringiana	93.3	33.3
Chiococca parvifolia	100.0	77.8
Chrysobalanus icaco	40.0	55.6
Chrysophyllum oliviforme	53.3	44.4
Citharexylum fruiticosum	26.7	11.1
Coccoloba diversifolia	13.3	11.1
Coccothrinax argentata	26.7	0.0
Colubrina arborescens	6.7	0.0
Conocarpus erectus	13.3	66.7
Corchorus siliquosus	0.0	11.1
Crossopetalum ilicifolium	80.0	11.1
Crossopetalum rhacoma	13.3	0.0
Crotolaria pumila	93.3	33.3
Croton linearis	80.0	77.8
Dodonea viscosa	100.0	88.9
Eugenia axillaris	66.7	77.8
Eugenia foetida	6.7	0.0
Eupatorium coelestinum	53.3	44.4
Eupatorium mikanoides	6.7	21.4
Exothea paniculata	13.3	22.2
Ficus aurea	26.7	22.2
Ficus citrifolia	86.7	77.8
Forestiera segregata	33.3	22.2
Guapira discolor	80.0	66.7
Guetarda elliptica	100.0	77.8
Guetarda scabra	86.7	88.9
Hypericum brachyphyllum	0.0	11.1
Hypericum hypericoides	13.3	66.7
Ilex cassine	73.3	77.8
Ilex krugiana	26.7	0.0
Iva microcephala	0.0	11.1

Species	High elevation (2–4 m above msl)	Low elevation (1–2 m above msl)
Jacquemontia curtisii	80.0	100.0
Janquinia keyensis	6.7	0.0
Kosteletzkya virginica	0.0	11.1
Lantana depressa	40.0	33.3
Lantana involucrata	66.7	33.3
Licania michauxii	33.3	11.1
Lysiloma latisiliquum	33.3	0.0
Magnolia virginiana	6.7	11.1
Metopium toxiferum	100.0	100.0
Morinda royoc	100.0	100.0
Mosiera longipes	73.3	44.4
Myrica cerifera	100.0	100.0
Myrsine floridana	100.0	88.9
Persea borbonia	60.0	100.0
Pinus elliottii var. *densa*	100.0	100.0
Psychotria nervosa	13.3	11.1
Quercus virginiana	53.3	33.3
Randia aculeata	100.0	100.0
Rhus copallinum	100.0	100.0
Sabal palmetto	100.0	100.0
Salix caroliniana	20.0	11.1
Schinus terebinthfolius	66.7	100.0
Senna chapmanii	46.7	55.6
Serenoa repens	100.0	100.0
Simarouba glauca	6.7	0.0
Solanum donianum	6.7	55.6
Stillingia sylvatica	40.0	44.4
Tetrazigia bicolor	100.0	100.0
Trema micrantha	46.7	44.4
Waltheria indica	40.0	88.9
Zamia integrifolia	80.0	33.3

Harcombe et al. 1993; Peet and Allard 1993). Larger changes in plant species composition occur over smaller differences in topography in southeastern savannas than in other North American savannas (Peet and Allard 1993).

A large number of plant communities occur within southeastern savannas. Peet and Allard (1993) delineated 23 plant communities based on ordination of data on vascular plant species collected in 216 longleaf pine stands in the Atlantic and eastern Gulf Coastal plains and fall-line sandhills. The strongest association occurred between species composition and local topographic gradients, which Peet and Allard (1993) interpreted as being related to soil moisture. Species composition also differed between sites with seasonally saturated and better-drained soils; within these two groups, communities also sorted along moisture gradients. Differences between species composition of sites in the inner and outer Coastal Plain were less pronounced. Likewise, using data on woody species from 74 sites in the Big Thicket region of Texas, Harcombe et al. (1993) discerned six forest types (including savannas) associated with local topography. The strongest relationship occurred between species composition and soil texture (percent sand in surface soils). Savannas had the highest sand content, which was associated with low moisture availability (Harcombe et al. 1993).

Restricted species distributions result partly from local changes in environmental conditions over very small distances along a topographic gradient. Thus, several hundred plant species may occur within an area smaller than a hectare (Platt, Evans and Davis 1988; Peet and Allard 1993; Stout and Marion 1993). In addition, there is high species richness of vascular plants at any given elevation along the topographic gradient. Frequent fires that remove aboveground biomass and litter, coupled with low availability of nutrients for regrowth following fires, appear to limit local aboveground growth and litter production by ground-cover plants (Peet, Glenn-Lewin and Wolf 1983). Slight differences in fire season also may favor different ground-cover species (Platt, Evans and Davis 1988). As a result, many small-statured plant species of a variety of growth forms co-occur within small areas: 30–40+ species/1 m^2, 70–90 species/100 m^2, and 100–150 species/1,000 m^2 (J. Walker and Peet 1983; Platt, Evans and Davis 1988; Platt, Glitzenstein and Streng 1991; Peet and Allard 1993; Streng, Glitzenstein and Platt 1993; R. K. Peet, unpublished data; W. J. Platt, unpublished data). This species richness, the highest reported for North American savannas (Peet and Allard 1993), rivals that of tropical forests, but at a very different scale.

Vertebrate Fauna

Avian and mammalian faunas of the southeast, which tend to be temperate in origin, are depauperate compared to those of higher latitudes in North America (Layne 1984; Stout and Marion 1993). In contrast, herpetofaunas of longleaf pine savannas (which tend to be associated with faunas of the southwestern United States; Guyer and Bailey 1993) and of south Florida slash pine savannas (which tend to be associated with

West Indian faunas; Snyder, Herndon and Robertson 1990; Stout and Marion 1993) are more diverse in the southeast than at higher latitudes in North America (Kiester 1971; Guyer and Bailey 1993; Means 1996). Analyses by Guyer and Bailey (1993) indicated that the herpetofaunas, especially amphibians of longleaf pine savannas, were more diverse and contained more specialized species than herpetofaunas of other pine-dominated habitats at similar latitudes in other parts of the world.

A number of vertebrate species are characteristically associated with southeastern pine savannas, especially those in the longleaf pine region. No mammals are restricted to pine savannas, but southeastern pocket gophers (*Geomys pinetis*) and southeastern fox squirrels (four subspecies of *Sciurus niger*) are largely sympatric with the range of longleaf pine (Engstrom 1993). Red-cockaded woodpeckers (*Picoides borealis*), brown-headed nuthatches (*Sitta pusilla*), rufous-sided towhees (*Pipilo erythropthalmus*), bobwhite quail (*Colinus virginianus*), and Bachman's sparrows (*Aimophila aestivalis*) are typical of southeastern pine savannas (Jackson 1988; Engstrom 1993; Stout and Marion 1993). Among the many characteristic reptiles are gopher tortoises (*Gopherus polyphemus*), eastern indigo snakes (*Drymarchon corais*), eastern diamondback rattlesnakes (*Crotalus adamanteus*), yellow-lipped snakes (*Rhadinaea flavilata*), Florida pine snakes (*Pituophis melanoleucus mugitus*), Miami black-headed snakes (*Tantilla oolitica*), mole skinks (*Eumeces egregius*), and fence lizards (*Sceloporus undulatus*) (Guyer and Bailey 1993; Stout and Marion 1993; Means 1996). Likewise, amphibians, such as the flatwoods salamanders (*Ambystoma cingulatum*), striped newts (*Notophthalmus perstriatus*), gopher frogs (*Rana areolata*), and oak toads (*Bufo quercicus*) have life cycle stages commonly associated with southeastern pine savannas (Guyer and Bailey 1993; Stout and Marion 1993; Means 1996).

Gopher tortoises influence faunal diversity

in southeastern pine savannas. They dig deep burrows in the sandy upland soils (Franz and Auffenberg 1978; Hermann 1993; Ware, Frost and Doerr 1993). These burrows serve as refuges for more than 300 other obligate and facultative commensal species, including gopher frogs, indigo snakes, pine snakes, diamondback rattlesnakes, and many arthropods (Jackson and Milstrey 1989; Myers 1990; Folkerts, Deyrup and Sisson 1993; Means 1996). At least 16 obligate arthropod inquilines occur in gopher tortoise burrows; many of these have characteristics (depigmentation, elongate appendages, and reduced eye size) that are associated with subterranean conditions (Folkerts, Deyrup and Sisson 1993). The burrows also disrupt the ground cover, producing local microhabitats that may be germination sites for plants (Kaczor and Hartnett 1990; Hermann 1993).

Red-cockaded woodpeckers symbolize dependence of endemic species on old-growth conditions that have largely disappeared in most second-growth stands (Doren, Platt and Whiteaker 1993; Noel, Platt and Moser 1998). This woodpecker depends on large (probably 100 years or older) old-growth pine trees for nest cavities and foraging, as well as open space in the midstory and understory (Ligon 1970; Jackson, Lennartz and Hooper 1979; Hovis and Labisky 1985; Porter and Labisky 1986; Myers 1990; James, Hess and Kufrin 1997). Thus, only savannas with at least some large, older trees (usually ones not logged earlier in the century) and which have no midstory of hardwoods have viable colonies of red-cockaded woodpeckers (Hooper, Robinson and Jackson 1980; Lennartz et al. 1983; Conner and Rudolph 1991; Loeb, Pepper and Doyle 1992). James, Hess and Kufrin (1997) recently have shown that size, density, and productivity of red-cockaded woodpecker social units were positively associated with those characteristics of ground cover composition and pine regeneration indicating frequent occurrence of local fires, thus suggesting that fire history also is important for this species.

Cavities excavated by red-cockaded woodpeckers in living pine trees also provide refuges for a large number of other species, including birds, mammals, snakes, lizards, and insects (Means 1996). Because old-growth savanna pine trees killed by lightning often remain standing for decades (Platt, Evans and Rathbun 1988; Platt and Rathbun 1993), nest cavities excavated by red-cockaded woodpeckers also may be used by other species for many years after host trees have died. Thus, red-cockaded woodpeckers are, indirectly, keystone species in pine savannas.

Arthropod Fauna

Folkerts, Deyrup and Sisson (1993) conservatively estimate the arthropod fauna associated with xeric longleaf pine savannas at 4,000–5,000 species. These authors have summarized the more common arthropods of xeric longleaf pine savanna habitats in the east Gulf Coastal Plain.

Among the best-known arthropods are longleaf pine herbivores. Forty-two species, fewer than the number that attack healthy trees of other species of southern pines, are known to feed on healthy, undamaged longleaf pine (Folkerts, Deyrup and Sisson 1993). For example, southern pine beetles (*Dendroctonus frontalis*; Scolytidae) and Nantucket pine tip moths (*Rhyacionia frustrana*; Tortricidae) do not attack healthy longleaf pines. Other species, for example plant-feeding gall midges (Cecidomyiidae) may rarely attack longleaf pine, but are more common on other species of southern pines (Folkerts, Deyrup and Sisson 1993).

Only a few herbivores are strongly linked to longleaf pine. Two monophagous insects are known to attack healthy longleaf pines. Larvae of longleaf pine seedworms (*Cydia ingens;* Tortricidae), which consume embryonic tissue of seeds, can cause sizable reductions in seed crops (Baker 1972). When small, larvae of Louisiana longleaf needleminers (*Coleotechnites chillcotti*; Gelechiidae) mine needles, but when large, they tie needles together. Another moth larva, not specific to longleaf pine, that attacks and webs the foliage and thus can be conspicuous is the pine webworm (*Tetralopha robustella*; Pyralidae). Two other species that attack seeds and cones of several pine species, including longleaf pine, and which can cause heavy losses, are nymphs and adults of the leaf-footed pine seed bug (*Leptoglossus corculus*; Coreidae) (DeBarr 1967) and larvae of southern pine coneworms (*Dioryctria amatella*; Pyralidae) (Coulson and Franklin 1970).

Resistance of longleaf pine to herbivory has been hypothesized to result from physical and chemical defenses. Total resin flow, flow rates, viscosity, and time to crystallization are related to resistance (Barbosa and Wagner 1989). Longleaf pine is noted for mobilizing large volumes of resin, which is slow to crystallize at attack sites (Hodges et al. 1979; Folkerts, Deyrup and Sisson 1993). Monoterpenes (alpha- and beta-pinene), especially in needles, may also reduce herbivory (Smith 1965). Reductions in seed predation have been hypothesized to result from masting, which has been suggested to satiate/starve invertebrate and vertebrate predators (Silvertown 1980; Folkerts, Deyrup and Sisson 1993; Means 1996).

Fire Ecology of Southeastern Savannas

Succession Models

Although the uniqueness of the flora and fauna of southeastern pine savannas has been recognized, changes in these communities occurred before the ecology was studied. The importance of fire in southeastern savannas became recognized when fire suppression resulted in habitat losses, both for pine and game species (Stoddard 1935; Croker and Boyer 1975; Croker 1987). Frequent dormant season fires reduced densities of hardwood shrubs and trees, but had little effect on pines (Chapman 1932a, b; Wahlenberg 1946;

Croker and Boyer 1975; Waldrop, White and Jones 1992). Fire return intervals of more than about a decade were noted as resulting in replacement of herbaceous ground cover species and small pines by hardwood shrubs and trees. Studies of fire exclusion led to the concept of southeastern pine savannas as subclimaxes (sensu Tansley 1935; also see Clements 1916, 1936; Whittaker 1953) maintained by frequent fires (Harper 1911, 1914, 1927, 1943; Wells and Shunk 1931; Heyward 1939; Laessle 1942, 1958; Monk 1960, 1967, 1968; Wade, Ewel and Hofstetter 1980; Myers 1985). In the absence of fire, pine savannas were hypothesized to be replaced by regional climaxes of mixed-species temperate or subtropical hardwood forests (Harper 1911, 1927; Small 1929, 1930; Davis 1943; Egler 1952; Alexander 1953, 1958, 1967; Quarterman and Keever 1962; Monk 1965; Myers 1985).

Characteristics of Fires

Data on frequencies of thunderstorms, lightning, and the resulting natural fires in the southeast became available after the middle of the century. Adiabatic thunderstorms occur almost daily during the wet summer season (Hela 1952; Robertson 1953; Komarek 1964; Changery 1981), resulting in one of the highest frequencies of lightning in the world (Goodman and Christian 1993); cloud-to-ground lightning flashes occur at annual rates of 1–10/km^2 in peninsular Florida and along the Gulf Coastal Plain (Maier, Boulanger and Sax 1979). As a result, fire frequency in southeastern pine savannas is typically more than once a decade (Chapman 1932a, b; Wahlenberg 1946; Vogl 1973; Christensen 1981, 1988; D. L. Taylor 1981; Whelan 1985, 1995; Platt, Glitzenstein and Streng 1991; Robbins and Myers 1992; Olson and Platt 1995). Moreover, thunderstorms are strongly seasonal, with a midsummer peak in June–July–August (Komarek 1964, 1974; Maier, Boulanger and Sax 1979; Snyder 1986, 1991; Goodman and Christian

Figure 2.3. Prescribed fire in a central Florida longleaf pine flatwoods savanna (Bee Island in Myakka River State Park, Sarasota and Manatee counties). The prescribed head fire with flame lengths of 2–5 m is burning through a ground cover dominated by *Serenoa repens, Quercus minima, Ilex glabra, Lyonia lucida, Aristida beyrichiana, Schizachyrium stoloniferum, Sorghastrum secundum,* and *Andropogon ternarius* in May, 1993. Photo taken by Maynard L. Hiss.

1993). Fire frequency increases during the spring (April–June) when intervals between successive rainfalls also are increasing; at these times, natural lightning-initiated fires that at least sometimes are of high intensity (Figure 2.3) burn large areas, especially in dry years (Komarek 1964, 1974; D. L. Taylor 1981; Huffman and Blanchard 1991; Snyder 1991; Gunderson and Snyder 1994). Summer fires occur more frequently than spring fires, but tend to burn smaller areas and appear usually to be of lower intensity (Huffman and Blanchard 1991; Glitzenstein, Platt and Streng 1995). In the fall, intervals between successive rainfalls increase, but the frequency of thunderstorms and lightning declines rapidly. As a result, natural lightning-initiated fires are uncommon in the fall and winter (Komarek 1964, 1974; Snyder 1991).

Ecology and Natural Fire Regimes

Studies focusing on the dynamics of pine savannas in relation to characteristics of natural fire regimes were initiated once natural fire regimes were delineated. Results have called into question the applicability of successional models. Study of the dynamics of old-growth stands of longleaf and south Florida slash pine have suggested that these populations, although not close to equilibrium, are likely to persist over many centuries; replacement by hardwoods is not likely, given high natural fire frequencies (Platt, Evans and Rathbun 1988; Doren, Platt and Whiteaker 1993; Platt and Rathbun 1993). Experimental studies have demonstrated that subtle differences in fire regimes can change the composition and characteristics of southeastern savannas (Platt, Evans and Davis 1988; Platt, Glitzenstein and Streng 1991; Waldrop, White and Jones 1992; Glitzenstein, Platt and Streng 1995; Olson and Platt 1995). In particular, fires during the transition from spring to summer, when intervals between successive rainfalls are increasing and fires may be more intense, result in much greater mortality of hardwoods, but not pines or ground-cover species, than do fires at other times of the year (Platt, Glitzenstein and Streng 1991; Glitzenstein, Platt and Streng 1995). In addition, flowering of many ground-cover species, including most grasses (Platt, Glitzenstein and Streng 1991; Streng, Glitzenstein and Platt 1993) are enhanced by these early growing season fires. Recognition of such patterns can be effectively used in restoration of fire-suppressed savannas (e.g., Huffman and Blanchard 1991). Fires occurring at those times of the year, when they most often occurred historically, produce open savannas, dominated by overstory pines and herbaceous ground-cover species, that are not replaced by hardwood forests.

An Evolutionary Fire Ecology Model

Data from the effects of fire on plant populations and communities in southeastern savannas suggest a nonsuccessional scenario in which a "dependable" environment (Chapman 1932a) involving predictable fire regimes could result from high frequencies of lightning strikes at certain times of the year, combined with pyrogenic vegetation (Platt, Glitzenstein and Streng 1991; Glitzenstein, Platt and Streng 1995). The high frequencies of fires that occur most often around the onset of the lightning season in the spring could mold characteristics of the indigenous flora if such fire regimes occurred predictably over long intervals of time (Mutch 1970; Platt, Glitzenstein and Streng 1991; Glitzenstein, Platt and Streng 1995). These adaptations fall into three groups: resistance to fire, adaptation to environmental conditions produced by fires, and modification of fire regimes (Platt 1994).

Resistance to fire is universal among plants indigenous to southeastern savannas. More than 95% of the ground-cover plant species are perennials, as expected where fires occur frequently (Keeley 1981; Olson and Platt 1995). Most herbaceous species resprout from buds on rhizomes or tap roots protected from fire. A few woody species produce lignotubers, swellings of stems containing many dormant buds (e.g., *Ilex coriacea*, *Byrsonima lucida*) (Robertson 1953; Olson and Platt 1995); palms (e.g., *Serenoa repens*) have enclosed terminal buds (Abrahamson 1984; Snyder 1986). Although fire-related mortality of shrubs tends to be very low in frequently burned savannas, (e.g., Rebertus, Williamson and Moser 1989a, b; Streng, Glitzenstein and Platt 1993; Brewer and Platt 1994a, b), resistance may increase with size of shrub genets (Olson and Platt 1995).

Longleaf and south Florida slash pines also are well known for their fire resistance (Chapman 1932a, b; Wahlenberg 1946; Robertson 1953; Croker and Boyer 1975; Platt, Evans and Rathbun 1988; Landers 1991; Doren, Platt and Whiteaker 1993). Juveniles become fire-resistant by the second year, when bark, lateral buds, and secondary needles are present (Grace and Platt 1995a, b). Once in the "grass stage" (where height growth is very slow compared to increases in stem diameter and root systems; Mohr 1897; Schwarz 1907; Wahlenberg 1946; Platt, Evans and Rathbun 1988), juveniles have terminal buds surrounded by scales and abundant needles containing resins that slow combustion, lowering the heat produced (Chapman 1932a, b; Croker and Boyer 1975). Indeterminate flushing also results in rapid refoliation following growing season fires (Chapman 1932a, b; Wahlenberg 1946; Platt, Glitzenstein and Streng 1991). Although young savanna pines are vulnerable to intense fires during height growth, rapid increases in height and bark thickness, as well as loss of lower branches, reduce the susceptible period (Heyward 1939; Croker and Boyer 1975; Platt, Evans and Rathbun 1988; Rebertus, Williamson and Platt 1993). Large savanna pines have thick, layered bark that protects the cambium from frequent fires, although extremely hot fires, such as those following long periods of fire suppression, can kill even large trees (Chapman 1932a, b; Abrahamson 1984; Croker and Boyer 1975; Platt, Evans and Rathbun 1988).

Adaptation to environmental conditions produced by fire also appears widespread in southeastern pine savannas. Most ground-cover species, including grasses and forbs, flower after fire, especially when it occurs during the growing season (Lemon 1949, 1967; Robertson 1953; Vogl 1973; Whelan 1985; Snyder 1986; Platt, Evans and Davis 1988; Landers 1991; Platt, Glitzenstein and Streng 1991; Peet and Allard 1993; Streng, Glitzenstein and Platt 1993). Some dominant species, such as *Aristida beyrichiana*, flower and produce viable seeds only following growing season fires (Parrott 1967; Seamon, Myers and Robbins 1989; Streng, Glitzenstein and Platt 1993). *Pityopsis graminifolia* (golden-leafed aster), a dominant forb in many coastal upland pine savannas, exhibits increased clonal growth and flowering following early growing season fires, but not fires at other times of the year (Brewer and Platt 1994a, b). This phenotypic plasticity may represent transient responses to increased light and nutrients after fire (Brewer and Platt 1994b; Brewer et al. 1996). Savanna pines also are adapted for postfire conditions. Germination and establishment are dependent on bare mineral soil (Wahlenberg 1946; Robertson 1953; Croker and Boyer 1975). Survival and growth of juveniles are increased following fires (Chapman 1932a, b; Wahlenberg 1946; Croker and Boyer 1975; Grelen 1975; Grelen 1978).

Modification of fire regimes by indigenous species also occurs in southeastern savannas. Savanna pines convert lightning strikes into more widespread, but lower-intensity, surface fires (Platt, Evans and Rathbun 1988; Landers 1991). Compared to other tree species, savanna pines have a higher probability of lightning strike (Myers 1990); upright snags also may be hit repeatedly, thereby increasing the chance of fire; the resinous heartwood may smolder and ignite a ground fire days after ignition of the wood by lightning (A. R. Taylor 1974; Platt, Evans and Rathbun 1988). In addition to facilitating ignition, savanna pines also facilitate fire spread. Needles of savanna pine species are the longest and most resin-filled in North America (Landers 1991); retention times are only two years (Mirov 1967); and some needles are shed annually (Landers 1991). Needlefall increases during dry periods, which often occur near the time of year when fires are most likely (Herndon and Taylor 1985; Snyder 1986). Decomposition of dry needles is slow (Herndon and Taylor 1985); the resins present (Joye, Proveaux and Lawrence 1972) are volatilized at low temperatures, but produce high heats of combustion (Johnson 1992).

Although pines appear to facilitate frequent fires, the ground-cover vegetation provides the bulk of fuel during fires. Streng and Harcombe (1982) have predicted that rates of fire spread are higher in savannas than in forests, especially at higher humidities during the growing season. The nature and structure of ground-cover fuels are predicted to influence both rate of spread and fire intensity (Streng and Harcombe 1982; Platt, Glitzenstein and Streng 1991). Pine needles and grasses, especially in open areas, generate high heats of combustion rapidly, but the flame residence times are short (Shafizadeh and DeGroot 1976; Johnson 1992). Maximum fire temperatures recorded near ground level during growing season fires in frequently burned savannas consistently reach 500–800 °C (Williamson and Black 1981; Platt, Glitzenstein and Streng 1991; Glitzenstein, Platt and Streng 1995; Olson and Platt 1995). Highly flammable species result in even greater fire intensities; for example, high densities of *Serenoa repens* can elevate fire intensities in flatwoods savannas (Figure 2.3).

The resistance–adaptation–modification model has important ramifications for plant populations, savanna communities, and regional landscapes (Platt 1994). The apparent key to such an evolutionary scenario, especially in a region of high annual rainfall, is the high (and predictable) rates of lightning strikes in the seasonally dry environment of the southeastern Coastal Plain. In such an environment, the presence of plants that enhance ignition could, in addition to influencing characteristics of fires, result in increased frequency and predictability of postfire environmental conditions. These features in turn would likely constitute strong selection pressures on indigenous populations (Mutch 1970; Williamson and Black 1981; Platt, Glitzenstein and Streng 1991; Platt 1994). However, differential mortality of individuals should decrease in importance as populations contain mostly "resistant" individuals. Differential fitness will more likely result from differential adaptation to postfire conditions. These hypothesized selection pressures thus differ from those postulated by Bond and Midgley (1995) and Bond and van Wilgen (1996). Although influenced by concepts expressed by Mutch (1970), these arguments for the evolution of ignition potential and flammability do not invoke group selection (see Whelan 1995; Bond and van Wilgen 1996). Instead, predictable fire regimes and, hence, postfire environments are predicted to be molded by combined effects of individuals of all species influencing the characteristics of fires. Interactions between seasonal hydrology, low nutrient availability, and predictable fire regimes also appear to influence the local distribution and abundances of plant species in ways that enhance further the predictability of postfire environments (J. Walker and Peet 1983; Harcombe et al. 1993). Fires in surrounding landscapes, although less frequent, may become more predictable (Means 1996).

Nonetheless, despite very predictable environmental conditions, neither pine savannas nor landscapes containing these pyrogenic communities are necessarily close to some equilibrium state.

Fire Ecology and Invasibility

Combinations of certain substrates and climatic characteristics that have been regionally consistent over long periods of time appear to have resulted in distinctive environmental conditions in the southeastern United States. These conditions apparently were conducive for evolution of a unique pine savanna biota that includes specialized, endemic species, as well as high regional and local diversity of most flora and fauna. As a result, the occurrence of these species, and especially longleaf or south Florida slash pine, has been commonly used to indicate that savannas were present historically on given sites.

The life history characteristics of southeastern savanna pines differ markedly from those of the other southern pines (Landers 1991). Despite resistance to and dependence on disturbance, southeastern savanna pines have not been invasive, either locally or in other continents, unlike many other pine species (see Rejmanek and Richardson 1996). Juveniles of savanna pines, which require large investments in protective "defenses" against fires (see Loehle 1987), do not grow rapidly and thus appear unable to invade most communities (Richardson and Bond 1991; Rejmanek and Richardson 1996).

I propose that the noninvasiveness of savanna pines in most landscapes reflects an absence in most landscapes of those environmental conditions to which savanna pines are adapted. The prerequisites for savanna-type environments like those in the southeast (i.e., very frequent lightning strikes during seasonally dry periods, and low-nutrient soils) are rare in most landscapes. Where they occur, such environments are being eliminated by combinations of fire suppression and habitat destruction. Moreover, pine savanna environments appear to be modified biotically by fire-adapted species, including the pines themselves; once destroyed, such environments cannot easily be reconstituted.

Southeastern savannas and open successional communities in non–fire frequented habitats should not be confused, even though they have a somewhat similar physiognomy. The former are persistent (given a natural fire regime), whereas the latter are transient. Although savanna pines might well be capable of invading most open habitats as a result of capability for considerable dispersal and establishment under harsh environmental conditions, juveniles have been selected for fire resistance, not rapid growth. Thus, specialized pine savanna species will not be an important component of any community in which open stages are transient in nature. However, persistently open habitats, especially those where fires occur frequently during dry periods of the growing season, should be highly invasible by pine savanna species, which may modify the fire regime once established, biotically altering environmental conditions. This comparison of pine savannas with more transient pine-dominated communities suggests that it is not disturbance per se that is critical for invasion (see Rejmanek 1989, 1996; Hobbs 1991; Richardson and Bond 1991; Whitmore 1991), but rather the type of disturbance that governs invasibility by a particular species. Delineation of environmental conditions resulting in successful invasion by indigenous species should be useful in guiding restoration of southeastern pine savannas.

Acknowledgments

I owe much to many people, beginning with my parents, Lucile and William J. Platt, Jr., and Dr. Howard V. Weems, Jr., who instilled a profound interest in the ecology of Florida. More recently, Bob Peet, Paul Harcombe, Bruce Means, and Maynard Hiss have provided many insights and fruitful discussions related to topics in this review. Ed and Roy Komarek and Larry Landers engaged me in many discussions regarding the role of fire in southeastern pine savannas – I dedicate this chapter to the memory of their intense interest in the role of fire in shaping the southeastern landscape. Much of my own work has benefited from the close collaboration of Jeff Glitzenstein, Donna Streng, Steve Rathbun, Jim Hamrick, Jim DeCoster, and Bob Doren. Graduate students at Florida State and Louisiana State universities – Greg Evans, Mary Davis, Steve Brewer, Sue Grace, Matt Olson, Paul Drewa, Charles Kwit, Rob Gottschalk, Jarrod Thaxton, Jean Huffman, and Susan Carr – have worked on topics important in this review. Margaret Platt, Joe Connell, Maynard Hiss, Latimore Smith, Jim DeCoster, Jean Huffman, Roger Anderson, Jim Fralish, and Jerry Baskin provided very useful comments on earlier drafts of this manuscript. Last, but not the least, I thank the conservation and state organizations – Tall Timbers Research Station, The Nature Conservancy, the Florida Department of Parks and Recreation, the U.S. National Park Service, U.S. Forest Service, and U.S. Fish and Wildlife Service – for logistical support and use of pine savannas in scientific study conducted over the past couple of decades. Support for this review has come from National Science Foundation Grants BSR 8718803 and DEB 9314265 (W. J. Platt, PI) and by the National Park Service, Everglades National Park, Cooperative Agreement 5280-4-9004 (W. J. Platt, PI).

REFERENCES

Abrahamson, W. G. 1984. Species responses to fire on the Florida Lake Wales Ridge. *American Journal of Botany* 71:35–43.

Abrahamson, W. G., and Hartnett, D. C. 1990. Pine flatwoods and dry prairies. In *Ecosystems of Florida*, ed. R. Myers and J. Ewel, pp 102–149. Orlando, Fla.: University of Florida Press.

Alexander, T. R. 1953. Plant succession on Key Largo, Florida, involving *Pinus caribaea* and *Quercus virginiana*. *Quarterly Journal of the Florida Academy of Sciences* 16:133–138.

Alexander, T. R. 1958. High hammock vegetation of the southern Florida mainland. *Quarterly Journal of the Florida Academy of Sciences* 21:293–298.

Alexander, T. R. 1967. A tropical hammock on Miami limestone – a twenty-five year study. *Ecology* 48:863–867.

Armentano, T. V., Doren, R. F., Platt, W. J., and Mullins, T. 1995. Effects of Hurricane Andrew on Coastal and interior forests of southern Florida. *Journal of Coastal Research* 21:111–144.

Baker, W. L. 1972. *Eastern Forest Insects*. Washington, D.C.: USDA Forest Service, Miscellaneous Publication 1175.

Barbosa, P., and Wagner, M. R. 1989. *Introduction to Forest and Shade Tree Insects*. New York: Academic Press.

Batista, W. B., and Platt, W. J. 1997. *An Old-Growth Definition for Southern Mixed Hardwood Forests*. USDA Forest Service, Southern Research Station, Gen. Tech. Rep. 10.

Bond, W. J., and Midgley, J. J. 1995. Kill thy neighbour: an individualistic argument for the evolution of flammability. *Oikos* 73:79–85.

Bond, W. J., and van Wilgen, B. W. 1996. *Fire and Plants*. New York: Chapman and Hall.

Brewer, J. S., and Platt, W. J. 1994a. Effects of fire season and herbivory on reproductive success in a clonal forb, *Pityopsis graminifolia* (Michx.) Nutt. *Journal of Ecology* 82:665–675.

Brewer, J. S., and Platt, W. J. 1994b. Effects of fire season and soil fertility on clonal growth in a pyrophilic forb, *Pityopsis graminifolia* (Asteraceae). *American Journal of Botany* 81:805–814.

Brewer, J. S., Platt, W. J., Glitzenstein, J. S., and Streng, D. R. 1996. Effects of fire-generated gaps on growth and reproduction of golden aster (*Pityopsis graminifolia*). *Bulletin of the Torrey Botanical Club* 123:295–303.

Bridges, E. L., and Orzell, S. L. 1989. Longleaf pine communities of the west Gulf Coastal Plain. *Natural Areas Journal* 9:246–263.

Bromley, S. W. 1939. Factors influencing tree destruction during the New England hurricane. *Science* 90:15–16.

Brouillet, L., and Whetstone, R. D. 1993. Climate and physiography. In *Flora of North America North of Mexico*, ed. Flora of North America Editorial Committee, pp. 15–46. Oxford: Oxford University Press.

Brown, C. A. 1944. Historical commentary on the distribution of vegetation in Louisiana, and some recent observations. *Proceedings of the Louisiana Academy of Sciences* 8:35–46.

Brown, R. B., Stone, E. L., and Carlisle, V. W. 1990. Soils. In *Ecosystems of Florida*, ed. R. Myers and J. Ewel, pp. 35–69. Orlando, Fla.: University of Florida Press.

Changery, J. J. 1981. *National Thunderstorm Frequencies for the Contiguous United States*. NUREG/CR-2552. Washington, D.C.: U.S. Nuclear Regulatory Commission.

Chapman, H. H. 1932a. Is the longleaf type a climax? *Ecology* 13:328–334.

Chapman, H. H. 1932b. Some further relations of fire to longleaf pine. *Journal of Forestry* 30:602–604.

Chen, E., and Gerber, J. F. 1990. Climate. In *Ecosystems of Florida*, ed. R. Myers and J. Ewel, pp. 11–34. Orlando, Fla.: University of Florida Press.

Christensen, N. L. 1981. Fire regimes in southeastern ecosystems. In *Proceedings of the Conference: Fire Regimes and Ecosystem Properties*, ed. H. A. Mooney, T. M. Bonnicksen, N. L. Christensen, J. E. Lotan, and W. A. Reiners, pp. 112–136. USDA Forest Service, Gen. Tech. Rep. WO-26.

Christensen, N. L. 1985. Shrubland fire regimes and their evolutionary consequences. In *The Ecology of Natural Disturbances and Patch Dynamics*, ed. S. T. A. Pickett and P. S. White, pp. 85–100. New York: Academic Press.

Christensen, N. L. 1988. Vegetation of the southeastern Coastal Plain. In *North American Terrestrial Vegetation*, ed. M. G. Barbour and W. D. Billings, pp. 317–363. Cambridge: Cambridge University Press.

Christensen, N. L. 1993. The effects of fire on nutrient cycles in longleaf pine ecosystems. *Proceedings Tall Timbers Fire Ecology Conference* 18:205–214.

Clements, F. E. 1916. *Plant Succession: An Analysis of the Development of Vegetation*. Carnegie Institute of Washington Publication 242.

Clements, F. E. 1936. Nature and structure of the climax. *Journal of Ecology* 24:252–284.

Conner, R. N., and Rudolph, D. C. 1991. Effects of midstory reduction and thinning in red-cockaded woodpecker cavity tree clusters. *Wildlife Society Bulletin* 19:63–66.

Coulson, R. N., and Franklin, R. T. 1970. The biology of *Dioryctia amatella* (Lepidoptera: Phycitidae). *Canadian Entomologist* 102:679–684.

Craighead, F. C., Sr. 1971. *The Trees of South Florida*. Vol. I. *The Natural Environments and their Succession*. Coral Gables, Fla.: University of Miami Press.

Croker, T. C., Jr. 1987. *Longleaf Pine: A History of Man and a Forest*. USDA Forest Service, Forestry Report R8-FR7.

Croker, T. C., Jr., and Boyer, W. D. 1975. *Regenerating Longleaf Pine Naturally*. USDA Forest Service, Southern Experiment Station, Research Paper SO-105.

Davis, J. H. 1943. *The Natural Features of Southern Florida, Especially the Vegetation and the Everglades*. Florida Geological Survey, Bulletin 25.

DeBarr, G. L. 1967. *Two New Sucking Insects in Southern Pine Seed Bed*. USDA Forest Service, Research Note SE-78.

Delcourt, P. A., and Delcourt, H. R. 1987. *Long-term Dynamics of the Temperate Zone: A Case Study of Late-Quaternary Forest History in Eastern North America*. Ecological Study Series 63. New York: Springer-Verlag.

Delcourt, P. A., Delcourt, H. R., Morse, D. F., and Morse, P. A. 1993. History, evolution, and organization of vegetation and human culture. In *Biodiversity of the Southeastern United States: Lowland Terrestrial Communities*, ed. W. H. Martin, S. G. Boyce, A. C. Echternacht, pp. 47–79. New York: John Wiley & Sons.

Doren, R. F., Platt, W. J., and Whiteaker, L. D. 1993. Density and size structure of slash pine stands in the everglades region of south Florida. *Forest Ecology and Management* 59:295–311.

Egler, F. E. 1952. Southeast saline Everglades vegetation, Florida, and its management. *Vegetatio* 3:213–265.

Eleuterius, L. N. 1968. Floristics and ecology of Coastal bogs in Mississippi. M.S. thesis, University of Southern Mississippi, Hattiesburg, Miss.

Engstrom, R. T. 1993. Characteristic mammals and birds of longleaf pine forests. *Proceedings Tall Timbers Fire Ecology Conference* 18:127–138.

Folkerts, G. W. 1982. The gulf coast pitcher plant bogs. *American Scientist* 70:260–267.

Folkerts, G. W., Deyrup, M. A., and Sisson, C. D. 1993. Arthropods associated with xeric longleaf pine habitats in the southeastern United States. *Proceedings Tall Timbers Fire Ecology Conference* 18:159–192.

Franz, R., and Auffenberg, W. 1978. The gopher tortoise: a declining species. In *Proceedings of the Rare and Endangered Wildlife Symposium*, ed. R. R. Odum and J. L. Landers, pp. 61–63. Athens, Ga.: Georgia Department of Natural Resources, Game and Fish Division, Technical Bulletin WL4.

Frost, C. C. 1993. Four centuries of changing landscape patterns in the longleaf pine ecosystem. *Proceedings Tall Timbers Fire Ecology Conference* 18:17–44.

Gilliam, F. S., Yurish, B. M, and Goodwin, L. M. 1993. Community composition of an old growth longleaf pine forest: relationship to soil texture. *Bulletin of the Torrey Botanical Club* 120:287–294.

Glitzenstein, J. S., and Harcombe, P. A. 1988. Effects of the December 1983 tornado on forest vegetation of the Big Thicket, Southeast Texas, U.S.A. *Forest Ecology and Management* 25:269–290.

Glitzenstein, J. S., Platt, W. J., and Streng. D. R. 1995. Effects of fire regime and habitat on tree dynamics in north Florida longleaf pine savannas. *Ecological Monographs* 65:441–476.

Goodman, S. J., and Christian, H. J. 1993. Global observations of lightning. In *Atlas of Satellite Observations Related to Global Change*, ed. R. J. Gurney, J. L. Foster, and C. L. Parkinson, pp. 191–219. Cambridge: Cambridge University Press.

Grace, S. L., and Platt, W. J. 1995a. Effects of adult tree density and fire on the demography of pre–grass stage juvenile longleaf pine (*Pinus palustris* Mill.). *Journal of Ecology* 95:75–86.

Grace, S. L., and Platt, W. J. 1995b. Neighborhood effects on juveniles in an old-growth stand of longleaf pine (*Pinus palustris* Mill.). *Oikos* 72:99–105.

Grazulis, T. P. 1984. *Violent Tornado Climatology, 1880–1982*. NUREG/CR-3670, PNL-5006RB. Washington, D.C.: U.S. Nuclear Regulatory Commission.

Grelen, H. E. 1975. *Vegetative Response to Twelve Years of Seasonal Burning on a Louisiana Longleaf Pine Site*. USDA Forest Service, Southern Experiment Station, Research Note SO-192.

Grelen, H. E. 1978. *May Burns Stimulate Growth of Longleaf Pine Seedlings*. USDA Forest Service, Research Note SO-234.

Greller, A. M. 1980. Correlations of some climate

statistics with distribution of broadleaved forest zones in Florida, U.S.A. *Bulletin of the Torrey Botanical Club* 107:189–219.

Greller, A. M. 1989. Correlation of warmth and temperatures with the distributional limits of zonal forests of eastern North America. *Bulletin of the Torrey Botanical Club* 116:145–163.

Gresham, C. A., Williams, T. M., and Lipscomb, D. J. 1991. Hurricane Hugo wind damage to southeastern U.S. Coastal forest tree species. *Biotropica* 23:420–426.

Gunderson, L. H., and Snyder, J. R. 1994. Fire patterns in the southern everglades. In *Everglades: The Ecosystem and its Restoration*, ed. S. M. Davis and J. C. Ogden, pp. 291–305. Delray Beach, Fla.: St. Lucie Press.

Guyer, C., and Bailey, M. A. 1993. Amphibians and reptiles of longleaf pine communities. *Proceedings Tall Timbers Fire Ecology Conference* 18:139–158.

Harcombe, P. A., Glitzenstein, J. S., Knox, R. G., Orzell, S. L., and Bridges, E. L. 1993. Vegetation of the longleaf pine region of the west Gulf Coastal Plain. *Proceedings Tall Timbers Fire Ecology Conference* 18:83–104.

Hardin, E. D., and White, D. L. 1989. Rare vascular plant taxa associated with wiregrass (*Aristida stricta*) in the southeastern United States. *Natural Areas Journal* 9:234–245.

Harper, R. M. 1911. The relation of climax vegetation to islands and peninsulas. *Bulletin of the Torrey Botanical Club* 38:515–525.

Harper, R. M. 1914. *Geography and Vegetation of Northern Florida*. Florida Geological Survey, 6th Annual Report, pp. 163–451.

Harper, R. M. 1927. *Natural Resources of Southern Florida*. Florida Geological Survey, 18th Annual Report, pp. 27–206.

Harper, R. M. 1943. *Forests of Alabama*. Alabama Geological Survey Monograph 10:1–230.

Harper, R. M. 1948. A preliminary list of the endemic flowering plants of Florida. Part I. Introduction and history of exploration. *Quarterly Journal of the Florida Academy of Sciences* 11:25–35.

Harshberger, J. W. 1914. The vegetation of south Florida, south of 27°30′ north, exclusive of the Florida keys. *Transactions of the Wagner Free Institute of Science, Philadelphia, Pennsylvania* 7:49–189.

Hela, I. 1952. Remarks on the climate of south Florida. *Bulletin of the Marine Society of the Gulf of Mexico and Caribbean Sea* 2:438–447.

Hermann, S. M. 1993. Small-scale disturbances in longleaf pine forests. *Proceedings Tall Timbers Fire Ecology Conference* 18:265–274.

Herndon, A., and Taylor, D. L. 1985. *Litterfall in Pinelands of Everglades National Park*. South Florida Research Center, Report SFRC-85/01, Everglades National Park, Homestead, Fla.

Heyward, F. 1939. The relation of fire to stand composition of longleaf pine forests. *Ecology* 20:287–304.

Hobbs, R. J. 1991. Disturbance as a precursor to weed invasion in native vegetation. *Plant Protection Quarterly* 6:99–104.

Hodges, J. D., Elan, W. W., Watson, W. F., and Nebecker, T. E. 1979. Oleoresin characteristics and susceptibility of four southern pines to southern pine beetle (Coleoptera: Scolytidae) attacks. *Canadian Entomologist* 111:889–898.

Hooper, R. G., Robinson, A. F., and Jackson, J. A. 1980. *The Red-cockaded Woodpecker: Notes on Life History and Management*. USDA Forest Service, Gen. Rep. SA-GR9.

Hovis, J. A., and Labisky, R. F. 1985. Vegetative associations of red-cockaded woodpecker colonies in Florida. *Wildlife Society Bulletin* 13:307–314.

Huffman, J. M., and Blanchard, S. W. 1991. Changes in woody vegetation in Florida dry prairie and wetlands during a period of fire exclusion, and after dry-growing season fire. In *Fire and the Environment: Ecological and Cultural Perspectives*, ed. S. C. Nodvin and T. A. Waldrop, pp. 75–83. USDA Forest Service, Gen. Tech. Rep. SE-69.

Jackson, D. R., and Milstrey, E. G. 1989. The fauna of gopher tortoise burrows. In *Gopher Tortoise Relocation Symposium Proceedings*, ed. J. E. Diemer, D. R. Jackson, J. L. Landers, J. N. Layne, and D. A. Wood, pp. 86–98. Tallahassee, Fla.: Florida Game and Fresh Water Fish Commission.

Jackson, J. A. 1988. The southeastern pine forest ecosystem and its birds: past, present, and future. In *Bird Conservation 3*, ed. J. A. Jackson, pp. 119–159. Madison, Wis.: University of Wisconsin Press.

Jackson, J. A., Lennartz, M. R., and Hooper, R. G. 1979. Tree age and cavity initiation by red-cockaded woodpeckers. *Journal of Forestry* 77:102–103.

James, F. C., Hess, C. A., and Kufrin, D. 1997. Species-centered environmental analysis: indirect effects of fire history on red-cockaded woodpeckers. *Ecological Applications* 7:118–129.

Johnson, E. A. 1992. *Fire and Vegetation Dynamics: Studies from the North American Boreal Forest.* Cambridge: Cambridge University Press.

Jordan. C. L. 1984. Florida's weather and climate: implications for water. In *Water Resources Atlas of Florida*, ed. E. A. Fernald and D. J. Patton, pp. 18–25. Tallahassee, Fla.: Florida State University.

Joye, N. M., Jr., Proveaux, A. T., and Lawrence, R. V. 1972. Composition of pine needle oil. *Journal of Chromatographic Science* 10:590–592.

Kaczor, S. A., and Hartnett, D. C. 1990. Gopher tortoise (*Gopherus polyphemus*) effects on soil and vegetation in a Florida sandhill community. *The American Midland Naturalist* 123:100–111.

Keeley, J. E. 1981. Reproductive cycles and fire regimes. In *Proceedings of the Conference: Fire Regimes and Ecosystem Properties*, ed. H. A. Mooney, T. M. Bonnicksen, N. L. Christensen, J. E. Lotan, and W. A. Reiners, pp. 231–277. USDA Forest Service, Gen. Tech. Rep. WO-26.

Kelly, J. F., and Bechtold, W. A. 1990. The longleaf pine resource. In *Proceedings: Symposium on the Management of Longleaf Pine, April 4–6, 1989*, ed. R. M. Farrar, pp. 11–22. USDA Forest Service, Gen. Tech. Rep. SO-75.

Kiester, A. R. 1971. Species density of North American amphibians and reptiles. *Systematic Zoology* 20:127–137.

Komarek, E. V., Sr. 1964. The natural history of lightning. *Proceedings Tall Timbers Fire Ecology Conference* 3:139–183.

Komarek, E. V., Sr. 1974. Effects of fire on temperate forests and related ecosystems: southeastern United States. In *Fire and Ecosystems*, ed. T. T. Kozlowski and C. E. Ahlgren, pp. 251–277. New York: Academic Press.

Laessle, A. M. 1942. The plant communities of the Welaka Area. *University of Florida Publications, Biological Sciences Series* 4:5–141.

Laessle, A. M. 1958. The origin and successional relationship of sandhill vegetation and sand-pine scrub. *Ecological Monographs* 28:361–387.

Landers, L. L. 1991. Disturbance influences on pine traits in the southeastern United States. *Proceedings Tall Timbers Fire Ecology Conference* 17:61–98.

Layne, J. N. 1984. The land mammals of south Florida. In *Environments of South Florida: Present and Past. Memoir 2*, ed. P. J. Gleason, pp. 269–296. Coral Gables, Fla.: Miami Geological Society.

Lemon, P. C. 1949. Successional responses of herbs in the longleaf–slash forest after fire. *Ecology* 30:135–145.

Lemon, P. C. 1967. Effects of fire on herbs of the southeastern United States and central Africa. *Proceedings Tall Timbers Fire Ecology Conference* 6:112–127.

Lennartz, M. R., Knight, H. A., McClure, J. P., and Rudis, V. A. 1983. Status of red-cockaded woodpecker nesting habitat in the south. In *Red-cockaded Woodpecker Symposium II. Proceedings*, ed. D. A. Wood, pp. 13–19. Tallahassee, Fla.: Florida Game and Fresh Water Fish Commission.

Ligon, J. D. 1970. Behavior and breeding biology of the red-cockaded woodpecker. *Auk* 87:255–278.

Lockett, S. M. 1874. *Louisiana As It Is*. Baton Rouge, La.: Louisiana State University Press (reprinted 1969).

Loeb, S. C., Pepper, W. D., and Doyle, A. T. 1992. Habitat characteristics of active and abandoned red-cockaded woodpecker colonies. *Southern Journal of Applied Forestry* 16:120–125.

Loehle, C. 1987. Tree life history strategies: the role of defense. *Canadian Journal of Forest Research* 18:209–222.

Maier, M. W., Boulanger, A. G., and Sax, R. I. 1979. *An Initial Assessment of Flash Density and Peak Current Characteristics of Lightning Flashes to Ground in South Florida*. U.S. Nuclear Regulatory Commission Report CR-1024.

Marks, P. L., and Harcombe. P. A. 1981. Forest vegetation of the Big Thicket, southeast Texas. *Ecological Monographs* 51:287–305.

Martin, D. L., and Smith, L. M. 1991. *A Survey and Description of the Natural Plant Communities of the Kisatchie National Forest: Winn and Kisatchie Districts*. Baton Rouge, La.: Louisiana Department of Wildlife and Fisheries.

Martin, P. S., and Harrell, B. E. 1957. The Pleistocene history of temperate biotas in Mexico and eastern North America. *Ecology* 38:468–480.

Martin, W. H., and Boyce, S. G. 1993. Introduction: the southeastern setting. In *Biodiversity of the Southeastern United States: Lowland Terrestrial Communities*, ed. W. H. Martin, S. G. Boyce, and A. C. Echternacht, pp. 1–46. New York: John Wiley & Sons.

McPherson, G. 1997. *Ecology and Management of North American Savannas*. Tucson, Ariz.: University of Arizona Press.

Means, D. B. 1996. Longleaf pine forests, going,

going. In *Eastern Old-growth Forests*, ed. M. E. Davis, pp. 210–229. Washington, D.C.: Island Press.

Means, D. B., and Grow, G. O. 1985. The endangered longleaf pine community. *ENFO*, September, pp. 1–12.

Means, D. B., and Mohler, P. E. 1978. The pine barrens treefrog: fire, seepage bogs, and management implications. In *Proceedings of Rare and Endangered Wildlife Symposium, August 3–4, 1978, Athens, Georgia*, ed. R. R. Odum and L. L. Landers, pp. 77–83. Georgia Department of Natural Resources, Game and Fish Division, Tech. Bull. WL4.

Meylan, P. A. 1982. The squamate reptiles of the Inglis 1A fauna (Irvingtonian: Citrus Co., Florida). *Bulletin of the Florida State Musuem* 27:1–85.

Meylan, P. A. 1995. Pleistocene amphibians and reptiles from the Leisey Shell Pit, Hillsborough County, Florida. *Bulletin of the Florida State Musuem* 37:273–297.

Mirov, N. T. 1967. *The Genus Pinus*. New York: Ronald Press.

Mohr, C. 1897. *The Timber Pines of the Southern United States*. Washington, D.C.: USDA, Division of Forestry.

Monk, C. D. 1960. A preliminary study of the relationships between the vegetation of a mesic hammock community and a sandhill community. *Quarterly Journal of the Florida Academy of Sciences* 23:1–12.

Monk, C. D. 1965. Southern mixed hardwood forest of north-central Florida. *Ecological Monographs* 35:335–354.

Monk, C. D. 1967. Tree species diversity in the eastern deciduous forest with particular reference to north central Florida. *American Naturalist* 101:173–187.

Monk, C. D. 1968. Successional and environmental relationships of the forest vegetation of north central Florida. *The American Midland Naturalist* 79:441–457.

Mutch, R. W. 1970. Wildland fires and ecosystems – a hypothesis. *Ecology* 51:1046–1051.

Myers, R. L. 1985. Fire and the dynamic relationship between Florida sandhill and sand pine scrub vegetation. *Bulletin of the Torrey Botanical Club* 112:241–252.

Myers, R. L. 1990. Scrub and high pine. In *Ecosystems of Florida*, ed. R. Myers and J. Ewel, pp. 150–193. Orlando, Fla.: University of Florida Press.

Nash, G. V. 1895. Notes on some Florida plants. *Bulletin of the Torrey Botanical Club* 22:141–161.

Neumann, C. J., Jarvinen, B. R., McAdie, C. J., and Elms, J. D. 1993. *Tropical Cyclones of the North Atlantic Ocean, 1871–1992 (with Storm Tracks Updated through 1993)*. Fourth Revision, November 1993. Asheville, N.C.: National Climatic Center, National Oceanic and Atmospheric Administration, Historical Climatology Series 6-2.

Noel, J. M., Maxwell, A., Platt, W. J., and Pace, L. 1995. Effects of Hurricane Andrew on cypress (*Taxodium distichum* var. *nutans*) in south Florida. *Journal of Coastal Research* 21:184–196.

Noel, J. M., Platt, W. J., and Moser, E. B. 1998. Structural characteristics of old- and second-growth stands of longleaf pine (*Pinus palustris*) in the Gulf Coastal region of the U.S.A. *Conservation Biology* 12:533–548.

Noss, R. F. 1989. Longleaf pine and wiregrass: keystone components of an endangered ecosystem. *Natural Areas Journal* 9:211–213.

Noss, R. F., LaRoe, III, E. T., and Scott, J. M. 1995. *Endangered Ecosystems of the United States: A Preliminary Assessment of Loss and Degradation*. U.S. Department of Interior, National Biological Survey Biological Report 28.

Olson, M. S., and Platt, W. J. 1995. Effects of habitat and growing season fires on resprouting of shrubs in longleaf pine savannas. *Vegetatio* 119:101–118.

Outcalt, K. W., and Outcalt, P. A. 1994. *The Longleaf Pine Ecosystem: An Assessment of Current Conditions*. Unpublished data on file. USDA Forest Service, Southern Research Station, Gainesville, Fla.

Parrott, R. T. 1967. A study of wiregrass (*Aristida stricta* Michx.) with particular reference to fire. M.S. thesis, Duke University, Durham, N.C.

Peet, R. K. 1993. A taxonomic study of *Aristida stricta* and *A. beyrichiana*. *Rhodora* 95:25–37.

Peet, R. K., and Allard, D. J. 1993. Longleaf pine vegetation of the southern Atlantic and eastern Gulf coast regions: a preliminary classification. *Proceedings Tall Timbers Fire Ecology Conference* 18:45–81.

Peet, R. K., Glenn-Lewin, D. C., and Wolf, J. W. 1983. Prediction of man's impact on plant species diversity: a challenge for vegetation science. In *Man's Impact on Vegetation*, ed. W. Holzner, M. J. A. Werger, and I. Ikusima, pp. 41–54. The Hague: Dr. W. Junk.

Phillips, W. S. 1940. A tropical hammock on the Miami (Florida) limestone. *Ecology* 21:166–175.

Platt, W. J. 1994. Evolutionary models of plant population/community dynamics and conservation of southeastern pine savannas. In *Proceedings of the North American Conference on Savannas and Barrens*, ed. J. S. Fralish, R. C. Anderson, J. E. Ebinger, and R. Szafoni, pp. 262–273. Chicago, Ill.: U.S. Environmental Protection Agency, Great Lakes National Program Office.

Platt, W. J., Evans, G. W., and Davis, M. M. 1988. Effects of fire season on flowering of forbs and shrubs in longleaf pine forests. *Oecologia* 76:353–363.

Platt, W. J., Evans, G. W., and Rathbun, S. L. 1988. The population dynamics of a long-lived conifer (*Pinus palustris*). *American Naturalist* 131:491–525.

Platt, W. J., Glitzenstein, J. S., and Streng, D. R. 1991. Evaluating pyrogenicity and its effects on vegetation in longleaf pine savannas. *Proceedings Tall Timbers Fire Ecology Conference* 17:143–161.

Platt, W. J., and Rathbun, S. L. 1993. Dynamics of an old-growth longleaf pine population. *Proceedings Tall Timbers Fire Ecology Conference* 18:275–297.

Platt, W. J., and Schwartz, M. W. 1990. Temperate hardwood forests. In *Ecosystems of Florida*, ed. R. Myers and J. Ewel, pp. 194–229. Orlando, Fla.: University of Florida Press.

Plummer, G. L. 1963. Soils of the pitcher plant habitats in the Georgia Coastal Plain. *Ecology* 44:727–734.

Porter, M. L., and Labisky, R. F. 1986. Home range and foraging habitat of red-cockaded woodpeckers in north Florida. *Journal of Wildlife Management* 50:239–247.

Quarterman, E., and Keever, K. 1962. Southern mixed hardwood forest: climax in the southeastern Coastal Plain: U.S.A. *Ecological Monographs* 32:167–185.

Rebertus, A. J., Williamson, G. B., and Moser, E. B. 1989a. Fire-induced changes in *Quercus laevis* spatial pattern in Florida sandhills. *Journal of Ecology* 77:638–650.

Rebertus, A. J., Williamson, G. B., and Moser, E. B. 1989b. Longleaf pine pyrogenicity and turkey oak mortality in Florida xeric sandhills. *Ecology* 70:60–70.

Rebertus, A. J., Williamson, G. B., and Platt, W. J. 1993. Impacts of temporal variation in fire regime on savanna oaks and pines. *Proceedings Tall Timbers Fire Ecology Conference* 18:215–225.

Rejmanek, M. 1989. Invasibility of plant communities. In *Biological Invasions: A Global Perspective*, ed. J. A. Drake, H. A. Mooney, F. di Castri, R. H. Groves, F. J. Kruger, M. Rejmanek, and M. Williamson, pp. 369–388. New York: John Wiley & Sons.

Rejmanek, M. 1996. Species richness and resistance to invasions. In *Diversity and Processes in Tropical Forest Ecosystems*, ed. G. H. Orians, R. Dirzo, and J. H. Cushman, pp. 153–172. New York: Springer-Verlag.

Rejmanek, M., and Richardson, D. M. 1996. What attributes make some plant species more invasive? *Ecology* 77:1655–1661.

Richardson, D. M., and Bond, W. J. 1991. Determinants of plant distribution: evidence from pine invasions. *American Naturalist* 137:639–668.

Robbins, L. E., and Myers, R. L. 1992. *Seasonal Effects of Prescribed Burning in Florida: A Review.* Tall Timbers Research Station, Miscellaneous Publication 8.

Robertson, W. B., Jr. 1953. *A Survey of the Effects of Fire in Everglades National Park.* Mimeographed report. Homestead, Fla.: U.S. Department of the Interior, National Park Service, Everglades National Park.

Scholes, R. J., and Archer, S. R. 1997. Tree–grass interactions in savannnas. *Annual Review of Ecology and Systematics* 28:517–544.

Schwartz, M. W. 1994. Natural distribution and abundance of forest species and communities in northern Florida. *Ecology* 75:687–705.

Schwarz, G. F. 1907. *The Longleaf Pine in Virgin Forest: A Silvical Study.* New York: John Wiley and Sons.

Seamon, P. A., Myers, R. L., and Robbins, L. E. 1989. Wiregrass reproduction and community restoration. *Natural Areas Journal* 9:264.

Shafizadeh, F., and DeGroot, W. F. 1976. Combustion characteristics of cellulosic fuels. In *Thermal Uses and Properties of Carbohydrates and Lignins*, ed. F. Shafizadeh, K. V. Sarkanen, and D. A. Tillman, pp. 1–17. New York: Academic Press.

Silvertown, J. A. 1980. The evolutionary ecology of mast seeding in trees. *Biological Journal of the Linnean Society* 14:235–250.

Simberloff, D. 1993. Species-area and fragmentation effects on old-growth forests: prospects for longleaf pine communities. *Proceedings Tall Timbers Fire Ecology Conference* 18:1–13.

Simpson, R. H., and Lawrence, M. B. 1971. *Atlantic Hurricane Frequencies along the U.S. Coastline*. National Oceanic and Atmospheric Administration, Technical Memo. NWS SR-58.

Small, J. K. 1921a. Historic trails by land and by water. *Journal of the New York Botanical Garden* 22:193–222.

Small, J. K. 1921b. Old trails and new discoveries. A record of exploration in Florida in the spring of 1919. *Journal of the New York Botanical Garden* 22:25–40; 49–64.

Small, J. K. 1929. The Everglades. *Scientific Monthly* 28:80–87.

Small, J. K. 1930. Vegetation and erosion on the Everglades Keys. *Scientific Monthly* 30:33–49.

Smith, R. H. 1965. Effect of monterpene vapors on the western pine beetle. *Journal of Economic Entomology* 58:509–510.

Snyder, J. R. 1986. *The Impact of Wet Season and Dry Season Prescribed Fires on Miami Rock Ridge Pineland, Everglades National Park*. South Florida Research Center Report SFRC- 86/06.

Snyder, J. R. 1991. Fire regimes in subtropical south Florida. *Proceedings Tall Timbers Fire Ecology Conference* 17:303–319.

Snyder, J. R., Herndon, A., and Robertson, W. B. 1990. South Florida rockland ecosystems: tropical hammocks and pinelands. In *Ecosystems of Florida*, ed. R. Myers and J. Ewel, pp. 230–274. Orlando, Fla.: University of Florida Press.

Soil Survey Staff. 1975. *Soil Taxonomy: A Basic System of Classification for Making and Interpreting Soil Surveys*. USDA Soil Conservation Service, Agriculture Handbook Number 436.

Stahle, D. W., Cleaveland, M. K., and Hehr, J. G. 1988. North Carolina climate changes reconstructed from tree rings: A.D. 372 to 1985. *Science* 240:1517–1519.

Stoddard, H. L. 1935. Use of controlled fire in southeastern upland game management. *Journal of Forestry* 33:346–351.

Stout, I. J., and Marion, W. R. 1993. Pine flatwoods and xeric pine forests of the southern (lower) Coastal Plain. In *Biodiversity of the Southeastern United States. Lowland Terrestrial Communities*, ed. W. H. Martin, S. G. Boyce, and A. C. Echternacht, pp. 373–446. New York: John Wiley & Sons.

Streng, D. R., Glitzenstein, J. S., and Platt, J. W. 1993. Evaluating season of burn in longleaf pine forests: a critical literature review and some results from an ongoing long-term study. *Proceedings Tall Timbers Fire Ecology Conference* 18:227–263.

Streng, D. R., and Harcombe, P. A. 1982. Why don't east Texas savannas grow up to forest? *The American Midland Naturalist* 108:278–294.

Tansley, A. G. 1935. The use and abuse of vegetational concepts and terms. *Ecology* 16:284–307.

Taylor, A. R. 1974. Ecological aspects of lightning in forests. *Proceedings Tall Timbers Fire Ecology Conference* 13:455–482.

Taylor, D. L. 1981. *Fire History and Fire Records for Everglades National Park, 1948–1979*. Report T-619. Homestead, Fla.: U.S. National Park Service, South Florida Research Center.

Tebo, M. 1985. The southeastern piney woods: describers, destroyers, survivors. M.A. thesis, Florida State University, Tallahassee, Fla.

Thornbury, W. D. 1965. *Regional Geomorphology of the United States*. New York: John Wiley & Sons.

Tomlinson, P. B. 1980. *The Biology of Trees Native to Tropical Florida*. Allston, Mass.: Harvard University Printing Office.

Vogl, R. J. 1973. Fire in the southeastern grasslands. *Proceedings Tall Timbers Fire Ecology Conference* 12:175–198.

Wade, D., Ewel, J., and Hofstetter, R. 1980. *Fire in South Florida Ecosystems*. USDA Forest Service, Gen. Tech. Rep. SE-17.

Wahlenberg, W. G. 1946. *Longleaf Pine: Its Use, Ecology, Regeneration, Protection, Growth and Management*. Washington, D.C.: Charles Lathrop Pack Forestry Foundation.

Waldrop, T. A., White, D. L., and Jones, S. M. 1992. Fire regimes for pine–grassland communities in the southeastern United States. *Forest Ecology and Management* 47:195–210.

Walker, H. J., and Coleman, J. M. 1987. Atlantic and gulf Coastal provinces. In *Geomorphic Systems of North America: Centennial Special Volume 2*, ed. W. L. Graf, pp. 51–110. Boulder, Colo.: Geological Society of America.

Walker, J. 1993. Rare vascular plant taxa associated with the longleaf pine ecosystems: patterns in taxonomy and ecology. *Proceedings Tall Timbers Fire Ecology Conference* 18:105–125.

Walker, J., and Peet, R. K. 1983. Composition and species diversity of pine–wiregrass savannas of the Green Swamp, North Carolina. *Vegetatio* 55:163–179.

Ware, S., Frost, C., and Doerr, P. D. 1993. Southern mixed hardwood forest: the former longleaf pine forest. In *Biodiversity of the Southeastern United States. Lowland Terrestrial Communities,* ed. W. H. Martin, S. G. Boyce, and A. C. Echternacht, pp. 447–493. New York: John Wiley & Sons.

Watts, W. A. 1992. Camel lake: a 40,000 year record of vegetational and forest history from northwest Florida. *Ecology* 73:1056–1066.

Webb, S. D. 1990. Historical biogeography. In *Ecosystems of Florida,* ed. R. Myers and J. Ewel, pp. 70–100. Orlando, Fla.: University of Florida Press.

Webb, S. D., and Wilkins, K. T. 1984. Historical biogeography of Florida Pleistocene mammals. In *Contributions in Quaternary Vertebrate Paleontology, Special Publication of the Carnegie Museum Number 8,* ed. H. H. Genoways and M. R. Dawson, pp. 370–383. Pittsburgh, Pa.: Carnegie Museum.

Wells, B. W. 1928. Plant communities of the Coastal Plain of North Carolina and their successional relations. *Ecology* 9:230–242.

Wells, B. W., and Shunk, I. V. 1931. The vegetation and habitat factors of coarser sands of the North Carolina Coastal Plain: an ecological study. *Ecological Monographs* 1:465–520.

Werner, P. A., ed. 1991. *Savanna Ecology and Management: Australian Prespectives and Intercontinental Comparisons.* Oxford: Blackwell Scientific Publications.

Whelan, R. J. 1985. Patterns of recruitment to plant populations after fire in western Australia and Florida. *Proceedings of the Ecological Society of Australia* 14:169–178.

Whelan, R. J. 1995. *The Ecology of Fire.* Cambridge: Cambridge University Press.

Whitmore, T. C. 1991. Invasive woody plants in perhumid tropical climates. In *Ecology of Biological Invasions in the Tropics,* ed. P. S. Ramakrisnan, pp. 35–40. New Delhi: International Scientific Publications.

Whittaker, R. H. 1953. A consideration of climax theory: the climax as population and pattern. *Ecological Monographs* 23:41–78.

Williams, J. L. 1827. *A View of West Florida.* Philadelphia: H. S. Tanner & J. L. Williams, Publishers.

Williams, J. L. 1837. *The Territory of Florida: or Sketches of the Topography, Civil and Natural History of the Country, the Climate, and the Indian Tribes from the First Discovery to the Present Time.* Gainesville, Fla.: University of Florida, Facsimile Edition, 1962.

Williamson, G. B., and Black, E. M. 1981. High temperatures of forest fires under pines as a selective advantage over oaks. *Nature* 293:643–644.

Winsberg, M. D. 1990. *Florida Weather.* Orlando: University of Central Florida Press.

Woolfenden, G. E. 1959. A Pleistocene avifauna from Rock Springs, Florida. *Wilson Bulletin* 71:183–187.

3 New Jersey Pine Plains

The "True Barrens" of the New Jersey Pine Barrens

DAVID J. GIBSON, ROBERT A. ZAMPELLA,
AND ANDREW G. WINDISCH

A snapshot of the Plains will often seem to take in huge expanses of forest,
as if the picture had been made from a low-flying airplane, unless a human
being happens to have been standing in the camera's range, in which case
the person's head seems almost grotesque and planetary, outlined in sky
above the tops of the trees.

John McPhee (1968)

Introduction

The New Jersey Pine Barrens, or Pinelands, comprise a 550,000-ha mosaic of upland and wetland vegetation on the outer Coastal Plain of southern New Jersey (Little 1979; McCormick and Forman 1979). As noted by Little (1979), the physiognomy of much of the area does not resemble the common perception of areas termed *barrens*; that is, areas relatively "bare" of tree growth or with only stunted trees (Heikens and Robertson 1994; Homoya 1994; Tyndall 1994). In the lowlands, Atlantic white cedar (*Chamaecyparis thyoides*) swamp forests approach 26,200 trees ha^{-1} with a basal area of 56–57 m^2 ha^{-1} (McCormick 1979). In the upland oak–pine forest, black oak (*Quercus velutina*), white oak (*Q. alba*), scarlet oak (*Q. coccinea*), chestnut oak (*Q. prinus*), pitch pine (*Pinus rigida*), and shortleaf pine (*P. echinata*) tree density can average 824 trees ha^{-1} with a basal area of 22 m^2 ha^{-1} (Gibson, Collins and Good 1988). Nevertheless, within the Pinelands there exist a number of distinct regions of dwarfed pitch pine forest known as the Pine Plains or pygmy forest (McCormick and Buell 1968; Good, Good and Andresen 1979; Windisch 1986).

Other areas of the upland forest outside the Pine Plains also have similar "barrens" characteristics. It is these areas of "true barrens" that form the focus of this chapter. We also limit our coverage to upland areas. Except for comparative purposes, the generally more diverse wetland areas bordering the stream channels that interdigitate the Pine Plains are excluded.

Historic and Current Geographic Distribution

Depending upon differences in definition, 4,000–9,000 ha of Pine Plains originally existed (Harshberger 1916; Good, Good and Andresen 1979; Windisch 1986; Buchholz and Zampella 1987). The two largest areas, the East Plains (2,368 ha) and the West Plains (2,467 ha), are separated by approximately 6 km (Good, Good and Andresen 1979). Both straddle the Ocean and Burlington County borders near Warren Grove (Figure 3.1). Smaller tracts known as Little Plains (268 ha), Spring Hill Plains (79 ha), Bass River Plains (49 ha), and the Westecunk Plains (31 ha), are found within 2–3 km of the East and West Plains, and scattered small occurrences are found in the Forked River Mountains 7–8

Roger C. Anderson, James S. Fralish, and Jerry M. Baskin, eds. *Savannas, Barrens, and Rock Outcrop Plant Communities of North America.* Copyright © 1999 Cambridge University Press. All rights reserved.

km northeast of the West Plains (Windisch 1986). Approximately 60,000 ha (10%–15%) of the Pinelands region surrounding, and northeast of, the Pine Plains is covered with stunted pitch pine – and blackjack oak (*Quercus marilandica*) – dominated forests with little or no cover by tree-sized oaks (woody plants >3 m height). These barrens (Mc-Cormick 1970, 1979; New Jersey Natural Heritage Program Database, New Jersey Department of Environmental Protection, Trenton, N.J.), along with the Pine Plains, make up the "true barrens" of the New Jersey Pinelands. Other communities in the Pinelands with tree-sized oaks and shortleaf pine tend to lack or poorly express the distinct fire regime, physiognomy, and characteristic species assemblages of the "true barrens" communities.

The barrens character of the area was noted in early accounts, including the earliest land surveys of 1832–33 (Lutz 1934). Early Dutch navigators did not hold a favorable view of the area for agricultural settlement and negotiated larger-than-normal individual land grants "because there is much barren land." Assessing the soils in 1685, John Reid, a Scottish landscape architect, noted, "There are also some barren land, viz white sandy land, full of Pine trees." George Keith, a land surveyor, in 1687 wrote of "a great tract of barren lands consisting of pine land and sand." (All quotes are cited by Wacker 1979.) An advertisement in 1823 for a coach trip refers to the "Grouse Plains" in reference to the now extinct, but then common, heath hen (*Tympanuchus cupido cupido*) (Good, Good and Andresen 1979).

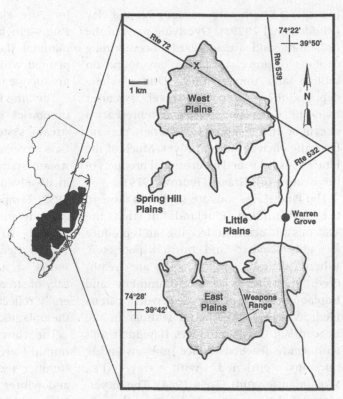

Figure 3.1. Distribution of Pine Plains vegetation in the Pinelands (shaded) of New Jersey (adapted from Buchholz and Zampella 1987). The approximate location of the former old Cedar Bridge fire tower (Figure 3.4) is shown by the X in the West Plains. The small Bass River Plains and Westecunk Plains are not shown, but occur 1 km to the east and south, respectively, of the East Plains (Windisch 1986).

Geology, Soils, and Climate

The geology, soils, and climate of the Pine Plains do not differ appreciably from the surrounding Pinelands. Indeed, this lack of a consistent environmental characteristic for the Pine Plains has led to much controversy and speculation regarding their origin (Lutz 1934; Good, Good and Andresen 1979).

The major surficial geological formation underlying the Pine Plains is the Cohansey sand, a deep (30–76 m) Miocene deposit of deltaic, beach, and tidal flat quartz-arenite, interspersed with carbonaceous silt and clay (Rhodehamel 1979a). Overlying parts of the Cohansey sand are localized unconforming deposits of gravel that often are found on hilltops and upper slopes. One of these deposits, the Beacon Hill Gravel, is coarser than the Cohansey Sand and represents a stream or fluvial deposit of materials, perhaps from the ancient Hudson River. Much of the Pine Plains is formed on erosional products of the Beacon Hill Gravel (Tedrow 1979)

The Pine Plains soils are similar to those in the surrounding Pinelands in that the siliceous and acidic underlying sands produce a highly leached and nutrient-poor soil. When Cohansey sand deposits are weathered, they form Quartzipsamment and Haplaquod soils of the Lakewood catena (Tedrow 1979). Weathered feldspar in the Beacon Hill Gravel produces Hapludult soils with sandy, leached surface horizons and B_t horizons enriched with clay (i.e., Woodmansie sand) (Trela 1984). The mineralogy of the parent materials contribute to low cation exchange capacity and pH in the Woodmansie, the latter ranging from 3.8 in the A_1 horizon to 4.1 in the C_1 horizon (Douglas and Trela 1979).

Several late-19th-century investigators believed that soils of the Pine Plains were lower in essential nutrients than those in the surrounding Pinelands (Good, Good, and Andresen 1979). Others have suggested that the common hardpan or ironstone, aluminum toxicity, strong winds, or the elevated land was a cause of the Pine Plains (Pinchot 1899; Harshberger 1916; Joffe and Watson 1933). Whittaker (1979) suggested that the Pine Plains represent an edaphic climax in which vegetational heterogeneity is the result of gradients in soil texture and nutrients. Buchholz and Motto (1981) showed that Pine Plains soils had more active soil mycorrhizae than soils collected in the surrounding Pinelands, and Buchholz (1980) suggested that soil calcium may be limiting

for Pine Plains pitch pine. However, Pine Plains–Pinelands differences in the soils are minimal (Lutz 1934; Andresen 1959), and planted white pine (*Pinus strobus*) is equally productive in both areas (Little 1972).

The Pine Plains are drained by streams and tributaries of three Atlantic Coastal basin stream systems; the Oswego, Wading, and Bass rivers. The surface waters of these stream systems are acidic (pH <4.5) and low in dissolved substances (Morgan and Good 1988; Zampella 1994). In Pinelands streams, groundwater discharge accounts for 89% of annual stream discharge (Rhodehamel 1979b). Because groundwater is the primary source of stream flow, surface water chemistry of streams draining the Pine Plains generally reflects that of water discharging from the uplands.

The climate of the region is temperate, humid, predominantly continental, with summer temperatures averaging 22–24 °C, and winter temperatures averaging 0–2 °C (Havens 1979; Rhodehamel 1979b). Average annual precipitation is 1,097 mm with the mid- to late-summer months receiving the highest levels. There is considerable year-to-year variation in precipitation, with both extended summer droughts and heavy rains and storms being frequent.

Composition and Structure of Vegetation

As part of the New Jersey Pinelands, the Pine Plains form part of Braun's Atlantic Slope section of the Oak–Pine Forest Region in the deciduous forest formation (Braun 1950). Christensen (1988) included these forests within the vegetation of the southeastern Coastal Plain. Dwarf pine plains similar to those considered here also occur on Long Island and on the summits of the Shawangunk Mountains, New York (McIntosh 1959; Olsvig, Cryan and Whittaker 1979).

Pine Plains communities are characterized

Figure 3.2. View of the West Plains off Stevenson Road on October 11, 1982 (photography by R. A. Zampella). Tree canopy height is 2–3 m.

by a single- to trilayered dwarf woodland dominated by 1–3-m-tall pitch pine and blackjack oak trees (Good, Good and Andresen 1979) (Figure 3.2). Scrub or bear oak (*Quercus ilicifolia*) is also common, along with several shrubs such as mountain laurel (*Kalmia latifolia*), inkberry (*Ilex glabra*), sweetfern (*Comptonia peregrina*), black huckleberry (*Gaylussacia baccata*), dangleberry (*G. frondosa*), and low blueberry (*Vaccinium pallidum*) (Table 3.1). Black huckleberry and low blueberry are the most abundant shrubs (Little 1979). Trees common in the surrounding Pinelands, such as shortleaf pine, black oak, scarlet oak, white oak, and chestnut oak, are absent (Good, Good and Andresen 1979), and they are thought to have been eliminated by fire when the Pine Plains formed hundreds or thousands of years ago (McCormick 1979). Bearberry (*Arctostaphylos*

uva-ursi), false heather (*Hudsonia ericoides*), little bluestem (*Schizachyrium scoparium*), and pyxie moss (*Pyxidanthera barbulata*) are recorded as being abundant herbs and shrubs (Table 3.1). Floristically, Lutz (1934) listed 52 vascular plant species (1 fern, 1 gymnosperm, and 50 angiosperms) and 30 nonvascular plants (27 lichens and 3 mosses) from the Pine Plains.

In a phytosociological survey of the Pinelands, Olsson (1979) categorized seven relevés from the Pine Plains in a separate unit, Entity A2-Pygmy Forest, within the pine–oak vegetation type. Detrended Correspondence Analysis (Hill and Gauch 1980) of the 12 vegetation entities of Olsson that contained pitch pine and blackjack oak showed the pygmy forest entity from the Pine Plains to be an extreme representative of Pinelands vegetation (Figure 3.3). In this

Table 3.1. *Common higher plants of the Pine Plains. Nomenclature follows Gleason and Cronquist (1991). Some rows sum to 101% because of rounding.*

Species	Frequency of occurrence				Average abundance	
	Abundant	Common(%)[1]	Rare		Absent	(%)[2]
Dwarfed trees and shrubs						
Gaylussacia baccata	100	-	-		-	>25
Pinus rigida	81	13	6		-	>25
Quercus ilicifolia	50	44	6		-	6.25–25
Quercus marilandica	25	44	25		6	>25
Vaccinium pallidum	13	75	13		-	6.25–25
Comptonia peregrina	13	25	56		6	-
Gaylussacia frondosa	-	25	44		31	-
Ilex glabra	-	13	44		44	-
Kalmia latifolia	-	13	25		63	<6.25
Smilax glauca	-	6	31		63	-
Quercus stellata	-	6	-		94	-
Subshrubs and herbaceous plants						
Hudsonia ericoides	25	38	19		19	<6.25
Arctostaphylos uva-ursi	19	38	25		19	-
Pyxidanthera barbulata	6	13	50		31	-
Schizachyrium scoparium	-	31	69		-	-
Gaultheria procumbens	-	25	44		31	<6.25
Tephrosia virginiana	-	25	38		38	-
Melampyrum lineare	-	19	56		25	-
Leiophyllum buxifolium	-	13	6		81	-
Aster linariifolius	-	6	81		13	-
Epigaea repens	-	6	44		50	-
Corema conradii	-	-	19		81	6.25–25

[1] From 16 stations surveyed by Lutz (1934).
[2] From a synopsis of 7 releves by Olsson (1979)

analysis, pitch pine, scrub oak, lichen (*Cladonia strepsilis*), and broom crowberry (*Corema conradii*) were assigned negative DCA axis 1 loadings to reflect their high average abundance and/or constancy in the Pine Plains vegetation entity. The principal vegetation gradient from Olsson's data, which reflects soil moisture as DCA axis 1, contrasts the pygmy forest and other oak–pine forests with marsh vegetation.

Lutz (1934) noted that, to an observer,

much of the pitch pine in the Pine Plains appeared always to be 12 years old; this observation is due in part to the small size of the trees as well as the young age of first reproduction. Based upon ring counts from stem sections cut at the soil surface, Buchholz and Good (1982) reported a mean stand age of 8.9 ± 2.6 years, and similar ages were reported by Lutz (1934). However, the maximum age of the genets remain unknown, but Pinchot (1899) aged the main root of a num-

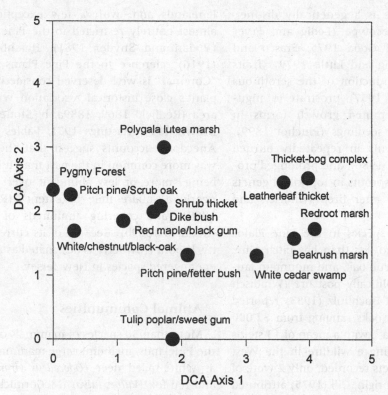

Figure 3.3. Detrended Correspondence Analysis of Vegetation Entities of Olsson (1979) containing woody vegetation. Eigenvalues and cumulative percentage variance accounted for in the data set were 0.65% and 23% for DCA axis 1, and 0.30% and 33% for DCA axis 2.

ber of stumps to be 40–100 years old. Buchholz and Good (1982) reported a mean density of 24,900 stems ha^{-1} of pitch pine from the Pine Plains and a mean net annual aboveground production of 9,020 kg ha^{-1}. These levels of density and production are higher than in the surrounding Pinelands or in the Long Island Pine Barrens. The density of pitch pine stools ranges from 2,500 to 10,000 ha^{-1} with density decreasing during the years following fire (Andresen 1959; Good, Good and Andresen 1979). Gill (1975) attributed the uniform spatial dispersion of pitch pine individuals to seed predation.

Pitch pine has a thick bark allowing relatively high stem survival during fire, and there is a well-developed basal crook that further protects dormant buds just below the soil surface (Ledig and Little 1979). Despite

prodigious seed production from serotinous cones, which open at temperatures ranging from 47 °C to 75 °C (Frasco and Good 1976; Ledig and Little 1979; Givnish 1981), regeneration following fire is almost entirely by root crown sprouting, with less than 1% of stems originating from seedlings (Buchholz and Good 1982; Buchholz 1983). Nearly 100% of the Pine Plains pitch pine have serotinous cones, with the frequency decreasing in the surrounding areas where fire frequencies decline (Givnish 1981). Seedlings that become established grow slowly, and Pine Plains seedlings are shorter, lighter, and have a more shrubby growth form than individuals derived from seed collected in the surrounding Pinelands (Good and Good 1975).

Comparative studies indicate that the Pine

Plains pitch pine is a genetically distinct, fire-maintained ecotype (Ledig and Fryer 1972; Good and Good 1975; Frasco and Good 1976; Ledig and Little 1979). Traits include early production of the serotinous cones (Andresen 1957), prostrate to angular, multiple-stemmed growth forms in many unburned seedlings (Pinchot 1899), poor growth forms in repeatedly burned genets (Windisch 1986), and prolonged production of basal sprouts in top-killed genets for 10–15 years after fire (Windisch and Good 1991).

Other canopy species in the Pine Plains have been studied less than has pitch pine. Blackjack oak, scrub oak, and mountain laurel all sprout prolifically post fire (Windisch and Good 1991). Buchholz (1983) reported stem densities of oaks ranging from 82,000 ha^{-1} to 239,000 ha^{-1}, with a mean of 13 stems per genet following a wildfire in the West Plains. Of 65 genets recorded, only 2 were of postfire seedling origin. Gill (1975) attributed the random dispersal of individual blackjack oak clumps within larger, asymmetric clumps (400 m^2) to fire, which prevented seedling recruitment.

There are no plant species endemic to the Pine Plains; however, broom crowberry or corema (*Corema conradii*) is characteristic of the area. Broom crowberry is a highly branched, dioecious shrub up to 5 dm, in height, with terminal heads of purplish flowers (Gleason and Cronquist 1991). The species was first described by John Torrey, who visited a West Plains location in 1833 (Redfield 1884). Most adult plants are killed by fire; however, broom crowberry reproduction appears to be dependent on fire to stimulate germination of seed accumulated in the soil seed bank (Dunwiddie 1990).

Broom crowberry is one of 14 northern species reaching the southern limit of their geographical range in the Pinelands (McCormick 1970). In New Jersey, this relict species (Fairbrothers 1979) is limited to the Pinelands and, with a few exceptions, is almost entirely restricted to the Pine Plains (Vivian and Snyder 1981). Harshberger's (1916) reference to the Pine Plains as the "Coremal" is well deserved considering the plant's close historical association with the area (Redfield 1869, 1889a, b; Stone 1911; Brown 1914; Hastings 1931; Fables 1948). Anecdotal accounts suggest that the plant was more common in the past than it is now, being "more or less abundant over a tract nearly a square mile in extent" (Redfield 1889a) and covering "hundreds of acres" (Redfield 1889b). Because of its current rarity, broom crowberry is designated as a state-endangered species in New Jersey.

Animal Communities

More than 30 species of mammals occur in the Pinelands, including large mammals such as white-tailed deer (*Odocoileus virginianus*) and red fox (*Vulpes fulva*) (McCormick 1970; Wolgast 1979). However, the only published accounts of mammals found in the Pine Plains deal with shrews and rodents. Connor (1953) trapped only the white-footed mouse (*Peromyscus leucopus*) in the Pine Plains. The white-footed mouse typically dominates dry Pinelands uplands (Craig and Dobkin 1993). A more diverse small mammal community is found in wetlands of the Pine Plains and non-Plains pitch pine forest (Connor 1953; Craig and Dobkin 1993).

Bird diversity in the central portions of the Pinelands is generally limited (McCormick 1970; Leck 1979; Kerlinger 1984), and this characterization of the ecosystem applies equally to the Pine Plains. Brush (1987) notes that the summer resident communities in the West Plains are dominated by the prairie warbler (*Dendroica discolor*), rufous-sided towhee (*Pipilo erythrophthalmus*), and brown thrasher (*Toxostoma rufum*). Carolina chickadees (*Parus carolinensis*) are the most common resident species. Migrants and winter residents are rare. Historically, the Pine Plains was the habitat and last New Jersey stronghold for the now-extinct heath hen. By 1870, the Pine Plains population of this once abundant and heavily hunted species was extinct (Stone 1909).

The New Jersey Natural Heritage Program Database includes numerous documented occurrences of the pine snake (*Pitouphis melanoleucus*) and the Pine Barrens treefrog (*Hyla andersoni*) in the Pine Plains and vicinity, and some widely scattered accounts of the timber rattlesnake (*Crotalus horridus*) and corn snake (*Elaphe guttata*) in adjacent pinelands. The Pine Barrens treefrog, corn snake, and timber rattlesnake are listed as state endangered, and the pine snake as state threatened.

Of the 25 or more rare Lepidoptera found in the New Jersey Pinelands, many have been documented in the Pine Plains and surrounding pitch pine–blackjack oak/scrub oak barrens (Schweitzer 1989; New Jersey Natural Heritage Program Database). Several of these rare Lepidoptera, such as *Agrotis buchholze* (Buchholz's dart moth), *Catocala herodias gerhardi* (pine barrens underwing moth), *Catocala jair* (jair underwing moth), *Heterocampa varia* (a notodontid moth), and *Itame* sp. (an unnamed species of geometrid moth), which were common in the West Plains, are generally absent in the nearby oak–pine forests (Madenjian 1979; species identifications revised by D. Schweitzer, The Nature Conservancy). Many of these rare Lepidoptera require food plants such as scrub and blackjack oak, and certain Pineland shrubs and herbs (Covell 1984), which are uncommon in infrequently burned oak–pine forests.

Ecological Relationships: The Role of Fire

Fire and its interaction with physical landscape features have played a major ecological role in structuring Pine Plains and associated community landscape patterns, species composition, physiognomy, and pitch pine population dynamics. At the landscape level, the Pine Plains are near the center of sandy uplands largely undissected by major streams. During the early part of the 20th century, anthropogenic fires burned into these sites from various distances and directions, at intervals of 4–15 years (Lutz 1934; Givnish 1981). Pitch pine–blackjack oak and scrub oak barrens occur in surrounding portions of these fire-prone uplands and on several other similarly large sandy upland divides (McCormick 1970). Frequent fires at less than 20–30 year intervals in both these communities have caused selective removal of all oak and shortleaf pine trees. The serotinous, shrubby ecotypes of pitch pine suggest that short fire intervals may have been typical in the Pine Plains for many centuries, or millennia.

Because of high flammability of mature communities in the Pine Plains, 90% or more of those burned are consumed by a high-intensity crown fire (Windisch and Good 1991). Following fire, canopy and surface fuel loads are drastically reduced, making stands highly resistant to burning for the next 4–5 years (Windisch and Good 1986). Crown fires produce near 100% stem mortality, but most genets quickly resprout. Immediately following fire, there is a pulse of sprout production that continues for at least 15 years (Buchholz and Good 1982). After 10–20 years of postfire growth, resprouting pine and oak genets begin to close the canopy and subcanopy, and surface litter covers the soil. During the first year after fire, combined pine and oak sprout densities may exceed 500,000 stems per ha (Buchholz 1983), but competition between sprouts within and among genets causes high mortality during the next several years. Canopy height increases and stem and genet density decreases with increasing stand age (i.e., time since last fire) as subcanopy and subordinate stems are

Figure 3.4. View of the former old Cedar Bridge fire tower (32 m tall) in the West Plains along State Highway Route 72, circa 1937 (photographer unknown). See Figure 3.1 for exact location. Height of the low-growing shrubby pitch pine and blackjack oak trees in the foreground is less than 2 m.

shaded out and several larger stems per genet occupy the canopy space (Andresen 1959; Good, Good and Andresen 1979; Buchholz and Good 1982; Windisch and Good 1991). These structural changes greatly increase canopy and surface fuel loads and stand flammability (Windisch and Good 1986, 1991).

Several headwater streams and intermittently wet swales and depressions that support wetland communities penetrate the Pine Plains. The wider, more saturated parts of these are surrounded by patches of arborescent pine barrens and upper transitional plains (Windisch 1986). These wetlands tend to burn less intensely than the surrounding uplands, and they often impede the spread or reduce the intensity of fires, creating a "fire shadow."

These "fire shadows" are analogous to the hardwood hammocks that develop adjacent to streams in sandhill or sandpine scrub regions in the southeastern United States (Laessle 1958), and to fire-sensitive *Pinus strobus–Pinus resinosa* or *Pinus–Abies* forests that develop next to lakes within relatively flammable *Pinus banksiana–Picea mariana*–sprout hardwood communities of Minnesota (Heinselman 1973).

As with other fire-dependent systems (Mushinsky and Gibson 1991), fire alters the habitat structure, which influences the species and density of birds, small mammals, amphibians, reptiles, arthropods, and soil microorganisms (Boyd and Marucci 1979; Conant 1979; Dindal 1979; Leck 1979; Wolgast 1979). For example, Brush (1987) noted a similar overall species composition in the avifauna of burned and unburned Pine Plains; however, downy woodpecker (*Picoides pubescens*), field sparrow (*Spizella pusilla*), and common nighthawk (*Chordeiles minor*) were found only on the burned plains. Tempel (1976) hypothesized that the opening of serotinous cones on pitch pine only after fire in the Pine Plains was, in part, an adaptive response to minimize seed predation by red squirrels (*Tamiasciurus hudsonicus*). Serotinous cones have fewer seeds per cone than nonserotinous cones. However, since they are retained on the plant until a fire, they then are subject to high levels of predation (Stiles 1979).

Historical descriptions and photographs (Figure 3.4) of the Pine Plains suggest the shrubland community was generally lower and much more open in the past, with abundant patches or networks of open sand and gravel (Redfield 1889b; Pinchot 1899; Stone 1911; Harshberger 1916). These accounts and Lutz (1934) suggest that the Pine Plains had much greater plant species diversity around the turn of the century than that noted in modern studies (Olsson 1979). Many shrub, subshrub, herb, and graminoid species found only sporadically, or not at all, in recent years (Olsson 1979) were far more frequent and/or abundant before fire suppression began in the 1940s (Forman and Boerner 1981). Historically, more frequent fires prevented the buildup of fuel loads and caused less intense fires than those recently observed (Windisch and Good 1991). The less intense crown fires and many sandy unburned gaps likely would have created "safe sites" where some fire-sensitive adult plants such as broom crowberry and pyxie moss (*Pyxidanthera barbulata*) survived. The greater abundance of graminoids and several flowering herbs and shrubs may have enhanced Lepidoptera populations. It is suggested that managing large portions of the current Pine Plains through prescribed burning at 5–15-year intervals will restore the historical richness in species diversity (Lutz 1934) and reduce catastrophic wildfire hazard.

Current Status, Management, and Threats

The chance that the Pine Plains would be lost through public or private real estate development decreased substantially with passage of the National Parks and Recreation Act of 1978 and the New Jersey Pinelands Protection Act of 1979. These laws created the Pinelands Commission, a state agency charged with developing and implementing a regional comprehensive management plan to protect the natural and cultural resources of the 405,000-ha Pinelands National Reserve (Good and Good 1984; Collins and Russell 1988). Protection of the Pine Plains by regulation and acquisition was a major element in the federal and state Pinelands legislation.

A major landmark in the development of the Pinelands preservation movement was a 1965 proposal by the Burlington County governing board to develop a major jetport and new city that would have eliminated the Pine Plains communities (Collins and Russell

1988). Opposition to this proposal heightened public concern and contributed to the momentum that eventually led to passage of Pinelands legislation.

Present land use in the Pine Plains is similar to that described by McCormick (1970). The major human developments include Coyle Airfield in the West Plains, the Warren Grove Weapons Range in the East Plains, and a Federal Aviation Administration (FAA) communication center in the Little Plains. Also, both the West Plains and East Plains are bisected by state highways (Figure 3.1). A major disturbance not described by McCormick (1970) is an upland site located along the western boundary of the West Plains where chemical wastes (volatile organics, semivolatile organics, and metals) were dumped during the 1950s and 1960s. A cleanup study of this site currently is being conducted through the federal superfund program.

Since 1979, more than 20,000 ha within the Plains and adjacent areas have come into public ownership through the New Jersey Pinelands acquisition program (Zampella 1988). Most of the Pine Plains are now publicly owned. During the 1980s, the Warren Grove Weapons Range extended its ownership and leasehold over approximately 3,400 ha of East Plains and adjacent areas. Use of both private and state lands located within and adjacent to the Plains area is strictly limited under regulations developed by the Pinelands Commission (Collins and Russell 1988). Under some circumstances, the regulations are applied to federal installations.

Although the fire-dependent Pine Plains is largely intact and preserved within publicly owned lands, its long term integrity is not assured. Fire suppression and the lack of a comprehensive prescribed burning program has eliminated the historical fire regime that maintained species and community diversity over large areas.

Research Needs

During this century, there has been an overall decrease in fire frequency within the Pinelands region (Forman and Boerner 1981; Buchholz and Zampella 1987). A detailed, long-term fire study would assess the ecological consequences of a changed fire regime on integrity of the Pine Plains communities; for example the effect of a loss of fire-adapted species and genotypes, and a decrease in fire patch size (Forman and Boerner 1981). Changes in land use within and adjacent to the Pine Plains must also be monitored to assess both direct (e.g., habitat destruction) and indirect (e.g., fireshed integrity) consequences.

The effect of wild and prescribed fire on the structure and composition of Pinelands vegetation has long been a topic of scientific research (Buell and Cantlon 1953; Little and Moore 1953; Forman 1979; Boerner 1981; Windisch and Good 1991; Zampella, Moore and Good 1992). However, the pre–European settlement fire regime is poorly understood, as all information is based upon recent fire records. Prescribed burns have been used in the Pine Plains for fire hazard reduction, but their ecological effects have not been fully assessed. Additional research on fire frequency and intensity, the relationship between recent fire and plant community attributes (e.g., species composition and physiognomy, pitch pine demography), and the effect of fire on the characteristic plant and animal species should be completed. The distribution and habitat relationships of broom crowberry should be documented further and used as a basis for management (Vivian 1978).

Research on reclamation and succession of disturbed sites is an important management need. For example, dwarf pitch pine and blackjack oak saplings grown from local seed were used to revegetate bulldozed areas at the Warren Grove Weapons Range (Fimbel and Kuser 1993). Recently reclaimed and disturbed areas provide an opportunity to monitor the long-term success of restoration efforts and to continue research on alternate revegetation approaches. The potentially negative effects of soil amendments (e.g., composted sludge, inorganic fertilizers) on species composition and groundwater also should be investigated.

Summary

The Pine Plains region of the New Jersey Pinelands occupies 4,000–9,000 ha in southern New Jersey. The sandy soils of this upland region are characterized by dwarfed and stunted pitch pine (*Pinus rigida*) and blackjack oak (*Quercus marilandica*) trees, along with a suite of fire-adapted shrubs and herbaceous plants. Historically, frequent fires prevented fuel buildup and maintained a dense shrub layer interspersed with open areas of bare sand and sparse plant cover. There are no species endemic to the region, but broom crowberry (*Corema conradii*), a relict species, generally is limited in its distribution in New Jersey to the Pine Plains. This fire-dependent landscape is largely intact and preserved within publicly owned lands. Nevertheless, the long-term integrity of the system will depend upon a better understanding, and the implementation, of a comprehensive prescribed burning program.

Acknowledgments

A tremendous debt of gratitude is owed to the late Ralph E. Good for sharing his knowledge and enthusiasm of the Pinelands with us. It was through his efforts that so much of the recent research into the Pinelands was conducted.

REFERENCES

Andresen, J. W. 1957. Precocity of *Pinus rigida* Mill. *Castanea* 22:130–134.

Andresen, J. W. 1959. A study of pseudo-nanism in *Pinus rigida*. *Ecological Monographs* 29:309–332.

Boerner, R. E. J. 1981. Forest structure dynamics following wildfire and prescribed burning in the New Jersey Pine Barrens. *The American Midland Naturalist* 105:321–333.

Boyd, H. P., and Marucci, P. E. 1979. Arthropods of the Pine Barrens. In *Pine Barrens: Ecosystem and Landscape*, ed. R. T. T. Forman, pp. 505–526. New York: Academic Press.

Braun, E. L. 1950. *Deciduous Forests of Eastern North America*. London: Collier Macmillan Pub. Co.

Brown, S. 1914. *Corema conradii* Torrey. *Bartonia* 6:1–7.

Brush, T. 1987. Birds of the central Pine Barrens: abundance and habitat use. *Bulletin of the New Jersey Academy of Science* 32:5–17.

Buchholz, K. 1980. Mineral nutrient accumulations in Plains and Barrens populations of *Pinus rigida* Mill., and an analysis of density, standing biomass and net annual aboveground productivity of Plains *P. rigida*. Ph.D. dissertation, Rutgers University, New Brunswick, N.J.

Buchholz, K. 1983. Initial responses of pine and oak to wildfire in the New Jersey Pine Barren Plains. *Bulletin of the Torrey Botanical Club* 110:91–96.

Buchholz, K., and Good, R. E. 1982. Density, age structure, biomass and net annual aboveground productivity of dwarfed *Pinus rigida* Mill. from the New Jersey Pine Barren Plains. *Bulletin of the Torrey Botanical Club* 109:24–34.

Buchholz, K., and Motto, H. 1981. Abundances and vertical distributions of mycorrhizae in Plains and Barrens forest soils from the New Jersey Pine Barrens. *Bulletin of the Torrey Botanical Club* 108:268–271.

Buchholz, K., and Zampella, R. A. 1987. A 30-year fire history of the New Jersey Pine Plains. *Bulletin of the New Jersey Academy of Science* 32:61–69.

Buell, M. F., and Cantlon, J. E. 1953. Effects of prescribed burning on ground cover in the New Jersey pine region. *Ecology* 34:520–528.

Christensen, N. L. 1988. Vegetation of the Southeastern Coastal Plain. In *North American Terrestrial Vegetation*, ed. M. Barbour and W. Billings, pp. 318–363. Cambridge: Cambridge University Press.

Collins, B. R., and Russell, E. W. B. 1988. *Protecting the New Jersey Pinelands*. New Brunswick, N.J.: Rutgers University Press.

Conant, R. 1979. A zoogeographical review of the amphibians and reptiles of southern New Jersey, with emphasis on the Pine Barrens. In *Pine Barrens: Ecosystem and Landscape*, ed. R. T. T. Forman, pp. 467–488. New York: Academic Press.

Connor, P. F. 1953. Notes on the mammals of a New Jersey Pine Barrens area. *Journal of Mammalogy* 34:227–235.

Covell, C. V. 1984. *A Field Guide to the Moths of Eastern North America*. Boston: Houghton Mifflin Company.

Craig, L. J., and Dobkin, D. S. 1993. *Community Dynamics of Small Mammals in Mature and Logged Atlantic White Cedar Swamps of the New Jersey Pine Barrens*. Albany,N.Y.: New York State Museum.

Dindal, D. L. 1979. Soil arthropod microcommunities of the Pine Barrens. In *Pine Barrens: Ecosystem and Landscape*, ed. R. T. T. Forman, pp. 527–539. New York: Academic Press.

Douglas, L. A., and Trela, J. J. 1979. Mineralogy of Pine Barrens soils. In *Pine Barrens: Ecosystem and Landscape*, ed. R. T. T. Forman, pp. 95–109. New York: Academic Press.

Dunwiddie, P. W. 1990. Rare plants in coastal heathlands: observations on *Corema conradii* (Empetraceae) and *Helianthemum dumosum* (Cistaceae). *Rhodora* 92:22–26.

Fables, D. 1948. Field trip reports, April 14, 1948, Pine Barrens. *Bulletin of the Torrey Botanical Club* 75:582.

Fairbrothers, D. E. 1979. Endangered, threatened, and rare vascular plants of the Pine Barrens and their biogeography. In *Pine Barrens: Ecosystem and Landscape*, ed. R. T. T. Forman, pp. 395–405. New York: Academic Press.

Fimbel, R. A., and Kuser, J. E. 1993. Restoring the pygmy pine forests of New Jersey Pine Barrens. *Restoration Ecology* 1:117–129.

Forman, R. T. T., ed.. 1979. *Pine Barrens: Ecosystem and Landscape*. New York: Academic Press.

Forman, R. T. T., and Boerner, R. E. 1981. Fire frequency and the Pine Barrens of New Jersey. *Bulletin of the Torrey Botanical Club* 108:34–50.

Frasco, B., and Good, R. E. 1976. Cone, seed and germination characteristics of pitch pine (*Pinus rigida* Mill.). *Bartonia* 44:50–57.

Gibson, D. J., Collins, S. L., and Good, R. E. 1988. Ecosystem fragmentation of oak–pine forest in the New Jersey Pinelands. *Forest Ecology and Management* 25:105–122.

Gill, D. E. 1975. Spatial patterning of pines and oaks in the New Jersey Pine Barrens. *Journal of Ecology* 63:291–298.

Givnish, T. 1981. Serotiny, geography and fire in

the Pine Barrens of New Jersey. *Evolution*
35:101–123.

Gleason, H. A., and Cronquist, A. 1991. *Manual of
Vascular Plants of Northeastern United States and
Adjacent Canada*. Bronx, New York: The New
York Botanical Garden.

Good, R. E., and Good, N. F. 1975. Growth char-
acteristics of two populations of *Pinus rigida*
Mill. from the Pine Barrens of New Jersey.
Ecology 56:1215–1220.

Good, R. E., and Good, N. F. 1984. The Pinelands
National Reserve: An ecosystem approach to
management. *BioScience* 34:169–173.

Good, R. E., Good, N. F., and Andresen, J. W.
1979. The Pine Barrens Plains. In *Pine Barrens:
Ecosystem and Landscape*, ed. R. T. T. Forman, pp.
283–295. New York: Academic Press.

Harshberger, J. W. 1916. *The Vegetation of the New
Jersey Pine Barrens, an Ecologic Investigation*.
Philadelphia: Christopher Sower Co.

Hastings, G. T. 1931. Weekend of September 25–27.
Bulletin of the Torrey Botanical Club 31:179–182.

Havens, A. V. 1979. Climate and microclimate of
the New Jersey Pine Barrens. In *Pine Barrens:
Ecosystem and Landscape*, ed. R. T. T. Forman, pp.
113–131. New York: Academic Press.

Heikens, A. L., and Robertson, P. A. 1994.
Barrens of the Midwest: A review of the literature. *Castanea* 59:184–194.

Heinselman, M. L. 1973. Fire in the virgin forests
of the Boundary Waters Canoe Area,
Minnesota. *Quaternary Research* 3:329–382.

Hill, M. O., and Gauch, H. G. 1980. Detrended
correspondence analysis: An improved ordination technique. *Vegetatio* 42:47–58.

Homoya, M. A. 1994. Barrens as an ecological
term: an overview of usage in the scientific lit-
erature. In *Proceedings of the North American
Conference on Barrens and Savannas: Living in the
Edge*, ed. J. S. Fralish, R. C. Anderson, J. E.
Ebinger, and R. Szafoni, pp. 295–303. Chicago,
Ill.: U.S. Environmental Protection Agency,
Great Lake National Program Office.

Joffe, C. F., and Watson, C. W. 1933. Soil profile
studies: V. Mature podzols. *Soil Science* 5:313–329.

Kerlinger, P. 1984. Avian community structure
along a successional gradient in the New Jersey
Pine Barrens. *Records of New Jersey Birds*
9:71–78.

Laessle, A. M. 1958. The origin and successional
relationships of sandhill vegetation and sand
pine scrub. *Ecological Monographs* 28:361–387.

Leck, C. F. 1979. Birds of the Pine Barrens. In
Pine Barrens: Ecosystem and Landscape, ed. R. T. T.
Forman, pp. 457–466. New York: Academic
Press.

Ledig, F. T., and Fryer, J. H. 1972. A pocket of
variability in *Pinus rigida*. *Evolution*
26:259–266.

Ledig, F. T., and Little, S. 1979. Pitch pine (*Pinus
rigida* Mill.): ecology, physiology, and genetics.
In *Pine Barrens: Ecosystem and Landscape*, ed. R.
T. T. Forman, pp. 347–371. New York:
Academic Press.

Little, S. 1972. Growth of planted white pine and
pitch pine seedlings in a South Jersey Plains
area. *Bulletin of the New Jersey Academy of Science*
17:18–23.

Little, S. 1979. The Pine Barrens of New Jersey. In
*Heathlands and Related Shrublands of the World. A.
Descriptive Studies*, ed. R. L. Specht, pp. 451–464.
Amsterdam: Elsevier Scientific Publishing
Company.

Little, S., and Moore, E. B. 1953. *Severe Burning
Treatment Tested on Lowland Pine Sites*. Station
Paper No. 64:1–11. U.S. Forest Service,
Northeastern Forest Experiment Station.

Lutz, H. J. 1934. Ecological relations in the pitch
pine Plains of southern New Jersey. *Yale
University School of Forestry Bulletin* 38:1–80.

Madenjian, J. 1979. The effects of vegetative
structure on Macroheteroceran species diversity at two sites in the New Jersey Pine
Barrens. M.S. thesis, Rutgers University, New
Brunswick, N.J.

McCormick, J. 1970. *The Pine Barrens: A Preliminary
Ecological Inventory*. Trenton, N.J.: New Jersey
State Museum. Museum Report No. 2.

McCormick, J. 1979. The vegetation of the New
Jersey Pine Barrens. In *Pine Barrens: Ecosystem
and Landscape*, ed. R. T. T. Forman, pp.
229–243. New York: Academic Press.

McCormick, J., and Buell, M. F. 1968. The
Plains: pygmy forests of the New Jersey Pine
Barrens. *Bulletin of the New Jersey Academy of
Science* 13:20–34.

McCormick, J., and Forman, R. T. T. 1979.
Introduction: Location and boundaries of the
New Jersey Pine Barrens. In *Pine Barrens:
Ecosystem and Landscape*, ed. R. T. T. Forman,
pp. xxxv–xli. New York: Academic Press.

McIntosh, R. P. 1959. Presence and cover in
pine–oak stands of the Shawangunk
Mountains, New York. *Ecology* 40:482–485.

McPhee, J. 1968. *The Pine Barrens.* New York: Farrar, Straus and Giroux.

Morgan, M. D., and Good, R. E. 1988. Stream chemistry in the New Jersey Pinelands: the influence of precipitation and watershed disturbance. *Water Resources Bulletin* 24:1091–1100.

Mushinsky, H. R., and Gibson, D. J. 1991. The influence of fire periodicity on habitat structure. In *Habitat Structure: The Physical Arrangement of Objects in Space,* ed. S. S. Bell, E. D. McCoy, and H. R., Mushinsky, pp. 237–259. London: Chapman & Hall.

Olsson, H. 1979. Vegetation of the New Jersey Pine Barrens: A phytosociological classification. In *Pine Barrens: Ecosystem and Landscape,* ed. R. T. T. Forman, pp. 245–263. New York: Academic Press.

Olsvig, L. S., Cryan, J. F., and Whittaker, R. H. 1979. Vegetational gradients of the pine plains and barrens of Long Island, New York. In *Pine Barrens: Ecosystem and Landscape,* ed. R. T. T. Forman, pp. 265–282. New York: Academic Press.

Pinchot, G. 1899. *A Study of Forest Fires and Wood Production in Southern New Jersey.* Annual Report of the State Geologist, 1898. New Jersey Geological Survey.

Redfield, J. H. 1869. *Corema conradii* (Torrey). *The American Naturalist* 3:327–328.

Redfield, J. H. 1884. *Corema conradii* and its localities. *Bulletin of the Torrey Botanical Club* 11:97–101.

Redfield, J. H. 1889a. Notes on *Corema conradii. Proceedings of the Academy of Natural Sciences of Philadelphia* 41:135–136.

Redfield, J. H. 1889b. *Corema* in New Jersey. *Bulletin of the Torrey Botanical Club* 16:193–195.

Rhodehamel, E. C. 1979a. Geology of the Pine Barrens of New Jersey. In *Pine Barrens: Ecosystem and Landscape,* ed. R. T. T. Forman, pp. 39–60. New York: Academic Press.

Rhodehamel, E. C. 1979b. Hydrology of the New Jersey Pine Barrens. In *Pine Barrens: Ecosystem and Landscape,* ed. R. T. T. Forman, pp. 147–167. New York: Academic Press.

Schweitzer, D. 1989. A progress report on the identification and prioritization of New Jersey's rare lepidoptera: 1981–1987. In *New Jersey's Rare and Endangered Plants and Animals,* ed. E. F. Karlin. Mahwah, N.J.: Institute for Environmental Studies, Ramapo College.

Stiles, E. W. 1979. Animal communities of the New Jersey Pine Barrens. In *Pine Barrens: Ecosystem and Landscape,* ed. R. T. T. Forman, pp. 541–553. New York: Academic Press.

Stone, W. 1909. The birds of New Jersey. *Annual Report of the New Jersey State Museum, 1908,* Part II: 11–347.

Stone, W. 1911. *Corema conradii* in Ocean County, NJ, East of the Plains. *Bartonia* 3:26.

Tedrow, J. C. F. 1979. Development of Pine Barrens soils. In *Pine Barrens: Ecosystem and Landscape,* ed. R. T. T. Forman, pp. 61–79. New York: Academic Press.

Tempel, A. S. 1976. Racial differences in cone characteristics of *Pinus rigida* in relation to seed predation by the red squirrel *(Tamiasciurus hudsonicus).* M.S. thesis, Rutgers University, New Brunswick, N.J.

Trela, J. J. 1984. Soil formation on Tertiary landsurfaces of the New Jersey coastal plain. Ph.D. dissertation, Rutgers University, New Brunswick, N.J.

Tyndall, R. W. 1994. Preface to the Barrens Symposium. *Castanea* 59:182–183.

Vivian, E. 1978. Habitat investigations on threatened plant species in the New Jersey Pine Barrens and their implications. In *Natural and Cultural Resources of the New Jersey Pine Barrens,* ed. J. W. Sinton, pp. 132–145. Pomona, N.J.: Stockton State College.

Vivian, E., and Snyder, D. B. 1981. *Rare and Endangered Vascular Plant Species in New Jersey.* Washington, D.C.: U.S. Fish and Wildlife Service.

Wacker, P. O. 1979. Human exploitation of the New Jersey Pine Barrens before 1900. In *Pine Barrens: Ecosystem and Landscape,* ed. R. T. T. Forman, pp. 3–23. New York: Academic Press.

Whittaker, R. H. 1979. Vegetational relationships of the Pine Barrens. In *Pine Barrens: Ecosystem and Landscape,* ed. R. T. T. Forman, pp. 315–332. New York: Academic Press.

Windisch, A. G. 1986. Delineation of the New Jersey Pine Plains and associated communities. *Skenectada* 3:1–16.

Windisch, A. G., and Good, R. E. 1986. *A Preliminary Analysis of Controlled Burning Effectiveness in Maintaining New Jersey Pine Plains Communities.* New Brunswick, N.J.: Rutgers University.

Windisch, A. G., and Good, R. E. 1991. Fire behavior and stem survival in the New Jersey

Pine Plains. *Proceedings of the Tall Timbers Fire Ecology Conference* 17:273–299.

Wolgast, L. J. 1979. Mammals of the New Jersey Pine Barrens. In *Pine Barrens: Ecosystem and Landscape*, ed. R. T. T. Forman, pp. 443–455. New York: Academic Press.

Zampella, R. A. 1988. Acquisition, management, and use of public lands. In *Protecting the New Jersey Pinelands, a New Direction in Land Use Management*, ed. B. R. Collins and E. W. B. Russell, pp. 214–231. New Brunswick, N.J.: Rutgers University.

Zampella, R. A. 1994. Characterization of surface water quality along a watershed disturbance gradient. *Water Resources Bulletin* 30:605–611.

Zampella, R. A., Moore, A. G., and Good, R. E. 1992. Gradient analysis of pitch pine (*Pinus rigida* Mill.) lowland communities in the New Jersey Pinelands. *Bulletin of the Torrey Botanical Club* 119:253–261.

4 Vegetation, Flora, and Plant Physiological Ecology of Serpentine Barrens of Eastern North America

R. WAYNE TYNDALL AND JAMES C. HULL

Introduction

The term *barren* is a historical one, used by settlers to refer to a landscape or landscape feature with little, if any, timber-sized trees, although tree species may have been present in smaller size classes. We use the term *serpentine barren* to refer to a serpentine outcrop and its associated vegetation. Prior to European settlement, serpentine barrens vegetation in much of eastern North America was composed predominately of fire-maintained communities of grassland and savanna (grassland with trees). This chapter considers the history, flora, vegetation, and physiological ecology of the serpentine barrens of eastern North America, with emphasis on the Mid-Atlantic states of Maryland and Pennsylvania. In this region, afforestation of nearly all undeveloped barrens, and more than 90% of undeveloped historic communities, has occurred during the past 50 years. The invading trees are relatively fire intolerant species such as *Pinus virginiana* (Virginia pine) and *Juniperus virginiana* (redcedar) (nomenclature follows Gleason and Cronquist [1991] unless authorities are given).

Serpentine soils are derived from ultramafic rocks, which occur in a discontinuous band along the eastern edge of the Appalachian mountain system from Newfoundland and Quebec, Canada, through New England, U.S.A., to Alabama on the Piedmont Plateau (Reed 1986; Brooks 1987). Ultramafic rocks are thermally altered and largely plutonic.

They are ferro-magnesium silicates high in magnesium and iron and low in aluminum, calcium, and silica. Serpentinite is a hydrated ultramafic rock, and it may be ultrabasic and contain heavy metals such as chromium and nickel (Reed 1986). The specific mineral composition of the rock and derived soil is highly variable, depending upon conditions (e.g., temperature and surrounding rocks) at the time of formation, and upon degree of metamorphosis and weathering of the rock following exposure (Proctor and Woodell 1975). The result is a complex mixture of substrates with numerous local characteristics. For example, in different locations, serpentine outcrops are associated with mining of chromite, talc, green marble, and asbestos (Reed 1986; Brooks 1987).

Serpentine soils are deficient in most mineral nutrients required by plants, have unbalanced ratios of required nutrients, contain toxic heavy metals, and have a pH different from surrounding soils (Brooks 1987). In addition, the soils often are poorly developed and rocky, but there may be deep clay soils (Robinson, Edgington and Byers 1935). The degree of soil formation is highly variable, depending upon climate, time of exposure, slope, and substrate. Physical characteristics of serpentine soils, especially those associated with formation of shallow, eroded, low clay soils, often are identified with reduced moisture (Proctor and Woodell 1975).

The environmental factors just discussed usually are cited as major contributors to the principal characteristics of serpentine vegeta-

Roger C. Anderson, James S. Fralish, and Jerry M. Baskin, eds. *Savannas, Barrens, and Rock Outcrop Plant Communities of North America.* Copyright © 1999 Cambridge University Press. All rights reserved.

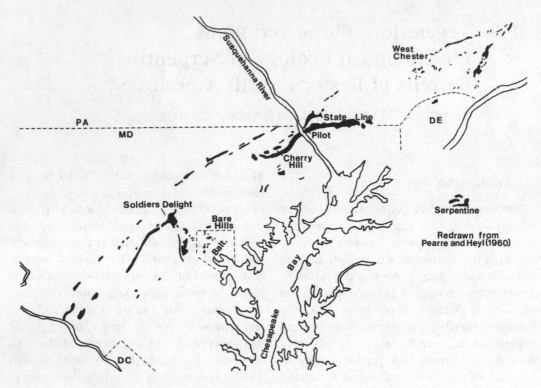

Figure 4.1. Distribution of serpentine bedrock in Maryland, Pennsylvania, and Delaware.

tion; that is, distinct physiognomy and community type, dominance of specific taxa, predominance of xerophytic species, low species richness, low productivity, and coexistence of low pH– and high pH–tolerant species (Whittaker 1954a; Proctor and Woodell 1975; Miller 1981; Proctor et al. 1981; Brooks 1987). In addition, use of fire by Amerindians has been documented as a primary factor on Maryland and Pennsylvania barrens (Figure 4.1) (Marye 1920, 1955a, b, c), where more than 90% of serpentine barren coverage in the eastern United States is concentrated (Latham 1993).

The Mid-Atlantic regional climate is temperate continental moderated by the Chesapeake Bay, Delaware Bay, and Atlantic Ocean. The growing season generally begins in April and continues through September or mid-October (U.S. Weather Bureau 1956; Reybold and Matthews 1976; Ruffner and Bair 1987).

Mean daily minimum and maximum temperatures during the growing season range from about 4 °C and 17 °C, respectively, for April to about 17 °C and 29 °C, respectively, for July. Mean rainfall varies from 72 mm to 108 mm per month during the growing season, with locally significant differences since most rainfall is produced by thunderstorms.

Landscape History

Presettlement Conditions in Maryland and Pennsylvania

Prior to settlement, circa 1750, the barrens in Maryland covered expansive areas in the upper parts of Baltimore, Harford, and Carroll counties (Marye 1920). In 1722, the barrens were described in correspondence as "a Vast Body of Barrens . . . what is called so, because there is no wood on it; besides Vast Quantities of Rockey Barrens . . . [and] the

Lands . . . all along the west side of Baltimore Co[unty], are cutt off and separated . . . by large Barrens, many miles over" (Marye 1955a). The following description occurs in a 1753 letter: "About thirty miles from Navigable Water is a Range of barren dry Land without Timber about nine miles [14 km] wide which keeps a Course about North East and South West parallel with the mountains thro this province Virginia & Pennsilvania" (Marye 1955a). In 1771, descriptions of extensive barrens with sparse vegetation were recorded in certificates of land surveys (Marye 1955a). For example, four opinions of a 620-acre tract in Baltimore County were "Verry poor bare Barrens"; "Barrense, hilly and stony"; "Poore hilly Barrance & much broke with stone & Verey scarce of Timber"; and "exceedingly poor & much broke with stone and Little or no Timber of any sort." The "Barrens of Patapsco" included Soldiers Delight (Marye 1955b), the largest remaining serpentine area in the eastern United States (700–800 ha).

Barrens in Pennsylvania were described in 1683 as "extensive treeless spaces in the wilderness" (Marye 1955a). In 1737, they comprised about 52,000 ha and their width at the Maryland border was reported to be 32 km (Marye 1955b). The latter estimate correlates well with the 37-km strip of ultramafic bedrock along this border (Pearre and Heyl 1960).

Between 1580 and 1652, ultramafic bedrock in Maryland was within the sole hunting domain of the Susquehannock Indians (Marye 1955c). Villages were centralized in Pennsylvania near the Maryland–Pennsylvania border on both sides of the Susquehanna River (Scharf 1881; Kinsey 1969; Witthoft 1969). (In 1570, this area was occupied by the Shenk's Ferry culture of unknown linguistic affiliation [Witthoft 1969].) The barrens were shared with other tribes after 1652, before hunting by Amerindians ceased about 1730 (Marye 1955c).

Marye (1955b) concluded that fire deliberately set by Native Americans was responsible for maintaining vegetative conditions on the barrens. The following quotations are from Marye's transcriptions of historical documents. According to one source, "It was the custome of the Indians in the autumn to set fire to and burn the barrens of York [Pennsylvania] and Baltimore [Maryland] Counties. . . . Grass covered the region, except for an occasional fringe of trees along the streams {cf. the timbered bottoms of the South Branch of Patapsco River, flanked by barrens}. When the Indians no longer set their fires, trees began to creep back." Another source described "narrow fringes of timber bordering the water courses" of this "vast prairie." Another, in reference to the York Barrens of Pennsylvania, stated "that the Indians for many years, and until 1730 or 1731, to improve this portion of their Great Park for the purpose of hunting, fired the copse or bushes as often as their convenience seemed to call for it; and thus, when the whites commenced settling here, they found no timber, hence they applied the term Barrens, a common appellation at that time, to such portions of the country, however fertile the soil."

"Fire hunting" was the primary technique for harvesting deer on Maryland serpentine (Marye 1955b; Porter 1983), as it was in a variety of habitats throughout the Southeast (Maxwell 1910; Swanton 1946). Access was provided by a system of "highways," especially the Old Indian Road that approached Soldiers Delight within 6 km to the north, east, and south (Marye 1920). Fire also may have been used to enhance game food production (Maxwell 1910) and for communication and warfare (Moore 1972).

Postsettlement Conditions in Maryland and Pennsylvania

European settlement of the barrens region was delayed until about 1750. Among factors causing this delay were the psychological

Figure 4.2. (a) Photograph of Soldiers Delight published in 1910. Trees are *Quercus marilandica* and *Q. stellata*. (Reprinted from Shreve et al. 1910.) **(b)** Soldiers Delight in 1996 dominated by *Pinus virginiana*.

Soldiers Delight." The barrens of Soldiers Delight were a favorite range for stock (Marye 1955c). Barrens that were not farmed or grazed became wooded (Marye 1955c). By 1800, timber was being advertised in Annapolis and Baltimore newspapers as growing in the barrens; for example, "Portions of the country that were sixty or seventy years ago [1775–85] without any timber are now [1845] thickly covered with sturdy oaks and large hickories" (Marye 1955b).

In the four principal serpentine areas in Maryland with remnant historic vegetation (Figure 4.1; Soldiers Delight, Bare Hills, Cherry Hill, and Pilot), grazing was still a factor in the early 1900s (Tyndall 1992a, b). Selective clearing of redcedar also was important in at least one of these areas. However, by the 1930s grazing and clearing had ceased in all four areas, based on aerial photography, and rapid conifer invasion and expansion ensued (Tyndall 1992a). By 1990, Virginia pine woodland and forest replaced 80% of grassland- and oak-dominated areas visible on 1938 aerial photographs in each of three areas. In the fourth area, Soldiers Delight, more than 50% of 1938 grassland and oak-covered areas was replaced by Virginia pine. In 1990, only one small opening remained at Pilot and several at Cherry Hill (Tyndall and Farr 1989, 1990). At all four sites, conifers established first in steep ravines and floodplains and then spread into shallow ravines and upper and lower ends of ridges. Conifers usually invaded the midsection of south-facing slopes last.

The most rapid rate of conifer expansion

effects of a vast barren landscape, scarcity of timber and water, and lack of an effective treaty with the Five Nations of Iroquois before 1744 (Porter 1975, 1979). After 1750, barrens were used immediately by settlers for livestock grazing, and "lands which lay within easy distance of the Barrens, were considered to be more valuable on that account" (Marye 1955c). For example, one parcel was advertised in the Maryland Gazette as "convenient for stock, there being an outlet to the 'Barrens of Patapsco' ....This 'outlet' . . . must have been by way of between 1938 and 1990 occurred in Robert E. Lee Park, the remaining undeveloped part of the original Bare Hills serpentine barren (Pearre and Heyl 1960; Tyndall 1992a). About 90% of the area covered by grassland in 1938 was invaded by Virginia pine, red-cedar, and the European buckthorn (*Rhamnus frangula*). Remnant grassland vegetation is fragmented into at least 15 openings over an area of about 3 ha.

In a photograph taken at ground level near the turn of the century at Soldiers Delight (Figure 4.2a), Virginia pine is not evident, nor is it mentioned in corresponding descriptions (Knox 1984). In 1914, the State Forester (Besley 1914) identified neither seedlings of Virginia pine nor merchantable stands of hardwood in Soldiers Delight. The nearest mapped stand of Virginia pine was 2.4 km away, and only two others were within 10 km. More than 80% of Virginia pine coverage in the county was east of Baltimore City, and most of it was along the Chesapeake Bay and tidal tributaries. "The desolation of the serpentine barrens around Soldiers Delight, with its rocky soil and stunted vegetation of cedar and meagre grass" was reported by the Maryland Geological Survey (1929). In 1937, aerial photographs show that conifers in upland areas were widely scattered. By 1980, however, stands of Virginia pine woodland and forest had developed, and pine was the most important tree species in many stands previously dominated by oaks (Hull and Wood 1984). Virginia pine was as much as 55 years younger than its codominant, *Quercus marilandica* (blackjack oak) (Knox 1984). Many openings have recently developed closed canopies of pine; others have succeeded to pine savannas and woodlands (Figure 4.2b). All remaining serpentine openings are expected to develop into conifer forests. Evidence for an edaphic climax community is lacking as even the driest southwest-facing slopes are being colonized.

Present-Day Vegetation: Eastern United States

Recently, plant communities were studied at the four principal serpentine areas in Maryland (Monteferrante 1973; Tyndall and Farr 1989, 1990; Tyndall 1992a, b). Recent reports of the vegetation on Pennsylvania serpentine are limited in the scientific literature. However, Pennell (1910, 1912, 1929) carefully described conditions of the vegetation in the early 1900s; a synopsis of Pennell's serpentine flora was prepared by Wherry (1963). Floristic and ecological information on Pennsylvania serpentine also is published in Miller (1977, 1981), Reed (1986), and Latham (1993).

Maryland

Schizachyrium scoparium is the most important characteristic species at each site examined in Maryland (Table 4.1), with the exception of Pilot, where it is a codominant with *Sporobolus heterolepis* (Tyndall and Farr 1990). Seasonally dominant species are *Carex umbellata, Panicum sphaerocarpon, Cerastium arvense* var. *villosum* Holl. & Britt., *Arabis lyrata, Viola sagittata, Potentilla canadensis*, and *Rubus flagellaris* in spring (April to mid-June); *Scleria pauciflora, Senecio anonymus, Panicum depauperatum, Oenothera fruticosa*, and *Talinum teretifolium* in summer (mid-June through August); and *Schizachyrium scoparium, Solidago nemoralis, Aristida dichotoma, A. purpurascens, Asclepias verticillata, Smilax rotundifolia, Polygonum tenue, Sorghastrum nutans, Saxifraga virginiensis*, and *Aster depauperatus* in fall (September and October). Three of these characteristic taxa are rare in Maryland: *C. arvense* var. *villosum, T. teretifolium*, and *A. depauperatus*.

Cerastium arvense var. *villosum* in Maryland occurs only on serpentine soil, and it may occur on all serpentine areas of Maryland and Pennsylvania (Pennell 1929; Wherry 1963; Wherry, Fogg and Wahl 1979; Hart

Table 4.1. *Mean Importance Values (MIV) for characteristic herbaceous layer species on Maryland serpentine. A characteristic species is defined as a taxon occurring in vegetation sampling plots in at least three of the four principal sites (Cherry Hill, Pilot, Soldiers Delight, and Bare Hills/Robert E. Lee). Mean importance values were calculated by averaging importance values across the four sites for each taxon. Importance values were based on relative cover and relative frequency results in Tyndall and Farr (1989 and 1990; results adjusted to exclude tree species) and Tyndall (1992b).*

Species	MIV (%)	Number of sites with sampled plants
Schizachyrium scoparium	52.0	4
Scleria pauciflora	13.8	4
Cerastium arvense var. *villosum*	11.9	4
Aristida purpurascens	11.9	3
Arabis lyrata	10.0	4
Aristida dichotoma	8.3	4
Panicum sphaerocarpon	8.0	4
Asclepias verticillata	6.6	3
Aster depauperatus	6.6	3
Carex umbellata	6.2	3
Senecio anonymus	5.8	3
Solidago nemoralis	5.8	3
Panicum depauperatum	5.7	4
Oenothera fruticosa	5.3	4
Viola sagittata	4.5	3
Smilax rotundifolia	4.2	3
Polygonum tenue	3.0	4
Sorghastrum nutans	2.8	3
Saxifraga virginiensis	2.3	3
Talinum teretifolium	1.4	4
Potentilla canadensis	1.2	3
Rubus flagellaris	0.7	4

1980; Reed 1986; Brooks 1987). Pennell (1929) reported it on "rocky banks and dry thin soil, abundant on all Serpentine Barrens of southeastern Pennsylvania and adjacent Delaware and northeastern Maryland; also on serpentine on Staten Island, New York, where it seems to be less marked."

Talinum teretifolium is restricted in Maryland to serpentine, except for one population on gneiss (Reed 1986; Maryland Department of Natural Resources 1996). *Talinum teretifolium* typically grows on granite outcrops of the Piedmont from Virginia to Georgia, but it also occurs on serpentine out-

crops at the far northern portion of its range in Maryland and Pennsylvania (Ware and Pinion 1990). Outlier populations also exist on sandstone on western and southern margins of its range; it does not occur on limestone (Ware and Pinion 1990).

Aster depauperatus in Maryland is restricted to serpentine, but the exact number of sites is uncertain due to its confusion with *A. pilosus* var. *pringlei*. Hart (1990) concluded that *A. depauperatus* may be the Midwestern prairie species, *A. parviceps*. Populations on serpentine in Maryland and Pennsylvania, and on diabase in North Carolina (Levy and

Wilbur 1990), could have migrated with prairie vegetation during the Hypsithermal period (6,000–8,000 yr BP), or earlier along the southern edge of the terminal moraine (Hart 1990).

At the four principal Maryland sites, plant community differences are greatest between sites at Cherry Hill and Pilot and between sites at Robert E. Lee and Pilot (Tyndall 1992b). Between Cherry Hill and Pilot, the major differences in vegetation are the absence of *Sporobolus heterolepis* at Cherry Hill, the absence of *Aristida purpurascens* in Pilot plots, and the difference in importance values (relative cover + relative frequency) for *Cerastium arvense* var. *villosum*. Between Robert E. Lee and Pilot, the major differences in vegetation are the absence of *S. heterolepis* at Robert E. Lee, and the difference in importance values for *Schizachyrium scoparium* and for *C. arvense* var. *villosum*. Species richness and diversity, based on the Shannon–Wiener index (H', Shannon and Weaver 1949), are greatest at Cherry Hill (richness = 33, H' = 2.49). Values are similar for remaining sites (Soldiers Delight: richness = 26, H' = 1.95; Pilot: 24, 1.88; Robert E. Lee: 23, 1.88).

Cherry Hill. Coniferous woodland and forest dominated primarily by Virginia pine, redcedar, and thickets of *Smilax rotundifolia* now occupy most of Cherry Hill. Skeletons of dead trees of open-grown *Quercus stellata* (post oak) and blackjack oak stand in a park-like distribution, and only a small portion of the original oak population has survived in the shade of the taller pine canopy. A few scattered pine/cedar savanna openings occur at this site, mostly on south- to southwest-facing slopes. The two largest openings (collectively about 3 ha) were sampled in 1987 (Tyndall and Farr 1989) prior to manual clearing of conifers in winter 1989/1990. Bare rock and bare soil accounted for 14.4% of the surface area, cryptogams for 10.6% (lichen = 8.6%, moss = 2.0%), herbaceous vegetation for 67.6% and tree/sapling trunks for 2.5%. Trees of *Q. marilandica* were not

observed in the openings, although seedlings and saplings were present.

The pine/cedar savanna community in the two openings consisted of 69 vascular plant taxa, with about 40% of herbaceous cover produced by *Schizachyrium scoparium* and *Aristida purpurascens*. Rare species included *Aster depauperatus, Talinum teretifolium, Cerastium arvense* var. *villosum, Panicum flexile,* and *Linum sulcatum. Panicum flexile* is "highly state rare" in Maryland (Maryland Department of Natural Resources 1996), and *Linum sulcatum* has been documented at only one other site in Maryland, Soldiers Delight.

Pilot. Nearly all of this serpentine area has succeeded to a *Pinus virginiana–Smilax rotundifolia* forest, and only a single 0.5-ha opening remained in 1988 before restoration efforts began. *Sporobolus heterolepis* and *Schizachyrium scoparium* were community dominants, comprising 56% of the total vegetation cover (Tyndall and Farr 1990). *Sporobolus heterolepis* was the most abundant species, forming dense mats on the steep-sloping perimeter of the opening in conifer savanna and woodland. *Schizachyrium scoparium* was most abundant in the nonperimeter region of the opening in grassland and conifer savanna. The opening had 45 vascular plant taxa, and rare species included *Aster depauperatus, Talinum teretifolium,* and *Sporobolus heterolepis. Sporobolus heterolepis* is very rare in Maryland (Maryland Department of Natural Resources 1996). It is known from only one other site in the state, a powerline right-of-way that crosses Goat Hill, one of the State Line Barrens (Figure 4.1). Bare rock and bare soil accounted for 66.2% of the surface area in 1988, lichens for 4.4%, mosses for 0.9%, and herbaceous vegetation for 28.5%. However, vegetation cover probably was lower than normal due to abnormally low rainfall from May to August during the study. This extended summer drought inhibited Virginia pine expansion and favored redcedar. Mortalities of Virginia pine seedling and sapling were about 66% and 50%, respectively, whereas redcedar seedling and sapling mortalities were negligible. Immediate effects were 24% and 44% gains in importance for redcedar trees and saplings, respectively.

Robert E. Lee. *Schizachyrium scoparium* dominated the herbaceous layer in openings at Robert E. Lee in 1991, with a total cover of about 52% and a frequency of 100% (Tyndall 1992b). Rare species included *Talinum teretifolium*, *Aster depauperatus*, and *Linum sulcatum* (Maryland Department of Natural Resources 1996). Bare rock and soil coverage was 22.9%.

An April 1991 fire of unknown origin at Robert E. Lee preceded sampling and thoroughly burned the ground layer of most remnant openings (Tyndall 1992b). Interestingly, *Talinum teretifolium* was found only in plots of burned openings, whereas *Potentilla canadensis* and *Sisyrinchium mucronatum* occurred only in plots of unburned openings. Fixed-point photographs, taken 13 days after the fire and during each sampling event, show that most conifer seedlings were killed, whereas tree mortality was negligible.

Soldiers Delight. Historic plant communities at Soldiers Delight are primarily *Schizachyrium scoparium* grassland and *Q. marilandica–Q. stellata* savanna. These oaks formed parklike stands (Figure 4.2a) in 1910 (Shreve et al. 1910). The savanna community was most abundant in mesic habitats, such as north-facing slopes, steep ravines, and relatively flat uplands above the steep ravines. Nearly all areas once dominated by oak savannas are now forested with Virginia pine. Composition of the herbaceous layer of the historic oak savanna is unknown.

The current forest vegetation on Soldiers Delight consists primarily of Virginia pine, *Q. marilandica*, *Q. stellata*, and *Sassafras albidum* (Hull and Wood 1984; Knox 1984; Wood 1984). *Quercus velutina* is present in a few stands (Knox 1984; Wood 1984), and redcedar is infrequent (Monteferrante 1973). Common understory shrubs include *Gaylussacia baccata*, *Rubus cuneifolius*, *Smilax*

rotundifolia, *Rhus radicans*, and *Vaccinium vacillans* (Wood 1984). Monteferrante (1973) reported *Schizachyrium scoparium*, *Hedyotis caerulea*, and *Viola sagittata* as common in the forest understory.

Schizachyrium scoparium–dominated (cover = 55% and frequency = 100%) grassland vegetation is most abundant on southwest-facing slopes, especially mid-slope sections in the early stage of pine expansion (Tyndall 1992b). Most important associated species are *Panicum sphaerocarpon*, *Cerastium arvense* var. *villosum*, and *Arabis lyrata* in spring; *Scleria pauciflora*, *Senecio anonymus*, and *Panicum depauperatum* in summer; and *Solidago nemoralis*, *Aristida dichotoma*, *A. purpurascens*, and the rare *Aster depauperatus* in fall. Other rare species in the sampling area include *Talinum teretifolium* and *Linum sulcatum*. Bare rock and soil coverage in 1989 was 30.8%. Short-term effects of conifer clearing and prescribed burning on grassland vegetation in this study area are discussed in Tyndall (1994).

The most notable rare and endangered plant species in Soldiers Delight is *Agalinis acuta* (sandplain gerardia), the only federally protected (threatened) plant species on eastern U.S. serpentine. Historically, its primary habitat was sandy soil of the coastal plain (U.S. Fish and Wildlife Service 1993). Of the 49 historical populations, ranging from Massachusetts to Maryland, only 10 remain, and less than 10% of the rangewide population occurs on sandy soil. The remaining 90% or more occurs in grassland and pine savanna openings on Soldiers Delight, where manual removal of conifers followed by prescribed burning are the primary tactics of habitat restoration.

Gentianopsis crinita, fringed gentian, is another rare and endangered species at Soldiers Delight, with a dwindling population primarily because of conifer expansion. Because *G. crinita* is shade intolerant and its primary habitats at Soldiers Delight are stream banks and wetlands, it probably was one of the first species affected by pine expansion. Other rare and endangered species with extant populations at Soldiers Delight include *Carex hystericina, C. richardsonii, Desmodium rigidum, Gentiana andrewsii, Panicum oligosanthes, P. flexile,* and *Pycnanthemum torreyi* (Maryland Department of Natural Resources 1996).

Pennsylvania

Pennell (1910, 1912, 1929) studied the flora of the State Line and (West) Chester serpentine districts (Figure 4.1). State Line Barrens are mostly in a single belt about 21 km long and 2 km wide paralleling the Maryland and Pennsylvania border (Pennell 1910; Pearre and Heyl 1960). The West Chester Serpentine group, about 30 km northeast of the State Line district, consists of 12–20 small, scattered areas up to 2 km long.

The Chester group was described by Pennell (1910) as covered mostly by a "sparse growth of timber of markedly xerophytic type . . . . The round bushy growth of the thick-leafed oaks, with open park-like spaces between, is the characteristic feature of this woodland." *Quercus stellata* was abundant on all the Chester barrens and *Q. marilandica* on most. Virginia pine was restricted to a few trees in one locality, and *P. rigida* was absent. *Smilax rotundifolia* was especially abundant in depressions where redcedar was the main tree. Grasses and sedges comprised most of the herbaceous vegetation in the numerous parklike openings; *Schizachyrium scoparium, Panicum philadelphicum, P. sphaerocarpon, Aristida dichotoma, A. longespica,* and *Scleria pauciflora* were abundant. Where the canopy was more closed than previously described, *Salix tristis, Corylus americana, Rhus glabra, Ceanothus americana, Gaylussacia baccata,* and *Vaccinium vacillans* formed "dry upland thicket" vegetation.

In the State Line Barrens, *Pinus rigida* was predominant, with *Q. stellata* and *Q. marilandica* as important associated species

(Pennell 1910). Redcedar occurred primarily in upland depressions with greenbrier thickets. *Sporobolus heterolepis* was reported from the State Line Barrens but not from the Chester group. *Campanula rotundifolia, Galium boreale, Lactuca canadensis, Aster concinnus,* and *Cerastium arvense* var. *villosissimum* grew on serpentine cliffs along Octoraro Creek (Pennell 1929). *Cerastium arvense* var. *villosissimum* was described by Pennell, who reported it as being restricted to Octoraro Creek (Pennell 1929). It is not known by the state rare and endangered species program as occurring in Maryland (Maryland Department of Natural Resources 1996).

The three most characteristic serpentine species in Maryland, *Schizachyrium scoparium, Scleria pauciflora,* and *Cerastium arvense* var. *villosum* (Table 4.1), were abundant in dry, open barrens of Pennsylvania according to Pennell (1910, 1912, 1929). The other species listed in Table 4.1 were described by Pennell as common, frequent, or present in all Pennsylvania serpentine barrens, except *Talinum teretifolium* ("local"); *Carex umbellata* ("occasional"); and *Smilax rotundifolia,* which was found in depressions of barrens, frequently growing on redcedar as a vine.

The serpentine diploid maidenhair fern, *Adiantum aleuticum* (Ruprecht) Paris (includes *A. pedatum* ssp. *calderi* Cody and *A. pedatum* var. *aleuticum* Rupr. [Paris 1991]) reportedly occurs in the State Line Barrens at Goat Hill. In eastern North America, it is restricted to serpentine in southeastern Canada and in the northeastern United States (Lellinger 1985; Paris 1991). *Adiantum aleuticum* is also the common maidenhair fern in a variety of habitats in western North America (Paris 1991). It is divergent genetically from the common eastern woodland maidenhair fern (*A. pedatum sensu stricto*) (Paris and Windham 1988; Paris 1991). The Green Mountain maidenhair fern, *A. viridimontanum* Paris, is a recently described allotetraploid derivative of a hybrid between *A. pedatum* and *A. aleuticum* and is known only from a few serpentine outcrops in Vermont (Paris and Windham 1988; Paris 1991).

North Carolina

Physiognomically, the *Pinus rigida* (pitch pine) savanna on an olivine–serpentine outcrop in western North Carolina (Buck Creek) (Mansberg and Wentworth 1984) is similar to pitch pine sites in the State Line Barrens of Pennsylvania. However, the herbaceous vegetation differs considerably from that found in Maryland. Of the five dominant herbaceous species at the North Carolina site, that is, *Senecio plattensis, Schizachyrium scoparium, Andropogon gerardii, Hexastylis arifolia* var. *ruthii,* and *Thalictrum macrostylum,* only *S. scoparium* is a characteristic species of Maryland serpentine vegetation. In addition, 37 of the 44 "prevalent" herbaceous species at the North Carolina site were not sampled in the four Maryland sites (Tyndall and Farr 1989, 1990; Tyndall 1992b). Small serpentine intrusions occur in Wake County that are characterized by stunted open canopies dominated by *Pinus echinata* and *Q. marilandica,* with *Schizachyrium scoparium* and *Stipa avenacea* in the understory (Levy and Wilbur 1990). Radford (1948) conducted a floristic study of the vascular flora of olivine deposits in North Carolina and Georgia, and correlated species and community distribution with mineral distribution.

Delaware and New York

In Delaware, more than 200 ha of serpentinite were present prior to settlement, but most of this area is now under Hoopes Reservoir near Mt. Cuba in New Castle County (W. A. McAvoy, State Botanist, Delaware Natural Heritage Program, Dover, Delaware, personal communication, 1997). Fewer than 6 ha of exposed bedrock remain in the state, including lawns, roadbanks, and early successional, weedy woodlands. State rare species on Delaware serpentine are *Cerastium arvense* var. *villosum, Aster laevis, Arabis lyrata, Panicum oligosanthes,* and *Senecio anonymus* (McAvoy,

personal communication, 1997). In New York, a few exposures of serpentine rock occur on Staten Island (Reed 1986). Probably none of the indigenous serpentine flora of these sites has survived, except for a few mosses and lichens (Reed 1986).

Vermont

In Vermont, Zika and Dann (1985) explored 19 dunite and serpentine areas for rare vascular plants. Twenty-one new locations or relocations for *Adiantum pedatum* var. *aleuticum, Arenaria macrophylla, Kalmia latifolia, Lycopodium selago,* and *Thelypteris simulata* were reported, and the distribution of ultramafic bedrock was mapped for the state. Four stations for *Adiantum pedatum* var. *aleuticum* subsequently were determined to be the newly described serpentine tetraploid maidenhair fern, *A. aleuticum* (Paris 1991).

Eastern Canada

Ultramafic rock outcrops continue beyond Vermont through the Gaspé Peninsula of eastern Quebec, to the northern tip of Quebec in the Ungava Peninsula, and into Newfoundland (Fernald 1911, 1926, 1933; Brooks 1987). In the southern part of the Gaspé Peninsula, serpentine vegetation in the eastern townships occurs in a mostly continuous strip in an area 100 km long and 25 km wide (Brooks 1987). Citing Legault (1976) and Legault and Blais (1968), Brooks (1987) reported vegetation on the serpentine talus slope at one site (Mt. Silver) to be dominated by a scrub association of *Juniperus communis* with scattered dwarfed *Betula papyrifera.* This site, and another steep, dry talus of serpentine, supported *Cheilanthes siliquosa* Maxon, a fern rare and endemic to serpentine in eastern North America, but common on serpentine in western North America (Legault and Blais 1968; Brooks 1987). *Cheilanthes siliquosa* also occurs on Mt. Albert toward the tip of the Gaspé Peninsula (Legault and Blais 1968).

Serpentine on Mt. Albert is completely devoid of trees, in sharp contrast to surrounding amphibolite, which is covered by a close growth of shrubby spruce and *Abies balsamea* (balsam fir) (Rune 1954). This vast, flat tableland of serpentine physiognomically resembles the Arctic tundra (Rune 1954). The most abundant of the characteristic serpentine vascular plants on Mt. Albert is *Arenaria marcescens* Fern. (*Minuartia marcescens* (Fern) House) (Rune 1954). On Mt. Albert and in Newfoundland, this species is restricted to serpentine (Brooks 1987).

In Newfoundland, ultramafic bedrock is distributed in three parallel bands, with the west coast band the largest and most important (Brooks 1987; Roberts 1991). On the west coast in the Bay of Islands complex, Table Mountain serpentine massif is one of the most prominent features of the alpine plateau (Bouchard, Hay and Rouleau 1978). It has the appearance of a rocky, denuded wasteland colonized by only scanty vegetation cover (Bouchard, Hay and Rouleau 1978) and appears as a yellow-orange mass in the greenery of surrounding nonserpentine hills (Dearden 1979). Serpentine barrens vegetation occurs on a variety of landforms, with the surface usually greater than 80% exposed bedrock interspersed with solitary wind-pruned shrubs and tufts of grass or other herbs (Bouchard, Hay and Rouleau 1978).

Five rare species were documented by Bouchard et al. (1986) on Table Mountain serpentine; *Arenaria marcescens* Fern. is quite common, and populations of *Danthonia intermedia*, *Eleocharis nitida* Fern., *Festuca altaica*, and *Salix arctica* Pallas are few in number. *Adiantum aleuticum* (*A. pedatum* var. *aleuticum*) is common in moist, sheltered crevices (Dearden 1979).

The White Hills Mountains of the Great Northern Peninsula (Hay, Bouchard and Brouillet 1994) occur about 250 km north of Table Mountain and form two adjoining plateau-like massifs of ultramafic, serpentinized peridotite and dunite bedrock. Three

species of Caryophyllaceae new to Newfoundland were recently documented by Hay, Bouchard and Brouillet (1994): *Minuartia biflora* (Wats.) Schinzl. & Thell, *Sagina caespitosa* (J. Vahl) Lange, and *S. saginoides* (L.) Karst. Significant range extensions were made for *Cerastium terrae-novae* Fern. & Wieg., *Danthonia intermedia*, and *Salix arctica* Pallas.

Important nutrient stresses on Newfoundland serpentine are confounded by physical stress factors of drought, wind erosion, snow abrasion, and cryoturbation (Bouchard, Hay and Rouleau 1978; Roberts 1980, 1991). Of the physical factors, Roberts (1991) concluded that cryoturbation had the greatest effect on vegetation.

Physiological Ecology
Mineral Nutrition of Serpentine Soils

Serpentine vegetation most frequently is controlled by two edaphic variables: minerals and water. Worldwide, the mineral status of serpentine soils is characterized by low availability of one or more essential nutrients (especially Ca, K, N, and P), by an imbalance of mineral nutrients (especially a low ratio of Ca to Mg), by toxicity of heavy metals (notably Ni and Cr), and by extreme vertical pH gradients (Robinson, Edgington and Byers 1935; Proctor and Woodell 1975; Koenigs et al. 1982; Hull and Wood 1984; Mansberg and Wentworth 1984; Brooks 1987). The relative importance of these factors is dependent upon the specific mineralogy of the substrate and upon soil features (e.g., pH and leaching) that influence retention and availability of nutrients. No single explanation satisfies all observations of the interactions between plants and soil (Proctor and Woodell 1975; Koenigs et al. 1982; Brooks 1987).

Exchangeable Ca:Mg ratios of serpentine of the Mid-Atlantic region of the United States range from 0.03 to 1.86, with a mean of 0.49 (Robinson, Edgington and Byers 1935). Many of these soils have limiting concentrations of both elements. In Newfoundland, serpentine communities most distinct from surrounding vegetation have soil with low calcium and high magnesium concentrations, whereas those most similar to surrounding vegetation have high calcium and low magnesium concentrations (Dearden 1979). Compared to nonserpentine species, serpentine plants typically have one or more of the following traits: greater tolerance of high Mg and low Ca levels, higher Mg requirement for maximum growth, lower Mg absorption, higher Ca absorption, and Mg exclusion from leaves (Kruckeberg 1954; Walker, Walker and Ashworth 1955; Main 1974; Marrs and Proctor 1976; Koenigs et al. 1982; Brooks 1987). Excessive amounts of soil Ni (Roberts 1991) and certain plants able to hyperaccumulate heavy metals such as nickel occur on some serpentine soils (Brooks 1987; Gabbrielli et al. 1990). However, there are no published reports from eastern North America regarding hyperaccumulation.

Wood (1984) examined the relationship between several elements (Ca, Mg, K, P, Cr, and Ni) in soil and tissues of five tree species on and adjacent to Soldiers Delight. Neither Cr nor Ni was present at detectable levels in soils or plant tissues, and thus heavy metals were not controlling factors at the sites analyzed. At sites dominated by serpentine oaks, *Q. stellata* and *Q. marilandica*, soil Ca was similar to, or greater than, that at sites dominated by nonserpentine oaks, *Q. alba* and *Q. velutina*. Conversely, soil Mg was considerably greater at serpentine sites. The Ca:Mg ratio at nonserpentine sites was 5.6, whereas the Ca:Mg ratio at serpentine sites was 0.21. Since Ca varied little between sites, Wood (1984) concluded that distribution of forest species on and off serpentine soils is dependent upon the Mg concentration of the soils.

Serpentine Water Relations

Vegetation on serpentine soils typically is more xerophytic than that on other soils of the same region (Whittaker 1954b). Serpentine soils of Maryland characteristically are rocky and shallow, with clays that provide an inadequate supply of water for plants (Reybold and Matthews 1976). However, physical measurements of water status (capillary water, available water, and permanent wilting percentage) show few differences between serpentine and nonserpentine soils (Miller 1977). For example, dawn xylem water potentials at Soldiers Delight were not lower in serpentine oaks (*Q. marilandica* and *Q. stellata*) than in nonserpentine oaks (*Q. alba* and *Q. velutina*) early in the growing season (May to July) (Hull and Wood 1984). In August, midday xylem potentials differed between species, but not consistently between substrates. In mid-July, dawn xylem potentials of *Q. stellata* and *Q. marilandica* (serpentine oaks) were significantly (p < 0.05) lower than those of *Q. velutina*. Dawn xylem potentials of *Q. alba* differed from those of *Q. marilandica*, but not from *Q. stellata*. However, none of these dawn xylem potentials were as low as those recorded in other studies of nonserpentine substrates (Hinckley and Bruckerhoff 1975; Reich and Hinckley 1980). Water potentials were greater in serpentine than in nonserpentine soils. Consequently, water relations as the sole causative factor for distribution of oak species onto serpentine cannot be supported (Hull and Wood 1984).

Present distribution of vegetation on serpentine soils may be affected by water availability as determined by soil depth. Grasslands occur on shallow, rocky soils with low moisture storage in the soil profile (Monteferrante 1973). Savanna occurs in slightly deeper soils, and forest presumably occurs in the deepest soils. There was a significant correlation between dawn xylem potentials of oaks and forest soil water potentials at 20–30-cm depths beneath crowns of the oaks (Hull and Wood 1984). Consequently, it is likely that water storage is a contributing factor in the rate of invasion and expansion of woody species on serpentine soils.

Summary

Serpentine soils are derived from serpentinite, a hydrated ultramafic rock. In eastern North America, serpentinite is distributed in a discontinuous band from Newfoundland, Canada, to Alabama, U.S.A. In the eastern United States, more than 90% of extant vegetation occurs in Maryland and Pennsylvania. Serpentine soils in Maryland and Pennsylvania are associated with limited soil nutrients and water and, prior to settlement (circa 1750), with frequent and widespread Amerindian fires. These ecological factors contributed to the maintenance of expansive xerophytic grassland and savanna communities until Amerindian extirpation. Settlers used the term *barrens* in the late 1600s and early 1700s to refer to the lack or paucity of timber-sized trees on the serpentine outcrops. Between 1750 and the early 1900s, livestock grazing and selective clearing of woody plants inhibited afforestation in some areas, whereas unaffected areas became forested. Today, very little grassland and oak savanna remain in Maryland and Pennsylvania, and the term *barrens* is sometimes confusingly used to refer to openings that have not yet become forested. Newfoundland barrens are not forested; graminoids and herbs with scattered shrubs characterize the vegetation. Effects of nutrient stress, that is, excessive Mg:Ca ratios, low levels of essential macronutrients and micronutrients, and possibly excessive amounts of nickel in some microsites, are confounded on Newfoundland serpentine with physical stress factors of drought, wind erosion, snow abrasion, and cryoturbation.

REFERENCES

Besley, F. W. 1914. Map of Baltimore County and Baltimore City showing the forest areas by commercial types. Baltimore, Md.: Maryland Board of Forestry.

Bouchard, A., Hay, S., and Rouleau, E. 1978. The vascular flora of St. Barbe South District, Newfoundland: an interpretation based on biophysiographic areas. *Rhodora* 80:228–308.

Bouchard, A., Hay, S., Gauvin, C., and Bergeron, Y. 1986. Rare vascular plants of Gros Morne National Park, Newfoundland, Canada. *Rhodora* 88:481–502.

Brooks, R. R. 1987. *Serpentine and its Vegetation, a Multidisciplinary Approach*. Portland, Ore.: Dioscorides Press.

Dearden, P. 1979. Some factors influencing the composition and location of plant communities on a serpentine bedrock in western Newfoundland. *Journal of Biogeography* 6:93–104.

Fernald, M. L. 1911. A botanical expedition to Newfoundland and southern Labrador. *Rhodora* 13:109–162.

Fernald, M. L. 1926. Two summers of botanizing in Newfoundland. *Rhodora* 28:49–63, 74–87, 89–111, 115–129, 145–155, 161–178, 181–204, 210–225, 234–241. Reprinted in *Contributions from the Gray Herbarium of Harvard University* 76:49–241.

Fernald, M. L. 1933. Recent discoveries in the Newfoundland flora. *Rhodora* 35:1–16, 47–63, 80–107, 120–140, 161–185, 203–223, 231–247, 265–283, 298–315, 327–346, 364–386, 395–403. Reprinted in *Contributions from the Gray Herbarium of Harvard University* 101:1–403.

Gabbrielli, R., Pandolfini, T., Vergnano, O., and Palandri, M. R. 1990. Comparison of two serpentine species with different nickel tolerance strategies. *Plant and Soil* 122:271–277.

Gleason, H. A., and Cronquist, A. 1991. *Manual of the Vascular Plants of Northeastern United States and Adjacent Canada*. Bronx, N.Y.: The New York Botanical Garden.

Hart, R. 1980. The coexistence of weeds and restricted native plants on serpentine barrens in southeastern Pennsylvania. *Ecology* 61:689–701.

Hart, R. 1990. *Aster depauperatus*: a midwestern migrant on eastern serpentine barrens? *Bartonia* 56:23–28.

Hay, S. G., Bouchard, A., and Brouillet, L. 1994. Additions to the flora of Newfoundland. III. *Rhodora* 96:195–203.

Hinckley, T. M., and Bruckerhoff, D. N. 1975. The effects of drought on water relations and stem shrinkage of *Quercus alba*. *Canadian Journal of Botany* 53:62–72.

Hull, J. C., and Wood, S. G. 1984. Water relations of oak species on and adjacent to a Maryland serpentine soil. *The American Midland Naturalist* 112:224–234.

Kinsey, W. F., III. 1969. Historic Susquehannock pottery. In *Susquehannock Miscellany*, ed. J. Witthoft and W. F. Kinsey, III., pp. 61–98. Harrisburg, Pa.: Pennsylvania Historical and Museum Commission.

Knox, R. G. 1984. Age structure of forests on Soldiers Delight, a Maryland serpentine area. *Bulletin of the Torrey Botanical Club* 111:498–501.

Koenigs, R. L., Williams, W. A., Jones, M. B., and Wallace, A. 1982. Factors affecting vegetation on a serpentine soil. II. Chemical composition of foliage and soil. *Hilgardia* 50:15–26.

Kruckeberg, A. R. 1954. The ecology of serpentine soils. III. Plant species in relation to serpentine soils. *Ecology* 35:267–274.

Latham, R. E. 1993. The serpentine barrens of temperate eastern North America: critical issues in the management of rare species and communities. *Bartonia* 57 (supplement):61–74.

Legault, A. 1976. Field trip to the serpentine area of Black Lake-Thetford Mines, Province of Quebec, Canada. Canadian Botanical Association Unpub. Rep.

Legault, A., and Blais, V. 1968. Le *Cheilanthes siliquosa* Maxon dans le nord-est Americain. *Naturaliste Canadien* 95:307–316.

Lellinger, D. B. 1985. *A Field Manual of the Ferns and Fern Allies of the United States and Canada*. Washington, D.C.: Smithsonian Institution Press.

Levy, F., and Wilbur, R. L. 1990. Disjunct populations of the alleged serpentine endemic, *Aster depauperatus* (Porter) Fern., on diabase glades in North Carolina. *Rhodora* 92:17– 21.

Main, J. L. 1974. Differential responses to magnesium and calcium by native populations of *Agropyron spicatum*. *American Journal of Botany* 61:931–937.

Mansberg, L., and Wentworth, T. R. 1984. Vegetation and soils of a serpentine barren in western North Carolina. *Bulletin of the Torrey Botanical Club* 111:273–286.

Marrs, R. H., and Proctor, J. 1976. The response of serpentine and non-serpentine *Agrostis stolonifera* to magnesium and calcium. *Journal of Ecology* 64:953–964.

Marye, W. B. 1920. The Old Indian Road. *Maryland Historical Magazine* 15:107–124, 208–229, 345–395.

Marye, W. B. 1955a. The great Maryland barrens. *Maryland Historical Magazine* 50:11–23.

Marye, W. B. 1955b. The great Maryland barrens: II. *Maryland Historical Magazine* 50:120–142.

Marye, W. B. 1955c. The great Maryland barrens: III. *Maryland Historical Magazine* 50:234–253.

Maryland Department of Natural Resources. 1996. Biological and conservation database. Annapolis, Md.: Heritage and Biodiversity Conservation Programs.

Maryland Geological Survey. 1929. *Baltimore County.* Baltimore, Md.: John Hopkins University Press.

Maxwell, H. 1910. The use and abuse of forests by the Virginia Indians. *William and Mary College Quarterly Historical Magazine* 19:73–103.

Miller, G. L. 1977. An ecological study of the serpentine barrens in Lancaster County, Pennsylvania. *Proceedings of the Pennsylvania Academy of Science* 51:169–176.

Miller, G. L. 1981. Secondary succession following fire on a serpentine barren. *Proceedings of the Pennsylvania Academy of Science* 55:62–64.

Monteferrante, F. J. 1973. A phytosociological study of Soldiers Delight, Baltimore County, Maryland. M.S. thesis, Towson State College, Towson, Md.

Moore, C. T. 1972. Man and fire in the Central North American Grassland 1535–1890: a documentary historical geography. Ph.D. thesis, University of California, Los Angeles.

Paris, C. A. 1991. *Adiantum viridimontanum*, a new maidenhair fern in eastern North America. *Rhodora* 93:105–122.

Paris, C. A., and Windham, M. D. 1988. A biosystematic investigation of the *Adiantum pedatum* complex in eastern North America. *Systematic Botany* 13:240–255.

Pearre, N. C., and Heyl, A. V., Jr. 1960. Chromite and other mineral deposits in serpentine rocks of the Piedmont Upland, Maryland, Pennsylvania, and Delaware. *U.S. Geological. Survey Bulletin* 1082–K:707–833.

Pennell, F. W. 1910: Flora of the Conowingo barrens of southeastern Pennsylvania. *Proceedings of the Academy of Natural Sciences of Philadelphia* 62:541–584.

Pennell, F. W. 1912. Further notes on the flora of the Conowingo or serpentine barrens of southeastern Pennsylvania. *Proceedings of the Academy of Natural Sciences of Philadelphia* 64:520–539.

Pennell, F. W. 1929. On some critical species of the serpentine barrens. *Bartonia* 12:1–23.

Porter, F. W., III. 1975. From backcountry to county: the delayed settlement of western

Maryland. *Maryland Historical Magazine* 70:329–349.

Porter, F. W., III. 1979. The Maryland frontier, 1722–1732: Prelude to settlement in western Maryland. In *Geographical Perspectives on Maryland's Past.* Occasional Paper No. 4, Geography Department, ed. R. D. Mitchell and E. K. Muller, pp. 90–107. College Park, Md.: University of Maryland.

Porter, F. W., III. 1983. *Maryland Indians Yesterday and Today.* Baltimore, Md.: Maryland Historical Society.

Proctor, J., and Woodell, S. R. J. 1975. The ecology of serpentine soils. In *Advances in Ecological Research,* ed. A. MacFadyen, 9:255–366. London: Academic Press.

Proctor, J. W., Johnston, R, Cottam, D. A., and Wilson, A. B. 1981. Field-capacity water extracts from serpentine soils. *Nature* 294:245–246.

Radford, A. E. 1948. The vascular flora of the olivine deposits of North Carolina and Georgia. *Journal of the Elisha Mitchell Scientific Society* 64:45–106, plate 8.

Reed, C. F. 1986. *Floras of the Serpentinite Formations in Eastern North America, with Descriptions of Geomorphology and Mineralogy of the Formations.* Baltimore, Md.: Reed Herbarium.

Reich, P. B., and Hinckley, T. M. 1980. Water relations, soil fertility, and plant nutrient composition of a pigmy oak ecosystem. *Ecology* 61:400–416.

Reybold, W. U., III, and Matthews, E. D. 1976. *Soil Survey of Baltimore County, Maryland.* Washington, D.C.: USDA, Soil Conservation Service.

Roberts, B. A. 1980. Some chemical and physical properties of serpentine soils from western Newfoundland. *Canadian Journal of Soil Science* 60:231–240.

Roberts, B. A. 1991. The serpentinized areas of Newfoundland, Canada: a brief review of their soils and vegetation. In *The Vegetation of Ultramafic (Serpentine) Soils: Proceedings of the First International Conference on Serpentine Ecology,* ed. A. J. M. Baker, J. Proctor, and R. D. Reeves, pp. 53–66. Andover, Hampshire: Intercept Ltd.

Robinson, W. O., Edgington, G., and Byers, H. G. 1935. *Chemical Studies of Infertile Soils Derived from Rocks High in Magnesium and Generally High in Chromium and Nickel.* Washington, D.C.: USDA, Technical Bulletin No. 471; 28. United States Government Printing Office.

Ruffner, J. A., and Bair, F. E. 1987. *Weather of U. S. Cities,* 3rd edition, Vol. 1. Detroit, Ill.: Gale Research Company.

Rune, O. 1954. Notes on the flora of the Gaspé Peninsula. *Svensk Botanisk Tidskrift* 48:117–138.

Scharf, J. T. 1881. *History of Baltimore City and County.* Philadelphia: Louis H. Everts. Reprinted 1971. Baltimore Md.: Regional Publishing Company.

Shannon, C. E., and Weaver, W. 1949. *The Mathematical Theory of Communication.* Urbana, Ill.: University of Illinois Press.

Shreve, F., Chrysler, M. A., Blodgett, F. H., and Besley, F. W. 1910. *The Plant Life of Maryland.* Baltimore, Md.: Johns Hopkins University Press.

Swanton, J. R. 1946. *The Indians of the Southeastern United States.* Washington, D.C.: Smithsonian Institution Press.

Tyndall, R. W. 1992a. Historical considerations of conifer expansion in Maryland serpentine "barrens." *Castanea* 57:123–131.

Tyndall, R. W. 1992b. Herbaceous layer vegetation on Maryland serpentine. *Castanea* 57:264–272.

Tyndall, R. W. 1994. Conifer clearing and prescribed burning effects to herbaceous layer vegetation on a Maryland serpentine "barren." *Castanea* 59:255–273.

Tyndall, R. W., and Farr, P. M. 1989. Vegetation structure and flora of a serpentine pine–cedar savanna in Maryland. *Castanea* 54:191–199.

Tyndall, R. W., and Farr, P. M. 1990. Vegetation and flora of the Pilot serpentine area in Maryland. *Castanea* 55:259–265.

U.S. Fish and Wildlife Service. 1993. *Sandplain Gerardia* (Agalinis acuta) *Population and Habitat Viability Assessment, Shelter Island, New York, 6–8 July 1993.* Briefing Book. Cortland, N.Y.

U.S. Weather Bureau. 1956. *Climatic Guide for Baltimore, Maryland.* Number 40–18. Washington, D.C.: U. S. Government Printing Office.

Walker, R. B., Walker, H. M., and Ashworth, P. R. 1955. Calcium–magnesium nutrition with special reference to serpentine soils. *Plant Physiology* 30:214–221.

Ware, S., and Pinion, G. 1990. Substrate adaptation in rock outcrop plants: eastern United States *Talinum* (Portulacaceae). *Bulletin of the Torrey Botanical Club* 117:284–290.

Wherry, E. T. 1963. Some Pennsylvania barrens and their flora. I. Serpentine. *Bartonia* 33:7–11.

Wherry, E. T., Fogg, J. M., Jr., and Wahl, H. A. 1979. *Atlas of the Flora of Pennsylvania.* Philadelphia: The Morris Arboretum, University of Pennsylvania.

Whittaker, R. H. 1954a. The ecology of serpentine soils. I. Introduction. *Ecology* 35:258–259.

Whittaker, R. H. 1954b. The ecology of serpentine soils. IV. The vegetational response to serpentine soils. *Ecology* 35:275–288.

Witthoft, J. 1969. Ancestry of the Susquehannocks. In *Susquehannock Miscellany,* ed. J. Witthoft and W. F. Kinsey, III, pp. 19–60. Harrisburg, Pa.: The Pennsylvania Historical and Museum Commission.

Wood, S. G. 1984. Mineral element composition of forest communities and soils at Soldiers Delight, Maryland. M.S. thesis, Towson State University, Towson, Md.

Zika, P. F., and Dann, K. T. 1985. Rare plants on ultramafic soils in Vermont. *Rhodora* 87:293–304.

5 The Mid-Appalachian Shale Barrens

SUZANNE H. BRAUNSCHWEIG, ERIK T. NILSEN,
AND THOMAS F. WIEBOLDT

Introduction

Shale barrens have been a source of fascination to naturalists for the past one hundred years. In the late 19th and early 20th century, botanists noticed sites in the Middle Appalachians supporting an unusual herbaceous flora distinct from the surrounding eastern deciduous forest. Steele (1911) formally introduced shale barren communities to botanists with the following passage:

. Several of the species considered are inhabitants of a type of land widely distributed through the mountains of middle Virginia which might well be denominated "shale barrens." . . . The barrenness is perhaps largely due to the constant washing away of fine particles of soil, but in some cases it seems as if it must be chargeable to chemical composition. . . . The variety of plant life is very considerable and together with many plants well known on other substrata, these barrens possess a number peculiar unto themselves.

Thus, the first description delineated shale barren communities on the basis of substrate and presence of a unique flora. In addition, Steele (1911) speculated that in at least some cases soil chemistry must be a factor creating barrens. According to Platt (1951), a surface layer of rock fragments and its rapid erosion are important in maintaining shale barrens. Therefore, it is the structural characteristics of the ground surface and root zone that set shale barren communities apart from the surrounding sandstone ridges and limestone valleys.

Nevertheless, soil features alone are not adequate to delineate a shale barren community. The site also must contain some indicator species endemic to shale barren communities (Keech 1996). Several workers have studied the natural history of shale barren flora to determine which species characterize shale barrens (Wherry 1935; Artz 1937, 1948; Allard 1946; Platt 1951; Keener 1983; Walck 1994). From these and other studies, a core of obligate shale barren species has been developed and used to indicate a shale barren community (Keech 1996; Rawinski et al. 1996). In some instances, indicator species were used to delineate shale barren communities on substrates not usually associated with the historical definition of shale barrens. Even though there is controversy about the delineation of shale barren communities, there are barrens with unique flora located in the Middle Appalachians on shale substrate. We present an overview of the current knowledge of their microclimate, soil characteristics, flora, vegetation, and conservation issues.

Geology and Physiography

Shale barrens occur predominately in the Ridge and Valley Physiographic Province, a long, relatively narrow belt of folded sedimentary strata between the Blue Ridge on the southeast and the Allegheny Plateau on the northwest. Long northeast–southwest trending ridges that are unusually uniform in elevation and separated by narrow valleys

Roger C. Anderson, James S. Fralish, and Jerry M. Baskin, eds. *Savannas, Barrens, and Rock Outcrop Plant Communities of North America.* Copyright © 1999 Cambridge University Press. All rights reserved.

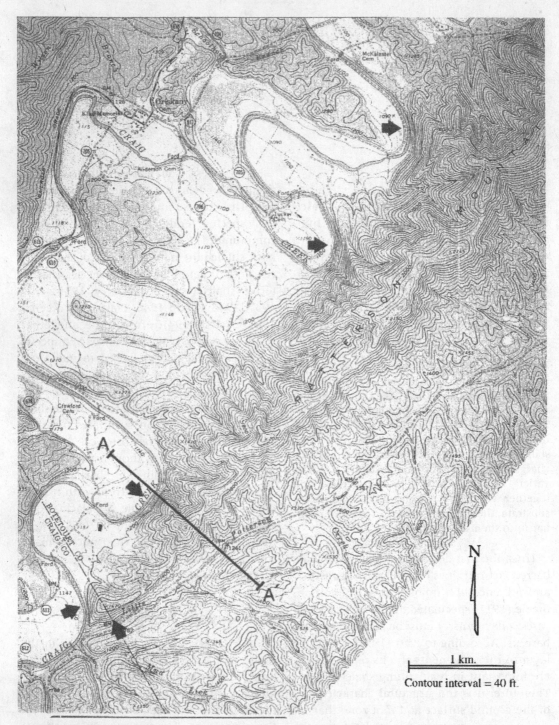

Figure 5.1. Northwest portion of the Oriskany Quadrangle, 7½–minute series, showing part of Craig Creek Valley, Botetourt Co., Va. Arrows indi-cate shale barren locations. Note the steep side slopes undercut by streams.

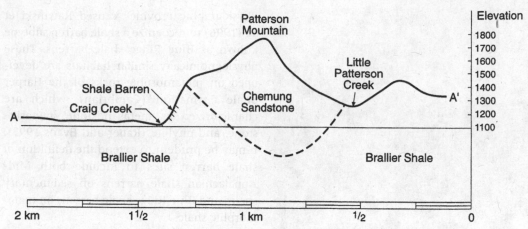

Figure 5.2. A geologic cross-section showing the position of a shale barren and topography relative to substrate along transect A–A' of Figure 5.1.

characterize the region, commonly called the Folded Appalachians. Folding and faulting have resulted in the same geologic formations being repeated several times from east to west. Shaly woodlands of oak and oak–pine occupy most slopes. Shale barrens form locally on dry exposures where a stream abuts the shale slope. Forests generally cover the landscape, except in valley bottoms and where limestone substrates provide arable land.

Different nomenclature used by geologists for shales in different regions of the Mid-Appalachians makes stratigraphic correlation with vegetation difficult. Names most commonly applied to shales in Virginia are Millboro and Brallier shales; in West Virginia, Brallier and Romney shales; and in Pennsylvania, Brallier and Trimmers Rock formations. Each formation varies in lithology on a geographic scale, with beds of varying texture and thickness in alternating layers. This, together with differences in strike and dip of the bedding, has dramatic effects on the steepness and stability of individual barrens, which vary from even-contoured slopes to shale cliffs. Most commonly, barrens are characterized by more-or-less even slopes broken by low ledges where thicker or coarser erosion-resistant strata crop out on the slope.

Topography is strongly related to bedrock geology. Shales, generally being the least resistant rocks, form the lowest ridgetops and drainage divides and the narrowest and most numerous interfluves. Shale country, consequently, is characterized by low, shaly hills or mountains of intermediate elevations and often strikingly flat valley bottoms (Figure 5.1). When present, erosion-resistant sandstone ridgecaps protect the more easily eroded shales and produce steep side slopes and narrow-backed spur ridges (Figure 5.2). Within this framework, streams undercut the slope and create unstable conditions in which shale barrens typically form.

Shale layers capable of developing shale barren communities occur on Ordovician, Silurian, and Devonian rocks. Individual geologic formations are massive and contain shale strata interbedded with sandstone and siltstone. As a result, shale barrens are not confined to rocks of a particular age and develop wherever strata occur in the right topographic setting. Slope aspect ranges from west facing to southeast facing, although barrens occasionally may have a slight northerly aspect. Elevation is generally within 300 to 575 m. Shale barrens are steep, occurring on slopes of more than 20%, and a stream (Figure 5.3) commonly undercuts them.

Figure 5.3. Cedar Creek shale barren, Frederick Co., Virginia. A typical shale barren. Note the stream at the base of the barren. Photograph by T. F. Wieboldt.

Shale barrens are most commonly associated with a thick (ca. 1,800 m) series of fissile shales of Devonian age, particularly Brallier shales (Figure 5.4), thin siltstones, and flaggy and brittle sandstones (Hack 1965). Within the Massanutten syncline, however, barrens are found on limy shales of the Martinsburg Formation of Ordovician age. Morse (1983) also described an area of typical shale barren vegetation on the Rose Hill Shale formation of Silurian age in Allegheny County, Maryland. Because of the great thickness of these formations, they crop out over broad horizontal distances, with shales predominantly occupying the valleys and lowlands, and sandstone members being important ridge formers.

The presence of shale barren endemics (see section on flora) in the Blue Ridge Physiographic Province caused Rawinski et al. (1996) to recognize a shale barren subtype known as Blue Ridge shale barrens. These physiognomically similar habitats are developed on metamorphic rocks of the Harper Shale Formation (Cambrian), which are characterized as metasandstone, metasiltstone, and phyllite (Rader and Evans 1993). It may be prudent to extend the definition of shale barren sites to include both Mid-Appalachian shale barrens on sedimentary shales and Blue Ridge shale barrens on metamorphic shales.

Soils

The shale barren soil profile is atypical, as it usually lacks an organic O horizon. The soil surface is covered with a thin layer of weather resistant shaly rock fragments, with little to no leaf litter present. The A horizon is thin, with little organic accumulation. The B horizon is absent and the C horizon, which may be up to 30 cm thick, is interspersed with pieces of fractured shale. Particle size distribution of the <2-mm fraction of shale barren soil from Ironto, Montgomery County, Virginia, is sand 60%, silt 30%, and clay 10%.

Several authors have proposed that shale barren soils must be deficient in some essential nutrient (Steele 1911; Allard 1946; Platt 1951) because of the low amount of organic matter and clay present in the soil. A deficiency of essential nutrients has been hypothesized as being the cause for the sparseness of vegetation on the barrens. However, while soil profile development may be less than that occurring on non-shale barren sites, aspects of the soil chemistry are not notably different. The pH of shale barren soil (4.0 to 5.0) and availability of some inorganic nutrients (Table 5.1) are comparable to that found in soil supporting nonbarrens vegetation on north and south slopes on various geologic formations in southwestern Virginia (Platt 1951; Schiffman 1990). Nevertheless, the reported availability of inorganic nutrients is extremely low and if availability of nitrogen is related to soil organic matter, as it is in other soils, soil nutrients may play a role in limiting nonendemic plant growth on shale barrens.

Climate and Microclimate

Soil surface temperature and solar irradiance are the most striking environmental differences between shale barren and non–shale barren sites. The shale barrens have a much higher soil surface temperature, and experience considerably higher irradiance, than other natural sites in the same region. Both of these factors can be attributed to the west- or south-facing aspect of the shale barrens and the sparseness of overstory canopy, which permit increased insolation to reach the ground surface and raise its temperature.

Midday air temperatures on shale barrens are comparable to those found in the four desert regions of North America (Barbour, Burk and Pitts 1987). Maximum air temperatures as high as 34–47 °C have been reported (Platt 1951; Braunschweig 1993). Soil surface temperatures on shale barrens in Virginia can reach as high as 50–60 °C (Platt 1951; Braunschweig 1993). Kaltenbach (1988) reports soil surface temperatures for a shale barren in western Maryland that are comparable to the Virginia studies. Platt (1951) states that the mean annual temperatures on shale barrens vary little throughout their geographic range.

Irradiance for herbaceous species on a shale barren is considerably higher than that in the surrounding deciduous forest (Figure 5.5).

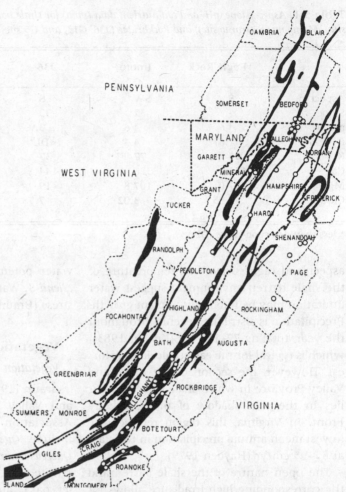

Figure 5.4. The shale barren range of the Mid-Appalachians. Black areas indicate the Brallier shale formation: open circles indicate known shale barren sites. Figure reprinted, with permission, from Platt (1951).

During the growing season, photosynthetically active radiation (PAR) in the subcanopy of a non–shale barren slope can reach a maximum of approximately 50 $\mu mol/m^2/s$ (ca. 2.5% full sun). On a shale barren, however, light intensity commonly remains above 1,000 $\mu mol/m^2/s$ PAR (50% full sun) during a summer day. Shale barren endemics appear to be obligate heliophytes that cannot acclimate to the low light conditions of a closed canopy forest (Braunschweig 1993).

The steep southerly- or westerly-facing

Table 5.1. *Aspect, slope, pH, and soil nutrient data (ppm) for shale barren sites in Botetourt (Eagle Rock site), Montgomery (Ironto site), and Rockbridge (J36, G13, and G5 sites) counties, Virginia.*

	Eagle Rock	Ironto	J36	G13	G5
Aspect	S	SW	S	S	S
Slope	-	32	33	37	36
pH	4.6	4.6	3.9	4.3	4.4
P	1.67	4.4	<DL*	1	10
K	55	77	37	48	103
Ca	453	326	144	48	228
Mg	57.0	107.8	19	14	44
Ca/Mg	7.95	3.02	7.58	3.43	5.18

* less than detection limits (DL)

aspect and high surface soil temperature of the shale barren environment suggest water limitation may be important to plant growth. Precipitation is distributed evenly throughout the year, totaling 72–110 cm (Keener 1983), which is typical for the eastern deciduous forest. However, the section of the Ridge and Valley Province in which shale barrens occur lies in the rain shadow of the Allegheny Front. In Virginia, this region receives the lowest mean annual precipitation in the state at 85–95 cm/yr (Hayden 1979).

The open nature of the shale barren and the corresponding high irradiance allows for high evaporation from the soil; however, the endemic plants apparently are not subjected to appreciable drought stress, even during the summer months (Steele 1911; Braunschweig, personal observation). The soil structure of the shale barren promotes water percolation into the C horizon, so runoff is minimized. Root systems of shale barren plants are extensive and extend deep into the fractured rock of the C horizon (Braunschweig and Nilsen, personal observation). The combination of good percolation, low evapotranspiration by trees, and deep root systems of endemic plants should make water resources nonlimiting. Suggestions by Baskin and Baskin (1988) that shale barren sites are characterized by low water availability have not been substantiated. Measurements of

water potential on the endemic *Eriogonum allenii* S. Wats. indicated no significant water stress (Braunschweig, personal observation.).

Vegetation and Flora

Vegetation

Braun (1950) included forests in the Ridge and Valley Province in her Oak–Chestnut Association. Dominant canopy species include *Quercus prinus* L. (chestnut oak), *Quercus rubra* L.(red oak), *Quercus alba* L. (white oak), *Quercus coccinea* Muenchh. (scarlet oak), and *Acer rubrum* L. (red maple) (Braun 1950; Adams and Stephenson 1983). *Castanea dentata* (Marsh.) Borkh. (American chestnut) remains as a persistent part of the forest community in the understory (Parker, Hill and Kuehnel 1993). Although the vegetation of the Oak–Chestnut Forest Region (sensu Braun) is largely consistent throughout its range, it is interspersed with various rock outcrop communities (Baskin and Baskin 1988) whose vegetation suggests a xeric environment (Allard 1946).

Braunschweig (1993) characterized the woody vegetation (stems >1.0 cm diameter at breast height) of two shale barren sites in southwestern Virginia (Table 5.2). Species with highest importance on shale barren slopes were *Pinus virginiana* P. Mill., *Quercus prinus*, *Quercus rubra*, *Quercus ilicifolia*

Figure 5.5. Canopy photographs for shale (A) and wooded, nonshale (B) sites in southwestern Virginia. Note the extremely open canopy of the shale barren site. Photographs by S. H. Braunschweig.

Table 5.2. *Importance values (IV = relative density + relative basal area + relative frequency) for all woody species for two shale barrens in southwestern Virginia. All individuals with a diameter >1.0 cm were measured.*

Species	Ironto	Craig Co.
Amelanchier arborea (Michx. f.) Fernald	22.1	-
Carya spp.	10.6	24.0
Celtis occidentalis L.	-	17.4
Cercis canadensis L.	5.24	8.7
Fraxinus americana L.	-	17.1
Juniperus virginiana L.	2.57	124.0
Nyssa sylvatica Marsh.	9.6	-
Pinus virginiana P. Mill.	126.2	25.2
Quercus ilicifolia Wangenh.	19.5	17.0
Quercus prinus L.	53.0	13.6
Quercus rubra L.	27.2	41.3
Quercus spp.	5.3	-
Rhus spp.	2.52	4.0
Sassafras albidum (Nutt.) Nees	2.54	-
Vitis spp.	2.61	7.9

Wangenh., *Juniperus virginiana* L., and *Quercus rubra*.

Non–shale barren, south-facing slopes in southwestern Virginia may be dominated by tree species such as *Pinus pungens* Lamb, *Quercus prinus*, and *Nyssa sylvatica* Marsh. (Williams 1989), or *Carya glabra* (P. Mill.) Sweet and *Quercus prinus* (Stephenson 1982). Additionally, nonbarren slopes have a well-developed shrub layer dominated by species such as *Gaylussacia baccata* (Wang.) K. Koch, *Kalmia latifolia* L., *Leucothoe recurva* (Buckley) Gray, and *Menzesia pilosa* (Michaux) Jussieu (M. V. Lipscomb and E. T. Nilsen, unpublished manuscript, 1991). These species generally are not present on shale barrens.

The differences in vegetation between shale barren and nonbarren slopes may result from variation in soil organic matter and available soil depth for rooting. Additionally, although shale barren endemics do not appear to experience water stress, it may be that typical forest tree species cannot tolerate the exposed nature of the barren environment.

Platt (1951) found that the woody vegetation associated with shale barrens was similar throughout the range of the barrens. However, dominant species may differ from one barrens to another. For example, the Ironto, Virginia, shale barren is dominated by *Pinus virginiana* (IV = 126.2) whereas the Craig County barren is dominated by *Juniperus virginiana* (IV = 124.0). This difference in dominant species indicates site-specific variation in vegetation. Such variability may be the result of localized differences in substrate, slope, aspect, the condition of surrounding vegetation, or patterns of seed dispersal.

Flora, with Particular Reference to Endemic Species

The general flora of the barrens has been discussed at length in only two papers. Allard (1946), in a thorough study of selected barrens of the Massanutten Mountains, lists 287 species in 172 genera, but his list appears to include species typical of adjacent Appalachian oak and oak–pine woodland areas. A broader survey of the shale barren region by Platt (1951) listed 128 taxa, of which 100 are vascular plants. The latter list includes those taxa more closely associated

Table 5.3. *Taxonomic list of shale barren endemics and near endemics. Species are listed in taxonomic order following the Engler–Prantl system.*

A. Endemics

> *Eriogonum allenii* S. Wats
> *Clematis albicoma* Wherry
> *Clematis coactilis* (Fernald) Keener
> *Clematis viticaulis* Steele
> *Arabis serotina* Steele
> *Trifolium virginicum* Small ex Small and Vail
> *Oenothera argillicola* Mackenzie
> *Senecio antennariifolius* Britt.

B. Near endemics

> *Allium oxyphilum* Wherry
> *Paronychia montana* (Small) Pax and K. Hoffman
> *Arabis laevigata* (Muhl. ex Willd.) Poiret var. *burkii* Porter
> *Astragalus distortus* Torr. and Gray var. *distortus*
> *Taenidia montana* (Mackenzie) Cronq.
> *Calystegia spithamaea* (L.) Pursh ssp. *purshiana* (Wherry) Brummitt
> *Scutellaria ovata* Hill ssp. *rugosa* (Wood) Epling var. *rugosa*
> *Antennaria virginica* Stebbins
> *Hieracium traillii* Greene
> *Solidago harrisii* Steele

C. Transitional

> *Phlox buckleyi* Wherry

with the high-light environments of classic shale barrens. Keener (1983) classifies 18 taxa as either strict endemic, preferential endemic, or disjunct racial endemic. The Blue Ridge shale barren subtype includes only two of the shale endemic species per site, *Senecio antennariifolius* Britt. and *Clematis coactilis* (Fernald) Keener. Numerous miscellaneous notes have been published on shale barren flora but, to date, there are no comprehensive species lists or quantitative studies of shale barren herbs. Information is largely observational, or documented in the form of specimen data in herbaria.

Interest in shale barrens has been derived largely from the endemic vascular plant taxa found in this natural community. A number of lists of shale barren endemics have been compiled, no two of which are in complete agreement (Wherry 1935; Core 1940, 1952, 1966; Keener 1970, 1983;

Bodkin 1973). The number of taxa has varied from as few as 7 strict endemics (Wherry 1935) to 21 principal species characteristic of shale barrens (Core 1940), depending on how different authors perceived shale barren flora.

An excellent discussion of the origin and taxonomy of the endemic flora may be found in Keener's (1983) review paper. Endemics are classified according to an hypothesized biohistory based on cytology, probable ancestry, and range. Endemics have also been called geographic endemics, near endemics, indicator species, and principal species. Here we refer to them as endemic, near-endemic, or transitional species. The present list (Table 5.3) largely follows Keener (1983) with slight modification. Compared to Keener's list, two species are omitted, three added, and the taxonomic status of two has been changed. Keener (1983) comments on a presumed hybrid origin for *Aster schistosus* Steele, which is omitted from the list since neither putative parent (*A. cordifolius* L. or *A. laevis* L.) is an endemic or otherwise restricted geographically. *Helianthus laevigatus* Torr. and Gray is likewise omitted since it occurs commonly in a wide variety of sunny or thinly wooded habitats well south of the region. Among the additions, *Arabis serotina* Steele has been shown to be distinct from *A. laevigata* (Muhl. ex Willd) Poiret var. *burkii* Porter (Wieboldt 1987), both of which are associated with shale barrens. The former is a strict endemic of the southern portion of the range, whereas *A. laevigata* var. *burkii* is most often on barrens but ranges farther north and northeast in thin, rocky woodlands of the shale barren region. *Scutellaria ovata* Hill ssp. *rugosa* (Wood) Epling var. *rugosa* is a species that previously has not appeared on lists of endemics but is a regular member of the midsummer flora of the barrens. As shown by Pittman (1988), it occupies an Appalachian range coincident with other near-endemic taxa and, being disjunct from the Ozarks, fits the pattern of having an affinity with a more western flora. The range of *Hieracium traillii* Greene coincides mostly with the shale barren region, and the species consistently occurs on shale barrens or in transitional shaly woodlands. *Pseudotaenidia montana* Mackenzie was transferred to *Taenidia* (Cronquist 1982) based on an earlier study by Guthrie (1968), who had demonstrated a very close relationship to *Taenidia integerrima* (L.) Drude. *Solidago harrisii* Steele is maintained as a species rather than as a variety of *S. arguta* Aiton based on studies of the *arguta–boottii* complex (Morton 1973) and a consistent morphology of plants in other (especially limestone) habitats well beyond the shale barren region. *Phlox buckleyi* Wherry (swordleaf phlox) is a species routinely included on lists of endemics, though it has long been known as atypical. Though predominately limited to a small portion of the Ridge and Valley Province in Virginia and West Virginia, swordleaf phlox occurs in shaly woods and roadbanks rather than on shale barrens. A few individuals of a few of the more frequent endemics or near endemics may occur with *Phlox buckleyi* in these marginal or transitional habitats.

The lack of strong edaphic limitations undoubtedly contributes to the lack of consensus as to what constitutes endemic status. Classification is further confounded by the fact that the so-called strict endemics occur occasionally on other substrates and are sometimes found well beyond the borders of the shale barren region. Taxa previously considered endemics occur in other habitats with sufficient frequency to stretch the term *endemic* beyond traditional concepts. For example, *Solidago harrisii* now is known to occur on limestone cliffs and barrens in southwestern Virginia and on the western edge of the Cumberland Plateau in Kentucky (Campbell et al. 1993). *Trifolium virginicum* Small ex Small and Vail recently has been found on diabase and ultramafic substrates on the northern and central Piedmont in Virginia, respectively (T. Rawinski, personal communication). *Clematis coactilis* occurs with some regularity on limestone and dolomite in the southwestern portion of its range.

The affinities of certain endemic species to western taxa have been noted by Keener (1983) and others. The connection between the shale barren flora and that of the western United States is evidenced further by the presence of disjunct populations of non-endemics on shale barrens. Since its discovery in the Appalachians as a disjunct from the southwestern United States, *Cheilanthes castanea* Maxon has been found infrequently on shale barrens and a few other rock outcrop types in the same region (Knobloch and Lellinger 1969). Emory (1970) reported a remarkable 2,000-km disjunction, from western Oklahoma, of *Asplenium septentrionale* (L.) Hoffman on a shale barren in Monroe County, West Virginia. A separate subspecies (ssp. *appalachiana* (T.M.C. Taylor)

Windham) of the western *Woodsia scopulina* D.C. Eaton, although found on a variety of rock types, is frequent on shale barrens in the southern part of the region. Among angiosperms, two taxa occur between the Appalachians and the vicinity of the Mississippi River or further west without any intervening population. One taxon, *Astragalus distortus* Torr. and Gray var. *distortus*, is considered a disjunct "racial" endemic by Keener (1983). The other, *Scutellaria ovata* ssp. *rugosa* var. *rugosa*, grows in a variety of open, rocky habitats in the Appalachians but is especially frequent on shale barrens.

Other species occur as disjuncts, but with intermediate populations between their point of origin and the shale barren region. *Paronychia virginica* Sprengel occurs in Arkansas, Oklahoma, Texas, and Alabama, with disjunct populations in the Appalachians primarily on limestone or dolomite but occasionally on shale barrens. *Erysimum capitatum* (Douglas ex Hooker) Greene, a primarily western species with sporadic occurrences eastward (Price 1987), is known in the Appalachians only from a few barrens in Allegheny and Bath counties, Virginia.

Cryptogams are a conspicuous part of the shale barren flora, but very few studies have been done on them. Twenty-one mosses and 2 liverworts were listed by Ammons (1943) in a study of one site nears Cabins, West Virginia. Platt (1951) listed 24 lichens and 4 mosses, and Allard (1946) mentions 4 lichens as being especially prevalent, as well as 3 species of reindeer lichen (*Cladonia*), and *Cetraria islandica* (L.) Ach, which he discussed at length in a separate paper (Allard 1940).

Autecology

Seedling Establishment

High soil surface temperatures may have a detrimental effect on seedling establishment of herbaceous plants on shale barrens. Most nonadapted species probably are prevented from establishing a population on shale barrens at the seedling stage. A great majority of young seedlings that germinate in the spring on shale barrens die before the summer, due to intolerance of the high temperature near the soil surface (Platt 1951; Kaltenbach 1988). Platt (1951) reported seedlings burned from contact with the soil surface. He suggested that seedlings of shale barren endemic species are capable of producing a deep root system quickly enough in the spring to establish water flow that will be critical as a buffer against the high temperatures and soil surface drying of the summer. Some endemics, such as the biennial *Arabis serotina*, establish a root system during one year and utilize that resource for flowering the next year. Others, such as the polycarpic perennial *Eriogonum allenii*, may utilize vegetative reproduction to circumvent the stresses of seedling survival on the shale barren.

Ecophysiology of Endemic Plants

Endemic plants that grow in shale barren environments experience a fragmented rock substrate, with little to no surface organic cover, which receives high irradiance. Since most shale barren endemics are small herbaceous perennials, they are strongly influenced by the microenvironment near the soil surface. Throughout the early part of the 20th century, most authors believed that physiological tolerance of belowground stresses such as an unusual soil-solution chemistry and drought were the main factors contributing to the growth of endemic plants on shale barrens (Wherry 1935; Core 1940; Allard 1946). However, there are several lines of evidence that reduce the importance of edaphic factors to shale barren plants.

Unlike many other barren environments considered in this book, there is no evidence that soil-solution chemistry of shale barren sites is any different than that of many other locales in the Mid-Appalachian region. Shale barren plants are easily grown in standard greenhouse growth medium (Baskin and

Table 5.4. *Photosynthetic characteristics derived from Light Response Curves (LRC) from shaded and unshaded* Eriogonum allenii *(data from Braunschweig 1993).*

LRC values	Unshaded plants		Shaded plants	
Maximum photosynthesis ($\mu mol/m^2/s$)	15.259*	(5.922)	6.738*	(4.84)
Light saturation ($\mu mol/m^2/s$)	1,094.38	(194.79)	937.88	(408.7)
Light compensation point ($\mu mol/m^2/s$)	41.41	(17.2)	32.05	(17.72)
Quantum yield ($\mu mol\ CO_2\ /\mu mol\ PAR$	0.063*	(0.025)	0.037*	(0.013)

* maximum photosynthesis and quantum yield are significantly different between shaded and unshaded treatments at the 0.05 level. Values are means (standard deviation).

Baskin 1988; Braunschweig, personal observation), thus they have no special requirement for a growth medium that mimics the soil conditions in the shale barren. The facts that (1) soil inorganic nutrients and water are not limiting to growth or stressful for shale barren plants, and (2) shale barren plants do not need special growth media in greenhouse situations, support the notion that aboveground factors are more important determinants of endemism than below ground factors in shale barren sites.

Due to the climatic conditions of shale barrens and morphological characteristics of the endemic plants, several authors have suggested that endemic species of shale barrens are "obligate heliophytes" (Platt 1951; Keener 1983; Braunschweig 1993), a characteristic that restricts the plants to high-light barrens and reduces their competitive ability in other disturbed habitats, where shading occurs from neighboring individuals.

To test the hypothesis that shale barren endemic plants are high-light–adapted species without the capacity to acclimate to low light, Platt (1951) exposed five shale barren endemic species (grown in pots) to high, intermediate, and heavily shaded light regimes, and observed their growth. All five species grew best in the highest light, and none grew normally in heavy shade. *Oenothera argillicola* Mackenzie was particularly sensitive to shaded conditions: All plants of this species died after only four weeks. These qualitative results were reexamined in a more natural way by manipulating light availability for in situ individuals of *Eriogonum allenii* (Braunschweig 1993). Shaded conditions severely inhibited photosynthetic capacity of *E. allenii* (Table 5.4). The high intercellular carbon dioxide concentration and low stomatal conductance of shaded plants indicated strong nonstomatal limitation to photosynthesis. The limited photosynthesis reduced growth and fitness of *E. allenii* in shaded conditions compared to natural light conditions. In fact, plants grown under 73% shade did not flower, or grow, the first year of treatment and they died during the second year.

Further evidence for the lack of shade acclimation in *E. allenii* was found by examining light response curves of photosynthesis under high light and shaded conditions. It generally is accepted that plants adapted to high light are capable of acclimating to low-light conditions by developing a lower light saturation point, lower light compensation point, and an unchanged or higher quantum yield. All of these adjustments increase photosynthetic efficiency under low-light conditions (Nilsen and Orcutt 1996). Plants of *E. allenii* maintained under 47% shade had a similar light saturation point and a lower quantum yield than plants exposed to natural light in a greenhouse (Braunschweig 1993). These data and those from in situ studies provide clear evidence that *E. allenii* is incapable of adjusting its net photosynthesis under low-light conditions to maintain a positive carbon balance.

Given the high summer temperatures of the shale barren environment (Platt 1951; Keener 1983; Kaltenbach 1988), it appears that the shale barren endemics have the capacity to tolerate temperature extremes. Thermotolerance of buds of several shale barren species was studied in a greenhouse experiment (Platt 1951). Bud mortality occurred between 42 °C and 45 °C, which is at least 10 °C below the maximum temperatures recorded on shale barren sites. The temperature response curve of photosynthesis for *Eriogonum allenii* showed a thermal optimum of 25 °C for plants grown at 20 °C or at 30 °C (Braunschweig 1993). However, membrane thermotolerance of mature leaves of *E. allenii* was maintained to 52 °C (Braunschweig 1993). These studies suggest that buds of shale barren endemic species have low thermal tolerance and that photosynthesis is inhibited at summer temperatures found on shale barren sites. However, once the leaves are formed, they are capable of maintaining membrane integrity at the high temperatures occurring on shale barren sites. Furthermore, leaves of many shale barren endemic species have morphological traits that serve to reduce leaf temperature under high-light conditions, such as a thick layer of trichomes, a waxy leaf surface, and a steep leaf angle (Nilsen and Orcutt 1996).

These results also suggest that most of the growth and carbon accumulation of shale barren endemic plants occurs in the early spring and the late fall, when temperatures are moderate. However, some shale barren species reach peak activity during the hottest time of the year. For example, *Oenothera argillicola* and *Eriogonum allenii* bloom during the middle of the summer. This discrepancy between field observation and laboratory results indicates that further study is required on thermotolerance of shale barren species.

Natural and Anthropogenic Disturbances

Shale barrens are unstable sites due to active erosion of the thinly bedded and easily fragmented shale strata. As habitats, however, they are very stable over long periods of time, being self-maintaining as long as the shale hills or ridges and undercutting stream are a part of the landscape. Natural disturbances other than erosion are largely insignificant. The absence or sparseness of fuel on shale barrens renders fire relatively unimportant, although Rawinski et al. (1996) suggested that it may play an important role in the life history of endemics such as *Clematis coactilis* on Blue Ridge shale barrens.

Certain human activities have affected shale barrens locally, but on a landscape scale these would likely have only minor effects on the frequency or abundance of the endemic flora. Road, railroad, and dam construction have all been noted as destructive activities (Bartgis 1987). Shale banks occasionally are quarried and the rock used for road repairs; the base of the slope of one site studied by Platt is now severely degraded. Lack of productivity perhaps is a saving grace in the case of shale barrens. The absence of merchantable timber has spared these areas from commercial logging, although individual trees of *Pinus strobus* L. and *Juniperus virginiana* sometimes are harvested locally for lumber and fence posts, respectively. Because of the treacherous terrain, shale barrens usually are fenced to keep livestock off them, although sheep and goats occasionally are allowed access. The principal effect of grazing is denudation of the site of its native flora and the introduction of a host of exotic weedy species. Deer browsing may adversely affect one endemic, *Arabis serotina*. Eight of 11 populations in West Virginia were browsed, with 15%–70% percent of the inflorescences in a population being partially or completely destroyed (Bartgis 1987). Although foot traffic can have a very noticeable short-term detrimental effect, few, if any, shale barrens receive sufficient visitation to have lasting effects. The occasional hunter or naturalist who visits these habitats traverses only the most accessible portion of a site, resulting in little to no impact on endemic populations.

Conservation and Management

Efforts to conserve and preserve shale barrens have occurred at federal and state levels. *Arabis serotina* is a federal endangered species and various other endemic or disjunct taxa are considered worthy of protection in one or more of the states in which they occur. In general, the impacts of real or apparent threats are unknown, underscoring the need for additional life history and monitoring studies. Since some endemic or disjunct plants occur in only a few sites, protection or management in these instances may be prudent.

In some cases, exotic weed species may pose a threat to shale barren endemics (Keech 1996). Barrens where this is the case are usually partially artificial, having been enlarged through overgrazing by livestock such as goats. In natural shale barrens, invasion of exotic weeds is not a problem, although awareness of the issue is of some interest because of the endangered status of several endemics. Generally, shale barrens require no particular management other than protection. Private landowners should be informed about these unique habitats and

encouraged to leave them undisturbed.

Faunistically, shale barrens also have attracted the attention of naturalists and conservationists. Two rare butterflies are now principally associated with shale barrens (J. C. Ludwig, personal communications). *Pyrgus wyandot* Edwards (the grizzled skipper) formerly ranged from New York south to North Carolina, and *Euchloe olympia* Edwards (the Olympia marblewing) is the only eastern representative of a western genus (Klots 1951). The disjunct distribution of the Olympia marblewing is similar to the disjunct distribution of several endemic plant species.

Future Research

Much remains to be learned about the shale barren community. Additional research on community structure, across a variety of shale barren sites, will provide information about variation among shale barrens. Considerable work needs to be done to better understand the life history and ecophysiology of most of the shale barren endemics. No work has been done on the pollination or dispersal strategies of the endemics or addressed the question of how, or if, endemics migrate from one barren to another. Such research will provide basic knowledge about the shale barren habitat and could provide useful information for future management needs.

REFERENCES

Adams, H. S., and Stephenson, S. L. 1983. A description of the vegetation on the south slopes of Peters Mountain, southwestern Virginia. *Bulletin of the Torrey Botanical Club* 110:18–22.

Allard, H. A. 1940. "Iceland moss," *Cetraria islandica*, in Virginia. *Virginia Journal of Science* (old series) 1:17–25.

Allard, H. A. 1946. Shale barren associations on Massanutten Mountain, Virginia. *Castanea* 11:71–120.

Ammons, N. 1943. Bryophytes of the Appalachian shale barrens. *Castanea* 8:128–131.

Artz, L. 1937. Plants of the shale banks of the Massanutten Mountains of Virginia. *Claytonia* 3:45–50; 4:10–15.

Artz, L. 1948. Plants of the shale barrens of the tributaries of the James River. *Castanea* 13:141–145.

Barbour, M. G., Burk, J. H., and Pitts, W. D. 1987. *Terrestrial Plant Ecology*. 2nd edition. Menlo Park, N.J.: Benjamin Cummings Publishing Company, Inc.

Bartgis, R. L. 1987. Distribution and status of *Arabis serotina* Steele populations in West Virginia. *Proceedings of the West Virginia Academy of Science* 59:73–78.

Baskin, J. M., and Baskin, C. C. 1988. Endemism in rock outcrop plant communities of unglaciated eastern United States: an evaluation of the roles of the edaphic, genetic, and light factors. *Journal of Biogeography* 15:829–840.

Bodkin, N. L. 1973. Shale barren species and endemic derivation. *Studies and Research* 31:80–100. Harrisonburg, Va.: James Madison University.

Braun, E. L. 1950. *Deciduous Forests of Eastern North America*. New York: Macmillan Publishing Co., Inc.

Braunschweig, S. H. 1993. The acclimation ability of the shale barren endemic *Eriogonum allenii* to light and heat. Ph.D. dissertation, Virginia Polytechnic Institute and State University, Blacksburg, Va.

Campbell, J. N. N., Cicerello, R. R., Kiser, J. D., Kiser, R. R., MacGregor, J. R., and Risk, A. C. 1993. *Cooperative Inventory of Endangered, Threatened, Sensitive, and Rare Species, Daniel Boone National Forest, Redbird Ranger District.* Frankfort, Ky.: Kentucky State Nature Preserves Commission.

Core, E. L. 1940. The shale barren flora of West Virginia. *Proceedings of the West Virginia Academy of Science* 14:27–36.

Core, E. L. 1952. Ranges of some plants of the Appalachian shale barrens. *Castanea* 17:105–116.

Core, E. L. 1966. *The Vegetation of West Virginia*. Parsons, W. Va.: McClain Printing Company.

Cronquist, A. 1982. Reduction of *Pseudotaenidia* to *Taenidia* (Apiaceae). *Brittonia* 34:365–367.

Emory, D. L. 1970. A major North American range extension for the forked spleenwort,

Asplenium septentrionale. American Fern Journal 60:129–134.

Guthrie, R. L. 1968. A biosystematic study of *Taenidia* and *Psuedotaenidia* (Umbelliferae). Ph.D. dissertation, West Virginia University, Morgantown, W. Va.

Hack, J. T. 1965. *Geomorphology of the Shenandoah Valley, Virginia and West Virginia, and the Origin of Residual Ore Deposits.* U.S. Geological Survey Professional Paper Number 484.

Hayden. B. P. 1979. *Atlas of Virginia Precipitation.* Charlottesville, Va.: University of Virginia Press.

Kaltenbach, N. 1988. An investigation of soil temperature, soil moisture, and seedling establishment at a shale barren site in Greenbriar State Forest, Maryland. Undergraduate senior thesis, Towson State University, Baltimore, Md.

Keech, D. 1996. *Managing Invasive Weeds on Shale Barrens.* Chevy Chase, Md.: The Nature Conservancy, MD/District of Columbia Office.

Keener, C. S. 1970. *The Natural History of the Mid-Appalachian Shale Barren Flora. The Distributional History of the Biota of the Southern Appalachians.* Part 2. *Flora.* Res. Div. Monograph 2. Blacksburg, Va.: Virginia Polytechnic Institute and State University.

Keener, C. S. 1983. Distribution and biohistory of the endemic flora of the mid-Appalachian shale barrens. *The Botanical Review* 49:65–115.

Klots, A. B. 1951. *A Field Guide to the Butterflies of North America, East of the Great Plains.* Boston: Houghton Mifflin.

Knobloch, I. W., and Lellinger, D. B. 1969. *Cheilanthes castanea* and its allies in Virginia and West Virginia. *Castanea* 34:59–61.

Morse, L. E. 1983. A shale barren on Silurian strata in Maryland. *Castanea* 48:206–208.

Morton, G. H. 1973. The taxonomy of the *Solidago–arguta-boottii* Complex. Ph.D. dissertation, University of Tennessee, Knoxville, Tenn.

Nilsen, E. T., and Orcutt, D. 1996. *The Physiology of Plants under Stress.* New York: John Wiley.

Parker, G. G., Hill, S. M., and Kuehnel, L. A. 1993. Decline of understory American Chestnut (*Castanea dentata*) in a southern Appalachian forest. *Canadian Journal of Forest Research* 23:259–265.

Pittman, A. B. 1988. Systematic Studies in *Scutellaria* Section *Mixtae* (Labiatae). Ph.D. dissertation, Vanderbilt University, Nashville, Tenn.

Platt, Robert. 1951. An ecological study of the

mid-Appalachian shale barrens and of the plants endemic to them. *Ecological Monographs* 21:269–300.

Price, R. A. 1987. Systematics of the *Erysimum capitatum* alliance (Brassicaceae) in North America. Ph.D. dissertation, University of California, Berkeley.

Rader, E. K., and N. H. Evans, eds. 1993. *Geologic Map of Virginia – Expanded Explanation.* Charlottesville, Va.: Division of Mineral Resources.

Rawinski, T. J., Hickman, K. N., Waller-Elling, J., Fleming, G. P., Austin, C. S., Helmick, S. D., Huber, C., Kappesser, G., Huber, F. C., Jr., Bailey, T., and Collins, T. K. 1996. *Plant Communities and Ecological Land Units of the Glenwood Ranger District, George Washington and Jefferson National Forests.* Virginia Natural Heritage Technical Report 96–20. Richmond, Va.: Virginia Department of Conservation and Recreation, Division of Natural Heritage.

Schiffman, P. M. 1990. Environmental determination and forest structure and composition: A naturally replicated experiment. Ph.D. dissertation, Virginia Polytechnic Institute and State University, Blacksburg, Va.

Steele, E. S. 1911. New or noteworthy plants from the eastern United States. *Contributions from the United States National Herbarium* 13:359–374.

Stephenson, S. L. 1982. Exposure induced differences in the vegetation, soils and microclimate of north- and south-facing slopes in southwestern Virginia. *Virginia Journal of Science* 33:36–49.

Walck, J. L. 1994. A contribution to the ecological life history of *Senecio antennariifolius*. *Castanea* 59:1–11.

Wherry, E. T. 1935. Fifteen notable shale barren plants. *Claytonia* 2:19–21.

Wieboldt, T. F. 1987. The shale barren endemic *Arabis serotina* (Brassicaceae). *Sida* 12:381–389.

Williams, C. E. 1989. Population ecology of *Pinus pungens* in pine–oak forests of Southwestern Virginia. Ph.D. dissertation, Virginia Polytechnic Institute and State University, Blacksburg, Va.

6 Granite Outcrops of the Southeastern United States

DONALD J. SHURE

Introduction

Granite outcrops of southeastern United States are a product of differential weathering of Precambrian metamorphic rock located within the Piedmont Plateau (McVaugh 1943). The Piedmont Plateau consists of a shield of granite schists and gneisses extending for almost 1,120 km from Virginia to Alabama. The more easily weathered materials are covered by deep soils, whereas the more resistant granite rocks remained exposed because the products of decomposition wash away as rapidly as they are formed (McVaugh 1943; Burbanck and Platt 1964). Consequently, long-term geological processes have left a multitude of relatively unbroken exposed granite, either as low domes rising as much as 200 m above the surrounding terrain or as flat rocks that vary in size from a few square meters to many hectares. Major outcrops within the southeastern Piedmont (Figure 6.1) extend over six states, but are most abundant in the upper Piedmont region east of Atlanta, Georgia (McVaugh 1943; Murdy 1968; Murdy, Johnson and Wright 1970; Quarterman, Burbanck and Shure 1993). These "flat rocks" or granite outcrops provided a unique habitat to plants that have evolved over a long evolutionary period.

Several distinct plant communities (Figure 6.2) occupy the microhabitats of the granite outcrops (Quarterman, Burbanck and Shure 1993). The open, exposed surfaces of the

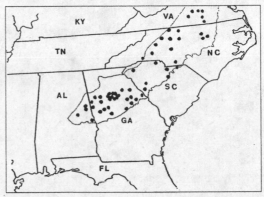

Figure 6.1. Location of the major granite outcrops within the Piedmont Plateau (delimited by dotted lines) of the southeastern United States. Adapted from Murdy (1966) and Quarterman, Burbanck and Shure (1993).

rock are quickly colonized by low-growing lichen and moss associations (Snyder 1971). Exfoliation depressions, which occur through long-term weathering processes, create "solution pits" of varying size that eventually form isolated soil island communities within the rock matrix. A few exfoliation depressions perched on the top of rock domes retain sufficient moisture to form shallow pools containing an exceptionally unusual plant assemblage (McVaugh 1943; Burbanck and Platt 1964). In addition, the edge or marginal zone between exposed rock and surrounding Piedmont fields or forests supports a mixture of outcrop and non-outcrop species. Each of these habitats will be examined to illustrate the uniquely adapted flora that occur on granite outcrops.

Roger C. Anderson, James S. Fralish, and Jerry M. Baskin, eds. *Savannas, Barrens, and Rock Outcrop Plant Communities of North America.* Copyright © 1999 Cambridge University Press. All rights reserved.

Figure 6.2. General view of Mt. Arabia, a granite outcrop in DeKalb County, Georgia, in late April 1994. Shown are a series of soil island communities at different stages of successional development.

Microenvironment

The plant taxa of granite outcrops include primarily species capable of withstanding environmental extremes. Microenvironmental conditions are considerably more severe compared to the surrounding Piedmont systems experiencing similar macroclimates. For example, granite outcrops are characterized by high solar irradiance and widely fluctuating temperatures (Shure and Ragsdale 1977; Lugo and McCormick 1981). The combination of high insolation, heat absorption at the rock surface, and limited evaporative cooling can produce rock surface temperatures of 50–55 °C (Cotter and Platt 1959; Phillips 1982; King 1985). Water availability also is relatively limited because of rapid runoff from the rock surface and low water-holding capacity of shallow sandy soils in the exfoliation depressions (Wiggs and Platt 1962). Rapid evapotranspiration during high daytime temperatures can intensify moisture stress. The combination of these hot xeric conditions produces a "microenvironmental desert" within a relatively mesic macroclimate; thus, outcrops support plants that have a life form more similar to deserts than to surrounding Piedmont habitats (Phillips 1982; Houle 1987). These species exhibit a variety of adaptations that permit survival in this xeric microenvironment (Lugo 1969; Lugo and McCormick 1981).

Endemism

Endemism is the consequence of the disjunct distribution of these extreme granite microenvironments. Twelve plant species (Table 6.1) and four varieties or subspecies of plants are considered strict endemics of south-

Table 6.1. *Habitat and distribution of plant species considered endemic to granite outcrops on the southeastern Piedmont. Distributional relationships include recorded occurrence on granite outcrops in 6 regional zones (Regional) along the Piedmont from Alabama to Virginia (Murdy 1968) and from 20 localities (Local) from North Carolina to Alabama (McVaugh 1943). Dash lines reflect species that were not included in particular distributional analyses. Compiled from McVaugh (1943), Murdy (1968), and Houle (1987), except data for* I. tegetiformans *was reported by Rury (1978).*

Species[a]	Major habitat	Distribution	
		Regional	Local
Isoetes melanospora	Inundated soil islands (pools)	1	4
Isoetes tegetiformans	Inundated soil islands (pools)	-	-
Amphianthus pusillus	Inundated soil islands (pools)	5	11
Cyperus granitophilus	Inundated soil islands, outcrop ecotone	6	10
Juncus georgianus	Inundated soil islands	6	10
Helianthus porteri	Successional soil islands	3	9
Rhynchospora saxicola	Inundated soil islands	2	4
Phacelia maculata	Outcrop ecotonal areas	5	–
Sedum pusillum	Shallow soil islands in outcrop ecotones	5	8
Portulaca smallii	Deeper soil islands, ecotonal areas	6	7
Quercus georgiana	Deeper soil islands	4	4
Panicum lithophilum	Inundated soil islands	-	2

[a] List does not include four endemic subspecies or varieties: *Oenothera linifolia* var. *glandulosa*, *Oenothera fruticosa* var. *subglobosa*, *Phacelia dubia* var. *georgiana*, and *Agalinis tenuifolia* subsp. *polyphylla*.

eastern granite outcrops (McVaugh 1943; Murdy 1968). A few additional granite outcrop species, including *Diamorpha cymosa* (synonyms: *Diamorpha smallii*; *Sedum smallii*), *Arenaria uniflora*, *Talinum mengesii*, and *T. teretifolium*, also occur on sandstone outcrops (Baskin and Baskin 1988). Several of the more widely distributed granite outcrop endemics (*Amphianthus pusillus*, *Isoetes melanospora*, and *Isoetes tegetiformans*) are associated with the shallow pools occurring in continuously flooded soil islands. Some species (*Cyperus granitophilus*, *Panicum lithophilum*, *Juncus georgianus*, and *Rhynchospora saxicola*) occupy temporary pools and moist areas found in deep, unevenly hollowed out soil island depressions. Other endemics (*Helianthus porteri*, *Phacelia maculata*, *Portulaca smallii*, *Sedum pusillum*, and *Quercus georgiana*) occur in soil island depressions at various stages of community development, particularly in marginal areas between the rock surface and

surrounding fields or forests (Quarterman, Burbanck and Shure 1993). The greatest concentration of these endemic plant species occurs in the central Piedmont of Georgia, with the number of endemics decreasing along southwest and northeast axes from this center (Murdy 1968; Harvill 1976). This "center of endemism" in Georgia is coincident with the area of greatest concentration of exposed rock surfaces within the Piedmont (Figure 6.1). Several cryptic as well as warning-colored arthropod species also are restricted to granite outcrop surfaces within the region (Quarterman, Burbanck and Shure 1993). No vertebrate species are endemic to southeastern granite outcrops.

The relatively high degree of endemism associated with granite outcrops suggests an extremely long-term habitation by plants that have been subject to active speciation processes (McVaugh 1943; Murdy 1968). Murdy (1968) suggests that most of the present endemic plant taxa evolved through either gradual or abrupt adaptations to outcrop habitats rather than representing remnants of more widespread species that evolved elsewhere. The Piedmont region marks the geographical limit of distribution for many Coastal Plain and Appalachian Plateau plant species. Thus, the environment found on the granite outcrop offers an opportunity for semi-isolated and geographically peripheral plant populations either (1) to become locally adapted, develop as endemic ecotypes, and subsequently undergo gradual ecogeographical speciation (Murdy 1966), or (2) to experience more abrupt saltational speciation processes in the extreme outcrop microenvironments (Murdy 1968). *Portulaca smallii, Rhynchospora saxicola, P. maculata*, and the recently evolved ecotype, *Phacelia dubia* var. *georgiana*, are endemics that may have evolved through a gradual ecogeographical origin. In contrast, genetic evidence suggests a more abrupt origin for the rock outcrop endemics *Diamorpha cymosa, Cyperus granitophilus*, and *Talinum teretifolium* (Murdy 1968).

Baskin and Baskin (1988) provide strong evidence suggesting that the adaptation to high photosynthetic photon flux density is an obligate requirement for rock outcrop plant endemics, and thus may be the proximal cause of species restriction to outcrop habitats. These authors maintain that these obligate heliophyte species thrive only in high-light microenvironments and are poor competitors in shaded situations (Cotter and Platt 1959; Mellinger 1972). In contrast, endemic species do not appear to be restricted to outcrop habitats because of a low genetic diversity (McCormick and Platt 1964; Murdy and Carter 1985) or a requirement for the chemical, physical, or biotic resources specific to the granite outcrop (Baskin and Baskin 1988).

Habitats and Communities

Exposed Rock Surfaces

Low-growing lichens and mosses are widely distributed on undisturbed granite surfaces. Crustose lichens often are major colonizers of bare rock (Whitehouse 1933). Additional colonizers include the foliose lichen *Xanthoparmelia conspersa* and the moss *Grimmia laevigata*. The presence of these colonizing species exerts a variable influence on further vegetation development on the rock surface. Lichens generally form a relatively stable community that plays no further role in vegetation development (Oosting and Anderson 1939; McVaugh 1943; Keever, Oosting and Anderson 1951; Winterringer and Vestal 1956; Quarterman, Burbanck and Shure 1993). The corrosive action of lichens may loosen granite particles, but the transport function of wind and water rapidly offsets this process (Whitehouse 1933).

Mosses represent the earliest pioneer plants that promote soil formation and vegetation mat development on exposed rock surfaces (Oosting and Anderson 1939; Keever, Oosting and Anderson 1951; Keever 1957). *Grimmia* germinates in the presence or absence of lichens and begins to trap blowing or washing mineral soil particles. Soil formation proceeds slowly within *Grimmia* mats until lichens (*Cladonia* spp.) and/or small vascular plants invade, usually at the center of the mat. However, the establishment of lichens within *Grimmia* mats on Georgia outcrops usually promotes die back and loss of *Grimmia* cover and a reversal of community development (Snyder 1971). In contrast, *Grimmia* mats established on outcrop surfaces in North Carolina and Virginia often initiate a long-term process of mat development (Quarterman, Burbanck and Shure 1993). This process involves a successional sequence of soil-building species, including *Cladonia leporina, Selaginella rupestris*, and *Polytrichum ohioense*, as well as annual and perennial herbs characteristic of soil island depressions elsewhere (Oosting and Anderson 1937, 1939; McVaugh 1943; Keever, Oosting and Anderson 1951; Quarterman, Burbanck and Shure 1993). These species generally invade at the deeper, central portion of the mat and expand outward in a series of concentric zones that reflect their sequential occurrence. Thus, in different regions, mat building on granite surfaces is relatively varied as to rate and extent of development (Palmer 1970; Leslie and Burbanck 1979; Quarterman, Burbanck and Shure 1993).

Shallow Pools

Shallow, semipermanent pool communities occasionally are present in larger, domed outcrops within the southeastern Piedmont. These pools form in shallow, flat-bottom depressions where the presence of a continuous, uneroded rim surrounding the depression prevents complete drainage following heavy rainfall (McVaugh 1943). The pools are characterized by shallow, sandy substrates, prolonged periods of inundation up to 10–15 cm during winter and spring, and complete desiccation over most of the summer.

Three uniquely adapted endemic plant species (*Isoetes melanospora*, *Isoetes tegetiformans*, and *Amphianthus pusillus*) thrive to varying degrees in these shallow pool communities. The two endemic perennial quillworts, *I. melanospora* and *I. tegetiformans*, often "carpet" the substrate of these shallow pools in several locations in Georgia (McVaugh 1943; Mathews and Murdy 1969; Rury 1978). *Isoetes* species utilize a CAM-like photosynthetic pathway involving the nocturnal fixation of net CO_2 as malic acid in their submerged leaves (Keeley 1982). The malic acid is subsequently broken down during the day, releasing CO_2, which is used in photosynthesis. *Amphianthus pusillus* is another endemic species with a unique growth form and reproductive habit. This winter annual has submerged and floating leaves, and bears flowers on elongated floating stems and at the plant base where they are submerged and cleistogamous (Baker 1956; Murdy 1968). It is scattered throughout the pools in relation to localized variations in depth and duration of flooding. Their unique adaptations suggest a long-standing period of endemism, especially since the species constitutes a monotypic genus of uncertain origin (Murdy 1968). Outcrop pools containing these three species occur throughout much of the Piedmont, except in North Carolina (Wyatt and Fowler 1977) and Virginia (Berg 1974).

Soil Islands

Large numbers of soil island depressions are scattered on the open rock surface of each granite outcrop (Figure 6.2). Winter cold and summer heat loosen and fragment the surface layers of granite, which slowly weather and form exfoliation depressions of varying diameter and depths (Hopson 1958). Rainfall and ice formation also play an important role in the weathering process. The development of these depressions on the granite surface sets the stage for the gradual accumulation of a sandy–gravelly mineral substrate.

Soil accumulation within the many exfoliation depressions ultimately initiates a process of primary succession (Burbanck and Phillips 1983) involving a fairly predictable series of changes in abiotic conditions and biotic populations (Shure and Ragsdale 1977; Figure 6.3). The initial accumulation of mineral soil in a depression allows colonization by pioneer plants highly adapted to extreme microenvironmental conditions (Table 6.2). With the passage of time, there is a gradual deepening of the soil, accompanied by an increase in soil organic matter and moisture-holding capacity. These changes promote a gradual reduction of microenvironmental fluctuations within the soil islands (McCormick, Lugo and Sharitz 1974; Lugo and McCormick 1981) and a concurrent shift to plant and animal components with substantially different adaptations compared to those found in earlier stages (Shure and Ragsdale 1977). This process of primary succession has been documented by long-term measurements in permanently marked communities (Phillips 1981; Burbanck and Phillips 1983). Succession to a relatively stable community (e.g., oak–hickory forest)

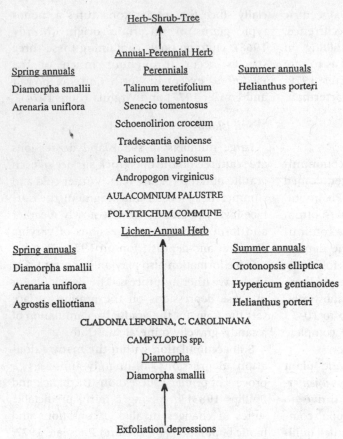

Figure 6.3. Plant community composition during early stages of primary succession in exfoliation depressions on granite outcrops of the Georgia Piedmont. Major species, including soil builders (names capitalized), are indicated. Diagram is modified from Quarterman, Burbanck and Shure (1993).

region of Georgia. Burbanck and Platt (1964) initially divided the island communities into diamorpha-dominated, lichen–annual herb, annual–perennial herb, and herb–shrub communities, based on correlations between maximum soil depth and characteristic flora. Rogers (1971) used the term *shrub–tree community* to describe large soil islands dominated by trees and having a shrub understory. The pioneer diamorpha community has been synonymously designated as *Sedum smallii* soil islands (Houle and Phillips 1989 a, b). The original herb–shrub stage has since been expanded to a more inclusive herb–shrub–tree community because of the broad overlap in soil depth between herb–shrub and shrub–tree islands (Burbanck and Phillips 1983; Quarterman, Burbanck and Shure 1993). These designations offer a convenient and appropriate reference point for identifying particular "stages" of successional development. Nevertheless, the numerous soil islands represent a continuum of biotic stages associated with an abiotic gradient.

These four community stages will be used to illustrate the complexity of interspecific interactions and adaptive responses that operate during long-term community development within exfoliation depressions. The objective is to provide a mechanistic interpretation of the process of successional development and to depict the nature of plant evolutionary adaptations that have developed in response to strong limiting factors within shallow soil islands on granitic outcrop surfaces.

Diamorpha Community. Numerous examples of the diamorpha community are present

may take over 1,000 years (Burbanck and Platt 1964) and may be reversed as a result of climatic extremes (Phillips 1981). The overall process may vary with size, depth, and configuration of the initial exfoliation depression (McCormick, Lugo and Sharitz 1974; Shure and Ragsdale 1977).

Four stages of community development are generally recognized during primary succession on granite outcrop surfaces (Burbanck and Platt 1964; Burbanck and Phillips 1983). This model of plant succession was developed primarily through extensive studies of outcrop communities from the central Piedmont

Table 6.2. *Abiotic variables from diamorpha (D), lichen–annual herb (LAH), annual–perennial herb (APH), and herb–shrub–tree (HST) stages of community development in soil island depressions on granite outcrops. Data sources are (1) Burbanck and Platt (1964); (2) Shure and Ragsdale (1977); (3) Rogers (1971), HST community only; (4) Houle (1990); and (5) Houle and Phillips (1989b).*

		Community stage			
Factor	D	LAH	APH	HST	Source
Community area (m²)	2.8	5.5	12.4	25.5	(1)
	2.7	4.0	16.9		(2)
	2.4	4.3	18.8	120.5	(4)
Maximum soil depth (cm)	5.3	10.5	24.5	44.0	(1)
	7.6	10.8	23.1	66.9	(4)
Soil bulk density (g cm⁻³)	1.03	0.83	0.44		(2)
	0.94	0.91	0.54		(5)
Cation exch. capacity (meq/100 g)	5.2	9.1	19.2	28.3	(2, 3)
Soil pH	4.0	4.2	4.5	5.8	(1, 3)
	4.4	4.5	4.6		(2)
Soil organic matter (%)	3.0	4.0	8.6	14–16	(1, 3)
Soil temperature fluctuation (°C)[a]	12.1	5.9	1.6		(2)
Soil inundation (no. incidents)[b]	77	73	10		(2)
Soil desiccation (no. incidents)[b]	83	28	0		(2)

[a] Based on a 24-hr period during midsummer.
[b] Based on number of occurrences of soil inundation or soil desiccation (<2% soil water) in 10 communities of each seral stage sampled 31 times over a 33-month period.

in shallow depressions on granite surfaces (Figure 6.4). This pioneer community develops as mineral soil accumulates through wind- and water-borne deposition (Wiggs and Platt 1962). Soil varies in depth from 2 to 9 cm, is acidic, high in coarse- and medium-sized sand particles (Wiggs and Platt 1962), and is infertile due to the low organic matter and cation exchange capacity (Table 6.2). These shallow soils are subject to extreme and rapidly fluctuating temperatures in summer and to periodic occurrences of soil inundation following heavy rainfall. The extreme nature of these abiotic factors limits plants to those adapted to these conditions.

A generally depauperate plant community is present in shallow soil islands. *Diamorpha cymosa* is the principal species. This near-endemic, winter annual germinates in the fall and overwinters in a frost-resistant rosette form that can survive up to 3–4 weeks of inundation. It flowers and completes its life cycle in early May before the onset of summer heat and desiccation (Wiggs and Platt 1962). Ants serve as key pollinators for the self-incompatible *Diamorpha* plants (Wyatt 1981; Wyatt and Stoneburner 1981), which exhibit a negative statistical correlation between density and fecundity (Clay and Shaw 1981). *Diamorpha* seedlings are capable of surviving 2–3 weeks of severe desiccation in early fall; the mature plants use water stored in succulent leaves (Wiggs and Platt 1962). Thus, this plant is a true drought-evading and drought-resistant species (Shantz 1927; Wiggs and Platt 1962) that is exceptionally well adapted to withstand the extreme moisture conditions of shallow granite substrates. The retention of seeds on capsules near the top of erect dead plants removes the seeds from summer soil surface temperature extremes and extends seed viability during an obligate 70–120-day after-ripening period (Wiggs and Platt 1962;

Figure 6.4. The diamorpha community occupying a shallow soil island depression on Mt. Arabia in late April 1994. The community consists of a dense population of *Diamorpha smallii*, which is in flower.

Baskin and Baskin 1992). The adaptations of *Diamorpha* plants permit them to thrive on extreme sites where interspecific competition is minimal or absent.

The depauperate nature of the diamorpha community is more a product of extreme limiting factors than of seed availability. Houle and Phillips (1988) recorded 21 plant species in the seed bank. Nevertheless, few species have the capacity to germinate and survive the moisture and temperature extremes of these shallow, sandy substrates. *Arenaria* (syn. *Minuartia*) *uniflora* is another small winter annual that occasionally germinates in areas of deeper, more organically rich soil of the diamorpha community during wet years (Burbanck and Platt 1964; Shure and Ragsdale 1977). The larger *Arenaria* plants can competitively isolate the smaller *Diamorpha* plants to shallower soils, where *Arenaria* does not survive (Wiggs and Platt 1962; Sharitz and

McCormick 1973). Additional annual species such as *Hypericum gentianoides, Crotonopsis elliptica, Bulbostylis capillaris,* and *Helianthus porteri* (formerly *Viguiera porteri*) also may germinate in favorable microsites during particularly wet years, although few survive to complete their life cycle (Shure and Ragsdale 1977; Houle and Phillips 1988, 1989a). The intense selective pressure of soil conditions in the diamorpha community results in large numbers of a physically small, but highly adapted, species; reduced vegetation cover and standing crop biomass; and extremely low plant species richness (Table 6.3; Houle 1990).

The rate of soil development is relatively slow in diamorpha communities (Oosting and Anderson 1939). Burbanck and Phillips (1983) found no discernable evidence of plant succession over a 22-year period. However, many small *Diamorpha* plants trap some wind- or water-borne materials and are

Table 6.3. *Vegetation density, biomass, and diversity in diamorpha (D), lichen–annual herb (LAH), and annual–perennial herb (APH) stages of successional development on southeastern granite outcrops. Data are presented from 10 replicate communities of each stage present on Panola Mountain, Georgia (Shure and Ragsdale 1977). Means and standard errors are indicated where appropriate. Plant diversity calculations are based on the Shannon–Weaver (1949) information theory formula (H') and were averaged for two growing seasons.*

Community characteristic	Community stage		
	D	LAH	APH
Plant density (peak number m^{-2})	10,871	3,060	7,736 (1,168)[a]
Standing crop biomass (g m^{-2})	40 ± 8.9	515 ± 75	1132 ± 171
Litter biomass (g m^{-2})	15 ± 4.5	182 ± 71	602 ± 107
Richness (total species)	5	16	24
Richness (no. species/comm.)	3.0 ± 0.42	8.2 ± 0.55	11.1 ± 0.50
Plant diversity (H')	0.16 ± 0.05	1.06 ± 0.13	0.97 ± 0.09

[a] Value in parentheses is for vascular plant species (mosses excluded).

an annual source of organic matter. Seedlings of other species that appear during favorable years also add organic matter. More importantly, the addition of several species of lichens (particularly *Cladonia leporina*) through runoff from upslope plays a prominent role in initiating the soil-building process (Burbanck and Platt 1964). These *Cladonia* lichens also are well adapted to survive on exposed, shallow soil through their ability to absorb and store water in their fungal elements and to obtain nutrients and food from their algal component (Burbanck and Platt 1964). The year-round presence of this intricately branched lichen cover effectively intercepts and traps debris while providing a favorable microsite for seed germination. The successional transition, which usually is initiated by the addition of lichens in the deeper portions of diamorpha soil islands, may last decades or even hundreds of years before reaching the next stage in community development. The presence of pioneer plants apparently facilitates the successional transition to later stages (Houle and Phillips 1989b).

Lichen–Annual Herb (LAH) Community. Substrate conditions are more favorable for plant establishment in the LAH community compared to the diamorpha stage of development (Figure 6.5). Maximum soil depth varies from 7 to 18 cm in the LAH community, which occupies larger and deeper soil islands (Table 6.2). The soil remains sandy; however, soil desiccation occurs less often because the soil moisture level and organic matter content are greater and flooding is still frequent. Soil temperature fluctuations are reduced under the greater protective cover of additional vegetation and litter (Tables 6.2 and 6.3). The combined reduction in these potential stresses sets the stage for successful colonization by additional annual species.

Plant community richness increases as the magnitude of environmental limitations diminishes in the LAH community (Shure and Ragsdale 1977; Houle 1990; Table 6.3). The presence of *Cladonia* lichens provides the cover needed to facilitate germination of seed. The seed bank (20 species) is similar in species composition to that of the diamorpha stage (Houle and Phillips 1988). However, germination of particular species and their distribution within the LAH community is highly dependent on the moisture and nutrient conditions present at different soil depths (Mellinger 1972; Sharitz and McCormick

Figure 6.5. Lichen–annual herb community on Mt. Arabia in late April 1994. Lichens (*Cladonia*) and mosses are present in the shallower, downslope portion of the soil island (lower left), whereas several annual species, including *Arenaria, Agrostis,* *Bulbostylis,* and small seedlings of *Helianthus,* have become established in the deeper, upslope (upper right) portion of the community. *Diamorpha* plants are scattered along the margin of the community, particularly in the shallow edge (upper left).

1973; McCormick, Lugo and Sharitz 1974; Meyer, McCormick and Wells 1975). Strong interspecific competition plays a critical role in determining the final patterns of species occurrence (Sharitz and McCormick 1973). *Diamorpha* plants survive in openings throughout the LAH soil islands, but usually are competitively restricted to the more shallow soil margins, where other species cannot tolerate the limiting conditions. The deeper soil toward the center of the LAH community favors germination and development of the spring-flowering annuals *Arenaria* and *Agrostis elliottiana,* and several summer- or fall-flowering annuals, including *Crotonopsis, Hypericum, Bulbostylis,* and *Helianthus*; a few perennial species may sporadically occur in deeper substrate.

The characteristic zonation pattern present along a soil depth gradient in certain lichen–annual and annual–perennial communities has been the subject of considerable experimental evaluation (Sharitz and McCormick 1973; McCormick, Lugo and Sharitz 1974). Efforts have focused on the response of three rock outcrop endemics (*Diamorpha, Arenaria,* and *Helianthus*) that exhibit differential tolerances and competitive abilities across the gradient. Asymmetrical competition creates zonal patterns whereby *Diamorpha, Arenaria,* and *Helianthus* may occur sequentially as soil becomes progressively deeper from the edge to the center of the community. A similar sequence occurs over time as soil development progresses within a particular community. Apparently, interspecific competition is most pronounced during the dry season, when soil moisture is at a critical level (Sharitz and McCormick 1973). However, Meyer, McCormick and Wells (1975) and McCormick, Lugo and Sharitz (1974) suggested that soil nutrients (particularly nitrogen) often act as a secondary regulatory factor in shaping the competitive response during the approximately 100 days of favorable moisture within the growing season. More recently, Baskin and Baskin (1988) hypothesized that shading imposed by a high density of progressively larger plant species in this zonal sequence may competitively hinder root development and limit nutrient uptake by smaller plant species.

Limitations imposed by moisture, nutrients, and light may occur concurrently or separately at different times of the year. These limitations shape the distributional patterns of species occurrence as soil depth varies within successional soil islands. The zonal patterns are relatively dynamic, and communities can shift considerably in composition or zonal sequence in response to seasonal and particularly year-to-year variations in precipitation and available soil moisture (Houle and Phillips 1989a). Interestingly, the loss of a species during one or more drought years may be compensated for by seeds that persist in the seed bank and germinate over several years (Mellinger 1972; Houle and Phillips 1988).

The gradual increase in soil depth in the LAH community thus exerts a major influence on system-level properties and species composition. The less restrictive conditions in LAH soil islands promote a sharp increase in vegetation biomass and diversity compared to the diamorpha stage (Table 6.3). Many examples of the LAH community exhibited successional changes in depth and species composition over 22 years, suggesting an accelerated rate of community development (Burbanck and Phillips 1983). In time, the gradual transition to a recognizable annual–perennial herb community is considered complete when soil islands contain at least two species of perennial herbs, mature *Helianthus* plants, and common to abundant patches of *Polytrichum* (Burbanck and Phillips 1983). The mosses *Polytrichum commune* and *Aulacomnium palustre* are the major soil builders in the transition to annual–perennial stages (Quarterman, Burbanck and Shure 1993).

Annual–Perennial Herb (APH) Community. Island area and depth continue to increase over time as soil development progresses in the APH community (Table 6.2). Maximum soil depth reaches 13–41 cm through the annual addition of organic matter from autochthonous sources and through ongoing weathering processes at the granite–soil interface at the bottom of island depressions. These processes produce a 40%–50% reduction in soil bulk density and an approximate doubling of soil organic matter and cation exchange capacity between the LAH and APH stages; soil pH increases only slightly. More important is the diminished fluctuation in potentially limiting conditions associated with temperature and moisture extremes on outcrop surfaces. Soil buildup over time produces a fourfold reduction in soil temperature fluctuations between the LAH and APH stages. Vegetation present in the deeper APH islands affords protective cover for the soil. Added vegetation and litter also prevent soil desiccation even during drought (Table 6.2). Eventually, the buildup of soil may create a convex structure in many APH communities as the soil surface rises above the rim of the exfoliation depression. Soil inundation is absent on these convex surfaces.

The gradual amelioration of abiotic stresses within APH soil islands enables successful establishment of many plant species (Figure 6.6). The soil seed bank (28 species) reflects the substantial increase in perennial species. There also is a closer correlation between species presence in the seed bank and presence in the existing plant community than in earlier successional stages (Houle and Phillips 1988). Patches of *Polytrichum* mosses within APH islands offer protection from drought

Figure 6.6. Annual–perennial herb community on Mt. Arabia in late April 1994. Vertically oriented leaves of *S. tomentosus*, which is in the final stage of flowering, are present in the center of the community, surrounded by standing dead plants and early spring growth of several annual and perennial herbs and grasses. *Diamorpha* plants primarily occur on the downslope margin (right) of the community.

(Leslie 1975) and provide a safe microsite for seed germination of perennial species. Lichens and small annual plants gradually are restricted to the outer margin of the islands.

Five perennial plants (*Polytrichum commune, Andropogon virginicus, Tradescantia ohiensis, Schoenolirion croceum, Senecio tomentosus*) and mature flowering individuals of the annual *Helianthus* are considered indicative of the APH stage (Burbanck and Phillips 1983). A number of other species also sporadically occur on soil islands on different outcrops (see Burbanck and Platt 1964). *Schoenolirion* and *Senecio* are two drought-adapted species that are among the first perennial plants to invade the APH community (Burbanck and Platt 1964). *Schoenolirion* possesses a deeply buried bulb that remains dormant through the summer drought. In contrast, *Senecio* has a woody underground stem and a deep root system that may obtain water from deeper soil layers to avoid drought. Many perennial species are restricted primarily to deeper portions of the APH islands because of abiotic restrictions or in response to the inhibitory effects of annual species that often are dominant competitors in shallower areas (Houle and Phillips 1989b). However, *Helianthus* is an annual species that frequently occupies large areas of the islands without displacing perennials (Quarterman, Burbanck and Shure 1993). The tendency of *Helianthus* seed to germinate and for seedlings to grow rapidly when moisture is favorable, combined with its ability to withstand prolonged wilting, helps ensure success in the highly variable year-to-year soil moisture regime (Cumming

1969; Mellinger 1972; McCormick, Lugo and Sharitz 1974). Nevertheless, many *Helianthus* plants often do not survive extreme drought. The species relies on a persistent soil seed bank for subsequent reestablishment when conditions are favorable (Mellinger 1972).

The combination of a large island area and a deep soil promotes a diverse assemblage of vascular as well as nonvascular (e.g., soil lichens) plant species in the APH community (Bostick 1968). Community standing crop and litter biomass increase 30–40-fold as soil develops from the diamorpha to APH stages (Table 6.3). The plant community present in APH islands may be approaching equilibrium conditions (Shure and Ragsdale 1977) as it has less seasonal and yearly turnover in species composition than the communities on shallower soil (Houle and Phillips 1989a). However, continued deepening of the soil permits the successful establishment of woody species, which facilitates the transition to the final stage in primary succession.

Herb–Shrub–Tree (HST) Community. Herb–shrub–tree communities are present on larger, deeper, and more organically rich soil islands compared to those of earlier stages (Table 6.2; Figure 6.7). This community represents a possible edaphic climax or end point of succession on granite outcrops, but species composition remains relatively dynamic (Phillips 1981; Burbanck and Phillips 1983). Phillips (1981) recognized pine, mixed pine–hardwood, and hardwood stands as distinct variants of the HST stage. *Pinus taeda* seedlings generally initiate woody species colonization in well-developed APH communities. Seedlings may germinate in LAH or early APH stages during especially wet years, but they invariably die within a short period. In contrast, seedlings on deeper APH soil islands reach maturity and initiate the HST stage of development. *Pinus taeda* usually is accompanied by perennial species characteristic of the annual–perennial stage. Species include sedge (*Carex communis, Scleria oligantha*) and grass (*Panicum* spp.), as well as

woody plants (*Smilax glauca* and *Vaccinium arboreum*) and, occasionally, trees (*Juniperus virginiana, Carya glabra,* or *Prunus injucunda*). Large numbers of seedlings of *Quercus nigra* and a granite outcrop endemic, *Q. georgiana,* are sometimes present (Burbanck and Platt 1964; Phillips 1981). Periodic droughts often kill many woody species present at the pine stage of development, and this may reset the successional process. *Pinus taeda* mortality in these stands is inversely statistically correlated to the soil depth beneath each tree (Phillips 1981).

Mixed pine–hardwood stands develop on deeper islands (maximum soil depth = 75–85 cm). *Pinus taeda* and *Carya glabra* are dominants, but *Rhus copallina, Bumelia lanuginosa, Quercus* spp., and *Vaccinium arboreum* also are relatively important (Phillips 1981). The near absence of *P. taeda* seedlings and saplings under the shade of a developing hardwood canopy indicates a future transition to hardwood stands. Hardwood stands (maximum soil depth 100–125 cm) contain *Carya glabra* and *Q. prinus* as dominants, with *Bumelia, Prunus serotina, Ptelea trifoliata, Vaccinium,* and several other species occasionally present (Phillips 1981). *Pinus taeda* is absent on the deeper soil of these hardwood stands. Drought-induced mortality is greatly reduced in both mixed and hardwood stands. Interestingly, of the four communities (diamorpha, LAH, APH, and HST) associated with various stages of soil development, plant species richness was significantly positively correlated with island area and available soil depth only in the HST community (Houle 1990).

Overview of Community Development

Primary succession on southeastern granite outcrops occurs through the reciprocal interactions between biota and substrate (Shure and Ragsdale 1977). The shallow pioneer diamorpha communities are subject to severe

Figure 6.7. Herb–shrub–tree community on Mt. Arabia in late April 1994. This view illustrates the presence of a thick growth of shrubs, vines, and trees in these deeper soil islands and the effects of a severe drought in late summer 1993 on loblolly pine survival. The two large pine trees in the community did not survive the drought, whereas the pine in the background (left) withstood the moisture deficit. A number of the long-standing pine trees on Mt. Arabia died as a result of the 1993 summer drought.

abiotic stresses and experience large nutrient and material fluxes during periodic flooding (Hay 1973). Plant communities shift over time as soil depth gradually increases in the soil islands of exfoliation depressions. As succession progresses, the number and size of plant species colonizing soil islands become larger. Density-dependent factors such as competition often act as intense regulators in shaping plant distribution patterns in relation to limiting factors of the substrate. The gradual increase in plant biomass and vegetative cover over time enhances soil development, thereby facilitating further plant establishment. The combination of increased plant cover and deeper soil ultimately reduces moisture and temperature fluctuations in the substrate. These processes continue until the soil level of the APH communities reaches or exceeds the rim of the exfoliation depression and the stress associated with flooding or desiccation is minimized or absent. Nutrient recycling becomes important within these deep, convex APH or HST communities in comparison to major nutrient fluxes in the shallow soil of pioneer stages. The extent of limiting conditions within each particular soil island sets the rate and extent of community development. However, as many soil island communities undergo long-term succession, they tend to converge in structure (Shure and Ragsdale 1977) and experience a reduced seasonal and annual fluctuation in species composition (Houle and Phillips 1989a).

Strong statistical correlations exist between vegetation composition and abiotic conditions throughout successional development on granite outcrop surfaces. The strength of this relationship suggests that plant species occurrence during primary succession often is related to specific features of substrate development (Chapin et al. 1994). Facilitative or competitive biotic interactions on granite outcrops appear to have played a major role in fine tuning the patterns of species response to the changing complex of abiotic limiting factors (Houle and Phillips 1989b).

The limiting conditions within soil island depressions also exert a major influence on ecosystem functioning (McCormick, Lugo and Sharitz 1974; Lugo and McCormick 1981). These conditions are closely related to soil depth within and among communities. The magnitude of temperature and moisture extremes clearly diminishes with increased soil depth, which also promotes an increase in structural complexity of the system, net biomass production, and plant species diversity (Lugo and McCormick 1981). Plant photosynthetic capacity potentially is relatively large given the high insolation on granite outcrops. Nevertheless, plant production is largely concentrated during periods of suitable growing conditions, when temperature is less limiting and water and nutrients are readily available. Plant growth is accelerated during these periods and conducive to rapid biomass accumulation. However, periodic summer moisture stress can promote a high plant respiration rate that becomes energetically wasteful (Lugo and McCormick 1981). This point is reached despite various adaptations (McCormick, Lugo and Sharitz 1974) minimizing transpirational losses of water through extended periods of temporary wilting (*Helianthus*); a midday depression in transpiration, photosynthesis, and respiration (*Sedum, Talinum*); inward folding of shoots (*Polytrichum*); or utilization of a grassy habit (*Bulbostylis*). Dormancy becomes adaptive when temperature and moisture stresses become maximal, dormancy has contributed to the establishment of separate spring and summer flora in lichen–annual and annual–perennial successional stages (Figure 6.3). Energy stored in outcrop plants during favorable periods subsequently is utilized to maintain various morphological adaptations that permit survival during recurring stress (Lugo and McCormick 1981).

Community development is relatively similar on granite outcrops of the southwestern United States (Uno and Collins 1987; Collins, Mitchell and Klahr 1989). In both areas, the successional patterns result from a strong positive relationship between species occurrence and island size and soil depth. Although the actual plant species involved in the two regions are relatively different, their life-form distributions are comparable because of similar desertlike microenvironments (Walters and Wyatt 1982). The principal regional differences are that southwestern outcrops have fewer endemic species (Walters and Wyatt 1982) and shallower soil islands (Uno and Collins 1987) and, consequently, succession is restricted to herbaceous stages of development (Uno and Collins 1987; Collins, Mitchell and Klahr 1989).

Marginal Zone

Many species characteristic of granite outcrops are relatively abundant in the narrow, often treeless, marginal zone between exposed rock and the surrounding forest ecosystem (McVaugh 1943). Spatially, this marginal zone may be relatively dynamic, extending as peninsulas of varying width onto the exposed rock or into surrounding habitats. Soil depth and composition in the marginal zones are similar to deeper soil islands on the outcrop surface, although increased shade and gradual seepage of groundwater from surrounding areas usually promote more mesic conditions. Nevertheless, the relatively rapid evaporative loss of water from these marginal zone soils during dry summers generally prevents the establishment and spread of non-outcrop species. The particular set of abiotic conditions dictated by shade, drainage, and soil depth conditions at a given location determines the specific mix of outcrop versus invading species. Thus, plant species composition may vary somewhat across the marginal zone from outcrop to surrounding forest.

The vegetational community associated with marginal zone areas is considered relatively stable (McVaugh 1943; Quarterman, Burbanck and Shure 1993). Soil input and losses are often balanced in marginal zone substrates, thereby precluding successional changes in species composition. Many endemic outcrop species, including *Cyperus granitophilus*, *Phacelia maculata*, *P. dubia* var. *georgiana*, *Portulaca smallii*, *Quercus georgiana*, and *Sedum pusillum*, thrive to varying degrees within these marginal zone habitats. Other near-endemic or outcrop specialists, such as *Minuartia brevifolia*, *Schoenolirion croceum*, and *Talinum teretifolium*, usually are abundant (McVaugh 1943). Peripheral zones along the edges of marginal communities may contain *Grimmia*, *Polytrichum*, *Selaginella*, and *Cladonia* (Quarterman, Burbanck and Shure 1993), although mat building further onto the rock surface is limited or absent (McVaugh 1943).

Future Research

Several decades of research on southeastern granite outcrops have provided a solid foundation for understanding plant habitat associations, physiological adaptations, successional relationships, and ecosystem functioning within this desertlike habitat. Future efforts should focus on nutrient uptake processes, root-to-shoot ratios, or plant carbon allocation to growth versus reproduction under different microenvironmental conditions or interspecific associations. Other areas of concentration could include more integrative assessments of the mechanism by which plant species adjust to climatic stress, the extent of genetic divergence in relation to cli-

matic conditions of different local outcrops or across the regional northeast-to-southwest distribution of outcrop occurrence, and, finally, further experimental evaluation of the mechanisms of plant species replacement during the primary successional process (Houle and Phillips 1989b).

Little research has concentrated on animal populations of granite outcrops (Quarterman, Burbanck and Shure 1993). A documentation of existing animal species in granite outcrop habitats would provide a strong foundation for a variety of topical approaches. Studies concentrating on plant–animal interactions are particularly needed to assess the potential importance of arthropod populations in the pollination and spread of outcrop plants, their role in seed consumption and dispersal mechanisms, and the consequence of insect herbivore damage on plant survival and fitness under various outcrop conditions. For example, the spring flowering of *Senecio tomentosus* attracts large numbers of a host-specific hemipteran herbivore (*Neacoryphus bicrucis*: Lygaeidae), which extracts nutrients from flowers, stems, and seeds. *Neacoryphus* sequesters pyrrolizidine alkaloids obtained from *Senecio* and uses them for defense against predators (McLain and Shure 1985). In return, the aggressive courtship behavior of *Neacoryphus* indirectly may benefit *Senecio* by excluding other insect herbivores (McLain and Shure 1987). There may be similar types of plant–animal interactions that alter the response of other outcrop plant species to abiotic and biotic stresses.

The presence of numerous soil island communities of varying size and degree of isolation on the granite surface offers an excellent opportunity to test hypotheses emerging in landscape ecology, patch dynamics, island biogeography, and biodiversity. Studies documenting the diversity of plants and animals in different soil islands could provide valuable information on species responses to specific patch variables, including size, configuration, stage of development, and the degree of isolation within the granite matrix. Animal responses to these variables could be examined to better appreciate the nature of consumer adjustments to particular patch attributes.

Community Threats and Conservation

The relatively fragile nature of the biotic communities on granite outcrops suggests their potential susceptibility to anthropogenic processes. Quarrying is perhaps the most destructive of these processes. For example, the 22 active commercial quarrying operations near Atlanta, Georgia, have caused varying degrees of damage to granite outcrops in this region (Quarterman, Burbanck and Shure 1993). Grazing, land development for commercial and residential use, and extensive hiking, picnicking, and off-road vehicle activity can cause severe damage to these special biotic communities. Human incursion has had a direct impact on lichen survival on the granite surface. Lichen growth is greatly reduced or absent where trampling has damaged the colony and has increased susceptibility to air pollution or other abiotic stresses (Quarterman, Burbanck and Shure 1993).

The need for conservation of granite outcrops is especially evident near metropolitan areas such as Atlanta. The increasing evidence of damage from human use has led to the establishment of several state or county parks to protect outcrops. This important step has permitted a small permanent park staff to educate the public and prevent further damage. Other means of protection have been afforded through the occurrence of outcrops in national forests or battlefield sites, as well as on private preserves. In addition, the current federal listing of *Amphianthus pusillus* and *Liatris helleri* as threatened and of *Isoetes melanospora* and *I. tegetiformans* as endangered species should provide habitat protection for

granite outcrops with resident populations of these species. Effort should be expended to maintain these populations while monitoring the fate of other species with small resident populations. Perhaps the best means of accomplishing preservation is through local efforts to purchase and set aside parcels of land containing outcrops of special significance. Education of the public concerning the ecological and evolutionary significance of these unique "desertlike" habitats is a first step in this process.

Summary

Granite outcrops occur sporadically throughout the Piedmont of the southeastern United States. These exposed rock outcrops have a desertlike microenvironment to which plants have developed unique morphological, physiological, and life-cycle adaptations over a long evolutionary period. Endemism is a product of the disjunct distribution of these outcrops and their extreme microenvironments. The plant communities consist primarily of lichen–moss associations on exposed rock surfaces, occasional shallow, semipermanent pools containing several uniquely adapted endemic plants, and various diamorpha, lichen–annual herb, annual–perennial herb, and herb–shrub–tree communities that represent seral stages of primary succession in numerous soil-filled exfoliation depressions on the granite surface. Primary succession is initiated when a sandy, inorganic substrate accumulates in the depressions. During succession, there is a gradual shift to larger and longer-lived plant species as soil depth and soil organic matter increase, and as soil temperature and moisture fluctuations are reduced through the interaction of biota and substrate. Facilitative or competitive biotic interaction during succession plays a major role in governing species responses to the changing complex of abiotic limiting factors. The timing of the successional trajectory may vary depending on the initial size, depth, and configuration of each exfoliation depression. Nevertheless, as succession progresses, the communities in the soil islands on each outcrop tend to converge and exhibit less seasonal and annual fluctuation in species composition.

REFERENCES

Baker, W. B. 1956. Some interesting plants on the granite outcrops of Georgia. *Georgia Mineral Newsletter* 9:10–19.

Baskin, J. M., and Baskin, C. C. 1988. Endemism in rock outcrop plant communities of unglaciated eastern United States: an evaluation of the roles of the edaphic, genetic and light factors. *Journal of Biogeography* 15:829–840.

Baskin, J. M., and Baskin, C. C. 1992. Germination characteristics of *Diamorpha cymosa* seeds and an ecological interpretation. *Oecologia* 10:17–28.

Berg, J. D. 1974. Vegetation and succession on Piedmont granitic outcrops of Virginia. M.S. thesis, College of William and Mary, Williamsburg, Va.

Bostick, P. E. 1968. The distribution of some soil fungi on a Georgia granite outcrop. *Bull. Georgia Academy of Science* 26:149–154.

Burbanck, M. P., and Phillips, D. L. 1983. Evidence of plant succession on granite outcrops of the Georgia Piedmont. *American Midland Naturalist* 109:94–104.

Burbanck, M. P., and Platt, R. B. 1964. Granite outcrop communities of the Piedmont Plateau in Georgia. *Ecology* 45:292–306.

Chapin, F. S., III, Walker, L. R., Fastie, C. L., and Sharman, L. C. 1994. Mechanisms of primary succession following deglaciation at Glacier Bay, Alaska. *Ecological Monographs* 64:149–175.

Clay, K., and Shaw, R. 1981. An experimental demonstration of density-dependent reproduction in a natural population of *Diamorpha smallii*, a rare annual. *Oecologia* 51:1–6.

Collins, S. L., Mitchell, G. S., and Klahr, S. C. 1989. Vegetation–environment relationships in a rock outcrop community in southern Oklahoma. *American Midland Naturalist* 122:339–348.

Cotter, D. J., and Platt, R. B. 1959. Studies on the ecological life history of *Portulaca smallii*. *Ecology* 40: 651–668.

Cumming, F. P. 1969. An experimental design for the analysis of community structure. M.S. thesis, University of North Carolina, Chapel Hill, N.C.

Harvill, A. M., Jr. 1976. Flat-rock endemics in Gray's Manual range. *Rhodora* 78:145–147.

Hay, J. D. 1973. An analysis of *Diamorpha* systems on granite outcrops. M.S. thesis, Emory University, Atlanta, Ga.

Hopson, C. A. 1958. Exfoliation and weathering at Stone Mountain, Georgia, and their bearing on disfigurement of the Confederate Memorial. *Georgia Mineral Newsletter* 11:65–79.

Houle, G. 1987. Vascular plants of Arabia Mountain, Georgia. *Bulletin of the Torrey Botanical Club* 114:412–418.

Houle, G. 1990. Species–area relationship during primary succession in granite outcrop plant communities. *American Journal of Botany* 77:1433–1439.

Houle, G., and Phillips, D. L. 1988. The soil seed bank of granite outcrop plant communities. *Oikos* 52:87–93.

Houle, G., and Phillips, D. L. 1989a. Seasonal variation and annual fluctuation in granite outcrop plant communities. *Vegetatio* 80:25–35.

Houle, G., and Phillips, D. L. 1989b. Seed availability and biotic interactions in granite outcrop plant communities. *Ecology* 70:1307–1316.

Keeley, J. E. 1982. Distribution of diurnal acid metabolism in the genus *Isoetes*. *American Journal of Botany* 69:254–257.

Keever, C. 1957. Establishment of *Grimmia laevigata* on bare granite. *Ecology* 38:422–429.

Keever, C., Oosting, H. J., and Anderson, L. E. 1951. Plant succession on exposed granite of Rocky Face Mountain, Alexander County, North Carolina. *Bulletin of the Torrey Botanical Club* 78:401–421.

King, P. S. 1985. Natural history of *Collops georgianus* (Coleoptera: Melyridae). *Annals Entomological Society of America* 78:131–136.

Leslie, K. A. 1975. Vegetative reproduction of *Polytrichum commune* in granite outcrop communities. M.S. thesis, Emory University, Atlanta, Ga.

Leslie, K. A., and Burbanck, M. P. 1979. Vegetation of granitic outcroppings at Kennesaw Mountain, Cobb County, Georgia. *Castanea* 44:80–87.

Lugo, A. E. 1969. Energy, water, and carbon budgets of a granite outcrop community. Ph.D. dissertation, University of North Carolina, Chapel Hill, N.C.

Lugo, A. E., and McCormick, J. F. 1981. Influence of environmental stressors upon energy flow in a natural terrestrial ecosystem. In *Stress Effects on Natural Ecosystems*, ed. G. W. Barrett and R. Rosenberg, pp. 79–102. New York: John Wiley & Sons.

Matthews, J. M., and Murdy, W. H. 1969. A study of *Isoetes* common to the granite outcrops of the southeastern Piedmont, United States. *Botanical Gazette* 130:53–61.

McCormick, J. F., Lugo, A. E. and Sharitz, R. R. 1974. Experimental analysis of ecosystems. In *Handbook of Vegetation Science VI: Vegetation and Environment*, ed. B. R. Strain and W. D. Billings, pp. 159–179. The Hague: Dr. W. Junk Publishers.

McCormick, J. F., and Platt, R. B. 1964. Ecotypic differentiation in *Diamorpha cymosa*. *Botanical Gazette* 125:271–279.

McLain, D. K., and Shure, D. J. 1985. Host plant toxins and unpalatability of *Neacoryphus bicrucis* (Hemiptera: Lygaeidae). *Ecological Entomology* 10:291–298.

McLain, D. K., and Shure, D. J. 1987. Pseudocompetition: interspecific displacement of insect species through misdirected courtship. *Oikos* 49:291–296.

McVaugh, R. 1943. The vegetation of the granitic flat-rocks of the southeastern United States. *Ecological Monographs* 13:119–166.

Mellinger, A. C. 1972. Ecological life cycle of *Viguiera porteri*. Ph.D. dissertation, University of North Carolina, Chapel Hill, N.C.

Meyer, K. A., McCormick, J. F., and Wells, C. G. 1975. Influence of nutrient availability on ecosystem structure. In *Mineral Cycling in Southeastern Ecosystems*, ed. F. G. Howell, J. B. Gentry, and M. H. Smith, pp. 756–779. Symposium Series CONF-740513.

Murdy, W. H. 1966. The systematics of *Phacelia maculata* and *P. dubia* var. *georgiana*, both endemic to granite outcrop communities. *American Journal of Botany* 53:1028–1036.

Murdy, W. H. 1968. Plant speciation associated with granite outcrop communities of the southeastern Piedmont. *Rhodora* 70:394–407.

Murdy, W. H., and Carter, E. M. B. 1985. Electrophoretic study of the allopolyploidal origin of *Talinum teretifolium* and *T. appalachianum* (Portulacaceae). *American Journal of Botany* 72:1590–1597.

Murdy, W. H., Johnson, T. M., and Wright, V. K. 1970. Competitive replacement of *Talinum men-gesii* by *T. teretifolium* in granite outcrop communities of Georgia. *Botanical Gazette* 131:186–192.

Oosting, H. J., and Anderson, L. E. 1937. The vegetation of a barefaced cliff in western North Carolina. *Ecology* 18:280–292.

Oosting, H. J., and Anderson, L. E. 1939. Plant succession on granite rock in eastern North Carolina. *Botanical Gazette* 100:750–768.

Palmer, P. G. 1970. The vegetation of Overton rock outcrop, Franklin County, North Carolina. *Journal of the Elisha Mitchell Society* 86:80–87.

Phillips, D. L. 1981. Succession in granite outcrop shrub–tree communities. *American Midland Naturalist* 106:313–317.

Phillips, D. L. 1982. Life forms of granite outcrop plants. *American Midland Naturalist* 107:206–208.

Quarterman, E., Burbanck, M. P., and Shure, D. J. 1993. Rock outcrop communities: limestone, sandstone and granite. In *Biodiversity of the Southeastern United States: Upland Terrestrial Communities*, ed. W. H. Martin, S. G. Boyce, and A. E. Echternacht, pp. 35–86. New York: John Wiley & Sons.

Rogers, S. E. 1971. Vegetational and environmental analysis of shrub–tree communities on a granite outcrop. M.S. thesis, Emory University, Atlanta, Ga.

Rury, P. M. 1978. A new and unique, mat-forming Merlin's grass (*Isoetes*) from Georgia. *American Fern Journal* 68:99–108.

Shannon, C. E., and Weaver, W. 1949. *The Mathematical Theory of Communication*. Urbana, Ill.: University of Illinois Press.

Shantz, H. L. 1927. Drought resistance and soil moisture. *Ecology* 8:145–157.

Sharitz, R. R., and McCormick, J. F. 1973. Population dynamics of two competing annual plant species. *Ecology* 54:723–740.

Shure, D. J., and Ragsdale, H. L. 1977. Patterns of primary succession on granite outcrop surfaces. *Ecology* 58:993–1006.

Snyder, J. M. 1971. Interactions within the weathering environment of lichen–moss ecosystems on exposed granite. Ph.D. dissertation, Emory University, Atlanta, Ga.

Uno, G. E., and Collins, S. L. 1987. Primary succession on granite outcrops in southwestern Oklahoma. *Bulletin of the Torrey Botanical Club* 114:387–392.

Walters, T. M., and Wyatt, R. 1982. The vascular flora of granite outcrops in the Central Mineral

Region of Texas. *Bulletin of the Torrey Botanical Club* 109:344–364.

Whitehouse, E. 1933. Plant succession on central Texas granite. *Ecology* 14:391–405.

Wiggs, D. N., and Platt, R. B. 1962. Ecology of *Diamorpha cymosa. Ecology* 43:654–670.

Winterringer, G. S., and Vestal, A. G. 1956. Rockledge vegetation in southern Illinois. *Ecological Monographs* 26:105–130.

Wyatt, R. 1981. Ant-pollination of the granite outcrop endemic *Diamorpha smallii* (Crassulaceae). *American Journal of Botany* 68:1212–1217.

Wyatt, R., and Fowler, N. 1977. The vascular flora and vegetation of the North Carolina granite outcrops. *Bulletin of the Torrey Botanical Club* 104:245–253.

Wyatt, R., and Stoneburner, A. 1981. Patterns of ant-mediated pollen dispersal in *Diamorpha smallii* (Crassulaceae). *Systematic Botany* 6:1–7.

7 High-Elevation Outcrops and Barrens of the Southern Appalachian Mountains

SUSAN K. WISER AND PETER S. WHITE

Introduction

Despite the lack of a climatic tree line, the southern Appalachian mountains support several treeless, high-elevation (>1,200 m) communities. The three principal open communities are heath balds, grassy balds, and rock outcrops. Other open communities include frequent, but small, rocky, steep streamsides and seeps, and, very rarely, mountain bogs (Schafale and Weakley 1990). The information on heath and grassy balds will be briefly reviewed, but the focus will be on outcrops because they support a highly distinctive flora and are the least studied of the three primary open communities. Notably, they support one of the richest floras of rare species of any regional habitat, including both rare endemics and northern alpine disjuncts.

Heath Balds

Heath balds are species-poor communities with only 10–20 vascular plant species on a site (White and Renfro 1984) and are dominated by ericaceous evergreen shrubs. They occur within a restricted elevation range (1,220–1,525 m) on narrow ridges and adjacent south and west slopes, and become larger and more frequent in localities disturbed by logging activity (White, Wilds and Stratton, unpublished data). In Great Smoky Mountains National Park, heath balds are more frequent (>400) than grassy balds (~30) but dominate less than 50% of apparently suitable topographic sites.

Because of dense shade, thick leaf litter, and high soil acidity (aluminum may reach levels toxic to tree roots), heath balds are stable or only slowly invaded by trees. Some heath balds created by logging remain treeless one hundred years later. Forest communities with a heath understory can dominate similar sites on narrow ridges and may convert to heath balds after the trees have been killed by wind or debris avalanches. On these sites, succession to forest is extremely slow. Dominance of ericaceous shrubs, low species diversity, and near absence of herbaceous plants distinguish heath balds from shrub communities (i.e., the "shrub balds" of Ramseur 1960) that dominate some successional stages on grassy balds.

Grassy Balds

Grassy balds are species-rich communities (often >200 vascular plant species on a site) dominated by grasses, sedges, and broad-leaved herbs. The origin of grassy balds has been much debated (Billings and Mark 1957; Gersmehl 1970, 1973; Lindsay and Bratton 1979; Saunders 1980; Stratton and White 1982). Although some evidence is circumstantial, scientists generally accept the following information.

1. Grassy balds were universally used as summer grazing pastures from the early 1800s to the 1930s, at the latest (Gersmehl 1970; Lindsay 1976; Lindsay and Bratton 1979; White 1984).

Roger C. Anderson, James S. Fralish, and Jerry M. Baskin, eds. *Savannas, Barrens, and Rock Outcrop Plant Communities of North America*. Copyright © 1999 Cambridge University Press. All rights reserved.

2. Soon after grazing ceased (due to a change in the farming pattern and local economy), all grassy balds became vulnerable to woody plant invasion (Bruhn 1964; Lindsay and Bratton 1979, 1980). Woody cover has increased on all balds not managed for the open condition, and some balds have a complete woody cover after 50 years.

3. No current natural processes maintain grassy balds or create new ones (Saunders 1980). Fires, either natural or human caused, do not create grassy balds. Fire may be useful in management, but only after removal of established woody plants by other means.

4. Grassy balds do not differ from forested sites in elevation, topography, or soil, and they occupy a very small percentage of potential sites based on the distribution of existing balds (Gersmehl 1970; Stratton and White 1982).

5. Rare species occur on balds, but most species also are found in other high-elevation, open habitats like rock outcrops and seeps (Stratton and White 1982). Rare species may have spread to grassy balds from open habitats nearby; for example, grassy balds adjacent to outcrops contain more rare species than those isolated from outcrops. Grassy balds have a high percentage (>13%) of exotic field weeds (Stratton and White 1982).

6. Oral histories and a few early photographs indicate that some grassy balds were created from hardwood forest through tree felling and girdling (Lindsay 1976). The dominant grass of the balds, *Danthonia compressa*, occurs in open, high-elevation forests. However, some balds were reported to be open before European settlement (K. Langdon, Great Smoky Mountain National Park, personal communication).

7. Few Native American artifacts have been found on the balds, and the few oral histories that refer to balds likely postdate, rather than predate, European immigrant arrival (Gersmehl 1973; Lindsay 1976).

These observations suggest that European immigrants created grassy balds, although the earliest Europeans described a few balds as open. It is possible that hot fires during the Hypsithermal, or during other periods of prolonged warmth and drought, created open habitats that were subsequently maintained by Native Americans or by large, now-extirpated, grazing animals (e.g., woodland buffalo and elk). However, this combination of factors rarely occurred and probably did not result in the vegetation patterns of the present landscape.

Because grassy balds are rich in wildflowers and rare species and provide scenic views of the surrounding landscape in a region where unobstructed views are rare, they are being managed for the open condition by several agencies and conservation groups (Bratton and White 1980). The best management practice includes hand cutting of woody plants, local use of herbicides to control sprouting in blueberries (*Vaccinium* species) and blackberries (*Rubus* species), and seeding with native plants, including *Danthonia compressa* (Lindsay and Bratton 1979). At Great Smoky Mountains National Park, managers attempted livestock grazing, but the animals did not discriminate between rare and common species and lost weight during the weeks they grazed on the balds. Experimental livestock grazing is still being studied at Roan Mountain.

Rock Outcrops

Distribution and Origin

High-elevation outcrops occur in the high peaks region of the southern Appalachians, primarily in western North Carolina and eastern Tennessee (Figure 7.1). This region includes Mt. Mitchell, the highest peak in eastern North America, at 2,037 m. The outcrops we studied ranged in elevation from 1,200 to 2,030 m to encompass the area thought to have been above tree line during the peak of the last glacial advance (Delcourt and Delcourt 1985). Notably, outcrops in this elevation range support over 80% of important rare and endemic species populations, most of which may have been derived from the past alpine tundra flora.

Although most outcrops in the region are naturally occurring landscape features, a few were created in recent times by severe soil erosion following devastating logging fires during the period 1880–1930 (e.g., Charlie's Bunion in the Great Smoky Mountains, Potato Knob in the Black Mountains) or by debris avalanches following intense local rain storms (Feldkamp 1984). Common outcrop species (e.g., *Saxifraga michauxii, Carex misera*) occur on recently created outcrops, but rare species occur only on the oldest landslide scars near populations on natural outcrops. The age of these scars has not been determined but certainly they predate 1930s aerial photography. Most outcrops remain open sites due to loss of individual rocks and boulders, soil and vegetation erosion, and exfoliation, rather than by large-scale debris avalanches.

Physical Environment

Above 1,200 m, mean annual precipitation ranges from 1,270 to 2,000 mm, increasing with elevation and toward the southwest (Ruffner 1985). Precipitation peaks in June and July. Mean July temperatures are cool for the region and decrease with increasing elevation (17 °C at 1,662 m and 15 °C at 1,743 m). Frosts occur from late September to mid-May. On Grandfather Mountain, one of the most exposed summits, mean wind speed ranges from 18.5 km hr^{-1} in summer to 33 km hr^{-1} in winter, with extreme gusts up to 270 km hr^{-1} (based on five year's data provided by Grandfather Mountain, Inc.).

Outcrop exposure varies widely from the summits of isolated peaks to outcrop bases that may be sheltered by surrounding forest. Because rainfall is high and temperature low, moisture stress usually is low, except near

Highlands, North Carolina (Figure 7.1), where outcrops occur primarily on south-facing slopes of Whiteside quartz–diorite plutonic formations. These rock "domes" generally lack fractures and crevices, thus runoff is rapid.

Outcrops occur on rocks ranging from felsic granites, gneisses, and schists, to mafic gabbros and amphibolites. Mafic rock predominates north of Grandfather Mountain (Figure 7.1). At Grandfather Mountain, mafic pockets and veins occur within the felsic, meta-arkose matrix. South of Grandfather Mountain, rock types are felsic but vary in origin (igneous vs. sedimentary), degree of metamorphism, and mineralogy (e.g., mica content).

Soils typically are <30 cm deep and are high in organic matter; pH averages 4.1 and ranges from 3.1 to 7.2 (Wiser 1993). Where persistent seepage occurs, soil water-holding capacity, pH, and Na are higher, and sulfate lower, than on dry sites. Soils on mafic bedrock have higher B, Cu, and P, but lower Na than those on felsic bedrock. Soil cations and pH are unrelated to bedrock type, perhaps because of high rainfall and rapid leaching on these steep sites.

Floristics and Biogeography

Our description of high-elevation outcrop floristics is based on compositional analysis of 154 plots, each 100 m^2 in size and enclosing two to seven 1-m^2 subplots (Wiser 1993, 1994; Wiser, Peet and White 1996). Plots were located on 42 peaks in 11 general localities (Figure 7.1).

A total of 281 vascular plant species, predominantly native, summer-flowering, herbaceous perennials, were recorded (Wiser 1994). Species restricted to open habitats comprise 18% of the flora; this group includes species restricted to outcrops (31 species, e.g., *Carex misera, Selaginella tortipila, Asplenium montanum*), and those restricted to outcrops and other open habitats at high elevations (21 species, e.g., *Sibbaldiopsis triden-*

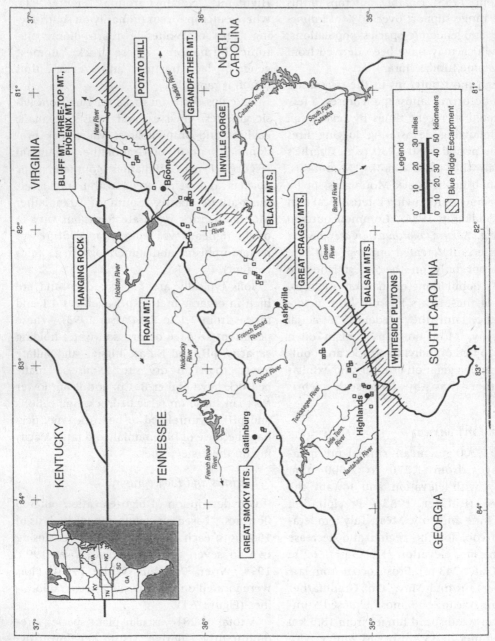

Figure 7.1. Study locations (rectangles) and positions (small squares) of high-elevation rock outcrop plots. (Reprinted from Wiser, Peet and White 1996.)

tata, Prenanthes roanensis). Although these species represent only a minor portion of the total outcrop flora, they have a higher mean frequency than others and average 42% of the total plant cover. They also are important biogeographically. A portion of the open-land species derives from past alpine tundra vegetation, including species disjunct from alpine and arctic environments far to the north (e.g., *Juncus trifidus, Scirpus cespitosus*), and endemics restricted to high-elevation outcrops (e.g., *Geum radiatum, Solidago spithamaea*) (Table 7.1). High-elevation outcrops provide the combination of low competition, high light, and low temperatures required by these species. Other open-land species (e.g., *Carex debilis, Carex umbellata, Danthonia spicata*, and *Paronychia argyrocoma*) probably were absent from the alpine tundra flora but emigrated from lower elevations as the climates warmed. This view is supported by the occurrence of these species in northern treeless habitats below tree line but not in the alpine zone (Wiser 1998).

The remaining 82% of the flora consists of herbs, shrubs, and stunted trees not restricted to open habitats but often common in nearby forest (e.g., *Chelone lyonii, Vaccinium* spp., *Abies fraseri*). Over half of these occurred in less than 5 of the 154 plots and so are only incidental on outcrops, most likely persisting because of spatial mass effect (sensu Shmida and Ellner 1984). The high proportion of the total outcrop flora composed of forest species indicates that the nearby forest strongly influences outcrop community composition.

Structural and Compositional Variation

The prevalence of bare or lichen-encrusted bedrock distinguishes outcrops from surrounding habitats; on average, 55% of the ground surface in the 100-m² plots was exposed bedrock (Wiser 1993). Herbaceous plants comprise 70% of the species and, on average, 67% of the total plant cover. In crevices with deep soil, stunted trees and shrubs produce a local structure and compo-

sition that forms a transition to heath balds and surrounding vegetation. Plant cover may be patchy and highly heterogeneous and is in large part determined by bedrock fracturing and crevice formation.

Three common species occur across all outcrop vegetation types: *Carex misera, Saxifraga michauxii*, and *Vaccinium corymbosum*. Other widespread species include *Scirpus cespitosus, Sorbus americana, Rhododendron catawbiense, Polypodium appalachianum*, and *Agrostis perennans*. Variation in community composition primarily is related to elevation and potential solar radiation gradients (Figure 7.2; Wiser, Peet and White 1996), which generally are important to southern Appalachian vegetation (Whittaker 1956; Callaway, Clebsch and White 1987). Composition is less strongly related to bedrock geology, soil chemistry, geographic position, and the presence of continuous seepage.

Within a mountain range (Figure 7.1), there are some consistent relationships between outcrop composition and environment (Wiser, Peet and White 1996). On 100-m² plots, elevation or potential solar radiation, or both, are strongly related to composition. Rank correlations with Detrended Correspondence Analyses (DCA) ordination axes range from 0.50 to 0.95, p < 0.05. At a smaller scale of resolution (1-m² subplots within individual ranges), soil depth and percentage exposed bedrock are consistently related to composition (p < 0.05 for rank correlation with DCA ordination axes). In contrast, relationships between composition and soil chemical gradients vary from range to range, for both plot sizes, presumably because of changing geology.

Dichanthelium acuminatum, Carex umbellata, Kalmia latifolia, and *Krigia montana* characterize the lower-elevation outcrops examined (1,200–1,600 m). *Carex biltmoreana*, a rare southern Appalachian endemic, is confined primarily to outcrops in this elevation band. On mafic bedrock or wet outcrops, vegetation is dominated by broad-leaved herbs and dis-

Table 7.1. *Vascular plant species of high-elevation outcrops in the southern Appalachians that are southern Appalachian endemics or disjunct from eastern North America. An asterisk (*) indicates that the southern Appalachian outcrop is the primary habitat for the species.*

Globally rare[1]	Regionally rare[2]	Not rare
Endemic species		
Abies fraseri[3]	*Carex ruthii*	*Angelica triquinata*
Cacalia rugelia	*Huperzia porophila*	*Chelone lyoni*
Calamagrostis cainii*	*Hypericum buckleyi**	*Diervilla sessilifolia*
Carex biltmoreana*	*Hypericum graveolens*	*Parnassia asarifolia*
Carex misera*	*Hypericum mitchellianum*	*Rhododendron catawbiense*
Geum radiatum*[4]	*Krigia montana**	*Saxifraga michauxii**
Glyceria nubigena	*Menziesia pilosa*	*Selaginella tortipila**
Houstonia purpurea var.	*Sedum rosea* var. *roanensis**	*Solidago glomerata*
montana*	*Zigadenus leimanthoides*	*Vaccinium erythrocarpum*
Liatris helleri*		
Prenanthes roanensis		
Rhododendron vaseyi		
Senecio millefolium*		
Solidago spithamaea*		
Northern species		
No species	*Alnus viridis* ssp. *crispa*	*Viola macloskeyi* ssp. *pallens*
	Campanula rotundifolia*	
	Clintonia borealis	
	Deschampsia cespitosa*	
	Gentiana linearis*	
	Huperzia appalachiana*	
	Juncus trifidus var.	
	carolinianus*	
	Minuartia groenlandica*	
	Muhlenbergia glomerata	
	Rubus idaeus ssp.	
	sachalinensis	
	Sanguisorba canadensis	
	Scirpus cespitosus*	
	Sibbaldiopsis tridentata	
	Trisetum spicatum*[5]	

[1] Includes species that have fewer than 100 occurrences globally, following Weakley (1990).
[2] Rare in either North Carolina or Tennessee according to Weakley (1990) or Tennessee Rare Plant Protection Program (1992).
[3] Closely related to the northern *Abies balsamea*.
[4] Closely related to *Geum peckii*, a northern Appalachian endemic.
[5] Presumed extirpated in the southern Appalachians.

tinguished by *Coreopsis major* and *Schizachyrium scoparium*. On dry, lower-elevation, felsic outcrops, composition varies with geographic position and subtle bedrock differences. On the southern Whiteside plutons, vegetation is characterized by dense *Selaginella tortipila* mats. These may be so extensive that less than 15% of the bedrock

surface on these gently sloping, south-facing outcrops is
visible (Figure 7.3). Felsic outcrops in the Great Craggy,
Balsam, and Black mountains
to the north lack large
Selaginella mats and are distinguished by a high frequency of
Aronia arbutifolia and *Kalmia
latifolia*. Herbaceous plants
comprise a low proportion
(24%) of the total cover, and
the vegetation grades into adjacent heath balds. The rare
Hypericum buckleyi is restricted
primarily to lower-elevation
felsic bedrock.

Characteristic of outcrops
above 1,600 m are *Aster acuminatus*, *Picea
rubens*, *Abies fraseri*, *Solidago glomerata*, *Menziesia
pilosa*, *Vaccinium erythrocarpum*, and *Carex brunnescens*, species commonly encountered in
nearby forests or balds. Compositional variation on outcrops adjacent to forest is less
marked than that observed below 1,600 m,
which primarily is a consequence of the relative abundance of dominant species. Rock
outcrops on Anakeesta slate are an exception, as they are the only high-elevation type
where *Rhododendron carolinianum* occurs with
R. catawbiense, the typical rhododendron of
such sites. Rare outcrop species most commonly occur above 1,600 m. Across the
region, individual rare species are unevenly
distributed; *Gentiana linearis* is restricted to
the Great Smoky Mountains, *Calamagrostis
cainii* occurs only from the Black Mountains
south, and *Houstonia purpurea* var. *montana*
and *Minuartia groenlandica* occur only on
mafic or intermediate bedrock.

Some of the higher-elevation outcrops are
surrounded and influenced by grass or
shrub balds that developed after past disturbance. *Angelica triquinata* and *Krigia montana*, species that are abundant on the
neighboring balds, are most frequent on
such outcrops. On most high-elevation out-

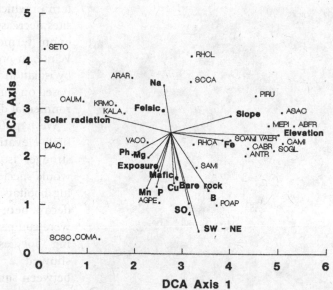

Figure 7.2. Detrended Correspondence Analysis
using data from 154 plots, each 100 m² in size.
Environmental relationships are shown with vectors, except for rock types, which are indicated by
their centroids. Scores for common species mentioned in the text also are shown. Codes for species
are ABFR = *Abies fraseri*, AGPE = *Agrostis perennans*,
ANTR = *Angelica triquinata*, ARAR = *Aronia arbutifolia*, ASAC = *Aster acuminatus*, CABR = *Carex brunnescens*, CAMI = *Carex misera*, CAUM = *Carex
umbellata*, COMA = *Coreopsis major*, DIAC =
Dichanthelium acuminatum, KALA = *Kalmia latifolia*,
KRMO = *Krigia montana*, MEPI = *Menziesia pilosa*,
PIRU = *Picea rubens*, POAP = *Polypodium appalachianum*, RHCA = *Rhododendron catawbiense*, RHCL =
Rhododendron minus, SAMI = *Saxifraga michauxii*,
SCSC = *Schizachyrium scoparium*, SCCA = *Scirpus
cespitosus*, SETO = *Selaginella tortipila*, SOGL =
Solidago glomerata, SOAM = *Sorbus americana*,
VACO = *Vaccinium corymbosum*, VAER = *Vaccinium
erythrocarpum*.

crops, exotic species are uncommon, but six
of the eight exotic species encountered (i.e.,
Polygonum aviculare, *Phleum pratense*, *Poa
compressa*, *Rumex acetosella*, *Taraxacum officinale*, *Trifolium pratense*) occur only on, or are
most frequent on, outcrops adjacent to
shrub or grass balds.

The most species-rich outcrop community
type (26 species/100 m²) is rare, occurring

Figure 7.3. Dense mats of *Selaginella tortipila* are characteristic of communities on the smooth, felsic outcrops on Whiteside quartz–diorite plutons in southwestern North Carolina.

only on constantly wet outcrops. Such sites have a dense layer of herbs (up to 94% of total cover) and often are partly shaded by adjacent forest. The vegetation is distinguished by *Chelone obliqua*, *Aster divaricatus*, *Oxypolis rigidior*, and *Houstonia serpyllifolia*. Several rare species, *Filipendula rubra*, *Huperzia porophila*, and *Tofieldia glutinosa* ssp. *glutinosa* occur only on these sites.

Geographic Position and Distance Decay

Although environment controls some compositional variation among outcrops, geography also may influence the composition of these highly isolated communities. Because rock outcrops are small habitat islands isolated in space, seed dispersal and plant migration among them may be restricted. This isolation may produce a pat-

tern in which community similarity between sites decreases with the distance, a phenomenon termed *distance decay* (Nekola and White, in press). The spatial pattern imposed by isolation of islandlike habitats is superimposed on community differences due to environmental differences among sites.

We hypothesized that species restricted to high-elevation outcrops would show a stronger isolation and distance effect than would species that also occurred in surrounding habitats. Therefore, the rates of distance decay between outcrop-restricted species were compared to those for more wide ranging species. Outcrop restricted species showed a steeply declining relationship between similarity (Jaccard's index) and distance; unrestricted species showed no relationship (Figure 7.4).

Distance-controlled variation among communities has important consequences regarding development of conservation strategies. Strong distance decay implies rapid turnover of species in space. Regardless of whether this variation is underlain by spatial influences or environment, many sites must be protected to encompass the geographic variation in the habitat and vegetation. Given the steep distance decay among outcrops, particularly for species that make the outcrop flora unique, a network of protected sites, well distributed across the southern Appalachian ranges, is necessary to adequately represent variation among outcrop communities.

Rare Plant Species

Forty species for which the high-elevation outcrop is a primary habitat are considered rare in Tennessee or North Carolina (Weakley 1990; Tennessee Rare Plant Protection Program 1992; Wiser 1994) (Figure 7.5). Of these, five are extremely rare endemics: *Geum radiatum*, *Calamagrostis cainii*, *Houstonia purpurea* var. *montana*, *Liatris helleri*, and *Solidago spithamaea*. Seven species occur in the Northeastern alpine and farther north in the Canadian arctic: *Huperzia appalachiana, Juncus trifidus, Minuartia groenlandica, Scirpus cespitosus, Sibbaldiopsis tridentata, Agrostis mertensii,* and *Campanula rotundifolia.*

Some outcrop species are uncommon in the region simply because their habitat is rare. For example, *Saxifraga michauxii,* a broadly distributed southern Appalachian endemic, was observed on all mountain ranges and occurred on 77% of plots; it is essentially as rare as its habitat. In contrast, some species will be rarer than their habitat because of extirpation on habitat patches and slow dispersal among outcrops. For example, the federally listed endangered species *Solidago spithamaea* occurred in only 8% of the plots, and these were restricted to two mountain ranges. Quantitative models based on current habitat characteristics indicate that apparently suitable habitat for this species exists beyond its current range (Wiser, Peet and White in press). Thus, rarity is a matter of degree determined by the presence or absence of a species on an already-rare outcrop habitat.

Gymnoderma lineare, an endemic, is the only globally rare lichen, and *Cephaloziella obtusilobula* is the only globally rare bryophyte occurring on these outcrops. Fifteen other bryophytes occur that are considered rare in North Carolina (Table 7.2).

Fauna

High-elevation outcrops also support rare fauna (A. Boynton, North Carolina Wildlife Commission, and S. Hall, North Carolina Natural Heritage Program, personal communication). Bird species include peregrine falcon (*Falco peregrinus*); raven (*Corvus corax,* regionally rare); and golden eagle (*Aquila*

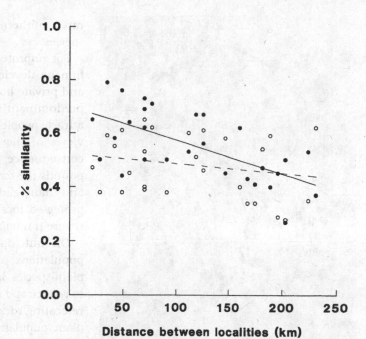

Figure 7.4. The relationship between distance-between-study-locations (Figure 7.1) and their floristic similarity, as measured by Jaccard's index, for outcrop-restricted species (filled circles; solid line is a fitted linear regression, $r^2 = 0.41$, $p < 0.001$) and all other species (open circles; dashed line is a fitted linear regression, $r^2 = 0.04$, $p = 0.29$). Species found in fewer than five plots were excluded.

chrysaetos), which roosts but does not breed on the outcrops. Mammals include the federally endangered Virginia big-eared bat (*Plecotus townsendii virginianus*) and wood rats (*Neotoma magister* and *Neotoma floridana*). The only rare amphibian is the green salamander (*Aneides aeneus*), which usually occurs at lower elevations but ascends to at least 1,250 m on the Whiteside plutons (M. L. Hafer, Pisgah National Forest, U.S. Forest Service, personal communication). Rare arthropods include the band-winged locust (*Trimerotropis saxitalis*); the ledge locust (*Spharagemon saxatile*); the lampshade spider (*Hypochilus coylei*), which is endemic to the southern part of the study area in North Carolina; and *H. sheari,* which is endemic to the area extending from the Balsam to the Black mountains, North Carolina.

Figure 7.5. *Liatris helleri, Solidago spithamaea* (both federally listed endangered species), and *Scirpus cespitosus,* which are disjunct from northern alpine areas and rare in the southern Appalachians, are among the unusual rare species inhabiting high-elevation outcrops.

Protection Status and Threats

Outcrops on 36 of the 42 peaks studied have some level of protection for their biological diversity and natural values. They occur on lands owned by the National Park Service, U.S. Forest Service, North Carolina State Parks, and North Carolina Nature Conservancy, or occur in local parks or private reserves. This ownership pattern protects most of the described vegetation types. Unprotected and threatened vegetation types occur in the 1,200–1,600-m range on mafic bedrock north of Grandfather Mountain. Only one of the five peaks studied by Wiser (1993) currently is protected; two unprotected peaks support some of the largest and least-disturbed populations of rare southern Appalachian outcrop plant species.

For unprotected outcrops, habitat destruction by development for resorts, ski slopes, and private homes is a serious threat. In this predominantly forested region, outcrops attract people because they provide good views of the surrounding landscape. As a consequence, trampling by hikers compounds the damage created by rock climbers and hang-gliding enthusiasts. Even many protected localities with public access have severe trampling problems.

On outcrops where trampling threatens populations of federally listed endangered plant species, land managers have taken steps to reduce and mitigate adverse impacts by (1) relocating rock-climbing routes to avoid rare plant populations, (2) obstructing unofficial trails criss-crossing outcrops, (3) installing interpretive signs to keep visitors on designated trails, (4) installing boardwalks and overlooks to keep visitors off rock surfaces, (5) rejecting potential trail routes that would cross sensitive habitats, and (6) restoring habitat.

In 1990, a habitat restoration project began at Craggy Pinnacle (Johnson 1996). Populations of five rare plants (*Geum radiatum, Calamagrostis cainii, Juncus trifidus, Scirpus cespitosus,* and *Carex misera*) and three Appalachian endemics (*Sibbaldiopsis tridentata, Krigia montana,* and the moss *Polytrichum appalachianum*) were supplemented with plants raised primarily from seed collected on site. Because trampling had completely removed soil from much of the outcrop, habitat modules were constructed from biodegradable fabric filled with soil. To better understand survival requirements, species were reintroduced onto both suitable and unsuitable sites based on microhabitat conditions and species associations. Depending on species, 38%–75% of individuals survived the first year; survivorship was higher the second year.

Although acquisition of critical sites and better management of protected areas poten-

Table 7.2. *Rare lichens and bryophytes of southern Appalachian outcrops >1,200 m elevation. Based on Weakley (1990).*

Liverworts and hornworts	Mosses	Lichens
Anastrophyllum saxicola	*Bartramidula cernua*	*Gymnoderma lineare*
Bazzania naudicaulis	*Campylopus atrovirens*	
Cephaloziella obtusilobula	*Campylopus paradoxus*	
Cephaloziella spinicaulis	*Leptodontium flexifolium*	
Lophozia barbata	*Rhytidium rugosum*	
Lophozia excisa	*Sphagnum pylaesii*	
Lophozia hatcheri	*Sphagnum tenellum*	
Mylia taylorii		
Nardia scalaris		

tially can curb or mitigate habitat destruction, several other threats are more insidious. An exotic insect, the balsam wooly adelgid (*Adelges piceae*), has caused intensive mortality of the endemic Fraser fir (Eagar 1984). Near outcrops >1,850 m, fir occurred in almost monospecific stands. Tree loss has resulted in increased insolation and drying of soil, and wind-throw of dead trees may increase soil erosion during heavy rainstorms. Occasionally, the open condition allows some outcrop species to spread into formerly forested habitats; for example, *Gentiana linearis, Krigia montana*, and *Calamagrostis cainii* on Mt. LeConte, Great Smoky Mountains (White and Wofford 1984). Whether Fraser fir mortality will induce sufficient changes in outcrop habitat to affect other species is unknown.

Air pollution is a major threat to the high peaks region. In spruce–fir forests, pollution may induce changes in soil fertility and cation cycling (Eagar and Adams 1992). Deposition of nitrate and sulfate is substantial, and ozone levels frequently are high. High-elevation outcrops often are immersed in fog and clouds more acidic than precipitation itself. Tests are underway to determine ozone and acid precipitation impacts on some high-elevation species (K. Langdon, personal communication). Outcrop species such as *Geum radiatum* proved to be resistant to ozone, whereas *Aster acuminatus* and *Krigia montana* exhibited spotted leaves and reduction in vigor when exposed to ozone (Dowsett, Anderson and Hoffard 1992; K. Langdon, personal communication).

Research Opportunities

Contributions to understanding high-elevation outcrop ecosystems can be made at all levels: documenting species distributions, understanding the autecology of dominant and rare species, and clarifying causes and consequences of current biogeographical patterns. Basic information is lacking on the distribution of groups of organisms on the outcrops, particularly nonvascular plants and most animals. Likewise, there is little understanding of the role played by these organisms in community dynamics or ecosystem processes.

There is little quantitative information on life history, demography, or ecophysiology of high-elevation outcrop plants, which is prerequisite to informed management decisions. For example, it is unclear what limits populations, whether populations are stable, and if so, whether they are static or undergo cycles of extinction and recolonization. Several rare species may be declining (e.g., *Geum radiatum*; U.S. Fish and Wildlife Service 1993); thus the influence of various disturbances and stresses on this decline must be determined.

If global climate warming occurs, species restricted to outcrops at mountain summits will be at risk because no higher-elevation habitat exists for colonization. Ecophysiological research is required to determine whether current species distributions reflect present temperature regimes. Information from this research will allow us to predict whether existing habitats will be sufficient to protect the species or whether biologists and managers will have to create *ex situ* collections and find new sites for establishing populations.

Although floristic links between southern Appalachian outcrops and northern alpine and treeless vegetation are well established, details concerning the biogeography and evolution of the flora are sketchy. Genetic research is required to reveal the degree of relatedness between northern and southern Appalachian species populations and among populations of both regions. Physiological comparisons will help determine selective pressures that have driven evolution in different environments of the two regions.

Finally, high-elevation outcrops provide an ideal opportunity to examine how vegetation and diversity patterns are influenced by the environment of insular habitats. An important endeavor would be to determine whether species richness is simply related to outcrop size and isolation, as suggested by island theory (MacArthur and Wilson 1967), or whether local environmental conditions and microsite heterogeneity better explain species richness patterns.

Summary

The principal southern Appalachian treeless communities are heath balds, grassy balds, and rock outcrops. Heath balds occur on undisturbed ridges and are more frequent and larger in areas disturbed by logging. They are dominated by ericaceous evergreen shrubs and are poor in total and rare species. Grassy balds appear to have developed from high-elevation summer pastures of the 1800s. Outcrops are the least common of the three habitats but support the richest flora of endemic and rare species. The presence of species disjunct from northern alpine and arctic areas suggests that the flora in part is a relict of a Pleistocene alpine flora.

The highly variable environment of high-elevation outcrops is reflected by community composition. Composition varies with elevation, potential solar radiation, geological and soil variability, presence of continuous seepage, and composition of surrounding forests and balds. Geographic position and isolation are likely to influence composition, particularly for species restricted to outcrops. Within an individual outcrop, compositional variation is strongly related to local variation in soil depth and chemistry.

To adequately preserve the high variability in species composition within and between outcrops, reserves must include several outcrop sites from within each mountain range found in the southern Appalachians. Currently, mafic outcrops north of Grandfather Mountain, North Carolina, are the least adequately protected. However, even on protected outcrops, trampling and air pollution may still threaten this unique ecosystem.

REFERENCES

Billings, W. D., and Mark, A. F. 1957. Factors involved in the persistence of montane treeless balds. *Ecology* 38:140–142.

Bratton, S. P., and White, P. S. 1980. Grassy balds management in parks and nature preserves: issues and problems. In *Status and Management of Southern Appalachian Mountain Balds*, ed. P. R. Saunders, pp. 96–114. Southern Appalachian Research/Resource Management Cooperative, Western Carolina University, Cullowhee. N.C.

Bruhn, M. E. 1964. Vegetation succession on three grassy balds of the Great Smoky Mountains. M.S. thesis, University of Tennessee, Knoxville, Tenn.

Callaway, R. M., Clebsch, E. E. C., and White, P. S. 1987. A multivariate analysis of forest communities in the Western Great Smoky Mountains National Park. *The American Midland Naturalist* 118:107–120.

Delcourt, H. R., and Delcourt, P. A. 1985. Quaternary palynology and vegetational history of the southeastern United States. In *Pollen Records of Late-Quaternary North American Sediments*, ed. V. M. Bryant, Jr., and R. G. Holloway, pp. 1–37. Dallas, Tex.: American Association of Stratigraphic Palynologists Foundation.

Dowsett, S., Anderson, R., and Hoffard, W. 1992. *A Review of Selected Articles on Ozone Sensitivity and Associated Symptoms for Plants Commonly Found in the Forest Environment.* Atlanta, Ga.: USDA, Forest Service, Research Report R8-TP-18.

Eagar, C. 1984. Review of the biology and ecology of the balsam woolly aphid in southern Appalachian spruce–fir forests. In *The Southern Appalachian Spruce–Fir Ecosystem, Its Biology and Threats*, ed. P. S. White, pp. 36–50. Atlanta,Ga.: U.S. Department of Interior, National Park Service, Research/Resource Management Report SER-71.

Eagar, C., and Adams, M. B. 1992. *The Ecology and Decline of Red Spruce in the Eastern United States.* New York: Springer Verlag.

Feldkamp, S. M. 1984. Revegetation of upper elevation debris slide scars on Mount LeConte in the Great Smoky Mountains National Park. M.S. thesis, University of Tennessee, Knoxville, Tenn.

Gersmehl, P. 1970. A geographic approach to a vegetation problem: The case of the southern Appalachian grassy balds. Ph.D. dissertation, University of Georgia, Athens,Ga.

Gersmehl, P. 1973. Pseudo-timberline: the southern Appalachian grassy balds. *Arctic and Alpine Research* 5:A137–A138.

Johnson, B. R. 1996. Southern Appalachian rare plant reintroductions on granite outcrops. In *Restoring Diversity: Strategies for Reintroduction of Endangered Plants*, ed. D. A. Falk, C. I. Millar, and M. Olwell, pp. 433–443. Washington, D.C.: Island Press.

Lindsay, M. M. 1976. *History of the Grassy Balds in Great Smoky Mountains National Park.* Atlanta, Ga.: U.S. Department of Interior, National Park Service, Research/Resource Management Report SER-4.

Lindsay, M. M., and Bratton, S. P. 1979. Grassy balds of the Great Smoky Mountains: their history and flora in relation to potential management. *Environmental Management* 3:417–430.

Lindsay, M. M., and Bratton, S. P. 1980. The rate of woody plant invasion on two grassy balds. *Castanea* 45:75–87.

MacArthur, R. H., and Wilson, E. O. 1967. *The Theory of Island Biogeography.* Princeton, N.J.: Princeton University Press.

Nekola, J. C., and White, P. S. In press. The distance decay of similarity in ecology and evolution. *Journal of Biogeography.*

Ramseur, G. S. 1960. The vascular flora of the high mountain communities of the southern Appalachians. *Journal of the Elisha Mitchell Science Society* 76:82–112.

Ruffner, J. A. (ed.). 1985. *Climates of the States.* Detroit, Mich.: Gale Research Co.

Saunders, P. R. (ed.). 1980. *Status and Management of Southern Appalachian Mountain Balds.* Southern Appalachian Research/Resource Management Cooperative, Western Carolina University, Cullowhee, N.C.

Schafale, M. P., and Weakley, A. S. 1990. *Classification of the Natural Communities of North Carolina: Third Approximation.* Department of Environment, Health, and Natural Resources, State of North Carolina, Raleigh.

Shmida, A., and Ellner, S. 1984. Coexistence of plant species with similar niches. *Vegetatio* 58:29–55.

Stratton, D. A., and White, P. S. 1982. *Grassy Balds of Great Smoky Mountains National Park: Vascular Plant Floristics, Rare Plant Distributions, and an Assessment of the Floristic Data Base.* U.S. Department of Interior, National Park Service, Atlanta, Ga. Research/Resource Management Report SER-58.

Tennessee Rare Plant Protection Program. 1992. Rare plant list of Tennessee. Unpublished.

U.S. Fish and Wildlife Service. 1993. *Spreading Avens Recovery Plan.* Atlanta, Ga.: U.S. Department of Interior, Fish and Wildlife Service.

Weakley, A. S. 1990. *Natural Heritage Program List of the Rare Plant Species of North Carolina.* State of North Carolina, Report to Department of Environment, Health, and Natural Resources, Raleigh, N.C.

White, P. S. 1984. Impacts of cultural and historic resources on natural diversity: lessons from Great Smoky Mountains National Park, North Carolina and Tennessee. In *Natural Diversity in Forested Ecosystems*, ed. J. L. Cooley and J. H.

Cooley, pp. 119–132. Athens, Ga:. Institute of Ecology, University of Georgia.

White, P. S., and Renfro, L. A. 1984. Vascular plants of southern Appalachian spruce–fir: annotated checklists arranged by habitat, growth form, and geographical relationships. In *The Southern Appalachian Spruce–Fir Ecosystem: Its Biology and Threats*, ed. P. S. White, pp. 235–246. Atlanta, Ga.: U.S. Department of Interior, National Park Service, Research/Resource Management Report SER-71.

White, P. S., and Wofford, B. E. 1984. Rare native Tennessee vascular plants in the flora of Great Smoky Mountains National Park. *Journal of the Tennessee Academy of Science* 59:61–64.

Whittaker, R. H. 1956. Vegetation of the Great Smoky Mountains. *Ecological Monographs* 26:1–80.

Wiser, S. K. 1993. Vegetation of high-elevation rock outcrops of the Southern Appalachians: composition, environmental relationships, and biogeography of communities and rare species. Ph.D. dissertation, University of North Carolina, Chapel Hill, N.C.

Wiser, S. K. 1994. High-elevation cliffs and outcrops of the Southern Appalachians: vascular plants and biogeography. *Castanea* 59:85–116.

Wiser, S. K. 1998. Comparison of Southern Appalachian high-elevation outcrop plant communities with their Northern Appalachian counterparts. *Journal of Biogeography* 25:501–513.

Wiser, S. K., Peet, R. K., and White, P. S. 1996. High-elevation rock outcrop vegetation of the southern Appalachian mountains. *Journal of Vegetation Science* 7:703–722.

Wiser, S. K., Peet, R. K., and White, P. S. In press. Prediction of rare plant occurrence: a Southern Appalachian example. *Ecological Applications* 8:908–920.

CENTRAL/MIDWEST REGION

8   Dry Soil Oak Savanna in the Great Lakes Region

SUSAN WILL-WOLF AND FOREST STEARNS

Introduction

Dry soil oak savanna in the Great Lakes region occurs in Minnesota; Wisconsin; Michigan; northeastern Iowa; the northern portions of Illinois, Indiana, and Ohio; and the Great Lakes plains of southeastern Ontario (Chapman et al. 1995). Dry soil oak savanna, where low moisture availability limits biomass production, includes vegetation with quite varied physiognomy; terms for variants include *woodland, barrens, sand savanna, scrub oak savanna*, and *brush prairie*. The term *barrens* has no standard usage (Curtis 1959; Heikens and Robertson 1994; Hutchinson 1994; J. White 1994); we use this term and other terms only as defined in this chapter.

Variation in savanna vegetation physiognomy and species composition in the region has been related to interaction of disturbance regime, especially fire, with broad edaphic gradients (Grimm 1984; Bowles and McBride 1994). Fire frequency and intensity are in turn influenced by landscape structure (Leitner et al. 1991; Will-Wolf and Montague 1994) interacting with climate (Grimm 1985). Vegetation classification, used to define units for mapping and for management, frequently incorporates the assumption that site environment determines vegetation composition and structure. However, in recent savanna classifications for the upper Midwest, dry soil savanna vegetation units distinguished by structure and species composition are described as a function of different disturbance regimes as much as of different site environments (Homoya 1994; Faber-Langendoen 1995; Haney and Apfelbaum 1995). These relationships are critical for understanding savanna dynamics and for successful site management.

Savanna vegetation associated with dry, low-productivity sites in the Great Lakes region falls into two groups (Figure 8.1). Dry sand savanna (sand barrens), the more common of the two, occurs on acidic (pH 4.3–6.5), sandy soils with low available water-holding capacity (AWC) and low inorganic nutrient availability (Curtis 1959; A. S. White 1983; R. C. Anderson and Brown 1986; Udvig 1986; Leach 1996). Dry calcareous (limestone) savanna occurs on thin or excessively well drained, finer-textured soils (pH 7.0–7.5) with low water availability and average nutrient status, over calcareous or impervious parent material (Catling and Catling 1993; Armstrong 1994). Oak species of the red and black oak subgenus (*Erythrobalanus*; vascular plant nomenclature follows Gleason and Cronquist 1991) are characteristic of dry sand savanna, whereas oak species of the white oak subgenus (*Lepidobalanus*) dominate the tree layer of dry calcareous savanna (Table 8.1). Both dry sand or calcareous savanna grade into bur oak (*Quercus macrocarpa*) and/or black (*Quercus velutina*) and white oak (*Quercus alba*) savanna on deeper, richer soils with higher water availability (Figure 8.2). Dry calcareous savanna and dry sand savanna

Roger C. Anderson, James S. Fralish, and Jerry M. Baskin, eds. *Savannas, Barrens, and Rock Outcrop Plant Communities of North America.* Copyright © 1999 Cambridge University Press. All rights reserved.

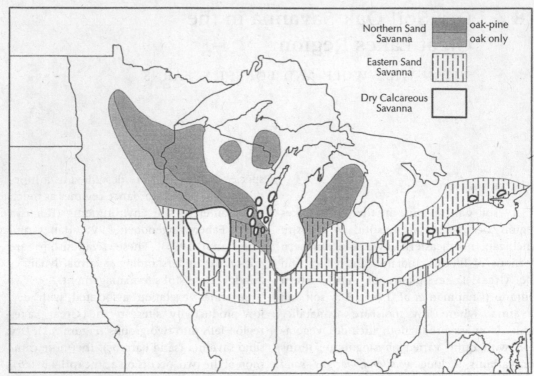

Figure 8.1. Distribution of dry soil oak savanna in the Great Lakes region, indicating the geographic area within which each vegetation type occurs. For dry calcareous savanna, sections delineated by a heavy line suggest the areas in which they occur. The largest and most western of these areas is the Driftless Area (see text). Scattered Eastern Sand Savanna sites occur in the Wisconsin River valley within the Driftless Area (not indicated on map).

remnants in southwestern Ontario were clearly separated based on species composition using principal coordinates ordination and UPGMA (unweighted pair-group method, arithmetic average) clustering (Catling and Catling 1993).

Several common ground-layer species, such as little bluestem (*Schizachyrium scoparium*), Pennsylvania sedge (*Carex pensylvanica*), and leadplant (*Amorpha canescens*), are found on most dry savanna sites regardless of soil type (Table 8.1). Ground-layer species characteristic (usually found at sites of one type; seldom found at sites of other types) of different types of dry sand savanna (Table 8.2) are relatively low in abundance (Will-Wolf and Stearns in press), though current species lists are preliminary.

Dry Sand Savanna

Distribution

Dry sand savanna is associated with sandy, low-nutrient, droughty soils of glacial moraines, outwash plains, sandy lake beds, and dune systems located relatively near former glacial margins. In the early 1800s, most of these areas were recorded as open barrens or savanna by government land surveyors. Their structure and composition before European settlement were maintained by natural and anthropogenic fire (Vestal 1936; Curtis 1959; Szeicz and MacDonald 1991). Curtis (1959) estimated that at the time of government land surveys in Wisconsin, over 1.7 million hectares could be characterized as barrens (dry sand savanna). Today, only scat-

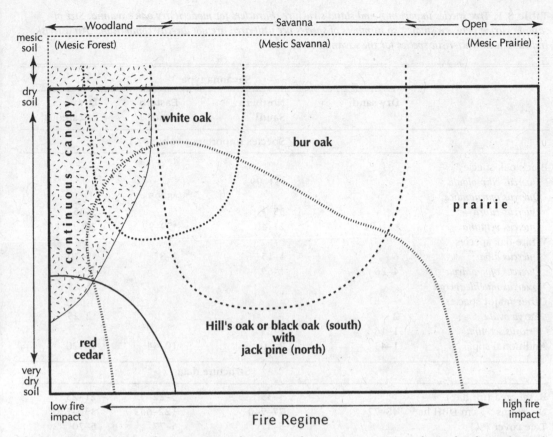

Figure 8.2. Distribution of tree species and vegetation physiognomy along soil moisture and fire impact gradients for Great Lakes region dry soil savanna. The relationships of dry sand savanna to other vegetation are emphasized; relationships of dry calcareous savanna are not represented. "Fire impact" includes fire frequency and fire intensity, which can vary independently.

tered remnants occur, and many have developed into closed-canopy oak woodland (Auclair 1976). Mossman, Epstein and Hoffman (1991) estimated that about 20,000 hectares (1.2% of the presettlement area) of dry sand savanna remain in Wisconsin, which is considerably more than the 0.02% remaining for Midwest wet, mesic, and dry soil oak savanna combined (Nuzzo 1986).

The region west of Lake Huron is divided, by a vegetation transition zone trending northwest to southeast, into a northern (and more eastern) part with mixed conifer–hardwood vegetation, cool summers, long winters, and continuous snow cover, and a southern (and more western) part with hard-

wood vegetation, warmer, drier summers, and winters with sporadic snow cover. In Wisconsin, this zone is known as the tension zone and is defined by the distributional boundaries of many plant species (Curtis 1959). A similar vegetation pattern occurs in Minnesota (U.S. Department of Agriculture 1974) and Michigan (Veatch 1959; McNab and Avers 1994)). The location of the transition zone coincides with the southern limit of oak–pine sand savanna (Figure 8.1). Jack pine (*Pinus banksiana*) often is present north of this zone.

Haney and Apfelbaum (1995) divided Great Lakes region dry sand savanna into Northern and Eastern Sand Savanna groups,

Table 8.1. *Tree species importance and stand structure characters for modern dry oak savanna. Size of sample varied between studies, so numbers of species per unit area are not directly comparable. An asterisk indicates a characteristic species for the savanna type.*

		Savanna type		
	Dry sand[1]	Northern Sand[2]	Eastern Sand[3]	Dry calcareous[4]
		Species importance value (%)		
Black oak species				
Quercus ellipsoidalis		*31–99		
Quercus marilandica			*8–95	
Quercus rubra	1	35		5–35
Quercus velutina	*31–92	21–40	*59–99	20
White oak species				
Quercus alba	6	1–15	1–8	*5–70
Quercus macrocarpa	4–26	1–69		40
Quercus muehlenbergii				*5–25
Other major species				
Carya ovata	2			*2–25
Prunus serotina	1–16	1–23		
Additional spp.	1–43	2–60	10–44	10–20
		Structure data		
Basal area (m² ha⁻¹)		4–35	5–22	21–23
No. stems >5 cm DBH ha⁻¹	48–72	77–940	122–643	330–730
Tree cover (%)		5–47	5–77	5–70
Shrub cover (%)		1–44	10–79	18–40
Herb cover (%)			42–61	10–66
No. species	49	21–71	28–86	34–105

[1] Data from Curtis (1959) and Leach (1996). For this column only, *Quercus velutina* includes *Q. ellipsoidalis* and hybrids.

[2] Data from A. S. White (1983), Udvig (1986), Tester (1989), Kline and McClintock (1994), Faber-Langendoen and Davis (1995), Kline and McClintock (1995), and Fralish (unpublished data).

[3] Data from Whitford and Whitford (1971), N. R. Henderson and Long (1984), R. C. Anderson and Brown(1986), and Catling and Catling (1993).

[4] Data from Catling and Catling (1993), Armstrong (1994), and M. Anderson, Will-Wolf and Howell (1996).

which differ in general geographic location (Figure 8.1), with overlap in southern Wisconsin. Northern Sand Savanna, in the northern and western parts of the region, has Hill's (northern pin oak; *Quercus ellipsoidalis*) and Hill's oak hybrids (Hill's oak × black oak; Hill's oak × northern red oak, *Q. rubra*), with jack pine present to the north. Eastern Sand Savanna, in the southern and eastern parts of the region, usually has black oak as the char-

acteristic tree species, with Hill's oak absent or a minor component. We favor this subdivision of dry sand savanna over others recently advocated because it recognizes the similarity of Northern Sand Savanna north and south of the vegetation transition zone.

The dry sand savanna groups share several common characteristic ground-layer species (Table 8.2), such as huckleberry (*Gaylussacia baccata*), bracken fern (*Pteridium aquilinum*),

Table 8.2. *Frequency of common shrub, vine, and herbaceous species. Most numbers are frequency in quadrats within a site; only Curtis (1959) included enough separate sites to report among-site frequency. Size and number of quadrats and size of study site differed between studies, so frequency values are not strictly comparable; ranges rather than averages are reported. Selected species have at least 10% within-site frequency from two reports, or they are both prevalent and modal in Wisconsin dry sand savanna (Curtis 1959). An asterisk indicates a characteristic species of that savanna type (Will-Wolf and Stearns in press). For instance, the shrubs Gaylussacia baccata and Vaccinium angustifolium are found occasionally in dry calcareous savanna, but are characteristic of dry sand savanna, albeit with low frequency.*

Species	Dry sand savanna[1]		Northern Sand Savanna[3]	Eastern Sand Savanna[4]	Dry calcareous savanna[5]
	Among sites[2]	Between sites			
Shrub and vine frequency (%)					
Ceanothus americanus			3	17–48	3–11
Cornus racemosa	57	6–30	27		5–33
Corylus americana	79	13–18	18–39	18	7
Gaylussacia baccata		*		4–28	1–7
Rhus spp. (shrubs)	57	2	3	5–36	2–30
Rosa spp. (mostly caroliniana)	64	14–16	–	14–80	8
Symphoriocarpos spp.					*4–60
Toxicodendron radicans			2–13	8–80	8–15
Vaccinium angustifolium		*	16	24–64	2–16
Vitis riparia			25	4–36	1
Herbaceous species frequency (%)					
Amorpha canescens	93	12–25	9	10	15–45
Amphicarpaea bracteata	43	13–21	2–39		8–11
Andropogon gerardii	57	9–18	6–44	4–10	5–35
Antennaria spp.	79	7–8	6	1–3	1–19
Asclepias spp.	50	2–7	13	20	3–10
Aster oolentangiensis				22	2–15
Aster linariifolius	43	7		*40	
Carex pensylvanica	50	12–46		12–76	10–19
Carex spp.			33–47	4–62	4–21
Comandra umbellata	86	11–15	5–26	12–52	8–15
Coreopsis palmata	71	7			15–36
Euphorbia corollata	10	32–36		20–36	6–46
Fragaria virginiana	71	10–15	5–6		1–40
Galium boreale	36	10	10		3–30
Helianthemum canadense	64	*13	3	5–20	1
Heuchera richardsonii	43	2			6–16
Krigia biflora	50	3	*		1–3
Lespedeza capitata	93	9		2–15	5–15
Lithospermum canescens	71	4			12–25
Lupinus perennis	43	*4	20	10	
Monarda fistulosa	50	10–19	2	1–10	2–40
Panicum oligosanthes			*9	1–20	
Panicum villosissimum	21		2–12		4–22
Pteridium aquilinum	57	*15		20–25	
Phlox pilosa		11		28–76	
Physalis spp.	64	4	6		3–35
Schizachyrium scoparium	64	6–22	12	28–48	9–52

(cont.)

Table 8.2 (cont.)

| Species | Dry sand savanna[1] | | Northern Sand Savanna[3] | Eastern Sand Savanna[4] | Dry calcareous savanna[5] |
	Among sites[2]	Between sites			
Smilacina racemosa	64	11–13	20	48–68	1–11
Smilacina stellata	71	13–21	3–16		8
Solidago nemoralis	57	9–11	8	1–35	2–25
Tradescantia ohiensis	64	12–16	2–14		1–2
Viola pedata	43	9			4–16

[1] Data from Gilbert and Curtis (1953), Curtis (1959), and Leach (1996). Both sand savanna types are combined in these studies.
[2] Data from Curtis (1959).
[3] Data from A, S. White (1983), Tester (1989), and Kline and McClintock (1995).
[4] Data from Whitford and Whitford (1971), N. R. Henderson and Long (1984), R. C. Anderson and Brown (1986), and Catling and Catling (1993).
[5] Data from Catling and Catling (1993), Armstrong (1994), M. Anderson, Will-Wolf and Howell (1996).

and blueberry (*Vaccinium angustifolium*). They also share several uncommon characteristic species, some of which are conservation listed in one or more states; for example, wild blue lupine (*Lupinus perennis*), sand fameflower (*Talinum rugospermum*), and goat's rue (*Tephrosia virginiana*). Most ground-layer species characteristic of either Northern or Eastern Sand Savanna (Will-Wolf and Stearns in press) are low in abundance.

Composition and History

Northern Sand Savanna. Northern Sand Savanna (Figure 8.1) occurs in Minnesota, Wisconsin, and Michigan (Western Lake states). The most characteristic tree, Hill's oak (Table 8.1), is closely related to scarlet oak (*Quercus coccinea*) and may be a subspecies of the latter (Gleason and Cronquist 1991). Hill's oak hybridizes extensively with black oak south of the vegetation transition zone (Curtis 1958) and with northern red oak (*Quercus rubra*) north of that zone (Gleason and Cronquist 1991). White oak and bur oak also are found occasionally. Ground-layer species characteristic of Northern Sand Savanna include beaked hazelnut (*Corylus cornuta*), two species of hawkweeds (*Hieracium kalmii* and *H. longipylum*), orange dwarf-dandelion (*Krigia biflora*), and a goldenrod (*Solidago ptarmicoides*), four species with primarily western North American distributions, and several state-listed species (Will-Wolf and Stearns in press).

Large, moderate- to high-quality remnants of Northern Sand Savanna north of the vegetation transition zone include Cedar Creek Natural Area (A. S. White 1983) in the Anoka Sand Plains of Minnesota, and Necedah Wildlife Refuge and Crex Meadows (Vogl 1964) in Wisconsin. Fort McCoy Military Reservation (Leach 1993; Maxwell and Givnish 1996) and the central sand plains (Leopold 1949; Curtis 1959) of Wisconsin straddle the vegetation transition zone.

North of the vegetation transition zone, oak-dominated Northern Sand Savanna occurs on sites similar in environment to jack pine barrens (see Chapter 21), but with different disturbance history. Much of what is now oak-dominated Northern Sand Savanna was described as jack pine barrens by 19th-century European visitors and settlers. "'The barrens' have a timber growth exclusively their own. The trees are either scrub pines, or black-jack oaks [Hill's oak], averaging in diameter about three or four inches and in height not over fifteen feet. In some places, as on the sandhill of the barrens, the trees are at considerable distances from each other" (Sweet 1880 in Curtis 1959). Presettlement fire regimes that resulted in jack pine dominance included stand-killing fires at 30–60-yr intervals (Curtis 1959; Vogl 1970; Whitney 1986).

Jack pine barrens were logged and abandoned, then burned during the late 1800s or early 1900s (sometimes farmed briefly and again abandoned in the 1930s) in central Minnesota (Grimm 1985), central and northern Wisconsin (Leopold 1949; Vogl 1964), and in large areas of the Lower Peninsula of Michigan (Whitney 1987). Although some of the Michigan sites reverted to jack pine barrens (Whitney 1987; also Chapter 21), many of them, and most of the Wisconsin and Minnesota sites, have become oak savanna and woodland. During the relatively unnatural combination of disturbances occurring with European settlement, the jack pine community component was often severely reduced or lost. Vogl (1964) noted that most of the herbaceous prairie species have persisted at Crex Meadows (with changes in abundance) in both burned and unburned stands despite major changes in overstory over more than 100 years.

Eastern Sand Savanna. Eastern Sand Savanna is found in northern and central Illinois, southern Wisconsin, southern Michigan (Madany 1981; Pruka and Faber-Langendoen 1995), northwestern Indiana (Homoya 1994), northwestern Ohio (Gordon 1966; Ohio Biological Survey 1966; Brewer 1989), and southeastern Ontario (Szeicz and MacDonald 1991; Catling and Catling 1993). Before European settlement, extensive areas of Eastern Sand Savanna occurred on lake plains and sandy outwash in southern Michigan and on the Lake Michigan shore- dune complexes of Indiana and Michigan (Cole and Taylor 1995; Rabe, Comer and Albert 1995).

Black oak is the characteristic tree, with other oak species occasionally present (Table 8.1). Dry sand savanna in central Illinois often has blackjack oak (*Quercus marilandica*), but

with ground-layer species similar to other Eastern Sand Savanna sites (Rodgers and Anderson 1979; R. C. Anderson and Brown 1986). Ground-layer composition includes a short list of characteristic species, including shining sumac (*Rhus copallina*), staghorn sumac (*Rhus typhina*), *Aster linariifolius*, two pinweeds (*Lechea mucronata* and *L. pulchella*), and three state-listed species (Will-Wolf and Stearns in press), in addition to species shared with other dry soil savanna types (Table 8.2).

Eastern Sand Savanna underwent postsettlement changes similar to that described for Northern Sand Savanna. For instance, in Fulton and Lucas counties southwest of Toledo, Ohio, savanna occurred on glacial lake sands where black oak and white oak, with average diameters of 38 cm and 47 cm respectively, accounted for 95% of savanna trees (Brewer 1989). After settlement, the area was farmed and then abandoned. Part of the area returned to savanna (250 ha remaining today), and part became forest dominated by black oak (Brewer and Grigore 1995).

Structure

Dry sand savanna, in contrast to deep-soil oak savanna (see Chapter 9) or dry calcareous savanna, often lacks scattered stately trees with wide crowns. The black oak group species characteristic of sand savanna are relatively more fire sensitive than white oak group species, grow slowly on low-nutrient sites (Figure 8.2), and are more prone than white oak group species to produce short-statured woody growth. With scrub oak, most stems are not "tree sized," so tree density is not an adequate measure of structure. In general, degree of canopy closure has been the most widely used structural description; for instance, savanna <50% canopy vs. forest >50% canopy (Curtis 1959; Bray 1960), or savanna 10%–25% canopy vs. woodland 25%–60% canopy (Faber-Langendoen 1995). However, dry sand savanna often has open areas interspersed with more-or-less closed-canopy patches, so for measurements made at scales near or below tree or open-area patch size (about 0.5 ha), percentage canopy values widely vary. "Brush prairie" physiognomy, with short shrubby growth, lacks a well-defined canopy and so does not fit either description.

Frequent or intense fire can produce scrub oak structure (short-statured, multiple-stemmed, small trees) or brush prairie structure (shrubby, small-stemmed patches of oak or shrub species) in clumps alternating with open patches of herbaceous (prairie) species. Such structure was common at the time of European settlement (Vestal 1936; Curtis 1959; Rodgers and Anderson 1979) but has since become considerably less common (Bowles and McBride 1994).

Fire suppression has resulted in oak woodland, often with the same tree species, but with a more uniform, tall canopy. Densities of 7–74 trees/ha and basal area of 1–9 m²/ha calculated from presettlement survey records (Rodgers and Anderson 1979; Cole and Taylor 1995; Rabe, Comer and Albert 1995) are much lower than values for modern stands (Table 8.1). One savanna at Indiana Dunes National Lakeshore with fewer, more intense burns, appears to be developing toward a scrub oak structure with few large trees, whereas a similar nearby site, with more frequent but less intense burns, has maintained a large tree canopy with little woody growth in the ground layer (N. R. Henderson and Long 1984). Grazing in post-European dry sand savanna maintained open understory by preventing growth of shrubs and tree seedlings (Curtis 1959; Auclair 1976; Brewer 1989; Catling and Catling 1993; Cole and Taylor 1995; Faber-Langendoen and Davis 1995, Chapman et al. 1995). In Ontario, shrub cover and herb species relative abundance vary widely between remnant examples of single dry soil savanna types (Catling and Catling 1993).

Studies of plant and animal communities on relatively large remnant sites (e.g., Mossman, Epstein and Hoffman 1991; Shuey

1995; Maxwell and Givnish 1996) have emphasized the innately patchy character of sand savanna structure. Linkage of site environment (topographic position, soil nutrients, etc.) to particular sand savanna physiognomy and composition on local (<1,000 ha: R. C. Anderson and Brown 1983; N. R. Henderson and Long 1984; Faber-Langendoen and Tester 1993) and regional (>1,000 ha: Grimm 1984; Leitner et al. 1991; Will-Wolf and Montague 1994; Bowles and McBride 1994) scales varies with site history and fire regime (and other disturbances) over short (years to decades) time scales. The effectiveness of topographic features as fire-breaks, as well as the nature of fire regimes, shifts with climate changes over long (decades to centuries) time scales (Grimm 1985; Cole and Taylor 1995). Thus, many investigators use a dynamic, nonequilibrium perspective adapted from Watt's (1947) model of patch dynamics to explain causes of variation in dry sand savanna structure. Situations in which site environment appears to determine vegetation (Roberts 1987) in dry sand savanna complexes (e.g., Chapman et al. 1995) may result from artificially uniform disturbance regimes operating over relatively short time scales (<50 yr) and at relatively small spatial scales (<10 ha).

Vegetation Dynamics

Vascular Plants. Differences in species composition occur for a variety of reasons, but as with structure, variation and changes in composition are best understood using a nonequilibrium model of vegetation dynamics. Shifts in ground-layer species composition are often related to changes in structure of woody vegetation. Overgrazing has been blamed for the loss of herbaceous plant diversity and the invasion of weedy exotics (Curtis 1959; Catling and Catling 1993; Ko and Reich 1993).

Some sand savanna sites have soil that is too nutrient poor and/or droughty for shade-tolerant tree species to thrive, so fire suppression does not lead to change in overstory composition (Habeck 1959; Whitford and Whitford 1971; Kotar, Kovach and Locey 1988: *Quercus/Gaultheria–Ceanothus*, *Quercus/Amopha*, and *Pinus/Amphicarpa* habitat types in northwestern Wisconsin). Low soil AWC in extremely dry years may limit the density and species composition of woody vegetation on these sites (Faber-Langendoen and Tester 1993).

On less extreme sites, fire suppression allows invasion of red maple (*Acer rubrum*) and other moderately shade-tolerant, dry-site tree species (Curtis 1959; Levy 1970; A. S. White 1983; Catling and Catling 1993; Brewer and Grigore 1995). Fralish (unpublished data) found during a 17-year study (1977–94) of northern Wisconsin dry sand savanna sites that sapling density increased from less than 1.0 ha^{-1} to over 12,000 ha^{-1} (90% red maple). During the same period, relative importance of Hill's oak trees decreased from 80%–90% to 50%–80%, and relative importance of red maple and aspen (*Populus* spp.) trees increased from 0% to 5%–20%.

Studies of ground-layer vegetation dynamics in dry sand savanna have focused on the impact of fire and its loss from the system, and the response of herbaceous plants to light gradients under tall tree savanna. The wide range of light environments available in savanna has long been associated with high ground-layer diversity (Bray 1958, 1960; Curtis 1959). Increase in overstory density in dry sand savanna following fire suppression decreases the proportion of high-light habitat available for ground-layer plants.

Loss of relatively shade-intolerant species through canopy closure has been extrapolated by observing recovery of species at several sites when fire was reintroduced. Kline and McClintock (1994, 1995) found increases in species diversity and cover of relatively shade-intolerant herbaceous species such as gray goldenrod (*Solidago nemoralis*) and ground-cherry (*Physalis* spp.) immediately after prescribed burns on the Grady Tract (University of Wisconsin Arboretum). They

distinguished light specialists (both high-light and low-light) from light generalists, a distinction supported by Leach's (1996) finding that 3–11 species are modal in each of five segments of a light gradient in Wisconsin sand savanna remnants. Grass species are more frequent and abundant in the high-light specialist group, often called "true prairie" species in savanna literature. Intermediate- and low-light categories include more forb species (e.g., Bray 1958, 1960; Ko and Reich 1993; Kline and McClintock 1995; Leach 1996) and most characteristic "savanna species." Many of the herbaceous plants described by Gilbert and Curtis (1953) as characteristic dry oak forest understory species now appear to be better described as dry savanna intermediate- and low-light specialist species and light generalist species that were able to persist when canopies closed after postsettlement fire suppression (Pruka 1994).

Cover of "true prairie" grasses such as big bluestem (*Andropogon gerardii*) and little bluestem (*Schizachryium scoparium*), forbs such as bastard toad-flax (*Comandra umbellata*) and several goldenrods (*Solidago* spp.), and density of "true prairie" shrubs such as smooth sumac (*Rhus glabra*) and *Rosa* spp. increased with burn frequency at Cedar Creek Natural Area (Minnesota), whereas density of nonprairie shrubs, such as dogwoods (*Cornus* spp.), and of trees declined (A. S. White 1983; Tester 1989). Annual burns enhanced growth of prairie shrubs at Hooper Branch Nature Preserve (central Illinois) while inhibiting growth of tree seedlings and saplings. Less-frequent burns allowed establishment and growth of black oak seedlings (Johnson and Ebinger 1992).

Light and moisture availability usually interact to constrain plant species distribution (Ko and Reich 1993; Leach 1996). Maxwell and Givnish (1996) found that the relative success of wild blue lupine in adjacent sunny vs. shady habitats differed between wet and dry years. The patchy mosaic of different habitats provided by a dry sand savanna complex may allow buffering against extremes in yearly weather variation for other dry sand savanna plant species as well. Investigation of the response of ground-layer composition has focused on open areas vs. tall-canopy understory; ground-layer composition in brushy or scrub oak patches has not been reported.

Cryptogams. From 15 to 30 common and widespread lichen species per site have been found on oak trunks in dry sand savanna (Will-Wolf 1980; Wetmore 1983). The foliose lichens *Punctelia rudecta*, *Physcia millegrana*, and *Candelaria concolor* and the crustose lichens *Candelariella xanthostigma* and *Rinodina papillata* were common at all sites. Number of species and percentage cover of lichens on trunks of bur oak in a Minnesota dry sand savanna declined with increased frequency of burns (Wetmore 1983). Declines were much greater near the ground than at 1.5 m on trunks, suggesting that burning had little short-term effect on lichens higher on trunks and in tree canopies. *Candelaria concolor* and *Physciella chloantha* increased in frequency or cover with increase in burn frequency, whereas a few species, including *Physcia millegrana* and *Rinodina papillata*, were little affected by burns. *Caloplaca flavorubescens* and *Phaeophyscia rubropulchra* declined drastically with annual burns. Slope exposure, direction of shading, tree species, and presence of the basidiomycete fungus *Aleurodiscus* sp. also affected composition and cover of lichens.

Dry soil savanna often has patches of cryptogamic or microbiotic soil crust composed of lichens, mosses, and cyanobacteria, a vegetation component more characteristic of arid western North America, where it is very important for soil stabilization (Brotherson, Rushforth and Johansen 1983). Soil lichens in selected quadrats declined from 95% cover before a sand prairie burn to 10% cover 2 years after the burn (Schulten 1985), although they were beginning to regrow. These results suggest that burns should be no more frequent than every 2–3 years to maintain microbiotic crust in open areas of dry sand savanna.

Plant Productivity and Nutrient Dynamics

At a dry sand savanna complex in east central Minnesota, the diversity of the herbaceous understory was inversely related to availability of nutrients, especially nitrogen, in low-nutrient sandy soils (Inouye and Tilman 1995; Tilman, Wedin and Knops 1996). On such sites, herbivory and shortage of nutrients other than nitrogen may limit legume density in the ground-layer communities (Ritchie and Tilman 1995).

On a dry sand savanna in southern Wisconsin, soil nutrients and water were more available under oak canopies, but light was less available than in open areas. Limited light was proposed as the major factor responsible for lower herbaceous plant biomass under oak canopies, since soil characteristics under canopies appeared more favorable for understory growth than did soil characteristics in the open (Ko and Reich 1993).

Soil pH increased with frequency of burning, and species richness peaked on sites burned about every 2 years on dry sand savanna in Minnesota (A. S. White 1983; Tester 1989; Faber-Langendoen and Davis 1995). Spring burns in a Wisconsin dry sand savanna increased summer productivity of herbaceous vegetation (Vogl 1965).

Grigore and Tramer (1996) reported that wild blue lupine (*Lupinus perennis*) had both positive (increased leaf production and seed set) and negative (increased seedling mortality) first-year productivity responses to prescribed burns. These authors recommended staggered burns at least 2 years apart, timed to occur before seedling emergence to encourage lupine growth.

Fauna

Mammals. Elk and bison, once relatively common on prairie and savanna in the upper Midwest (Schorger 1930, 1954), were extirpated by the late 1800s; their impact on the dry sand savanna system and the effect of their loss are not known. Several mammal species presently are found in dry sand savanna (Table 8.3), but limited knowledge of their needs and impacts restricts the development of management strategies (Chapman et al. 1995). In a Wisconsin "brush prairie" dry sand savanna, boreal red-backed voles and white-footed mice (woodland species) decreased in density with increasing burn frequency, whereas prairie deer mice and thirteen-lined ground squirrels (prairie species) increased in density with increasing burn frequency (Beck and Vogl 1972).

In a Minnesota dry sand savanna, large beard-tongue (*Penstemon grandiflorus*) grows on mounds made by the plains pocket gopher. Both species are generally confined to openings and are absent from closed canopy areas. Experiments by Davis et al. (1995) showed that competing herbaceous vegetation limited the success of large beard-tongue more than shade from above. These authors concluded that in unburned, closed-canopy woodland, absence of pocket gophers and their bare soil mounds ("competition-free" microhabitats) was a more likely cause for the loss of large beard-tongue than was shade from woody canopy.

Herptiles. Seven species of reptiles and amphibians (Table 8.3: Common) have been reported from several dry sand savanna sites; five species (Occasional) are relatively uncommon (Vogt 1981; Botts et al. 1994; Eckstein and Moss 1995). More study is needed on their status and management needs.

Birds. Birds of large dry sand savanna sites in Wisconsin (Table 8.3) include both species characteristic of these vegetation types and species formerly found in dry prairies. Dry savanna sites are apparently the only areas sufficiently large for them in most of the region (Mossman, Epstein and Hoffman 1991; Sample and Mossman 1994). Several common species, such as the eastern kingbird, brown thrasher, and American goldfinch, use both tall tree clumps and brush prairie areas;

other species are more specific. For example, the eastern bluebird, American robin, and eastern meadowlark prefer tall tree clumps, whereas the gray catbird, Brewer's blackbird, and the song sparrow prefer brushy savanna (Table 8.3). Ruffed grouse and sharp-tailed grouse need very large areas of brushy or scrub oak savanna to maintain viable populations (Kubisiak 1985; Gregg 1987). Bird communities of Wisconsin dry sand savanna are similar to those of other dry savanna types in the eastern United States (e.g., Atlantic Coastal Plain pine barrens; Brush and Stiles 1990; Morimoto and Wasserman 1991).

Arthropods. Arthropods are an important, but insufficiently documented, component of dry savanna systems. Many invertebrates, unlike birds, appear to have relatively restricted distributions among savanna vegetation types (Panzer 1988; Swengel 1993; Hess and Sedman 1994; Maxwell and Givnish 1996). In contrast, Panzer et al. (1995) reported that only 256 (<25%) of 1,100 insect species found on prairie and savanna remnants in the Chicago region were limited to these remnants of natural vegetation. Flower moths and root-boring moths showed the highest degree of remnant dependence. This survey was for a single year only and in an area with extremely fragmented natural vegetation, so the most restricted species may have been either missed or absent.

The most species and highest numbers of bark-inhabiting invertebrates were found on trees in a Minnesota dry sand savanna with a moderate level of fire disturbance compared to sites with annual or infrequent fires (Nicolai 1991). Arthropods from many different families showed complex responses to fire frequency, with some being specialist inhabitants of heavily disturbed sites.

Several species of moths and butterflies appear to be characteristic of dry sand savanna (Botts et al. 1994; Swengel 1994; Eckstein and Moss 1995); most of them are listed for conservation in at least one state

Table 8.3. *Animals found in dry sand savanna. See text for definition of common and occasional herptiles. Only characteristic moths and butterflies are reported.*

Mammals[1]

Canis latrans, coyote	*Peromyscus leucopus*, white-footed mouse
Citellus franklinii, Franklin's ground squirrel[5]	*Peromyscus maniculatus*, prairie deer mouse
Citellus tridecemlineatus, thirteen-lined ground squirrel	*Procyon lotor*, raccoon
Geomys bursarius, plains pocket gopher[5]	*Sciurus carolinensis*, gray squirrel
Mephites mephites, striped skunk	*Sciurus niger*, fox squirrel
Odocoileus virginianus, white-tailed deer	*Taxidea taxus*, badger[5]

Herptiles[2]

Common	Occasional
Bufo americanus, American toad	*Cnedimophorus sexlineatus*, six-lined racerunner
Bufo woodhousei, Fowler's toad	*Coluber constrictor*, racer
Eumeces fasciatus, five-lined skink[5]	*Ophisaurus attenuatus*, slender glass lizard[5]
Heterodon platirhinos, hognose snake	*Terrapene ornata*, ornate box turtle[5]
Hyla chrysoscelis, Cope's gray treefrog	*Thamnophis proximus*, western ribbon snake[5]
Opheodrys vernalis, green snake	
Pituophis melanoleucus, bullsnake	

Birds[3]

Ammodramus savannarum, grasshopper sparrow T[6]	*Pedioecetes phasionellus*, sharp-tailed grouse[5] B[6]
Bartramia longicauda, upland sandpiper[5] T	*Pipilo erythrophthalmus*, rufous-sided towhee B
Bonasa umbellus, ruffed grouse TB	*Pooecetes gramineus*, vesper sparrow TB
Carduelis tristis, American goldfinch TB	*Riparia riparia*, bank swallow T
Coccyzus erythropthalmus, black-billed cuckoo B	*Sialia sialis*, eastern bluebird T
Cyanocitta cristata, blue jay T	*Spizella pucilla*, field sparrow TB
Dumetella carolinensis, gray catbird B	*Sturnella magna*, eastern meadowlark T
Euphagus cyanocephalus, Brewer's blackbird B	*Toxostoma rufum*, brown thrasher TB
Icterus galbula, Baltimore oriole T	*Turdus migratorius*, American robin T
Melospiza melodia, song sparrow B	*Tyrannus tyrannus*, eastern kingbird TB
Molothrus ater, brown-headed cowbird TB	*Zenaida macroura*, mourning dove T
Passerina cyanea, indigo bunting TB	

Moths and butterflies[4]

Atrytonopsis hianna, dusted skipper[5]	*Hesperia ottoe*, ottoe skipper
Chlosyne gorgone, gorgone checkerspot	*Incisalia irus*, frosted elfin[5]
Erynnis martialis, mottled dusky wing[5]	*Oenis chryxus*, chryxus arctic
Erynnis persius, persius dusky wing[5]	*Lycaeides melissa samuelis*, Karner blue[5]
Euchloe olympia, Olympia marblewing[5]	*Schinia indiana phlox*, phlox flower moth[5]
Hesperia leonardus, Leonard's skipper[5]	*Speyeria idalia*, regal fritillary[5]
Hesperia metea, cobweb skipper[5]	*Strymon edwardsii*, Edward's hairstreak

[1] List from Botts et al. (1994) and Eckstein and Moss (1995).
[2] List from Vogt (1981), Botts et al. (1994), and Eckstein and Moss (1995).
[3] List from Mossman, Epstein and Hoffman (1991) and Botts et al. (1994).
[4] List from Botts et al. (1994), Swengel (1994), and Eckstein and Moss. (1995).
[5] Listed for conservation in at least one state in the region.
[6] T = savanna with tree clumps; B = brush prairie physiognomy.

(Table 8.3). Four species, the frosted elfin, Karner blue, regal fritillary, and persius dusky wing, are subjects of concern over most of their range (Chapman et al. 1995).

The federally endangered Karner blue butterfly is the best-known symbol for the rare lepidopterans of the dry sand savanna community (Andow, Baker and Lane 1994). Eggs of this butterfly overwinter on wild blue lupine, the only larval food plant; adults subsist briefly on nectar from various herbaceous perennial plants while they reproduce. Stable populations of the Karner blue remain in Minnesota, Wisconsin (Bleser 1994), Michigan (Rabe, Comer and Albert 1995), and the northeastern United States (see Chapter 3), but many populations have become extinct within the last 20 years.

Both historical evidence and modern observation confirm that the Karner blue butterfly and wild blue lupine thrive in the context of frequent, but patchy, burns. All Karner blue life cycle stages are fire sensitive, yet wild blue lupine is a high-light specialist that thrives in recently burned patches (Shuey 1994). The Karner blues need both shady and open patches, but they cannot disperse far, and the habitat value of patches changes with yearly climate variation (Maxwell and Givnish 1996). Large landscapes foster long-term survival of the Karner blue and wild blue lupine via metapopulation dynamics in patches of dry sand savanna with varied structure (Bleser and Leach 1994; Shuey 1995), an exemplary application for Watt's (1947) "flickering mosaic" model of patch dynamics (Bormann and Likens 1979; Shuey 1994). One of the largest and most stable population complexes of the Karner blue is found on the Fort McCoy Military Reservation, Wisconsin, where it is associated with irregular patchy fire and other disturbance (Leach 1993).

Conserving the Karner blue simultaneously conserves the frosted elfin and the persius dusky wing; the latter two species also require lupine as larval food, but are less sensitive to local extinction (Eckstein and Moss 1995). The Karner blue butterfly and other rare lepidopterans also link midwestern dry sand savanna with Atlantic Coast pine barrens. The overstory structure of jack pine barrens (Chapter 21), eastern pine barrens (Chapter 3), and sand savanna provides the needed shade patches, and wild blue lupine and other needed herbaceous nectar species are found in the understory.

Current Management and Future Needs

Many dry sand savanna sites continue to support savanna vegetation because the poor soil makes agriculture unprofitable. However, the vegetation of many remnants has grown too dense for maintenance of savanna structure, for oak reproduction, or for persistence of relatively shade-intolerant understory plant species with their associated invertebrates. Fragmentation has also probably led to loss of species, although (by analogy with prairies), equivalent fragmentation may result in fewer species lost from dry soil savanna than from mesic soil savanna (Leach and Givnish 1996).

Recent management and recovery plans for dry sand savanna (Botts et al. 1994; Leach and Ross 1995; Pruka and Faber-Langendoen 1995; Chapman et al. 1995) recommend thorough inventories of existing sites followed by expansion and/or improvement using prescribed burning. However, burning too frequently and/or too thoroughly can be deleterious to arthropods (Panzer 1988; Swengel 1994, 1995) and possibly cryptogams (Schulten 1985). The most successful management systems have included burns at irregular intervals and on only parts of an area, as applied at Fort McCoy Military Reservation.

High-quality dry sand savanna vegetation and healthy populations of the Karner blue butterfly and other rare arthropods persist at Fort MCoy Military Reservation, Wisconsin. Apparently, military operations have been excellent substitutes for the presettlement disturbance regime. After tank maneuvers or other severe disturbances occur, operations move elsewhere and the site is left to recover; the permanent artillery firing range has frequent random burns and little or no other management (Leach 1993). Controlled experiments on the short-term recovery of plant and insect communities from military practices and prescribed burns are being conducted (Maxwell and Givnish 1996).

Management of very large areas of brushy or scrub oak savanna for ruffed grouse and sharp-tailed grouse (Kubisiak 1985; Gregg 1987) also maintains habitat for other animal and plant species that require less space. Large dry sand savanna remnants also can serve as control areas to test models of the effects of habitat fragmentation on dry sand savanna species.

If areas north of the vegetation transition zone are managed more aggressively to restore pine barrens habitat (Borgerding, Bartelt and McCown 1995), the area of oak-dominated dry sand savanna will be reduced. Differences between oak-dominated savanna and jack pine barrens are linked to disturbance regime (Curtis 1959); sites should probably be managed as oak–jack pine complexes.

Dry Calcareous Savanna

Distribution and History

Dry calcareous savanna is scattered throughout the region (Figure 8.1), mostly as small (0.5–5 ha) areas interspersed with woodland and/or dry prairie. In the large, unglaciated Driftless Area of Wisconsin, Illinois, Minnesota, and Iowa, upper and ridgetop shoulder slopes support dry lime prairie/savanna areas (thin soil over dolomite) where white oak, chinquapin or yellow oak (*Quercus muehlenbergii*), and some black oak are the dominant species (Faber-Langendoen 1995; M. Anderson, Will-Wolf and Howell 1996). In the Baraboo Hills of southern Wisconsin, stunted oak–hickory woodland (Lange 1989) and isolated oak

glades (Clark, Isenring and Mossman 1993; Armstrong 1994) occur on thin (<30 cm) calcareous loess-derived soils over insoluble quartzite bedrock. Well-drained, gravelly, calcareous morainal soils support dry savanna in the Kettle Moraine area of eastern Wisconsin (R. Henderson 1995). Thin-soil ridgetop savannas are often adjacent to eastern redcedar (*Juniperus virginiana*) glades (Curtis 1959), which occur on ridges and ledges with thin, discontinuous soil pockets. In southeastern Ontario, thin soils over limestone on flat lowlands support remnants of dry calcareous savanna (Catling and Catling 1993). Dry calcareous savanna productivity is limited by moisture regime and soil volume (Catling and Catling 1993); bare soil or rock is common.

Dry calcareous oak savanna is not distinguished in reconstructions of pre-European settlement vegetation, so it is difficult to estimate its historic extent. Current remnants suggest that patches were widely scattered across the Great Lakes region. A similar community has been reported as far west as eastern Nebraska (Hanson 1922).

Edaphically stressful conditions have allowed degraded remnants to persist as small islands. For example, on a west-facing ridgetop site, the canopy never completely closed, even when fire was absent for over 75 years (M. Anderson, Will-Wolf and Howell 1996). The present condition of vegetation adjacent to remnants indicates that savanna area has been reduced and the density and cover of woody vegetation (Table 8.1) have increased as fire frequency decreased (M. Anderson, Will-Wolf and Howell 1996); savanna quality has been degraded by grazing (Catling and Catling 1993). With fire suppression and grazing, redcedar has expanded into dry lime prairie/savanna complexes.

Composition and Structure

White oak, bur oak, and shagbark hickory (*Carya ovata*) are characteristic tree species (Table 8.1), and chinquapin oak occurs at either end of the geographic range of the community (Mississippi River valley and southeastern Ontario). The ground layer is often diverse (Table 8.1). Most of the characteristic ground-layer species (Will-Wolf and Stearns in press), like the tick-trefoils (*Desmodium glutinosum* and *D. nudiflorum*), the horse-gentians (*Triosteum aurantiacum* and *T. perfoliatum*), and kittentail (*Besseya bullii*: conservation listed in six states), occur in low density, whereas snowberry (*Symphoriocarpos* spp.) is common. The Wisconsin sites have many herbaceous species found on dry lime prairies (O. Anderson 1954); several rare species occur also in cliff communities and cedar glades (Will-Wolf and Stearns in press). Ontario dry calcareous savanna (Catling and Catling 1993) has 40% of the prevalent species of Wisconsin oak openings (Curtis 1959) and also shares many species with Great Lakes alvar communities (see Chapter 23).

Dry calcareous savanna is more likely than dry sand savanna to have classic savanna structure; the oak species are relatively fire tolerant, and the thin soil slows buildup of herbaceous biomass that could fuel intense fires.

Management and Research Needs

Dry calcareous savanna should be burned periodically to keep the ground layer relatively open and the canopy relatively thin (M. Anderson, Will-Wolf and Howell 1996). Isolated sites should be divided and burned at different times; remnants too small to divide into burn units should be mowed or cut until sites are larger. A base survey is needed to determine the distribution of this savanna type and to better characterize its composition.

Summary

Two distinct types of edaphically limited dry soil oak savanna occur in the Great Lakes region: dry calcareous savanna and dry sand savanna. Dry sand savanna occurs on acidic, low-nutrient soils over sandy substrates associated with glacial lake beds and outwash plains and, thus, has a wide distribution. Dominant tree species are mostly from the black oak group. Northern Sand Savanna, with Hill's oak, and Eastern Sand Savanna, with black oak, are distinguished based on species composition and geography. Dry calcareous savanna, dominated by white oak group tree species, occurs on thin soils over calcareous substrate; it is widely distributed, but uncommon and insufficiently documented.

Fire frequency and intensity as affected by landscape structure interact with local site environment to influence physiognomy of dry sand savanna, which varies from brushy and scrub oak savanna to savanna with scattered clumps of tall trees. In the absence of fire or other disturbance, vegetation on all but the most edaphically extreme sites loses much of its savanna physiognomy, becoming structurally more dense and changing in composition through succession. Large remnants of dry sand savanna are being managed to restore natural processes, composition, and structure.

Successful management of dry sand savanna must simultaneously maintain or restore savanna structure and populations of herbaceous plants and of animals such as sharp-tailed grouse and Karner blue butterfly. This approach has focused on patch dynamics and nonequilibrium ecosystem processes. Ground-layer plants specialize on combinations of light and moisture that vary with climate from year to year. Invertebrates are killed by fire, but need plants that thrive in burned patches, and they need patches of vegetation with different structure and composition within close proximity. Lessons learned from the management of large dry sand savanna sites can contribute much to the management and restoration of other kinds of savanna for which no large remnants exist.

Acknowledgments

Much of the research and inventory work on dry soil savanna has been published in workshop and symposium publications and in private circulation reports; the "gray literature" of Minnesota and Wisconsin has been better represented here than that for other areas. We thank the editors for much patience and constructive criticism, Virginia Kline for helpful comments, and John Wolf for editorial assistance. We thank Kristen Westad, who helped compile information for tables, and Kandis Elliot, who drew the figures.

REFERENCES

Anderson, M., Will-Wolf, S., and Howell, E. A. 1996. *Eagle Valley Nature Preserve. Part 1. Site Analysis and Master Plan;* Part 2. *Appendices;* Part 3. *Implementation and Management Plan.* Madison, Wis.: University of Wisconsin, Department of Landscape Architecture.

Anderson, O. 1954. The phytosociology of dry lime prairies of Wisconsin. Ph.D. thesis, University of Wisconsin, Madison, Wis.

Anderson, R. C., and Brown, L. E. 1983. Comparative effects of fire on trees in a midwestern savanna and an adjacent forest. *Bulletin of the Torrey Botanical Club* 110:87–90.

Anderson, R. C., and Brown, L. E. 1986. Stability and instability in plant communities following fire. *American Journal of Botany* 73:364–368.

Andow, D. A., Baker, R. J., and Lane, C. P. 1994. *Karner Blue Butterfly: A Symbol of a Vanishing Landscape.* St. Paul, Minn.: Minnesota Agricultural Experiment Station. Miscellaneous Publication Series 84.

Armstrong, P. K. 1994. Vegetation survey of a quartzite glade in Auk County, Wisconsin. In *Proceedings of the North American Conference on Barrens and Savannas,* ed. J. S. Fralish, R. C. Anderson, J. E. Ebinger, and R. Szafoni, pp. 355–363. Chicago, Ill.: U.S. Environmental Protection Agency, Great Lakes National Program Office.

Auclair, A. N. 1976. Ecological factors in the development of intensive-management ecosystems in the Midwestern United States. *Ecology* 57:431–444.

Beck, A. M., and Vogl, R. J. 1972. The effects of spring burning on rodent populations in a brush prairie savanna. *Journal of Mammalogy* 53:336–346.

Bleser, C. A. 1994. Karner blue butterfly survey, management, and monitoring activities in Wisconsin: 1990–Spring 1992. In *Karner Blue Butterfly: A Symbol of a Vanishing Landscape,* ed. D. A. Andow, R. J. Baker, and C. P. Lane. St. Paul, Minn.: Minnesota Agricultural Experiment Station. Miscellaneous Publication Series 84.

Bleser, C. A., and Leach, M. K. 1994. Protecting the Karner blue butterfly in Wisconsin: shifting focus from individuals to populations and processes. In *Proceedings of the North American Conference on Barrens and Savannas,* ed. J. S. Fralish, R. C. Anderson, J. E. Ebinger, and R. Szafoni, pp. 139–144. Chicago, Ill.: U.S. Environmental Protection Agency, Great Lakes National Program Office.

Borgerding, E. A., Bartelt, G. A., and McCown, W. M. 1995. *The Future of Pine Barrens in Northwest Wisconsin: A Workshop Summary.* Madison, Wis.: Wisconsin Department of Natural Resources, PUBL-RS-913–94.

Bormann, F. H., and Likens, G. E. 1979. *Pattern and Process in Forested Ecosystem.* New York: Springer-Verlag.

Botts, P, Haney, A., Holland, K., and Packard, S. 1994. *Midwest Oak Ecosystems Recovery Plan.* Chicago, Ill.: U.S. Environmental Protection Agency, Great Lakes National Program Office.

Bowles, M. L., and McBride, J. L. 1994. Presettlement barrens in the glaciated prairie region of Illinois. In *Proceedings of the North American Conference on Barrens and Savannas,* ed. J. S. Fralish, R. C. Anderson, J. E. Ebinger, and R. Szafoni, pp. 75–83. U.S. Environmental Protection Agency, Great Lakes National Program Office, Chicago, Ill.

Bray, J. R. 1958. The distribution of savanna species in relation to light intensity. *Canadian Journal of Botany* 36:671–681.

Bray, J. R. 1960. The composition of savanna vegetation in Wisconsin. *Ecology* 41:721–732.

Brewer, L. G. 1989. *Vegetation Changes in the Oak Savannas and Woodlands of Northwestern Ohio.* Final Report to Division of Natural Areas and Preserves, Ohio Department of Natural Resources, Columbus, Ohio.

Brewer, L. G., and Grigore, M. T. 1995. Restoring oak savanna in northwestern Ohio: monitoring the progress. In *Proceedings of the Midwest Oak Savanna Conference,* ed. F. Stearns and K. Holland. Chicago, Ill.: U.S. Environmental Protection Agency, Great Lakes National Program Office. Internet document. Address: http://www.epa.gov/glnpo/oak/

Brotherson, J. D., Rushforth, S. R., and Johansen J. R. 1983. Effects of long-term grazing on cryptogamic crust cover in Navajo National Monument, Arizona. *Journal of Range Management* 36:579–581.

Brush, T., and Stiles, E. W. 1990. Habitat use by breeding birds in the New Jersey (USA) Pine Barrens. *Bulletin of the New Jersey Academy of Science* 35:13–16.

Catling, P. M., and Catling, V. R. 1993. Floristic composition, phytogeography and relationships of prairies, savannas, and sand barrens along the Trent River, eastern Ontario. *Canadian Field-Naturalist* 107:24–45.

Chapman, K. A., White, M. A., Huffman, M. R., and Faber-Langendoen, D. 1995. Ecology and stewardship guidelines for oak barrens landscapes in the upper midwest. In *Proceedings of the Midwest Oak Savanna Conference*, ed. F. Stearns and K. Holland. U.S. Environmental Protection Agency, Great Lakes National Program Office, Chicago, Ill. Internet document. Address: http://www.epa.gov/glnpo/oak/

Clark, F., Isenring, B., and Mossman, M. 1993. *The Baraboo Hills Inventory Final Report*. Madison, Wis.: The Nature Conservancy, Wisconsin Chapter.

Cole, K. L., and Taylor, R. S. 1995. Past and current trends of change in a dune prairie/oak savanna reconstructed through a multiple-scale history. *Journal of Vegetation Science* 6:399–410.

Curtis, J. T. 1958. *Native Woody Plants of the Arboretum*. Arboretum Pamphlet. Madison, Wis.: University of Wisconsin.

Curtis, J. T. 1959. *The Vegetation of Wisconsin*. Madison, Wis.: University of Wisconsin Press.

Davis, M. A., Ritchie, B., Graf, N., and Gregg, K. 1995. An experimental study of the effects of shade, conspecific crowding, pocket gophers and surrounding vegetation on survivorship, growth and reproduction in *Penstemon grandiflorus*. *The American Midland Naturalist* 134:237–243.

Eckstein, R., and Moss, B. 1995. Oak and pine barren communities. In *Wisconsin's Biodiversity as a Management Issue*, ed. Wisconsin Department of Natural Resources, pp. 98–114. Madison, Wis.: Wisconsin Department of Natural Resources.

Faber-Langendoen, D., ed. 1995. The Nature Conservancy's natural community classification and its application to midwest oak savannas and woodlands. In *Midwest Oak Ecosystems Recovery Plan: A Call to Action*, ed. M. K. Leach and L. Ross, Appendix A, pp. 65–76. Chicago, Ill.: U.S. Environmental Protection Agency, Great Lakes National Program Office.

Faber-Langendoen, D., and Davis, M. A. 1995. Effects of fire frequency on tree canopy cover at Allison Savanna, eastcentral Minnesota, USA. *Natural Areas Journal* 15:319–328.

Faber-Langendoen, D., and Tester, J. 1993. Oak mortality in sand savannas following drought in eastcentral Minnesota. *Bulletin of the Torrey Botanical Club* 120:248–256.

Gilbert, M. L., and Curtis, J. T. 1953. Relation of the understory to the upland forest in the prairie–forest border region of Wisconsin. *Transactions of the Wisconsin Academy of Sciences, Arts and Letters* 42:183–195.

Gleason, H. A., and Cronquist, A. 1991. *Manual of Vascular Plants of Northeastern United States and Adjacent Canada*. 2nd ed. Bronx, N.Y.: The New York Botanical Garden.

Gordon, R. B. 1966. The natural vegetation of Ohio in pioneer days. *Bulletin of the Ohio Biological Survey*, Vol. III, No. 2. The Ohio State University, Columbus.

Gregg, L. 1987. *Recommendations for a Program of Sharptail Habitat Preservation in Wisconsin*. Madison, Wis.: Wisconsin Department of Natural Resources, Research Report 141.

Grigore, M. T., and Tramer, E. J. 1996. The short-term effect of fire on *Lupinus perennis* (L.). *Natural Areas Journal* 16:41–48.

Grimm, E. C. 1984. Fire and other factors controlling the Big Woods vegetation of Minnesota in the mid-nineteenth century. *Ecological Monographs* 54:291–311.

Grimm, E. C. 1985. Vegetation history along the prairie–forest border in Minnesota. In *Archaeology, Ecology, and Ethnohistory of the Prairie–Forest Border Zone of Minnesota and Manitoba*, ed. J. Spector and E. Johnson, pp. 9–29. Reprints in *Archaeology*, vol. 31. J & L Reprint Co., Lincoln, Neb.

Habeck, J. R. 1959. A phytosociological study of the upland forest communities in the central sand plain area. *Transactions of the Wisconsin Academy of Sciences, Arts, and Letters* 48:31–48.

Haney, A., and Apfelbaum, S. I. 1995. Characterization of midwestern oak savannas. In *Proceedings of the Midwest Oak Savanna Conference*, ed. F. Stearns and K. Holland. Chicago, Ill.: U.S. Environmental Protection

Agency, Great Lakes National Program Office. Internet document. Address: http://www.epa.gov/ glnpo/oak/

Hanson, H. C. 1922. Prairie inclusions in the deciduous forest climax. *American Journal of Botany* 9:330–337.

Heikens, A. L., and Robertson, P. A. 1994. Barrens of the midwest: a review of the literature. *Castanea* 59:184–194.

Henderson, N. R., and Long, J. N. 1984. A comparison of stand structure and fire history in two black oak woodlands in northwestern Indiana. *Botanical Gazette* 145:222–228.

Henderson, R. 1995. Oak savanna communities. In *Wisconsin's Biodiversity as a Management Issue*, ed. Wisconsin Department of Natural Resources, pp. 88–96. Madison, Wis.: Wisconsin Department of Natural Resources.

Hess, D. F., and Sedman, Y. 1994. Butterfly (*Rhopalocera*) fauna of the oak barrens and adjacent habitats of Lake Argyle State Park and McDonough County, Illinois. In *Proceedings of the North American Conference on Barrens and Savannas*, ed. J. S. Fralish, R. C. Anderson, J. E. Ebinger, and R. Szafoni, pp. 179–184. Chicago, Ill.: U.S. Environmental Protection Agency, Great Lakes National Program Office.

Homoya, M. A. 1994. Indiana barrens: classification and description. *Castanea* 59: 204–213.

Hutchinson, M. A. 1994. The barrens of the midwest: an historical perspective. *Castanea* 59:195–203.

Inouye, R. S., and Tilman, D. 1995. Convergence and divergence of old-field vegetation after 11 years of nitrogen addition. *Ecology* 76:1872–1887.

Johnson, K. C., and Ebinger, J. E. 1992. Effects of prescribed burns on the woody vegetation of a dry sand savanna, Hooper Branch Nature Preserve, Iroquois County, Illinois. *Transactions of the Illinois State Academy of Science* 85:105–111.

Kline, V. M., and McClintock, T. 1994. Changes in a dry oak forest after a third prescribed burn. In *Proceedings of the North American Conference on Barrens and Savannas*, ed. J. S. Fralish, R. C. Anderson, J. E. Ebinger, and R. Szafoni, pp. 279–284. Chicago, Ill.: U.S. Environmental Protection Agency, Great Lakes National Program Office.

Kline, V. M., and McClintock, T. 1995. The ground layer of an oak forest in transition under prescribed burning. In *Proceedings of the Midwest Oak Savanna Conference*, ed. F. Stearns

and K. Holland. U.S. Environmental Protection Agency, Great Lakes National Program Office, Chicago, Ill. Internet document. Address: http://www.epa.gov/glnpo/oak/

Ko, L. J., and Reich, P. B. 1993. Oak tree effects on soil and herbaceous vegetation in savannas and pastures in Wisconsin. *The American Midland Naturalist* 130:31–42.

Kotar, J., Kovach, J. A., and Locey, C. T. 1988. *Field Guide to Forest Habitat Types of Northern Wisconsin*. Madison, Wis.: Department of Forestry, University of Wisconsin, and Wisconsin Department of Natural Resources.

Kubisiak, J. F. 1985. *Ruffed Grouse Habitat Relationships in Aspen and Oak Forests of Central Wisconsin*. Madison, Wis.: Department of Natural Resources, Technical Bulletin No. 151.

Lange, K. L. 1989. *Ancient Rocks and Vanished Glaciers: A Natural History of Devil's Lake State Park, Wisconsin*. Stevens Point, Wis.: Worzalla Publishing Company.

Leach, M. K. 1993. *Status and Distribution of the Karner Blue Butterfly at Fort McCoy, Wisconsin: Final Report of a Two Year Study*. Unpublished report to Natural Resources Management Division, Fort McCoy Military Reservation, Tomah, Wis., and The Nature Conservancy, Wisconsin Chapter, Madison, Wis.

Leach, M. K. 1996. Gradients in groundlayer composition, structure and diversity in remnant and experimentally restored oak savannas. Ph.D. dissertation, University of Wisconsin, Madison, Wis.

Leach, M. K., and Givnish, T. J. 1996. Ecological determinants of species loss in remnant prairies. *Science* 273:1555–1558.

Leach, M. K., and Ross, L., eds. 1995. *Midwest Oak Ecosystems Recovery Plan: A Call to Action*. U.S. Environmental Protection Agency, Great Lakes National Program Office, Chicago, Ill.

Leitner, L. A., Dunn, C. P., Guntenspergen, G. R., Stearns, F., and Sharpe, D. M. 1991. Effects of site, landscape features, and fire regime on vegetation pattern in presettlement southern Wisconsin. *Landscape Ecology* 5:203–217.

Leopold, A. 1949. *A Sand County Almanac*. New York: Oxford University Press, Inc.

Levy, G. F. 1970. Phytosociology of northern Wisconsin upland openings. *The American Midland Naturalist* 83:213–237.

Madany, M. H. 1981. A floristic survey of savannas in Illinois. In *Proceedings of the Sixth North American Prairie Conference*, ed. R. L. Stuckey and K. J. Reese, pp. 177–181. Ohio Biological Survey. Biological Note No. 15.

Maxwell, J. A., and Givnish, T. J. 1996. *Research on the Karner Blue Butterfly at Fort McCoy, Wisconsin: Progress Report for the 1995 Field Season*. Report submitted to the U.S. Fish and Wildlife Service and the Department of the Army, Fort McCoy Military Reservation, Tomah, Wis.

McNab, W. H., and Avers, P. E.,eds. 1994. *Ecological Subregions of the United States: Section Descriptions*. Washington, D.C.: USDA, Forest Service, Publication WO-WSA-5.

Morimoto, D. C., and Wasserman, F. E. 1991. Dispersion patterns and habitat associations of rufous-sided towhees, common yellowthroats, and prairie warblers in the southeastern Massachusetts (USA) pine barrens. *The Auk* 108:264–276.

Mossman, M. J., Epstein, E., and Hoffman, R. M. 1991. Birds of Wisconsin pine and oak barrens. *Passenger Pigeon* 53:137–163.

Nicolai, V. 1991. Reactions of the fauna on the bark of trees to the frequency of fires in a North American savanna. *Oecologia* 88:132–137.

Nuzzo, V. A. 1986. Extent and status of midwest oak savanna: presettlement and 1985. *Natural Areas Journal* 6:6–36.

Ohio Biological Survey. 1966. *Natural Vegetation of Ohio at the Time of the Earliest Land Surveys* (map). Adapted from Gordon 1966 by Ohio Department of Natural Resources, Columbus, Ohio.

Panzer, R. 1988. Managing prairie remnants for insect conservation. *Natural Areas Journal* 8:83–90.

Panzer, R., Stillwaugh, D., Gnaedinger, R., and Derkovitz, G. 1995. Prevalence of remnant dependence among the prairie- and savanna-inhabiting insects of the Chicago region. *Natural Areas Journal* 15:101–116.

Pruka, B. 1994. Distribution of understory plant species along light and soil depth gradients in an upland oak savanna in southern Wisconsin. In *Proceedings of the North American Conference on Barrens and Savannas*, ed. J. S. Fralish, R. C. Anderson, J. E. Ebinger, and R. Szafoni, pp. 213–216. Chicago, Ill.: U.S. Environmental Protection Agency, Great Lakes National Program Office.

Pruka, B., and Faber-Langendoen, D. 1995. Sample conservation and recovery plans for three midwestern oak savanna and woodland types. In *Midwest Oak Ecosystems Recovery Plan: A Call to Action*, ed. M. K. Leach and L. Ross, Appendix B, pp. 77–97. Chicago, Ill.: U.S. Environmental Protection Agency, Great Lakes National Program Office.

Rabe, M. L., Comer, P. J., and Albert, D. 1995. Enhancing habitat for the Karner blue butterfly: restoration of the oak–pine barrens in southwest Michigan. In *Proceedings of the Midwest Oak Savanna Conference*, ed. F. Stearns and K. Holland. U.S. Environmental Protection Agency, Great Lakes National Program Office, Chicago, Ill. Internet document. Address: http://www.epa.gov/glnpo/oak/

Ritchie, M. E., and Tilman, D. 1995 Responses of legumes to herbivores and nutrients during succession on a nitrogen-poor soil. *Ecology* 76:2648–2655.

Roberts, D. W. 1987. A dynamical system perspective on vegetation theory. *Vegetatio* 69:27–33.

Rodgers, C. S., and Anderson, R. C. 1979. Presettlement vegetation of two prairie peninsula counties. *Botanical Gazette* 140:232–240.

Sample, D. W., and Mossman, M. J. 1994. Birds of Wisconsin oak savannas: past, present, and future. In *Proceedings of the North American Conference on Barrens and Savannas*, ed. J. S. Fralish, R. C. Anderson, J. E. Ebinger, and R. Szafoni, pp. 155–159. Chicago, Ill.: U.S. Environmental Protection Agency, Great Lakes National Program Office.

Schorger, A. W. 1930. The range of the bison in Wisconsin. *Transactions of the Wisconsin Academy of Sciences* 30:122–132.

Schorger, A. W. 1954. The elk in early Wisconsin. *Transactions of the Wisconsin Academy of Sciences, Arts, and Letters* 43:5–23.

Schulten, J. A. 1985. The effects of burning on the soil lichen community of a sand prairie. *The Bryologist* 88:110–114.

Shuey, J. A. 1994. Dancing with fire: oak barrens/savanna patch dynamics, management, and the Karner blue butterfly. In *Proceedings of the North American Conference on Barrens and Savannas*, ed. J. S. Fralish, R. C. Anderson, J. E. Ebinger, and R. Szafoni, pp. 185–189. Chicago, Ill.: U.S. Environmental Protection Agency, Great Lakes National Program Office.

Shuey, J. A 1995. Dancing with fire: ecosystem dynamics, management, and the Karner blue (*Lycaeides melissa samuelis* Nabokov). In *Proceedings of the Midwest Oak Savanna Conference*, ed. F. Stearns and K. Holland. U.S. Environmental

Protection Agency, Great Lakes National Program Office, Chicago, Ill. Internet document. Address: http://www.epa.gov/ glnpo/oak/

Sweet, E. T. 1880. Geology of the western Lake Superior district; climate, soils and timber. *Geology of Wisconsin* 3:323–329.

Swengel, A. B. 1993. *Observations of Karner Blues and the Barrens Butterfly Community in Wisconsin 1987–1993*. Report to National Biological Survey and U.S. Fish and Wildlife Service, Baraboo, Wis.

Swengel, A. B. 1994. Conservation of the prairie–savanna butterfly community. In *Proceedings of the North American Conference on Barrens and Savannas*, ed. J. S. Fralish, R. C. Anderson, J. E. Ebinger, and R. Szafoni, pp. 133–138. Chicago, Ill.: U.S. Environmental Protection Agency, Great Lakes National Program Office.

Swengel, A. B. 1995. Effects of fire and hay management on abundance of prairie butterflies. *Biological Conservation* 76:73–85.

Szeicz, J. M., and MacDonald, G. M. 1991. Postglacial vegetation history of oak savanna in southern Ontario (Canada). *Canadian Journal of Botany* 69:1507–1519.

Tester, J. R. 1989 Effects of fire frequency on oak savanna in eastcentral Minnesota. *Bulletin of the Torrey Botanical Club* 116:134–144.

Tilman, D., Wedin, D., and Knops, J. 1996. Productivity and sustainability influenced by biodiversity in grassland ecosystems. *Nature* 379:718–720.

Udvig, T. T. 1986. Acid sensitive soils: the effects of red pine (*Pinus resinosa*) on converted oak sites. M.S. thesis, Southern Illinois University, Carbondale, Ill.

United States Department of Agriculture. 1974. *The Original Vegetation of Minnesota* (map redrafted from 1930 F. J. Marschner original). St. Paul, Minn.: North Central Forest Experiment Station, Forest Service

Veatch, J. O. 1959. *Presettlement Forest in Michigan* (map). East Lansing, Mich.: Michigan State University, Department of Resource Development.

Vestal, A. 1936. Barrens vegetation in Illinois. *Transactions of the Illinois State Academy of Science* 11:122–126.

Vogl, R. J. 1964. Vegetational history of Crex Meadows, a prairie savanna in northwestern Wisconsin. *The American Midland Naturalist* 72:157–175.

Vogl, R. J. 1965. Effects of spring burning on yields of brush prairie savanna. *Journal of Range Management* 18:202–205.

Vogl, R. J. 1970. Fire and the northern Wisconsin pine barrens. *Proceedings of the Annual Tall Timbers Fire Ecology Conference* 10:175–209.

Vogt, R. C. 1981. *Natural History of Amphibians and Reptiles in Wisconsin*. Milwaukee Public Museum, Milwaukee, Wis.

Watt, A. S. 1947. Pattern and process in the plant community. *Journal of Ecology* 35:1–22.

Wetmore, C. M. 1983. Lichen survival in a burned oak savanna. *Michigan Botanist* 22:47–52.

White, A. S. 1983. The effects of thirteen years of annual prescribed burning on a *Quercus ellipsoidalis* community in Minnesota. *Ecology* 64:1081–1085.

White, J. 1994. How the terms *savanna, barrens,* and *oak openings* were used in early Illinois. In *Proceedings of the North American Conference on Barrens and Savannas*, ed. J. S. Fralish, R. C. Anderson, J. E. Ebinger, and R. Szafoni, pp. 65–74. Chicago, Ill.: U.S. Environmental Protection Agency, Great Lakes National Program Office.

Whitford, P. B., and Whitford, K. 1971. Savannas in central Wisconsin, U.S.A. *Vegetatio* 23:77–87.

Whitney, G. G. 1986. The relation of Michigan's presettlement pine forests to substrate and disturbance history. *Ecology* 67:1548–1559.

Whitney, G. G. 1987. An ecological history of the Great Lakes Forest of Michigan. *Journal of Ecology* 75:667–684.

Will-Wolf, S. 1980. Structure of corticolous lichen communities before and after exposure to emissions from a "clean" coal-fired power generating station. *The Bryologist* 83:281–295.

Will-Wolf, S., and Montague, T. G. 1994. Landscape and environmental constraints on the distribution of presettlement savannas and prairies in southern Wisconsin. In *Proceedings of the North American Conference on Barrens and Savannas*, ed. J. S. Fralish, R. C. Anderson, J. E. Ebinger, and R. Szafoni, pp. 97–102. Chicago, Ill.: U.S. Environmental Protection Agency, Great Lakes National Program Office.

Will-Wolf, S. and Stearns, F. In press. Characterization of dry oak savanna in the upper Midwest. In *Proceedings of the Midwest Oak Savanna and Woodland Conference: Opportunities for Community*, ed. M. Boyce. Madison, Wis.: Wisconsin Academy of Sciences, Arts and Letters.

9 Deep-Soil Savannas and Barrens of the Midwestern United States

ROGER C. ANDERSON AND MARLIN L. BOWLES

Introduction

Eastern Prairie–Forest Transition

Midwestern savannas occupied a transitional area between eastern deciduous forest and tallgrass prairie. These savannas were part of the eastern prairie–forest transition (Curtis 1959; Anderson 1983; Nuzzo 1986) that extended as a broad arc along the eastern edge of the northern mixed and tallgrass prairies from the Canadian provinces of Alberta, Saskatchewan, and Manitoba southward into Texas (Figure 9.1). We discuss deep-soil savannas (also called black soil, mesic, and tallgrass savannas and barrens) that occurred in the glaciated landscapes of Minnesota, southern Michigan and Wisconsin, Ohio, Indiana, and Illinois. These deep-soil savannas occurred on sites with fine-textured soils, where growth of trees was not severely limited by edaphic factors. Savannas with shallow soil profiles over bedrock, and those with sandy soil with low fertility and water-holding capacity, are considered in other chapters (see Chapters 8 and 21). Essentially all of the original Midwest, mesic, deep-soil savanna vegetation was lost to fire protection and agricultural activities, including overgrazing (Curtis 1959; Nuzzo 1986). These savannas are among the rarest natural vegetation types in the world.

Savannas of the Midwest occupied the eastern edge of a large, triangular-shaped grassland that extended from the Rocky Mountains into the Midwest (Risser et al. 1981; Anderson 1990). The grassland narrowed eastward, producing the well-known prairie peninsula (Transeau 1935). Across the long west–east axis of this grassland, the climate becomes progressively less suitable for growth of grass and more favorable for trees, as periods of periodic drought and low humidity during summer decrease and annual precipitation and its reliability increase (Borchert 1950; Risser et al. 1981). During major drought years, moisture stress can be pronounced in the eastern part of this peninsula (Borchert 1950). Low winter precipitation can result in incomplete recharge of subsoil moisture, which is used by deep-rooted trees in soils without drainage restriction. Grasses generally have shallower roots than trees and do not rely on this source of moisture. Thus, summer droughts preceded by winters without deep-soil moisture recharge are more detrimental to trees than to grasses. For example, during the drought of 1933–34, the region of incomplete recharge of deep-soil moisture in the Midwest generally corresponded to the prairie peninsula (Britton and Messenger 1969).

Consequently, at its eastern end, the grassland became increasingly fragmented and interspersed with forest and savanna, forming a broad transition zone to the deciduous forest of the east and the conifer forest in the north. Interaction of climate, topography, and fire produced a mosaic of vegetation types, including prairie, forest, and savanna, that varied depending upon the relative importance of these interacting factors (Anderson 1991; Robertson, Anderson and Schwartz 1997).

Roger C. Anderson, James S. Fralish, and Jerry M. Baskin, eds. *Savannas, Barrens, and Rock Outcrop Plant Communities of North America.* Copyright © 1999 Cambridge University Press. All rights reserved.

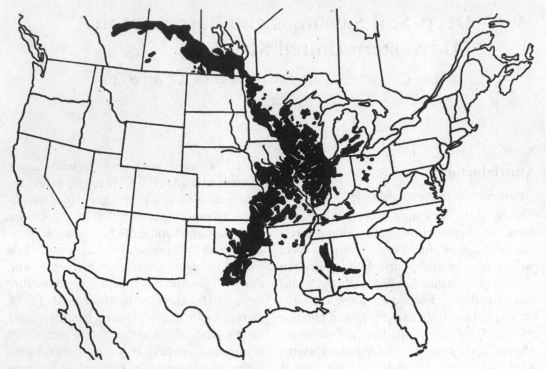

Figure 9.1. The distribution of the eastern prairie–forest transition (adapted from Anderson 1982 and Nuzzo 1986).

Defining the Savanna Community

As a result of varied tree canopy structure and density, the savanna vegetation matrix is transitional to forest and prairie (Curtis 1959; Anderson 1990; Leach 1996; Taft 1997). Fire regimes structure the savanna canopy (Faber-Langendoen and Davis 1995), which in turn creates a solar irradiance gradient and a corresponding distribution of ground-layer species (Gilbert and Curtis 1953; Bray 1958; Curtis 1959; Leach 1994; Pruka 1994a, b; Leach 1996; Bowles and McBride 1998). Shaded areas in savanna experience a shifting mosaic of light and shade to which the shade-tolerant species apparently are adapted. Areas beneath tree crowns receive high levels of solar irradiance before leaves develop on the tree canopy. This irradiance allows spring-flowering species, such as shooting star (*Dodecatheon meadia*) and wild hyacinth (*Camassia scilloides*), which are also found in prairies, to complete their relatively short aboveground growth cycles before a marked reduction in solar irradiance occurs (Gustafson and Anderson 1994; Gustafson 1996). Variation in soil depth and moisture, temporal change in fire regime and canopy structure, plant competition, and plant–animal interactions occur across the light gradient. The individualistic ecological responses of plant species across these multiple gradients produce a complex continuum of savanna vegetation (Bray 1958; Skarpe 1992; Belsky and Canham 1994; Leach 1994, 1996; Pruka 1994a; Packard and Mutel 1997).

The transitional vegetation types containing elements of prairies and forests have been called a variety of names, but most often, perhaps, they have been referred to as barrens, oak openings, or savannas. Heikens and Robertson (1994) note that the term *barrens* has generated considerable confusion in the botanical literature. *Barrens* has been used

Figure 9.2. Artist's conception of the prairie–savanna–woodland/forest continuum that characterized portions of the Midwestern U.S.A. prior to settlement by Europeans. (From Packard and Mutel 1997.)

synonymously with prairie, savanna, oak opening, glade, scrub prairie, brush prairie, grassland, and others. Similarly, the term *savanna* has been defined differently by various investigators, and use of it has changed historically. J. White (1994) indicated that in Illinois in the 1700s and 1800s, the word *savanna* referred to grassland with few or no trees and was used as a substitute for *prairie* by emigrants from Great Britain or New England.

Elsewhere, savanna has been defined as a vegetation type having a tree canopy cover of 10%–50% in Missouri (Nelson 1985), to complete canopy closure in Ohio (Nuzzo 1986) (see Figure I.1, Introduction, this volume). These varied definitions likely result from dynamic relationships occurring among forest, savanna, and prairie, and reflect the difficulty of separating a vegetation continuum into discrete classification units. There is also regional variation in vernacular usage of terms that describe vegetation. Use of these different definitions of savanna creates difficulties when comparisons are being made between studies.

We adopt Curtis's (1959) conceptual definition of savanna (i.e., as being part of a vegetation continuum between forest and prairie, Figure 9.2) as well as his classification for savannas. He arbitrarily defined savanna as a vegetation type having more than one mature tree per acre (2.5 trees ha^{-1}), but less than 50% tree canopy cover. Curtis divided the savanna into oak openings and barrens. Oak openings were characterized by scattered open-grown, broad crown oaks (usually bur oak [*Quercus macrocarpa*])

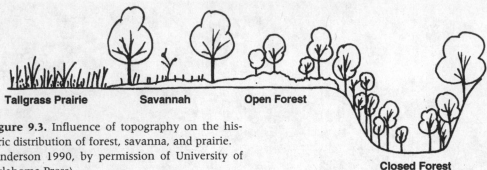

Figure 9.3. Influence of topography on the historic distribution of forest, savanna, and prairie. (Anderson 1990, by permission of University of Oklahoma Press)

with a mixed understory of prairie species and species adapted to shady woodland habitats. Barrens included brush prairies, sand barrens (similar to sand prairies in terms of herbaceous species composition), and the Hill's oak (*Quercus ellipsoidalis*) and jack pine (*Pinus banksiana*) barrens that occupied sandy soil in the central and northern portions of Wisconsin. Barrens on deep, fine-textured soils tended to have more shrubs and coppices of trees, which were maintained by reoccurring fires, than did oak openings. Nomenclature follows Mohlenbrock (1986).

Origin of Midwest Savanna

Several workers described changes in vegetation during the Holocene based on analysis of fossil pollen deposits (King 1981; Winkler, Swain and Kutzbach 1986; Baker et al. 1992; Griffin 1994). Pollen deposits indicate that mesic forests occupied much of the Midwest following Wisconsinan glaciation (9,000–10,000 yr BP). Many species of oaks probably were restricted to sites with low availability of soil nutrients and water and that were not occupied by mesophytic species.

Midwest savannas originated during the relatively warm and dry postglacial period that peaked about 5,500–8,000 yr BP (McAndrews 1966; King 1981; Winkler, Swain and Kutzbach 1986; Baker et al. 1992; Griffin 1994). However, there was regional variation within the Midwest regarding when

oak savannas became established during the Holocene (King 1981; Baker et al. 1996; Winkler 1997). In Illinois, the drying trend during the Holocene began about 8,700–7,900 yr BP, and prairie began to replace deciduous forest in southern Illinois. Prairie influx into central Illinois occurred a few hundred years later (about 8,300 yr BP) and oak forests began to displace mesic forests in northern Illinois at the same time. During the hottest and driest part of the Holocene in Illinois (8,000–6,000 yr BP), prairies occupied most of the state (King 1981). Oak openings appeared in northern Illinois and southern Wisconsin and Minnesota about 5,500 yr BP (McAndrews 1966; Griffin 1994). In Illinois, the climate became cool and moist about 5,000–3,500 yr BP, and the area of prairie decreased (Delcourt and Delcourt 1981; King 1981).

More recently, however, Baker et al. (1996) reported that in northeastern Iowa, forest dominated from about 8,000 to 5,100 yr BP. After 5,100 yr BP, forest was replaced by prairie, probably resulting from an increased flow of arid Pacific air and from fire, which these authors do not attribute to aboriginal burning. Oak savanna appeared about 3,000 yr BP in northeast Iowa, perhaps indicating the climate was becoming cooler and more moist, and less favorable for prairie than it had been previously. Similarly, Winkler (1997) found that in southcentral Wisconsin a Holocene peak in charcoal in lake and bog sediments (6,500–3,500 yr BP) indicated that the highest frequency of regional fires occurred during that portion of the Holocene. The peak of charcoal in sediments was associated with a fire-adapted oak savanna landscape (Winkler 1994, 1997). After 3,500 yr BP, the vegetation in southcentral Wisconsin was dominated by closed oak forest. This change in vegetation is suggested by a decrease in charcoal found in sediments and a decline in charcoal stable carbon isotope data (δ^{13}C below -26 per mil), indicating domination by C3 plants, which is typical of arboreal landscapes (Winkler 1997).

Despite some regional variation in the timing of vegetational change during the Holocene, maintenance of vegetation patterns immediately prior to European settlement is attributed to aboriginal burning of the landscape under a climatic regime that could support forest, prairie, or savanna (Curtis 1959; Grimm 1984; Anderson 1990). Occurrence of the three vegetation types on the landscape was a function of fire frequency, which was largely controlled by topography. Level to gently rolling landscapes had frequent fires and supported tallgrass prairies. In dissected landscapes, spread of fire was reduced, permitting establishment of trees. Closed forests were associated with areas sheltered from fires, such as ravines or along waterways that served as firebreaks. These sheltered locations supported shade-tolerant, fire-sensitive, mesophytic forests of sugar maple (*Acer saccharum*), basswood (*Tilia americana*), and, in the eastern portion of the prairie–forest transition, beech (*Fagus grandifolia*). Woodlands and savannas dominated by fire-tolerant oaks occurred in areas where fires occurred less frequently than in prairies, but with shorter return time than in closed mesophytic forests (Figure 9.3) (Gleason 1922; Curtis 1959; Anderson and Anderson 1975; Rodgers and Anderson 1979; Grimm 1984; Anderson 1990, 1991; Abrams 1992). Curtis (1959) proposed that during the Holocene, sites supporting mesic forest species were converted to prairie because the combination of drought and fires eliminated these species from all but the most sheltered sites. However, fires did not eliminate oaks because of the fire-resistant bark in some species, such as bur oak, and the ability to resprout after being top-killed, an adaptive feature in all oaks (Stearns 1991; Abrams 1992). Mesic forests experienced occasional fires with fire return times as long as centuries. In these forests, the shade-intolerant red oak was a pioneer species. Even with long fire intervals (centuries), the intolerant red oak apparently persisted on mesic sites, because of its long life span. However, shade-tolerant mesophytes could dominate smaller-tree size classes on mesic sites for most of the time between fires (Curtis 1959; Adams and Anderson 1980; Abrams 1992; Will-Wolf and Roberts 1993).

Rodgers and Anderson (1979) reported that in central Illinois, prairies were associated positively with Mollisols, but that forests and savannas tended to be negatively associated with these soils. Generally, savanna and forest tended to have the same associations with soils, and both were positively associated with Alfisols. These results suggest that savannas developed as a result of forest degradation. Nevertheless, in the Big Woods area of southeastern Minnesota, some presettlement savannas may have originated by trees invading prairie (Grimm 1984). Occurrence of aspen (*Populus tremuloides* and some *P. grandidentata*) and, to some extent, bur oak on prairie soils resulted from trees of these species invading prairie around the periphery of forested areas and forming savannas (Grimm 1984).

Types of Deep-Soil Savannas: Oak Openings and Barrens

Oak Openings

Oak openings occurred on fine-textured soil and, as described by European settlers in the 1800s, had an orchardlike appearance with scattered broad crown white and bur oaks with an understory of prairie species (Curtis 1959; J. White 1994). Soils of oak openings were considered to be fertile and suitable for the growth of crops (J. White 1994). Latrobe (1835) described the oak opening of Michigan and Illinois as having rich vegetable soils. Similarly, Little (1861) described the soil of the oak opening in Illinois as being "better adapted to production of fruit than our prairie soils." Shepherd (1850) described soils of the oak openings as being the best wheat lands.

In contrast to dominance by coppices of shrubs and oak grubs in barrens, as we describe later, Curtis (1959) indicated that the thick bark of bur oak helped it to withstand prairie fires and to colonize prairie. This allowed development of classic savanna structure associated with oak openings on mesic sites: scattered, large, open-grown trees with a prairie understory (i.e., with an orchard- or parklike appearance). Nearly annual fires that were set by native Americans and by occasional lightning strikes (Curtis 1959; Vogl 1977) maintained this savanna, which more closely resembled a grassland with trees than barrens. Drought-driven summer fires also may have been important in structuring this vegetation (Anderson 1982).

Oak openings had fewer shrubs than barrens, or in some cases oak openings appeared to lack woody understory, enhancing their parklike appearance. Nevertheless, there was a rapid woody succession of oak openings to closed oak forests following fire protection associated with European settlement (Cottam 1949; McCune and Cottam 1985; Bowles and McBride 1994; J. White 1994). Thus few, if any, intact savannas survive on deep silt loam soil in the Midwest (Curtis 1959; Madany 1981; Apfelbaum and Haney 1991; Packard 1991; Bowles and McBride 1996). Because this vegetation type disappeared soon after settlement, the original species composition and structure of savannas are poorly described. Burning often is presumed to have maintained a graminoid fuel matrix in savanna (Curtis 1959; Apfelbaum and Haney 1991; Packard 1991, 1993). However, oak sprouts and shrubs such as American hazel (*Corylus americana*) were frequent in oak openings, demonstrating a strong similarity to brush prairie (Cottam 1949; McAndrews 1966; Bowles, Hutchison and McBride 1994).

In Wisconsin, Bray (1960) found mostly grazed mesic savannas, formerly the most widespread savanna type. However, Bray sampled 103 species across 59 stands and established the distribution of herbaceous species and trees across a moisture gradient. The 103 species were distributed almost equally in their occurrence in prairie and forest, demonstrating that oak openings were floristically intermediate between the two vegetation types.

Bray (1958) also identified a strong relationship between variation in species composition and canopy gaps that was controlled by irradiance (Table 9.1). With a 10%–50% canopy cover definition of savanna, prairie herbaceous species are an important component of the savanna vegetation matrix, with an admixture of shade-tolerant savanna species found in woodlands with intermediate irradiance levels (Bray 1958). The eight silt loam savannas studied by Betz (1992) also reflect a strong floristic affinity to tallgrass prairies (Table 9.2), and he considered them to be most closely allied to oak openings. The transitional soils on these eight sites had A horizons shallower than in prairie soils. Betz (1992) suggests that the sites may have originated as the result of trees invading prairie, although transitional soils could result from prairie invading forest that was degraded by repeated fires.

Deep-Soil Barren/Brush Prairies

In the prairie-dominated, glaciated landscapes of central and northern Illinois, early settlers and surveyors used the term *barrens* to describe habitats perceived as being unproductive because they lacked canopy trees and their vegetation was intermediate between

Table 9.1. *Relation of selected common oak opening species to illuminance (after Bray 1955) from Curtis (1959). Values are average quadrat frequency in quadrats occurring within the indicated range of illuminance in lux.*

	Average frequency in quadrats with indicated illuminance		
	0–10,760 lux.	10,760–100,760 lux	100,760+ lux
Light species			
Amorpha canescens	28	61	77
Helianthus laetiflorus	24	38	77
Stipa spartea	21	37	63
Andropogon gerardii	21	35	61
Schizachyrium scoparium	8	13	56
Panicum leiburgii	16	35	55
Aster sericeus	2	20	47
Bouteloua curtipendula	9	18	47
Petalostemum purpureum	3	5	38
Aster ericoides	19	22	29
Intermediate species			
Euphorbia corollata	37	60	49
Galium boreale	43	55	23
Monarda fistulosa	37	49	22
Coreopsis palmata	37	47	45
Solidago ulmifolia	40	43	7
Aster azureus	37	41	4
Commandra richardsiana	17	41	34
Anemone cylindrica	28	31	7
Rosa sp.	27	30	23
Smilacina stellata	21	25	17
Shade species			
Amphicarpaea bracteata	61	50	15
Cornus racemosa	59	38	7
Helianthus strumosus	53	47	22
Aster laevis	40	29	0
Corylus americana	41	26	38
Aralia nudicaulis	36	17	0
Carex pensylvanica	33	30	5
Ratibida pinnata	29	24	4
Geranium maculatum	23	6	2
Apocynum androsaemifolium	19	14	7

grassland and forest (Bowles and McBride 1994). According to J. White (1994), the single unifying characteristic of barrens was the presence of fire-maintained brush and tree sprouts. Barrens often had tallgrass, but were characterized by underwood or brushwood of fire-resistant shrubs or oak grubs, rather than the open-grown trees of the oak openings (Bowles and McBride 1994; J. White 1994).

Gleason (1922) indicated that barrens in Illinois represented a late stage of fire-caused forest degradation. They were characterized by 1.2–1.5-m-high sprouts of scrub oak, apparently shingle oak (*Quercus imbricaria*),

Table 9.2. *Species occurring in >50% of the eight silt loam pioneer cemetery savannas studied by Betz (1992) in northern Illinois and northwestern Indiana.*

Species	Percentage presence
Amorpha canescens	50
Andropogon gerardii	88
Antennaria neglecta	63
Antennaria plantaginifolia	63
Asclepias verticillata	50
Aster ericoides	50
Baptisia leucantha	50
Carex bicknelli	63
Ceanothus americanus	63
Commandra richardsiana	75
Desmodium canadense	50
Desmodium illinoense	50
Dodecatheon meadia	50
Erigeron strigosus	75
Euphorbia corollata	88
Fragaria virginiana	50
Helianthus grosseserratus	50
Helianthus laetiflorus	50
Heuchera richardsonii	63
Hypoxis hirsuta	50
Lespedeza capitata	75
Lithospermum canescens	59
Lobelia spicata	75
Monarda fistulosa	100
Oxalis violacea	75
Panicum implicatum	50
Parthenium integrifolium	50
Pycnanthemum virginianum	75
Ratibida pinnata	50
Rosa carolina	88
Rudbeckia hirta	75
Salix humilis	50
Schizachyrium scoparium	75
Silphium integrifolium	88
Silphium terbinthinaceum	63
Sisyrinchium albidum	75
Solidago juncea	50
Solidago rigida	75
Sorghastrum nutans	100
Stipa spartea	50
Tradescantia ohiensis	63
Viola papilionacea	63
Viola pedatifida	63

black oak (*Q. velutina*), and Hill's oak, and American hazel and wild plum (*Prunus americana*). Similar vegetation in Wisconsin was called "brush prairie" by Curtis (1959). Grimm (1984) also indicated that part of the savanna habitat he studied in eastern Minnesota "was scrub, a dense thicket of fire-stunted oak and brush." According to Curtis, from a distance, brush prairies were indistinguishable from tallgrass prairies because nearly annual fires reduced trees to "brush" or "grubs." Oak grubs formed when nearly annual prairie fires killed the tops of young oaks but shoots would sprout from the persistent root system the next year. Over a period of years, possibly as long as centuries, a massive root system developed, often with a large, surface root plate 0.6–0.9 m in diameter. The term *grub* is from the German *gruben*, to dig, in reference to the method used by European settlers to laboriously remove these massive root systems from their agricultural fields (Curtis 1959). The shrub New Jersey tea or red root (*Ceanothus americanus*) also produced a large underground root burl as a result of repeated top-killing and sprouting.

The vegetation structure and ecological features of former barrens are poorly known. Barrens were most common on uneven or rolling topography or along stream drainages (e.g., Peck 1834), which reduced fire effects. They also developed on the west sides of forests penetrated by eastward-moving prairie fires driven by prevailing westerly winds (Gleason 1913). In addition to landscape fire effects, the thin bark of black, shingle, and Hill's oak probably facilitated top-killing by fire and resprouting, which contributed to formation of barrens. Because barrens lacked canopy trees, their spatial vegetational patterns may not have been related to a light gradient. However, the clonal structure of shrubs such as hazel may have resulted in a mosaic of low, woody vegetation with few grasses, and openings dominated by grasses and prairie forbs. Without fire, these communities were unstable. They probably disappeared more rapidly than oak openings, but left no old open-grown trees as markers of their former presence. According to Gleason (1922), large areas of barrens were converted into forest "as by magic" when the anthropogenic fires that maintained them were stopped and the oak sprouts became trees. As a result, even-aged forest stands with hazel understories may have been derived from barrens. Forests of this kind, however, often have trees with multiple stems that originated from basal root plates.

The floristic composition of former barrens can be reconstructed, to some extent, from historic descriptions. Bowles and McBride (1994) indicate that more than 30 shrub species may have characterized barrens, including hazel, New Jersey tea, dogwoods (Drummonds dogwood [*Cornus drummondii*], pale dogwood [*C. obliqua*], gray dogwood [*C. racemosa*], and red osier dogwood [*C. stolonifera*]), wild crab (*Malus coronaria* and *M. ioensis*), wild plum, sumac (*Rhus* spp.), rose (*Rosa* spp.), prairie willow (*Salix humilis*), and prickly ash (*Zanthoxylum americanum*). Barrens that formed along the western flanks of forests were dominated by hazel, with black oak and shingle oak forming the interior of the forest margin (Gleason 1913).

Bowles and McBride (1994) prepared a composite barrens flora using four historic annotated species lists compiled by Mead (1846), Higley and Raddin (1891), Brendel (1887), and Hus (1908) from areas that included barrens vegetation at the time of the Government Land Surveys. The list of 247 taxa includes 41 woody, 191 herbaceous, and 15 graminoid species thought to occur in deep-soil barrens. The most remarkable of the annotated lists included 108 "barrens" species published by S. B. Mead, a medical doctor, as part of a larger annotated list of Illinois plants from Augusta (Hancock

County), Illinois (Mead 1846). Although Packard (1988) suggested that the species list represents tallgrass savanna, it has little similarity to extant silt loam savannas. For example, only 18 (23%) of the 77 nonwoody species on Mead's 1846 list occur in the silt loam cemetery savannas (Betz 1992).

There is no way of knowing exactly the kinds of habitats from which historic species listed for barrens were derived. However, the historic barrens flora lists contain a mixture of herbs, grasses, shrubs, and trees from habitats other than barrens (Curtis 1959; Steyermark 1963; Mohlenbrock 1986; Swink and Wilhelm 1994). The diversity of species and the variety of extant habitats that these species currently occupy suggest that barrens occurred across a wide range of landscape and moisture conditions. It is likely that barrens, brush prairies, and oak openings were part of a vegetation continuum on the presettlement landscape.

Extant savannas, although degraded, provide some insight into the nature of this vegetation continuum. For example, in a savanna restoration site in McLean County, central Illinois, three of the four species of *Carex* (*C. blanda*, *C. cephalophora*, and *C. jamesii*) on the composite barrens list prepared by Bowles and McBride (1994) occur on the restoration site. Government Land Office (GLO) survey records indicate that this restoration site was likely an oak opening at the time of the survey (1836) (Anderson et al. 1994). The site has been prescribed burned in February or March each year since 1989. No plants were introduced during the restoration. *Carex* is the dominant ground cover. Currently, there are no C4 prairie grasses on the site, owing to a past history of cattle grazing and the development of a woody understory canopy after cattle were removed from the site in 1967. Grasses on the site are C3 species associated with open woodlands or forest edges (e.g., bottlebrush grass [*Elymus hystrix*], *Bromus pubescens*,

Table 9.3. *Frequency (>15%) in 25-m² quadrats of herbaceous species sampled on two dates in 1994 at the ParkLands Foundation Savanna Restoration site (McLean County, Illinois).*

	Sample date	
	May 12	June 22
Species	(% freq.)	(% freq.)
Grasses		
Dicanthelium acuminatum	72	92
Poa pratensis	68	88
Bromus pubescens	52	60
Danthonia spicata	16	32
Elymus villosus	16	32
Festuca obtusa	28	20
Sedges		
Carex blanda	72	36
Carex hirsutella	0	80
Carex pensylvanica	64	16
Carex cephalophora	20	48
Carex sparganoides	0	28
Forbs		
Geum canadense	80	88
Tradescantia virginiana	80	0
Taenidia integerrima	76	72
Dodecatheon meadia	72	0
Solidago ulmifolia	20	80
Viola sororia	68	56
Sanicula gregaria	64	76
Erigeron strigosus	36	56
Phlox divaricata	48	16
Claytonia virginica	48	0
Galium circaezans	24	44
Camassia scilloides	44	0
Solidago nemoralis	36	24
Potentilla simplex	16	32
Ambrosia artemisiifolia	8	28
Ranunculus fasicularis	28	0
Melilotus alba	28	0
Sisyrinchium albidum	24	20
Viola spp.	24	20
Erythronium albidum	20	0
Oxalis sp.	16	24
Plantago rugelii	0	24
Taraxacum officinale	20	20
Rudbeckia hirta	16	12
Antennaria plantaginifolia	8	16
Oxalis violaceae	16	4

Festuca obtusa, and *Dicanthelium acuminata*) (Table 9.3).

In spring, the restored savanna supports a number of species that distinguish it from the adjacent woodland, either because the species are somewhat ephemeral (e.g., the shooting star, wild hyacinth, and spiderwort [*Tradescantia virginiana*]) or because they flower in spring (e.g., yellow pimpernell [*Taenidia integerrima*] and early buttercup [*Ranunculus fasicularis*]). The frequency of several species differed markedly between May and June (Table 9.3). These differences reflect the ephemeral nature of some species, the progressive development of the vegetation from spring into summer, sample variation, and the difficulty in identifying some *Carex* spp. in a vegetative condition.

Current Status

Rare, Threatened, and Endangered Species

Apparently because of the transitional nature of savanna and the occurrence of woodland and prairie species in savanna, no rare, threatened, or endangered species are restricted to this habitat. For example, at least 20 tallgrass prairie species attributed to savanna have been listed as threatened or endangered at the state level because of concern for their conservation. However, most of these species are at their eastern geographic range limits, where prairie habitat is fragmented, rare, and intergrades with savanna. Because prairie is part of the savanna matrix, many of these regionally rare prairie plants also are present in savanna habitat. In contrast, only a few non-prairie species recorded for deep-soil savanna habitat have been listed at the state level. Some of these species may have strong savanna affinities because of their tolerance of intermediate light levels. For example, *Liatris scariosa* var. *nieuwlandii*, an Illinois threatened species, appears to be primarily a savanna species in Illinois (Bowles, Wilhelm

and Packard 1988), but not elsewhere (e.g., Voss 1996).

Preservation

Because deep-soil savannas were maintained only with recurrent fires (Curtis 1959; Anderson and Brown 1983; Cole and Taylor 1995; Bowles and McBride 1998), few examples survived intact into the 20th century. Fire cessation resulted in rapid conversion of many deep-soil savannas to forest, as oak sprouts and grubs quickly grew into closed oak forests (Cottam 1949; Curtis 1959). Overgrazing following European settlement altered savanna composition by reducing the abundance of, or eliminating, grazing-intolerant species. Agriculture and land development fragmented and destroyed remnants. As a result, Bray (1960) examined only a few examples of mesic (deep-soil) savanna in Wisconsin that were not subjected to grazing (Curtis 1959). In a survey for remnant natural areas in Illinois in the 1970s, only two high-quality deep-soil savanna remnants were located, both occurring in small pioneer cemeteries (J. White 1978; Madany 1981). In a 1985 survey of savannas known in the midwestern states, no additional high-quality deep-soil savanna remnants were found (Nuzzo 1986). Shrub, or brush prairies, on deep soils apparently are even more imperiled. Curtis (1959) did not provide data on this vegetation type in Wisconsin, and none was identified by recent statewide surveys. In a reassessment of the status of northern Illinois savannas, two barrens remnants and two additional oak openings, totaling less than 50 hectares, were found (Bowles and McBride 1996). In the absence of fire on sites where edaphic features limit tree growth, conversion of savannas to forest occurs more slowly than on deep-soil savannas. Consequently, in Illinois and surrounding midwestern states, almost all savanna remnants occur on droughty sand and gravel deposits, and no high-quality deep-soil savanna remnants are known (Leach and Ross 1995).

Restoration and Management

As a result of their open structure and dependence on fire, savanna remnants are susceptible to rapid successional conversion to woodlands and invasion by exotic species. Most remnants are small, thus it often is difficult to restore landscape level fire. The small scale of the fires may reduce their effectiveness in suppressing woody vegetation (A. S. White 1983; Anderson and Brown 1986; Tester 1989; Kline and McClintock 1994). Invasive shrubs such as buckthorns and honeysuckles resprout and thus might not be killed by fire. Furthermore, fire intensity often is low where exotic shrub cover is high because shade depresses or eliminates graminoid species that provide finely divided fuels (Heidorn 1991). Thus, exotic woody species often need to be controlled by artificial cutting and treatment with herbicides (Luken and Mattimiro 1991). However, management efforts to reduce the abundance of native shrubs, such as gray dogwood, should be tempered to (1) ensure maintenance of nesting habitat for understory birds (Whelan and Dilger 1992) and (2) restore pre-European woody understory structure (Bowles, Hutchison and McBride 1994; Bowles and Spravka 1994).

In savanna remnants, many late-successional savanna species are poorly represented as mature plants and in seed banks (Johnson and Anderson 1986). Thus, management to restore deep-soil savannas must consider introduction of seed or plants of species that may have been lost to grazing, fire protection, and canopy closure. Determining which species to introduce is confounded by a lack of baseline data on species composition from intact savannas. This information has been supplemented to some extent by historic lists and original land survey notes (Bowles and McBride 1994), knowledge of the ecology of the species (Pruka 1995), and data from rem-

nants that retain some original species (Anderson et al. 1994; Pruka 1994a, b; Leach 1996; Bowles and McBride 1998).

Summary

Deep-soil savannas of the Midwest occupied a transitional area between eastern deciduous forest and tallgrass prairie. These savannas included oak openings and barrens/brush prairie that tended to merge floristically and structurally with each other and with forest and prairie communities. Oak openings tended to have an open, parklike appearance with widely scattered trees and an understory of high floristic similarity to tallgrass prairies. Barrens/brush prairies apparently were derived from forested areas and had a woody undergrowth of trees and shrubs that was maintained in scrubby condition by repeated burning. Because of the transitional nature of barrens, brush prairie, and oak openings, there are relatively few species of plants that characterize these vegetation types. Maintenance of these savannas was strongly dependent upon periodic fire. In the later part of the 19th century, barrens and oak openings were rapidly converted to closed forest following cessation of fires associated with European settlement. Agricultural activities, including overgrazing, degraded most of the remaining savannas, and consequently there are few extant examples of this vegetation type.

REFERENCES

Adams, D. E., and Anderson, R. C. 1980. Species response to a moisture gradient in Central Illinois forests. *American Journal of Botany* 67:381–392.

Abrams, M. D. 1992. Fire and the development of oak forests. *Bioscience* 42:346–353.

Anderson, R. C. 1982. An evolutionary model summarizing the roles of fire, climate, and grazing animals in the origin and maintenance of grasslands: an end paper. In *Grasses and Grasslands: Systematics and Ecology,* ed. J. R. Estes,

R. J. Tyrl, and J. N. Brunken, pp. 297–308. Norman, Okla.: University of Oklahoma Press.

Anderson, R. C. 1983. The eastern prairie–forest transition – an overview. In *Proceedings of the Eighth North American Prairie Conference*, ed. R. Brewer, pp. 86–92. Kalamazoo, Mich.: Western Michigan University.

Anderson, R. C. 1990. The historic role of fire in the North American Grassland. In *Fire in North American Tallgrass Prairies*, ed. S. Collins and L. Wallace, pp. 8–18. Norman, Okla.: University of Oklahoma Press.

Anderson, R. C. 1991. Presettlement forests of Illinois. In *Proceedings of the Oak Woods Management Workshop*, ed. G. V. Burger, J. E. Ebinger, and G. S. Wilhelm, pp. 9–19. Charleston, Ill.: Eastern Illinois University.

Anderson, R. C., and Anderson M. R. 1975. The presettlement vegetation of Williamson County, Illinois. *Castanea* 40:345–363.

Anderson, R.C., and Brown, L. E. 1983. Comparative effects of fire on trees in a midwestern savannah and an adjacent forest. *Bulletin of the Torrey Botanical Club* 110:87–90.

Anderson, R. C., and Brown, L. E. 1986. Stability and instability in plant communities following fire. *American Journal of Botany* 75:364–368.

Anderson, R. C., Schmidt, D., Anderson, M. R., and Gustafson, D. 1994. ParkLands Savanna Restoration. In *Proceedings of the North American Conference on Savannas and Barrens*, ed. J. S. Fralish, R. C. Anderson, J. E. Ebinger, and R. Szafoni, pp. 275–278. Chicago, Ill.: U.S. Environmental Protection Agency, Great Lakes National Program Office.

Apfelbaum, S. I. and Haney, A. W. 1991. Management of degraded oak savanna remnants in the upper Midwest: preliminary results from three years of study. In *Proceedings of the Oak Woods Management Workshop*, ed. G. V. Burger, J. E. Ebinger, and G. S. Wilhelm, pp. 81–90. Charleston, Ill.: Eastern Illinois University.

Baker, R. G., Maher, L. J., Chumbley, C. J., and Van Zant, K. L. 1992. Patterns of Holocene environmental change in Midwestern United States. *Quaternary Research* 37:379–389.

Baker, R. G., Bettis, E. A., III, Schwert, D. P., Horton, D. G., Chumbley, C. A., Gonzalez, L. A., and Regan, M. K. 1996. Holocene paleoenvironments of northeast Iowa. *Ecological Monographs* 66.203–234.

Belsky, A. J., and Canham, C. D. 1994. Forest gaps and isolated savanna trees. *BioScience* 44:77–84.

Betz, R. F. 1992. Species composition of old settler savanna and sand prairie cemeteries in northern Illinois and Northwestern Indiana. In *Proceedings of the Twelfth North American Prairie Conference*, ed. D. D. Smith and C. A. Jacobs, pp. 79–87. Cedar Falls, Iowa: University of Northern Iowa.

Borchert, J. R. 1950. The climate of the central North American Grassland. *Annals of the Association of American Geographers* 40:1–29.

Bowles, M. L., Hutchison, M. D., and McBride, J. L. 1994. Landscape pattern and structure of oak savanna, woodland, and barrens in northeastern Illinois at the time of European settlement. In *Proceedings of the North American Conference on Savannas and Barrens*, ed. J. S. Fralish, R. C. Anderson, J. E. Ebinger, and R. Szafoni, pp. 65–73. Chicago, Ill:. U.S. Environmental Protection Agency, Great Lakes National Program Office.

Bowles, M. L., and McBride, J. L. 1994. Presettlement barrens in the glaciated prairie region of Illinois. In *Proceedings of the North American Conference on Savannas and Barrens*, eds. J. S. Fralish, R. C. Anderson, J. E. Ebinger, and R. Szafoni, pp. 75–85. Chicago, Ill.: U.S. Environmental Protection Agency, Great Lakes National Program Office.

Bowles, M. L., and McBride, J. L. 1996. *Evaluation and Classification of Savanna,.Woodland, and Barrens Natural Areas in Northern Illinois*. Lisle, Ill.: The Morton Arboretum.

Bowles, M. L., and McBride, J. L. 1998. Vegetation structure and chronological change in a Midwest North American Savanna. *Natural Areas Journal* 18:14–27.

Bowles, M., and Spravka, M. 1994. American hazelnut, an overlooked native shrub in northeastern Illinois. *Morton Arboretum Quarterly* 30:42–48.

Bowles, M. L., and Wilhelm, G. and Packard, S. 1988. The Illinois status of *Liatris scariosa* Willd. var. *nieuwlandii* Lunell: a new threatened species for Illinois. *Erigenia* 10:1–26.

Bray, J. R. 1955. The savanna vegetation of Wisconsin and an application of the concept order and complexity to the field of ecology. Ph.D. dissertation, University of Wisconsin.

Bray, J. R. 1958. The distribution of savanna species in relation to light intensity. *Canadian Journal of Botany* 36:671–681

Bray, J. R. 1960. The composition of savanna vegetation in Wisconsin. *Ecology* 41:721–732.

Brendel, F. 1887. *Flora Peoriana*. Peoria, Ill.: J. W. Franks and Sons.

Britton, W., and Messenger, A. 1969. Computed soil moisture patterns in and around the prairie peninsula during the great drought of 1933–34. *Transactions of the Illinois State Academy of Science* 62:181–187.

Cole, K. L., and Taylor, R. S. 1995. Past and current trends of change in a dune prairie/oak savanna reconstructed through multiple-scale history. *Journal of Vegetation Science* 6:399–410.

Cottam, G. 1949. The phytosociology of an oak woods in southwestern Wisconsin. *Ecology* 30:271–287.

Curtis, J. T. 1959. *The Vegetation of Wisconsin: An Ordination of Plant Communities*. Madison, Wis.: University of Wisconsin Press.

Delcourt, P., and Delcourt, H. A 1981. Vegetation maps for eastern North America: 40,000 yr BP to the present. In *Geobotany II*, ed. R. Romans, pp.123–165. New York: Plenum Press.

Faber-Langendoen, D., and Davis, M. A. 1995. Effects of fire frequency on tree canopy cover at Allison Savanna, East central Minnesota, USA. *Natural Areas Journal* 15:319–328.

Gilbert, M. L., and Curtis, J. T. 1953. Relation of the understory vegetation of the upland forest in the prairie–forest border region of Wisconsin. *Transactions of the Wisconsin Academy of Science, Arts, and Letters* 42:183–195.

Gleason, H. A. 1913. The relation of forest distribution and prairie fires in the middle west. *Torreya* 13:173–181.

Gleason, H. A. 1922. The vegetational history of the Middle West. *Annals of the Association of American Geographers* 12:39–85.

Griffin, D. 1994. Pollen analog dates for midwestern oak savannas. In *Proceedings of the North American Conference on Savannas and Barrens*, ed. J. S. Fralish, R. C. Anderson, J. E. Ebinger, and R. Szafoni, pp. 91–95. Chicago, Ill.: U.S. Environmental Protection Agency, Great Lakes National Program Office.

Grimm, E. C. 1984. Fire and other factors controlling the Big Woods vegetation of Minnesota in the mid-nineteenth century. *Ecological Monographs* 54:291–311.

Gustafson, D. J. 1996. Phenological, physiological, and germination responses of shooting star (*Dodecatheon meadia*: Primulaceae) occupying prairie and savanna habitats. M.S. thesis, Illinois State University, Normal, Ill.

Gustafson, D. J., and Anderson, R. C. 1994. Phenological and physiological response of shooting star (*Dodecatheon media* L.) in prairie and savanna habitats. In *Proceedings of the North American Conference on Savannas and Barrens*, ed. J. S. Fralish, R. C. Anderson, J. E. Ebinger, and R. Szafoni, pp. 217–222. Chicago, Ill.: U.S. Environmental Protection Agency, Great Lakes National Program Office.

Heidorn, R. 1991. Vegetation management guidelines: exotic buckthorns. *Natural Areas Journal* 11:216–217.

Heikens, A. L., and Robertson, P. A. 1994. Barrens of the Midwest: A review of the literature. *Castanea* 59:184–194.

Higley, W. K., and Raddin, C. S 1891. The flora of Cook County, Illinois, and a part of Lake County, Indiana. *Bulletin of the Chicago Academy of Science* 2:i–xxiii, 1–168.

Hus, H. 1908. An ecological cross section of the Mississippi River in the region of St. Louis, Missouri. *Nineteenth Annual report of the Missouri Botanical Garden* 19:128–258.

Johnson, R., and Anderson, R. C. 1986. The seed bank of a tallgrass prairie in Illinois. *The American Midland Naturalist* 115:123–130.

King, J. 1981. Late quaternary vegetational history of Illinois. *Ecological Monographs* 51:43–62.

Kline, V. M., and McClintock, T. 1994. Effect of burning on a dry oak forest infested with woody exotics. In *Spirit of the Land, Our Prairie Legacy, Proceedings of the Thirteenth North American Prairie Conference*, ed. R. G. Wickett, P. D. Lewis, A. Woodliffe, and P. Pratt, pp. 207–213. Windsor, ON, Canada: Department of Parks and Recreation.

Latrobe, C. J. 1835. *The Rambler in North America*. London: R. B. Seeley and W. Burnside.

Leach, M. K. 1994. Savanna plant species distributions along gradients of sun–shade and soils. In *Proceedings of the North American Conference on Savannas and Barrens*, ed. J. S. Fralish, R. C. Anderson, J. E. Ebinger, and R. Szafoni, pp. 223–227. Chicago, Ill.: U.S. Environmental Protection Agency, Great Lakes National Program Office, .

Leach, M. K. 1996. Gradients in groundlayer composition, structure and diversity in remnant and experimentally restored oak savanna. Ph.D. dissertation, University of Wisconsin, Madison, Wis.

Leach, M. K., and Ross, L. 1995. *Oak Ecosystem*

Recovery Plan. Chicago, Ill.: U.S. Environmental Protection Agency, Great Lakes National Program Office, and The Nature Conservancy.

Little, J. T. 1861. Orchards in Illinois. *Journal of the Illinois State Agricultural Society* 1:26–30.

Luken, J. O., and Mattimiro, D. T. 1991. Habitat-specific resilience of the invasive shrub amur honeysuckle (*Lonicera maackii*) during repeated clipping. *Ecological Applications* 1:104–109.

Madany, M. H. 1981. A floristic survey of savannas in Illinois. In *The Prairie Peninsula – In the Shadow of Transeau: Proceedings of the Sixth North American Prairie Conference*, ed. R. L. Stuckey and K. J. Reese, pp. 177–181. Columbus, Ohio: Ohio Biological Survey Biological Notes 15.

McAndrews, J. H. 1966. Postglacial history of prairie, savanna and forest in northwestern Minnesota. *Memoirs of the Torrey Botanical Club* 22:1–72.

McCune, B., and Cottam, G. 1985. The successional status of a southern Wisconsin oak woods. *Ecology* 66:1270–1278.

Mead, S. B. 1846. Catalogue of plants growing spontaneously in the State of Illinois, the principal part near Augusta, Hancock County. *Prairie Farmer* 6:35–36, 60, 93, 119–122.

Mohlenbrock, R. H. 1986. *Guide to the Vascular Flora of Illinois*. Carbondale, Ill.: Southern Illinois University Press.

Nelson, P. W. 1985. *The Terrestrial Natural Communities of Missouri*. Jefferson City, Mo.: Missouri Department of Natural Resources.

Nuzzo, V. 1986. Extent and status of Midwest oak savanna: presettlement and 1985. *Natural Areas Journal* 6:6–36.

Packard, S. 1988. Rediscovering the tallgrass savanna of Illinois. Article 01.14 In *The Prairie: Roots of our Culture: Foundation of our Economy. Proceedings of the 10th North American Prairie Conference*, ed. A. Davis and G. Stanford. Dallas, Tex.: The Native Prairie Association of Texas.

Packard, S. 1991. Rediscovering the tallgrass savanna of Illinois. In *Proceedings of the Oak Woods Management Workshop*, ed. G. V. Burger, J. E. Ebinger, and G. S. Wilhelm, pp. 31–54. Charleston, Ill.: Eastern Illinois University.

Packard, S. 1993. Restoring oak ecosystems. *Restoration and Management Notes* 11:5–16.

Packard, S., and Mutel, C. F. 1997. The *Tallgrass Restoration Handbook*. Washington, D.C.: Island Press.

Peck, J. M. 1834. *A Gazetteer of Illinois*. Jacksonville, Ill.: J. M. Goudy.

Pruka, B. W. 1994a. Distribution of understory plant species along light and soil depth gradients in an upland oak savanna remnant in southern Wisconsin. M.S. thesis, University of Wisconsin, Madison, Wis.

Pruka, B. W. 1994b. Distribution of understory plant species along light gradients in an upland oak savanna remnant in southern Wisconsin. In *Proceedings of the North American Conference on Savannas and Barrens*, ed. J. S. Fralish, R. C. Anderson, J. E. Ebinger, and R. Szafoni, pp. 213–215. Chicago, Ill.: U.S. Environmental Protection Agency, Great Lakes National Program Office, .

Pruka, B. 1995. Lists indicate recoverable oak savannas and open oak woodlands in southern Wisconsin. *Restoration and Management Notes* 13:124–126.

Risser, P., Birney, O., Blocker, H., Parton, W., and Weins, J. 1981 *The True Prairie Ecosystem*. Stroudsburg, Pa.: Hutchinson Ross Publishing Company.

Robertson, K., Anderson, R. C., and Schwartz, M. 1997. The tallgrass prairie mosaic. In *Conservation in Highly Fragmented Landscapes*, ed. M. Schwartz, pp. 55–87. New York: Chapman and Hall.

Rodgers, C., and Anderson, R. C. 1979. Presettlement vegetation of two prairie peninsula counties. *Botanical Gazette* 140:232–240.

Shepherd, C. L. 1850. Machine harvesting. *Prairie Farmer* 10:34.

Skarpe, C. 1992. Dynamics of savanna ecosystems. *Journal of Vegetation Science* 3:293–300.

Stearns, F. 1991. Oak woods: an overview. In *Proceedings of the Oak Woods Management Workshop*, ed. G.V. Burger, E. E. Ebinger, and G. S. Wilhelm, pp. 1–7. Charleston, Ill.: Eastern Illinois University.

Steyermark, J. A. 1963. *Flora of Missouri*. Ames, Iowa: Iowa State University Press.

Swink, F., and Wilhelm, G. 1994. *Plants of the Chicago Region*. 4th ed. Indianapolis, Ind.: Indiana Academy of Science.

Taft, J. 1997. Savanna and open woodland communities. In *Conservation in Highly Fragmented Landscapes*, ed. M. Schwartz, pp. 24–54. New York: Chapman and Hall.

Tester, J. R. 1989. Effects of fire frequency on oak savanna in east-central Minnesota. *Bulletin of the Torrey Botanical Club* 116:134–144.

Transeau, E. N. 1935. The prairie peninsula. *Ecology* 16:423–437.

Vogl, R. J. 1977. Fire: a destructive menace or a

natural process? In *Recovery of Damaged Ecosystems*, ed. J. Cairns, Jr., K. Dickson, and E. Herricks, pp. 262–289. Charlottesville, Va.: University Press of Virginia.

Voss, E. G. 1996. *Michigan Flora*. Part III, *Dicots (Pyrolaceae-Compositae)*. Ann Arbor, Mich.: Cranbrook Institute of Science Bulletin 61 and University of Michigan Herbarium.

Whelan, C. L., and Dilger, M. L. 1992. Invasive, exotic shrubs: a paradox for natural area managers? *Natural Areas Journal* 12:109–110.

White, A. S. 1983. The effects of thirteen years of annual burning on a *Quercus ellipsoidalis* community in Minnesota. *Ecology* 64:1081–1085.

White, J. 1978. *Illinois Natural Areas Inventory Technical Report*. Urbana, Ill.: Illinois Natural Areas Inventory.

White, J. 1994. How the terms savanna, barrens and oak openings were used in early Illinois. In *Proceedings of the North American Conference on Savannas and Barrens*, ed. J. S. Fralish, R. C. Anderson, J. E. Ebinger, and R. Szafoni, pp. 25–63. Chicago, Ill.: U.S. Environmental Protection Agency, Great Lakes National Program Office.

Will-Wolf, S., and Roberts, D. W. 1993. Fire and succession in oak–maple upland forests: a modeling approach based on vital attributes. In *John T. Curtis: Fifty Years of Plant Ecology*. ed. J. Fralish, R. McIntosh, and O. Loucks, pp. 217–236. Madison, Wis.: The Wisconsin Academy of Sciences, Arts, and Letters.

Winkler, M. G. 1994. Sensing plant community and climate change by charcoal–carbon isotope analysis. *Ecoscience* 1:340–345.

Winkler, M. G. 1997. Later quaternary climate, fire, and vegetation dynamics. In *Sediment Records of Biomass Burning and Global Change*. ed. J. Clark, H. Cachier, J. Goldammer, and B. Stocks, pp. 329–346. Berlin: Springler-Verlag.

Winkler, M. G., Swain, A. M., and Kutzbach, J. E. 1986. Middle Holocene dry period in the northern Midwestern United States: lake levels and pollen stratigraphy. *Quaternary Research* 25:235–250.

10 Open Woodland Communities of Southern Illinois, Western Kentucky, and Middle Tennessee

JAMES S. FRALISH, SCOTT B. FRANKLIN, AND DAVID D. CLOSE

Introduction

Scattered within the oak-hickory forest of southern Illinois and at Land Between The Lakes (LBL) in western Kentucky and northwest middle Tennessee are a variety of woodlands that have evolved from a barrens or savanna community of pre-European settlement time (before ca. 1820). In the present context, a *barren* is a community of widely scattered, short trees growing on relatively thin but continuous cover of extremely rocky soil; a *savanna* is dominated by scattered trees interspersed with prairie vegetation and located on somewhat deeper soil that may not have a large rock component. Trees commonly found in these communities are drought-tolerant scrub oak species (post oak, *Quercus stellata*; chestnut oak, *Q. prinus*; scarlet oak, *Q. coccinea*; southern red oak, *Q. falcata*; blackjack oak, *Q. marilandica*).

Woodland represents a late (successional) stage in the development of barrens and savanna communities and results from an extended period of protection (i.e., absence of disturbance, specifically fire). Like a barrens or savanna, it is a community dominated by extremely slow-growing, usually short, trees. The community has a higher tree density and a more closed (although still relatively open) canopy structure than a barrens or savanna. The generally gnarled, large, twisted limbs and crowns result from a combination of the open character of the overstory canopy, extreme temperatures (40 °C; Fralish, unpublished data) and low soil water-holding capacity, which produce high stress. Most woodland types usually have a distinct midcanopy stratum of ericaceous shrubs and a variety of herbs and bryophytes, many of which are restricted to this environment.

Before settlement (ca. 1820), fire had a major influence on the character of barrens and savanna in Missouri (Ladd 1991), southern Illinois (Fralish et al. 1991), western Kentucky (Bryant and Martin 1988), northwest middle Tennessee (DeSelm 1988), and LBL (McCrain and Grubb 1987; Franklin 1994). From the time of settlement into the 1930s, when state fire control laws generally were enacted, wildfire became an infrequent occurrence and burned considerably smaller areas (Guyette and Cutter 1991; Cutter and Guyette 1994). However, in some regions, settlers continued to use fire to clear land for homes and agriculture, and these fires occasionally escaped into the woods.

This chapter describes the woodland types of several natural regions of southern Illinois, and the types at Land Between The Lakes as representative of western Kentucky and northwest middle Tennessee. In each of these natural divisions, soil and topographic conditions are distinctly different, as are the associated plant communities. Also considered with these woodlands is a redcedar (*Juniperus*

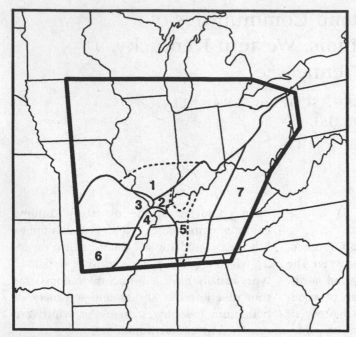

Figure 10.1. Physiographic provinces and natural divisions in Illinois, Kentucky, Tennessee, and adjacent states. Numbers indicate natural divisions: (1) Southern Till Plain, Central Lowland Province; (2) Shawnee Hills, Interior Low Plateaus; (3) Ozark Plateau Province; (4) Gulf Coastal Plain; (5) Western Highland Rim and Pennyroyal, Interior Low Plateaus; (6) Ouachita Mountains Province; (7) Appalachian Mountain Provinces.

virginiana)–dominated community generally referred to locally as a sandstone cedar glade. Following the criteria of Heikens (see Chapter 13), we distinguish a *glade* by the large areas of exposed bedrock and by soil isolated in depressions or lenses on the bedrock.

Natural Divisions

Within the tri-state area of southern Illinois, western Kentucky, and northwest middle Tennessee, woodland communities are found in five natural divisions (sensu Schwegman 1973) (Figure 10.1). There are three natural divisions in the southern one-third of Illinois, each representing a physiographic province: Southern Till Plain (East Central Lowlands), Ozark Hills (Ozark

Plateaus), and Shawnee Hills (western extension of the Interior Low Plateaus) (Fenneman 1938). LBL is located where the Eastern Highland Rim/Pennyroyal section of the Interior Low Plateau (Smalley 1980) interfaces with the Mississippi Embayment Section of the Gulf Coastal Plain (Braun 1950).

Regional Climate

The climate is continental. Average annual precipitation varies from 100 cm at Effingham, Illinois, at the northern terminus of Till Plain to 122 cm at Dover, Tennessee, at the south end of LBL (United States Department of Agriculture 1941). Although precipitation generally is evenly distributed throughout the year, a drought of 4–6 weeks may occur in July and August, particularly near the northern boundary. Temperature and the prevailing southwest wind interact to reduce the soil moisture level on middle and high west-, southwest-, and south-facing slopes (Downs 1976), where most woodlands and glades are located.

Post Oak Flatwoods of the Illinois Till Plain

The Southern Till Plain extends from the Grand Prairie Division of central Illinois to the forested Shawnee Hills Division in southern Illinois. The presettlement vegetation was a mixture of forest and prairie (Schwegman 1973). An analysis of witness tree data from the original land survey records indicates that presettlement vegetation ranged from open oak forest to oak savanna to prairie (Anderson and Anderson 1975; Phelps 1998). The post oak "flatwoods" is the only remaining savanna-related community in the division and, prior to settlement, was estimated to cover approximately 2.25 million ha (Taft, Schwartz and Loy 1995).

Prior to settlement, fire moved rapidly, frequently, and relatively unrestricted through the region due to the absence of a major river functioning as a fire break, interspersion of prairie with forest, and relatively low annual precipitation. Anderson and Anderson (1975) attributed the extensive areas of savanna and open forest to frequent fire. One important ignition source may have been the extensive Native American (Mississippian) settlement of more than 20,000 people in the Mississippi River east bank floodplain near Cahokia mounds, their spiritual and political center. Delcourt and Delcourt (1997) reached a similar conclusion regarding the use of fire by Native Americans in pre-Columbian times for the southern Appalachian highlands in the eastern deciduous forest region. Fire may also have come from the Grand Prairie region to the north, where precipitation is lower (80–90 cm yr^{-1}; United States Department of Agriculture 1941) and where fire was common.

Environment

Because of glaciation, the topography is gently to moderately rolling, and bedrock generally is buried below Illinoian glacial till and loess. At present, flatwood stands are relatively uncommon because the level topography (<2% slope) is ideal for agriculture. Wynoose, Bluford, Cisne, and Hoyleton silt loams (loess) are some of the major soil types under flatwoods (Fehrenbacher et al. 1984; Miles 1996). Wynoose soil, for example, has a clayey (30%–50% clay), strongly compacted (bulk density = 1.6–1.7 g/cc), distinctly mottled (ashy/reddish-brown) subsoil similar to a fragipan beginning at about a 15-cm depth and continuing downward for about a meter (Katerere 1979). Wynoose and related soil types have an extended period of soil saturation (~35% water) during the spring months because drainage is restricted by the level land surface and percolation is

restricted by the fragipan. However, during the late-summer drought period, soil moisture drops to below 5% (Fralish, unpublished data), which is near the permanent wilting point for most woody plant species. The combination of these extreme soil moisture conditions and compactness limits community richness to a small number of species.

Community Composition

An excellent example of a flatwoods on Wynoose soil is at Posen Woods Nature Preserve near Nashville, Illinois. Post oak is the major flatwoods dominant (Table 10.1) not only because it is drought tolerant (Burns and Honkala 1990b), but probably because its roots penetrate the compacted subsoil to a depth of nearly a meter or more. Thus, unlike other species, whose tree roots are restricted to near-surface horizons, post oak roots have access to soil water during the summer drought. Additional important species found in stands include black oak (*Quercus velutina*), white oak (*Q. alba*), and hickory (shagbark, *Carya ovata*; pignut, *Carya glabra-ovalis* complex; mockernut, *Carya tomentosa*).

Variation in flatwoods composition on Wynoose soil appears to be influenced by drainage. At Posen Woods, the slope is <1.0% within the post oak–reed grass (*Q. stellata–Cinna arundinacea*) habitat type (Fralish 1988b), which generally is isolated from a stream drainage system. Reed grass, considered a species of wet habitats, is indicative of the inundated soil condition during the spring months. Post oak is the single major tree species with an importance value (IV) (relative basal area) of 90%. Shagbark hickory (IV = 4%) is the next most important species, followed by black oak (IV = 3%) and mockernut hickory (IV = 3%).

Small drainage channels can be traced into the post oak–black oak–poison ivy (*Quercus stellata–Quercus velutina–Toxicodendron radicans*) habitat type that occupies a large area of Posen Woods. These channels apparently improve soil drainage and aeration; how-

Table 10.1. *Tree species importance values (relative basal area) for sandstone cedar glades and post oak woodland in the Till Plain, Ozark Hills, and Shawnee Hills sections of Illinois, and for post oak and chestnut oak woodlands at Land Between The Lakes (Western Highland Rim/Mississippi Embayment interface) of western Kentucky and northwest middle Tennessee.*

	Region/Community type					
	Till Plain	Ozark Hills	Shawnee Hills		Highland Rim	
Species	Post oak[1]	Post oak[2]	Redcedar[3]	Post oak[3]	Chestnut oak[4]	Post oak[4]
	Importance value (% basal area)					
Carya glabra	12.0	8.5		7.0	1.2	6.0
Carya ovata	4.4	0.2		0.6		0.3
Carya texana				11.6		
Carya tomentosa	2.0			1.8	0.2	0.2
Juniperus virginiana			85.4		0.1	
Nyssa sylvatica					1.2	0.3
Oxydendrum arboreum					3.0	
Quercus alba	4.2	5.8		13.2	10.1	11.8
Quercus coccinea				1.7	6.0	7.1
Quercus falcata				0.5	0.5	2.0
Quercus marilandica	1.6		4.3	15.5	1.7	8.5
Quercus prinus	0.9				61.3	3.8
Quercus stellata	56.0	50.0	1.8	39.5	6.9	51.1
Quercus velutina	18.6	34.5		7.8	7.6	5.2
Ulmus alata			5.8	0.4		
Other species	1.8 (8)[5]	0.1 (1)[5]	0 (0)	0.4 (1)[5]	0.2 (3)[5]	4.8 (5)[5]

Species	Region/Community type					
	Till Plain	Ozark Hills	Shawnee Hills		Highland Rim	
	Post oak[1]	Post oak[2]	Redcedar[3]	Post oak[3]	Chestnut oak[4]	Post oak[4]
			Site Environment			
Density (trees ha⁻¹)	242	255	140	340	429	404
Basal area (m² ha⁻¹)	22.4	19.79	5.09	14.97	20.96	19.19
Soil texture	Silt loam	Silt	Silt loam	Silt loam	Clay loam	Silty clay
Effective soil depth (cm)	49.0	56.5	<15.0	58.0	53.9	50.4
Percent rock (max. %)	0.0	38.0 (45)	0.0 (0.0)	50.0 (60)	34.0 (70)	19.0 (60)
Soil available water (cm)	31.0	15.2	2.0–6.0	5–10	8.4	10.8
Slope aspect	Flat	SW–S	W, SW, S, SE	W, SW, S	All	SW

[1] Data from Fralish (unpublished) collected at Posen Woods Nature Preserve; data based on three 0.04-ha plots.

[2] Data from Brennan (1983); data from five 0.04-ha plots in each of 3 stands.

[3] Data from Fralish (1976); data from five 0.04-ha plots in each of 5 redcedar stands and 5 post oak stands.

[4] Data from Franklin (1990); two 0.06-ha plots in each of 18 chestnut oak and 14 post oak stands.

[5] Number in parentheses indicates number of additional species in the community.

ever, slope remains low at less than 5%. Compared to the post oak–reed grass habitat type, black oak, white oak, pignut hickory, shagbark hickory, and mockernut hickory have greater importance (Table 10.1). Occasionally, shingle oak (*Quercus imbricaria*) and pin oak (*Quercus palustris*) are present (Taft, Schwartz and Loy 1995). In both habitat types, undisturbed stands have a cathedral-like overstory canopy because the straight, 20–30-m-tall trees have few low branches (Fralish 1988b); this tree and stand structure is unlike those of post oak stands in the Ozark Hills, Shawnee Hills, or at LBL. Menges, Dolan and McGrath (1987) studied a similar post oak community in southwestern Indiana.

The major midcanopy species in flatwoods are black cherry (*Prunus serotina*), sassafras, (*Sassafras albidum*), poison ivy, red mulberry (*Morus rubra*), and blackhaw (*Viburnum prunifolium*) (Fralish, unpublished data).

Summer/fall herbaceous plants that have >5% frequency (based on twenty-five 0.25 m^2 quadrats in each of six stands) include *Helianthus divaricatus* (12%), *Cinna arundinacea* (12%), *Agrostis perennans* (9%), *Dichanthelium acuminatum* (14%), *Carex pensylvanica* (7%), *Eleocharis verrucosa* (7%), *Carex festucacea* (6%), and *Carex hirsutella* (5%) (Taft, Schwartz and Loy 1995); 14 additional herbs were recorded. Only 5 herb species of the post oak flatwoods community in the Till Plain region were common to the post oak communities in the Illinois Ozark and Shawnee hills or at LBL in Kentucky and Tennessee: *Helianthus divaricatus*, *Parthenocissus quinquefolia* (36% frequency), *Toxicodendron radicans* (6%), *Danthonia spicata* (2%), *Porteranthus stipulatus* (2%).

The effect of fire on structure can be observed in Posen Woods, which contains a low (<6 m) stratum of hickory saplings. Close inspection indicates that these saplings are stump sprouts that originated after a 1937 fire killed the parent trees. Also, at the base of a number of black oak trees are old fire scars of similar size that apparently date from the same fire. A dendrochronological analysis of post oak increment cores indicated an exceptionally narrow growth ring during 1937, probably resulting from a substantially below average rainfall that produced a severe fire season. The woods apparently was not burned after 1937.

Threats to Conservation

Today, probably <0.1% of the original flatwoods remain. Forest remnants are widely separated and found primarily as small farm woodlots or in state forest, parks, and preserves. This fragmentation pattern is similar to that of the agricultural areas of Illinois (Iverson et al. 1989), southern Wisconsin (Sharpe et al. 1987; Dunn and Stearns 1993), and Ohio (Simpson et al. 1994), where most remnant stands are vulnerable to elimination at the whim of the landowner. Flatwoods continue to be cut and left in a degraded condition, or converted to agriculture. The herbaceous component probably has been impacted by grazing of domestic livestock.

Post Oak Woodland of the Illinois Ozark Hills

The Illinois Ozark Hills area is an extension of the Ozark Plateau of southeastern Missouri. Although post oak woodland is relatively common west of the Mississippi River, it is rare in the Ozark Hills (Brennan 1983). Stands are small (~2 ha) and isolated because they are restricted to a specific environment. No post oak stems were recorded as witness trees during the 1806–07 original land survey (McArdle 1991), although they may have been included with the white oak trees. However, in the Shawnee Hills land survey, post oak was identified as a separate species (Crooks 1988). In a study of circular plots superimposed on section and quarter-section corners, McArdle (1991) reported post oak importance values in the present forest at about 4%.

Prior to 1820 and settlement of the region, fire may have been less important than after 1850. Comparison of witness tree data from the 1806–07 original land survey with data collected from the same section corners indicated a higher component of oak and hickory and a reduced importance of sugar maple (*Acer saccharum*) and American beech (*Fagus grandifolia*) in 1990 (McArdle 1991; Fralish 1997). This composition change is attributed to an increase in fire frequency after 1850. Because the Mississippi River acted as a natural break, and due to the presence of deep, mesic drainages that characterize the Illinois Ozarks, fire had a long return time and a low intensity. Fire, as well as timber harvesting, continued to be used by landowners to clear woods for cattle and other purposes into the 1930s. During the first forest survey of Illinois, Miller (1920) reported that from the top of Bald Knob in the Illinois Ozark Hills he could not see "the mother of all rivers" (the Mississippi River 10 km away) because of smoke from at least six forest wildfires. Miller had been sent to southern Illinois because of reports that the forest was being destroyed by fire. Telford (1926) reported numerous fires and extensively burned woods, and Miller and Fuller (1922) reported fire as well as extensive logging. Therefore, from about 1850 to about 1930, when fire control laws were enacted, the Ozark woodland communities were heavily burned, considerably less dense, and referred to as "barrens." Environmentally, the shallow soil or exposed bedrock conditions of the present post oak woodlands meet the criteria of a barrens. Following fire control, the barrens slowly grew into more closed-canopy woodland.

Environment

The Illinois Ozark Hills region is characterized by rugged topography and a dendritic drainage pattern typical of a mature, dissected landscape. The area is underlain by cherty Devonian limestone; a loess deposit up to 10 m deep caps the limestone. However, erosion has removed most of the loess on isolated, high south and southwest slopes and has partially exposed the cherty limestone bedrock typical of post oak stands (Table 10.1) (Brennan 1983). The south and southwest slopes are hot, xeric sites because of the thin rocky soil, and because afternoon solar radiation produces the maximum impact on evapotranspiration. It is these xeric site conditions that have allowed post oak barrens to persist but also have restricted its development to a woodland condition after fire ceased in the 1930s. The surrounding oak–hickory forest on deeper soil rapidly has progressed toward a mesophytic community primarily dominated by sugar maple and American beech (Fralish 1997; Helmig 1997).

Community Composition

Post oak is the dominant overstory species (IV = ~50%), followed closely by black oak (Table 10.1), but white oak and several hickory species also are components. A basal area level of ~20 m^2 ha^{-1} indicates a moderately open community compared to a basal area of 25–33 m^2 ha^{-1} for moderately dense black and white oak stands, or 30–40 m^2 ha^{-1} for dense mixed stands of sugar maple, American beech, and other mesophytic species (Brennan 1983). The major midcanopy species include sparkleberry (*Vaccinium arboreum*), low blueberry (*Vaccinium vacillans*), serviceberry (*Amelanchier arboreum*), flowering dogwood (*Cornus florida*), eastern redbud (*Cercis canadensis*), and ironwood (*Ostrya virginiana*) (Warner 1982). Virginia creeper (*Parthenocissus quinquefolia*) and poison ivy also are common woody plants that seldom grow taller than the herb stratum.

The summer herb guild includes approximately 40 species, the most important of which are, in order of declining frequency, pussytoes (*Antennaria plantaginifolia*), poverty grass (*Danthonia spicata*), dittany (*Cunila origanoides*), panic grass (*Panicum dichotomum*), woodland sunflower (*Helianthus divaricatus*), and spreading aster (*Aster patens*) (Table 10.2). Species of moss, lichen (including *Cladina*), sedge (*Carex* spp.), and grass (mostly *Panicum* spp.) occur in all stands.

The relatively large number of herbs in post oak stands, compared to black oak–white oak and oak–sugar maple stands, results in part from the higher light conditions and reduced litter cover (Fralish 1997). The relatively thin post oak canopy (low leaf-to-area ratio) permits considerable light penetration and produces a low amount of surface organic matter after leaf drop occurs in the fall. Photosynthetically active radiation (PAR) at 70 cm above ground level was measured at approximately 234 μmol m^{-2} s^{-1} between 11:00 A.M. and 1:00 P.M. under a clear sky. PAR in an adjacent black oak stand was recorded at 138 μmol m^{-2} s^{-1}; in a white oak–black oak stand with a heavy midcanopy of sugar maple, PAR averaged <9.0 μmol m^{-2} s^{-1}.

Threats to Conservation

The potential for loss is low. Much of the forested land in the Ozark Hills is in federal or state ownership, including the Shawnee National Forest, Trail of Tears State Forest, and Union County Wildlife Refuge (state owned). Although timber harvesting in the National Forest is a potential threat, extensive forested areas are designated as wilderness, research natural areas, or interior management units for neotropical migrant birds. Most remnants are in need of periodic fire to reduce the density of woody stems and to increase the number and cover of herbs.

Redcedar Glades and Post Oak Woodlands of the Shawnee Hills

The Shawnee Hills is a 400,000-ha unglaciated east–west escarpment bisecting southern Illinois (Figure 10.1). Analysis of witness tree data from the 1806–07 original land survey records indicates that post oak was the dominant species of rocky southwest sites, and that there was a concentration of

this community toward the east side of the escarpment near areas known as Wildcat Hills, Cave Hill, Stone Face, Karbers Ridge, Rimrock Trail, Pounds Hollow, and Williams Tower Hill (Crooks 1988; Fralish et al. 1991). The sites supporting post oak stands were isolated within a forest composed primarily of white oak. Survey records provide no information on the sandstone cedar glades.

Elevation in the Shawnee Hills ranges from approximately 110 m above sea level where the region interfaces with the Till Plain on the north, to a general level of 160–210 m with a few hills extending above 240 m. Bedrock is massive Mississippian-age sandstone but with some interbedded limestone outcrops along the southern boundary with the Coastal Plain. Bedrock is covered by a loess deposit that varies locally in thickness from a few centimeters on outcrops to 3–5 m in stream terraces and north slopes. Topography varies from rolling to steep and hilly with no major fire breaks in the region.

Fire appears to have been a major factor prior to settlement. Given the general absence of mesophytic (fire-intolerant) species and the exceptionally high occurrence of oak as witness trees in the 1806–07 land survey records (Crooks 1988; Fralish et al. 1991), fire probably was not only frequent but also intense, due to the presence of relatively dry site conditions (low fuel moisture) over large areas, gentle topography, and absence of fire breaks.

The compositionally stable forest communities dominated by redcedar, post oak, white oak, northern red oak, and sugar maple have been shown to form a vegetational continuum along a soil moisture gradient from xeric to mesic (Chambers 1972; Cerretti 1975; Fralish 1976, 1988a, 1994; Fralish et al. 1978, 1991). The environment of the redcedar community positions it at the xeric end point of the gradient, whereas post oak woodland is in close juxtaposition toward the xeric–mesic section.

Sandstone Cedar Glade
Environment and Composition

Redcedar glades are widely scattered within the landscape but may be moderately large (2–3 ha). The community occurs only where thin soil lenses cover sandstone outcrops and cliff edges and, thus, differs considerably from the "limestone glades" considered by Baskin and Baskin (see Chapter 12). Available water-holding capacity is extremely low and ranges from 2 to 6 cm. (Table 10.1); thus, available soil water may be depleted below permanent wilting point several times during the growing season (Downs 1976). Aspect varies from southeast to southwest to west, but some sites are relatively level. Bedrock may be exposed over areas of 0.5 ha; here, foliose lichens, reindeer lichen (*Cladina* spp.), and cushion mosses cover the surface, and scattered redcedar seedlings often are rooted in cracks where some soil and organic material have accumulated.

The glade overstory has low species richness because only redcedar, post oak, blackjack oak, and winged elm (*Ulmus alata*) tolerate the xeric conditions (Fralish 1976; Table 10.1). Redcedar is the strong dominant, with an IV of 85%. The extreme openness of stands (basal area = 5.1 m² ha⁻¹) combined with the inverted cone architecture of redcedar permits considerable light penetration to the forest floor. Low richness characterizes the midcanopy, as sparkleberry, serviceberry, and bullbrier (*Smilax bona-nox*) generally are the only woody species; biomass averages 2,100 kg ha⁻¹ and varies from 500 kg ha⁻¹ to as high as 5,600 kg ha⁻¹ (O'Dell 1978).

However, the herb guild has high species richness, with 57 species identified in spring and 73 in August and September; various species of a few genera were not separated, so these numbers are conservative (J. Fralish and S. Jones, unpublished data). Although the spring and summer/fall herb flora are not completely mutually exclusive (lichens, mosses, ferns, and some sedges and grasses overlap), the total number sharply contrasts with the limited number of the woody species. The major spring herb species, in order of decreasing frequency, are little bluestem (*Schizachyrium scoparium*; 42% frequency), false dandelion (*Krigia dandelion*; 28%), poverty grass (*Danthonia spicata*; 15%), small-flowered bittercress (*Cardamine parviflora*; 12%), false garlic (*Nothoscordum bivalve*; 10%), American agave (*Manfreda virginica*; 6%), bluet (*Hedyotis pusilla*; 9%), wood sorrel (*Oxalis* spp; 7%), and rock-moss (*Sedum pulchellum*; 6%) (Jones 1974; Harty 1978; J. Fralish and S. Jones, unpublished data). The presence of succulent and extremely fleshy species such as opuntia cactus (*Opuntia humifusa*; 2%), false dandelion, American agave, and rock-moss indicate the extreme xeric nature of the glade environment.

Important species of the summer/fall herb guild (Table 10.2) in order of decreasing frequency, from 46% to 6%, are little bluestem, pineweed (*Hypericum gentianoides*), poverty grass, crotonoptis (*Crotonoptis elliptica*), sedge (*Carex* spp.), buttonweed (*Diodia teres*), American agave, and false dandelion. Frequencies for moss and lichens were greater than 50% during both the spring and summer/fall periods (Jones 1974; Harty 1978).

Post Oak Woodland
Environment and Composition

The post oak woodland is supported by deeper soil than that found in the sandstone cedar glade community, and soil continuously covers the bedrock. However, rock content (mostly sandstone flagstones) may be up to 60% by volume, so that there is only a modest water-holding capacity of 5–10 cm. Total depletion of available soil water also may occur at least twice during the growing season, but the thin soil in the redcedar community has longer drought periods. Quantitatively, the redcedar community has 500–700 cm-days of

Table 10.2. *Frequency (%) for common summer/fall herbaceous species in post oak and redcedar woodlands in the Ozark Hills and Shawnee Hills of Illinois, and at Land Between The Lakes (Western Highland Rim/Mississippi Embayment interface) in western Kentucky and northwest middle Tennessee. Values are rounded to the nearest whole number.*

Species	Ozark Hills Post oak[1]	Shawnee Hills Redcedar[2]	Shawnee Hills Post oak[2]	Highland Rim Chestnut oak[3]	Highland Rim Post oak[3]
Antennaria plantaginifolia	80	2	34	0	0
Asplenium platyneuron	20	5	3	0	2
Aster paters	36	1	7	0	0
Carex spp.	37	15	8	8	36
Crotonopsis elliptica	0	16	1	0	0
Cunila origanoides	50	2	48	8	18
Danthonia spicata	60	17	54	6	24
Euphorbia corollata	13	1	8	0	0
Galium pilosum	4	0	15	0	2
Galium spp.	3	1	1	0	1
Gerardia flava	1	0	7	0	12
Helianthus divaricatus	36	4	32	0	0
Hieraceum gronovii	16	1	13	2	1
Krigia spp.	10	8	6	0	0
Lespedeza spp.	2	1	9	8	2
Lichen (including Cladina)	33	95	72	18	6
Moss	87	52	70	58	56
Panicum boscii	10	0	0	10	6
Panicum dichotomum	46	4	6	8	32
Panicum spp.	1	13[4]	21[4]	6	2
Parthenocissus quinquefolia	43	3	4	0	26
Schizachyrium scoparium	10	45	29	2	2
Solidago spp.	53	1	7	2	10
Tephrosia virginiana	16	0	6	2	8

	Region/community type				
	Ozark Hills	Shawnee Hills		Highland Rim	
Species	Post oak[1]	Redcedar[2]	Post oak[2]	Chestnut oak[3]	Post oak[3]
Toxicodendron radicans	50	3	5	2	18
Vulpia octoflora	0	14	3	0	0
Total number of species	43	77	59	22	34

[1] Unpublished data collected by Fralish and Stephensen (1994); frequency based on 3 stands; ten 1.0-m² quadrats/stand.
[2] Data from Jones (1974); frequency based on 5 redcedar and 5 post oak stands; ten 1.0-m² quadrats/stand.
[3] Data from Close (1996); frequency based on 5 chestnut oak and 6 post oak stands; ten 2.0-m² quadrats/stand.
[4] Includes P. laxiflorum; may include P. boscii.

available soil water (amount of available water × length of storage), whereas the post oak community has approximately 800–900 cm-days (Downs 1976). Aspect varies from south to southwest to west; the sites generally have a slope of 15%–30%.

Post oak woodland has a higher average basal area (15 m² ha⁻¹) and a greater number of overstory species than the sandstone cedar glade due to the larger available soil water supply (Table 10.1) (Fralish 1987, 1994). Post oak, blackjack oak, white oak, black oak, pignut hickory, and Texas hickory (*Carya texana*) are the major components; the remaining six species have low importance. The well-developed midcanopy is dominated by sparkleberry and serviceberry, but flowering dogwood, ironwood, and eastern redbud occur in some stands. Midcanopy biomass averages 2,137 kg ha⁻¹, with a range of 169 kg ha⁻¹ to 4,200 kg ha⁻¹ in unburned stands (O'Dell 1978).

Approximately 60 species occur in the diverse spring herb stratum of post oak woodland. The stratum is dominated by poverty grass (53% frequency), false dandelion (40%), pussytoes (30%), dittany (30%), sedge (27%), little bluestem (21%), woodland sunflower (21%), panic grass (20%), wild bergamot (*Monarda bradburiana*; 16%), wood rush (*Luzula multiflora*; 14%), toothwort (*Dentaria laciniata*; 14%), false foxglove (*Aureolaria flava*; 11%), hairy hawkweed (*Hieracium gronovii*; 11%), short spiderwort (*Tradescantia virginiana*; 11%), and bristly buttercup (*Ranunculus hispidus*; 10%) (J. Fralish and S. Jones, unpublished data). A total of 54 species were recorded in the summer/fall flora. Many species are common both to the redcedar glades and to the post oak woodland of the Ozark Hills (Table 10.2).

Threats to Conservation

Although timber harvesting of post oak stands remains a potential threat, succession to white or black oak forest or to mesophytes

does not appear to be a problem because of the limited supply of available soil water (Fralish 1988a). Strong similarity in the composition of presettlement, disturbed, and stable forest indicates that post oak stands on dry, rocky southwest slopes are compositionally stable (Fralish et al. 1991).

Today, Forest Service restoration procedures call for periodic burning to restore the "original character" of barrens and the habitat of endangered species such as Mead's milkweed (*Asclepias meadii*). Various restoration projects are being carried out on federal, state, and private woodlands.

Post Oak and Chestnut Oak Woodlands at Land Between the Lakes

Land Between The Lakes (LBL) is a 68,800-ha forested peninsula between Kentucky Lake (Tennessee River) and Lake Barkley (Cumberland River). This interfluve, which is 64 km long and up to 13 km wide, is owned by the Tennessee Valley Authority (Thach and Doyle 1988). The area is highly dissected, and elevation ranges from 110 m at lake level to approximately 190 m on the Tennessee Ridge which bisects LBL for its entire length. Upland soil types have formed in (a) weathered Mississippian limestone (Western Highland Rim/Pennyroyal section), (b) white gravel (Tuscaloosa Formation of the Cretaceous system), (c) brown gravel (Lafayette Gravel of the Tertiary-Quaternary system) deposits of the Coastal Plain (Mississippi Embayment) (Braun 1950; Harris 1988), and (d) in a deposit of loess (United States Department of Agriculture 1953, 1981; Smalley 1980; Fralish and Crooks 1988, 1989; Franklin et al. 1993). Brown McNairy sand overlies the Tuscaloosa Gravel on some ridges.

Historic descriptions indicate that barrens were a common presettlement community that was maintained in a relatively open condition due to fire (Franklin 1994). Scrub oak (post, blackjack, chestnut, black, and southern red) were commonly recorded as witness trees in the 1820 general land survey of the Jackson Purchase (Bryant and Martin 1988) located directly west of LBL. DeFriese (1884), Richards and Berry (1885), and Loughridge (1888) as cited by Bryant and Martin (1988) reported that extensive prairie communities occurred in the Jackson Purchase. Loughridge (1888) indicated that with the cessation of fire after settlement, scrub oak and other shrubby plants rapidly invaded and replaced prairie within 30 years. Haywood (1959) also reported that extensive prairie was found in Stewart County (the southern one-third of LBL). A few barren remnants dominated by little bluestem and other prairie species still exist on the Western Highland Rim (DeSelm 1988).

During the early to mid-1800s, large areas of forest/barrens were severely disturbed by iron smelting. There were eight iron furnaces and numerous kilns that required charcoal for smelting. Near each kiln, the land was denuded of forest for several miles in all directions, and it was estimated that as much as 19% of LBL was heavily cut. The Center furnace continued operation until 1912 using scrub oak that could not be used as crossties (Franklin 1994). Livestock grazing, wildfire, and timber harvesting for railroad ties, lumber, and moonshine stills were continuing disturbances into the 1940s (Henry 1975; Wallace 1988; Franklin 1994).

A coenocline, developed to examine the location of compositionally stable communities along a moisture gradient, shows chestnut oak stands at the extreme xeric end of the gradient with post oak stands in close

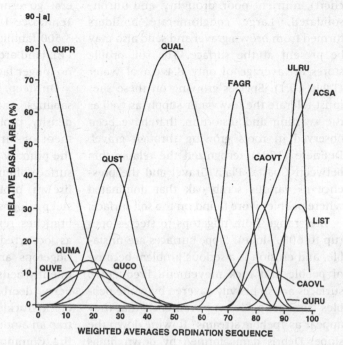

Figure 10.2. Coenocline developed from weighted averages ordination of compositionally stable stands at Land Between The Lakes in Kentucky and Tennessee. Only species with a relative basal area of >10% are shown. Acronyms are QUPR, *Quercus primus*, chestnut oak; QUST, *Q. stellata*, post oak; QUVE, *Q. veutina*, black oak; QUMA, *Q. marilandica*, blackjack oak; QUCO, *Q. coccinea*, scarlet oak; QUAL, *Q. alba*, white oak; FAGR *Fagus grandifolia*, American beech; ULRU, *Ulmus rubra*, red elm; ACSA, *Acer saccharum*, sugar maple; CAOVT, *Carya ovata*, shagbark hickory; QURU, *Q. rubra*, northern red oak, LIST, *Liquidambar styraciflua*, sweetgum; CAOVL = *Carya ovalis*, red hickory.

juxtaposition but nearer the xeric–mesic section (Figure 10.2) (Franklin 1990; Fralish et al. 1993; Franklin et al. 1993).

Chestnut Oak Woodland

This former gravel barrens community generally is located on extensive areas of white and brown gravel deposits near Kentucky Lake on the west side of the Tennessee Ridge, but with isolated pockets on the east side. The soil is extremely rocky (up to 65% by volume; white pebbles, mostly chert), nutrient poor, droughty, and unconsolidated. Large conglomerate boulders formed from brown gravel and sand also may be present at the surface. The soil profile stores an average of only 8.4 cm of water (Table 10.1). Species growing on these sites must tolerate the low water supply as well as the twisting and distortion that have been observed in roots growing through gravel. DeFriese (1884) recognized the relationship between Coastal Plain Gravel and the presence of various scrub oak that dominated where pebbles were found on the soil surface.

Sites range from ridgetops to steep slopes (up to 60% slope). Slope surfaces are unstable, and erosion is a serious problem because of pebble and rock movement. Frequently, surfaces are so heavily covered by white pebbles (similar to "desert pavement") that they appear as "pebble streams" flowing down the slope. Debris dams formed by down limbs and smaller branches are common and tend to stabilize steep slopes by preventing rock and organic matter movement.

Chestnut oak is the major dominant (IV = 61.6%; Franklin 1990; Table 10.1). Fourteen other tree species are present but, except for white oak (IV = 10.1%), have low importance values (<7.8%). Sourwood (*Oxydendrum arboreun*) is consistently present but has low importance values because of its small size.

Mature chestnut oak trees often are moderately large (DBH [diameter at breast height] = 30–40 cm), with an average age of 90 years for all stands. However, the age differential among the larger mature trees in a stand averaged 44 years (J. Fralish and S. Franklin, 1990 unpublished data). These data suggest that because of the xeric site conditions, stands slowly progressed from a barren community in the early 1900s to the present woodland. Moreover, 50%–75% of the trees are in the lowest 10–20-cm class, indicating a young stratum beneath the older overstory. The initiation of the younger stratum coincides with the 1944 survey of land by the fed-

eral government for impoundment of the Tennessee River (Kentucky Lake). In 1945, 3,500 families were moved from the present LBL land area, and cattle, pigs, and fire were no longer factors.

On steep, unstable gravel slopes, seedlings, saplings, and shrubs generally have a low density, probably from the excessive movement of surface rock. Here, sparkleberry is the primary midcanopy component. The soil surface often lacks leaf cover, suggesting that fire was probably not an important factor on steep slopes except when tree tops and branches remained after timber harvesting, as occurred in the 1800s. However, on ridgetops and gentle, stable slopes, a frequently dense understory of sweet low blueberry, deerberry (*Vaccinium stamineum*), and black huckleberry (*Gaylussacia baccata*) may trap an amount of leaf litter sufficient to carry fire. Common vines include muscadine grape (*Vitis rotundifolia*), summer grape (*Vitis aestivalis*), and catbrier (*Smilax glauca*).

The herb stratum includes a relatively small number of summer/fall species (18) but a few unidentified species were combined into genera such as *Carex* spp., *Lespedeza* spp., and *Panicum* spp. (Close 1996). Various species of cushion moss (primarily *Leucobryum glaucum*), other mosses (e.g., *Dicranum scoparium*), and lichen, particularly *Cladina rangiferina*, are common and often cover large areas (Table 10.2).

Post Oak Woodland

Unlike chestnut oak woodland, which dominates large contiguous areas of forest on steep slopes, post oak woodland usually occurs as isolated pockets on rolling topography. Thus, unlike the chestnut oak community, post oak stands were subject to more frequent fire, timber harvesting, and grazing. The lack of rock in the soil precludes classifying this community as a former barrens; in presettlement time, it was probably savanna.

This community is restricted to southwest slopes of moderate steepness (<25%) and shallow (<1 m) loess deposits. The loessal soil is droughty due to the presence of a fragipan (Lax soil type) at a depth of approximately 50 cm (Francis and Loftus 1977; Harris 1988; Fralish et al. 1993; Franklin et al. 1993). Available water-holding capacity indicates that the chestnut oak community occurs on more xeric sites than the post oak community (Table 10.1).

Post oak is the single major dominant (IV = 51%) followed by white oak (11.8%) and blackjack oak (8.6%). A variety of other oak species occur, including chestnut, black, southern red, and scarlet oak, as well as shagbark, pignut, and mockernut hickory. The basal area is moderate (19 m² ha⁻¹; Franklin 1990) and various canopy openings are present.

Post oak stands consistently have a mid-canopy of sparkleberry, serviceberry, flowering dogwood, sweet low blueberry, bullbrier, and muscadine grape. Thirty-one species were recorded in the summer/fall flora (Close 1996). The herbaceous component of the LBL stands is extremely similar to that of the Shawnee Hills and Ozark Hills post oak communities; however, the LBL stands have a substantially lower species richness (Table 10.2). Grazing of domestic livestock at LBL, particularly pigs, probably had a major impact on the herbaceous understory (Close 1996).

Threats

The greatest threat to continued existence of these communities is timber harvesting. Trees continue to be harvested at LBL but managers are aware that chestnut and post oak communities have low productivity because of the soil conditions. Examples of these community types are included in the Biosphere Reserve core area. Such protection is consistent with management of LBL as a demonstration area for recreation and forest resources. Succession to other forest types is not a threat as composition appears to be extremely stable.

Summary

The post oak and chestnut oak woodlands of southern Illinois and at Land Between The Lakes in western Kentucky and northwest middle Tennessee are the result of the removal of fire from the barrens and savannas of presettlement and early settlement periods. Although accounts describe the barrens and prairies as occurring across the landscape, only those remnants located on extreme sites survive to the present day. These sites are characterized by a near lack of soil (Shawnee Hills cedar glades), highly rocky soil of low water-holding capacity (Shawnee Hills post oak, and LBL post and chestnut oak woodlands), and moisture conditions that vary from ponded in the spring to permanent wilting point in midsummer (Till Plain post oak flatwoods).

A comparison of present stands with data generated from General Land Office survey records indicates that although density may have increased in 200 years, the overstory composition remained similar (percent similarity = 65%–76%) (Fralish et al. 1991). It would appear that presettlement-type post oak savanna could be restored with occasional burning, perhaps once every 3–5 years. In the nearby southeast Missouri Ozarks, Guyette and Cutter (1991) and Cutter and Guyette (1994) estimated that the fire return period for post oak savanna varied from 2.8 to 4.3 years until about 1850, but a major change in the mosaic pattern and tree density occurred when the return period increased to as high as 24 years after 1850.

In the Missouri Ozarks, removal of fire from the ecosystem has permitted redcedar to invade and degrade barrens and savanna; it is regarded as a highly fire-intolerant species (Burns and Honkala 1990a). Beilman and Brenner (1951) and Hall (1955) cite the absence of fire as the chief cause for the recent rapid redcedar expansion. The species is climax on sandstone outcrops in the Illinois Shawnee Hills (Fralish 1988a), and paradoxically, removal of the trees, either mechanically or by fire, possibly would destabilize the glade system by eliminating the root system and exposing soil to increased erosional forces.

Because of low fuel loading most of the year, it is doubtful that fire could be successfully used to open the overstory canopy of chestnut oak woodland at Land Between The Lakes. At present, there are few small oak trees or saplings; thus, as the large, old, and frequently diseased overstory trees die, canopy gaps will develop and the community probably will return to a more barrens structure. At the moment, the stands on the Cretaceous gravel of the Mississippi Embayment represent a most unusual community.

Acknowledgments

The research was supported by the Department of Forestry, Southern Illinois University, Carbondale, Illinois; McIntire-Stennis Cooperative Forest Research, The Center for Field Biology, Austin Peay State University, Clarksville, Tennessee; and Land Between The Lakes (TVA), Golden Pond, Kentucky. The authors acknowledge the Tennessee Valley Authority (Land Between The Lakes); Illinois Department of Conservation; Nature Preserves Commission; and U.S. Forest Service (Shawnee National Forest) for kindly permitting the collection of data from land under their jurisdiction.

REFERENCES

Anderson, R. C., and Anderson, M. R. 1975. The presettlement vegetation of Williamson County, Illinois. *Castanea* 40:345–363.

Beilman, A. P., and Brenner, L. G. 1951. Changing forest flora of the Ozarks. *Annals of the Missouri Botanical Garden* 38:283–291.

Braun, E. L. 1950. *Deciduous Forests of Eastern North America*. New York: Hafner Press.

Brennan, M. E. 1983. Site characteristics for major tree species in the Ozark region of southern Illinois. M.S. thesis, Department of Forestry, Southern Illinois University, Carbondale, Ill.

Bryant, W. S., and Martin, W. H. 1988. Vegetation of the Jackson Purchase of Kentucky based on the 1820 General Land Office survey records. In *Proceedings of the First Annual Symposium on the Natural History of Lower Tennessee and Cumberland River Valleys*, ed. D. H. Snyder, pp. 264–276. Clarksville, Tenn.: Center for Field Biology, Austin Peay State University.

Burns. R. M., and Honkala, B. H. 1990a. *Silvics of North America: Conifers*, Vol. 1. U.S. Forest Service, Agriculture Handbook 654.

Burns, R. M., and Honkala, B. H. 1990b. *Silvics of North America: Hardwoods*, Vol. 2. U.S. Forest Service Agriculture Handbook 654.

Cerretti, D. S. 1975. Vegetation and soil-site relationships for the Shawnee Hills Region, Southern Illinois. M.S. thesis, Department of Forestry, Southern Illinois University, Carbondale, Ill.

Chambers, J. L. 1972. The compositional gradient for undisturbed upland forests in Southern Illinois. M.S. thesis, Southern Illinois University, Carbondale, Ill.

Close, D. D. 1996. Evaluation of herbaceous diversity and differential species in mature forest stands at Land Between The Lakes, Kentucky and Tennessee. M.S. thesis, Southern Illinois University, Carbondale, Ill.

Crooks, F. B. 1988. Comparison of presettlement and old growth forest communities in the Shawnee Hills, Illinois. M.S. thesis, Department of Forestry, Southern Illinois University, Carbondale, Ill.

Cutter, B. E., and Guyette, R. P. 1994. Fire frequency on an oak–hickory ridgetop in the Missouri Ozarks. *The American Midland Naturalist* 132:393–398.

DeFriese, L. H. 1884. Timbers of the district west of the Tennessee River. In *Geological Survey of Kentucky: Timber and Botany*, pp. 141–170. Frankfort, Ky.: U.S. Geological Survey.

Delcourt, H. R., and Delcourt, P. A. 1997. Precolumbian Native American use of fire on southern Appalachian landscapes. *Conservation Biology* 11:1010–1014.

DeSelm, H. R. 1988. The barrens of the Western Highland Rim of Tennessee. In *Proceedings of the First Annual Symposium on the Natural History of Lower Tennessee and Cumberland River Valleys*, ed.

D. H. Snyder, pp. 199–219. Clarksville, Tenn.: Center for Field Biology, Austin Peay State University.

Downs, J. M. 1976. Soil water regimes for undisturbed forest communities in the Shawnee Hills, Southern Illinois. M.S. thesis, Department of Forestry, Southern Illinois University, Carbondale, Ill.

Dunn, C. P., and Stearns, F. 1993. Landscape ecology in Wisconsin: 1830–1990. In *John T. Curtis: Fifty Years of Wisconsin Plant Ecology*, ed. J. S. Fralish, R. P. McIntosh, and O. L. Loucks, pp. 197–216. Madison, Wis.: Wisconsin Academy of Science, Arts and Letters.

Fehrenbacher, J. B., Alexander, J. D. , Jansen, I. J., Darmody, R. G., Pope, R. A., and Flock, M. A.. 1984. *Soils of Illinois*. Bulletin 778. Champaign, Ill.: University of Illinois.

Fenneman, N. M. 1938. *Physiography of Eastern United States*. New York: McGraw-Hill Book Co.

Fralish, J. S. 1976. Forest site–community relationships in the Shawnee Hills Region, southern Illinois. In *Proceedings of the First Central Hardwood Forest Conference*, ed. J. S. Fralish, G. T. Weaver, and R. C. Schlesinger, pp. 65–87. Carbondale, Ill.: Southern Illinois University.

Fralish, J. S. 1987. Forest stand basal area and its relationship to individual soil and topographic factors in the Shawnee Hills. *Transactions of the Illinois State Academy of Science* 80:183–194.

Fralish, J. S. 1988a. Predicting potential stand composition from site characteristics in the Shawnee Hills forest of Illinois. *The American Midland Naturalist* 120:79–101.

Fralish, J. S. 1988b. Diameter–height–biomass relationships for *Quercus* and *Carya* in Posen Woods Nature Preserve. *Transactions of the Illinois State Academy Science* 81:31–38.

Fralish, J. S. 1994. The effect of site environment on forest productivity in the Illinois Shawnee Hills. *Ecological Applications* 4:134–143.

Fralish, J. S. 1997. Community succession, diversity and disturbance in the central hardwood forest. In *Conservation in Highly Fragmented Landscapes*, ed. M. W. Schwartz, pp. 234–266. New York: Chapman & Hall.

Fralish, J. S., and Crooks, F. B. 1988. Forest communities of the Kentucky portion of Land Between The Lakes: A preliminary assessment. In *Proceedings of the First Annual Symposium on the Natural History of Lower Tennessee and Cumberland River Valleys*, ed. D. H. Snyder, pp. 164–175. Clarksville, Tenn.: Center for Field Biology, Austin Peay State University.

Fralish, J. S., and Crooks, F. B. 1989. Forest composition, environment and dynamics at Land Between The Lakes in northwest middle Tennessee. *Journal of the Tennessee Academy of Science* 64:107–111.

Fralish, J. S., Crooks F. B., Chambers J. L., and Harty F. M . 1991. Comparison of presettlement, second-growth and old-growth forest on six site types in the Illinois Shawnee Hills. *The American Midland Naturalist* 125:294–309.

Fralish, J. S., Jones, S. M., O'Dell, R. K., and Chambers, J. L. 1978. The effect of soil moisture on site productivity and forest composition in the Shawnee Hills of southern Illinois. In *Proceedings of the Soil Moisture–Site Productivity Symposium*, ed. W. E. Balmer, pp. 263–285. Atlanta, Ga.: U.S. Forest Service.

Fralish, J. S., Franklin, S. C., Robertson, P. A., Kettler, S. M., and Crooks F. B. 1993. An ordination of compositionally stable and unstable forest communities at Land Between The Lakes, Kentucky and Tennessee. In *John T. Curtis: Fifty Years of Wisconsin Plant Ecology*, ed. J. S. Fralish, R. P. McIntosh, and O. L. Loucks, pp. 247–267. Madison, Wis.: Wisconsin Academy of Science, Arts and Letters.

Francis, J. K., and Loftus, N. F. 1977. *Chemical and Physical Properties of the Cumberland Plateau and Highland Rim Forest Soils*. USDA Forest Service, Southern Forest Experiment Station, New Orleans, La. Service Research Paper SO-138, pp. 33–44.

Franklin, S. B. 1990. The effect of soil and topography on forest community composition at Land Between The Lakes, KY and TN. M.S. thesis, Department of Forestry, Southern Illinois University, Carbondale, Ill.

Franklin, S. B. 1994. Late pleistocene and holocene vegetation history of Land Between The Lakes, Kentucky and Tennessee. *Transactions of the Kentucky Academy Science* 55:6–19.

Franklin, S. B., Robertson, P. A., Fralish, J. S., and Kettler, S. M. 1993. Overstory vegetation and successional trends of Land Between The Lakes, U.S.A. *Journal of Vegetation Science* 4:1–12.

Guyette, R. P., and Cutter, B. E. 1991. Tree-ring analysis of fire history of a post oak savanna in the Missouri Ozarks. *Natural Areas Journal* 11:93–99.

Hall, M. T. 1955. Comparison of juniper populations on an Ozark glade and old fields. *Annals of the Missouri Botanical Garden* 62:171–194.

Harris, S. E., Jr. 1988. Summary review of geology of Land Between The Lakes, Kentucky and Tennessee. In *Proceedings of the First Annual Symposium on the Natural History of Lower Tennessee and Cumberland River Valleys*, ed. D. H. Snyder, pp. 26–83. Clarksville, Tenn.: Center for Field Biology, Austin Peay State University.

Harty, F. M. 1978. Tree and herb species distribution, and herbaceous vegetation as an indicator of site quality for disturbed upland forest stands in the Greater Shawnee Hills of southern Illinois. M.S. thesis, Department of Forestry, Southern Illinois University, Carbondale, Ill.

Haywood, J. 1959. *Natural and Aboriginal History of Tennessee*. Kingsport, Tenn.: Kingsport Press.

Helmig, L. M. 1997. Predicting the threshold time of conversion from *Quercus–Carya* to mesophytic forest in the Illinois Ozark Hills. M.S. thesis, Department of Forestry, Southern Illinois University, Carbondale, Ill.

Henry, J. M. 1975. *The Land Between the Rivers*. Clarksville, Tenn.: Austin Peay State University, and Tennessee Valley Authority.

Iverson, L. R., Oliver, R. L., Tucker, D. P., Risser, P. G., Burnett, C. D., and Rayburn, R. G. 1989. *The Forest Resources of Illinois: An Atlas and Analysis of Spatial and Temporal Trends*. Special Publication 11. Champaign, Ill.: Illinois Natural History Survey.

Jones, S. M. 1974. Herbaceous vegetation as an indicator of site quality and potential stand composition for upland forest in southern Illinois. M.S. thesis, Department of Forestry, Southern Illinois University, Carbondale, Ill.

Katerere, Y. M. S. 1979. The influence of site characteristics on forest stand composition and basal area in the Southern Till Plain Division of Illinois. M.S. thesis, Department of Forestry, Southern Illinois University, Carbondale, Ill.

Ladd, D. 1991. Reexamination of the role of fire in Missouri oak woodlands. In *Proceedings of Oak Woods Management Workshop*, ed. J. E. Ebinger and G. S. Wilhelm, pp. 67–80. Charleston, Ill.: Eastern Illinois University.

Loughridge, R. N. 1888. *Report of the Geological and Economic Features of the Jackson Purchase Region*. Frankfort, Ky.: Kentucky Geological Survey.

McArdle, T. G. 1991. Comparison of presettlement and present forest communities by site type in the Illinois Ozark Hills. M.S. thesis, Department of Forestry, Southern Illinois University, Carbondale, Ill.

McCrain, G. R., and Grubb, A. L. 1987. *An Analysis of Past and Present Vegetation Patterns and Historic Parameters at Fort Donelson National Battlefield with Recommendations for Restoration and Future Management*. Raleigh, N.C.: Resource Management Co.

Menges, E. S., Dolan, R. W., and McGrath, D. J. 1987. *Vegetation, Environment, and Fire in a Post Oak Flatwoods/Barrens Association in Southwestern Indiana*. Indianapolis, Ind.: Holcomb Research Institute, Butler University.

Miles, C. C. 1996. *Soil Survey of Marion County, Illinois*. Washington, D.C.: USDA Natural Resources Conservation Service.

Miller, R. B. 1920. *Fire Prevention in Illinois*. Springfield, Ill.: Illinois Natural History Survey, Forestry Circular No. 2.

Miller, R. B., and Fuller, G. D. 1922. Forest conditions in Alexander County, Illinois. *Transactions of the Illinois State Academy of Science* 14:92–108.

O'Dell, R. K. 1978. A study of mid-canopy strata in upland forest communities of the Shawnee Hills regions of southern Illinois. M.S. thesis, Department of Forestry, Southern Illinois University, Carbondale, Ill.

Phelps, T. R. 1998. Comparison of presettlement and present vegetation cover of Marion County, Illinois, using a geographic information system. M.S. thesis, Department of Forestry, Southern Illinois University, Carbondale, Ill.

Richards, J. F., and Berry, G. N. 1885. The Jackson Purchase. In *Kentucky, A Brief History of the State*, vol. 2, ed. J. H. Battle, W. H. Perrin, and G. C. Kniffen. Louisville, Ky.: F. A. Battey.

Schwegman, J. 1973. *Natural Divisions of Illinois*. Rockford, Ill.: Illinois Nature Preserves Commission.

Sharpe, D. M., Guntenspergen, G. R., Dunn, C. P., Leitner, L. A., and Stearns, F. 1987. Vegetation dynamics in a southern Wisconsin agricultural landscape. In *Landscape Heterogeneity and Disturbance*, ed. M. G. Turner, pp. 137–55. New York: Springer-Verlag.

Simpson, J. W., Boerner, R. E., DeMers, M. N., Artigas, F. J., and Silva, A. 1994. Forty-eight years of landscape change on two contiguous Ohio landscapes. *Landscape Ecology* 9:261–270.

Smalley, G. W. 1980. *Classification and Evaluation of*

Forest Sites on the Western Highland Rim and *Pennyroyal*. New Orleans, La.: USDA Forest Service,.Gen. Tech. Rep. SO-30.

Taft, J. B., Schwartz, M. W., and Loy, R. P. 1995. Vegetation ecology of flatwoods on the Illinoian Till Plain. *Journal of Vegetation Science* 6:647–666.

Telford, C. J. 1926. *Third Report on a Forest Survey of Illinois*. Champaign, Ill.: Illinois Natural History, Survey Bulletin No. 16.

Thach, E. E., and Doyle, L. M. 1988. *Natural Resources Management, an Integral Part of Intensive Multiple Use: Land Between The Lakes, a Case History*. Golden Pond, Ky.: Tennessee Valley Authority.

United States Department of Agriculture. 1941. *Yearbook of Agriculture: Climate and Man*. Washington, D.C.

United States Department of Agriculture. 1953. *Soil Survey of Stewart County, Tennessee*. Washington, D.C.: Soil Conservation Service.

United States Department of Agriculture. 1981. *Soil Survey of Lyon and Trigg Counties, Kentucky*. Washington, D.C.: Soil Conservation Service.

Wallace, B. J. 1988. History of Land Between The Lakes. In *Proceedings of the First Annual Symposium on the Natural History of Lower Tennessee and Cumberland River Valleys*, ed. D. H. Snyder, pp. 84–144. Clarksville, Tenn.: Center for Field Biology, Austin Peay State University.

Warner, D. A. 1982. A study of mid-canopy strata in upland forest communities in the Ozark Hills regions of southern Illinois. M.S. thesis, Department of Forestry, Southern Illinois University, Carbondale, Ill.

11 The Big Barrens Region of Kentucky and Tennessee

JERRY M. BASKIN, CAROL C. BASKIN,
AND EDWARD W. CHESTER

Introduction

At the time of settlement by Europeans, ca. 1780, a large portion of the Kentucky Karst Plain (KKP, sensu Baskin, Baskin and Chester 1994; Figure 11.1) supported a type of vegetation known as barrens, that is, a deep-soil grassland with stunted trees (in particular, *Quercus* spp.) and shrubs interspersed with groves of timber (Gorin 1876; Austin 1904; Sauer 1927; Dicken 1935a). In Kentucky, this area became known as the Big Barrens Region, which extended into northwestern middle Tennessee (Transeau 1935; Shanks 1958; Chester 1988). Today, much of this region is used for crop and livestock production, and there are no "original" presettlement barrens on the KKP. Essentially all of the remaining forest is second growth (Smalley 1980); we know of only one old-growth forest on the KKP (Chester et al. 1995).

In Kentucky, historic deep-soil barrens also occurred on the Eastern Highland Rim (Haywood 1823; Lesley 1861; Dicken 1935a) and in the Jackson Purchase (Loughridge 1888; Dicken 1935a; Transeau 1935), and these will be discussed briefly.

Physical Setting

Boundaries

Based on geology and topography (including field examination of boundaries), our delimitations of the KKP agree, for the most part, with those defined by Quarterman and Powell (1978). The KKP (Figure 11.1, sensu

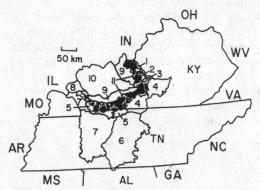

Figure 11.1. Historical occurrence of barrens (black) in the Big Barrens Region of Kentucky and Tennessee (modified from Baskin, Baskin and Chester 1994). Subsections (or section) of the Interior Low Plateaus Physiographic Province are 1 = Knobstone Escarpment and Knobs; 2 = Elizabethtown Plain; 3 = Brush Creek Hills; 4 = Greensburg Upland; 5 = Pennyroyal Plain; 6 = Central Basin (a section); 7 = Western Highland Rim; 8 = Marion; 9 = Mammoth Cave Plateau; and 10 = Ohio River Hills and Lowlands. The Dripping Spring Escarpment = 11. (Barrens on the Kentucky portion of the KKP are from Dicken 1935a, and those on the Tennessee portion are from Transeau 1935. Used with permission from the Southern Appalachian Botanical Society.)

Quarterman and Powell 1978) includes the Elizabethtown Plain (EP) and the Pennyroyal Plain (PP) subsections of the Highland Rim Section of the Interior Low Plateaus Physiographic Province (ILPP). These two subsections are separated by the Brush Creek Hills, a narrow and rugged subsection of the Shawnee Hills Section of the ILPP that extends northeastward from western Hart

County across northern Hart, southwestern Larue, northern Green, northwestern Taylor, and extreme southwestern Marion counties, Kentucky, to the Knobstone Escarpment and Knobs Subsection of the Bluegrass Section of the ILPP (Figure 11.1; Quarterman and Powell 1978).

In Kentucky, the boundaries of the EP and PP of Quarterman and Powell (1978) are generally similar to the Elizabethtown Area and the Pennyroyal Plain Area (PPA), respectively, of Sauer (1927). However, there is a notable difference in Quarterman and Powell's versus Sauer's boundaries of the PP (PPA) in its northwestern portion, as well as more minor differences elsewhere. The boundaries of the KKP in Kentucky are similar, in most respects, to those of Bailey and Winsor's (1964) Western Pennyroyal–Limestone Area. The Tennessee portion of the KKP is nearly equivalent to the Southern Pennyroyal of Fullerton et al. (1977).

Bedrock Geology

Information relating to bedrock geology and physiography was obtained from 7.5-minute geologic quadrangle maps of Kentucky (U.S. Geological Survey) and Tennessee (Division of Geology, State of Tennessee). Bedrocks on most of the KKP are the St. Louis or St. Genevieve, Upper Mississippian limestones of the Meramecian Series (Sable and Dever 1990). However, along some portions of its boundary, the KKP (Figure 11.1) extends onto bedrocks of the Chesterian (Upper Mississippian) or Osagean (Lower Mississippian) Series, and in other places the boundary is reached before leaving the St. Louis or St. Genevieve. Dip of the strata is westward (EP) and northwestward (PP) along the flank of the Cincinnati Arch into the Illinois Basin (Quarterman and Powell 1978).

Along a strip extending from southeast of Bowling Green in Warren County, Kentucky, to Horse Cave in Hart County, Kentucky, the boundary of the PP often is reached before leaving the St. Genevieve, which in some of

this area forms the base of the Dripping Spring Escarpment (see Figure 11.1). Small areas of the EP near its eastern boundary in Hardin County, Kentucky, are underlain by Salem or Harrodsburg limestones, and in Kentucky small areas at the eastern edge of the PP in Metcalfe and Simpson counties are underlain by Salem–Warsaw (undivided) limestones and the Fort Payne Formation. In addition to the St. Louis and St. Genevieve, the Warsaw and Fort Payne formations are bedrocks on the Pennyroyal Plain in Tennessee.

Physiography

The EP and PP subsections are level to steeply rolling karst plains (Sauer 1927; Dicken 1935b; Dicken and Brown 1938). Elevation on the KKP ranges from about 150 to 300 m above mean sea level; much of it is 150 to 200 m above mean sea level. According to Sauer (1927), there are some slight variations in physiography between the two subsections. Thus, compared to the PP, the EP (Sauer 1927, p. 46) "has fewer stretches [of land] that are conspicuously smooth and fewer flatwoods." Further, "in contrast to the Pennyroyal Plain, the region [EP] is drained relatively freely by or into streams and surface run-off is fairly great."

Sinkholes are numerous on the sinkhole plains (sensu Quinlan 1970) of the KKP (e.g., see Jillson 1924; Sauer 1927), but some areas are nearly sinkhole free. Most of the sinkholes are less than 200 m wide and 10 m deep; however, width and depth can be >1,600 m and 60 m, respectively (Quinlan 1970; Davies and LeGrande 1972). Jillson (1924) mentioned one sinkhole on the PP that had an area of 3,114 acres (1,261 ha). Several large natural sinkhole/depressional ponds and lakes occur on the KKP.

Soil erosion is a problem in the agricultural landscape of the KKP, and up to several meters of sediment have accumulated in the bottoms of the sinkholes (Sauer 1927, Dicken and Brown 1938). According to Sauer

(1927), many of the numerous small sink-hole ponds on the KKP did not exist prior to cultivation. They originated as a result of soil and detritus washing into the depressions from surrounding cultivated fields and blocking drainage. Further, even the floors of the flat-bottomed depressions present at settlement have been enlarged greatly. These basin floors may be the most fertile areas on the farm, and thus the primary (or only) place where row crops, such as tobacco, are grown (Sauer 1927; Dicken and Brown 1938).

Upland flats and depressions are distinctive features of the PP in southern Kentucky and in Tennessee. Some of the depressions are associated with sinkholes. Many of the wetlands have been cleared of their forests, drained (tiles and ditches), and converted to farmland. Floodplains, terraces, and/or rocky cliffs are associated with some of the streams of the region.

Soils

Soils of the KKP can be divided into three broad categories: (1) deep, moderately well drained to well-drained soils on level to steeply rolling uplands; (2) deep, well-drained to poorly drained soils on (in) floodplains, upland flats, and depressions; and (3) shallow to moderately deep, well-drained soils of broad ridges, side slopes, knobs and knoblike hills, benches, and nearly level to rolling karst uplands that often are associated with limestone rock outcrops. Parent materials for these (mostly) Alfisols and Ultisols include loess, residuum weathered from high-grade and cherty limestones, old alluvium, recent alluvium, and unconsolidated and weakly consolidated sandstone and shale. The great majority of soils on the KKP fit into category 1. Barrens were primarily on deep soils.

The major soil series on the KKP are the Crider, Pembroke, Nicholson, and Baxter. Crider, Pembroke, and Baxter soils are well drained, whereas the Nicholson is moderately well drained; all are on level to moder-

ately steep sloping uplands. Depth to bedrock is >1.5 m, and reaction is slightly (pH = 6.1–6.5) to very strongly (pH = 4.5–5.0) acidic, depending on soil series and horizon. The Crider–Pembroke and Pembroke–Crider are the most important agricultural soil associations on the KKP. Other important deep upland soils include Bewleyville, Dickson, Montview, and Ventrees. Bewleyville, Montview, and Ventrees soils are well drained, whereas those of the Dickson Series have a fragipan and thus are only moderately well drained. In the slumped sandstone/shale region of Hart, Hardin, and Larue counties, Kentucky, Gatton, Sonora, and Riney are the most important deep upland soils. Sonora and Riney soils are well drained, and Gatton soils, with a fragipan, are moderately well drained (various county soil surveys for Kentucky and Tennessee; Elder and Springer 1978; USDA-SCS 1994).

Drainage

Because of numerous sinkholes and sinking streams on the KKP, much of the drainage (all in some areas) is via complex underground drainage systems to the hydrologic base-level streams in the region, where the water emerges as springs. The Nolin River, Rough River, Otter Creek, Doe Run Creek, and Sinking Creek are hydrologic base-level streams on the EP (George 1976). The Green, Barren, and Little Barren rivers are the principle hydrologic base-level streams of the PP in the Central Kentucky Karst Region (Miotka and Pappenberg 1972; Hess, Wells and Brucker 1974; Hess et al. 1989; Quinlan et al. 1983). Drakes Creek, several forks of the Red River, and Little River and its Sinking and Muddy forks are the principal hydrologic base-level streams of the PP in the Western Kentucky Karst Region (Dyas 1985).

Climate

According to Thornthwaite (1948), the KKP has a humid mesothermal climate with intermediate thermal efficiency and with adequate precipitation for plant growth at all seasons ($B_2B'_2b'_3r$). In the Köppen system (Ackerman 1941), it has a mild, temperate, rainy climate without a distinct dry season and with a hot summer (Cfa). The climate of the KKP falls into region II of Borchert (1950), who described this climatic type as follows: "rainy winter; most of the winter rainfall of Anglo America east of the Rocky Mountains occurs in this area. The region receives as much or somewhat more rain in summer than in winter, but summers are sunnier." Figure 11.2 is a Walter-type climatic diagram for Mammoth Cave National Park, located immediately north of the Pennyroyal Plain in Barren, Edmonson, and Hart counties, Kentucky.

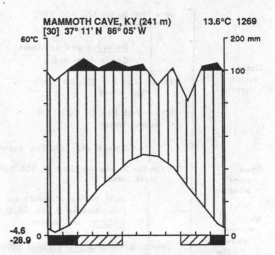

Figure 11.2. A Walter-type climatic diagram for Mammoth Cave National Park, which is immediately north of the Pennyroyal Plain in Barren, Edmonson, and Hart counties, Kentucky.

Paleovegetation

A vegetation chronosequence for the Interior Low Plateaus Region of Kentucky from 20,460 to ca. 220 yr BP has been constructed using fossil pollen recovered from a sediment core in Jackson Pond on the KKP in Larue County, Kentucky (Wilkins 1985; Wilkins et al. 1991). The vegetation on the KKP during this interval has changed in sequence from (1) a closed boreal forest of spruce and jack pine through (2) an open taiga-like boreal spruce woodland, (3) northern hardwood–conifer woodland, (4) mesic, open deciduous woodland, and (5) dry oak–hickory woodland, to (6) an oak–Poaceae "barrens" (at settlement) (Figure 11.3).

Pollen of neither grass nor "prairie" forbs shows a dramatic increase in percentage composition of nonarboreal pollen during the Hypsithermal Interval, as occurs in pollen diagrams when forest succeeds to grassland (e.g., Van Zant 1979; King 1981). However, there was a fairly dramatic rise in input of grass pollen at Jackson Pond after about 2,000 yr BP, suggesting that barrens vegetation on the KKP did not come into existence until after the Hypsithermal (Wilkins et al. 1991).

Potential Natural Vegetation: General Overview

Braun (1950) included almost all of the KKP in the Mississippian Plateau Section of her Western Mesophytic Forest Region; a small area in the northern part of the EP (Figure 11.1) is in the Hill Section of this forest region. She did not think the barrens on the KKP were "relics of the postglacial xerothermic period," but that their origin was more recent. Further, Braun stated that "the rapid changes which have taken place since disturbance of the Kentucky `barrens' point to a short period of dominance of prairie vegetation" (Braun 1950, p. 483).

Küchler (1964), on the other hand, mapped the potential natural vegetation of areas shown as barrens on Dicken's (1935a) map and as tallgrass prairie on Transeau's (1935) map as a mosaic of oak–hickory forest and bluestem prairie. He shows oak–hickory forest on the rest of the KKP. Thus, following Küchler, the climax vegetation of parts of the KKP on which barrens occurred is the same as that of the Prairie Peninsula of the Midwest. Interestingly, most vegetation maps

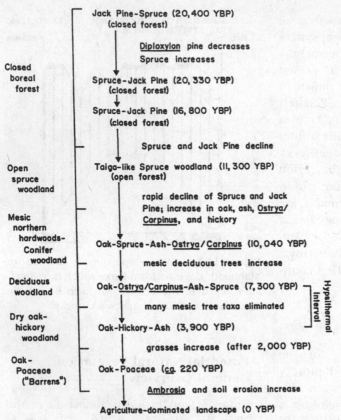

Figure 11.3. Paleovegetation sequence (20,460 to ca. 220 yr BP) on the KKP. Constructed from information in Wilkins (1985) and Wilkins et al. (1991).

of the tallgrass prairie published since 1935 include the Big Barrens in this vegetation type (e.g., Weaver 1954; Knapp 1965; Farney 1980; Anderson 1991; Chadwick 1993). An exception is Kucera's (1992) map of the North American tallgrass prairie. Vegetation maps of the United States published before 1935 do not show any prairies or barrens on the KKP (Sargent 1884; Shreve 1917; Shantz and Zon 1924).

Flora and Vegetation

The Fort Campbell Barrens

Chester et al. (1997) studied the floristic ecology of 22 stands of deep-soil barrens (sensu Baskin, Baskin and Chester 1994)

vegetation on the southwestern portion of the Pennyroyal Plain near the Kentucky–Tennessee border (Figure 11.4). The stands are in the area mapped as barrens or barrenlike vegetation by Shanks (1958) but west and southwest of the area mapped as barrens on the PP by Dicken (1935a), tallgrass prairie by Transeau (1935, 1956; also see Baskin, Stuckey and Baskin 1995), and oak–hickory forest/bluestem prairie by Küchler (1964). The nearly level to gently rolling karstic topography in this study area is nearly identical to nearby areas in Kentucky and Tennessee mapped as barrens by Dicken (1935a) and Transeau (1935). Thus, historically, barrens are likely to have been present here also. We know of no other good deep-soil barrens on the KKP.

The 22 barrens ranged from 1 to 40 ha, and averaged 12.2 ha. They are on the 42,510-ha Fort Campbell Military Reservation (FCMR) in parts of Trigg and Christian counties, Kentucky, and Montgomery and Stewart counties, Tennessee. The most important soils are Dickson silt loam, 1%–4% slopes (15 sites) and Montview silt loam, 5%–12% slopes, eroded (9 sites). Other important soils of these barrens are Hammack silt loam, 2%–6% slopes; Crider silt loam, 2%–8% slopes; Dickson silt loam, 4%–8% slopes; and Pickwick silt loam, 5%–12% slopes, eroded. Bedrock at 17 of the sites is Ste. Genevieve Limestone; part of each of the other 5 sites is underlain by Ste. Genevieve Limestone and part by St. Louis Limestone (see Chester et al. 1997).

USDA aerial photographs show that all 22 barrens once were cultivated fields. Since this

Figure 11.4. Portion of a 10-ha (top) and of a 30-ha (bottom) deep-soil barren on the Fort Campbell Military Reservation, Montgomery County, Tennessee. Soil is Dickson silt loam, 1%–4% slopes, and bedrock is Ste. Genevieve Limestone. The dominant plant is little bluestem. (Photographs taken by Carol C. Baskin on 18 September 1993.)

land was acquired by the Department of Defense in 1942, these barrens presumably have been created and/or maintained as prairie-grass barrens by prescribed, or sometimes accidental, burning by the military for use in various types of training exercises. Physiognomically (Figure 11.4), they appear to be quite similar to some of the presettlement barrens described by early travelers on the KKP (see Baskin, Baskin and Chester 1994). Throughout the growing seasons of 1993 and 1994 and in summer–autumn 1995, the authors and Dr. Eugene Wofford (Department of Botany, The University of Tennessee, Knoxville) did a floristic analysis of these 22 stands of barrens vegetation (Chester et al. 1997).

The flora of the FCMR barrens consists of (at least) 342 vascular taxa, of which 91% are native. The number of species observed on the 22 barrens ranged from 117 to 173, and averaged 145. Hemicryptophyte (52.3% of the taxa) is the dominant life form, and 90.6% of the taxa have the C3 photosynthetic pathway of carbon fixation; all but 2 of the 30 C4 taxa are perennial grasses. Nearly 88% of the barrens flora has an intraneous or extraneous-southern geographic distribution pattern; only about 5% are western taxa, and several of these are introduced. Most of the 311 native taxa occur in a variety of open to semi-open habitats. Nearly half of them occur in dry to moist woodlands, fields, or thickets; none is restricted to barrens and/or prairies, and none is endemic to the Big Barrens Region (Chester et al. 1997).

Little bluestem, *Schizachyrium scoparium*, is the most abundant species on these barrens. Sorensen's index values (I_s) for all pairwise comparisons of species composition among the 22 stands of barrens vegetation ranged from 61% to 87%, and averaged 73.5% (Chester et al. 1997), indicating membership in the same community type (Barbour, Burk and Pitts 1987). Eighty-two taxa occurred on 18–22 barrens (presence class 5), and 63 taxa, including *S. scoparium* and 7 other C4 peren-

nial grasses with an intraneous or extraneous-southern distribution, were present on all 22 barrens (Table 11.1).

Many plant species that occur in the KKP barrens also grow in the Midwest prairies (Baskin and Baskin 1981; Chester 1988). For the most part, however, these are geographically widespread eastern North American taxa of open habitats that are not restricted to prairies, for example, *Andropogon gerardii, Coreopsis tripteris, Eupatorium altissimum, Euphorbia corollata, Helenium autumnale, Helianthus hirsutus, Kuhnia eupatorioides, Liatris aspera, Lobelia spicata, Pycnanthemum tenuifolium, Schizachyrium scoparium, Solidago altissima* (= S. *canadensis* var. *scabra*), and *Sorghastrum nutans,* (Small 1933; Steyermark 1963; Radford, Ahles and Bell 1968; Gleason and Cronquist 1991). Further, the dominant grasses of the tallgrass prairie, such as *A. gerardii, S. scoparium,* and *S. nutans,* have intraneous distributions on the KKP. Nomenclature follows Wofford and Kral (1993).

The only stand of barrens vegetation on the KKP that has been sampled quantitatively is the Oakwood Barren, Montgomery County, Tennessee (DeSelm and Chester 1993), which was included in our floristic survey. Little bluestem was the dominant species (IV = 80.55/200) in this barren (Table 11.2). However, it was quite evident from our field observations that *S. scoparium* also was the dominant species in the other 21 stands of barrens vegetation we surveyed floristically on the FCMR (see Figure 11.4).

The Warfield Barren

The only stand of wet or wet–mesic barrens vegetation we know of on the KKP, the Warfield Barren, is on the Pennyroyal Plain in Montgomery County, Tennessee, about 1.0 km southeast of Guthrie, Kentucky. Most of this 1.4-km-long strip is in a depression between the Louisville and Nashville Railroad and U.S. Highway 41. Soils on 60% of the site are somewhat poorly drained (Taft, 42%) to poorly drained (Guthrie, 18%) silt loam soils.

Table 11.1. *Vascular plant taxa in presence class 5 (Cain and Castro 1959) in 22 stands of barrens vegetation on the southwestern portion of the Pennyroyal Plain in Christian and Trigg counties, Kentucky, and Montgomery and Stewart counties, Tennessee. Taxa marked with a superscript a were present on all 22 barrens. PP = photosynthetic pathway (C3, C4, CAM); G = geographical distribution pattern for native taxa (I = intraneous distribution, and EN, ES, and EW = extraneous northern, extraneous southern, and extraneous western distributions, respectively); LF = life-form (Ph = phanerophyte, H = hemicryptophyte, Cr = cryptophyte, and Th = therophyte). Information on photosynthetic pathway, geographical distribution, and life-form was obtained from various literature references.*

Taxon	PP	G	LF
Forbs			
[a]*Achillea millefolium*	C3	I	H
Agalinus fasciculata	C3	ES	Th
[a]*A. tenuifolia*	C3	I	Th
[a]*Allium vineale*	C3	nonnative	Cr
Ambrosia artemisiifolia	C3	I	Th
A. bidentata	C3	EW	Th
[a]*Apocynum cannabinum*	C3	I	H
[a]*Asclepias amplexicaulis*	C3	I	H
[a]*A. tuberosa*	C3	I	H
[a]*Aster dumosus*	C3	ES	H
[a]*A. pilosus*	C3	I	H
[a]*Chamaecrista fasciculata*	C3	I	Th
[a]*Chrysanthemum leucanthemum*	C3	nonnative	H
Cirsium discolor	C3	EN	H
[a]*Coreopsis major*	C3	ES	H
[a]*Daucus carota*	C3	nonnative	H
Desmodium canescens	C3	I	H
[a]*D. ciliare*	C3	I	H
D. paniculatum	C3	I	H
[a]*D. sessilifolium*	C3	I	H
Diodia teres	C3	ES	Th
[a]*Erigeron strigosus*	C3	I	H
[a]*Eupatorium altissimum*	C3	ES	H
[a]*E. hyssopifolium*	C3	ES	H
[a]*E. rotundifolium*	C3	ES	H
[a]*Euphorbia corollata*	C3	I	Cr
Euthamia graminifolia	C3	EN	H
[a]*Fragaria virginiana*	C3	EN	H
[a]*Galium pilosum*	C3	I	H
[a]*Gaura biennis*	C3	EN	H
[a]*Gnaphalium obtusifolium*	C3	I	Th
[a]*Hedyotis purpurea*	C3	ES	H
[a]*Helianthus hirsutus*	C3	I	Cr
[a]*H. mollis*	C3	I	Cr
[a]*H. occidentalis*	C3	I	H
Hypericum drummondii	C3	ES	Th
H. gentianoides	C3	I	Th
H. punctatum	C3	I	H
Kummerowia stipulacea	C3	nonnative	Th
[a]*Lespedeza procumbens*	C3	I	H
[a]*L. virginica*	C3	I	H
[a]*Linum striatum*	C3	I	H
Lobelia puberula	C3	ES	H

(cont.)

Table 11.1 (cont.)

Taxon	PP	G	LF
[a]Oxalis stricta	C3	I	H
Polygala sanguinea	C3	I	Th
[a]Potentilla simplex	C3	I	H
Prunella vulgaris	C3	nonnative	H
[a]Pycnanthemum pilosum	C3	ES	H
[a]P. tenuifolium	C3	ES	H
[a]Rudbeckia hirta	C3	I	H
Sabatia angularis	C3	I	H
[a]Sericocarpus linifolius	C3	I	H
Solidago altissima	C3	I	H
[a]S. juncea	C3	EN	H
[a]S. nemoralis	C3	I	H
[a]Stylosanthes biflora	C3	ES	H
Graminoids			
[a]Andropogon gyrans	C4	ES	H
[a]A. ternarius	C4	ES	H
[a]Carex complanata	C3	ES	Cr
[a]Elymus virginicus	C3	I	H
[a]Eragrostis spectabilis	C4	I	H
[a]Erianthus alopecuroides	C4	ES	H
Panicum anceps	C4	ES	Cr
[a]Schizachyrium scoparium	C4	I	H
[a]Setaria parviflora	C4	ES	H
[a]Sorghastrum nutans	C4	I	Cr
[a]Tridens flavus	C4	I	H
Shrubs			
Prunus angustifolia	C3	ES	Ph
[a]Rhus copallina	C3	I	Ph
[a]Rosa carolina	C3	I	Ph
[a]Rubus argutus	C3	I	H
[a]R. flagellaris	C3	I	H
Trees			
[a]Acer rubrum	C3	I	Ph
[a]Cornus florida	C3	I	Ph
[a]Diospyros virginiana	C3	I	Ph
[a]Nyssa sylvatica	C3	I	Ph
[a]Prunus serotina	C3	I	Ph
[a]Quercus falcata	C3	ES	Ph
[a]Sassafras albidum	C3	I	Ph
Vines			
[a]Lonicera japonica	C3	nonnative	Ph
[a]Smilax glauca	C3	ES	Ph
[a]Strophostyles umbellata	C3	ES	Cr

Table 11.2. *Percentage frequency, mean percentage cover, and importance value (IV) of vascular plant taxa in the Oakwood Barren on the southwestern portion of the Pennyroyal Plain, Montgomery County, Tennessee, for species with an IV ≥1.0. Percentage cover and importance values have been recalculated from data in DeSelm and Chester (1993), so cover is mean percentage cover per m² by species. IV = (Relative frequency + Relative cover)/2.*

Species	% frequency	% cover	IV
Schizachyrium scoparium	100	58.00	80.55
Pycnanthemum tenuifolium	75	6.75	16.74
Helianthus angustifolius	65	1.30	9.10
Solidago juncea	65	1.30	9.10
Diospyros virginiana	20	4.60	7.80
Lespedeza virginica	55	1.10	7.71
Smilax glauca	50	1.00	7.00
Sorghastrum nutans	30	2.40	6.34
Eupatorium rotundifolium	40	0.80	5.60
Aster solidagineus	30	1.50	5.27
Aster dumosus	40	0.40	5.13
Vernonia gigantea	30	1.20	4.92
Acer rubrum	30	0.90	4.56
Panicum commutatum	30	0.30	3.85
Eupatorium hyssopifolium	20	0.20	2.56
Panicum lanuginosum	20	0.20	2.56
Solidago nemoralis	20	0.20	2.56
Andropogon gyrans	15	0.60	2.45
Gerardia tenuifolia	15	0.15	1.92
Polygala verticillata	15	0.15	1.92
Scleria pauciflora	15	0.15	1.92
Rhus copallina	10	0.20	1.40
Boltonia diffusa	10	0.10	1.28
Diodia teres	10	0.10	1.28
Polygala sanguinea	10	0.10	1.28
8 other species	–	–	5.12
Total			199.92

The well-drained silt loam Pembroke (2%–5% slopes) and Crider (2%–5% slopes) soils cover 23% and 12% of the area, respectively. Bedrock at the site is Ste. Genevieve Limestone.

DeSelm and Chester (1993) reported 147 native and 19 nonnative plant taxa from the Warfield Barren. It is the only barren on the KKP in which prairie cordgrass, *Spartina pectinata*, has been collected. Specimens at APSC and TENN vouch for the presence of this species at the Warfield site in the early 1940s.

In the early 1980s, big bluestem, *Andropogon gerardii*, and Indian grass, *Sorghastrum nutans*, were the apparent dominants of this barren, which since has been converted to a pasture-like grassland with agronomic species (mostly fescue and orchard grass) (E. W. Chester, personal observation).

Historical Evidence

The only portions of Kentucky and Tennessee for which General Land Office surveys were conducted are west of the Tennessee River, that is, Jackson Purchase (sensu Lobeck 1928). However, ample evidence exists in the writings of persons who saw the Big Barrens Region in the late 18th and early 19th centuries to show that at the time of settlement by Europeans much of it was a grassland with forbs and shrubs interspersed with stunted oaks and hickories, that is, barrens (see Baskin, Baskin and Chester 1994).

Origin

Various theories have been proposed for the origin (and maintenance) of the barrens on the KKP (see Baskin, Baskin and Chester 1994). However, the one with the most support, and certainly the most reasonable one, is that these grasslands resulted from burning of forest by Native Americans, perhaps starting about 2,000 yr BP (Wilkins et al. 1991). Keith (1983) also concluded that fire was responsible for the origin and maintenance of barrens on the Mitchell Plain, a continuation of the KKP north of the Ohio River in southern Indiana (Quarterman and Powell 1978).

KKP Barrens Not Outlier of Tall Grass Prairie

Since publication of Transeau's (1935) paper on the Prairie Peninsula, the barrens of the KKP traditionally have been considered to be an outlier of the Midwestern tallgrass prairie. However, there is considerable evidence to the contrary. Baskin and Baskin (1981) and Baskin, Baskin and Chester (1994) gave eight reasons why the KKP barrens do not belong to the tallgrass prairie biome. Of these, the most convincing arguments for not including them in the Prairie Peninsula are that (1) the KKP remained forested throughout the Hypsithermal Interval (Delcourt and Delcourt 1981; Wilkins et al. 1991) and (2) soils of the presettlement KKP barrens developed under forest vegetation (Austin et al. 1953; Bailey and Winsor 1964).

Other Presettlement Deep-Soil Barrens in Kentucky

Historical evidence indicates that prairie-like barrens also occurred on the karstic (Upper Mississippian) limestone topography of the Eastern Highland Rim (sensu Quarterman and Powell 1978) and on the unconsolidated Quaternary–Tertiary deposits (Olive 1980) in the Jackson Purchase.

Eastern Highland Rim

According to Haywood (1823, p. 89), the barrens of the Upper Cumberland River (Eastern Highland Rim) were visited in 1769 by a company of 20 or more hunters from North Carolina. They established a base camp "in an open country called *barrens* six to seven miles from where the Wayne County [Kentucky] courthouse now stands." Then, "they dispersed in different directions to different parts of the country, the whole company still travelling to the southwest.... All the country through which these hunters passed was *covered with tall grass*, which seemed inexhaustible."

Lesley (1861, p. 492) wrote the following about the barrens of Wayne County (Eastern Highland Rim):

'The Barrens' have a gently undulating surface of the red calcareous soils of the *Lithostrotion* [a genus of fossil corals particularly characteristic of the St. Louis Limestone] division of the subcarboniferous [Mississ-ippian] limestone, and, at the period of the early settlement of this country, were simply destitute of trees, being represented in the records of the day as having the appearance of open prairies covered with long rank grass. But at present, where not under cultivation they luxuriate beneath a heavy cover of timber, principally of black oak and black walnut, though which hickory, dogwood, and blackgum are common.

The following information accompanied a soil sample collected by Joseph Lesley, Jr., in northern Wayne County, Kentucky, "one mile from the forks of the main Cumberland

[River] and Big South Fork" on the "Sub-carboniferous limestone formation" and sent to Robert Peter (1880, p. 71) for analysis.

The 'Barrens' form a strip of the first great terrace above and south of the Cumberland river, averaging five miles in width and extending lengthwise from the forks of the Cumberland to and beyond Monticello. Fifty years ago they were open prairie, with only occasional high swells, covered with black oak timber; now they are covered, where not cultivated, with a fine 'second growth,' mostly of black oak and hickory, with scattering dogwood and black gum.

Deep-soil, grass-dominated areas on the Eastern Highland Rim of Tennessee, none of which are shown on Transeau's (1935) map, have been described by DeSelm (1989, 1990).

Jackson Purchase

The lands of the Jackson Purchase were opened for settlement in 1820. Under an act of the Kentucky General Assembly of 1820, William T. Henderson was appointed surveyor to lay off the land into townships and sections. In the front of his 540 pages of field notes, Henderson wrote, "A considerable portion of the country being open barrens where trees could not be found, posts are fixed at the corners." The many references to corner posts in Henderson's notes indicate that barrens were extensive. Also, when he described the quality of the land at the end of each mile, he mentioned barrens a number of times. Unfortunately, Henderson failed to record distances to trees or their diameters, and thus his notes cannot be used to reconstruct the presettlement vegetation of the Jackson Purchase (Bryant and Martin 1988). Of the 10,330 trees Henderson recorded in this survey, 61.3% were oaks and 11% were hickories. Post oak (*Quercus stellata*), black oak (*Q. velutina*), blackjack oak (*Q. marilandica*), white oak (*Q. alba*), and red oak (*Q. rubra*) accounted for 95.9% of the oaks; species of hickories

(*Carya* spp.) were not distinguished (Bryant and Martin 1988).

Loughridge's (1888) *Report on the Geologic and Economic Features of the Jackson Purchase Region* [of Kentucky] includes a number of references to barrens, among which is the following (Loughridge 1888, p. 163):

There is within the Purchase counties a broad region which was formerly an open treeless prairie, but which, within the past thirty years, has been covered with a low growth of red and black jack oaks, and is still known as "Barrens." In the country west of Mayfield to Obion Creek it is said that twenty-five years ago the prairie grass was as high as the head of a man on horseback.

An agricultural map showing the locations of barrens, as well as other types of lands, is included in the report.

Transeau (1935) included the extensive areas of barrens in the Jackson Purchase in Kentucky, and a few small areas of barrens in the northern portion of the Jackson Purchase in Tennessee, on his map of the Prairie Peninsula. Küchler (1964) (following Transeau) mapped these barrens as a mosaic of bluestem prairie and oak–hickory forest. However, we suggest that they were not an outlier of the tallgrass prairie, for essentially the same reasons that the barrens of the KKP were not part of this vegetation type (see Baskin, Baskin and Chester 1994). Davis (1923, p. 55) stated, "It is the almost unanimous opinion of early observers competent to judge that the treeless character of the prairies [of the Jackson Purchase] was due to firing of grass by the Indians." As in the barrens of the KKP, those of the Jackson Purchase quickly grew up into forest after European settlement. Loughridge (1888, p. 299) stated, "Since the opening up of the country to civilization, a growth of red oak and black jack oak has sprung up over the entire region."

Further, the soils of the barrens are not prairie soils. According to Loughridge (1888, p. 282), the soil of the barrens "differs but little from that of the oak and hickory uplands already described, and has a growth of blackjack and red oaks." And Loughridge also writes (p. 246) that "the soil [of the barrens] is a light brown loam three to four feet in depth over a clayey bed of loam." In some places in the barrens, there were "tracts of badly drained lands on which water stands, and which are whitish, stiff and crawfishy" (Loughridge 1888, p. 262). Modern county soil surveys also show that the soils of the former Jackson Purchase barrens were not developed under prairie vegetation.

Transeau's Maps of the Prairie Peninsula

The map of the Prairie Peninsula published by Transeau (1935) is a modified version of a pre-1935 map he prepared (Baskin, Stuckey and Baskin 1995). A different version of Transeau's pre-1935 map is published in Weaver's *North American Prairie* (1954). The map in Weaver (1954) and another map left by Transeau (1956) with "(revised 1956)" written on it, are nearly identical. In comparison to Transeau's 1935 map, these two versions have been trimmed so that the western and southwestern boundaries of the Prairie Peninsula are about 165 km east and 160 km north, respectively, of their locations on the 1935 published map (Baskin, Stuckey and Baskin 1995). The barrens of the KKP and the Jackson Purchase, as published by Transeau (1935), are on all versions of the map. However, since there is overwhelming evidence that these barrens were never an outlier of the tallgrass prairie, they should not be included on further revisions of the Prairie Peninsula map.

Summary

During the past 20,460 years, the climatic climax vegetation of the KKP has changed (through several stages) from boreal forest to temperate deciduous forest, which is the potential natural (climatic climax) vegetation of the region. The present landscape of this region, however, is dominated by agricultural crops, cattle pastures, and second-growth forests.

Deep-soil barrens that covered a large portion of the KKP in Kentucky at the time of European settlement came into existence after the Hypsithermal Interval as a result of the burning of the forests by Native Americans. The only good deep-soil barrens remaining on the KKP are on the Fort Campbell Military Reservation (FCMR); they are a perennial-grass stage in secondary succession and are maintained by burning and/or bush-hogging. The FCMR barrens appear to be physiognomically quite similar to those described by visitors to the region in the late 1700s and early 1800s. Their flora consists of (at least) 342 vascular taxa, 91% of which are native. Little bluestem is the dominant species of these barrens.

In Kentucky, historic barrens also occurred on the Upper Mississippian karstic limestone topography on the Eastern Highland Rim and on unconsolidated Quaternary–Tertiary deposits in the Jackson Purchase. Like the KKP barrens, those on the Eastern Highland Rim and in the Jackson Purchase (1) originated by burning of forests by Native Americans, (2) quickly grew up into oak-dominated forests after settlement by Europeans, and (3) occurred on soil that developed under forest vegetation.

Transeau (1935), and later other plant ecologists and plant geographers, included the barrens of the KKP and of the Jackson Purchase on their maps of the Prairie Peninsula. However, evidence from paleovegetation, soils, and other fields of study is overwhelming that neither the barrens of the KKP nor those of the Jackson Purchase were outliers of the tallgrass prairie.

REFERENCES

Ackerman, E. A. 1941. The Köppen classification of climates in North America. *The Geographical Review* 31:105–111.

Anderson, R. C. 1991. Illinois prairies: A historical perspective. In *Our Living Heritage*, ed. L. M. Page and M. R. Jeffords, pp. 384–391. Urbana, Ill.: Illinois Natural History Survey.

Austin, M. E., Beadles, C. B., Moore, R. W., Matzek, B. J., and Jenkins, C. 1953. *Soil Survey of Stewart County, Tennessee*. USDA- SCS in cooperation with the Tennessee Agricultural Experiment Station and the Tennessee Valley Authority.

Austin, S. F. 1821 [1904]. Journal of Stephen F. Austin on his first trip to Texas, 1821. *The Quarterly of the Texas State Historical Association (The Southwestern Historical Quarterly)* 7:286–307.

Bailey, H. H., and Winsor, J. H. 1964. *Kentucky Soils*. University of Kentucky Agricultural Experiment Station Miscellaneous Publication Number 308.

Barbour, M. G., Burk, J. H., and Pitts, W. D. 1987. *Terrestrial Plant Ecology*. 2nd ed. Menlo Park, Calif.: The Benjamin/Cummings Publishing Company, Inc.

Baskin, J. M., and Baskin, C. C. 1981. The Big Barrens of Kentucky not part of Transeau's Prairie Peninsula. In *The Prairie Peninsula – In the "Shadow" of Transeau*, ed. R. L. Stuckey and K. J. Reese, pp. 43–48. Proceedings of the Sixth North American Prairie Conference, The Ohio State University, Columbus. Ohio Biological Survey Biological Notes Number 15.

Baskin, J. M., Baskin, C. C., and Chester, E. W. 1994. The Big Barrens Region of Kentucky and Tennessee: Further observations and considerations. *Castanea* 59: 226–254.

Baskin, J. M., Stuckey, R. L., and Baskin, C. C. 1995. Variations in Transeau's maps of the Prairie Peninsula. In *Proceedings of the Fourteenth North American Prairie Conference*, ed. D. C. Hartnett, pp. 227–231. Manhattan, Kans.: Kansas State University.

Borchert, J. R. 1950. The climate of the central North American grassland. *Annals of the Association of American Geographers* 40:1–39.

Braun, E. L. 1950. *Deciduous Forests of Eastern North America*. Philadelphia: Blakiston.

Bryant, W. S., and Martin, W. H. 1988. Vegetation of the Jackson Purchase of Kentucky based on the 1820 General Land Office Survey. In *Proceedings of the First Annual Symposium on the Natural History of Lower Tennessee and Cumberland River Valleys*, ed. D. H. Snyder, pp. 264–276. Clarksville, Tenn.: The Center for Field Biology of Land Between the Lakes, Austin Peay State University.

Cain, S. A., and Castro, G. M. 1959. *Manual of Vegetation Analysis*. New York: Harper and Brothers Publishers.

Chadwick, D. H. 1993. The American prairie – roots of the sky. *The National Geographic Magazine* 184(4): 90–119.

Chester, E. W. 1988. The Kentucky prairie barrens of northwest middle Tennessee: An historical and floristic perspective. In *Proceedings of the First Annual Symposium on the Natural History of Lower Tennessee and Cumberland River Valleys*, ed. D. H. Snyder, pp. 145–163. Clarksville, Tenn.: The Center for Field Biology of Land Between the Lakes, Austin Peay State University.

Chester, E. W., Noel, S. M., Baskin, J. M., Baskin, C. C., and McReynolds, M. L. 1995. A phytosociological analysis of an old-growth upland wet woods on the Pennyroyal Plain, southcentral Kentucky, USA. *Natural Areas Journal* 15:297–307.

Chester, E. W., Wofford, B. E., Baskin, J. M., and Baskin, C. C. 1997. A floristic study of barrens on the southwestern Pennyroyal Plain, Kentucky and Tennessee. *Castanea* 62:161–172.

Davies, W. E., and LeGrande, H. E. 1972. Karst of the United States. In *Important Karst Regions of the Northern Hemisphere*, ed. M. Herak and V. T. Springfield, pp. 467–505. New York: Elsevier.

Davis, D. H. 1923. The geography of the Jackson Purchase of Kentucky. *Kentucky Geological Survey Series* 6, Vol. 9.

Delcourt, P. A., and Delcourt, H. R. 1981. Vegetation maps of eastern North America: 40,000 yr BP to the present. In *Geobotany* II, ed. R. Romans, pp. 123–166. New York: Plenum Press.

DeSelm, H. R. 1989. The barrens of Tennessee. *Journal of the Tennessee Academy of Science* 64:89–95.

DeSelm, H. R. 1990. Flora and vegetation of some barrens on the Eastern Highland Rim of Tennessee. *Castanea* 55:187–206.

DeSelm, H. R., and Chester, E. W. 1993. Further studies on the barrens of the northern and western Highlands Rims of Tennessee. In *Proceedings of the First Annual Symposium on the Natural History of Lower Tennessee and*

Cumberland River Valleys, ed. S. W. Hamilton, E. W. Chester and A. F. Scott, pp. 137–160. Clarksville, Tenn.: Center for Field Biology, Austin Peay State University.

Dicken, S. N. 1935a. The Kentucky barrens. *Bulletin of the Geographical Society of Philadelphia* 43:42–51.

Dicken, S. N. 1935b. Kentucky karst landscapes. *Journal of Geology* 43:708–728.

Dicken, S. N., and Brown, H. B. Jr. 1938. *Soil Erosion in the Karst Lands of Kentucky*. USDA Circular Number 490.

Dyas, M. D. 1985. The western Kentucky speleological survey: A progress review. *The National Speleological Society Bulletin* 47(1):1–11.

Elder, J. A., and Springer, M. E. 1978. *General Soil Map of Tennessee*. USDA Soil Conservation Service in cooperation with the Tennessee Agricultural Experiment Station.

Farney, D. 1980. The tallgrass prairie: Can it be saved? *The National Geographic Magazine* 157(1):36–61.

Fullerton, R. O., McDowell, H. O., McMillion, M., Ray, B, and Terrell, R. P. 1977. Tennessee Geographical Patterns and Regions. Dubuque, Iowa: Kendall/Hunt.

George, A. I. 1976. Karst and cave distribution in north-central Kentucky. *The National Speleological Society Bulletin* 38(4):94–98.

Gleason, H. A., and Cronquist, A. 1991. *Manual of Vascular Plants of Northeastern United States and Adjacent Canada*. 2d ed. Bronx, N.Y.: The New York Botanical Garden.

Gorin, F. 1876. Barren County, Kentucky. (A series published in the Glasgow [Kentucky] Times). Reprinted in 1926 under the title of "Times of long ago" by J. P. Morton Co., Louisville, Ky.

Haywood, J. 1823. *The Civil and Political History of the State of Tennessee, from its Earliest Settlement up to the Year 1796, Including the Boundaries of the State*. Printed for the author by Heiskell Brown, Knoxville, Tenn.

Hess, J. W., Wells, S. G., and Brucker, T. A. 1974. A survey of springs along the Green and Barren rivers, Central Kentucky Karst. *The National Speleological Society Bulletin* 36(3):1–7.

Hess, J. W., Wells, S. G., Quinlan, J. F., and White, W. B. 1989. Hydrology of the South-central Karst. In *Karst Hydrology: Concepts from the Mammoth Cave Area*, ed. W. B. White and E. L. White, pp. 15–63. New York: Van Nostrand Reinhold.

Jillson, W. R. 1924. American karst country. *The Pan-American Geologist* 42(1):37–44.

Keith, J. H. 1983. Presettlement barrens of Harrison and Washington counties, Indiana. In *Proceedings of the Seventh North American Prairie Conference*, ed. C. L. Kucera, pp. 17–25. Springfield, Mo.: Southwest Missouri State University.

King, J. E. 1981. Late Quaternary vegetational history of Illinois. *Ecological Monographs* 51:43–62.

Knapp, R. 1965. *Die Vegetation von Nord- und Mittelamerika und der Hawaii-Inseln*. Stuttgart: Gustav Fischer Verlag.

Kucera, C. L. 1992. Tall grass prairie. In *Ecosystems of the World*. Vol. 8A. *Natural Grasslands. Introduction and Western Hemisphere*. Section II. *Grasslands of North America*, ed. R. T. Coupland, pp. 227–268. New York: Elsevier.

Küchler, A. W. 1964. *Potential Natural Vegetation of the Conterminous United States*. (Map and accompanying manual.) American Geographical Society Publication Number 36.

Lesley, J., Jr. 1861. Topographical and geological report of the country along the outcrop baseline, following the western margin of the Eastern Coalfield of the state of Kentucky through the counties of Carter, Rowan, Morgan, Bath, Montgomery, Powell, Estill, Owsley, Jackson, Rockcastle, Pulaski, Wayne, and Clinton, from a survey made during the years 1858–9. *Kentucky Geological Survey Series 1*, 4:438–494.

Lobeck, A. K. 1928. *The Geology and Physiography of the Mammoth Cave National Park*. Kentucky Geological Survey Series 6, Pamphlet Number 21.

Loughridge, R. N. 1888. *Report on the Geological and Economic Features of the Jackson Purchase Region, Embracing the Counties of Ballard, Calloway, Fulton, Graves, Hickman, McCracken, and Marshall*. Kentucky Geological Survey Reports of Special Subjects. Series 2, Vol. 6F.

Miotka, F.-D., and Pappenberg, H. 1972. Geomorphology and hydrology of the Sinkhole Plain and Glasgow Upland, Central Kentucky Karst: Preliminary report. *Caves and Karst* 14(4):25–32.

Olive, W. W. 1980. *Geologic Maps of the Jackson Purchase*. United States Geological Survey Miscellaneous Investigation Series Map I–1217 (with 11 pages of accompanying text).

Peter, R. 1880. Chemical report of the soils, coals, ores, clays, marls, mineral waters, rocks, etc., of Kentucky. Kentucky Geological Survey Report of Progress 5, New Series. Reprinted in *Kentucky Geological Survey Chemical Analyses* A2:1–93 [1885].

Quarterman, E., and Powell, R. L. 1978. *Potential Ecological/Geological Natural Landmarks on the Interior Low Plateaus.* Washington, D.C.: U.S. Department of the Interior.

Quinlan, J. F. 1970. The central Kentucky karst. *Études et Travaux de Mediterranée* 7:235–253.

Quinlan, J. F., Ewers, R. O., Ray, J. A., Powell, R. L., and Krothe, N. C. 1983. Ground-water hydrology and geomorphology of the Mammoth Cave Region, Kentucky, and of the Mitchell Plain, Indiana (Field Trip 7). In *Field Trips in Midwestern Geology*, Vol. 2, ed. R. H. Shaver and J. A. Sunderman, pp. 1–85. Geological Society of America and Indiana Geological Survey.

Radford, A. E., Ahles, H. E., and Bell, C. R. 1968. *Manual of the Vascular Flora of the Carolinas.* Chapel Hill, N.C.: University of North Carolina Press.

Sable, E. G., and Dever, G. R., Jr. 1990. *Mississippian Rocks in Kentucky.* U.S. Geological Survey Professional Paper Number 1503.

Sargent, C. S. 1884. *Report on the Forests of North America.* Tenth census of the United States. Vol. 9.

Sauer, C. O. 1927. *Geography of the Pennyroyal.* Kentucky Geological Survey Series 6, Vol. 25.

Shanks, R. E. 1958. Floristic regions of Tennessee. *Journal of the Tennessee Academy of Science* 33:195–210.

Shantz, H. L., and Zon, R. 1924. *Natural Vegetation.* Part 1, Sec. E. Atlas of American Agriculture. Washington, DC.: U.S. Department of Agriculture.

Shreve, F. 1917. A map of the vegetation of the United States. *The Geographical Review* 3:119–125 + map.

Small, J. K. 1933. *Manual of the Southeastern Flora.*

Chapel Hill, N.C.: University of North Carolina Press.

Smalley, G. W. 1980. *Classification and Evaluation of Forest Sites on the Western Highland Rim and Pennyroyal.* USDA Forest Service Gen. Tech. Rep. SO-30.

Steyermark, J. A. 1963. *Flora of Missouri.* Ames, Iowa: Iowa State University Press.

Thornthwaite, C. W. 1948. An approach toward a rational classification of climate. *The Geographical Review* 38:55–94 + map.

Transeau, E. N. 1935. The Prairie Peninsula. *Ecology* 16:423–437 + foldouts of figures 1, 9, 10, and 11. (Figure 1 is the map of the Prairie Peninsula.)

Transeau, E. N. 1956. Prairie Peninsula. Revised, unpublished black and white line drawing map, 11 x 15 inches. 1 p. (Reproduced on pages 176–177 in J. E. Weaver 1954, *North American Prairie.* Lincoln, Neb.: Johnsen Publishing Company.

USDA-SCS. 1994. *General Soils Map of Kentucky.* Fort Worth, Tex.: USDA, SCS National Cartography & GIS Center.

Van Zant, K. 1979. Late glacial and postglacial pollen and plant macrofossils from Lake West Okoboji, northwestern Iowa. *Quaternary Research* 12:358–380.

Weaver, J. E. 1954. *North American Prairie.* Lincoln, Neb.: Johnsen Publishing Company.

Wilkins, G. R. 1985. Late Quaternary vegetational history at Jackson Pond, Larue County, Kentucky. M.S. thesis, University of Tennessee, Knoxville, Tenn.

Wilkins, G. R., Delcourt, P. A., Delcourt, H. R., Harrison, F. W., and Turner, M. R. 1991. Paleoecology of central Kentucky since the last glacial maximum. *Quaternary Research* 36:224–239.

Wofford, B. E., and Kral, R. 1993. Checklist of the vascular plants of Tennessee.*Sida, Botanical Miscellany* No.10.

12 Cedar Glades of the Southeastern United States

JERRY M. BASKIN AND CAROL C. BASKIN

Introduction

Cedar or limestone glades (Figure 12.1) are open areas of rock pavement, gravel, flagstone, and/or shallow soil in which occur natural, long-persisting (edaphic climax) plant communities dominated by herbaceous angiosperms and/or cryptogams (J. M. Baskin and Baskin 1985a; Quarterman, Burbanck and Shure 1993). They may, or may not, be surrounded by forest (Galloway 1919). Cedar glades may support low densities of woody plants, which become established in deep soil-filled cracks in the bedrock (e.g., Picklesimer 1927 [see J. M. Baskin and Baskin 1996a]; Quarterman 1950b). The dominant plants are C4 summer annual grasses; C3 winter annual, summer annual, and/or perennial herbaceous dicots; mosses (primarily *Pleurochaete squarrosa*); the cyanobacterium *Nostoc commune*; and crustose, foliose, and fruticose lichens (Quarterman 1948, 1950a, b; Finn 1968; Mahr and Mathis 1981; Somers et al. 1986; Drew 1991; Dubois 1993)

Historically, in the Central Basin the term *cedar glades* has referred to rocky limestone openings ("glades") and the adjacent red-cedar–redcedar/hardwood–hardwood forest complex, that is, Quarterman's (1989) glade complex or ecosystem (e.g., Safford 1851; Galloway 1919; Freeman 1933; Quarterman 1950b; Mahr and Mathis 1981). However, in the past 20–30 years botanical studies in cedar glades of the southeastern United States (e.g., Ware 1969; J. M. Baskin and Baskin 1985a; Somers et al. 1986; Quarterman, Burbanck and Shure 1993; J. M. Baskin, Webb and Baskin 1995) have used *cedar glades* (or *limestone glades*) in reference to the rocky openings only, that is, *glades* or *open glades*, sensu Quarterman 1950b), and this is how we use the term in the present chapter.

Geography of Cedar Glades

As defined in this chapter, most cedar glades in the eastern United States are in the Southeast. Here, they occur in nine physiographic areas or regions (Figure 12.2) as defined by various authors (Safford 1869; Bailey and Winsor 1964; Sapp and Emplaincourt 1975; Stearns 1975; Clark and Zisa 1976; Quarterman and Powell 1978; Woodward and Hoffman 1991) and three physiographic provinces (sensu Fenneman 1938), that is, Interior Low Plateau, Appalachian Plateaus, and Ridge and Valley. A few cedar glade–like areas also have been reported in the Ridge and Valley of Alabama (Rollins 1963; Allison 1994). However, cedar glades are most numerous and best developed in the Central Basin of Tennessee; most are in the Inner (vs. Outer) Basin (sensu Edwards, Elder and Springer 1974).

Outside the Southeast in the unglaciated eastern United States, we have seen small open areas on limestone that more or less fit our definition of cedar glades in southern Ohio, southern Indiana, southwestern Missouri, southeastern Kansas, and southeastern and northeastern Oklahoma. All cedar glades we are aware of are south of

Roger C. Anderson, James S. Fralish, and Jerry M. Baskin, eds. *Savannas, Barrens, and Rock Outcrop Plant Communities of North America.* Copyright © 1999 Cambridge University Press. All rights reserved.

the boundary of Pleistocene glaciation. However, some of the alvars of the Great Lakes region (see Chapter 23) are similar to the rock pavement cedar glades of the Southeast.

Physical Environment of Cedar Glades

Bedrock Geology

Cedar glades of the southeastern United States occur primarily on Ordovician and Mississippian limestones, but some are on Ordovician and Silurian dolomites. Bedrock for cedar glades in the southeastern United States (obtained from numerous sources) is as follows: (1) Outer Bluegrass (Kentucky), primarily Lower and Middle Silurian dolomites and dolomitic limestones; (2) Pennyroyal Plain (Kentucky), Upper Mississippian Ste. Genevieve (primarily) and Girkin limestones; (3) Central Basin (Tennessee), Middle Ordovician Ridley and Lebanon (primarily) limestones of the Stones River Group; (4) Western Valley (Tennessee), Silurian limestones of the Dixon and other formations; (5) Tennessee Valley (Alabama), Upper Mississippian Tuscumbia Limestone; (6) Little Mountain (Alabama), Upper Mississippian Bangor Limestone; (7) Moulton Valley (Alabama), Upper Mississippian Bangor Limestone; (8) Sequatchie Valley (Alabama), Middle Ordovician Chickamauga Limestone; and (9) Ridge and Valley (Georgia, Tennessee, Virginia), Middle Ordovician limestones of the Stones River Group (northwest Georgia), Lower Ordovician Chelpultepec Dolomite of the Knox Group,

Figure 12.1. Photographs of flagstone/gravel (A) and solid-rock pavement (B, C) cedar glades in Cedars of Lebanon State Forest in the Inner Central Basin of Tennessee. A and B, cedar glades on Lebanon Limestone; C, cedar glade on Ridley Limestone. (Photographs by Carol C. Baskin. Fig. 12.1A taken on 2 May 1972 and Figs. 12.1B and C on 24 November 1994.)

Middle Ordovician limestones of the Stones River Group, Middle Ordovician Lenoir Limestone of the Chickamauga Group (east Tennessee), and Middle Ordovician Hurricane Bridge Limestone (southwestern Virginia).

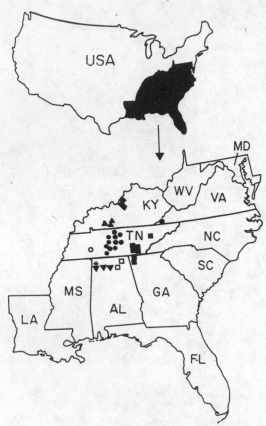

Figure 12.2. Distribution map of cedar glades in southeastern United States. Each symbol represents all cedar glades in a county: (♦) Outer Bluegrass, (▲) Pennyroyal Plain, (●) Central Basin, (○) Western Valley, (★) Tennessee Valley and Little Mountain, (▼) Moulton Valley, (□) Sequatchie Valley, (■) Ridge and Valley.

Soils

In the Inner Central Basin of Tennessee, soil on Lebanon Limestone flagstone glades (Figure 12.1A) is the Gladeville flaggy silty clay loam, 5%–15% slopes (True et al. 1977; North 1981), which is the soil type on about 14.8% of the area of this physiographic subregion (calculated from data in Springer and Elder 1980). Thin "flags" or fragments of limestone are scattered over the surface and throughout the soil profile. Depth to bedrock is 8–30 cm, and soil reaction is neutral to moderately alkaline. The Gladeville soil is a Clayey-skeletal, mixed, thermic Lithic

Rendoll. Sixty to eighty percent of this soil map unit is Gladeville flaggy silty clay loam, but outcrops of solid flat-rock "pavement" also occur on the Lebanon Limestone (Figure 12.1B) and on the Ridley Limestone (Figure 12.1C; see Wilson 1980). Within the boundaries of the Cedars of Lebanon State Park and Forest, Wilson County, Tennessee, a large portion of the Gladeville–Rock Outcrop Complex (USDA-NRCS unpublished soil survey field sheets) supports forests of redcedar, redcedar–hardwoods, and hardwoods. Cedar glades are natural openings on Lebanon Limestone and on Ridley Limestone within this forest matrix.

Flat-rock cedar glades are characterized by nearly level exposures of bare rock surrounded by shallow soil. Cedar glade flora and vegetation on these flat-rock–shallow soil complexes occur where depth to bedrock is <1.0 to ca. 25 cm (J. M. Baskin and Baskin 1985a). Soil also may accumulate in slight depressions on the flat-rock exposures, and this leads to formation of vegetation mats surrounded by bare rock (i.e., rock glades sensu Quarterman 1950b). Unlike the Gladeville, soil associated with flat-rock cedar glades contains fewer and smaller rocks. Deeper soils associated with the Gladeville–Rock Outcrop Complex in the Central Basin include, among others, the Talbott (Fine, mixed, thermic Typic Hapludalfs) and Barfield (Clayey, mixed, thermic Lithic Hapludolls). Depth to bedrock of Talbott soils is 50–100 cm and of Barfield soils 25–50 cm.

In the Moulton Valley (Alabama), cedar glades occur primarily in association with rock outcrops and shallow rocky soils of the Colbert (Fine, montmorillonitic, thermic Vertic Hapludalfs) and Talbott series. Areas where limestone boulders and/or bedrock cover a high percentage of the surface are mapped as Rockland Limestone or as Stony rolling land (Sherard, Wesson and Edwards 1959; Sherard et al. 1965). Cedar glades with flagstone/gravel, solid-rock pavement, or a mixture of flagstone/gravel and solid-rock pavement develop on the Bangor Limestone (see Figure 2 in J. M. Baskin, Webb and Baskin 1995).

In the other seven physiographic regions, where they are fewer and less well developed than in the Central Basin and Moulton Valley, cedar glades are associated with rock outcrop complexes of various soil series or with soil units mapped as Limestone Rockland; Limestone Rockland, rolling and hilly; Rockland Limestone; Stony smooth land, limestone; or Rolling stony land (information from county soil surveys). As in the Central Basin and Moulton Valley, the topography of cedar glades in these provinces mostly varies from nearly level to gently sloping.

In "The Cedars," Lee County, Virginia, limestone outcrops cover 90% or more of the surface, which generally is "gently undulating to rolling" (Henry et al. 1953). In some places, bedrock projects up to a few feet above the ground surface. A large portion of The Cedars landscape supports a cutover redcedar–hardwood forest. Cedar glades are small, (presumably) natural openings within this scrubby forest, where bedrock is at ground level or is covered with up to a few centimeters of soil. Gravel/flagstone may be present on the surface. Cedar glade–like areas also occur in some parts of The Cedars that have been cleared for pasture.

Studies of cedar glades on Gladeville soil in the Central Basin have reported mean soil depths in the range of 4.8–12.4 cm (Freeman 1933; Hemmerly 1976; Somers et al. 1986; Drew 1991; Baskauf 1993). However, these measurements may not represent the depth of soil in which plants are rooted (Quarterman 1989). Depth was measured by pushing a probe into the soil until it struck rock. However, since flagstone is scattered throughout the profile of the Gladeville soil and, in fact, makes up 35%–65% of the volume of this soil (True et al. 1977; North 1981), it is unlikely that the probe would be pushed to the bottom of the profile. Thus,

depth measurements reported for the Gladeville may be too shallow, and they may not represent the functional rooting depth for plants. For example, Hemmerly (1976) found that roots of mature plants of the federal endangered cedar glade endemic *Echinacea tennesseensis* growing in a Lebanon Limestone glade were up to 50 cm in length and extended to a depth of 30 cm into the soil. Further: "Roots often had grown through clefts in rocks and in some instances branched horizontally after reaching rock which they could not penetrate" (Hemmerly 1976, p. 12). Thus, it seems that the longest roots of *E. tennesseensis* penetrate the entire soil profile of the Gladeville soil, which can be up to 30 cm (North 1981).

Climate

The macroclimate of the cedar glades region of the southeastern United States is in Thornthwaite's (1948) $B_2B'_2b'_3r$ type (i.e., humid, mesothermal region with intermediate thermal efficiency and with adequate precipitation for plant growth at all seasons); Köppen's (Ackerman 1941) Cfa type (i.e., mild temperate rainy climate without a distinct dry season, and with a hot summer), and Borchert's (1950) region II (i.e., a rainy winter but as much or more rainfall in summer than in winter, summer sunnier than winter). A Walter-type climatic diagram for Columbia, Tennessee, in the Central Basin, is shown in Figure 12.3.

However, microclimate and soil moisture conditions of cedar glades are harsher for plant growth than the general meteorological data suggest. For example, soil moisture is more extreme for the shallow glade soils than for the deeper soils of the region. Soils often are saturated with water from late autumn to early spring, whereas in summer and early autumn they sometimes are below the permanent wilting point (Freeman 1933; C. C. Baskin, Baskin and Quarterman 1972). In his study of a cedar glade in the Inner Basin, Freeman (1933) found that soil moisture was

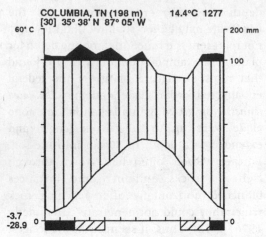

Figure 12.3. A Walter-type climate diagram for Columbia, Tennessee, in the Central Basin.

below the wilting coefficient 11 times between 13 June and 24 October.

Insolation at or near the soil surface on clear days in summer is high (ca. 1.3–1.4 cal cm^{-2} min^{-1}; J. M. Baskin and Baskin 1973, 1978), and photosynthetic photon irradiance in full sun during summer can reach levels of >2,000 µmol m^{-2} sec^{-1} (Baskauf 1993). Soil surface temperatures may be considerably higher than those recorded at nearby weather stations (Freeman 1933; Dubois 1993). Dubois (1993) reported a temperature of 43 °C at a depth of 2–3 cm on 4 September, and we recorded a temperature of 50 °C on the bare soil surface at midday in June (Baskin and Baskin unpublished).

Flora and Phytogeographical Relationships

Flora

A checklist of plants in the cedar glade flora of the southeastern United States compiled by J. and C. Baskin (unpublished) includes 406 dicots, 124 monocots, 4 gymnosperms, and 7 ferns, distributed among 279 genera and 74 families (59 dicot, 8 monocot, 2 gymnosperm, and 5 fern). Four hundred and forty-four of these 541 taxa are indigenous to the cedar glade region, and 97 are nonindige-

nous. The largest families are the Asteraceae (67 native, 19 nonnative taxa), Poaceae (46, 15), Fabaceae (23, 12), Cyperaceae (30, 0), and Brassicaceae (17, 6). *Panicum* (including *Dichanthelium*, 16 taxa), *Carex* (14), *Hypericum* (11), *Aster* (10), and *Leavenworthia* (10) are the largest genera in the flora. Nineteen of the taxa are endemic to cedar glades, and two are nearly endemic (Table 12.1). The physiographic region with the largest cedar glade flora is the Central Basin of Tennessee (328 native, 80 nonnative), followed by the Ridge and Valley (210, 33) and Moulton Valley of Alabama (209, 24).

Based on presence and abundance in southeastern cedar glades as reported in numerous studies, the characteristic species are *Arenaria patula, Asclepias verticillata, Croton capitatus, C. monanthogynus, Celtis* spp., *Cyperus aristatus, Dalea gattingeri, Diodia teres, Eleocharis compressa, Erigeron strigosus, Forestiera ligustrina, Hedyotis* spp., *Heliotropium tenellum, Hypericum dolabriforme/H. sphaerocarpum, Isoetes butleri, Juniperus virginiana, Leavenworthia* spp., *Nothoscordum bivalve, Ophioglossum engelmannii, Opuntia humifusa, Oxalis priceae* subsp. *priceae, Panicum capillare, P. flexile, Pediomelum subacaule, Rhus aromatica, Ruellia humilis, Scutellaria parvula/S. leonardii, Sedum pulchellum, Senecio anonymous, Sisyrinchium albidum, Solidago nemoralis, Sporobolus vaginiflorus, Talinum calcaricum, Trichostema brachiatum, Verbena simplex,* and *Ulmus alata.* Plant nomenclature primarily follows Wofford and Kral (1993), or specialists in groups with taxa endemic to southeastern cedar glades.

Life-Forms

Raunkiaer life-form was determined for each of the 541 taxa primarily from Gibson (1961) and from personal observations by the authors on plants in the field. Life-form distribution among the native, nonnative, and total flora, respectively, is as follows: phanerophytes (native = 11.5%, nonnative = 4.1%, total = 10.2%); chamaephytes (0.5%,

Table 12.1. *Presence of the 19 cedar glade endemics and 2 near-endemics in the nine physiographic regions in which southeastern cedar glades occur. OBGK = Outer Bluegrass (Kentucky), PPK = Pennyroyal Plain (Kentucky), CBT = Central Basin (Tennessee), WVT = Western Valley (Tennessee), TVA = Tennessee Valley (Alabama), LIMA = Little Mountain (Alabama), MOVA = Moulton Valley (Alabama), SQVA = Sequatchie Valley (Alabama), and RV = Ridge and Valley (Georgia, Tennessee, Virginia).*

Taxa	OBGK	PPK	CBT	WVT	TVA	LIMA	MOVA	SQVA	RV	Total
Endemics										
Astragalus bibullatus			x							1
Dalea gattingeri[a]			x		x		x	x	x	5
Delphinium alabamicum							x			1
D. carolinianum ssp. *calciphilum*		x	x				x		x	4
Echinacea tennesseensis			x							1
Leavenworthia alabamica (including vars. *alabamica* and *brachystyla*)					x	x	x			3
L. crassa (including vars. *crassa* and *elongata*)								x		1
L. exigua var. *exigua*			x	x					x	3
L. exigua var. *laciniata*	x									1
L. exigua var. *lutea*			x					x	x[b]	3
L. stylosa			x							1
L. torulosa		x	x		[x][c]				x	4
Lesquerella lyrata					x		x			2
Lobelia appendiculata var. *gattingeri*		x	x							2
Onosmodium molle		[x][c]	x				x			3
Pediomelum subacaule			x			x	x		x	4
Phacelia dubia var. *interior*			x							1
Talinum calcaricum		x	x				x			3
Trifolium calcaricum			x						x	2
Near-Endemics										
Astragalus tennesseensis			x		[x][c]		x			3
Dalea foliosa			x				x			2
Totals	1	5	16	1	5	2	10	3	7	50

[a] also known from one site in the Ozarks of Missouri (Summers, Skinner and Yatskicvych 1996)

[b] also on limestone (Rollins 1963, 1993) and dolomite (Allison 1994) gladelike areas in Ridge and Valley of Alabama

[c] [ ] = apparently extirpated from this physiographic region

2.1%, 0.7%); hemicryptophytes (52.0%, 29.9%, 48.1%); cryptophytes (12.6%, 6.2%, 11.5%); therophytes (23.0%, 57.7%, 29.2%); stem succulents (0.2%, 0.0%, 0.18%); and epiphytes (0.2%, 0.0%, 0.18%). Included in the 51 phanerophytes are 23 trees, 13 shrubs, and 15 (woody) vines. *Opuntia humifusa* is the only stem succulent, and *Polypodium polypodioides* is the only epiphyte. Of the 19 endemics (Table 12.1), 9 are therophytes, (all winter annuals), 9 are hemicryptophytes, and 1 (*Pediomelum subacaule*) is a cryptophyte. The 2 near-endemics (Table 12.1) are hemicryptophytes.

Photosynthetic Pathways

A photosynthetic pathway (C3, C4, CAM) was assigned to each of the 541 taxa based on information published in numerous papers. Photosynthetic pathways were inferred from information at the familial and/or generic level(s) for species whose pathways had not been determined. All woody plants were assumed to have the C3 pathway (see J. M. Baskin and Baskin 1981, 1985b; Eickmeier 1986; Ross and Van Horn 1988; J. M. Baskin, Webb and Baskin 1995).

The distribution of photosynthetic pathways among the 444 native taxa is C3, 90.5%; C4, 8.8%; CAM, 0.5%; and one (0.2%) holoparasite (*Cuscuta campestris*). Twenty-eight of the 39 C4 taxa are grasses. The 19 endemics and 2 near-endemics are C3 plants. Ninety-six of the 97 nonnative taxa use the C3 pathway; only *Portulaca oleracea* is a C4 plant. *Manfreda virginica* and *Opuntia humifusa* are the only 2 taxa that use the CAM pathway. Thus, 92.3% of the 541 taxa use the C3 pathway, 7.2% the C4 pathway, and 0.4% the CAM pathway; one taxon is holoparasitic.

Phytogeographical Relationships

Of 270 native taxa examined in the Central Basin cedar glade flora, 81.1% are intraneous (center of distribution includes middle Tennessee) and 18.9% are extraneous (center of distribution does not include Middle Tennessee) (J. M. Baskin and Baskin 1986; Bridges and Orzell 1986). The largest intraneous group includes species with their center of distribution in temperate eastern North America, and the largest extraneous group includes those whose center of distribution is west and northwest of Tennessee. The great majority of the native plant taxa in cedar glades in the southeastern United States are

widely distributed geographically and grows in a variety of habitats.

The geographical distribution of the 19 cedar glade endemics and 2 near-endemics is shown in Table 12.1. The highest number of endemics (14) are in the Central Basin of Tennessee, with the second highest number (8) in the Moulton Valley of Alabama. The 2 near-endemics occur in each of these two physiographic regions. *Dalea gattingeri* occurs in more physiographic regions (5) than any of the other 18 endemics (Table 12.1). Four of the endemics are restricted to the Central Basin, 2 to the Moulton Valley, and 1 to the Outer Bluegrass (Kentucky) physiographic region. Outside the southeastern cedar glades, an extant population of *Astragalus tennesseensis* is known only from a gravel hill prairie in central Illinois (Tazewell County; McFall 1984), and the only populations of *D. foliosa* known to be extant are in shallow-soil dolomite prairies in northern (Will County) Illinois (U.S. Fish and Wildlife Service 1993).

Ninety of the 97 introduced taxa are from Asia, Eurasia, or Europe; 1 from the Old World Tropics; 2 from Tropical America; and 4 from the United States west or south of the cedar glade region.

Plant Communities

Central Basin

Quarterman (1948, 1950a, b) described for the Central Basin of Tennessee what she thought were successional pathways from bare rock (or cracks in the rocks) to a climax mixed hardwood forest or to a subclimax redcedar forest. Her presumed successional pathways, however, probably represent vegetation sequences along xeric to dry to mesic environmental gradients that are unrelated to succession. The plant communities sampled by Quarterman were gravel glade, grass glade, glade–shrub, shrub–cedar, 88-year-old cedar forest, 103-year-old cedar forest, and cedar hardwood forest. She referred to gravel and grass glades, along with rock glades, as open glades, the "cedar glades" as defined in this chapter. Characteristic species of 22 open glades, based on constancy, were *Sporobolus vaginiflorus*, *Erigeron strigosus*, *Nostoc commune*, *Dalea gattingeri*, *Croton capitatus*, *C. monanthogynus*, *Ruellia humilis*, and the lichen *Dermatocarpon hepaticum*.

Based on an importance value (IV) calculated from Quarterman's (1948) data, the C4 summer annual grass *Sporobolus vaginiflorus* is the dominant in both gravel and grassy glades (Tables 12.2 and 12.3, respectively). The other species in Quarterman's samples in open glades are C3 summer annuals, winter annuals, biennials, and perennials. In terms of IV, two cedar glade endemics, *Pediomelum subacaule* and *Dalea gattingeri*, ranked second and sixth, respectively, in gravel glades and seventh and second, respectively, in grassy glades (Tables 12.2 and 12.3). Important shrubs and trees (woody transgressives) in gravel and grassy glades were *Forestiera ligustrina*, *Rhus aromatica*, *Symphoricarpos orbiculatus*, *Rosa carolina*, *R. setigera*, *Celtis laevigata*, *Juniperus virginiana*, and *Ulmus alata*.

Somers et al. (1986) used 313 1-m² plots to sample open glades in the Inner Basin in June and again in September–October; 132 taxa were identified in the sample plots. Using CLUSTER analysis, they identified seven plant community types: four xeric (*Panicum capillare*, foliose lichen, *Nostoc commune–Sporobolus vaginiflorus*, and *Dalea gattingeri*) and three subxeric (*Sporobolus vaginiflorus*, *Pleurochaete squarrosa*, and *Panicum flexile–Pleurochaete squarrosa–Sporobolus vaginiflorus*). Overall, *Sporobolus vaginiflorus* had the highest frequency/cover values, and the cedar glade endemic *Dalea gattingeri* the second highest.

Drew (1991; also see Drew and Clebsch 1995) sampled five Central Basin cedar glades with populations of the federal endangered plant species *Echinacea tennesseensis*. Using Drew's (1991) cover and frequency data, we calculated importance values in percentage (% IV) for taxa at each of the five sites. *Grindelia lanceolata* (% IV = 15.9) had

the highest importance value at one site, and *Hedyotis nigricans* (23.5 and 20.2) and *Sporobolus vaginiflorus* (28.1 and 17.1) at two sites each. *Sporobolus vaginiflorus* had the highest % IV averaged over all five sites (14.7), followed by *Echinacea tennesseensis* (11.9). Other important species included *Hedyotis nigricans*, *Schizachyrium scoparium*, *Ruellia humilis*, *Grindelia lanceolata*, *Dalea gattingeri*, *Hypericum sphaerocarpum*, and *Trichostema brachiatum*.

In a *Grindelia lanceolata* plant community type on Lebanon Limestone in the Central Basin, we identified 54 vascular plant taxa (51 native, 3 nonnative) in 85 1-m² sample plots (J. M. Baskin and Baskin 1996b). Species with the highest frequencies were *G. lanceolata* (100%), *Sporobolus vaginiflorus* (96.5%), the cedar glade endemic *Dalea gattingeri* (91.8%), *Ruellia humilis* (64.7%), *Isanthus brachiatus* (63.3%), the cedar glade endemic *Pediomelum subacaule* (56.5%), *Heliotropium tenellum* (45.9%), and *Croton capitatus* (36.5%).

Moulton Valley

In Alabama, cedar glades comparable to those in the Central Basin of Tennessee mostly are restricted to the Moulton Valley (J. M. Baskin, Webb and Baskin 1995); none has been sampled quantitatively. However, percentage presence was determined using the checklist of vascular plants for each of 16 glades (J. M. Baskin, Webb and Baskin 1995; unpublished). Species in presence class 5 (present in 81%–100% of the 16 glades) include the C4 summer annual grasses *Sporobolus vaginiflorus* (100% presence) and *Panicum capillare* (87.5%); the C3 winter annuals *Arenaria patula* (100%) and *Sedum pulchellum* (81.3%); the C3 summer annuals *Croton monanthogynus* (87.5%), *Heliotropium tenellum* (87.5%), and *Trichostema brachiatum* (87.5%); the C3 biennials *Erigeron strigosus* (100%) and *Rudbeckia triloba* (93.8%); the C3 perennial herbs *Hypericum sphaerocarpum* (100%), *Verbena simplex* (87.5%), *Dalea gat-*

Table 12.2. *Frequency, cover, and importance value (IV) of each herbaceous plant taxon in ten 16-m² plots in gravelly glades in the Central Basin of Tennessee. IV ([Relative frequency + Relative cover]/2) was calculated from data on % cover and % frequency in Quarterman (1948). Taxa marked with an asterisk (*) are endemic to cedar glades. PP = photosynthetic pathway (C3, C4, CAM); LC = life-cycle type (sA = summer annual, wA = winter annual, B = biennial, and Per = herbaceous polycarpic perennial).*

Species	PP	LC	Frequency (%)	Cover (%)	IV (%)
Sporobolus vaginiflorus	C4	sA	100	47.0	34.87
Pediomelum subacaule*	C3	Per	80	7.0	8.77
Arenaria patula	C3	wA	70	5.0	6.97
Euphorbia dentata	C3	sA	40	6.0	5.95
Nothoscordum bivalve	C3	Per	60	3.5	5.49
Hypericum sphaerocarpum	C3	Per	70	2.5	5.40
Dalea gattingeri*	C3	Per	70	2.0	5.10
Trichostema brachiatum	C3	sA	40	2.5	3.77
Erigeron strigosus	C3	B	60	+ᵃ	3.30
Isoetes butleri	C3	Per	20	3.5	3.29
Croton monanthogynus	C3	sA	50	+	2.75
Leavenworthia stylosa*	C3	wA	40	+	2.20
Euphorbia spathulata	C3	sA	40	+	2.20
Malvastrum hispidum	C3	sA	10	1.0	1.18
Plantago virginica	C3	wA	30	+	1.65
Croton capitatus	C3	sA	20	+	1.10
Sedum pulchellum	C3	wA	20	+	1.10
Galium virgatum	C3	wA	20	+	1.10
Leavenworthia uniflora	C3	wA	20	+	1.10
Talinum calcaricum*	C3	Per	10	+	0.55
Asclepias viridis	C3	Per	10	+	0.55
Oxalis violacea	C3	Per	10	+	0.55
Delphinium carolinianum ssp. calciphilum*	C3	Per	10	+	0.55
Scutellaria parvula	C3	Per	10	+	0.55
Total					100.04

ᵃ + = present

tingeri (81.3%), and *Ruellia humilis* (81.3%); and the CAM succulents *Manfreda virginica* (87.5%) and *Opuntia humifusa* (81.3%). Cedar glade endemics with the highest presence values were *D. gattingeri* (81.3%) and the winter annual *Leavenworthia alabamica* (75%).

Woody plants with the highest presence values in cedar glades in the Moulton Valley were the trees *Diospyros virginiana* (68.8%), *Juniperus virginiana* (68.8%), *Ulmus alata* (68.8%), *Fraxinus americana* (56.3%), *Celtis* spp. (56.3%), *Bumelia lycioides* (50%), and *Cercis canadensis* (50%); the shrubs *Forestiera*

ligustrina (75%), *Rhus aromatica* (75%), *Rosa carolina* (56.3%), *Hypericum frondosum* (50%), *Callicarpa americana* (43.8%), and *Rhamnus caroliniana* (43.8%); and the vines *Smilax bona-nox* (62.5%) and *Berchemia scandens* (43.8%).

Vascular cryptogams with the highest presence value in the Moulton Valley cedar glades were *Ophioglossum engelmannii* (68.6%) and *Isoetes butleri* (43.8%).

Thus, floristically the Moulton Valley cedar glades are quite similar to those in the Central Basin. Undoubtedly, quantitative

Table 12.3. *Frequency, cover, and importance value (IV) of each herbaceous plant taxon in ten 16-m² plots in grassy glades in the Central Basin of Tennessee. IV was calculated from data on % cover and % frequency in Quarterman (1948). Species marked with one asterisk (*) are endemic to cedar glades, and those marked with two asterisks (**) are nonnative. PP = photosynthetic pathway (C3, C4, CAM); LC = life-cycle type (sA = summer annual, wA = winter annual, swA = both summer and winter annual, B = biennial, and Per = herbaceous polycarpic perennial).*

Species	PP	LC	Frequency (%)	Cover (%)	IV (%)
Sporobolus vaginiflorus	C4	sA	100	52.0	38.31
Dalea gattingeri*	C3	Per	80	6.5	7.92
Erigeron strigosus	C3	B	70	2.5	4.86
Arenaria patula	C3	wA	40	4.5	4.77
Euphorbia dentata	C3	sA	60	3.0	4.73
Pediomelum subacaule*	C3	Per	50	3.5	4.58
Hypericum sphaerocarpum	C3	Per	40	3.0	3.80
Ophioglossum engelmannii	C3	Per	60	1.0	3.43
Croton monanthogynus	C3	sA	50	+ª	2.32
Galium virgatum	C3	wA	50	+	2.32
Oxalis violacea	C3	Per	40	+	1.85
Plantago virginica	C3	wA	40	+	1.85
Grindelia lanceolata	C3	B	20	1.0	1.58
Nothoscordum bivalve	C3	Per	30	+	1.39
Leavenworthia stylosa*	C3	wA	30	+	1.39
Hedyotis purpurea	C3	wA	30	+	1.39
Sisyrinchium albidum	C3	Per	30	+	1.39
Ruellia humilis	C3	Per	20	0.2	1.04
Viola egglestonii	C3	Per	20	+	0.93
Lobelia appendiculata var. gattingeri*	C3	swA	20	+	0.93
Trichostema brachiatum	C3	sA	20	+	0.93
Croton capitatus	C3	sA	10	+	0.47
Medicago lupulina**	C3	variable	10	+	0.47
Sedum pulchellum	C3	wA	10	+	0.47
Euphorbia spathulata	C3	sA	10	+	0.47
Isoetes butleri	C3	Per	10	+	0.47
Asclepias viridis	C3	Per	10	+	0.47
Scutellaria parvula	C3	Per	10	+	0.47
Panicum spp. (including Dichanthelium?)	C4(?)	sA(?)	10	+	0.47
Oxalis stricta	C3	Per	10	+	0.47
Acalypha gracilens	C3	sA	10	+	0.47
Chaerophyllum tainturieri	C3	wA	10	+	0.47
Bupleurum rotundifolium**	C3	wA	10	+	0.47
Physalis heterophylla	C3	Per	10	+	0.47
Ambrosia artemisiifolia	C3	sA	10	+	0.47
Hedyotis crassifolia	C3	wA	10	+	0.47
H. nuttalliana	C3	Per	10	+	0.47
Krigia virginica	C3	wA	10	+	0.47
Veronica officinalis**	C3	wA	10	+	0.47
Total					100.19

ª + = present

sampling would show that the plant communities of these cedar glades also are very similar to those of cedar glades in the Central Basin.

Ridge and Valley

In an open glade in Knox County, Tennessee, the very shallow soil zone was dominated by cryptogams, that is, a crustose lichen, *Verrucaria nigrescens*; two mosses, *Pleurochaete squarrosa* and *Tortella humilis;* and cyanobacteria (Finn 1968). The only vascular plants in this zone (all of minor importance) were *Fragaria virginiana*, *Hypericum dolabriforme*, *Manfreda virginica*, and *Pellaea atropurpurea*. Dominant plants in the herb zone were *Ruellia humilis*, *Aster oblongifolius*, *Sporobolus vaginiflorus*, and the moss *Tortella humilis*; and in the herb–moss zone, *Aster oblongifolius*, *Ruellia humilis*, *Sporobolus clandestinus*, and the moss *Pleurochaeta squarrosa*.

In a general way, the physical and vegetational features of the Mascot glade studied by Finn (1968) resemble those of the Central Basin cedar glades. However, there are some notable differences in the vegetation: (1) Overall, *Sporobolus vaginiflorus* is a much more important component of open glades in the Central Basin; (2) winter annuals (*Leavenworthia* spp., *Sedum pulchellum*, *Arenaria patula*) are not an important component of any vegetation zone in the Mascot glade; (3) none of the endemics, near-endemics, or Ozark–western disjuncts in the cedar glades of the Central Basin occur in the Mascot glade; (4) the Mascot glade does not have a shrub zone; and (5) pines do not occur in glade woods in the Central Basin.

DeSelm (1993) sampled two cedar glades in east Tennessee and one in northwest Georgia. Percentage cover given in DeSelm's paper was recalculated, so that it was mean percentage cover per m^2 by species, and this value was used in recalculating an importance value (IV = [relative frequency + relative cover]/2). The most important plants were the C3 perennial forbs *Aster oblongifolius* (IV = 58.35/200) and *Hypericum dolabriforme* (21.96) and the C3 perennial grass *Danthonia spicata* (20.69) at the Burnett Creek Road site (Knox County, Tennessee); *Juniperus virginiana* (IV= 29.31/200), the C4 summer annual grasses *Sporobolus neglectus* (17.95) and *Panicum flexile* (9.68), and the C4 perennial grass *Schizachyrium scoparium* (16.43) at the Henry Harris site (Rhea County, Tennessee); and the C4 summer annual *Sporobolus* spp. (IV = 65.49/200) and the C3 perennial forbs *Hypericum dolabriforme* (29.23), *Aster dumosus* (19.03), and *Dalea gattingeri* (11.93) at the Georgia Route 218 at 146 site (Catoosa County, Georgia).

Of the four cedar glades sampled in the Ridge and Valley (Finn 1968; DeSelm 1993), the vegetation of the Route 218 at 146 site in Catoosa County, Georgia, is most similar to that of the cedar glades of the Central Basin. At both sites, annual *Sporobolus* is dominant, and *Dalea gattingeri*, *Hypericum* (*dolabriforme* in the Ridge and Valley of Georgia, *sphaerocarpum* in the Central Basin of Tennessee), *Erigeron strigosus*, and *Croton monanthogynus* are relatively important species.

Summary

The most thorough study of cedar glade vegetation in the Southeast was by E. Quarterman, in the Central Basin of Tennessee, in the 1940s. Her "cedar glades" included the natural rocky openings ("glades" or "open glades")–glade/shrub thicket–shrub/redcedar thicket–redcedar forest–redcedar/hardwood forest–hardwood forest complex. However, most botanists now use *cedar glades* for the open glades only, and this is how we define the term. Thus, cedar glades are natural, rocky-shallow soil openings with a C4 annual grass–C3 annual/perennial forb-cryptogam-dominated vegetation. They consist of several more or less distinct "xeric" and "subxeric" plant communities, which are edaphic climaxes. Overall, the geographically widespread eastern North American species *Sporobolus vaginiflorus*, a C4 summer annual grass, is the most important dominant in southeastern cedar glades.

In the southeastern United States, cedar glades are developed best in the Central Basin of Tennessee, but they also occur in other physiographic regions, in Alabama, Georgia, Kentucky, Tennessee, and Virginia. The majority of the approximately 450 native vascular plant taxa are C3 therophytes or hemicryptophytes that have an intraneous distribution and occur in a variety of habitats in eastern/southeastern North America. However, the flora also includes 19 endemics, 2 near-endemics and a strong extraneous element from west and northwest of the cedar glade region.

REFERENCES

Ackerman, E. A. 1941. The Köppen classification of climates in North America. *The Geographical Review* 31:105–111.

Allison, J. R. 1994. A botanical "lost world" in central Alabama. In *Proceedings of the North American Conference on Savannas and Barrens*, ed. J. S. Fralish, R. C. Anderson, J. E. Ebinger, and R. Szafoni, pp. 323–327. Chicago, Ill.: U.S. Environmental Protection Agency, Great Lake National Program Office.

Bailey, H. H., and Winsor, J. H. 1964. *Kentucky Soils*. University of Kentucky Agricultural Experiment Station Miscellaneous Publication Number 308.

Baskauf, C. J. 1993. Comparative population genetics and ecophysiology of a rare and a widespread species of *Echinacea* (Asteraceae). Ph.D. dissertation, Vanderbilt University, Nashville, Tenn.

Baskin, C. C., Baskin, J. M., and Quarterman, E. 1972. Observations on the ecology of *Astragalus tennesseensis*. *The American Midland Naturalist* 88:167–182.

Baskin, J. M., and Baskin, C. C. 1973. Pad temperatures of *Opuntia compressa* during daytime in summer. *Bulletin of the Torrey Botanical Club* 100:56–59.

Baskin, J. M., and Baskin, C. C. 1978. Leaf temperatures of *Heliotropium tenellum* and their ecological implications. *The American Midland Naturalist* 100:488–492.

Baskin, J. M., and Baskin, C. C. 1981. Photosynthetic pathways indicated by leaf anatomy in fourteen summer annuals of cedar glades. *Photosynthetica* 15:205–209.

Baskin, J. M., and Baskin, C. C. 1985a. Life cycle ecology of annual plant species of cedar glades of southeastern United States. In *The Population Structure of Vegetation*, ed. J. White, pp. 371–398. Dordrecht: Dr. W. Junk.

Baskin, J. M., and Baskin, C. C. 1985b. Photosynthetic pathway in 14 southeastern cedar glade endemics, as revealed by leaf anatomy. *The American Midland Naturalist* 114:205–208.

Baskin, J. M., and Baskin, C. C. 1986. Distribution and geographical/evolutionary relationships of cedar glade endemics in southeastern United States. *Association of Southeastern Biologists Bulletin* 33:138–154.

Baskin, J. M., and Baskin, C. C. 1996a. Bessey Picklesimer's little-known quantitative study on the vegetation of a cedar glade in the Central Basin of Tennessee. *Castanea* 61:25–37.

Baskin, J. M., and Baskin, C. C. 1996b. The *Grindelia lanceolata* plant community type in cedar glades of the Central Basin of Tennessee. *Castanea* 61:339–347.

Baskin, J. M., Webb, D. H., and Baskin, C. C. 1995. A floristic plant ecology study of the limestone glades of northern Alabama. *Bulletin of the Torrey Botanical Club* 122:226–242.

Borchert, J. R. 1950. The climate of the central North American grassland. *Annals of the Association of American Geographers* 40:1–39.

Bridges, E. L., and Orzell, S. L. 1986. Distribution patterns of the non-endemic flora of middle Tennessee limestone glades. *Association of Southeastern Biologists Bulletin* 33:155–166.

Clark, W. Z., Jr., and Zisa, C. 1976 [reprinted 1988]. *Physiographic Map of Georgia*. Atlanta, Ga.: Georgia Department of Natural Resources.

DeSelm, H. R. 1993. Barrens and glades of the southern Ridge and Valley. In *Proceedings of the Fifth Annual Symposium on the Natural History of Lower Tennessee and Cumberland River Valleys*, ed. S. W. Hamilton, E. W. Chester, and A. F. Scott, pp. 81–135. Clarksville, Tenn.: Center for Field

Biology, Austin Peay State University.

Drew, M. B. 1991. The role of Tennessee cone-flower, *Echinacea tennesseensis*, in its native habitat: The vegetation and a demographic analysis. M.S. thesis, University of Tennessee, Knoxville, Tenn.

Drew, M. B., and Clebsch, E. E. C. 1995. Studies on the endangered *Echinacea tennesseensis* (Asteraceae): Plant community and demographic analysis. *Castanea* 60:60–69.

Dubois, J. D. 1993. Biological dinitrogen fixation in two cedar glade communities of middle Tennessee. *Journal of the Tennessee Academy of Science* 68:101–105.

Edwards, M. J., Elder, J. A., and Springer, E. 1974. *Soils of the Nashville Basin*. The University of Tennessee Agricultural Experiment Station Bulletin Number 499.

Eickmeier, W. G. 1986. The distribution of photosynthetic pathways among cedar glade plants. *Association of Southeastern Biologists Bulletin* 33:200–205.

Fenneman, N. M. 1938. *Physiography of the Eastern United States*. New York: McGraw-Hill.

Finn, L. L. 1968. Vegetation of a cedar glade area near Mascot, Tennessee, and observations on the autecology of three Arenaria taxa. M.S. thesis, University of Tennessee, Knoxville.

Freeman, C. P. 1933. Ecology of the cedar glade vegetation near Nashville, Tennessee. *Journal of the Tennessee Academy of Science* 8:143–228.

Galloway, J. J. 1919. *Geology and Natural Resources of Rutherford County, Tennessee*. Tennessee Geological Survey Bulletin Number 22.

Gibson, D. 1961. The life forms of Kentucky flowering plants. *The American Midland Naturalist* 66:1–60.

Hemmerly, T. E. 1976. Life cycle strategy of a highly endemic cedar glade species: *Echinacea tennesseensis* (Compositae). Ph.D. dissertation, Vanderbilt University, Nashville, Tenn.

Henry, E. F., Baisden, A. M., Conner, P. C., Perry, H. H., and Mason, D. D. 1953. *Soil Survey of Lee County, Virginia*. USDA Soil Conservation Service, in cooperation with the Virginia Agricultural Experiment Station and the Tennessee Valley Authority.

Mahr, W. P., and Mathis, P. M. 1981. Foliose and fruticose lichens of the cedar glades in Stones River National Battlefield (Rutherford County, Tennessee). *Journal of the Tennessee Academy of Science* 56: 66–67.

McFall, D. W. 1984. Vascular plants of the Manito Gravel Prairie, Tazewell County, Illinois.

Transactions of the Illinois State Academy of Science 77:9–14.

North, O. L. 1981. *Soil Survey of Davidson County, Tennessee*. USDA Soil Conservation Service, in cooperation with the University of Tennessee Agricultural Experiment Station.

Picklesimer, B. C. 1927. A quantitative study of cedar glades in middle Tennessee. M.A. thesis, George Peabody College for Teachers, Nashville, Tenn. (now part of Vanderbilt University).

Quarterman, E. 1948. Plant communities of cedar glades in middle Tennessee. Ph.D. dissertation, Duke University, Durham, N.C.

Quarterman, E. 1950a. Ecology of cedar glades. I. Distribution of glade flora in Tennessee. *Bulletin of the Torrey Botanical Club* 77:1–9.

Quarterman, E. 1950b. Major plant communities of Tennessee cedar glades. *Ecology* 31:234–254.

Quarterman, E. 1989. Structure and dynamics of the limestone cedar glade communities in Tennessee. *Journal of the Tennessee Academy of Science* 64:155–158.

Quarterman, E., Burbanck, M. P., and Shure, D. J. 1993. Rock outcrop communities: Limestone, sandstone, and granite. In *Biodiversity of Southeastern United States/Upland Terrestrial Communities*, ed. W. H. Martin, S. G. Boyce, and A. C. Echternacht, pp. 35–86. New York: John Wiley and Sons.

Quarterman, E., and Powell, R. L. 1978. *Potential Ecological/Geological Natural Landmarks on the Interior Low Plateaus*. Washington, D.C.: U.S. Department of the Interior.

Rollins, R. C. 1963. The evolution and systematics of *Leavenworthia* (Cruciferae). *Contributions of the Gray Herbarium* 192:3–98.

Rollins, R. C. 1993. *The Cruciferae of Continental North America. Systematics of the Mustard Family from the Arctic to Panama*. Stanford, Calif.: Stanford University Press.

Ross, J. S., and Van Horn, G. S. 1988. A survey of C_4 photosynthesis in plants of two Georgia cedar glades and a new state record for Georgia. *Castanea* 53:286–289.

Safford, J. M. 1851. The Silurian Basin of Tennessee, with notices of strata surrounding it. *American Journal of Science and Arts, 2nd Series* 12:352–361 + map.

Safford, J. M. 1869. *Geology of Tennessee*. Nashville, Tenn.: S. C. Mercer Printer.

Sapp, C. D., and Emplaincourt, J. 1975. *Physiographic Regions of Alabama*. Geological Survey Alabama Special Map 168.

Sherard, H., Wesson, H. J., and Edwards, M. J. 1959. *Soil Survey of Lawrence County, Alabama*. USDA Soil Conservation Service, in cooperation with the Alabama Agricultural Experiment Station, the Alabama Department of Agriculture and Industries, and the Tennessee Valley Authority.

Sherard, H., Young, R. A., Bryant, J. P., Smith, S. T., and Gibbs, J. A. 1965. *Soil Survey of Franklin County, Alabama*. USDA Soil Conservation Service, in cooperation with the Alabama Agricultural Experiment Station and the Alabama Department of Agriculture and Industries.

Somers, P., Smith, L. R., Hammel, P. B., and Bridges, E. L. 1986. Preliminary analysis of plant communities and seasonal changes in cedar glades in middle Tennessee. *Association of Southeastern Biologists Bulletin* 33:178–192.

Springer, M. E., and Elder, J. A. 1980. *Soils of Tennessee*. The University of Tennessee Agricultural Experiment Station Bulletin Number 596.

Stearns, R. G. 1975. Introduction. In *Field Trips in West Tennessee*, ed. R. G. Stearns, pp. 1–7. Tennessee Division of Geology Report of Investigations Number 36.

Summers, B., Skinner, M., and Yatskievych, G. 1996. *Dalea gattingeri*, a cedar glade endemic new to Missouri. *Missouriensis* 16(2):4–9.

Thornthwaite, C. W. 1948. An approach toward a rational classification of climate. *The Geographical Review* 38:55–94 + map.

True, J. C., Jackson, W. C., Davis, E. P., Wharton, C. F., and Sprouse, O. G. 1977. *Soil Survey of Rutherford County, Tennessee*. USDA Soil Conservation Service, in cooperation with the University of Tennessee Agricultural Experiment Station.

U.S. Fish and Wildlife Service. 1993. *Leafy Prairie Clover Technical Draft Plan*. Atlanta, Ga.: U.S. Fish and Wildlife Service.

Ware, S. A. 1969. Ecological role of *Talinum* (Portulacaceae) in cedar glade vegetation. *Bulletin of the Torrey Botanical Club* 96:163–175.

Wilson, C. W., Jr. 1980. *Geology of Cedars of Lebanon State Park and Forest*. Tennessee Division of Geology State Park Series Number 1.

Wofford, B. E., and Kral, R. 1993. Checklist of the vascular plants of Tennessee. *Sida, Botanical Miscellany* Number 10.

Woodward, S. L., and Hoffman, R. L. 1991. The nature of Virginia. In *Virginia's Endangered Species*, Coord. K. Terwilliger, pp. 23–48 + plates 1–24. Blacksburg, Va.: The McDonald and Woodward Publ. Co. Figure. 3 is a map of the physiographic provinces of Virginia drafted by S. L. Woodward (personal communication).

13 Savanna, Barrens, and Glade Communities of the Ozark Plateaus Province

ALICE LONG HEIKENS

Introduction

The Ozark Plateaus (Figure 13.1) supports a mosaic of prairies, forests, glades, barrens, and savannas depending on such factors as topography, bedrock, soils, fire, and native herbivores (Nelson 1985; Rebertus and Jenkins 1994). These plant communities vary widely in their composition and size, and they appear to have changed significantly since settlement by Europeans (Palmer 1921). Open, grassy communities, such as savannas, barrens, and glades, were more common in the northern and particularly the northwestern sections of the Ozark Plateau than in the southern Boston Mountains region, which was more densely forested (Palmer 1921).

In Missouri, *savanna* is defined as grassland with 10%–50% tree cover, little to no shrub component (Nelson 1985), and a parklike appearance. Fire appears to be important in maintaining savanna, which was considered by some to be a transitional or ecotonal community between forest and grassland (Nuzzo 1986).

Barrens were more common than savannas in the Missouri Ozarks, and early use of the term denoted a plant community with shallow, infertile, or rocky soil that supported an herbaceous understory of prairie species and no trees (Schroeder 1981; Nelson 1985) or only scattered, stunted trees (Steyermark 1951; Nelson 1985; Rebertus and Jenkins 1994). On occasion, brushy grasslands were called barrens (Schroeder 1981). Although

Figure 13.1. Location of the Ozark Plateaus Province. Note outlying areas in southwestern Illinois. (From Cozzens 1939; Schwegman 1973; Thom and Wilson 1980; Smith et al. 1984.)

Schwegman (1973) suggested that barrens probably occurred in all the natural divisions of Illinois, they were not included in his review of plant communities.

Although glades are common in Missouri (Nelson and Ladd 1983), quantitative studies that clearly distinguish glades, barrens, and savannas are lacking. Generally, glades are similar to barrens in several ways: (1) Soils are extremely shallow (0–50 cm) and contain many rock fragments; (2) bedrock often crops out; (3) exposure usually is south or west; and (4) vegetation is dominated by herbaceous species with or without scattered, stunted woody species (Nelson and Ladd 1983; Nelson 1985). Large areas of exposed bedrock further distinguish glades from barrens, which often have shallow (0–100 cm; Nelson 1985), rocky soil with less exposed bedrock. Also, soils on glades primarily occur in rock depressions, which contribute to the mosaic vegetation characteristic of glades (White and Madany 1978).

The terms *barrens* and *glades* commonly were used to describe presettlement communities in Missouri (Schroeder 1981) and Illinois (Hutchison 1994; White 1994). Although the term *savanna* was used extensively in Illinois (White 1994), it was not used in presettlement Missouri (Schroeder 1981). Instead, early botanists and land surveyors in Missouri described these communities as barrens and/or oak openings. Today, these barrens and oak openings would be classified as savannas or glades, and the term *barrens* is not used in Missouri (Nelson 1985).

It is apparent that qualitative descriptions of these plant community types overlap and not all botanists use the terms *glade*, *barrens*, and *savanna* similarly. Furthermore, the problem is complicated by the gradation among barrens, savanna, and glade types (Nelson and Ladd 1983). Currently, there is no consistent use of the terms *savanna* or *barrens* in the Ozarks, although there is less confusion with the delineation of *glades*.

Savanna, glade, and barrens are important plant communities in that some support rare species, serve as refuges for prairie-associated species (Logan 1992), represent remnants of presettlement communities, and/or provide habitats for numerous wildlife species.

In this chapter, *savanna* and *barrens* refer to oak openings with an herbaceous understory. The term *savanna* includes all oak openings, and *barrens* is a type of savanna having thin, rocky soil. Glades are distinguished from savannas by extensive bedrock outcrops, a dominate herbaceous vegetation, and shallow, rocky soil.

Geology, Soil, and Topography

The Ozark Plateaus Province, popularly called the Ozark Mountains (Braun 1950), occurs in southern Missouri, northern Arkansas, northeastern Oklahoma, extreme southeastern Kansas, and southwestern Illinois (Figure 13.1). The Ozark Plateaus Province consists of a northern portion, the Ozark Plateau, including the Springfield and Salem plateaus, and the rugged Boston Mountains to the south (Braun 1950). The Ozarks have been a continuous land mass since the end of the Paleozoic Era and are one of the oldest geological regions in North American (Palmer 1921; Beilmann and Brenner 1951; Steyermark 1959). They were uplifted at least twice; however, after the Cretaceous uplift, the region was eroded to a relatively flat plain, which existed until the second uplift in the late Tertiary (Steyermark 1959).

Bedrock types in the Ozark Plateau include limestone, chert, and dolomite of Cambro–Ordovician and Mississippian ages; bedrock in the Boston Mountains is sandstone and shale of Pennsylvanian age (Braun 1950; Steyermark 1951; Thom and Wilson 1980). In Illinois, the northern and southern Ozarkian sections are underlain by relatively pure limestone and cherty limestone, respectively; the central section is underlain by sandstone (Schwegman 1973). Precambrian igneous rocks, including felsite, rhyolite, and dillenite, are exposed in southeastern Missouri (Nelson and Ladd 1983). The Missouri and Arkansas Ozarks are unglaciated (Thom and Wilson 1980); however, the central section and part of the northern section of the Illinois Ozarks were glaciated during the Illinoisan stage of the Pleistocene (Schwegman 1973). Bedrock often is exposed along streams, on river bluffs, and in ravines (Schwegman 1973; Thom and Wilson 1980). Topography ranges from very steep to nearly level, with elevation of 120–550 m (Thom and Wilson 1980). Most of the Illinois Ozark soils were developed in deep loess; however, on steep slopes, loess is thin or absent due to erosion. On these slopes, bedrock and bedrock residuum influenced soil development (Fehrenbacher, Walker and Wascher 1967). Throughout the Missouri and Illinois Ozark regions, erosion has resulted in thin, rocky soils too shallow for intensive agriculture (Beilmann and Brenner 1951; Schwegman 1973; Thom and Wilson 1980; Guyette and Cutter 1991). Soil depth ranges from areas of exposed bedrock and no soil, to areas with soil 10–40 cm deep, to pockets of relatively deep soil (>1 m) (Palmer 1910; Nelson 1985). In general, barrens soils have a low moisture-holding capacity and undergo periodic droughts (Ray and Lawson 1955). Soil moisture in barrens is limited by shallow depths, high amounts of coarse rock fragments, and high permeability rates (Ray and Lawson 1955; Nelson 1985).

Presettlement Vegetation

Nelson (1985) estimated presettlement Missouri had 5.3 million hectares of savanna. Presettlement savannas also were quite abundant in Illinois; however, the extent of presettlement savanna that occurred in Arkansas or specifically in the Ozark region is unknown. Some botanists have proposed that savanna was extensive in presettlement times (Beilmann and Brenner 1951; Schroeder 1981; Ladd 1991), but others believe the area primarily was forest of varying tree densities (Steyermark 1959; Schwegman 1973). In addition, use of General Land Office notes to determine Missouri presettlement vegetation is problematic because some surveys were conducted 100 years after settlement (Nuzzo 1986).

Various accounts of presettlement vegetation described plant communities with scattered, open-grown trees, primarily xeric oak (Quercus) species, and a dense, herbaceous layer dominated by prairie species (Beilmann and Brenner 1951; Schroeder 1981). Although a brushy understory was present on some sites (Schroeder 1981; Rebertus and Jenkins 1994), surveyors noted the general absence of undergrowth and brush throughout much of the Ozark region (Schroeder 1981).

Swallow (1859) noted the beauty of the oak savannas, and Schoolcraft (1819) described savannas: "A tall, thick and rank growth of grass covers the whole country, in which the oaks are standing interspersed, like fruit trees in some well cultivated orchard, and giving to the scenery the most novel, pleasing, and picturesque appearance." Marbut (1911) described the Ozark Dome as a region of open woods with large treeless openings especially in areas of gentle topography. When the United States obtained the Louisiana Territory in 1803, the Ozarks were described as having an open aspect with numerous barrens, parklike groves, and a scattering of trees that did not inhibit the growth of grasses (Beilmann and Brenner 1951).

However, tree density increased when the area was settled between 1820 and 1850 (Sauer 1920). By 1884, Sargent noted the "improved forest condition" as a result of the fence law enactment, which prevented livestock from grazing in woodlands. Forest encroachment into openings was evidenced by prairie soils supporting forest (Braun 1950; Howell and Kucera 1956). Increases in American elm (Ulmus americana), eastern redcedar (Juniperus virginiana), and oak species have been reported (Marbut 1911; Howell and Kucera 1956; Rebertus and Jenkins 1994). This increase in woody cover has been attributed to several factors, including soil erosion, prairie deterioration, gully formation, and, most commonly, fire suppression (Howell and Kucera 1956).

Steyermark (1959) believed the presettlement accounts of Beilmann and Brenner (1951) were erroneous in that DeSoto and Coronado, who described the open nature of the vegetation, apparently never reached the Ozark region (Johnson and Malone 1958,

1959; Fernandez-Armesto 1991). Instead, Steyermark (1959) suggested that European settlers cleared well-developed forests for homesteads. He indicated that early travelers used established Native American paths and animal trails. Presumably, these routes provided an atypical view of the landscape, with emphasis on the open character of some areas that were not indicative of the forest as a whole. Also, he suggested that early travelers were accustomed to forest and overemphasized the grassy openings because they were unusual. However, Steyermark's views are not widely accepted. Most ecologists believe that before European settlement, the Ozark region supported a low-density forest of scattered trees or groves of trees intermixed with prairie (Howell and Kucera 1956; Schroeder 1981; Nelson 1985; Ladd 1991).

Community Composition and Structure

Glades and barrens often occur on south or west slopes; however, savannas tend to be situated in less harsh environments than barrens and glades (Rebertus and Jenkins 1994). Savannas occur on gentle slopes, level plains, and north and east exposures (Nelson 1985). At some locations, relationships among forest development, drainage systems, bedrock, and soils are evident (Palmer 1921; Cozzens 1939; Howell and Kucera 1956; Rebertus and Jenkins 1994). However, occasionally the absence of a well-developed forest on a site occupied by savanna defies explanation (Palmer 1921). Fire and herbivory may be important factors on these sites (Nelson 1985).

Currently, oak–hickory or oak–pine dominates the Ozark forest, but more xeric and mesic forest community types occur in areas of diverse topography. Extant savannas are small remnants or are restored, which may not be representative of the former large, presettlement communities (Jenkins and Rebertus 1994) because the majority of deep-

soil, fertile sites were destroyed by agricultural developments (Nelson 1985). Six natural savanna communities based on soil moisture content (e.g., dry, mesic, and wet–mesic) and substrate (e.g., sand, sandstone, and chert) are defined in the Ozark Plateaus Province by Nelson (1985). Limestone/dolomite, chert, sandstone, and igneous savannas occupy gentle to steep slopes with shallow, well-drained soils and with exposed bedrock, whereas mesic and wet–mesic savanna occur on deep soils with little exposed rock (Nelson 1985). Xeric oak species, eastern redcedar, and shortleaf pine (*Pinus echinata*) dominate the tree canopy on dry savannas, which have an understory of big bluestem (*Andropogon gerardii*), little bluestem (*Schizachyrium scoparium*), and Indian grass (*Sorghastrum nutans*) (Table 13.1). Mesic savannas are dominated by white oak (*Quercus alba*), bur oak (*Q. macrocarpa*), pin oak (*Q. palustris*), big bluestem, Indian grass, and prairie cordgrass (*Spartina pectinata*) (Nelson 1985).

In Missouri, Nelson (1985) did not describe barrens as a distinct plant community type, but classified them as savannas. He defined savannas as having 0%–50% woody canopy cover with little or no shrub layer and a dominant herbaceous layer. In Illinois, White and Madany (1978) used a somewhat similar definition for savannas, except that woody canopy ranged between 10% and 80% and shrubs might be present.

Illinois barrens are classified as a specific type of savanna that occurs in otherwise forested regions of the southern portion of the state (White and Madany 1978). These communities are further subdivided on the basis of soil moisture, for example, dry, dry–mesic, and mesic.

Ozark glades are classified by substrate, for example, limestone, dolomite, chert, sandstone, shale, and igneous (White and Madany 1978; Nelson 1985). Little bluestem is the dominant species on these glade types except for igneous glades, where Indian grass and dropseed (*Sporobolus clandestinus*) domi-

Table 13.1. *Summary of savanna and glade types in Missouri from Nelson (1985). Note that no dominant tree species are listed for glades because Nelson defines glades as herbaceous-dominated communities.*

Community type	Slope position	Soil	Dominant tree species	Dominant herb species
Limestone/dolomite savanna	Steep, upper slopes	Well-drained; shallow: max. depth 100 cm	*Quercus prinoides* var. *acuminata* *Fraxinus americana* *Juniperus virginiana*	*Andropogon gerardii* *Sorghastrum nutans* *Bouteloua curtipendula*
Chert savanna	Gentle–steep slopes	Well-drained; shallow: max. depth 100 cm	*Quercus stellata* *Quercus marilandica* *Quercus velutina* *Quercus alba*	*Sorghastrum nutans* *Andropogon gerardii*
Sandstone savanna	Gentle–steep slopes	Rapidly drained; variable depths: very shallow (max. depth 38 cm) to deep (max. depth >100 cm)	*Quercus marilandica* *Quercus stellata* *Pinus echinata*	*Schizachyrium scoparium* *Sorghastrum nutans* Mosses and lichens
Igneous savanna	Gentle–moderate slopes	Rapidly drained; very shallow: max. depth 38 cm	*Quercus marilandica* *Carya texana*	*Schizachyrium scoparium* *Sorghastrum nutans*
Mesic savanna	Lower north and east slopes	Deep: max. depth >100 cm	*Quercus macrocarpa*	*Andropogon gerardii* *Sorghastrum nutans*
Wet–mesic savanna	Gentle plains	Poorly drained: deep: max. depth >100 cm	*Quercus macrocarpa* *Quercus palustris*	*Spartina pectinata* *Andropogon gerardii*
Limestone glade	Moderate–steep, south and west slopes	Very rapidly drained; 0–38 cm		*Schizachyrium scoparium* *Bouteloua curtipendula*
Dolomite glade	Moderate–steep, often south and west slopes	Rapidly drained; 0–38 cm		*Schizachyrium scoparium* *Bouteloua curtipendula* *Sporobolus heterolepis*
Chert glade	Level–moderately steep slopes	Very rapidly drained; 0–38 cm		*Schizachyrium scoparium* *Sporobolus neglectus* *Coreopsis lanceolata* *Selaginella rupestris*
Sandstone glade	Gentle–moderately steep, often south and west slopes	Very rapidly drained; 0–38 cm		*Schizachyrium scoparium* *Crotonopsis elliptica*
Igneous glade	Gentle–moderately steep, often south and west slopes	Very rapidly drained; 0–38 cm		*Sorghastrum nutans* *Sporobolus clandestinus* *Ambrosia bidentata* *Diodia teres* *Cladina* spp.

nate (Nelson 1985). Species composition, especially characteristic species, varies considerably among the types of glades (Nelson and Ladd 1983; Nelson 1985).

Variation in vegetation between sites was studied at sandstone glades at Devil's Den State Park in northwestern Arkansas and at Calico Rock in north-central Arkansas (Jeffries 1985, 1987). Devil's Den glades occur on steep, primarily west-facing slopes, whereas Calico Rock glades occur on relatively level sites (Table 13.2). Soil pH was variable, especially at Calico Rock. Cryptogams, which significantly decreased with higher soil pH (Jeffries 1985), occurred more frequently at Calico Rock than at Devil's Den. Although both sandstone regions were described as dominated by herbaceous vegetation, little bluestem dominated Devil's Den, and forbs dominated at Calico Rock. Similarities between the two regions include shallow soils, a relatively large percentage of areas devoid of vegetation, and thin soils supporting sparse, xerophytic vegetation (Jeffries 1985, 1987).

Logan (1992) also noted differences among the sandstone glades in the Arkansas Ozark region. Flat-rock and relatively deep-soil sandstone glades were identified. However, at some locations the relatively deep-soil glades were difficult to delineate from limestone glades due to thin limestone strata and/or limestone-derived colluvium overlying the sandstone. Species composition and vegetational cover varied depending on slope angle and presence of limestone (Logan 1992).

Endemic and Rare Species

The age and physiographic diversity of the Ozark Plateaus Province contribute to its high species richness and to the presence of endemic species, which include *Echinacea paradoxa* var. *paradoxa*, *Rudbeckia missouriensis*, *Oenothera missouriensis* var. *missouriensis*, and *Trillium pusillum* var. *ozarkanum* (Thom and

Wilson 1980; Nelson 1985). Some of the rare species, such as *Trillium pusillum* var. *ozarkanum*, *Phlox pilosa*, and *Onosmodium subsetosum* (Steyer-mark 1963) probably evolved in the region. Other species that occur in Ozark barrens are common on the western plains; for example, *Andrachne phyllanthoides*, *Lesquerella gracilis*, *Evolvulus pilosus*, and *Stenosiphon linifolius* (Palmer 1921).

Disturbance Factors

Native Americans annually burned openings to increase grasses for grazers (Sauer 1920; Cutter and Guyette 1994). Although the frequency of fire generally was reduced after settlers moved into an area (Beilmann and Brenner 1951), some settlers continued to periodically burn barrens, savannas, and glades to control woody vegetation and enhance herbaceous vegetation (Read 1951). Before and after European settlement, the majority of fires probably were started by humans, because most Ozark thunderstorms with lightning strikes have sufficient rainfall to limit wildfires (Cutter and Guyette 1994).

Fire suppression resulted in loss of natural openings, including prairies, barrens, savannas, and glades, due to forest encroachment (Beilmann and Brenner 1951; Jenkins and Rebertus 1994). Presettlement fire interval was 2.8 and 4.3 years on two Ozark savannas, but after Native Americans left the area (ca. 1850), the interval increased to a 24-year average (Cutter and Guyette 1994). This reduction in fire frequency was attributed to a reduced cover of herbaceous vegetation as a result of woody plant invasion, grazing, and absence of regular burn regimes.

By the late 1800s and early 1900s, rapid degradation of barrens and savannas was evident, and few openings remained. The last areas into which woody plants encroached were the xeric, thin-soil openings in northern Arkansas and southwestern Missouri (Beilmann and Brenner 1951). It appears

Table 13.2. *Biotic and abiotic characteristics of two sandstone glade regions in the Arkansas Ozarks (Jeffries 1985, 1987). Devil's Den State Park sandstone glades are located in northwestern Arkansas, and Calico Rock sandstone glades are located in north-central Arkansas.*

Characteristic	Devil's Den	Calico Rock
Slope angle (degrees)	20–40	0–5
Mean soil depth (cm)	5.2	4.2
Soil pH	4.6–7.0	4.7–8.88
Percent occurrence of herbaceous vegetation	52.6	29.4
Percent occurrence of cryptogamic vegetation	7.8	32.0
Percent occurrence of bare soil, bare rock, and litter	39.6	38.6
Mean basal area for all trees (m^{-2} ha^{-1})	3.6	2.8
Most commonly occurring woody species	*Juniperus virginiana* *Quercus stellata*	*Juniperus virginiana*
Most commonly occurring herbaceous species	*Schizachyrium scoparium*	*Coreopsis grandiflora* *Crotonopsis elliptica*

that fire had less influence on thin-soil sandstone glades than on other natural forest openings due to the large amount of exposed bedrock and to the sparse vegetation that limited fuel loading on sandstone glades. Limestone and other glade types with deeper soils and more vegetation were more fire dependent and had more rapid woody encroachment in the absence of fire (Logan 1992).

Today, savannas, barrens, and glades are threatened by increased densities of eastern redcedar, blackjack oak (*Quercus marilandica*), and black hickory (*Carya texana*) that reduce the abundance and diversity of herbaceous species (Rebertus and Jenkins 1994). In addition, sugar maple (*Acer saccharum*) has encroached on some Missouri glades, barrens, and thinly wooded dry ridges (Rochow 1972; Nigh, Pallardy and Garrett 1985; Pallardy, Nigh and Garrett 1988, 1991). Although sugar maple typically occurs on cool slopes with deep, moist soils, this species can remain in the understory of xeric sites for years, and some ecologists question whether it will replace oak as a dominant canopy species (Rochow 1972; Nigh, Pallardy and Garrett 1985). Such succession is most probable on barrens and savannas that have

undergone substantial woody encroachment and provide some shade.

In addition to fire suppression, logging has contributed to the loss of some savannas (Nelson 1985) in that the removal of large, scattered trees has resulted in a dense growth of small trees (Cozzens 1939). Logging in the Ozarks peaked in the 1900s, and within 50 years most of the valuable timber was cut (Beilmann and Brenner 1951). As dense stands of small trees matured, the cover of prairielike herbaceous vegetation decreased as shade increased.

Overgrazing of native species and planting of nonnative species also contributed to the degradation of these communities (Nelson 1985). By the 1860s, many of the native Ozark savannas, barrens, and glades had deteriorated greatly, leaving stands of native grasses only in remote rugged areas (Read 1951). In contrast to dry savannas, wet and wet–mesic savannas were settled and farmed (Palmer 1921; Nelson 1985), leaving thin-soil xeric barrens as isolated remnants (Beilmann and Brenner 1951).

Fauna

Restoration projects have focused primarily on plant species, and thus little is known

Table 13.3. *Characteristic animal species of Missouri glades (Nelson 1985).*

Common name	Scientific name
Bachman's sparrow*	*Aimophila aestivalis*
Roadrunner*	*Geococcyx californianus*
Brush mouse*	*Peromyscus boylei*
Ornate box turtle	*Terrapene ornata ornata*
Western slender glass lizard	*Ophisaurus attenuatus attenuatus*
Eastern collared lizard	*Crotaphytus collaris collaris*
Six-lined racerunner	*Cnemidophorus sexlineatus sexlineatus*
Eastern narrowmouth toad	*Gastrophryne carolinensis*
Osage copperhead	*Agkistrodon contortrix phaeogaster*
Western worm snake	*Carphophis vermis*
Prairie ringneck snake	*Diadophis punctatus arnyi*
Prairie kingsnake	*Lampropeltis calligaster calligaster*
Eastern coachwhip snake	*Masticophis flagellum flagellum*
Bullsnake	*Pituophis melanoleucus sayi*
Western pigmy rattlesnake	*Sistrurus miliarius streckeri*
Flathead snake	*Tantilla gracilis*
Rough earth snake	*Virginia striatula*
Missouri tarantula	*Dugesiella hentzi*
Black widow spider	*Latrodectus mactans*
Plains scorpion	*Centruroides vittatus*
Lichen grasshopper	*Trimerotropis saxatilis*
Missouri woodland swallowtail	*Papilio joanae*
Metea skipper	*Hesperia metea*

* restricted to dolomites glades

about the effects of management, specifically fire, on wildlife (Robinson 1994). In addition, due to degradation and limited research on Ozark savanna, the effects of wildlife on these communities are difficult to identify (Nelson 1985). Species such as the bison (*Bison bison*), elk (*Cervus canadensis*), gray wolf (*Canis lupus*), and coyote (*Canis latrans*) probably inhabited Ozark savanna (Nelson 1985). Also, it appears that game was abundant, because Native Americans in the Ozarks did not cultivate maize (*Zea mays*). Instead, they lived on bison, elk, deer (*Odocoileus virginianus*), turkey (*Meleagris gallopavo*), grouse (*Bonasa umbellus*), and quail (*Colinus virginianus*) (Beilmann and Brenner 1951), as well as various plant species.

Nelson (1985) identified species characteristic of Missouri glades (Table 13.3). However, many animals seen in these communities also inhabit adjacent forest.

Studies have shown that Ozark glades, barrens, and savannas produce more grazing herbage than does hardwood forest (Read 1951; Segelquist and Green 1968). This suggests that these open communities may support more wildlife than mature forest, although acorns are more abundant in the hardwood forest and provide substantial food (Segelquist and Green 1968).

Birds characteristic of southern Illinois barrens include the eastern wood-pewee (*Contopus virens*), great crested flycatcher (*Myiarchus crinitus*), blue-gray gnatcatcher (*Polioptila caerulea*), northern cardinal (*Cardinalis cardinalis*), summer tanager (*Piranga rubra*), and indigo bunting (*Passerina cyanea*), all common species capable of occupying other habitats (Robinson 1994). Other species, which decline in abundance due to fire, typically are associated with dense understories and low canopy layers. These species include the acadian flycatcher (*Empidonax virescens*), wood thrush (*Hylocichla mustelina*), red-eyed vireo (*Vireo olivaceus*), worm-eating warbler (*Helmitheros vermivorus*), hooded warbler (*Wilsonia citrina*), and ovenbird (*Seiurus aurocapillus*). Therefore, Robinson (1994) suggested that species diversity may change with barrens restoration projects that use fire as a management tool. However, adjacent forests may serve as a refuge for these birds because they are common species often found in forests. Also, because barrens intergrade with forest, managing barrens with a landscape ecology approach to restore the original distribution of forest and barrens may not negatively influence these avian species.

In addition to changes in species abundance, Robinson (1994) determined that fire also increased brown-headed cowbirds (*Molothrus ater*) in southern Illinois. He suggested that barrens restoration efforts may adversely affect bird populations by increasing brood parasitism. In contrast, Hahn and Hatfield (1995) found that in New York, cowbird parasitism was not higher at the forest edge. Furthermore, they suggested that cowbird parasitism patterns were not consistent within the forest or across the landscape. They concluded that effects of savanna management on the rate and pattern of cowbird parasitism are difficult to predict due to regional variation in habitat and host preference. It is clear that additional research is needed to determine the impact of cowbird parasitism in savannas.

Summary

It appears that Ozark savannas, barrens, and glades have undergone substantial degradation since settlement due to fire suppression, overgrazing, agricultural practices, and logging. The once widespread and picturesque oak openings currently are represented by fragmented remnants on primarily xeric areas. These xeric communities have persisted because the droughty nature of the sites slowed forest encroachment, and these sites were not plowed because soils were unsuitable for agriculture.

The lack of distinct characteristics delineating savanna, barrens, and glade communities in the Ozarks has resulted in much confusion. Changes in terminology since European settlement, degradation of many of these communities, and variation of each community type complicate the development of clear community definitions. Quantitative research to delineate these plant community types is needed, as is additional research on the management and restoration of these areas, especially the effects of fire on wildlife species.

REFERENCES

Beilmann, A. P., and Brenner, L. G. 1951. The recent intrusion of forests in the Ozarks. *Annals of the Missouri Botanical Garden* 38:261–282.

Braun, E. L. 1950. *Deciduous Forests of Eastern North America*. Philadelphia: Blakiston Co.

Cozzens, A. B. 1939. Analyzing and mapping natural landscape factors of the Ozark Providence. *Transactions of the Academy of Science of St. Louis* 30:37–63.

Cutter, B. E., and Guyette, R. P. 1994. Fire frequency on an oak–hickory ridgetop in the Missouri Ozarks. *The American Midland Naturalist* 132:393–398.

Fehrenbacher, J. B., Walker, G. O., and Wascher, H. L. 1967. *Soils of Illinois*. Illinois Agricultural Experiment Station Bulletin 725. Urbana, Ill.: University of Illinois.

Fernandez-Armesto, F. 1991. *Times: Atlas of World Exploration*. New York: Harper Collins Publishers.

Guyette, R. P., and Cutter, B. E. 1991. Tree-ring analysis of fire history of a post oak savanna in the Missouri Ozarks. *Natural Areas Journal* 11:93–99.

Hahn, D. C., and Hatfield, J. S. 1995. Parasitism at the landscape scale: cowbirds prefer forests. *Conservation Biology* 9:1415–1424.

Howell, D. L., and Kucera, C. L. 1956. Composition of pre-settlement forests in three counties of Missouri. *Bulletin of the Torrey Botanical Club* 83:207–217.

Hutchison, M. D. 1994. Using the Public Land Survey fieldnotes to determine the presettlement character of the Midwestern landscape with particular reference to southern Illinois barrens. In *Proceedings of the North American Conference on Savannas and Barrens*, ed. J. S. Fralish, R. C. Anderson, J. E. Ebinger, and R. Szafoni, pp. 87–90. Chicago, Ill.: U.S. Environmental Protection Agency. Great Lakes National Program Office.

Jeffries, D. L. 1985. Analysis of the vegetation and soils of glades on calico rock sandstone in northern Arkansas. *Bulletin of the Torrey Botanical Club* 112:70–73.

Jeffries, D. L. 1987. Vegetation analysis of sandstone glades in Devil's Den State Park, Arkansas. *Castanea* 52:9–15.

Jenkins, S. E., and Rebertus, A. J. 1994. Spatial demography of an oak savanna in the Ozarks. In *Proceedings of the North American Conference on Savannas and Barrens*, ed. J. S. Fralish, R. C. Anderson, J. E. Ebinger, and R. Szafoni, pp. 107–111. Chicago, Ill.: U.S. Environmental Protection Agency. Great Lakes National Program Office.

Johnson, A., and Malone, D. 1958. *Dictionary of American Biography,* Vol. II, part I. New York: Charles Scribner Publishers.

Johnson, A., and Malone, D. 1959. *Dictionary of American Biography,* Vol. III. New York: Charles Scribner Publishers.

Ladd, D. 1991. Reexamination of the role of fire in Missouri oak woodlands. In *Proceedings of the Oak Woods Management Workshop*, ed. G. V. Burger, J. E. Ebinger, and G. S. Wilhelm, pp. 67–80. Charleston, Ill.: Eastern Illinois University.

Logan, J. M. 1992. The glades of the Buffalo National River, Arkansas. M.S. thesis, Iowa State University, Ames.

Marbut, C. F. 1911. *Soil Reconnaissance of the Ozark Region of Missouri and Arkansas.* Field Operations of the Bureau of Soils, U.S. Dept. of Agriculture. Washington, D.C.: U.S. Government Printing Office. (Cited in Schroeder 1981.)

Nelson, P. W. 1985. *The Terrestrial Natural Communities of Missouri.* Jefferson City, Mo.: Missouri Department of Natural Resources.

Nelson, P., and Ladd, D. 1983. Preliminary report on the identification, distribution and classification of Missouri glades. In *Proceedings of the Seventh North American Prairie Conference*, ed. C. L. Kucera, pp. 59–76. Springfield, Mo.: Southwest Missouri State University,

Nigh, T. A., Pallardy, S. G., and Garrett, H. E. 1985. Sugar maple–environment relationships in the river hills and central Ozark Mountains of Missouri. *The American Midland Naturalist* 114:235–251.

Nuzzo, V. A. 1986. Extent and status of Midwest oak savanna: presettlement and 1985. *Natural Areas Journal* 6:6–35.

Pallardy, S. G., Nigh, T. A., and Garrett, H. E. 1988. Changes in forest composition in central Missouri: 1968–1982. *The American Midland Naturalist* 120:380–390.

Pallardy, S. G., Nigh, T. A., and Garrett, H. E. 1991. Sugar maple invasion in oak forests of Missouri. In *Proceedings of the Oak Woods Management Workshop*, ed. G. V. Burger, J. E. Ebinger, and G. S. Wilhelm, pp. 21–30. Charleston, Ill.: Eastern Illinois University.

Palmer, E. J. 1910. Flora of the Grand Falls Chert Barrens. *Transactions of the Academy of Science of St. Louis* 19:97–112.

Palmer, E. J. 1921. The forest flora of the Ozark region. *Journal of the Arnold Arboretum* 2:216–232.

Ray, H. C., and Lawson, M. 1955. Site characteristics as a guide to forest and grazing use in the Ozarks. *Journal of Range Management* 8:69–73.

Read, R. A. 1951. Woodland forage in the Arkansas Ozarks. *Journal of Range Management* 4:391–396.

Rebertus, A. J., and Jenkins, S. 1994. *Savanna and Glade Vegetation of Turkey Mountain, Arkansas.* Omaha, Neb.: National Park Service, Midwest Regional Office.

Robinson, S. K. 1994. Bird communities of restored barrens and burned forests of southern Illinois. In *Proceedings on the North American Conference on Savannas and Barrens*, ed. J. S. Fralish, R. C. Anderson, J. E. Ebinger, and R. Szafoni, pp. 147–150. Chicago, Ill.: U.S. Environmental Protection Agency. Great Lakes National Program Office.

Rochow, J. J. 1972. A vegetational description of a mid-Missouri forest using gradient analysis techniques. *The American Midland Naturalist* 87:377–396.

Sargent, C. S. 1884. *Report on the Forests of North America.* U.S. Dept. of Interior, 10th Census,

Forestry. (Cited in Beilmann and Brenner 1951.)

Sauer, C. O. 1920. *The Geography of the Ozark Highland of Missouri.* Geographical Society of Chicago Bulletin No. 7. (Cited in Beilmann and Brenner 1951.)

Schoolcraft, H. R. 1819. *A View of the Lead Mines of Missouri.* New York: C. Wiley and Co. (Cited in Beilmann and Brenner 1951.)

Schroeder, W. A. 1981. *Presettlement Prairie of Missouri.* Jefferson City, Mo.: Missouri Department of Conservation, Natural History Series No. 2.

Schwegman, J. E. 1973. *The Natural Divisions of Illinois. Comprehensive Plan for the Illinois Nature Preserves System, Part 2.* Springfield, Ill.: Illinois Nature Preserves Commission.

Segelquist, C. A., and Green, W. E. 1968. Deer food yields in four Ozark forest types. *Journal of Wildlife Management* 33:330–337.

Smith, K. L., Pell, W. F., Rettig, J. H., Davis, R. H., and Robinson, H. W. 1984. *Arkansas's Natural Heritage.* Little Rock, Ark.: August House.

Steyermark, J. A. 1951. Botanical areas in the Missouri Ozarks. *Missouri Botanical Garden Bulletin* 39:126–133.

Steyermark, J. A. 1959. Vegetational history of the Ozark forest. *University of Missouri Studies* 31:1–138.

Steyermark, J. A. 1963. *Flora of Missouri.* Ames, Iowa: Iowa State University Press.

Swallow, G. C. 1859. *Geological Report of the Country along the Line of the Southwestern Branch of the Pacific Railroad.* St. Louis, Mo.: George Knapp and Co. (Cited in Beilmann and Brenner 1951.)

Thom, R. H., and Wilson, J. H. 1980. The natural divisions of Missouri. *Transactions of the Missouri Academy of Science* 14:9–23.

White, J. 1994. How the terms savanna, barrens, and oak openings were used in early Illinois. In *Proceedings of the North American Conference on Savannas and Barrens*, ed. J. S. Fralish, R. C. Anderson, J. E. Ebinger, and R. Szafoni, pp. 25–63. Chicago, Ill.: U.S. Environmental Protection Agency. Great Lakes National Program Office.

White, J., and Madany M. H. 1978. Classification of natural communities in Illinois. In *Illinois Natural Areas Inventory Technical Report.* Urbana, Ill.: Illinois Natural Areas Inventory.

14 The Cross Timbers

B. W. HOAGLAND, I. H. BUTLER, F. L. JOHNSON,
AND S. GLENN

Introduction

The cross timbers are a mosaic of forest, woodland, savanna, and prairie vegetation located in portions of Kansas, Oklahoma, and Texas (Figure 14.1). Nonetheless, two woody species characterize the cross timbers: post oak (*Quercus stellata*; Figure 14.2a) and blackjack oak (*Quercus marilandica*; Figure 14.2b) (Dyksterhuis 1948; Hale 1955; Rice and Penfound 1955). There are an estimated 4.8 million ha of cross timbers located between 38° N latitude in southeastern Kansas and 32° N latitude in north central Texas (Küchler 1964; Engle and Stritzke 1992). Approximately half (2.5 million ha) of the cross timbers are in Oklahoma (Rice and Penfound 1959; D. D. Dwyer and Santelman 1964). The cross timbers form two distinct bands of vegetation in Texas, known as the western and eastern cross timbers (Figure 14.1). These formations were referred to as the upper and lower cross timbers, respectively, by early European settlers due to their location along the Red River (Foreman 1947).

Josiah Gregg, an entrepreneur who was active in the region during the 1840s, left an apt description of the cross timbers that is relevant today:

The celebrated cross timbers, of which frequent mention has been made, . . . vary in width from five to thirty miles, and entirely cut off the communication betwixt the interior prairies and those of the Great Plains. They may be considered as a "fringe" of the great prairies, being a continuous brushy strip, composed of various kinds of undergrowth; such as blackjacks, post-oaks, and in some places hickory, elm, etc. intermixed with a very diminutive dwarf oak. (Fulton 1941)

Contemporary vegetation classifications describe the cross timbers as oak–hickory forest or post oak savanna (Rice and Penfound 1959; Sanders 1980; Diamond, Riskind and Orzell 1987; Lauver 1989; Hoagland 1996). However, classifying *cross timber* sites as savanna can be problematic due to the propensity of post oak and blackjack oak to root sprout and produce mottes of trees with interlocking crowns (Tharp 1926; Penfound 1962; Powell and Lowry 1980).

It is believed that the term *cross timbers* refers to the "timber" that had to be "crossed" as expeditions and settlers traveled west, although its true origins are obscure (Foreman 1947; Dyksterhuis 1948). The earliest European account of the cross timbers was penned by Athanase De Mezieres in a report from east Texas in 1772 (Bolton 1914). He noted that the native Americans referred to the cross timbers as the Grand Forest and that the vegetation was dense and difficult to traverse. Thomas Nuttall, in 1821, described the cross timbers as a "pathless thicket of somber timber" (Bruner 1931). The Ellsworth expedition entered the cross timbers in 1832 near present-day Tulsa (Luckhardt and Barclay 1938; Foreman 1947). Washington Irving accompanied the expedition and wrote: "I shall not easily forget the mortal toil, and the vexations of flesh and spirit, that we

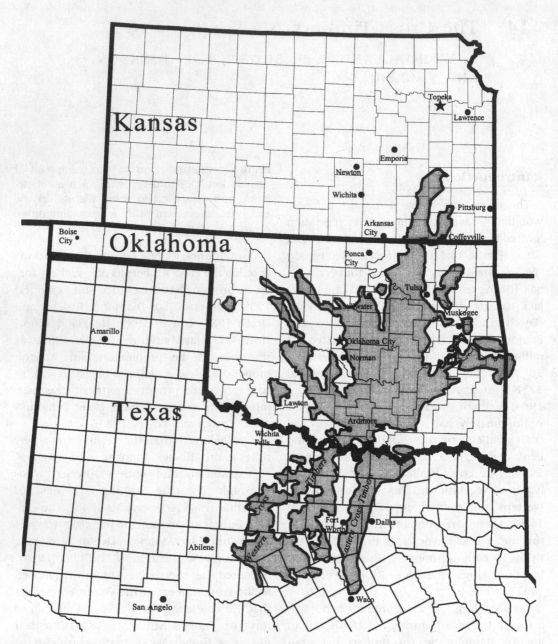

Figure 14.1. Location of the cross timbers region. (Adapted from Küchler 1964.)

underwent occasionally, in our wanderings through the Cross Timber. It was like struggling through forests of cast iron" (Irving 1956).

Elsewhere in the region, Captain R. B. Marcy reported little difficulty in traversing cross timbers vegetation. During the expedi-

tion in search of the Red River's source, Marcy wrote: "At six different points where I have passed through it [the cross timbers], I have found it characterized by these peculiarities; the trees, consisting primarily of post-oak and black-jack, standing at such intervals that

Figure 14.2. Aspects of the cross timbers prairie, savanna, forest vegetation mosaic in Osage County, Oklahoma: (a) a forested, south-facing slope dominated by post oak (*Quercus stellata*), and (b) savanna vegetation dominated by little bluestem (*Schizachyrium scoparium*) with blackjack oak (*Quercus marilandica*).

wagons can without difficulty pass between them in any direction" (Foreman 1947).

Physical Environment

Climate

The cross timbers region is located in the subtropical humid climate zone, which is characterized by hot summers and mild winters (Trewartha 1968). Precipitation varies longitudinally in the region and is highest in the east (Table 14.1; Figure 14.3; Court 1974; Corcoran 1982). Annual precipitation averages 102 cm in the eastern Oklahoma cross timbers, but only 71 cm in the west (Sutherland 1977). Approximately 35%–49% of the annual precipitation falls during the spring months (Harrison 1974). Temperature, on the other hand, varies latitudinally (Court 1974; Corcoran 1982). Regionally, annual average

Table 14.1. *Species composition, basal area (m²ha⁻¹), and annual precipitation (cm) for six cross timbers sites in Oklahoma. Sites are arranged by Oklahoma county along an east–west axis. (Compiled from original data collected by E. L. Rice and W. Penfound.)*

Species	Cherokee	Muskogee	Love	McClain	Grady	Major
	Basal area (m² ha⁻¹)					
Quercus stellata	2.19	5.64	9.91	7.26	7.03	0.58
Quercus marilandica	6.22	1.67	3.69	2.25	1.79	4.02
Quercus velutina	5.52	0.0	0.0	0.0	0.0	0.0
Carya texana	0.46	0.35	0.58	0.69	0.0	0.0
Ulmus alata	0.0	1.09	0.46	0.0	0.0	0.0
Juniperus virginiana	0.0	0.0	0.0	0.0	0.0	0.12
Other trees	0.58	0.06	1.38	0.35	0.06	0.06
Total stand basal area	14.97	8.81	16.02	10.55	8.88	4.78
	Precipitation (cm)					
	107	104	91	86	79	66

temperature is 15 °C in the north and 18.8 °C in the south (Bell and Hulbert 1974; Court 1974; Sutherland 1977). Growing season length ranges from 180 days in the north to 240 days in the southern cross timbers (Figure 14.4; Harrison 1974).

Midtropospheric pressure and winds above the cross timbers affect the spatial distribution of weather elements (Corcoran 1982). Prevailing winds are from the west (Court 1974). Budbreak and flowering are initiated in the eastern cross timbers when warm, moisture-laden air arrives from the Gulf of Mexico in spring (McCluskey 1972). The region also experiences periodic drought and a high frequency of large hail and tornadoes (Court 1974).

Geology and Soils

The cross timbers are located on the Osage Plains of the Central Lowland Physiographic Province (Hunt 1974). The Osage Plains are characterized as irregular plains with local relief from 30 to 90 m (National Atlas 1968; Hunt 1974). Major landforms in the region include low, east-facing cuestas, river bluffs, tablelands, gentle slopes, and deep ravines (Hunt 1974).

The surface geology of the region consists of Pennsylvanian, Permian, and Cretaceous sedimentary formations that dip gently westward and strike north-to-south (Hunt 1974). Pennsylvanian formations are exposed in a series of plains and intervening ridges in Oklahoma and Kansas. Permian shales and sandstones in central Oklahoma form gently rolling hills and broad plains that extend westward (Curtis and Ham 1972). Substantial Quaternary deposits occur along major streams in the region, including the Arkansas, Brazos, Canadian, Cimarron, South Canadian, Red, Trinity, and Washita rivers (National Atlas 1968).

The close association between cross timbers vegetation and coarse, arenaceous soils has been recognized for many years (Hill 1887; Bruner 1931; Dyksterhuis 1948). Cretaceous sandstones in Texas and Pennsylvanian sandstones in Oklahoma and Kansas are the parent materials of these soils (Dyksterhuis 1948; Branson and Johnson 1979). Deep, coarse-textured soils support larger trees than shallow, coarse-textured soils (Engle and Stritzke 1992). Heavy-textured, shale-derived soils produce savannas with trees in dense mottes (Powell and Lowry 1980). Prairie and glade vegetation occur on fine-textured clay soils derived from limestone and shale (Dyksterhuis 1948; Gray and Galloway 1959; Bell and Hulbert 1974; Rhodes 1980).

Alfisols are the predominant soil order throughout the cross timbers, with a minor component of Inceptisols. Two major soil series, Windthorst (moderately well drained) and Stephenville (well drained), are low-fertility Alfisols that belong to the Paleustalf and Haplustalf Great Groups, respectively. The well-drained Darnell soil series is an Inceptisol with low fertility and a member of the Ustochrept Great Group (Aandahl 1982). The Stephenville (yellowish red to light brown) and Darnell (grayish brown to light brown) soils are common in the northern cross timbers, and Windthorst (reddish brown to yellowish brown) soils are prominent in the southern cross timbers. Soil depth is shallowest for the Darnell (<10–50 cm), intermediate for the Stephenville (50–90 cm), and deepest for the Windthorst (>90cm) (Gray and Galloway 1959; Godfrey, McKee and Oakes 1973; Aandahl 1982).

Vegetation and Flora

Paleovegetation

The origins of the cross timbers are linked closely to the emergence of the North American grasslands. Events such as the Miocene uplift of the Rocky Mountains, growth of the Antarctic ice sheet, and the cooling of oceanic waters all contributed to increasing aridity in central North America, which promoted the development of grass-lands (Axelrod 1985). Oak and hickory species migrated into the eastern Oklahoma flora some 12,000 yr BP, displacing jack pine, spruce, and fir (Delcourt and Delcourt 1981). Prairie vegetation established on the south-ern plains as early as 12,000 yr BP (Axelrod 1985), and oak savannas were present in the region about 10,000 yr BP (Delcourt and Delcourt 1981). The regional climate became warmer and drier during the next 2,000 years, facilitating the spread of prairie, oak

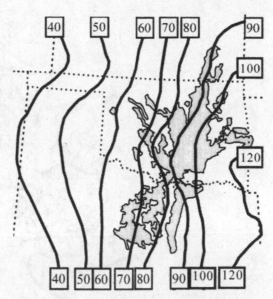

Figure 14.3. Average annual precipitation (cm) in the cross timbers region. (Adapted from Court 1974.)

savannas, and oak–hickory vegetation into the Interior Highlands (Delcourt and Delcourt 1981). What percentage of this Late Holocene vegetation was composed of post oak or blackjack oak cannot be discerned. Paleobotanical evidence suggests that hickory species were more abundant in northeastern Oklahoma approximately 1,000 yr BP and that they subsequently declined as the abun-dance of oak species increased (Hall 1982). Weaver and Clements (1929) concluded that the cross timbers formed during a period of climatic amelioration and now represent a relict of the oak–hickory forest. Indeed, floristic affinities between the cross timbers and the eastern oak–hickory forest are strong (Smeins and Diamond 1986).

Vegetation Composition

Although post oak and blackjack oak have come to characterize the cross timbers, it must be reiterated that the region is a mosaic of prairie, woodland, and savanna vegetation (Dyksterhuis 1948, 1957; Rice and Penfound 1959). Post oak and blackjack oak may con-stitute up to 90% of the canopy cover (Rice

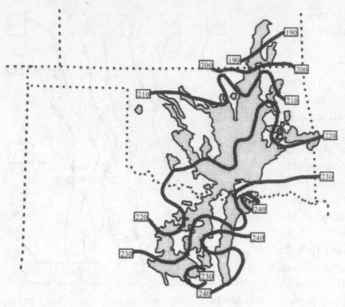

Figure 14.4. Latitudinal trends in length of growing season (days) in the cross timbers region. (Adapted from Harrison 1974.)

and Penfound 1959) and 50% of the basal area in cross timber stands (Table 14.1; Kennedy 1973). In fact, post oak and blackjack oak account for 70% of the total woody basal area in Oklahoma (Rice and Penfound 1959). Both species are slow growing, low in stature, and intolerant of shade and competition, and both frequently reproduce from root sprouts and have limited commercial value (Liming 1942; Rice and Penfound 1959; Stransky 1990). Post oak produces root sprouts more quickly than blackjack oak following disturbance (Powell and Lowry 1980; Stransky 1990). Self- and intraspecific root grafting is common in both species, but interspecific root grafts are rare (Love 1971).

These two species also differ in many life history and physiological traits. The abundance of both species is correlated strongly with soil sand content, but the abundance of post oak is greater on soils with high organic content (Klahr 1989; Collins and Klahr 1991). Post oak grows in a wide range of nutrient and moisture conditions (Johnson and Risser 1972, 1973). Post oak growth rates

are positively correlated with precipitation and become increasingly climate sensitive in the western cross timbers (Stahle and Hehr 1984). Blackjack oak is susceptible to periodic drought and has a higher mortality rate than post oak. Blackjack oak often thrives at sites unfavorable for other species due to its ability to tolerate low soil fertility and moisture stress (Rice and Penfound 1959; Johnson and Risser 1972). In addition, there is evidence that reproduction in cross timbers blackjack oak populations is declining (Johnson and Risser 1975).

Black hickory (*Carya texana*), black oak (*Quercus velutina*), and redcedar (*Juniperus virginiana*) are woody species of secondary importance in the cross timbers (Coppock et al. 1955; Rice and Penfound 1955, 1959; Penfound 1963; Johnson and Risser 1972). Black hickory responds to soil fertility and moisture gradients much as does post oak (Johnson and Risser 1972). Black oak can tolerate low soil fertility but requires relatively high soil moisture (Johnson and Risser 1972).

The regional abundance of redcedar has increased dramatically since European settlement, due primarily to the suppression of fire (Dyksterhuis 1948; Snook 1985; Engle and Kulbeth 1992). However, individuals in excess of 250 years in age have been reported from the region (Stahle and Chaney 1994). Redcedar is involved in a chain of events that expedites the process of canopy closure in savanna vegetation. First, birds deposit redcedar seeds. Redcedar seeds germinate in large numbers, and, once established, facilitate further invasion of grassland and savanna vegetation by other woody species (Rykiel and Cook 1986). Again, suppression of fire is key in this process.

Prominent members of the woody understory in the cross timbers region include chittamwood (*Bumelia lanuginosa*), redbud (*Cercis canadensis*), roughleaf dogwood (*Cornus drummondii*), Mexican plum (*Prunus mexicana*), buckbrush (*Symphoriocarpos orbiculatus*), winged sumac (*Rhus copallina*), smooth sumac (*R. glabra*), winged elm (*Ulmus alata*), and fox grape (*Vitis vulpina*) (Gould 1902; Lathrop 1958; Harrison 1974; Johnson and Risser 1974, 1975). The actual composition and density of the woody understory vary regionally and from site to site. For example, stem densities of winged elm are higher in the eastern, versus the western, cross timbers of Texas (McCluskey 1972). Roughleaf dogwood, winged sumac, and smooth sumac are ecotonal species (Petranka and McPherson 1979). Winged sumac, aggressively invades prairie vegetation from the ecotone by means of shallowly rooted rhizomes (Petranka and McPherson 1979). Roughleaf dogwood and Mexican plum occur in woodlands on sandy soil (Schnell, Risser and Helsel 1981; Ewing, Stritzke and Kulbeth 1984). Chittamwood and winged elm are also abundant in the woody seedling layer (Penfound 1963; Dooley and Collins 1984).

The herbaceous understory of the cross timbers is similar in composition to the surrounding prairie (Dyksterhuis 1948; Küchler 1964, 1974). However, understory development and species richness often are suppressed allelopathically by post oak and blackjack oak, particularly in closed-canopy stands (McPherson and Thompson 1972). Little bluestem (*Schizachyrium scoparium*) is the dominant grass in most cross timbers stands, but big bluestem (*Andropogon gerardii*) and Indiangrass (*Sorghastrum nutans*) may be codominant (Küchler 1964). Other commonly encountered grasses include threeawn (*Aristida purpurascens*), sideoats grama (*Bouteloua curtipendula*), hairy grama (*B. hirsuta*), poverty grass (*Danthonia spicata*), small panicgrass (*Panicum oligosanthes*), switchgrass (*P. virgatum*), tall dropseed (*Sporobolus asper*), and purpletop (*Tridens flavus*) (Dyksterhuis

1948; Rice and Penfound 1955, 1959; Küchler 1964; Johnson and Risser 1975; Smeins and Diamond 1986). In savannas, sideoats grama and tall dropseed are most abundant on clay soils, and small panicgrass is most abundant on sandy soils (Dyksterhuis 1948; Ewing, Stritzke and Kulbeth 1984).

Regional Variation

Local topo-edaphic gradients and regional climatic variation (Lathrop 1958; Penfound 1962; Harrison 1974) affect the flora and vegetation of the cross timbers region. Dyksterhuis (1948) wrote that ecologists could understand the cross timbers only in the context of regional environmental gradients. For example, along a north–south axis, cross timbers vegetation ranges from closed forests of post oak, blackjack oak, and pignut hickory (*Carya glabra*) in Kansas to post oak–blackjack oak–mesquite (*Prosopis juliflora*) savannas in Texas (Dyksterhuis 1948). The codominant status of black hickory decreases from east to west (Risser and Rice 1971; Little 1981). Black hickory is replaced by bitternut hickory (*C. cordiformis*) and pignut hickory (*C. glabra*) as the dominant hickories in the Kansas cross timbers (Hale 1955; Lathrop 1958). Likewise, the abundance of black oak declines in the western cross timbers due to physiological constraints (Hall and McPherson 1980). Species with affinities to the shortgrass steppe and southwestern United States, such as buffalograss (*Buchloë dactyloides*), *Opuntia* spp., and mesquite, become increasingly abundant in the western cross timbers of Texas (Dyksterhuis 1948).

Topographic complexity in the eastern cross timbers region (National Atlas 1968) produces an intricate interplay of vegetation types ranging from bottomland, mesic, and closed-canopy post oak forests, to oak savannas, and prairie. In this area, cross timbers vegetation is developed best on south-facing sandstone slopes and ridgelines (Rice and Penfound 1959). Sugar maple (*Acer saccharum*), bitternut hickory, pecan (*C. illinoensis*), green ash (*Fraxinus pennsylvanica*), black walnut (*Juglans nigra*), eastern cottonwood (*Populus deltoides*), chinquapin oak (*Quercus muhlenbergii*), Shumard oak (*Quercus shumardii*), and American elm (*Ulmus americana*) are common trees in bottomland and mesic forests (Barclay 1948; Dyksterhuis 1948; Hale 1955; Buck 1964; Rice 1965). Cedar elm (*Ulmus crassifolia*) is prominent in bottomland forests of the eastern cross timbers in Texas (Marcy 1982).

Several species in the cross timbers have relictual or disjunct distributions. The most striking example is the occurrence of seaside alder (*Alnus maritima*) on gravel bars and stream banks in the Blue River drainage of Oklahoma (Johnston and Pontotoc counties). The principal populations of seaside alder are found on the Delmarva Peninsula of Maryland and Delaware (Little 1981). The history of this disjunction has yet to be elucidated (Zanoni et al. 1979). Relict populations of sugar maple occur in the Witchita Mountains and Caddo Canyons of west-central Oklahoma (Clements 1936; Little 1939, 1981). The Caddo Canyons were eroded from Permian sandstone and provide an ameliorated microclimatic hospitable to sugar maple (Rice 1960; Penfound1962).

Localized vegetation types dominated by plateau live oak (*Quercus fusiformis*) and Ashe juniper (*Juniperus asheii*) occur in the Oklahoma cross timbers. These two species are sympatric on the Edwards Plateau of Texas, but not in Oklahoma. Patchy forests of plateau live oak occur on rocky granite slopes in the Witchita and Quartz mountains of southwestern Oklahoma (Little 1981). Ashe juniper grows on shallow limestone soils and slopes in the Arbuckle Mountains of south-central Oklahoma, an area with strong floristic affinities to the Edwards Plateau (Hopkins 1941; Dale 1956).

Vegetation Structure

The ratio of post oak to blackjack oak in a cross timbers stand ranges from 2:1 to 3:1,

depending on slope, aspect, and/or geographic location (Luckhardt and Barclay 1938; Kennedy 1973; Sims 1988). Occasionally, stem density of blackjack oak may surpass that of post oak on south-facing slopes, but, because blackjack oak rarely exceeds 30 cm in diameter, basal area values are roughly equivalent (Luckhardt and Barclay 1938; Rice and Penfound 1955, 1959). An inverse relationship between black hickory and post oak and blackjack oak basal areas has been noted at some sites (Luckhardt and Barclay 1938). In addition to an east–west decline in the total basal area of cross timbers woody species, the height of blackjack oak and post oak decreases from an average of 15 m in the east to 6 m in the west (Table 14.1; Dyksterhuis 1948; Rice and Penfound 1959; Penfound 1962).

Biomass and Nutrient Cycling

Productivity, biomass, and nutrient dynamics are the least-studied aspects of cross timbers ecology. A study of the central Oklahoma cross timbers found net primary productivity averaged approximately 15,155 kg ha^{-1} yr^{-1} (Table 14.2). Aboveground perennial material accounts for 79% of the biomass in the cross timbers. Leaves account for 2%, and roots 18%, of the live biomass. Woody overstory species contribute 95% of the 5,386 kg ha^{-1} yr^{-1} annual average litterfall. Since post oak and blackjack oak retain some leaves throughout the winter, litterfall peaks both in November and March (Johnson 1973; Johnson and Risser 1974).

Nutrient concentration and cycling in central Oklahoma cross timbers are comparable to other deciduous forests (Table 14.3; Johnson and Risser 1974). Nutrient concentration is highest in leaves, with the exception of calcium, which is sequestered in large quantities in the bark. The concentration of nitrogen, phosphorus, and potassium in leaves decreases from May to September. However, the leaf concentration of calcium increases throughout the growing season.

Table 14.2. *Net annual primary production in a central Oklahoma cross timbers stand. Units are kg ha⁻¹ yr⁻¹. Production of minor tree species was less than 1% of the total and is included with post oak.*

Component	Post oak	Blackjack oak	Stand	% of total
Leaves	3,472	1,287	4,759	31.4
Current twigs	201	133	334	2.3
Branches	2,967	861	3,828	25.2
Trunks	2,100	1,594	3,694	24.4
Roots			2,240	14.8
Understory			300	2.0
Total			15,155	

Total nutrient concentration is lower in blackjack oak than in post oak.

Trials of Fire and Water

Fire plays an important role in the maintenance of the cross timbers vegetation mosaic. Native Americans employed fire as a management tool throughout the cross timbers and prairie (Foreman 1947; Irving 1956; Blinn 1958; Anderson 1990). Ellsworth (1937; Blinn 1958) noted that fires moved quickly through the prairie, but slowed or halted at the margins of cross timbers vegetation. Fire suppression became an active policy following European settlement of the region, thus allowing woody vegetation to expand at the expense of savannas and prairies (Rice and Penfound 1959; Bell and Hulbert 1974; Sims 1988). As fire frequency decreases, the rate of canopy closure increases in oak savannas (Johnson and Risser 1975; Henderson and Epstein 1995).

Increased number of woody stems per hectare (Dooley 1983; Dooley and Collins 1984), and development of even-aged blackjack oak understories (Liming 1942), are two responses of cross timbers vegetation to burning. Fire-induced mortality is highest for trees under 4 cm in diameter, but few trees of greater size are destroyed (Johnson and Risser 1975). The greater number of woody seedlings on burned, versus unburned, plots may be the result of reduced litter depth (McPherson and Thompson 1972). Among understory species, the biomass of little bluestem increases following burns (Engle and Stritzke 1992).

The season in which a cross timbers stand burns has important consequences for stand structure. Anderson (1972) hypothesized that fire during the growing season would adversely impact the survival of woody species. Early settlers recorded that dry-season fires eliminated woody understory species and allowed tallgrass prairie species to luxuriate (Dyksterhuis 1948). Dormant-season fire stimulates prolific root sprouting in blackjack oak and post oak (Dyksterhuis 1948). Woody species in a cross timbers community respond differently to late-winter versus summer burns (Adams, Anderson and Collins 1982). Roughleaf dogwood, winged sumac, and post oak were eliminated by late-winter burns. The abundance of persimmon (*Diospyros virginiana*) and smooth sumac increases following burns regardless of season. The diversity of woody species decreases following both late-winter and summer burns (Adams, Anderson and Collins 1982).

Major droughts in the region occur at approximately 20-year intervals (Rice and Penfound 1959; Johnson and Risser 1975). Major droughts occurred in 1930–33, 1952–54, and 1980, as a result of eastward shifts of the Rocky Mountain midtropospheric ridge (Corcoran 1982). Woody plants are more susceptible to drought than herbaceous plants,

Table 14.3. *Annual mean concentrations of six mineral elements in various components of a central Oklahoma cross timbers stand. Units are $\mu g\ g^{-1}$ dry weight.*

Component	Mineral element					
	N	P	K	Ca	Mg	Mn
Quercus stellata	47,562	3,880	36,363	179,569	9,189	4,078
Quercus marilandica	36,860	2,989	38,712	110,759	7,683	3,306
Understory species (combined)	7,306	796	6,250	14,724	2,110	405
Roots	4,330	514	2,636	11,251	1,446	588
Litterfall	7,902	698	5,479	16,754	2,662	938
Litter	5,357	467	4,393	14,346	1,697	1,013
Soil depth: 0–15 cm	429	88	62	325	104	108
Soil depth: 30–45 cm	189	51	50	149	74	83

particularly when deep-soil moisture is not recharged and available to trees following winter precipitation (Britton and Messenger 1969).

Drought can profoundly affect the composition of cross timbers vegetation (Rice and Penfound 1959; Johnson and Risser 1975). A drought in the years 1886–87 led to a replacement of mixed grass prairie vegetation with short grass steppe species in the Texas cross timbers (Dysterhuis 1948). Sixty to 80% of the redcedar in Kansas succumbed during the Dust Bowl of the 1930s (Albertson 1940). There also was a loss of understory species such as smooth sumac and buckbrush (Albertson 1940).

Rice and Penfound (1959) reported a lag between the onset of the 1950s drought in Oklahoma and the period of highest woody plant mortality. Many trees were lost in the last year of the drought, 1954, but widespread mortality did not occur until 1956. Trees in western Oklahoma were more strongly affected than those in the east, and mortality was higher for blackjack oak than for post oak (Rice and Penfound 1959). Mortality in populations of black oak also exhibited a lag response to the mid-1950s drought (J. P. Dwyer, Cutter and Wetteroff 1995).

Herbivory

Vertebrate and invertebrate herbivory apparently influence the vigor and structure of cross timbers vegetation, although few studies have directly addressed the issue. Post oak is susceptible to predation by many insects, including defoliators, leaf rollers, tent caterpillars, sawflies, leaf miners and skeletonizers, aphids, lace bugs, scales, gall wasps, and mites (Stransky 1990). Extreme infestations of larval *Cecropia* species are common in drought years (McCluskey 1972). Fungal pathogens, such as oak wilt (*Ceratocystis fagacerum*) and chestnut blight (*Cryphonectria parasitica*), also parasitize post oak rangewide (Stransky 1990).

More is known about the effects of domestic cattle grazing than about the effects of browsing by native ungulates on vegetation in the cross timbers region. Spanish explorers and settlers introduced cattle to the upper Trinity River region as early as 1788 (Dysterhuis 1948). Many of these animals escaped from captivity and contributed to the decline of grazing capacity by 1880 (Dysterhuis 1948; Smeins and Diamond 1986). Heavy grazing reduces fuel loads and thus the severity of fires (Dysterhuis 1957; Smeins and Diamond 1986; Archer 1995). An indirect effect of grazing is the conversion of cross timbers and prairie vegetation to exotic pasture grasses such as king ranch bluestem (*Andropogon ischaemum*), bermudagrass (*Cynodon dactylon*), and Johnsongrass (*Sorghum halepense*) (Dysterhuis 1948; Johnson 1986).

Rare, Threatened, or Endangered Species

The decline of oak savannas throughout the Midwest has had a negative impact on many species (Henderson and Epstein 1995). However, there are no species restricted to the cross timbers region that are listed as federally threatened or endangered. The eastern prairie fringed (*Platanthera praeclara*) and western prairie fringed (*P. leucophea*) orchids inhabit high-quality prairie sites and are federally listed threatened species. Although these orchids occur within the cross timbers, their geographic range extends beyond the region (Zanoni et al. 1979; U.S. Fish and Wildlife Service 1993).

Likewise, several threatened species of birds migrate through the cross timbers but their geographic ranges extend beyond the region; examples include black-capped vireo (*Vireo atricapillus*) and least tern (*Sterna antillarium*), both of which are federally listed as endangered species. The black-capped vireo provides a striking example of the interactive effects of savanna canopy closure on associated animal species. Black-capped vireos prefer to nest in open savanna vegetation. But the decrease in open savanna vegetation, fostered by the increased abundance of redcedar, has led to rising populations of brown-headed cowbirds (*Molothrus ater*). Consequently, cowbird parasitism of black-capped vireo nests has increased and is hampering vireo conservation efforts (U.S. Fish and Wildlife Service 1993; Grzybowski, Tazik and Schnell 1994). Also, the use of pesticides for range management in the cross timbers region can have negative impacts on breeding bird populations (Schulz et al. 1992).

Several species once common in the cross timbers were extirpated following European settlement. The passenger pigeon (*Ectopistes migratorius*) and bison (*Bison bison*) are probably the most familiar examples (Sutton 1967; Caire et al. 1989). The Carolina parakeet (*Conuropsis carolinensis*) was a migratory visitor to the cross timbers, and possibly a resident (Sutton 1967). Black bears (*Ursus americanus*) are no longer a member of the cross timbers fauna, but Josiah Gregg reported that they were common in the

1840s (Caire et al. 1989). Prairie dog towns were once prevalent in the western cross timbers, but their numbers have declined dramatically as a result of eradication programs (Caire et al. 1989). Prairie dog towns have a pronounced effect on vegetation composition and structure, and they host species-rich vertebrate assemblages (Tyler 1968). The existence of a 1928 collection from central Oklahoma reveals that the black-footed ferret (*Mustela nigripes*) also may have been a member of the cross timbers fauna (Caire et al. 1989).

Current Status

The cross timbers have played a profound role in the natural history and cultural identity of the region (Brown 1992). However, to the authors' knowledge, no effort has been made to evaluate the conservation status of the cross timbers at either the state or the federal level. Likewise, there is no state or national park dedicated to the interpretation of the cultural or natural history of the cross timbers. Nonetheless, excellent examples of cross timbers vegetation can be found at several state parks, state wildlife management areas, national recreation areas, and national wildlife refuges in Oklahoma and Texas (Smeins and Diamond 1986).

Although this chapter has focused on natural history, the cross timbers have a rich cultural context as well. Captain Marcy noted: "In the early days, traders, trappers, and other travelers in the country employed the Cross Timbers as a datum line for location, and measured distances of places from this well known landmark, as in populated parts of the world reference is made to the meridian of Greenwich" (Foreman 1947). The modest impact of human settlement in the Oklahoma cross timbers changed dramatically with the land run of 1889, when, virtually overnight, the area was occupied by 50,000 settlers. Interestingly, people staking claims sought prairie vegetation and avoided cross timbers stands. Human settlement patterns in the cross timbers were again profoundly affected by the Dust Bowl era of the 1930s (Brown 1992).

Like most temperate and tropical savannas, the regional heterogeneity of the cross timbers is threatened by the encroachment of woody species and the resulting canopy closure (Rice and Penfound 1959; Johnson and Risser 1975; Johnson 1986; Archer 1995). Woody plant invasion has been rapid over a short time span (50–100 years); accentuated by extreme climatic events such as drought; associated with livestock grazing and with elimination of native browsers and/or fire suppression; influenced by topo-edaphic factors; and driven by an influx of unpalatable, stress-tolerant species (i.e., redcedar). In addition, the effects of woody plant invasion are irreversible over time frames relevant to ecosystem management (Archer 1995).

Certainly, there are extensive research needs in the cross timbers. Although the upland forests have been well studied, landscape-level patterns and processes need to be evaluated. To understand the cross timbers, we must understand regional heterogeneity and community organization along large-scale and local gradients (Dyksterhuis 1957; Butler 1995).

REFERENCES

Aandahl, A. R. 1982. *Soils of the Great Plains: Land Use, Crops, and Grasses.* Lincoln, Neb.: University of Nebraska Press.

Adams, D. E., Anderson, R. C., and Collins, S. L. 1982. Differential response of woody and herbaceous species to summer and winter burning in an Oklahoma grassland. *The Southwestern Naturalist* 27:55–61.

Albertson, F. W. 1940. Studies of native cedars in west central Kansas. *Transactions of the Kansas Academy of Science* 43:85–95.

Anderson, R. C. 1972. Prairie history, management and restoration in southern Illinois. In *Proceedings of the Second Midwest Prairie Conference,* ed. J. Zimmerman, pp. 15–22. Madison, Wis. n.p.

Anderson, R. C. 1990. The historic role of fire in North American grasslands. In *Fire in North American Tallgrass Prairie,* ed. S. L. Collins and

L. L. Wallace, pp. 9–18. Norman, Okla.: University of Oklahoma Press.

Archer, S. 1995. Tree–grass dynamics in a *Prosopis*–thornscrub savanna parkland: reconstructing the past and predicting the future. *Ecoscience* 2:83–99.

Axelrod, D. I. 1985. Rise of the grassland biome, central North America. *The Botanical Review* 51:163–201.

Barclay, H. G. 1948. The woody vegetation of Bear's Glen, a Washington Irving stopover. *Proceedings of the Oklahoma Academy of Science* 28:39–57.

Bell, E. L., and Hulbert, L. C. 1974. Effect of soil on occurrence of Cross Timbers and Prairie in Southern Kansas. *Transactions of the Kansas Academy of Science* 77:203–209.

Blinn, W. C. 1958. The short-grass plains and post oak–blackjack woodland of Oklahoma in historical perspective. M.S. thesis, Oklahoma State University, Stillwater, Okla.

Bolton, H. E. 1914. *Athanase De Mezieres and the Louisiana–Texas Frontier: 1768–1780.* New York: Arthur H. Clark Co., Inc.

Branson, C. C., and Johnson, K. S. 1979. Generalized geologic map of Oklahoma (1:2,000,000 scale color map). In *Geology and Earth Resources of Oklahoma,* ed. K.S. Johnson. Norman, Okla.: Oklahoma Geological Survey.

Britton, W., and Messenger, A. 1969. Computed soil moisture patterns in and around the prairie peninsula during the great drought of 1933–34. *Transactions of the Illinois State Academy of Science* 62:181–187.

Brown, B. J. 1992. Cultural ecology of the Garber-Wellington cross timbers in eastern Cleveland county, Oklahoma. Ph.D. dissertation, University of Oklahoma, Norman, Okla.

Bruner, W. E. 1931. The vegetation of Oklahoma. *Ecological Monographs* 1:100–188.

Buck, P. 1964. Relationships of the woody vegetation of the Wichita Mountains Wildlife Refuge to geological formations and soil types. *Ecology* 45:336–344.

Butler, I. H. 1995. *Element Stewardship Abstract for Central Cross Timber.* Norman, Okla.: Oklahoma Biological Survey, Oklahoma Natural Heritage Inventory.

Caire, W., Tyler, J. D., Glass, B. P., and Mares, M. A. 1989. *Mammals of Oklahoma.* Norman, Okla.: University of Oklahoma Press.

Clements, F. E. 1936. The origin of the desert climax and climate. In *Essays in Geobotany in Honor of William Albert Setchell,* ed. T. H.

Goodspeed, pp. 87–139. Berkeley: University of California Press.

Collins, S. L., and Klahr, S. C. 1991. Tree dispersion in oak-dominated forests along an environmental gradient. *Oecologia* 86:471–477.

Coppock, R. K., Ely, C. A., Ficken, R. W., and Smith, M. G. 1955. An evaluation of the quadrat method in the blackjack–post oak forest. *Proceedings of the Oklahoma Academy of Science* 36:49–50.

Corcoran, W. T. 1982. Moisture stress, mid-tropospheric pressure patterns, and the forest grassland transition in the south central states. *Physical Geography* 3:148–159.

Court, A. 1974. The climate of the conterminous United States. In *Climates of North America*, ed. R. A. Bryson and F. K. Hare, pp. 193–343. New York: Elsevier Scientific Pub. Co.

Curtis, N. M., and Ham, W. E. 1972. *Geomorphic Provinces of Oklahoma* (1:2,000,000 scale map). Norman, Okla.: Oklahoma Geological Survey.

Dale, E. E., Jr. 1956. A preliminary survey of the flora of the Arbuckle Mountains, Oklahoma. *Texas Journal of Science* 8:41–73.

Delcourt, P. A., and Delcourt, H. R. 1981. Vegetation maps for eastern North America: 40,000 yr B.P. to the present. In *Geobotany II*, ed. R. C. Romans, pp. 123–165. New York: Plenum Pub. Corp.

Diamond, D. D., Riskind, D. H., and Orzell, S. L. 1987. A framework for plant community classification and conservation in Texas. *Texas Journal of Science* 39:203–221.

Dooley, K. L. 1983. Description and dynamics of some western oak forests in Oklahoma. Ph.D. dissertation, University of Oklahoma, Norman, Okla.

Dooley, K. L., and Collins, S. L. 1984. Ordination and classification of western oak forests in Oklahoma. *American Journal of Botany* 71:1221–1227.

Dwyer, D. D., and Santelmann, P. W. 1964. *A Comparison of Post Oak–Blackjack Oak Communities on Two Major Soil Types in North Central Oklahoma*. Stillwater, Okla.: Oklahoma Agricultural Experiment Station, Oklahoma State University.

Dwyer, J. P., Cutter, B. E., and Wetteroff, J. J. 1995. A dendrochronological study of black and scarlet oak decline in the Missouri Ozarks. *Forest Ecology and Management* 75:69–75.

Dyksterhuis, E. J. 1948. The vegetation of the western cross timbers. *Ecological Monographs* 18:235–376.

Dyksterhuis, E. J. 1957. The savannas concept and its use. *Ecology* 38:435–442.

Ellsworth, H. L. 1937. *Washington Irving on the Prairie, or A Narrative of a Tour of the Southwest in the Year 1832*; ed. S. T. Williams and B. D. Simison New York: American Book Co.

Engle, D. M., and Kulbeth, J. D. 1992. Growth dynamics in crowns of eastern redcedar at 3 locations in Oklahoma. *Journal of Range Management* 45:301–305.

Engle, D. M., and Stritzke, J. F. 1992. Vegetation management in the Cross Timbers. In *Range Research Highlights 1983–1991, Circular E-905*, pp. 1–4. Stillwater, Okla.: Cooperative Extension Service, Division of Agricultural Sciences and Natural Resources, Oklahoma State University.

Ewing, A. L., Stritzke, J. F., and Kulbeth, J. D. 1984. *Vegetation of the Cross Timbers Experimental Range, Payne County, Oklahoma*. Stillwater, Okla.: Oklahoma Agricultural Experimental Station, Oklahoma State University.

Foreman, C. 1947. *The Cross Timbers*. Muskogee, Okla.: The Star Printery.

Fulton, M. G. 1941. *Diary and Letters of Josiah Gregg; Southwestern Enterprises 1840–1847*. Norman, Okla.: University of Oklahoma Press.

Godfrey, C., McKee, G., and Oakes, H. 1973. *General Soils Map of Texas* (1:1,555,000 scale color map). Miscellaneous Publication MP-1304. College Station, Tex.: Texas Agricultural Experiment Station.

Gould, C. N. 1902. Notes on trees, shrubs and vines in the Cherokee Nation. *Transactions of the Kansas Academy of Science* 18:145–146.

Gray, F., and Galloway, H. M. 1959. *Soils of Oklahoma*. Stillwater, Okla.: Oklahoma Agricultural Experiment Station, Oklahoma State University.

Grzybowski, J. A., Tazik, D. H., and Schnell, G. D. 1994. Regional analysis of black-capped vireo breeding habitats. *Condor* 96:512–544.

Hale, M. E. 1955. A survey of upland forests in the Chautauqua Hills, Kansas. *Transactions of the Kansas Academy of Science* 58:165–168.

Hall, S. A. 1982. Late Holocene paleoecology of the southern plains. *Quaternary Research* 17:391–407.

Hall, S. L., and McPherson, J. K. 1980. Geographic distribution of two species of oaks in Oklahoma in relation to seasonal water potential and transpiration rates. *The Southwestern Naturalist* 25:283–295.

Harrison, T. P. 1974. A floristic study of the woody vegetation of the North American Cross

Timbers. Ph.D. dissertation, North Texas State University, Denton, Tex.

Henderson, R. A., and Epstein, E. J. 1995. Oak savannas in Wisconsin. In *Our Living Resources: A Report to the Nation on the Distribution, Abundance, and Health of U.S. Plants, Animals, and Ecosystems*, ed. E. T. LaRoe, G. S. Farris, C. E. Puckett, P. D. Doran, and M. J. Mac, pp. 230–232. Washington, D.C.: U.S. Department of the Interior, National Biological Service.

Hill, R. T. 1887. The topography and geology of the cross timbers and surrounding regions in northern Texas. *American Journal of Science* 33:291–303.

Hoagland, B. W. 1996. *Preliminary Plant Community Classification for Oklahoma*. Norman, Okla.: Oklahoma Biological Survey, Oklahoma Natural Heritage Inventory.

Hopkins, M. 1941. The floristic affinities of the Arbuckle Mountains in Oklahoma. *American Journal of Botany* 28:16 (abstract).

Hunt, C. B. 1974. *Natural Regions of the United States and Canada*. San Francisco: W. H. Freeman and Company.

Irving, W. 1956. *A Tour on the Prairies*, ed. J. F. McDermott. Norman, Okla.: University of Oklahoma Press.

Johnson, F. L. 1986. Oak–hickory savannas and transition zones: preservation status and management problems. In *Wilderness and Natural Areas in the Eastern United States: A Management Challenge*, ed. D. L. Kulhavy and R. N. Conner, pp. 345–347. Nacogdoches, Tex.: Stephen F. Austin University.

Johnson, F. L. 1973. Biomass, annual net primary production, and dynamics of six mineral elements in a post oak–blackjack forest. Ph.D. dissertation, University of Oklahoma, Norman, Okla.

Johnson, F. L., and Risser, P. G. 1972. Some vegetation–environment relationships in the upland forests of Oklahoma. *Journal of Ecology* 60:655–663.

Johnson, F. L., and Risser, P. G. 1973. Correlation analysis of rainfall and annual ring index of central Oklahoma blackjack and post oak. *American Journal of Botany* 60:475–478.

Johnson, F. L., and Risser, P. G. 1974. Biomass, annual net primary production, and dynamics of six mineral elements in a post oak–blackjack oak forest. *Ecology* 55:1246–1258.

Johnson, F. L., and Risser, P. G. 1975. A quantitative comparison between an oak forest and an oak savanna in central Oklahoma. *The*

Southwestern Naturalist 20:75–84.

Kennedy, R. K. 1973. An analysis of selected Oklahoma upland forest stands including both overstory and understory components. Ph.D. dissertation, University of Oklahoma, Norman, Okla.

Klahr, S. C. 1989. Spatial pattern in post oak–blackjack oak forests along an environmental gradient in Oklahoma. M.S. thesis, University of Oklahoma, Norman, Okla.

Küchler, A. W. 1964. *Potential Natural Vegetation of the Conterminous United States* (1:3,168,000 scale map). Washington D.C.: American Geographical Society. Special Publication Number 36.

Küchler, A. W. 1974. A new vegetation map of Kansas. *Ecology* 55:586–604.

Lathrop, E. W. 1958. The flora and ecology of the Chautauqua Hills in Kansas. *University of Kansas Science Bulletin* 39:79–209.

Lauver, C. L. 1989. *Preliminary Classification of the Natural Communities of Kansas*. Lawrence, Kans.: Kansas Natural Heritage Program, Kansas Biological Survey.

Liming, F. G. 1942. Blackjack oak in the Missouri Ozarks. *Journal of Forestry* 40:249–252.

Little, E. L. 1939. The vegetation of the Caddo County canyons, Oklahoma. *Ecology* 20:1–10.

Little, E. L. 1981. *Forest Trees of Oklahoma*. Oklahoma City, Okla.: Oklahoma Forestry Division.

Love, H. S., Jr. 1971. The detection and nature of functional underground connections between certain woody species of the Oklahoma cross timbers. M.S. thesis, Oklahoma State University, Stillwater, Okla.

Luckhardt, R. L., and Barclay, H. G. 1938. A study of the environment and floristic composition of an oak–hickory woodland in northeastern Oklahoma. *Proceedings of the Oklahoma Academy of Science* 18:25–32.

Marcy, L. E. 1982. Habitat types of the eastern cross timbers of Texas. M.S. thesis, Texas A&M University, College Station, Tex.

McCluskey, R. L. 1972. Some population parameters of *Quercus stellata* in the Texas Cross Timbers. Ph.D. dissertation, North Texas State University, Denton, Tex.

McPherson, J. K., and Thompson, G. L. 1972. Competitive and allelopathic suppression of understory by Oklahoma oak forests. *Bulletin of the Torrey Botanical Club* 99:293–300.

National Atlas. 1968. Map of topographic relief, Sheet #59 (1:17,000,000 scale color map). Washington, D.C.: U.S. Geological Survey.

Penfound, W. T. 1962. The savanna concept in Oklahoma. *Ecology* 43:774–775.

Penfound, W. T. 1963. The composition of a post oak forest in south-central Oklahoma. *The Southwestern Naturalist* 8:114–115.

Petranka, J. W., and McPherson, J. K. 1979. The role of *Rhus copallina* in the dynamics of the forest–prairie ecotone in north-central Oklahoma. *Ecology* 60:956–965.

Powell, J., and Lowry, D. P. 1980. Oak (*Quercus* spp.) sprouts growth rates on a central Oklahoma shallow savannah range site. *Journal of Range Management* 33:312–313.

Rhodes, L. K. 1980. Correlation between vegetation and geologic formations in Oklahoma. *Oklahoma Geology Notes* 40:47–62.

Rice, E. L. 1960. The microclimate of a relict stand of sugar maple in Devil's Canyon in Canadian County Oklahoma. *Ecology* 41:445–453.

Rice, E. L. 1962. The microclimate of sugar maple stands in Oklahoma. *Ecology* 46:19–25.

Rice, E. L. 1965. Bottomland forests of north-central Oklahoma. *Ecology* 46:708–714.

Rice, E. L., and Penfound, W. T. 1955. An evaluation of the variable-radius and paired-tree methods in the blackjack–post oak forest. *Ecology* 36:315–320.

Rice, E. L., and Penfound, W. T. 1959. The upland forests of Oklahoma. *Ecology* 40:593–608.

Risser, P. G. and Rice, E. L. 1971. Phytosociological analysis of Oklahoma upland forest species. *Ecology* 52:940–945.

Rykiel, E. J., and Cook, T. L. 1986. Hardwood–cedar clusters in the post oak savanna of Texas. *The Southwestern Naturalist* 31:73–78.

Sanders, I. L. 1980. Post oak – blackjack oak. In *Forest Cover Types of the United States and Canada*, ed. F. H. Eyre, pp. 38–39. Washington, D.C.: Society of American Foresters.

Schnell, G. D., Risser, P. G, and Helsel, J. F. 1981. Shrub distribution patterns in Oklahoma with comparisons to those exhibited by trees. *Bulletin of the Torrey Botanical Club* 108:54–66.

Schulz, C. A., Leslie, D. M., Lochmiller, R. L., and Engle, D. M. 1992. Herbicide effects on cross timbers breeding birds. *Journal of Range Management* 45:407–411.

Sims, P. L. 1988. Grasslands. In *North American Terrestrial Vegetation*, ed. M. G. Barbour and W. D. Billings, pp. 266–286. New York: Cambridge University Press.

Smeins, F. E., and Diamond, D. D. 1986. Grasslands and savannas of east central Texas:

ecology, preservation status and management problems. In *Wilderness and Natural Areas in the Eastern United States: A Management Challenge*, ed. D. L. Kulhavy and R. N. Conner, pp. 381–394. Nacogdoches, Tex.: Stephen F. Austin University.

Snook, E. C. 1985. Distribution of eastern red-cedar on Oklahoma rangelands. In *Eastern Red-cedar in Oklahoma*, ed. R. F. Wittwer and D. M. Engle, pp. 45–52. Stillwater, Okla.: Cooperative Extension Service Division of Agriculture, Oklahoma State University.

Stahle, D. W., and Chaney, P. L. 1994. A predictive model for the location of ancient forests. *Natural Areas Journal* 14:151–158.

Stahle, D. W., and Hehr, J. G. 1984. Dendroclimatic relationships of post oak across a precipitation gradient in the Southcentral United States. *Annals of the Association of American Geographers* 74:561–573.

Stransky, J. J. 1990. *Quercus stellata* Wangenh. Post Oak. In *Silvics of North America*. Vol. 2, *Hardwoods*, ed. R. M. Burns and B. H. Honkala, pp. 738–743. Washington, D.C.: USDA, Forest Service.

Sutherland, S. M. 1977. The climate of Oklahoma. In *Geography of Oklahoma*, ed. J. W. Morris, pp. 40–53. Oklahoma City, Okla.: Oklahoma Historical Society.

Sutton, G. M. 1967. *Oklahoma Birds*. Norman, Okla.: University of Oklahoma Press.

Tharp, B. C. 1926. Structure of Texas vegetation east of the 98th meridian. Ph.D. dissertation, University of Texas, Austin, Tex.

Trewartha, G. T. 1968. *An Introduction to Climate*, 4th ed. New York: McGraw-Hill Book Co., Inc.

Tyler, J. D. 1968. Distribution and vertebrate associates of the black-tailed prairie dog in Oklahoma. Ph.D. dissertation, University of Oklahoma, Norman, Okla.

U.S. Fish and Wildlife Service. 1993. Endangered and threatened species of Oklahoma. Tulsa, Okla: Oklahoma Ecological Services Field Office of the U.S. Fish and and Wildlife Services and other agencies.

Weaver, J. E., and Clements, F. E. 1929. *Plant Ecology*. New York: McGraw-Hill Book Co., Inc.

Zanoni, T. A., Gentry, J. L., Tyrl, R. J., and Risser, P. G. 1979. Endangered and threatened plants of Oklahoma. Norman, Okla.: Department of Botany and Microbiology, University of Oklahoma, and Department of General and Evolutionary Biology, Oklahoma State University, Stillwater.

WESTERN/SOUTHWESTERN REGION

15   Ponderosa and Limber Pine Woodlands

DENNIS H. KNIGHT

Introduction

The term *woodland* is used for a variety of shrub- and tree-dominated communities on sites that are not favorable for the development of a forest, whereas the term *savanna* implies a grass/forb-dominated community with trees at a lower density than expected in the forests of the area. Tree-dominated woodlands are rare across the western Great Plains of North America, occurring mostly along rivers and occasionally on rocky escarpments. Various broad-leaved trees dominate the riparian zone, especially plains cottonwood (*Populus deltoides*) (plant nomenclature follows Kartesz 1994), boxelder (*Acer negundo*), green ash (*Fraxinus pennsylvanica*), and peachleaf willow (*Salix amygdaloides*). On escarpments, ponderosa pine (*Pinus ponderosa* var. *scopulorum*) is common (Figure 15.1). Associated species include Rocky Mountain juniper (*Juniperus scopulorum*), skunkbush sumac (*Rhus trilobata*), big sagebrush (*Artemisia tridentata*), mountain mahogany (*Cercocarpus montanus*), little bluestem (*Schyzachyrium scoparium*), and sideoats grama (*Bouteloua curtipendula*) (Hansen and Hoffman 1988). Bur oak (*Quercus macrocarpa*) is a common associate in some areas, such as in the Black Hills of South Dakota (Hoffman and Alexander 1987). Near the Rocky Mountains, ponderosa pine often occurs beyond the rocky escarpments, forming parklike savannas on the deeper, finer-textured soils of the adjacent grassland. Limber pine (*Pinus flexilis*) sometimes occurs with ponderosa pine, but usually it appears at higher elevations where the climate is drier and colder (Peet 1988).

Similarly, desert shrublands of the Southwest grade into woodlands near the mountains or on plateaus. Woodlands dominated by various species of oak (*Quercus* spp.), especially Gambel oak (*Q. gambelii*), are common in some areas (Dick-Peddie 1993). Elsewhere, pinyon pine (*Pinus edulis*) and juniper (*Juniperus scopulorum* or *J. osteosperma*) are widespread. Ponderosa pine often invades such woodlands above 1800 m, or it forms extensive savannas with an understory of various forbs and grasses, including blue grama (*Bouteloua gracilis*), screwleaf muhly (*Muhlenbergia virescens*), muttongrass (*Poa fendleriana*), and Arizona fescue (*Festuca arizonica*) (Muldavin, DeVelice and Ronco 1996). Limber pine is absent from the foothills of the southern Rocky Mountains, occurring only at much higher elevations at the edge of the alpine tundra. As on the western Great Plains, pine-dominated woodlands in the Southwest typically occur on coarse-textured soils.

Ponderosa pine, limber pine, and pinyon pine also occur west of the Rocky Mountains, on the Colorado Plateau and in the Great Basin. Most of the Great Basin is too dry for woodlands. However, pinyon pine (*P. edulis* and *P. monophylla*) and Utah juniper (*J. osteosperma*) form woodlands over extensive areas. Ponderosa pine occurs in isolated groves on coarse-textured soils in the Great Basin (Figure 15.2), and in the foothills of the

Figure 15.1. Ponderosa pine woodlands are common on escarpments and rocky soils in the western Great Plains, such as in eastern Montana, where this photo was taken (elevation 850 m). Common associates of the pine are Rocky Mountain juniper and skunkbush sumac. Mixed-grass prairie and shrublands dominated by big sagebrush (*Artemisia tridentata*) and silver sagebrush (*A. cana*) are found on the deeper soils in the vicinity of such woodlands.

Rocky Mountains to the east and the Sierra Nevada and Cascade Mountains to the west (Barbour 1987; Franklin and Dyrness 1988). Often it occurs with Gambel oak and western juniper (*J. occidentalis*), or in savannas with an understory of grasses, forbs, and low shrubs, especially bluebunch wheatgrass (*Agropyron spicatum*), big sagebrush (*Artemisia tridentata*), low sagebrush (*A. arbuscula*), bitterbrush (*Purshia tridentata*), serviceberry (*Amelanchier utahensis*), and curlleaf mountain-mahogany (*Cercocarpus ledifolius*). In the Sierra Nevada, Jeffrey pine (*Pinus jeffreyi*) forms savannas similar to those of ponderosa pine, although Jeffrey pine is more abundant at higher elevations (Barbour 1987). Limber

pine is restricted to near the tops of the islandlike mountain ranges that characterize the Great Basin, where it occurs commonly with intermountain bristlecone pine (*Pinus longaeva*) (Billings 1990).

The association of pine woodlands with rocky escarpments and coarse-textured soils is a common observation. Daubenmire (1968) suggested that rock outcrops have woody plants because they provide above-average moisture compared to fine-textured soils at the same elevation. This explanation has led to a principle known as the inverse texture effect (Noy-Meir 1973; Sala et al. 1988). Coarse soils (including rocky escarpments with surficial cracks) in semi-arid cli-

Figure 15.2. Isolated woodlands dominated by ponderosa pine and Jeffrey pine occur in the big sagebrush–dominated basins of western Nevada, south of Reno, and are associated with coarse, nutrient-deficient soils caused by hydrothermal activity that occurred in the Miocene (DeLucia and Schlesinger 1990). Elevation 1615 m. Ponderosa pine and Jeffrey pine also dominate the forests in the distant foothills of the Sierra Nevada (Vasek and Thorne 1977). The woodlands in the foreground are in the foothills of the Virginia Mountains and are dominated by pinyon pine and Utah juniper. Photo by Evan DeLucia. Similar stands of ponderosa pine occur on sandy soils in eastern Oregon (Franklin and Dyrness 1988).

mates allow for deeper infiltration of water than fine-textured soils, which reduces the amount of water evaporated from the soil surface and increases the amount of moisture available for plants whose roots extend deep into the coarser soil. Rocky substrates appear dry, but the hard surfaces funnel the available water into cracks that, over long periods of time, accumulate sufficient amounts of fine particles, nutrients, and organic matter to sustain the growth of trees. Nutrient availability undoubtedly is lower than on the fine-textured soils of adjacent grasslands. However, water appears to be the factor limiting tree

establishment in western North America, and pines, along with other conifers, tolerate infertile soils (Knight et al. 1994).

The distribution of pine-dominated woodlands in the Rocky Mountain region has changed rapidly during the last 10,000 years. During the Pleistocene, the climate of the Southwest and Great Basin was cooler, and limber pine was widespread over much of the area now dominated by big sagebrush, pinyon pine, and juniper (Betancourt 1990; Thompson 1990). At that time, limber pine occurred at an elevation 500–1,000 m lower than at present. Today, woodlands of this nature occur only far to the north, such as in Wyoming and Montana. Ponderosa pine, now so abundant in the Southwest, is thought to have been rare in that area about 10,000–11,000 years ago, possibly because of excessively dry summers and fewer fires than later in the Holocene (Betancourt 1990). Frequent burning seems to give ponderosa pine a competitive edge because of its thick bark. On the Colorado Plateau, the earliest record of ponderosa pine is 10,300 years old (Betancourt 1990).

Ponderosa Pine

Ponderosa pine is common on lowland escarpments and in the foothills of most major mountain ranges in the western United States (Figure 15.3). Isolated populations evolved into at least two varieties during the Pleistocene: the Pacific variety (*P. ponderosa* var. *ponderosa*) in California, Oregon, and Washington, and the Rocky Mountain variety (*P. ponderosa* var. *scopulorum*) elsewhere. Ponderosa pine hybridizes with Jeffrey pine in California, where the two species are sympatric. Less-frequent hybridization has been reported between *P. ponderosa* and four predominately Mexican species (*P. montezumae, P. arizonica, P. engelmannii,* and *P. washoensis* (Oliver and Ryker 1990).

Site Relationships

Over part of its range, ponderosa pine forms extensive forests, such as in the Black Hills of South Dakota and on the Mogollon Plateau of Arizona. The trees can become sufficiently dense that many grassland and shrubland species cannot survive. Elsewhere ponderosa pine forms intermittent escarpment woodlands and foothill savannas (Veblen and Lorenz 1991). On escarpments, it occurs on shale as well as on sandstone, limestone, granite, and basalt. Where ponderosa pine occurs, the mean annual precipitation typically is 280–760 mm and the frost-free period is 90 to >200 days. Winters

may be cold in the northern part of its range, but mild in the south. The growing season temperature is warm with frequent spring and summer rains. Ponderosa pine seedlings are less tolerant of cold temperatures than lodgepole pine seedlings (Cochran and Berntsen 1973), an observation that may explain why ponderosa pine is widespread at lower elevations and why, at higher elevations, it is restricted to south slopes.

A common associate of ponderosa pine is Douglas-fir (*Pseudotsuga menziesii*). Although this tree is better adapted to higher elevations than ponderosa pine, it also can form open, foothill savannas, with many of the understory species associated with ponderosa pine (Houston 1973; Peet 1988; Knight 1994). Like ponderosa pine, mature Douglas-fir have thick bark and are tolerant of some surface fires (Crane and Fischer 1986).

Ponderosa pine establishment appears to be dependent on adequate summer precipitation. This conclusion is based on its abundance on the western Great Plains, where summer rains are frequent, and its absence at the same elevation in nearby intermountain basins, where summer rainfall is much less frequent. Similarly, ponderosa pine is abundant in the Southwest only where summer rainfall is comparatively high. DeLucia and Schlesinger (1990) found that the stomata of ponderosa pine close with less water stress (<-2.0 MPa) than those of juniper (-3.0 MPa) and big sagebrush (-4.5 MPa) growing on the same site. They concluded that ponderosa pine has a high water-use efficiency, but that it is not competitive for limited moisture with juniper and big sagebrush on fine-textured soils.

Effect of Fire

The density of ponderosa pine is affected by fire in many areas. However, because of the lack of fuel continuity, fire is less likely to occur in escarpment woodlands than in savannas. Indeed, Wells (1965) argued that the presence of woodlands on escarpments is an indication of little or no burning. Escarpments usually provide marginal conditions for tree growth. Seed dispersal and the distribution of safe sites for seedling establishment, such as rock crevices, are probably the most important factors influencing vegetation structure, not the length of time between fires.

In contrast, fire is an extremely important factor in the ponderosa pine savannas adjacent to escarpment woodlands and at higher elevations in the foothills. Here the understory is dominated by a nearly continuous cover of flammable grasses and shrubs that historically have led to surface fires every 5 to 20 years (Barrett and Arno 1982; Higgins 1984; Gruell 1985; Fisher, Jenkins and Fisher 1987; Covington et al. 1994). Whether started by lightning or humans, such fires have little effect on species composition because the understory plants usually sprout from root crowns and the thick bark of older trees prevents high levels of tree mortality. With surface fire suppression by Euro-Americans, the density of ponderosa pine in savannas increased rapidly (Habeck and Mutch 1973; Covington and Moore 1994a, b; Morgan 1994; Arno, Smith and Krebs 1997). Moreover, as tree density increases, abundance of understory vegetation declines (Moir 1966; McPherson 1992). In addition, the understory changes from species requiring nearly full sunlight to shade-tolerant species; and the leaf area of the evergreen conifers increases, which increases evapotranspiration and reduces stream flow (Orr 1975). Without fire, fuels continue to accumulate and the forests can be subjected to what many consider unnaturally intense canopy fires that kill nearly all of the trees.

The high levels of flammability and other ecosystem changes that develop with surface-fire suppression have led to the conclusion that prescribed burning should be used to keep fuel levels down and maintain savanna structure. Fuel loads are now so high, how-

 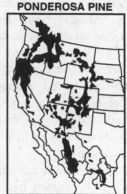

LIMBER PINE **PONDEROSA PINE**

Figure 15.3. Maps showing the general distribution of limber pine and ponderosa pine (after Critchfield and Little 1966). Limber pine occurs at higher elevations, where the climate is cooler and drier, such as in western Wyoming and the mountains of central Nevada.

ever, that harvesting seems necessary to reduce tree density before surface fires can be prescribed, a practice that is presently applied in some areas. Although nutrients are limiting in most pine forests (Knight et al. 1994), research on prescribed burning suggests that nitrogen losses through volatilization during fires are minimal (Covington and Sackett 1992). A more severe problem may be the accumulation of leaf litter during a long period of fire suppression, causing forest floor fuels to burn with sufficient intensity to kill some mature trees despite their thick bark (Swezy and Agee 1991).

Currently, there is some controversy about whether harvesting is necessary to restore the structure of ponderosa pine woodlands prior to prescribed burning. For example, in the Black Hills of South Dakota there is evidence to suggest that heavy fuel accumulation and crown fires were characteristic of ponderosa pine forests (Shinneman and Baker 1997). Forests of ponderosa pine are common in the Black Hills, as they are in other areas where environmental conditions are optimal for this tree. However, savannas predominate on the fringes of the forests, such as in nearby Devil's Tower National

Park, where an examination of 90 fire scars by Fisher, Jenkins and Fisher (1987) suggested an average surface-fire return interval of 19 years prior to Euro-American settlement in 1900, and 42 years since that time. The preferred management practices for forests and savannas probably differ; generalized guides for fire management in all woodlands dominated by ponderosa pine may not be possible.

There also is uncertainty about the causes of the increased tree density that has been observed. Madany and West (1983) found that livestock grazing reduces competition from understory plants, thereby facilitating the establishment of young trees. They concluded that livestock grazing was more important than fire suppression. Heavy grazing reduces the flammability of a savanna by reducing fuel loads, thus making the decline in fire return interval a contributing factor rather than the primary factor. Savage and Swetnam (1990) found evidence, however, that livestock grazing in the 1800s was so severe that most seedlings did not survive. This level of grazing pressure was reduced in the early 1900s, and a decline in livestock abundance, coupled with a wet climatic cycle, favored seedling establishment. Dense forests became established at this time, long before fire suppression was effective, but also at a time when the spread of fires may have been less likely due to increased precipitation (Savage and Swetnam 1990).

Tree density in ponderosa pine savannas has undoubtedly fluctuated through time with changing climatic conditions and with different levels of herbivory. Density also varies spatially. Some early explorers described the savannas as open parklands, whereas others noted patches of dense tree growth interspersed with open grasslands essentially devoid of trees. Where trees occurred, several age classes typically were represented (Cooper 1960, 1961), sometimes all within a single stand, but elsewhere with different age classes found in different patches (West 1969). Ponderosa pine seedling establishment commonly is described as episodic, presumably the result of a periodic, abundant seed crop coinciding with favorable conditions for establishment and a sufficiently long fire-free period for seedlings to develop resistance to fire (White 1985; Oliver and Ryker 1990). Notably, surface fires are thought to reduce competition and to expose mineral soil, thereby providing favorable conditions for seedling establishment.

Damaging Agents

The animals of ponderosa pine forests have been the subject of numerous investigations. Although the seeds are adapted for wind dispersal, rodent caches facilitate seedling establishment (West 1968), and porcupines (*Erethizon dorsatum* L.) frequently slow tree growth by feeding on the bark, as do Abert squirrels (*Sciurus aberti* True; six subspecies) in the Southwest (States and Wettstein 1997). Two native bark beetles, the mountain pine beetle (*Dendroctonus ponderosae* Hopkins) and western pine beetle (*D. brevicomis* LeConte), are especially important in causing tree mortality (Schmid 1988; Veblen and Lorenz 1991; Lundquist 1995). Most trees are protected against bark beetle invasion through the production of sticky resins, but this protection can be lost as the trees age or as the resources required for resin production become more limiting, possibly due to increased stand density following fire suppression (Larrson et al. 1983; Waring and Pitman 1985; Olsen, Schmid and Mata 1996). The beetle larvae feed on the phloem of the inner bark, typically girdling the tree and thereby reducing the flow of carbohydrates to the roots. Interestingly, some trees are more susceptible than others, suggesting genetic variation within the population. Experimental girdling leads to death after a 4–6-year period, but beetle-infested trees die within one year. Such rapid tree death has been attributed to the introduction of blue-stain fungi (*Ceratocystis* and *Europhium* spp.) by the beetles. The fungi quickly colonize the sapwood and impede water and nutrient flow to the leaves. The beetle–fungus association may be mutualistic, with the beetle serving as a vector for the dispersal of the fungi and benefiting by facilitating a fungal infection that creates an innerbark environment favorable for beetle larvae.

Perhaps the best-known vertebrate associated with ponderosa pine is the Abert squirrel, also known as the tassel-eared squirrel (Kieth 1965; States et al. 1988; Austin 1990; Snyder 1993). The squirrel is essentially confined to ponderosa pine forests in the central and southern Rocky Mountains (primarily south of Wyoming and Central Utah). Unlike other forest squirrels, which feed on seeds and fruits, the Abert squirrel feeds year-round on the inner bark of twigs and small branches. Notably, the squirrel avoids feeding on some trees, possibly because of genetically controlled variation in bark characteristics (Pederson and Welch 1985; Linhart 1988; Snyder 1993; Linhart, Snyder and Gibson 1994). Abert squirrels also consume the sporocarps of various species of hypogeous and epigeous fungi (truffles and mushrooms, respectively), two kinds of Basidiomycetes that form mycorrhizae with ponderosa pine. It has been suggested that the relationship is mutualistic because the squirrels help disperse the spores of the fungi, thereby increasing the range of sites over which the pine can become established (Kotter and Farentinos 1994). The squirrels also cache some of the pine seed in the soil, facilitating seedling establishment.

A common parasite of ponderosa pine is dwarf mistletoe (*Arceuthobium vaginatum*), an angiosperm with explosive fruits that expel their single seed up to 10 m or more. As with bark beetles and squirrels, the mistletoe obtains a large portion of its nutrition from the innerbark. Populations of dwarf mistletoe may have been controlled by periodic fires prior to settlement by Euro-Americans,

which would explain why this parasite appears to be more common now than in presettlement times (Hessburg, Mitchell and Flip 1993). Also, with fire suppression, mistletoe probably spreads more rapidly as the distance between potential hosts declines. Although mistletoe slows the growth of trees, Bennetts et al. (1996) found that bird abundance and diversity were higher in stands with this parasite.

Increasing tree density, high levels of flammability, and increasing populations of bark beetles and mistletoe have led some managers to the conclusion that many ponderosa pine forests and woodlands are in "poor health." Such problems are complicated by the many people who build homes where ponderosa pine is found, thereby creating severe logistical problems for possible solutions such as prescribed burning and tree harvesting (Covington and Wagner 1996; Hardy and Arno 1996). In environments less favorable for ponderosa pine establishment and growth, such as on escarpments at lower elevations, the problems are less critical.

Limber Pine

Limber pine often occurs with ponderosa pine in Wyoming and Montana, but the two species differ taxonomically and ecologically in various ways. Phylogenetically, limber pine is a white pine (subgenus *Strobus*), with 5 needles per fascicle and a seed adapted for bird dispersal. In contrast, ponderosa pine is in subgenus *Pinus*, with 2–3 needles per fascicle and wind-dispersed seed. Limber pine usually is found at somewhat higher elevations than ponderosa pine and is more tolerant of colder climates. However, it occurs on the Great Plains at elevations of 870 m (e.g., at Pinebluffs on the Wyoming/Nebraska border and in western North Dakota). Limber pine also occurs as high as 3,810 m in Colorado, where it grows at the edge of alpine tundra and apparently fills the same ecological niche as three other pines: whitebark pine (*P. albicaulis*) in the northern Rocky Mountains, Colorado bristlecone pine in the southern Rocky Mountains (*P. aristata*), and intermountain bristlecone pine in the mountains of the Great Basin (*P. longaeva*). No other conifer in North America spans such a broad elevational gradient as limber pine, a distribution that suggests considerable genetic diversity. Schuster, Alles and Mitton (1989) concluded that gene flow occurs between the high- and low-elevation populations.

Limber and ponderosa pine also differ in longevity, with some limber pine trees living for more than 1,500 years, whereas ponderosa pine rarely live to be 500 years old (Oliver and Ryker 1990; Steele 1990). An adaptation that may contribute to the longevity of limber pine is its unusually "limber" branches, which minimize breakage in the windy environments where it often is found. Limber pine occurs in the subalpine forests of the northern Rocky Mountains, but not as a dominant. A relatively cool optimum temperature for germination and seedling photosynthesis (Lepper 1974) reflects the tendency for the species to occur at higher elevations.

Common associates of limber pine are Rocky Mountain juniper, big sagebrush, curl-leaf mountain-mahogany, common juniper (*Juniperus communis*), snowberry (*Symphoricarpos oreophilus*), Oregon grape (*Mahonia repens*), and foothill grasses such as bluebunch wheatgrass and king spikefescue (*Festuca kingii*). In some areas limber pine grows with ponderosa pine, Douglas-fir, or aspen (*Populus tremuloides*), occurring on shale, limestone, and sandstone, as well as on granite, quartzite, volcanic, and badland soils. As with ponderosa pine, limber pine usually is found on coarse-textured soil (Figure 15.4).

The local distribution of limber pine is strongly influenced by birds, the primary seed dispersal agent. The Clark's nutcracker (*Nucifraga columbiana* Wilson) is cited most often in this regard, but the Steller's jay (*Cyanocitta stelleri* Gmelin) and pinyon jay (*Gymnorhinus cyanocephalus* Wied.) disperse the seed as well (Lanner and Vander Wall 1980; Benkman, Balda and Smith 1984; Tomback and Linhart 1990; Vander Wall 1990; Benkman 1995). The nutcracker has been observed to carry seeds up to 23 km, with as many as 125 seeds in a sublingual pouch. The birds bury 1–5 seeds in caches 2–3 cm deep, typically on south slopes or windswept ridges and escarpments where they often can access the seeds during winter due to discontinuous snow cover. Groves of limber pine often occur in such habitats, and it is common to see clusters of limber pine seedlings emerging from the ground. Red squirrels (*Tamiasciurus hudsonicus* Erxleben) also cache the seeds, along with those of other "stone" pines, such as the bristlecone, whitebark, and pinyon pines. Black and grizzly bears (*Ursus americanus* Audubon and Bachman and *U. arctos horribilis* Ord, respectively) feed on the seeds as well, especially in autumn as they build fat reserves for the winter (McCutchen 1996). Large seed crops are produced every 2–4 years.

Feeding by red squirrels has led to the evolution of specialized seeds and cones that minimize herbivory. Benkman (1995) measured reproductive characteristics of limber pine in two areas, one with squirrels and the other without. He found that the pine without squirrels allocated more than two times as much energy to seeds and much less to protective cone scales, seed coats, and resin. Squirrels often clip the cone-bearing branches of whitebark pine, which fall to the ground and become more accessible to bears. The same phenomenon may occur with limber pine.

Like ponderosa pine, limber pine can be attacked by the mountain pine beetle, at least at lower elevations, and commonly it is a host for mistletoe. Porcupines also feed on the bark. In addition, budworms (*Choristoneura lambertiana ponderosana* Obraztsov) sometimes feed on the buds and leaves, and the introduced white pine blister rust (*Cronartium*

Figure 15.4. Limber pine and Rocky Mountain juniper occur only along fracture planes of this sandstone escarpment north of Rawlins, Wyoming, providing a striking illustration of the inverse texture effect (see text for explanation). Shrublands dominated by big sagebrush and greasewood are found on the deeper, fine-textured soils. Elevation 2128 m.

ribicola Fisch.) can occur where limber pine grows in association with the intermediate host, *Ribes* spp. (Steele 1990).

Limber pine occurs in habitats where growth is very slow and where understory biomass often is low and discontinuous. Not surprisingly then, fires are infrequent (Crane and Fischer 1986). Older trees have thick bark, which probably functions to protect the tree from surface fires just as it does for ponderosa pine. Dense clusters of needles often develop around the terminal buds of both pines, which also may protect the apical meristem. When fires occur, the community is dominated for many years by grasses, forbs, and shrubs capable of sprouting. Trees invade slowly, even with the assistance of the Clark's

nutcracker and other jays (Tomback and Linhart 1990). In contrast to fire suppression with ponderosa pine, fire suppression usually does not lead to rapid increases in limber pine density.

Research Needs

For many years research has been underway on ponderosa pine and the diverse array of woodlands it dominates. However, unlike in the Southwest, the climatically driven contractions and expansions of the northern half of its range during the Pleistocene have not been determined. Some research suggests that, with the projected climate warming of the 21st century, ponderosa pine could expand into areas where it is now absent (such as in Yellowstone National Park; Bartlein, Whitlock and Shafer 1997). A major cause of change in ponderosa pine woodlands during the 1900s is the continued suppression of fires. The long-term effects of this human influence, aside from increased tree density, have not been determined. Timber harvesting will be used as an alternative to fire in some areas, but the various effects of silvicultural practices on plant and animal species composition, plant–animal interactions, the spread of pathogens, and other community characteristics are unknown.

With regard to limber pine, one of the greatest concerns is the spread of white pine blister rust, a disease that is not native to North America. The epidemiology of blister rust on limber pine is poorly known. For both ponderosa and limber pine, much is yet to be learned about the ways in which the species are adapted to such a wide range of environments, and how scattered woodlands dominated by these trees contribute to the biodiversity of the landscapes in which they occur.

Summary

Limber pine and ponderosa pine typically occur on escarpments and in the foothills of mountain ranges, environments that are cooler and more mesic than the adjacent grasslands and shrublands below and warmer and drier than the forests above. The irregular topography creates windbreaks. On south slopes where snow often melts in midwinter, food for a variety of wildlife is available in a comparatively warm environment. Moreover, the inverse texture effect of coarse soils, combined with sparse tree cover, promotes a comparatively high level of forage production. Foothill and escarpment woodlands comprise a small proportion of the landscape, but like riparian zones, they provide critical habitat for many species of animals. Plant diversity also may be high due to the numerous geologic substrata that commonly are exposed and that create variability in soil characteristics.

Ponderosa pine and limber pine woodlands are quite different ecologically, and they present different management/restoration challenges. With fire suppression, ponderosa pine becomes denser in some areas and is more likely to expand into adjacent grasslands. Forage production declines and the probability of a severe crown fire increases. Typically, a large number of people live and work in the same environment as ponderosa pine, which makes fuel management extremely difficult. In contrast, tree density in limber pine woodlands increases less rapidly with fire suppression. The growing season is cooler and drier in limber pine woodlands, and consequently human occupancy is lower. Unless the introduced white pine blister rust becomes a severe problem, the management and restoration of limber pine woodlands should be much easier than for ponderosa pine woodlands.

Acknowledgments

I gratefully acknowledge the helpful suggestions of Roger C. Anderson, Jerry M. Baskin, James S. Fralish, Stephen T. Jackson, and Jack S. States on an earlier draft of this chapter.

REFERENCES

Arno, S. F., Smith, H. Y., and Krebs, M. A. 1997. *Old Growth Ponderosa Pine and Western Larch Stand Structures: Influences of Pre-1900 Fires and Fire Exclusion.* USDA, Forest Service, Res. Pap. INT-RP-495.

Austin, W. J. 1990. The foraging ecology of Abert squirrels. Ph.D. dissertation, Northern Arizona University, Flagstaff, Ariz.

Barbour, M. G. 1987. California upland forests and woodlands. In *North American Terrestrial Vegetation*, ed. M. G. Barbour and W. D. Billings, pp. 131–164. New York: Cambridge University Press.

Barrett, S. W., and Arno, S. F. 1982. Indian fires

as an ecological influence in the northern Rockies. *Journal of Forestry* 80:647–651.

Bartlein, P. J., Whitlock, C., and Shafer, S. L. 1997. Future climate in the Yellowstone National Park region and its potential impact on vegetation. *Conservation Biology* 11:782–792.

Benkman, C. W. 1995. The impact of tree squirrels (*Tamiasciurus*) on limber pine seed dispersal adaptations. *Evolution* 49:585–592.

Benkman, C. W., Balda, R. P., and Smith, C. C. 1984. Adaptations for seed dispersal and the compromises due to seed predation in limber pine. *Ecology* 65:632–642.

Bennetts, R. E., White, G. C., Hawksworth, F. G., and Severs, S. E. 1996. The influence of dwarf mistletoe on bird communities in Colorado ponderosa pine forests. *Ecological Applications* 6:899–909.

Betancourt, J. L. 1990. Late Quaternary biogeography of the Colorado Plateau. In *Packrat Middens: The Last 40,000 Years of Biotic Change*, ed. J. L. Betancourt, T. R. Van Devender, and P. S. Martin, pp. 259–292. Tucson, Ariz.: University of Arizona Press.

Billings, W. D. 1990. The mountain forests of North America and their environments. In *Plant Biology of the Basin and Range*, ed. C. B. Osmund, L. F. Pitelka, and G. M. Hidy, pp. 47–86. New York: Springer-Verlag.

Cochran, P. H., and Berntsen, C. M. 1973. Tolerance of lodgepole and ponderosa pine seedlings to low night temperatures. *Forest Science* 19:272–280.

Cooper, C. F. 1960. Changes in vegetation, structure, and growth of southwestern pine forests since white settlement. *Ecological Monographs* 30:129–164.

Cooper, C. F. 1961. Pattern in ponderosa pine forests. *Ecology* 42:493–499.

Covington, W. W., Everett, R. L., Steele, R., Irwin, L. L., Daier, T. A., and Auclair, A. N. D. 1994. Historical and anticipated changes in forest ecosystems of the Inland West of the United States. In *Assessing Forest Ecosystem Health in the Inland West*, ed. R. N. Sampson and D. L. Adams, pp. 13–63. New York: The Haworth Press, Inc.

Covington, W. W., and Moore, M. M. 1994a. Postsettlement changes in natural fire regimes and forest structure: ecological restoration of old-growth ponderosa pine forests. In *Assessing Forest Ecosystem Health in the Inland West*, ed. R. N. Sampson and D. L. Adams, pp. 153 181. New York: The Haworth Press, Inc.

Covington, W. W., and Moore, M. M. 1994b. Southwestern ponderosa forest structure: changes since Euro-American settlement. *Journal of Forestry* 92:39–47.

Covington, W. W., and Sackett, S. S. 1992. Soil mineral nitrogen changes following prescribed burning in ponderosa pine. *Forest Ecology and Management* 54:175–191.

Covington, W. W., and Wagner, P. K., coordinators. 1996. *Conference on Adaptive Ecosystem Restoration and Management: Restoration of Cordilleran Conifer Landscapes of North America*. USDA Forest Service, Gen. Tech. Rep. RM-GTR-278.

Crane, M. F., and Fischer, W. C. 1986. *Fire Ecology of the Forest Habitat Types of Central Idaho*. USDA Forest Service, Gen. Tech. Rep. INT-218.

Critchfield, W. B., and Little, E. L., Jr. 1966. *Geographic Distribution of the Pines of the World*. USDA, Forest Service, Misc. Publ. 991.

Daubenmire, R. 1968. Soil moisture in relation to vegetation distribution in the mountains of northern Idaho. *Ecology* 49:431–438.

DeLucia, E. H., and Schlesinger, W. H. 1990. Ecophysiology of Great Basin and Sierra Nevada vegetation on contrasting soils. In *Plant Biology of the Basin and Range*, ed. C. B. Osmund, L. F. Pitelka, and G. M. Hidy, pp. 143–178. New York: Springer-Verlag.

Dick-Peddie, W. A. 1993. *New Mexico Vegetation: Past, Present and Future*. Albuquerque, N.Mex.: University of New Mexico Press.

Fisher, R. F., Jenkins, M. J., and Fisher, W. F. 1987. Fire and the prairie-mosaic of Devils Tower National Monument. *The American Midland Naturalist* 117:250–257.

Franklin, J. F., and Dyrness, C. T. 1988. *Natural Vegetation of Oregon and Washington*. Corvallis, Ore.: Oregon State University Press

Gruell, G. E. 1985. Fire on the early western landscape: an annotated record of wildland fires 1776–1900. *Northwest Scientist* 59:97–107.

Habeck, J. R., and Mutch, R. W. 1973. Fire-dependent forests in the Northern Rocky Mountains. *Journal of Quaternary Research* 3:408–424.

Hansen, P. L., and Hoffman, G. R. 1988. *The Vegetation of the Grand River/Cedar River, Sioux, and Ashland Districts of the Custer National Forest: A Habitat Type Classification*. USDA Forest Service, Gen. Tech. Rep. RM-157.

Hardy, C. C., and Arno, S. F. 1996. *The Use of Fire in Forest Restoration*. USDA Forest Service, Gen. Tech. Rep. INT-GTR-341.

Hessburg, P. F., Mitchell, R. G., and Flip, G. M. 1993. Historical and current roles of insects and pathogens in eastern Oregon and Washington forest landscapes. In *Eastside Forest Ecosystem Health Assessment*. Vol. III: *Assessment*, compiled by P. F. Hessburg, pp. 485–536. Portland, Ore.: USDA Forest Service, Pacific Northwest Research Station.

Higgins, K. F. 1984. Lightning fires in North Dakota grasslands and pine-savanna lands of South Dakota and Montana. *Journal of Range Management* 37:100–103.

Hoffman, G. R., and Alexander, R. R. 1987. *Forest Vegetation of the Black Hills National Forest of South Dakota and Wyoming: A Habitat Type Classification*. USDA Forest Service, Res. Pap. RM-276.

Houston, D. B. 1973. Wildfires in northern Yellowstone National Park. *Ecology* 54:1111–1117.

Kartesz, J. T. 1994. *A Synonymized Checklist of the Vascular Flora of the United States, Canada, and Greenland*. Portland, Ore.: Timber Press.

Kieth, J. O. 1965. The Abert squirrel and its dependence on ponderosa pine. *Ecology* 46:150–163.

Knight, D. H. 1994. *Mountains and Plains: The Ecology of Wyoming Landscapes*. New Haven, Conn.: Yale University Press.

Knight, D. H., Vose, J., Baldwin, C., Ewel, K., and Grodzinska, C. 1994. Contrasting patterns in pine forest ecosystems. *Ecology Bulletins* (Sweden) 43:9–19.

Kotter, M. M., and Farentinos, R. C. 1994. Tassel-eared squirrels as spore dispersal agents of hypogeous mycorrhizal fungi. *Journal of Mammalogy* 65:684–687.

Lanner, R. M., and Vander Wall, S. B. 1980. Dispersal of limber pine seed by Clark's nutcracker. *Journal of Forestry* 78:637–639.

Larrson, S., Oren, R., Waring, R. H., and Barrett, J. W. 1983. Attacks of mountain pine beetle as related to tree vigor of ponderosa pine. *Forest Science* 29:395–402.

Lepper, M. G. 1974. *Pinus flexilis* James and its environmental relationships. Ph.D. dissertation, University of California, Davis, Calif.

Linhart, Y. B. 1988. Ecological and evolutionary studies of ponderosa pine in the Rocky Mountains. In *Ponderosa Pine: The Species and Its Management*, ed. D. M. Baumgartner and J. E. Lotan, pp.77–89. Pullman, Wash.: Coop. Extension, Washington State University.

Linhart, Y. B., Snyder, M. C., and Gibson, J. P. 1994. Differential host utilization in a population of ponderosa pine. *Oecologia* 98:117–120.

Lundquist, J. E. 1995. Pest interactions and canopy gaps in ponderosa pine stands in the Black Hills, South Dakota, USA. *Forest Ecology and Management* 74:37–48.

Madany, M. H., and West, N. E. 1983. Livestock grazing–fire regime interactions within montane forests of Zion National Park, Utah. *Ecology* 64:661–667.

McCutchen, H. E. 1996. Limber pine and bears. *Great Basin Naturalist* 56:90–92.

McPherson, G. R. 1992. Comparison of linear and non-linear overstory–understory models for ponderosa pine: a conceptual framework. *Forest Ecology Management* 55:31–34.

Moir, W. H. 1966. Influence of ponderosa pine on herbaceous vegetation. *Ecology* 47:1045–1048.

Morgan, P. 1984. Dynamics of ponderosa and Jeffrey pine forests. In *Flammulated, Boreal, and Great Grey Owls in the United States: A Technical Conservation Assessment*, ed. G. H. Hayward and J. Verner, pp. 41–73. USDA Forest Service, Gen. Tech Rep. RM-253.

Muldavin, E. H., DeVelice, R. L., and Ronoco, F., Jr. 1996. *A Classification of Forest Habitat Types: Southern Arizona and Portions of the Colorado Plateau*. USDA Forest Service, Gen. Tech. Rep. RM-GTR-287.

Noy-Meir, I. 1973. Desert ecosystems: environment and producers. *Annual Review of Ecology and Systematics* 4:25–51.

Oliver, W. W., and Ryker, R. A. 1990. *Pinus ponderosa* Dougl. ex Laws. In *Silvics of North America*, Vol. 1, *Conifers*, coordinators, R. M. Burns and B. H. Honkala, pp. 413–424. USDA, Agriculture Handbook 654.

Olsen, W. K., Schmid, J. M., and Mata, S. A. 1996. Stand characteristics associated with mountain pine beetle infestations in ponderosa pine. *Forest Science* 310–327.

Orr, H. K. 1975. *Watershed Management in the Black Hills: The Status of our Knowledge*. USDA Forest Service, Res. Paper RM-141.

Pederson, J. C., and Welch, B. L. 1985. Comparison of ponderosa pines as feed and non-feed trees for Abert squirrels. *Journal of Chemical Ecology* 11:149–157.

Peet, R. K. 1988. Forests of the Rocky Mountains. In *North American Terrestrial Vegetation*, ed. M. G. Barbour and W. D. Billings, pp. 63–101. New York: Cambridge University Press.

Sala, O. E., Parton, W. J., Joyce, L. A., and

Lauenroth, W. K. 1988. Primary production of the central grassland region of the United States. *Ecology* 69:40–45.

Savage, M., and Swetnam, T. W. 1990. Early 19th-century fire decline following sheep pasturing in a Navajo ponderosa pine forest. *Ecology* 71:2374–2378.

Schmid, J. M. 1988. Insects of ponderosa pine: impacts and control. In *Ponderosa Pine: The Species and Its Management*, ed. D. M. Baumgartner and J. E. Lotan, pp.93–97. Pullman, Wash.: Coop. Extension, Washington State University.

Schuster, W. S., Alles, D. L., and Mitton, J. B. 1989. Gene flow in limber pine: evidence from pollination phenology and genetic differentiation along an elevational transect. *American Journal of Botany* 76:1395–1403.

Shinneman, D. J., and Baker, W. L. 1997. Nonequilibrium dynamics between catastrophic disturbances and old-growth forests in ponderosa pine landscapes of the Black Hills. *Conservation Biology* 11:1276–1288.

Snyder, M. A. 1993. Interactions between Abert's squirrel and ponderosa pine: the relationship between selective herbivory and host plant fitness. *American Naturalist* 141:866–879.

States, J. S., Gaud, W. S., Allred, W. S., and Austin, W. J. 1988. Foraging patterns of tassel-eared squirrels in selected ponderosa pine stands. In *Management of Amphibians, Reptiles, and Small Mammals in North America*, ed. R. Szaro, K. Severson, and D. Patton, pp. 425–431. USDA Forest Service, Gen. Tech. Rep. RM-166.

States, J. S., and Wettstein, P. J. 1998. Food habitats and evolutionary relationships of the tassel-eared squirrel (*Sciurus aberti*). In *Ecology and Evolutionary Biology of Tree Squirrel*, ed. M. A. Steele, J. F. Merritt, and D. A. Zegers, pp. 185–194. Marinsville, Pa.: Virginia Museum of Natural History.

Steele, R. 1990. *Pinus flexilis* James. In *Silvics of North America*, Vol. 1, *Conifers*, coordinators,

R. M. Burns and B. H. Honkala, pp. 348–354. USDA, Agriculture Handbook 654.

Swezy, D. M., and Agee, J. K. 1991. Prescribed-fire effects on fine root and tree mortality in old-growth ponderosa pine. *Canadian Journal of Forest Research* 21:626–634.

Thompson, R. S. 1990. Late Quaternary vegetation and climate in the Great Basin. In *Packrat Middens: The Last 40,000 Years of Biotic Change*, ed. J. L. Betancourt, T. R. Van Devender, and P. S. Martin, pp. 200–239. Tucson, Ariz.: University of Arizona Press.

Tomback, D. F., and Linhart, Y. B. 1990. The evolution of bird-dispersed pines. *Evolutionary Ecology* 4:185–219.

Vander Wall, S. B. 1990. *Food Hoarding in Animals*. Chicago, Ill.: University of Chicago Press.

Vasek, F. C., and Thorne, R. F. 1977. Transmontane coniferous vegetation. In *Terrestrial Vegetation of California*, ed. M. G. Barbour and J. Major, pp. 797–832. New York: John Wiley & Sons.

Veblen, T. T., and Lorenz, D. C. 1991. *The Colorado Front Range: A Century of Ecological Change*. Salt Lake City, Utah: University of Utah Press.

Waring, R. H., and Pitman, G. B. 1985. Modifying lodgepole pine stands to change susceptibility to mountain pine beetle attack. *Ecology* 66:889–897.

Wells, P. V. 1965. Scarp woodlands, transported grassland soils, and the concept of grassland climate in the Great Plains region. *Science* 148:246–249.

West, N. E. 1968. Rodent-influenced establishment of ponderosa pine and bitterbrush seedlings in central Oregon. *Ecology* 49:1009–1011.

West, N. E. 1969. Tree patterns in central Oregon ponderosa pine forests. *The American Midland Naturalist* 81:584–590.

White, A. S. 1985. Presettlement regeneration patterns in a southwestern ponderosa pine stand. *Ecology* 66:589–594.

16 The Sand Shinnery Oak (*Quercus havardii*) Communities of the Llano Estacado
History, Structure, Ecology, and Restoration

SHIVCHARN S. DHILLION AND MICHELE H. MILLS

These lands [shinnery oak communities] are perhaps the most fragile of all ecosystems on the southern High Plains of Texas and the landowner cannot afford to abuse them.

(Pettit 1979)

Introduction

Sand shinnery oak communities are some of the least known and most poorly described communities in the southwestern United States and until recently were given little attention. Historically they have been subjected to degradation and eradication, specifically destructive grazing, herbiciding, and other large-scale disturbances. Early observations by biologists and naturalists suggested that shinnery oak communities may be refugia for a variety of plants, mammals, insects, and birds. More recent studies indicate a rich flora and fauna, including three federally endangered and one locally threatened species. Today, there is a strong need for conservation and restoration, especially because eradication schemes are used widely (e.g., Test 1972; Pettit 1979; Sears et al. 1986). Unfortunately, no conservation or formal restoration is planned for these threatened communities. This is the first review of the ecology and future conservation of the sand shinnery oak communities.

Historical Background

Little has been written about shinnery oak communities. Before the late 1800s, the Llano Estacado (panhandle of Texas and high plains of Texas and New Mexico) was occupied by Native Americans, causing Europeans to avoid settling the area until the late 1870s (Biggers 1991). Early explorers described the area as "exceedingly monotonous and uninteresting, being a continuous succession of barren sand-hills" covered with a "dense growth of dwarf oak bushes" (Marcy 1854, quoted in Rowell 1967). Early surveyors considered shinnery oak land to be worthless, but the first European settlers of the Llano Estacado were cattlemen, who soon found shinnery oak to be beneficial to cattle during periods of drought, when grasses were scarce (Holden 1970).

The Dominant Species: Shinnery Oak and Its Hybrids

Sand shinnery oak (*Quercus havardii* Rydb.) originally was named by Rydberg for Dr. Valery Havard, who collected the type specimen of this species on the Llano Estacado of Texas (Rydberg 1901). The name *shinnery* oak comes from the Louisiana French word *chênière*, an oak grove (Oklahoma Forestry Division 1981). The continuous sea of oak consists of shinnery oak clones differing only slightly in leaf and acorn morphology. Each clone may range in diameter from 10 to 70 m (Mayes 1994). However, differences in depth of the sandy horizon and in the amount of precipitation cause clones to be taller and denser in the eastern than in the western portion of the range (Sullivan 1980). Stems arising from rhizomes average 1 m in height (Mayes 1994). *Quercus havardii* is a deciduous white oak with vegetative and reproduction characteristics typical of those of the genus *Quercus* (Rydberg 1901; Muller 1951; Correll and Johnston 1979; Elias 1980; Vines 1982). Shinnery oak clones live an average of 11–15 years (Pettit 1986), but some may live much longer (Wiedeman 1960); rhizomes up to 25 years old were excavated by S. Dhillion and C. Friese (unpublished) at Beasley Ranch, Yoakum County, Texas.

Whereas shinnery oak averages 1 m in height, hybrids between *Q. havardii* and other oak species reach heights of 2–4 m (Jones 1982), including those between *Q. havardii* and post oak (*Q. stellata*) (Muller 1951; Correll and Johnston 1979; Vines 1982). These two species were in contact over a wide range "in comparatively ancient times" (Muller 1951) but, due to increased climatic aridity becoming unfavorable for post oak, they are now only slightly sympatric at the eastern edge of the range of shinnery oak (Muller 1951). In the western part of its range, shinnery oak hybridizes with Mohr's oak (*Q. mohriana*) (Muller 1951; Correll and Johnston 1979; Vines 1982) and possibly with *Q. undulata* and *Q. gambelii* (Muller 1951).

Geographic Range of Shinnery Oak Communities

Shinnery oak communities frequently occur on dunes extending from eastern New

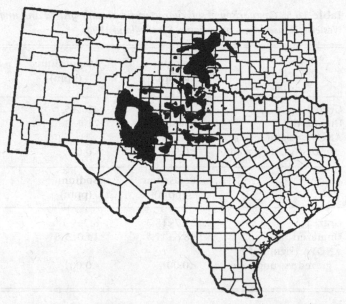

Figure 16.1. Present geographical range of shinnery oak communities.

Mexico across Texas high plains and into western Oklahoma (Muller 1951; Test 1972) at elevations below 1,200 m (Gould and Carlton 1970). The most recent estimate of the area occupied by these communities in Texas, New Mexico, and Oklahoma (Bóo 1974) is 3 million ha (Figure 16.1), which is lower than earlier estimates of 5 million ha, resulting from loss of these habitats to croplands or grasslands (Smith and Rechenthin 1964; Robison and Fisher 1968). In New Mexico, the geographic range of shinnery oak may have spread since the late 1800s (York and Dick-Peddie 1969); however, in Texas the most recent estimate is that shinnery oak has declined from 3.6 million ha to 1.2 million ha (Allred 1949; Deering and Pettit 1971). These primarily privately owned lands have been converted to cropland or grassland due to the poor grazing quality of shinnery oak ranges, reduction of herbaceous plant biomass associated with the spread of shinnery oak, and toxicity of shinnery oak to livestock in the spring (although it is palatable to cattle at other times) (Pettit 1977).

Table 16.1. *Comparison of soil characteristics of cattle-grazed and ungrazed regions of sand shinnery oak in Western Texas. Values are mean ± SD (Holland 1994).*

	Calcium (ppm)	Magnesium (ppm)	Nitrogen (ppm)	Phosphorous (ppm)
Grazed	289.8±63.4	42.4±5.1	1.6±0.3	1.1±0.1
Ungrazed	530.4±258.0	71.4±36.7	1.4±0.4	1.6±1.0
ANOVA (signif. of F) grazed vs. ungrazed	<0.001	<0.001	<0.01	<0.001
	Potassium (ppm)	Sodium (ppm)	Organic Matter (%)	Mycorrhizae (#/25g soil)
Grazed	21.7±18.3	8.2±9.1	1.0±0.0	15.1±13.1
Ungrazed	98.8±47.5	14.0±5.8	1.3±0.0	22.1±45.4
ANOVA (signif. of F) grazed vs. ungrazed	<0.001	<0.001	<0.01	0.5

Soil Types, Blowouts, and Dunes

Shinnery oak grows on approximately 50 soil series, all of which have a sandy, nutrient-poor surface horizon (Table 16.1; Small 1975; Secor et al. 1983; Pettit 1986; Holland 1994). Runoff is minimal and rainwater percolates quickly through the sandy soil (Wilhite 1960; Sullivan 1980). Prevailing winds often form sand dunes, which shinnery oak helps stabilize (Gould and Carlton 1970). Dune height can reach 6 m, with an average slope of 40% (Wilhite 1960). In many shinnery oak communities, blowouts form on the southwest sides of dunes (Wilhite 1960), which is the general direction of the prevailing winds. Blowouts are caused by overgrazing or other disturbances that remove dune-stabilizing vegetation. Without the vegetation, winds blow sand out of these unstable areas until shinnery oak roots are exposed (Sullivan 1980). This process kills shinnery oak, resulting in more sand being blown out, eventually exposing the clay layer beneath the sand (Melton 1940; Green 1951; Sullivan 1980). Depth of sand above the clay layer varies from a few millimeters to 3 m (Sullivan 1980). No research has been conducted on these soils to determine effects of the clay layer on water movement or vegetation growth. However, compared to a sandy soil, a soil with a high clay content will retain more water, which is then available for plants with shallow roots (Sullivan 1980).

Climate

The climate where shinnery oak grows is semiarid (Wiedeman 1960; Sullivan 1980), with annual rainfall decreasing from 76 cm in the eastern portion of the range to 35 cm or less in the southwestern portion; approximately 80% of precipitation falls between April and October (NOAA 1985). Minimum temperatures average -4.4 °C in January, and maximum temperatures reach or exceed 37.7 °C in July (NOAA 1985). Frost-free days range from 181 to 223 per year (Sullivan 1980). Snow and ice storms occur between October and April, with high winds in March and April, followed by violent summer storms (Bednarz, Hayden and Fischer 1990).

Vegetation Composition and Dynamics

Shinnery oak associations frequently are named after the dominant grass species in the community. The climax shinnery oak community has been described both as a tall-grass–shinnery oak community (Graff 1983; Dickerson 1985) and as a mixed-grass prairie (Holechek, Pieper and Herbel 1989; Sims 1989). Sullivan (1980) defined the shinnery oak associations along the Texas–New Mexico border and grouped them into three categories: (1) shinnery oak–tall-grass, (2) shinnery oak–mid-grass, and (3) mesquite–shinnery oak–mid-grass. These associations varied in dominant species due to differences in percentage sand in the soil, the northeast-to-southwest gradients of decreasing annual precipitation, and increasing temperature and evapotranspiration (Sullivan 1980). Percentage ground cover of shinnery oak has not been published for Oklahoma sites, but it ranges from 77% in Yoakum and Cochran counties along the Texas–New Mexico border to 6%–20% near southeastern New Mexico, where the climate is arid (Sullivan 1980). Shinnery oak also decreases in stature along this gradient, apparently due to the decreasing sand content in soils (Sullivan 1980; Jones 1982).

In Oklahoma, shinnery oak commonly is associated with sand sagebrush grasslands or with mixed-grass prairies (Wiedeman 1960). Three abundant species in most shinnery oak communities are red three-awn (*Aristida longiseta*), little bluestem (*Schizachyrium scoparium*), and sand sagebrush (*Artemisia filifolia*). Other important perennial plants in shinnery oak communities are *Yucca* spp., sand paspalum (*Paspalum setaceum*), sand dropseed (*Sporobolus cryptandrus*), and mesquite (*Prosopis glandulosa*) (Tables 16.2 and 16.3).

Holland (1994) provided the most comprehensive characterization of shinnery oak vegetation. The west Texas shinnery oak community she studied was not treated with herbicide but it had been grazed by cattle. Ten percent of the ground cover was herbaceous plants, with a density of approximately 10 individuals per 4 m² (Figure 16.2a). The remaining ground cover was approximately 27% leaf litter, and 50% woody species, 48% being shinnery oak with 13% bare ground. Although low in abundance, herbs represented 84% of the species present in the community; total species richness was composed of 63% forbs and 21% grasses (Holland 1994).

Shinnery oak communities are habitat for at least one known federally endangered plant, *Callirhoë scabriuscula*. In 1858, a United States Army Surgeon, Dr. Sutton Hayes, discovered this red-flowered mallow in a shinnery oak community near the Colorado River in Coke County, Texas (Dorr 1994). Its roots grow more than a meter into the soil, thus it is well adapted to growing on sand dunes. Today, this species is known from only 12 populations, several of which have only 10 plants, making the species one of the rarest in North America (Dorr 1994).

In communities dominated by shinnery oak, the density and richness of other species is low (e.g., Figure 16.2b). Density of annuals increases in wet years, but survival of forb seedlings is low (Pettit 1977, 1979; Sullivan 1980; Graff 1983). Timing of precipitation is also important, because heavy rains in late summer can cause changes in species composition and abundances, with the greatest effect on annuals (Martin 1982). For example, species richness sampled in belt transects was significantly (P = 0.0039) higher in 1993 than 1994 (Figure 16.2b). In 1994, there was heavy precipitation in May and July, and low precipitation during the remainder of the year, whereas spring and summer precipitation was more consistent in 1993 (NOAA 1993, 1994).

Animals

Mammals

Little is known about the species of mammals inhabiting shinnery oak communities, although many of the same species of rodents occur in Texas and southeastern New Mexico (Gennaro 1982). In west Texas, density of these rodents, such as Ord's kangaroo rat (*Dipodomys ordii*) and southern grasshopper

Table 16.2. *Forb species reported in shinnery oak communities. Nomenclature follows Correll and Johnston (1979).*

Scientific name	Common name	% Studies
Eriogonum annuum	Annual wild buckwheat	62
Commelina erecta	Erect dayflower	57
Ambrosia psilostachya	Western ragweed	48
Dithyrea wislizenii	Spectaclepod	48
Thelesperma megapotamicum	Prairie greenthread	48
Hymenopappus flavescens	Yellow woollywhite	43
Palafoxia sphacelata	Rayed palafoxia	43
Paronychia jamesii	James' nailwort	43
Heterotheca latifolia	Camphorweed	33
Linum rigidum	Stiff-stem flax	38
Calylophus serrulatus	Yellow evening primrose	33
Cryptantha jamesii	James' cryptantha	33
Erigeron modestus	Plains fleabane	33
Euphorbia fendleri	Fendler euphorb	33
Hedyotis humifusa	Mat bluet	33
Lithospermum incisum	Narrowleaf gromwell	29
Mirabilis linearis	Linearleaf four o'clock	33
Solanum elaeagnifolium	Silverleaf nightshade	33
Opuntia phaeacantha	Brownspine pricklypear	32
Chenopodium leptophyllum	Slimleaf goosefoot	29
Croton texensis	Texas croton	29
Evolvulus nuttallianus	Hairy evolvulus	29
Heliotropium convolvulaceum	Bindweed heliotrope	29
Oenothera rhombipetala	Fourpoint evening primrose	29
Polygala alba	White milkwort	29
Salsola iberica	Russian thistle	29
Senecio spartioides fremontii	Broom groundsel	29

Forbs mentioned in 10%–25% of studies: *Abronia fragrans, Amaranthus retroflexus, Artemisia caudata, Asclepias arenaria, A. pumila, Astragalus mollissimus, Conyza canadensis, Cassia fasciculata rostrata, Comandra pallida, Coryphantha macromeris, Croton dioicus, C. lindheimerianus, C. pottsii, Cryptantha minima, Dalea lanata, D. nana, Euphorbia missurica, Eurytaenia texana, Ferocactus wislizenii, Froelichia floridana, F. gracilis, Gaillardia pulchella, Gaura coccinea, G. villosa, Helianthus annuus, H. petiolaris, Heterotheca villosa, Hoffmanseggia glauca, H. jamesii, Hymenoxys scaposa, Ipomopsis longiflora, Krameria lanceolata, Kuhnia eupatorioides, Lesquerella* spp., *Liatris punctata, Linaria texana, Machaeranthera australis, M. pinnatifida, M. tanacetifolia, Maurandya wislizenii, Melampodium leucanthum, Mentzelia nuda, Mollugo verticillata, Monarda pectinata, M. punctata, Nerisyrenia camporum, Opuntia leptocaulis, Opuntia* spp., *Pectis angustifolia, Penstemon buckleyii, Petalostemum purpureum, P. villosum, Phacelia integrifolia, Phyllanthus abnormis, Physalis viscosa, Plantago patigonica, Polanisia jamesii, Portulaca mundula, P. oleracea, Ratibida columnaris, Reverchonia arenaria, Senecio douglasii jamesii, S. multicapitatus, Stillingia sylvatica, Talinum calycinum, Townsendia exscapa, Tradescantia occidentalis, Verbesina enceloides, Xanthisma texanum, Zinnia grandiflora.*

Sources: Wiedeman 1960; Test 1972; Small 1975; Pettit 1979; Biondini 1980; Best and Jackson 1982; Gennaro 1982; Jones 1982; Martin 1982; Sears 1982; Webb and Guthery 1982; Jacoby, Slosser and Meadors 1983; Secor et al. 1983; Plumb 1984; Dickerson 1985; Villena and Pfister 1990; Olawsky and Smith 1991; Dick-Peddie 1992; Dhillion et al. 1994; Holland 1994.

Table 16.3. *Graminoids and woody plants reported in shinnery oak communities.*

Graminoids	Common name	% Studies
Paspalum setaceum	Thin paspalum	76
Sporobolus cryptandrus	Sand dropseed	76
Aristida longiseta	Red three-awn	76
Munroa squarrosa	False buffalograss	71
Schizachyrium scoparium	Little bluestem	71
Bouteloua hirsuta	Hairy grama	67
Leptoloma cognatum	Fall witchgrass	62
Cenchrus incertus	Common sandbur	57
Eragrostis oxylepis	Red lovegrass	57
Andropogon gerardii	Big bluestem	48
Yucca campestris	Plains yucca	43
Bouteloua curtipendula	Sideoats grama	38
Sporobolus giganteus	Giant dropseed	38
Cyperus spp.	Sedges	33
Yucca angustifolia	Narrow-leaf yucca	33
Chloris cucullata	Hooded windmill grass	29
Cyperus uniflorus	Oneflower flatsedge	29
Eragrostis sessilispica	Tumble lovegrass	29
Cycloloma atriplicifolium	Tumble ringwing	29
Woody plants		
Quercus havardii	Sand shinnery oak	100
Artemisia filifolia	Sand sagebrush	90
Prosopis glandulosa	Mesquite	48
Gutierrezia sarothrae	Broom snakeweed	43
Caesalpinia jamesii	James' rushpea	33

Grasses mentioned in less than 25% of studies: *Andropogon hallii, Aristida wrightii, Aristida* spp., *Bothriochloa barbinodis, B. saccharoides, Bouteloua eriopoda, B. gracilis, Brachiaria ciliatissima, Buchloe dactyloides, Calamovilfa gigantea, Chloris verticillata, Cyperus schweinitzii, Digitaria sanguinalis, Eleocharis montevidensis, Eragrostis curtipedicellata, E. curvula, E. lehmanniana, E. trichodes, Erioneuron pulchellum, Juncus scirpoides, Lycurus phleoides, Muhlenbergia porteri, Panicum capillare, P. havardii, P. obtusum, P. ramisetum, P. virgatum, Poa* spp., *Schedonnardus paniculatus, Scirpus acutus, Scleropogon brevifolius, Setaria leucopila, S. macrostachya, Sporobolus contractus, S. flexuosus, Stipa comata, Triplasis purpurea.*

Woody plants mentioned in less than 25% of studies: *Baccharis texana, Celtis reticulata, Chrysothamnus pulchellus, Dalea formosa, Desmanthus cooleyi, Ephedra antisyphilitica, Juniperus virginiana, Larrea tridentata, Lycium berlandieri, Mimosa biuncifera, Prunus gracilis, Rhus aromatica, Salix goodingii, Sapindus saponaria, Schrankia occidentalis, S. uncinata, Tamarix* spp., *Vitis* spp., *Ziziphus obtusifolia.*

mouse (*Onychomys leucogaster*), was higher where grass cover increased after shinnery oak cover had been reduced by herbicides (Willig et al. 1993). Rodents consume large quantities of seeds both before and after dispersal, and thus they may have a significant effect on plant densities (Sullivan 1980; M. H. Mills personal observations). Prairie dogs occasionally occupy these habitats, but quickly remove shinnery oak from areas they occupy (Osborn 1942). Other small mammals observed in shinnery oak communities

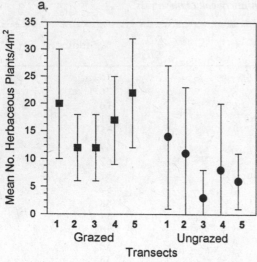

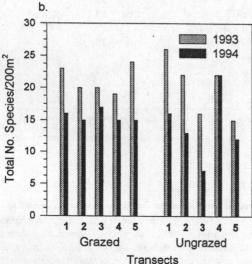

Figure 16.2. (a) Mean numbers of herbaceous plants per 4 m² sampled in ten 2 m × 250 m long belt transects in 1993 for cattle-grazed and ungrazed sites in a shinnery oak community located in Yoakum Co., Texas (Holland 1994), and (b) species richness for each belt transect (2 m × 100 m) sampled once during August or September in 1993 and 1994. (Holland 1994 and M. H. Mills unpublished data.)

include the Mexican and spotted ground squirrels, plains pocket gopher, deer mouse, white-footed mouse, northern grasshopper mouse, house mouse, hispid cotton rat, southern plains woodrat, and several species

of pocket mice and kangaroo rats (Gennaro 1982; Willig et al. 1993; Dhillion et al. 1994). Larger mammals include the desert cottontail, blacktailed jackrabbit, striped skunk, porcupine, coyote, swift fox, badger, bobcat, mule deer, and pronghorn (Gennaro 1982; Dhillion et al. 1994; M. H. Mills personal observations).

Birds

Many species of birds are residents in shinnery oak habitat, for example, the 22 raptor species observed in a shinnery oak habitat in New Mexico (Bednarz, Hayden and Fischer 1990). Of these raptors, the peregrine falcon and aplomado falcon are federally endangered species, and the ferruginous hawk and Swainson's hawk are considered rare. Despite the absence of tall trees and the homogeneous cover, shinnery oak communities are important habitat for these birds, primarily due to the abundance of mammals and reptiles that serve as prey for them (Bednarz, Hayden and Fischer 1990).

Other birds of interest include the lesser prairie chicken (*Tympanuchus pallidicintus*) and quail (*Colinus virginianus* and *Callipepla squamata*) (e.g., Donaldson 1969; Taylor and Guthery 1980; Webb and Guthery 1982). Because prairie chickens prefer to nest in grasses surrounded by shrubs approximately 1 m tall, shinnery oak is the prime habitat on the plains for this species. However, herbiciding and intense grazing of shinnery oak greatly reduce (1) production of acorns, a primary food source for prairie chickens (Jackson and DeArment 1963; Riley, Davis and Smith 1993), and (2) vertical screening cover that prairie chickens need for nesting (Haukos and Smith 1989). As a result, the number of prairie chickens and of quail has decreased in this region (Haukos and Smith 1989; Giesen 1994). Resident birds of shinnery oak communities include the barn swallow, burrowing owl, Chihuahuan white-necked raven, Cooper's hawk, golden eagle, lark bunting, loggerhead shrike,

Mississippi kite, mourning dove, northern Bullock's oriole, northern harrier, northern mockingbird, osprey, prairie falcon, red-tailed hawk, scissor-tailed flycatcher, sharp-shinned hawk, turkey vulture, western kingbird, western screech owl, and white-crowned sparrow (Jackson and DeArment 1963; Test 1972; Taylor and Guthery 1980; Ligon and Haydock 1981; Webb and Guthery 1982; Best 1986; Haukos and Smith 1989; Olawsky and Smith 1991; S. G. Mayes, personal observations; Wildlife Diversity Program, Oklahoma).

Other Animals

There are few detailed studies of small animals in shinnery oak communities, but species lists have been compiled for beetles in New Mexico (Whitford 1982) and ants in New Mexico and western Texas (Whitford 1982; Dhillion et al. 1994; McGinley, Dhillion and Newmann 1994). Social insects influence the soil by transporting as much as a metric ton of deep soil per hectare to the surface during the growing season (Whitford 1982). Soil disturbances by ants could greatly influence vegetation because colony densities are as high as 1,600 colonies per hectare of shinnery oak habitat (Whitford 1982). Beetles have been estimated to have a biomass of 5 kg per hectare between July and September (Whitford 1982), but because they have not been studied adequately in these habitats their significance in the system is not known.

In New Mexico, Gennaro (1982) found that of amphibians and reptiles, the western whiptail lizard, side-blotched lizard, and six-lined racerunner were the most common species in shinnery oak habitats. In addition to the Texas horned lizard, which is listed as threatened in Texas, other reptiles occurring in the habitat are the western box turtle, western hognose snake, coachwhip, glossy snake, bullsnake, longnosed snake, night snake, massasauga, western diamondback rattlesnake, and western rattlesnake. With the exception of this reptile study and a few ecological studies of birds and mammals, animals have not been adequately studied in shinnery oak habitats. Therefore, it is neither known what effect the reduction of shinnery oak cover or conversion of habitats to grassland or cropland has had on these species, nor how this affects other species interacting with them.

Animal–Plant–Soil Interactions

Up to 15 cm of leaf litter accumulates on the soil surface in the shinnery oak community during the winter, and it usually remains throughout the year. Tannins from shinnery oak leaves can prevent germination of herbaceous plant seeds, and litter can mechanically impede growth of seedlings of some species (J. Jeffery, unpublished data). The effect of leaf litter on plant growth was apparent in a western Texas shinnery oak community grazed by cattle until the early 1970s. Herbaceous seedlings were more abundant in areas where there was bare ground or small animal–created disturbances than where leaf litter was deep or there was little bare ground (Dhillion et al. 1994). Dhillion et al. (1994) reported that the density of herbaceous plant seedlings was independent of the percentage cover of shinnery oak. However, the density of mature herbaceous plants was negatively correlated with the cover and height of oak at the same site (Holland 1994). The results of these studies suggest that seedling establishment is strongly affected by availability of germination sites (Dhillion et al. 1994; Dhillion in press), whereas long-term survival may be influenced by competition with the dominant plant species (Holland 1994).

The spatial dispersion patterns of the disturbances also influence the distribution of seedlings (Dhillion et al 1994), and thus small animals influence the distribution of mature plants in the community. Small animal–disturbances have clumped spatial distributions, and so herbaceous plants also tend to be clumped. In Holland's (1994) comparison of this ungrazed site with an adjacent site that had been continuously grazed by cattle since the late 1800s, plant clumps were larger and less distinct where cattle grazed the shinnery oak community than at the ungrazed site (Holland 1994). Because number and distribution of small animal disturbances was similar on both sites, this difference in clump size is due to cattle's trampling leaf litter and exposing bare ground, providing colonization sites for herbaceous species that are unavailable at the ungrazed site. Consequently, where cattle grazing was light, the density of herbaceous plants was higher than at ungrazed sites (p = 0.0317; Figure 16.2a; Holland 1994), and the mean number of species per 4-m² quadrat was significantly (p < 0.01) higher than on the ungrazed site in both 1993 and 1994.

Grazing by domestic livestock also can alter soil properties beneath shinnery oak communities. For example, goat grazing causes an increase in nitrogen, phosphorus, and potassium (Escobar et al. 1995), whereas light-intensity cattle grazing indirectly decreases the amount of calcium, magnesium, phosphorous, potassium, sodium, and organic matter in soil (Holland 1994; Table 16.1). Small animals also alter soil by creating patches that are high in moisture and soil nutrients, and so contribute to the success of many plant species in shinnery oak communities (S. Dhillion, M. McGinley, C. Friese and J. Zak, unpublished data). Disturbance sites created by ants, rabbits, and gophers favor germination and establishment of different grass species in the shinnery oak community (Dhillion et al. 1994; McGinley, Dhillion and Newmann 1994; Dhillion in press; N. Mehdiabadi, unpublished data). Disturbance mounds are richer in inorganic nutrients, litter, and mycorrhizal propagules, and they support grass seedlings with more nutrient uptake, root length, tillering, and mycorrhizal colonization than seedlings growing on undisturbed sites (McGinley, Dhillion and

Newmann 1994; Dhillion in press). Soil displacement presumably increases nutrient availability via inducing enhancement of microbial activity in concert with rapid nutrient turnover rates and increased infection of mycorrhizal fungi (Dhillion in press).

Management and Eradication: Grazing, Burning, and Herbicides

Goats occasionally are used to control shinnery oak (McIlvain 1954; Escobar et al. 1995), which provides good forage for these animals (Villena and Pfister 1990). Burning and shredding are not used today to control shinnery oak, because they stimulate the rhizomes to resprout, resulting in higher stem density the following year (McIlvain and Armstrong 1966; Pettit 1986). Since the 1940s, shinnery oak has been reduced or eradicated by herbicides due to its toxicity to cattle in the spring and to the belief that it competes with forage grasses (Deering 1972; Scifres 1972; Pettit 1979; Scifres, Stuth and Bovey 1981; Sears 1982; Jones 1982; Galbraith 1983; Pettit 1986; Sears et al. 1986). Herbiciding shinnery oak is considered beneficial because of the increase in the herbaceous component and the nutritive value of vegetation in the community (e.g., Robison and Fisher 1968; Scifres and Mutz 1978; Biondini 1980; Jacoby, Slosser and Meadors 1983). However, many workers have reported that eradication of shinnery oak is undesirable because it serves as cover and as a necessary forage for livestock and wildlife in times of drought (e.g., Rechenthin et al. 1964; Pettit 1979), and it prevents wind erosion of the soil (Test 1972).

The Future: Restoration, Conservation, and Management

Although shinnery oak communities are refugia for a number of animals, including some that are threatened or endangered, and are habitat to at least one federally endangered plant species, little has been done toward establishing a management plan. Lack of research has left the status of rare or endangered species uncertain. Presently, there is no regional or state conservation or restoration plan for shinnery oak communities.

Nevertheless, these habitats are regarded as extremely vulnerable to anthropogenic and natural disturbances (Bednarz, Hayden and Fischer 1990). Shinnery oak must be recognized for its ability to stabilize the erosion and large-scale land degradation that is occurring at an alarming rate on the southern high plains in Texas, New Mexico, and Oklahoma. These dune habitats provide poor agricultural landscapes and are inadequate for long-term heavy grazing. In contrast, these habitats are critical for the survival of wildlife dependent on its intact structure of rolling sandy plains with blowouts, dominant woody plants, and high diversity of herbaceous plants. It is important to note that the maintenance of the herbaceous flora has been linked to the presence of burrowing animals and their activities associated with soil displacement. Research also suggests that disturbances within the communities operate to maintain species diversity by providing open patches rich in biotic and abiotic resources for establishment of herbaceous plants. Not surprisingly, the habitat's survival, conservation, and restoration are linked to sustaining the relationships and interactions of its components.

Future Research Needs

Shinnery oak communities are unique and deserve international attention, but first they have to be acknowledged locally as worthy of protection. Considerable basic research is needed on this system, including vegetation, animal, and microbial surveys. Studies on species interactions are necessary to determine how these affect species diversity in shinnery communities. From the few studies

conducted to date, it is obvious that these interactions are very important in structuring and regulating the communities. Studies have examined the effect of anthropogenic reductions in shinnery oak cover on abundance of birds, small rodents, and some plants (e.g., Jacoby, Slosser and Meadors 1983; Haukos and Smith 1989; Willig et al. 1993); yet, the impact on the majority of the species is unclear. Surveys of plant and animal species composition and abundances have not been adequate to determine the status of all species in the community.

Accurate mapping is needed to determine the amount of shinnery oak present and the rate at which it is being destroyed over time due to land use. This information would be essential to determine the effects of climate change on the geographic distribution of shinnery oak. Studies also are needed to examine geographic and genetic variation in the shinnery oak communities from New Mexico to Oklahoma, and to determine the potential for hybridization between this species and other oak species.

Acknowledgments

The authors thank Mr. Johnnie Fitzgerald for allowing us to conduct research on his property in Yoakum County, Texas. He also graciously provided information that helped us gain insight into the history and dynamics of shinnery oak communities. Institutional support from Texas Tech University to M. H. M. and from the University of Oslo and the Agricultural University of Norway to S. S. D. is acknowledged.

REFERENCES

Allred, B. W. 1949. Distribution and control of several woody plants in Texas and Oklahoma. Journal of Range Management 2:17–29.

Bednarz, J. C., Hayden, T., and Fischer, T. 1990. The raptor and raven community of the Los Medaños Area in Southeastern New Mexico: A unique and significant resource. In Ecosystem management: Rare Species and significant habitats, ed. R. S. Mitchell, C. J. Sheviak, and D. J. Leopold, pp. 92–101. Proceedings of the 15th Annual Natural Areas Conference. Albany, N.Y.: New York State Museum Bulletin No. 471.

Best, T. L. 1986. Feeding ecology of mourning doves (*Zenaida macroura*) in southeastern New Mexico. *Southwestern Naturalist* 31:33–38.

Best, T. L., and Jackson, D. W. 1982. Statistical evaluation of plant density data collected at the Los Medaños Site, New Mexico (1978–1980). In *Ecosystem Studies at the Los Medaños Site, Eddy County, New Mexico*, ed. J. Braswell and J. S. Hart, Chap. 8, pp. 1–374. Albuquerque, N. Mex.: U.S. Department of Energy, WTSD-TME 3141.

Biggers, D. H. 1991. *Buffalo Guns and Barbed Wire. A Combined Reissue of Pictures of the Past and History That Will Never Be Repeated, by Don Hampton Biggers (1902)*. Lubbock, Tex.: Texas Tech University Press.

Biondini, M. E. 1980. Nutritive value of sandy land forages treated with different rates of tebuthiuron. M.S. thesis, Texas Tech University, Lubbock, Tex.

Bóo, R. M. 1974. Root carbohydrates in sand shinnery oak (*Quercus havardii* Rydb.). M.S. thesis, Texas Tech University, Lubbock, Tex.

Correll, D. S., and Johnston, M. C. 1979. *Manual of the Vascular Plants of Texas*. Renner, Tex.: Texas Research Foundation.

Deering, D. W. 1972. Effects of selected herbicides and fertilization on a sand shinnery oak community. M.S. thesis, Texas Tech University, Lubbock, Tex.

Deering, D., and Pettit, R. 1971. Sand shinnery oak acreage survey. *Research Highlights, Texas Tech University* 2:14.

Dhillion, S. S. In press. Environmental heterogeneity, animal disturbances, microsite characteristics and seedling establishment in a semi-arid community. *Restoration Ecology*.

Dhillion, S. S., McGinley, M. A., Friese, C. F., and Zak, J. C. 1994. Construction of sand shinnery oak communities of the Llano Estacado: animal disturbances, plant community structure, and restoration. *Restoration Ecology* 2:51–60.

Dickerson, R. L., Jr. 1985. Short duration versus continuous grazing on sand shinnery oak range. M.S. thesis, Texas Tech University, Lubbock, Tex.

Dick-Peddie, W. A. 1992. *New Mexico Vegetation:*

Past, Present, and Future. Albuquerque, N.Mex.: University of New Mexico Press.

Donaldson, D. D. 1969. Effect on lesser prairie chickens of brush control in western Oklahoma. Ph.D. dissertation, Oklahoma State University, Stillwater.

Dorr, L. J. 1994. Plants in peril, 21. *Callirhoe scabruiscula*. *Kew Magazine* 11:146–154.

Elias, T. S. 1980. *The Complete Trees of North America: Field Guide and Natural History*. New York: Van Nostrand Reinhold Co.

Escobar, E. N., McKinney, T. S., Moseley, M., Blackwell, R., Inglish, S., and Rose, M. 1995. Impact on sand shinnery (*Quercus havardii*) after three years of goat browsing in Western Oklahoma. *Journal of Animal Science* 73:134 (abstract).

Galbraith, J. M. 1983. Plant and soil water relationships following sand shinnery oak control. M.S. thesis, Texas Tech University, Lubbock, Tex.

Gennaro, A. L. 1982. Ecology of amphibians, reptiles, and mammals at the Los Medaños Waste Isolation Pilot Plant (WIPP) Project Area of New Mexico. In *Ecosystem Studies at the Los Medaños Site, Eddy County, New Mexico*, ed. J. Braswell and J. S. Hart, Chap. 6, pp. 1–42. Albuquerque, N.Mex.: U.S. Department of Energy, WTSD-TME 3141.

Giesen, K. M. 1994. Movements and nesting habitat of lesser prairie-chicken hens in Colorado. *Southwestern Naturalist* 39:96–98.

Gould, W. L., and Carlton, H. H. 1970. *Control of Shinnery Oak, Mesquite, and Cresosotebush in New Mexico*. New Mexico Inter-Agency Range Committee, Rep. No. 4., Agriculture Research Service, USDA.

Graff, P. S. 1983. Trampling effects on seedling production and soil strength under short duration and continuous grazing. M.S. thesis, Texas Tech University, Lubbock, Tex.

Green, F. E. 1951. Geology of sand dunes, Lamb and Hale counties [Texas]. M.S. thesis, Texas Technological College, Lubbock, Tex.

Haukos, D. A., and Smith, L. M. 1989. Lesser Prairie Chicken nest site selection and vegetation characteristics in tebuthiuron-treated and untreated sand shinnery oak in Texas. *Great Basin Naturalist* 49:624–626.

Holden, W. C. 1970. *The Espuela Land and Cattle Company: A study of a Foreign-Owned Ranch in Texas*. Austin, Tex.: Texas State Historical Association.

Holechek, J. L., Pieper, R. D., and Herbel, C. H. 1989. *Range Management: Principles and Practices.*

Englewood Cliffs, N.J.: Prentice Hall.

Holland, M. 1994. Disturbance, environmental heterogeneity, and plant community structure in a sand shinnery oak community. M.S. thesis, Texas Tech University, Lubbock, Tex.

Jackson, A. S., and DeArment, R. 1963. The lesser prairie chicken in the Texas panhandle. *Journal of Wildlife Management* 27:733–737.

Jacoby, P. W., Slosser, J. E., and Meadors, C. H. 1983. Vegetational responses following control of sand shinnery oak with tebuthiuron. *Journal of Range Management* 36:510–512

Jones, V. E. 1982. Effects of tebuthiuron on a sand shinnery oak (*Quercus havardii*) community. Ph.D. dissertation, Texas Tech. University, Lubbock, Tex.

Ligon, J. D., and Haydock, J. 1981. Densities and species composition of the avifauna of the Los Medaños WIPP Site, Southeastern New Mexico. In *Ecosystem Studies at the Los Medaños Site, Eddy County, New Mexico*, ed. J. S. Hart, Chap. 1, pp. 1–42. Albuquerque, N.Mex.: U.S. Department of Energy, WTSD-TME 3106.

Martin, W. C. 1982. Floristic studies at the Los Medaños Site. In *Ecosystem Studies at the Los Medaños Site, Eddy County, New Mexico*, ed. J. Braswell and J. S. Hart, Chap. 3, pp. 1–117. Albuquerque: U.S. Department of Energy, WTSD-TME 3141.

Mayes, S. G. 1994. Clonal population structure of *Quercus havardii* (sand-shinnery oak). M.S. thesis, Texas Tech University, Lubbock, Tex.

McGinley, M. A., Dhillion, S. S., and Newmann, J. 1994. Environmental heterogeneity and seedling establishment: ant–plant–microbe interactions. *Functional Ecology* 8:607–615.

McIlvain, E. H. 1954. Interim report on shinnery oak control studies in the Southern Great Plains. In *Proceedings 11th North Central Weed Control Conference*, pp.125–126, Fargo, N.D.

McIlvain, E. H., and Armstrong, C. G. 1966. A summary of fire and forage research on shinnery oak rangelands. *Proceedings 5th Annual Tall Timbers Fire Ecology Conference*, pp.127–129, Tall Timbers Research Station, Tallahassee, Fla.

Melton, F. A. 1940. A tentative classification of sand dunes, its application to dune history in the Southern High Plains. *Journal of Geology* 48:113–174.

Muller, C. H. 1951. The oaks of Texas. *Renner: Contributions from the Texas Research Foundation* 1:21–323.

National Oceanic and Atmospheric Administration.

1985. *Climates of the States*. 3rd ed. Detroit: Gale Research Company.

National Oceanic and Atmospheric Administration. 1993. *Climatological Data, United States by Sections* 98(13):3–4.

National Oceanic and Atmospheric Administration. 1994. *Climatological Data, United States by Sections* 99(13)3–4.

Oklahoma Forestry Division. 1981. *Forest Trees of Oklahoma*. State Department of Agriculture Publication No. 1. Oklahoma City.

Olawsky, C. D., and Smith, L. M. 1991. Lesser prairie-chicken densities on tebuthiuron-treated and untreated sand shinnery oak rangelands. *Journal of Range Management* 44:364–368.

Osborn, B. 1942. Prairie dogs in shinnery (oak scrub) savannah. *Ecology* 23: 110–115.

Pettit, R. D. 1977. The ecology and control of sand shin oak. In *Proceedings of the 15th Range Management Conference, Lubbock, Texas*, pp. 6–11.

Pettit, R. D. 1979. Effects of picloram and tebuthiuron pellets on sand shinnery oak communities. *Journal of Range Management* 32:196–200.

Pettit, R. D. 1986. *Sand Shinnery Oak: Control and Management*. Management Note No. 8, Contribution No. T-9-431, College of Agricultural Sciences, Texas Tech University, Lubbock, Tex.

Plumb, G. E., Jr.1984. Grazing management following sand shin oak control. M.S. thesis, Texas Tech University, Lubbock, Tex.

Rechenthin, C. A., Bell, H. M., Pederson, R. J., and Polk, D. B. 1964. *Grassland Restoration*. Part II. *Brush Control*. USDA Soil Conservation Service Unnumbered Bulletin. Temple, Tex.

Riley, T. Z., Davis, C. A., and Smith, R. A. 1993. Autumn–winter habitat use of lesser prairie-chickens (*Tympanuchus pallidicinctus*, Tetraonidae). *Great Basin Naturalist* 53:409–411.

Robison, E. D., and Fisher, C. E. 1968. Chemical control of sand shinnery oak and related forage production. In *Brush Research in Texas–1968*. College Station, Tex.: Texas Agricultural Experiment Station Bulletin PR-2583-2609:5-8.

Rowell, C. M. 1967. Vascular plants of the Texas Panhandle and South Plains. Ph.D. dissertation, Oklahoma State University, Stillwater. Okla.

Rydberg, P. A. 1901. The oaks of the Continental Divide north of Mexico. *Bulletin of the New York Botanical Garden* 2:187–233.

Scifres, C. J. 1972. Sand shinnery oak response to silvex sprays of varying characteristics. *Journal of Range Management* 25:464–466.

Scifres, C. J., and Mutz, J. L. 1978. Herbaceous vegetation changes following applications of tebuthiuron for brush control. *Journal of Range Management* 31:375–378.

Scifres, C. J., Stuth, J. W., and Bovey, R. W. 1981. Control of oaks (*Quercus* spp.) and associated woody species on rangeland with tebuthiuron. *Weed Science* 29:270–275.

Sears, W. E. 1982. Biomass and nitrogen dynamics of a herbicide converted sand shinnery oak community. M.S. thesis, Texas Tech University, Lubbock, Tex.

Sears, W. E., Britton, C. M., Webster, D. B., and Pettit, R. D. 1986. Herbicide conversion of a sand shinnery oak (*Quercus havardii*) community: Effects on nitrogen. *Journal of Range Management* 39:403–407.

Secor, J. B., Shamash, S., Smeal, D., and Gennaro, A. L. 1983. Soil characteristics of two desert plant community types that occur in the Los Medaños area of southeastern New Mexico. *Soil Science* 136:133–144.

Sims, P. L. 1989. Grasslands. In *North American Terrestrial Vegetation*, ed. M. G. Barbour and W. D. Billings, pp. 265–286. New York: Cambridge University Press.

Small, M. W. 1975. Selected properties of contiguous soils supporting and devoid of sand shinnery oak (*Quercus havardii* Rydb.). M.S. thesis, Texas Tech University, Lubbock, Tex.

Smith, H. N., and Rechenthin, C. A. 1964. *Grassland Restoration*. Part I. *The Texas Brush Problem*. Temple, Tex.: USDA Soil Conservation Service.

Sullivan, J. C. 1980. Differentiation of sand shinnery oak communities in West Texas. M.S. thesis, Texas Tech University, Lubbock, Tex.

Taylor, M. A., and Guthery, F. S. 1980. Fall–winter movements, ranges, and habitat use of lesser prairie chickens. *Journal of Wildlife Management* 44:521–524.

Test, P. S. 1972. Soil moisture depletion and temperature affected by sand shinnery oak (*Quercus havardii* Rydb.) control. M.S. thesis, Texas Tech University, Lubbock, Tex.

Villena, F., and Pfister, J. A. 1990. Sand shinnery oak as forage for Angora and Spanish goats. *Journal of Range Management* 43:116–122.

Vines, R. A. 1982. *Trees of North Texas*. Austin, Tex.: University of Texas Press.

Webb, W. M., and Guthery, F. S. 1982. Response of bobwhite to habitat management in northwest Texas. *Wildlife Society Bulletin* 10:142–146.

Whitford, W. G. 1982. Arthropod and decomposition studies at the WIPP Site. In *Ecosystem Studies at the Los Medaños Site, Eddy County, New Mexico*, ed. J. Braswell and J. S. Hart, Chap. 5, pp. 1–17. Albuquerque, N.Mex.: U.S. Department of Energy, WTSD-TME 3141.

Wiedeman, V. E. 1960. Preliminary ecological study of the shinnery oak area of western Oklahoma. M. S. thesis, University of Oklahoma, Norman, Okla.

Wildlife Diversity Program. *Birds of the Shinnery Oak–Grasslands*. Oklahoma City, Okla.: Oklahoma Department of Wildlife Conservation.

Wilhite, A. T., Jr. 1960. Range Management. In *Texas Soil Survey*, Soil Survey Staff, USDA Soil Conservation Service, Yoakum Co. Washington D.C.: U.S. Government Printing Office.

Willig, M. R., Colbert, R. L., Pettit, R. D., and Stevens, R. D. 1993. Response of small mammals to conversion of a sand shinnery oak woodland into a mixed mid-grass prairie. *Texas Journal of Science* 45:29–44.

York, J. C., and Dick-Peddie, W. A.. 1969. Vegetation changes in southern New Mexico during the past hundred years. In *Arid Lands in Perspective*, ed. W. G. McGinnies and B. J. Goldman, pp. 157–166. Tucson, Ariz.: University of Arizona Press.

17 Oak Savanna in the American Southwest

MITCHEL P. McCLARAN AND GUY R. McPHERSON

Introduction

Oak savanna in the southwestern United States and northern Mexico represents a diverse, widely distributed vegetation type. The northern Sierra Madre of Mexico has 41 species of oak (*Quercus* L.) (Nixon 1993), and in southeastern Arizona, plant species richness is greater in open oak woodland than in any other plant community (Whittaker and Niering 1975). In this chapter we focus on the southwestern United States, because there is little information about oak savanna in Mexico.

Current and Past Geographic Distribution

Oak savanna generally is considered a subset of the more extensive evergreen oak woodland, or encinal (D. E. Brown 1982; Rzedowski 1983). The savanna is characterized by tree canopy cover between 1% and 30% and is restricted to relatively low (1,100–2,200 m) and dry elevations, where encinal grades into desert grassland or mattoral (shrubland) (White 1948; Gentry 1957; Rzedowski 1983; McPherson 1992; McClaran 1995) (Figure 17.1). Savanna tends to be located at lower elevations as the latitude increases. In Arizona, New Mexico, and northern Sonora, the savanna generally has a discontinuous distribution in the foothills of isolated mountain ranges that are separated by lowlands covered by desert grassland and shrubland. In the northern part of the Sierra Madre Occidental, in Chihuahua, Durango, Sinaloa, and central-southern Sonora, the savanna forms a somewhat more continuous distribution in large areas of rolling topography above desert grasslands or mattoral (Rzedowski 1983) (Figure 17.2).

Encinal occupies over 7 million ha of wildlands in northern Mexico (Sharman and Ffolliott 1992) and about 0.5 million ha in the southwestern United States (McClaran, Allen and Ruyle 1992). Descriptions generally do not differentiate between the relatively closed-canopy encinal and more open savanna, but we estimate that savanna comprises about 20%, or 1.5 million ha, of the encinal.

Analysis of ground-based, repeat photography suggests that since 1890 the savanna in southern Arizona has retreated upslope approximately 100 m (Hastings and Turner 1965). Analysis of more extensive evidence from land survey records and repeat aerial photography by Bahre (1991) found no consistent, permanent change in the encinal distribution since 1872. Bahre noted that encinal was temporarily removed near mines and settlements between 1880 and 1910 but has since reestablished.

Stable carbon isotope analysis of soil organic carbon suggests that between 700 and 1,700 yr BP, savanna trees moved downslope about 50 m in the Huachuca Mountains of southern Arizona (McClaran and McPherson 1995). However, the distribution of oak savanna during geologic time is poorly understood because the fossil record cannot describe the density or cover of trees.

Figure 17.1. Emory oak (*Quercus emoryi* Torr.) and perennial grasses form a southwestern oak savanna on the southern slopes of the Huachuca Mountains in southeastern Arizona.

Post-Cretaceous development of regional floras indicates that the Madro-Tertiary geoflora was ancestral to the current Sierra Madrean flora as well as to most floras of the Great Basin and California (Axelrod 1958). Many species associated with southwestern oak savanna have been present in southwestern North America for at least 70 million years.

Climate, Geology, and Soils

Climate in the oak savanna is hot and dry, with summer daytime air temperatures often in excess of 35 °C. However, freezing temperatures occur each winter, with an average frost-free period of about 200 days in northern areas and over 300 days in southern areas (D. E. Brown 1982).

Average annual precipitation varies from 350 to 600 mm and is bimodally distributed (White 1948; Gentry 1957; National Oceanic and Atmospheric Administration 1995). In the north, about one-half of the annual precipitation occurs with brief, high-intensity convective storms during July, August, and September. Most of the remaining precipitation is from longer, low-intensity frontal storms in winter. Some snow falls nearly every year. From north to south, there is a pattern of decreasing winter precipitation, increasing summer precipitation, and warmer winter temperatures (Rzedowski 1983).

In the north, historic changes in the spatial distribution and extent of oak savanna have been associated with variability in the seasonal distribution of precipitation. Davis (1994) used fossil pollen to suggest that a period of greater summer precipitation than is currently received occurred from 700 to 1,350 yr BP. Downslope movement of oaks into desert grassland from 700 to 1,700 yr BP (McClaran and McPherson 1995) corresponds with this period of higher summer precipitation. Recently, oak top-kill or mortality was associated with periods of reduced summer precipitation (Leopold 1924; Marshall 1957; Hastings and Turner 1965).

Changes in the seasonal distribution of precipitation are associated with atmospheric circulation patterns at the global level. For example, strengthening of the El Niño Southern Oscillation contributes to increased winter precipitation. Increased summer precipitation occurs when a strong Bermuda high pressure cell (Sellers and Hill 1974; Carleton 1985; Harrington, Cerveny and Balling 1992), or subtropical ridge, is displaced slightly north of its usual location, thereby contributing to greater advection of moisture from the Pacific Ocean (Carleton, Carpenter and Weser 1990).

Oak savanna occurs on level areas and slopes up to 80%, on all aspects, and on several substrates. Oak savanna is associated with a large variety of geologic substrates and geomorphic surfaces in the Sierra Madre Occidental and southern Rocky Mountains, including granite, limestone, volcanic alluvial deposits, alluvial fans, and valley bottoms (Ortlieb and Roldan-Q. 1981).

Most oak savanna soils are shallow and rocky, although occasionally depth may exceed 6 m. The typical soils are classified as Aridisols (Hendricks 1985), which indicates they are dry more than 6 months each year and contain little organic matter in the A horizon (Birkelund 1974). Some oak savanna may occur on Mollisols (Medina 1987), which are relatively well developed soils commonly associated with grassland.

Vegetation Structure and Function

On a regional scale, the encinal and savanna in the southwestern United States represent a depauperate form of the more *Quercus*-rich vegetation in Mexico (D. E. Brown 1982). For example, the Sierra Madre flora includes over 150 species of oak (Rzedowski 1983), but only 14 species are

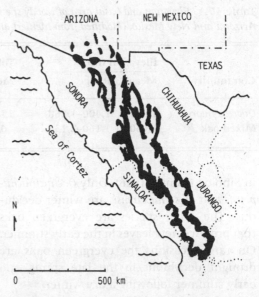

Figure 17.2. Distribution of encinal in the American Southwest. (Adapted from D. E. Brown and Lowe 1980 and Rzedowski 1983.)

found in the United States (Wooten and Standley 1915; Kearney and Peebles 1951).

On a landscape scale, the number of oak species increases with elevation in all areas of the savanna and encinal (Wallmo 1955; Marshall 1957; Whittaker and Niering 1964, 1965; Goldberg 1982; Niering and Lowe 1984; Rzedowski 1983). In addition, soil conditions may influence species distributions. Goldberg (1982) found that evergreen oaks can be restricted to nutrient-poor soils in Mexican encinal.

Emory oak (*Quercus emoryi* Torr.) is present nearly throughout the savanna in the United States and Mexico. In the United States and northern Mexico, it is associated with Mexican blue oak (*Q. oblongifolia* Torr.), Arizona white oak (*Q. arizonica* Sarg.), and grey oak (*Q. grisea* Liebm.) (Lasueur 1945; White 1948; D. E. Brown 1982; McPherson 1992). Further south in Chihuahua (Shreve 1939; Lasueur 1945) and Durango (Gentry 1957), associated oaks are Chihuahua oak (*Q. chihuahensis* Trel.), *Q. chuchuichupensis* C.H. Mull., and *Q. santaclarensis* C.H. Mull.; cusi (*Q. albocincta* Trel.) is a common associate

Table 17.1. *Elevation and basal area of woody species for two southwestern oak savanna communities in Arizona and New Mexico. (Modified from Mehlert and McPherson 1996.)*

Community	Elevation (m)		Total basal area (m²/ha)		*Quercus* basal area (m²/ha)	
	Mean	Range	Mean	Range	Mean	Range
Quercus emoryi	1,620	1,400–1,860	15.3	7.5–27.8	13.5	7.5–24.1
Mixed oak	1,660	1,160–2,130	6.9	0.6–19.3	5.1	0.6–18.9

in Sinoloa (Gentry 1946). Only *Q. chuchuichupensis* and *Q. santaclarensis* are winter deciduous. The other species are evergreen oaks that produce new leaves in the early summer. On rare occasions the evergreen oaks are drought deciduous in the late spring and early summer following a dry winter.

Between 11 and 15 encinal and savanna communities were identified in Arizona and New Mexico (Mehlert and McPherson 1996). Two of the communities represented southwestern oak savanna (Table 17.1): a nearly pure Emory oak savanna, and a mixed-oak community characterized by some combination of Emory oak, Arizona white oak, Mexican blue oak, grey oak, junipers (*Juniperus deppeana* Steud., *J. monosperma* (Engelm.) Sarg., *J. scopulorum* Sarg., or *J. osteosperma* (Torr)) and pinyon pine (*Pinus cembroides* Zucc.).

The diameter size-class distribution of Arizona white and Emory oaks in the United States savanna has the greatest number of individuals in the 10–30-cm classes, and there is a negative exponential decline of individuals in progressively larger size classes (Sanchini 1981). There is no evidence of major gaps in size or age classes (Sanchini 1981; Borelli, Ffolliott and Gottfried 1994) such as those reported for the Californian oak savanna (McClaran and Bartolome 1989). Absence of gaps in size class distribution may result from prolific resprouting from the trunk base when trees are cut or burned (Babb 1992; Caprio and Zwolinski 1992) and from establishment of new plants from acorns (Nyandiga and McPherson 1992).

Herbaceous species in oak savannas include perennial bunchgrasses and several forb species (White 1948; Gentry 1957; D. E. Brown 1982; McPherson 1992, 1994). Nonnative species are nearly absent. Most grasses have a C4 photosynthetic pathway. The common grasses are Texas bluestem (*Andropogon cirratus* Hack.), threeawns (*Aristida* L. spp.), sideoats grama (*Bouteloua curtipendula* (Michx.) Torr.), blue grama (*B. gracilis* (H.B.K.) Lag. ex Stend.), hairy grama (*B. hirsuta* Lag.), purple grama (*B. radicosa* (Fourn.) Griffiths), plains lovegrass (*Eragrostis intermedia* Hitchc.), Mexican lovegrass (*E. mexicana* (Hornem.) Link), green sprangletop (*Leptochloa dubia* (H.B.K.) Ness.), bullgrass (*Muhlenbergia emersleyi* Vasey), pinyon ricegrass (*Piptochaetium fimbriatum* (H.B.K.) Hitchc.), squirreltail (*Sitanion hystrix* (Nutt.) J.G. Smith), crinkleawn (*Trachypogon montufari* (H.B.K.) Ness.), and Mexican gamagrass (*Tripsacum lanceolatum* Rupr.). Several species of annual forbs (e.g., *Helianthus petiolaris* Nutt.) emerge in the early spring, following the cessation of freezing temperatures. A second group of forbs (e. g., *Portulaca cornata* Small) emerges coincident with the summer rains (McPherson 1994).

Species composition of herbaceous plants does not differ between areas beneath trees and those away from them in the savanna, but herbaceous biomass is about 40% less under trees (Haworth and McPherson 1994). Soil type and slope significantly influence herbaceous biomass production: sandy loam soils with 1%–15% slopes have the greatest production (1,100–2,800 kg/ha), whereas rocky soils on steep slopes have the lowest production (220–350 kg/ha) (McClaran, Allen and Ruyle 1992).

We estimated the relative contribution of trees and herbaceous plants to total biomass in an oak savanna using stable carbon isotope analysis of soil organic carbon (SOC). The relative isotopic composition (^{12}C and ^{13}C isotopes, expressed as δ^{13}C) in plant biomass is related to the photosynthetic pathway: C3 plants average -27‰ δ^{13}C and C4 plants average -12‰ δ^{13}C. The δ^{13}C value of SOC is approximately equal to the plant carbon deposited on and in the soil layers (Boutton 1991). In the oak savanna, 92% of the herbaceous biomass has a C4 pathway (Haworth and McPherson 1994), and the oak trees have a C3 pathway. Therefore, the δ^{13}C value provides an estimate of the relative contribution of tree and herbaceous biomass to SOC. Values of -13‰, -20‰, and -27‰ suggest 100% herbaceous biomass, equal biomass of herbaceous and tree forms, and 100% tree biomass, respectively.

In the winter of 1992, we collected soil samples in an oak savanna in San Rafael Valley, Arizona, with 15% tree cover and a herbaceous layer of C4 perennial grasses and only scattered C3 herbaceous plants. Soil samples were collected from five depths (0–5, 20–25, 45–50, 70–75, and 95–100 cm) at 20 sites in the savanna. Sites were selected in a stratified-random manner so that 15% of the sites (n = 3) were beneath trees, and the remainder were more than 5 m from all tree canopies. All sites were approximately 50–75 m apart. Methods of sample preparation and isotopic analysis in this study followed those used by McClaran and McPherson (1995).

There was no difference (ANOVA, P>0.20) in the δ^{13} values among soil depths. The mean (± standard deviation) of the δ^{13}C values for each depth were -19.2‰ (±2.0) at 0–5 cm, -17.3‰ (±1.7) at 20–25 cm, -17.4‰ (±1.8) at 45–50 cm, -18.1‰ (±3.0) at 70-75 cm, and -18.6‰ (±3.4) at 95–100 cm. These results suggest that there is (1) nearly equal contri-bution of tree (about a 55% contribution) and herbaceous (about a 45% contribution) biomass to SOC at the soil surface, and (2) deep infiltration of herbaceous SOC to soil depths, where tree roots presumably are the only living plant matter.

Ecological Relationships

The stability of grass–tree ratios is critical to ecological relations because the proportion of trees and grasses within a savanna has broad implications for management and land use. Life-form changes affect virtually all resources, including wildlife, water, livestock, fuelwood, and recreation. Therefore, our discussion focuses on factors controlling the grass–tree ratio, particularly (1) abiotic factors, (2) grass–tree interactions, and (3) animal activities.

Abiotic Factors

Resource partitioning often is suggested to explain the stability of grass–tree ratios in tropical and subtropical savannas (e.g., Walter 1954, 1979; Knoop and Walker 1985; Sala et al. 1989; J. R. Brown and Archer 1990; Bush and Van Auken 1991). Bimodal precipitation distribution is thought to allow stable coexistence of woody plants and grasses – woody plants use moisture that percolates through surface layers when grasses are dormant, whereas grasses exploit growing-season precipitation (Neilson 1986; Archer 1989; Lauenroth et al. 1993). Thus, shifts in the seasonality of precipitation could have profound impacts on grass–tree ratios in savannas (Weltzin and McPherson 1995; Weltzin 1998). In southwestern oak savanna, oak seedling emergence is constrained during most years by low soil moisture; seedlings rarely emerge without supplemental water, regardless of soil type (Germaine et al. 1997).

Savanna genesis and maintenance also have been interpreted on the basis of soil morphology. It has been suggested that argillic (clay-rich) horizons in the soil constrain establishment and development of woody plants in the American Southwest, thereby contributing to scattered, low-statured woody plants and trees (McAuliffe 1994,. 1995). McAuliffe hypothesized that water-impermeable argillic horizons (1) reduce water availability to woody plants in summer below thresholds necessary for survival (clay particles bond tightly with water), or (2) result in perched water tables near the surface in the winter, which may contribute to woody plant mortality (the clay layer impedes infiltration). Argillic soils generally are found in widely dispersed patches of 5–500 ha. In southeastern Arizona, about 10% of Emory oak seedlings persist for more than 2 years in shallow (<0.5 m), clayey soils, whereas 20% persist in deep (>2.0 m), loamy soils (Germaine 1997).

Effects of soil–water relations on oak savanna may be mediated by the fire regime. Less than 25% of Arizona white oak trees, and 5%–15% of Emory oak and Mexican blue oak trees, were killed by a single fire. Trees less than 10 cm in diameter were more likely to resprout than larger trees (Babb 1992; Caprio and Zwolinski 1992; Caprio 1994). In addition, Mexican blue oak was less likely to be top-killed by fire than Emory oak (Caprio and Zwolinski 1995). In contrast to the negative effects of fire on oak, grass cover and productivity were minimally affected (Bock and Bock 1992; McPherson 1995). If fire-induced mortality of oak trees is approximately balanced by recruitment, then periodic fires may contribute to stable grass–tree mixtures in oak savanna over long temporal scales. The hypothesis that oak savannas are maintained, at least partially, by periodic fires is supported by the increased density of oak trees within the savanna following the abrupt cessation of periodic fires subsequent to Anglo settlement (Bahre 1991).

Grass–Tree Interactions

Interference from herbaceous plants can constrain oak seedling recruitment in savan-

nas (Weltzin and McPherson 1994, 1995). The primary effects of interference appear to occur belowground (McPherson 1993). Herbaceous biomass is reduced beneath oak trees compared to open areas, but as previously noted, there is no change in species composition beneath trees (Haworth and McPherson 1995).

The relationship between woody and herbaceous plants in oak savanna is mutually detrimental (i.e., competitive), at least in the seedling stage (Figure 17.3). Sapling-sized oaks have significant impacts on their physical environment, and these impacts increase as trees become larger. Compared to treeless areas, the environment under mature trees receives 90% less incident light (Jake Weltzin, University of Arizona, personal communication), soil temperatures are warmer in winter and cooler in summer, and soil nitrogen (total) and organic carbon are greater (McPherson, Boutton and Midwood 1993). Stem-flow redistributes precipitation from the canopy to the main stems, resulting in a concentration of soil water near the tree base (Haworth and McPherson 1995). These differences in microclimate and nutrient concentrations beneath savanna trees compared to tree-free areas are common throughout the world (Belsky and Canham 1994).

Animal Activities

Demographic studies indicate that invertebrate herbivory is a common source of oak seedling mortality (McPherson 1993; Peck and McPherson 1994). However, climatic and edaphic factors probably interact with interference from neighboring plants and herbivory to constrain oak recruitment (Weltzin and McPherson 1995). Vertebrates, especially birds, consume large numbers of acorns. Acorn predation from migratory gray-breasted jays (*Aphelocoma ultramarina*) and white-winged doves (*Zenaida asiatica*) is particularly high during summer. However, jays also cache acorns in locations suitable for germination, thereby providing the primary mechanism for long-distance dispersal for oaks (Hubbard and McPherson 1997).

Role of Humans

Humans have influenced the oak savanna for more than 11,000 years (Propper 1992). For example, hunting reportedly was intensive enough to drive the Pleistocene megafauna to extinction by 11,000 yr BP (Martin 1963, 1984). In some limited areas of the savanna, domesticated plants were used 2,500 yr BP; significant cultural developments, including relatively permanent settlements, were present by 2,000 yr BP; and irrigation technology was present 700–1,400 yr BP (Propper 1992; Huckell 1995).

In southeastern Arizona, the role of humans in the historic period includes the activities of three Native American groups: Sobaipuri, Tohono O'Odam, and Apache (Propper 1992). The agrarian–hunter–gatherer Sobaipuri and Tohono O'Odam ventured into the savanna from lower elevations to gather acorns for food and wood for fuel, and to hunt for game animals. The hunter–raider Apache resided primarily in the savanna and woodland.

Raids by the Apache greatly limited use and settlement of the savanna by other groups during the Spanish and Mexican settlement periods and in the initial years of the United States settlement period (Officer 1987; Bahre 1991). Spanish exploration began in 1540 with the Coronado Expedition, and settlements were established in the mid-1700s. Settlement in the Mexican period began in 1821. The United States period of settlement started in 1854 with the Gadsden Purchase of land south of the Gila River to the current International Border.

Following the elimination of Apache raids in the late 1880s, the extent and intensity of human influences, particularly mining and livestock grazing, increased greatly. The development of silver and copper mining opera-

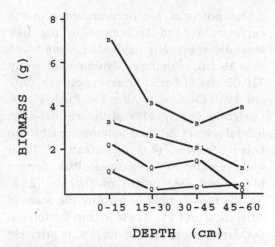

Figure 17.3. Belowground biomass allocation of *Quercus emoryi* Torr. and *Bouteloua curtipendula* (Michx.) Torr. grown together and separately. Q = *Quercus* grown alone; Q' = *Quercus* grown with *Bouteloua*; B = *Bouteloua* grown alone; and B' = *Bouteloua* grown with *Quercus*. Plants were grown for 14 months in pots of native soil (20 cm diameter, 60 cm deep). The pots were placed in the treeless areas of a *Q. emoryi* savanna.

tions resulted in considerable wood cutting for fuel and timber. Between 1875 and 1900, there were 27 active mines in southeastern Arizona, all of which used oak and other trees to fuel smelting furnaces and stamp mills. Bahre and Hutchinson (1985) estimated that between 1878 and 1886, nearly 48,000 cords were used for industrial purposes and another 31,000 cords were used domestically in the Tombstone area. Evidence of harvesting intensity during the mining period is still present: 43% of existing oak trees within 40 km of Tombstone originated as sprouts from the cut stumps (Bahre and Hutchinson 1985).

Livestock numbers in southeastern Arizona burgeoned to more than 200,000 in 1890, diminished due to extreme drought in 1891–92, and increased again by the first decade of the 20th century (Bahre 1995). Use of the savanna by livestock during this period was limited because of the limited availability of drinking water (McClaran, Allen and Ruyle 1992).

The potential for uncontrolled resource exploitation and settlement in the oak savanna was greatly reduced in Arizona and New Mexico with the establishment of nearly 320,000 ha of Forest Reserves between 1902 and 1910 (McClaran, Allen and Ruyle 1992). Currently, nearly 85% of the encinal is in federal ownership and administered by the Forest Service (73%), Bureau of Land Management (6%), National Park Service (4%), and Department of Defense (2%). Nearly 11% is held in trust by the State of Arizona; about 1% is held in trust for Tohono O'Odam people; and about 3% is privately owned (McClaran, Allen and Ruyle 1992). In comparison to the encinal, it is likely that private ownership of the savanna is on the order of 10%–15% and Forest Service ownership is about 60%.

Fuelwood harvest on public lands has been negligible for the last 50 years (Bennett 1992, 1995), and it pales in comparison to the cutting during the mining era. A similar pattern of fuelwood harvesting has taken place in Mexico. However, in Mexico the mining period started earlier in some areas, and reduced cutting has occurred only since the 1970s because alternative fuels have only recently become available (Rzedowski 1983).

Livestock use in the savanna and encinal administered by the Forest Service generally has declined since the establishment of the Forest Reserves. Grazing has spread from areas near natural water sources (streams and springs) to more remote areas following the construction of livestock watering developments, fences, and other infrastructures between 1930 and 1950 (Allen 1989). Grazing occurs on about 220,000 ha of savanna and woodland administered by the Forest Service, and the average stocking rate of 1 cow ha^{-1} mo^{-1} results in about 35%–55% utilization of aboveground herbaceous biomass (McClaran, Allen and Ruyle 1992). Cattle graze these areas all year, or they are rotated seasonally among different areas. Between November and January, oak can account for up to 25% of the cattle diet, because the dormant herbaceous biomass is low in protein and phosphorus (McClaran, Allen and Ruyle 1992). Nevertheless, in Arizona, there is no consistent browseline created by cattle (McClaran, Allen and Ruyle 1992). Furthermore, there is no difference in seedling recruitment between ungrazed and grazed areas, except beneath tree canopies, where seedling density is greater in ungrazed areas (Weltzin and McPherson 1994).

Prehistorically, fires occurred every 1–2 decades on a site (McPherson 1992); currently, fires rarely or never occur on most sites. Livestock-induced reductions in fine fuel, combined with construction of firebreaks (e.g., roads, pastures, housing developments), and effective fire suppression during and after the settlement period have contributed to reduced fire frequency.

The increase in recreational use of public and some private savanna that has occurred in the past 20 years has stimulated an ecotourism economy in southern Arizona. Deer and quail hunting long has been the dominant recreational use. However, an increasing number of people flock to the savanna to enjoy cooler temperatures, observe a variety of uncommon bird species, and escape the stress of urban life (McClaran, Allen and Ruyle 1992).

Residential development has increased in the savanna and adjacent desert grassland in the last 20 years. Sierra Vista, Sonoita, Nogales, and Patagonia are some of the fastest growing towns in Arizona and they are located in, or adjacent to, the savanna (Bahre 1995).

Threats to Preservation

Federal ownership of over 60% of the area largely ensures preservation of the oak savanna in Arizona. Forest Service activities are judged in reference to conservation of the native vegetation (Allen 1989), and that practice will probably continue. However, residential development on privately owned savanna, the possibility of land sales by the State of Arizona, and residential development adjoining the savanna will threaten the integrity of the remaining savanna. These activities will further constrict landscape-level processes such as fire, water drainage patterns, animal movements, and seed dispersal.

The increasing residential development and growing market for 5–20-ha ranchette parcels is reducing the stability of the livestock grazing industry, and this reduction could have a positive feedback on the rate of residential development (McClaran, Allen and Ruyle 1992). Greater residential development results in higher property values and property taxes, and increases logistical problems from land use conflicts. The increased land values reduce the ability to pay a mortgage with income from cattle sales, and the lure of land-sale profits can result in more rapid parcelization. Increased taxes can preclude profitability of a livestock operation, but some tax relief is available through differential property taxation for agricultural land in Arizona (McClaran, Allen and Ruyle 1992).

Future Research Needs

Understanding what constrains the establishment of woody plants in oak savannas is fundamental to predicting the response to management because the grass–tree ratio influences many elements of the savanna. To this end, future research should describe how the grass–tree interactions are affected by native herbivores, land use and land tenure, and potential changes in global atmospheric chemistry and climate. Limited research suggests that vertebrates and invertebrates form seasonally important constraints on oak seedling survival (McPherson 1993; Peck and McPherson 1994). However, the spatial extent and importance of these mortality vectors, as well as the identity of the particular herbivore species responsible, for the observed mortality, have not been determined. Understanding how land use and tenure changes can influence the dynamics of landscape-level processes controlling grass–tree ratios will become increasingly important as human numbers and behaviors change. Increased atmospheric CO_2 may become an important factor controlling grass–tree ratios in the savanna, particularly if the increase is accompanied by shifts in seasonal precipitation or other climatic factors (Weltzin and McPherson 1995). Predicting changes in grass–tree ratios will be improved if experiments are undertaken that focus on interactions between various atmospheric factors (e.g., concentrations of greenhouse gases) and climatic factors (e.g., temperature, precipitation).

Finally, increasing the amount of information available about the oak savanna in Mexico is critical, because the majority of the savanna is in Mexico.

Summary

The southwestern oak savanna covers about 1.5 million ha at the lower and drier elevations of the encinal or mixed evergreen woodland in the southwestern United States and northern Mexico. The evergreen Emory oak is the most common and widely distributed oak species, and C4 perennial grasses dominate the understory and grassy interspaces in the savanna.

The ratio of grass and tree life-forms in the savanna is influenced by amount and seasonal distribution of precipitation, clay concentration in the soil, fire, negative influences of grass on tree seedlings and trees on grass production, acorn predation by birds and invertebrates, acorn dispersal by birds, and browsing of seedlings by a variety of herbivores.

In the United States, the establishment of Forest Reserves by the federal government between 1902 and 1910 and the continued administration of the land by the U.S. Forest

Service have reduced the level of human impact on more than 60% of the savanna. Continued administration by the Forest Service should restrict future residential development to private lands at the lowest elevations of the savanna, but this may create significant discontinuities in the distribution of the savanna.

Future research to improve our understanding of the savanna in Mexico should be a priority.

Acknowledgments

Isotopic analysis was supported by the USGS Biological Research Division Global Change Research Program. The clarity and depth of this chapter were enhanced by contributions from Drs. Barbara Allen-Diaz and James W. Bartolome and graduate students in the Oak Savanna Seminar at the University of California, Berkeley, during fall 1995. Phil Jenkins assisted with scientific names and authorities. Susan Jorstad drafted Figure 17.2.

REFERENCES

Allen, L. S. 1989. Livestock and the Coronado National Forest. *Rangelands* 11:9–13.

Archer, S. A. 1989. Have southern Texas savannas been converted to woodlands in recent history? *The American Naturalist* 134:545–561.

Axelrod, D. I. 1958. Evolution of the Madro-Tertiary geoflora. *The Botanical Review* 24:433–509.

Babb, G. D. 1992. Sprouting response of *Quercus arizonica* and *Quercus emoryi* following fire. M.S. thesis, University of Arizona, Tucson, Ariz.

Bahre, C. J. 1991. *A Legacy of Change: Historic Human Impact on Vegetation in the Arizona Borderlands*. Tucson, Ariz.: University of Arizona Press.

Bahre, C. J. 1995. Human impacts on the grasslands of southeastern Arizona. In *The Desert Grassland,* ed M. P. McClaran and T. R. Van Devender, pp. 230–264. Tucson, Ariz.: University of Arizona Press.

Bahre, C. J., and Hutchinson, C. F. 1985. The impact of historical fuelwood cutting on the semidesert woodlands of southeastern Arizona. *Journal of Forest History* 29:175–186.

Belsky, A. J., and Canham, C. D. 1994. Forest gaps and isolated savanna trees: an application of patch dynamics in two ecosystems. *BioScience* 44:77–84.

Bennett, D. A. 1992. Fuelwood extraction in southeast Arizona. In *Proceedings of the Symposium on Ecology and Management of Oak and Associated Woodlands: Perspectives in the Southwestern United States and Northern Mexico,* tech. coords. P. F. Ffolliott, G. J. Gottfried, D. A. Bennett, V. M. Hernandez C., A. Ortega-Rubio, and R. H. Hamre, pp. 96–97. USDA Forest Service, Gen. Tech. Rep. RM-218.

Bennett, D. A. 1995. Fuelwood harvesting in the sky islands of southeastern Arizona. In *Biodiversity and Management of the Madrean Archipelago: The Sky Islands of Southwestern United States and Northwestern Mexico,* tech. coords. L. F. DeBano, G. J. Gottfried, R. H. Hamre, C. B. Edminster, P. F. Ffolliott, and A. Ortega-Rubio, pp. 519–523. USDA Forest Service, Gen. Tech. Rep. RM-264.

Birkelund, P. W. 1974. *Pedology, Weathering, and Geomorphological Research*. New York: Oxford University Press.

Bock, J. H., and Bock, C. E. 1992. Short-term reductions in plant densities following prescribed fire in an ungrazed semidesert shrub–grassland. *The Southwestern Naturalist* 37:49–53.

Borelli, S., Ffolliott, P. F., and Gottfried, G. J. 1994. Natural regeneration in encinal woodlands of southeastern Arizona. *The Southwestern Naturalist* 39:179–183.

Boutton, T. W. 1991. Stable carbon isotope ratios of natural materials: II. Atmospheric, terrestrial, marine, and freshwater environments. In *Carbon Isotope Techniques,* ed. D. C. Coleman and B. Fry, pp.173–185. New York: Academic Press.

Brown, D. E., ed. 1982. Biotic communities of the American Southwest – United States and Mexico. *Desert Plants* 4:1–342.

Brown, D. E., and Lowe, C. H. 1980. *Biotic Communities of the Southwest*. USDA Forest Service, Gen. Tech. Rep. RM-78.

Brown, J. R., and Archer, S. 1990. Water relations of a perennial grass and seedlings versus adult woody plants in a subtropical savanna, Texas. *Oikos* 57:366–374.

Bush, J. K., and Van Auken, O. W. 1991.

Importance of time of germination and soil depth on growth of *Prosopis glandulosa* (Leguminosae) seedlings in the presence of a C$_4$ grass. *American Journal of Botany* 78:1732–1739.

Caprio, A. C. 1994. Fire effects and vegetation response in a Madrean oak woodland, southeastern Arizona. M.S. thesis, University of Arizona, Tucson, Ariz.

Caprio, A. C., and Zwolinski, M. J. 1992. Fire effects on two oak species, *Quercus emoryi* and *Q. oblongifolia*, in southeastern Arizona. In *Proceedings of the Symposium on Ecology and Management of Oak and Associated Woodlands: Perspectives in the Southwestern United States and Northern Mexico*, tech. coords. P. F. Ffolliott, G. J. Gottfried, D. A. Bennett, V. M. Hernandez C., A. Ortega-Rubio, and R. H. Hamre, pp. 150–154. USDA Forest Service, Gen. Tech. Rep. RM-218.

Caprio, A. C., and Zwolinski, M. J. 1995. Fire and vegetation in a madrean oak woodland, Santa Catalina Mountains, southeastern Arizona. In *Biodiversity and Management of the Madrean Archipelago: The Sky Islands of Southwestern United States and Northwestern Mexico*, tech. coords. L. F. DeBano, G. J. Gottfried, R. H. Hamre, C. B. Edminster, P. F. Ffolliott, and A. Ortega-Rubio, pp. 389–398. USDA Forest Service, Gen. Tech. Rep. RM-264.

Carleton, A. M. 1985. Synoptic and satellite aspects of the southwestern U.S. summer 'monsoon.' *Journal of Climatology* 5:389–402.

Carleton, A. M., Carpenter, D. A., and Weser, P. J. 1990. Mechanisms of interannual variability of the Southwest United States summer rainfall maximum. *Journal of Climate* 3:999–1015.

Davis, O. K. 1994. The correlation of summer precipitation in the southwestern U.S.A. with isotopic records of solar activity during the Medieval Warm Period. *Climatic Change* 26:271–287.

Gentry, H. S. 1946. Sierra Tauicharmona – a Sinoloa plant locale. *Bulletin of the Torrey Botanical Club* 73:356–362.

Gentry, H. S. 1957. *Los Pastizales de Durango, Estudio Ecologico, Fisiografico y Floristico*. Mexico City: Ediciones del Instituto Mexicano de Recursos Naturales Renovables.

Germaine, H. L. 1997. Constraints on establishment of Emory oak at lower treeline. M.S. thesis, University of Arizona, Tucson, Ariz.

Germaine, H. L., McPherson, G. R., Rojahn, K., Nicholas, H., and Weltzin, J. 1997. Constraints on germination and emergence of Emory oak. In *Oak Woodland Ecology, Management, and Urban Interface Issues*, tech. coords. N. H. Pillsbury, J. Verner, and W. D. Tieje, pp. 225–230. USDA Forest Service, Gen. Tech. Rep. PSW-160.

Goldberg, D. E. 1982. The distribution of evergreen and deciduous trees relative to soil type: an example from the Sierra Madre, Mexico, and a general model. *Ecology* 63:942–951.

Harrington, J. A., Jr., Cerveny, R. S., and Balling, R. C., Jr. 1992. Impact of the Southern Oscillation on the North American Southwest monsoon. *Physical Geography* 13:318–330.

Hastings, J. R., and Turner, R. M. 1965. *The Changing Mile*. Tucson, Ariz.: University of Arizona Press.

Haworth, K., and McPherson, G. R. 1994. Effects of *Quercus emoryi* on herbaceous vegetation in a semi-arid savanna. *Vegetatio* 112:153–159.

Haworth, K., and McPherson, G. R. 1995. Effects of *Quercus emoryi* trees on precipitation distribution and microclimate in a semi-arid savanna. *Journal of Arid Environments* 31:153–170.

Hendricks, D. M. 1985. *Arizona Soils*. Tucson, Ariz.: University of Arizona, College of Agriculture.

Hubbard, J. A., and McPherson, G. R. 1997. Acorn selection by Mexican jays: a test of a tri-trophic relationship hypothesis. *Oecologia* 110:143–146.

Huckell, B. B. 1995. *Of Marshes and Maize: Preceramic Agricultural Settlement in the Cienega Valley, Southeastern Arizona*. Anthropological Papers of the University of Arizona Number 59. Tucson, Ariz.: University of Arizona Press.

Kearney, T. H., and Peebles, R. H. 1951. *Arizona Flora*. Berkeley: University of California Press.

Knoop, W. T., and Walker, B. H. 1985. Interactions of woody and herbaceous vegetation in a southern African savanna. *Journal of Ecology* 73:235–253.

Lasueur, H. 1945. *The Ecology of the Vegetation of Chihuahua, Mexico, North of Parallel Twenty-Eight*. The University of Texas Publication No. 4521.

Lauenroth, W. K, Urban, D. L., Coffin, D. P., Parton, W. J., Shugart, H. H., Kirchner, T. B., and Smith, T. 1993. Modeling vegetation structure–ecosystem process interactions across sites and ecosystems. *Ecological Modeling* 67:49–80.

Leopold, A. 1924. Grass, brush, timber and fire in southern Arizona. *Journal of Forestry* 22:1–10.

Marshall, J. T. 1957. *Birds of the Pine–Oak Woodland in Southern Arizona and Adjacent Mexico*.

Pacific Coast Avifauna Number 32. Berkeley, Calif.: Cooper Ornithological Society.

Martin, P. S. 1963. *The Last 10,000 Years*. Tucson, Ariz.: University of Arizona Press.

Martin, P. S. 1984. Pleistocene overkill: the global model. In *Quaternary Extinctions*, ed. P. S. Martin and R. G. Klein, pp. 354–403. Tucson, Ariz.: University of Arizona Press.

McAuliffe, J. R. 1994. Landscape evolution, soil formation, and ecological patterns and processes in Sonoran Desert bajadas. *Ecological Monographs* 64:111–148.

McAuliffe, J. R. 1995. Landscape evolution, soil formation, and ecological patterns and processes in desert grasslands. In *The Desert Grassland*, ed. M. P. McClaran and T. R. Van Devender, pp. 100–129. Tucson, Ariz.: University of Arizona Press.

McClaran, M. P. 1995. Desert grasslands and grasses. In *The Desert Grassland*, eds. M. P. McClaran and T. R. Van Devender, pp. 1–30. Tucson, Ariz.: University of Arizona Press.

McClaran, M. P., Allen, L. S., Ruyle, G. B. 1992. Livestock production and grazing management in the encinal oak woodlands of Arizona. In *Proceedings of the Symposium on Ecology and Management of Oak and Associated Woodlands: Perspectives in the Southwestern United States and Northern Mexico*, tech. coords. P. F. Ffolliott, G. J. Gottfried, D. A. Bennett, V. M. Hernandez C., A. Ortega-Rubio, and R. H. Hamre, pp. 57–64. USDA Forest Service, Gen. Tech. Rep. RM-218.

McClaran, M. P., and Bartolome, J. W. 1989. Fire related recruitment in stagnant *Quercus douglasii* populations. *Canadian Journal of Forest Research* 19:580–585.

McClaran, M. P., and McPherson, G. R. 1995. Can soil organic carbon isotopes be used to describe grass–tree dynamics within a savanna and at the savanna–grassland ecotone? *Journal of Vegetation Science* 6:857–862.

McPherson, G. R. 1992. Ecology of oak woodlands in Arizona. In *Proceedings of the Symposium on Ecology and Management of Oak and Associated Woodlands: Perspectives in the Southwestern United States and Northern Mexico*, tech. coords. P. F. Ffolliott, G. J. Gottfried, D. A. Bennett, V. M. Hernandez C., A. Ortega-Rubio, and R. H. Hamre, pp. 24–33. USDA Forest Service, Gen. Tech. Rep. RM-218.

McPherson, G. R. 1993. Effects of herbivory and herbs on oak establishment in a semi-arid tem-perate savanna. *Journal of Vegetation Science* 4:687–692.

McPherson, G. R. 1994. Response of annual plants and communities to tilling in a semi-arid temperate savanna. *Journal of Vegetation Science* 5:415–420.

McPherson, G. R. 1995. The role of fire in desert grasslands. In *The Desert Grassland*, eds. M. P. McClaran and T. R. Van Devender, pp. 130–151. Tucson, Ariz.: University of Arizona Press.

McPherson, G. R., Boutton, T. W., and Midwood, A. J. 1993. Stable carbon isotope analysis of soil organic matter illustrates vegetation change at the grassland/woodland boundary in south-eastern Arizona, USA. *Oecologia* 93:95–101.

Medina, A. L. 1987. Woodland communities and soils of Fort Bayard, southwestern new Mexico. *Journal of the Arizona–Nevada Academy of Sciences* 21:99–112.

Mehlert, S., and McPherson, G. R. 1996. Effects of basal area or density as sampling metrics on oak woodland cluster analyses. *Canadian Journal of Forest Research* 26:38–44.

National Oceanic and Atmospheric Administration. 1995. *Climatological Data, Arizona*. Asheville, N.C.: National Climatic Data Center.

Neilson, R. P. 1986. High resolution climatic analysis and southwest biogeography. *Science* 232:27–34.

Niering, W. A., and Lowe, C. H. 1984. Vegetation of the Santa Catalina Mountains: community types and dynamics. *Vegetatio* 58:3–28.

Nixon, K. C. 1993. The genus *Quercus* in Mexico. In *Biological Diversity in Mexico*, ed. T. P. Ramamoorthy, R. Bye, A. Lot, J. Fa, pp. 447–458. New York: Oxford University Press.

Nyandiga, C. O., and McPherson, G. R. 1992. Germination of two warm-temperate oaks, *Quercus emoryi* and *Q. arizonica*. *Canadian Journal of Forest Research* 22:1395–1401.

Officer, J. E. 1987. *Hispanic Arizona, 1536–1856*. Tucson, Ariz.: University of Arizona Press.

Ortlieb, L., and Roldan-Q., J., eds. 1981. *Geology of Northwestern Mexico and Southern Arizona*. Hermosillo, Sonora, Mexico: Estacion Regional del Noroeste, Instituto de Geologia, U.N.A.M.

Peck, R. A. B., and McPherson, G. R. 1994. Shifts in lower treeline: the role of seedling fate. *United States Section, International Association of Landscape Ecology Annual Meeting Abstracts* 9:105–106.

Propper, J. G. 1992. Cultural, recreational, and

esthetic values of oak woodlands. In *Proceedings of the Symposium on Ecology and Management of Oak and Associated Woodlands: Perspectives in the Southwestern United States and Northern Mexico,* tech. coords. P. F. Ffolliott, G. J. Gottfried, D. A. Bennett, V. M. Hernandez C., A. Ortega-Rubio, and R. H. Hamre, pp. 98–102. USDA Forest Service, Gen. Tech. Rep. RM-218.

Rzedowski, J. 1983. *Vegetacion de Mexico.* 2d ed. Mexico City: Editorial Limusa.

Sala, O. E., Golluscio, R. A., Lauenroth, W. K., and Soriano, A. 1989. Resource partitioning between shrubs and grasses in the Patagonian steppe. *Oecologia* 49:101–110.

Sanchini, P. J. 1981. Population structure and fecundity patterns in *Quercus emoryi* and *Q. arizonica* in southeastern Arizona. Ph.D. dissertation, University of Colorado, Boulder, Colo.

Sellers, W. D., and Hill, R. H. 1974. *Arizona Climate* 1931–1972. Tucson, Ariz.: University of Arizona Press.

Sharman, J. W., and Ffolliott, P. F. 1992. Structural diversity in oak woodlands of southeastern Arizona. In *Proceedings of the Symposium on Ecology and Management of Oak and Associated Woodlands: Perspectives in the Southwestern United States and Northern Mexico,* tech. coords. P. F. Ffolliott, G. J. Gottfried, D. A. Bennett, V. M. Hernandez C., A. Ortega-Rubio, and R. H. Hamre, pp. 132–136. USDA Forest Service, Gen. Tech. Rep. RM-218.

Shreve, F. 1939. Observations on the vegetation of Chihuahua. *Madrono* 5:1–13.

Wallmo, O. C. 1955. Vegetation of the Huachuca Mts., Arizona. *The American Midland Naturalist* 54:466–480.

Walter, H. 1954. Die verbuschung, eine erscheinung der subtropischen savannengebiete, und ihre ökologischen urscachen. *Vegetatio* 5/6:6–10.

Walter, H. 1979. *Vegetation of the Earth and Ecological Systems of the Geo-Biosphere.* New York: Springer-Verlag.

Weltzin, J. F. 1998. Biotic and abiotic constraints on shifts in temperate savanna ecotones at lower treeline. Ph.D dissertation, University of Arizona, Tucson, Ariz.

Weltzin, J. F., and McPherson, G. R. 1994. Distribution of Emory oak (*Quercus emoryi* Torr.) seedlings at lower treeline. *United States Section, International Association of Landscape Ecology Annual Meeting Abstracts* 9:130.

Weltzin, J. F., and McPherson, G. R. 1995. Potential effects of climate change on lower treelines in the southwestern United States. In *Biodiversity and Management of the Madrean Archipelago: The Sky Islands of Southwestern United States and Northwestern Mexico,* tech. coords. L. F. DeBano, G. J. Gottfried, R. H. Hamre, C. B. Edminster, P. F. Ffolliott, and A. Ortega-Rubio, pp. 180–193. USDA Forest Service, Gen. Tech. Rep. RM-264.

White, S. S. 1948. The vegetation and flora of the Rio de Bavispe in northeastern Sonora, Mexico. *Lloydia* 11:229–302.

Whittaker, R. H., and Niering, W. A. 1964. Vegetation of the Santa Catalina Mountains, Arizona. I. Ecological classification and distribution of species. *Journal of the Arizona Academy of Science* 3:9–34.

Whittaker, R. H., and Niering, W. A. 1965. Vegetation of the Santa Catalina Mountains, Arizona. II. A gradient analysis of the south slope. *Ecology* 46:429–452.

Whittaker, R. H., and Niering, W. A. 1975. Vegetation of the Santa Catalina Mountains, Arizona. V. Biomass, production and diversity along the elevation gradient. *Ecology* 56:771–790.

Wooten, E. O., and Standley, P. C. 1915. *Flora of New Mexico.* Washington D.C.: United States Government Printing Office.

18 Juniper–Piñon Savannas and Woodlands of Western North America

NEIL E. WEST

Introduction

Lands with semiarid climates west of 103° W longitude, currently occupied by at least one drought-tolerant juniper (section Sabina) and/or one drought-tolerant pine (subsection *cembroides* = the piñons), encompass close to 72 million acres (Figure 18.1) in the United States (West, Rea and Tausch 1975) and unknown additional acres in Mexico. This huge area has many characteristic trees, which have been geographically subdivided in Table 18.1.

Probably because they have more tolerance to drought and cold (DeLucia and Schlesinger 1991; Lajtha and Getz 1993), junipers are more widespread (Miller and Wigand 1994) than the piñons (Malusa 1992). This is why *juniper–piñon* is used here rather than the more commonly used *piñon–juniper* to describe this vegetation type. The furthest north that any self-sown piñon occurs is in extreme southern Idaho. Although pure juniper savanna or woodland is common on dry and/or cold sites throughout the extent of the type (Figure. 18.1), pure single needle piñon stands are found only on a few mountain ranges in extreme western Nevada and adjacent California.

Although junipers and piñons are found mixed, they also are segregated across elevational gradients. Piñons typically dominate at middle to relatively higher elevations, whereas most junipers dominate the woodland belt at lower elevations (West, Rea and Tausch 1975; Padien and Lajtha 1992). An

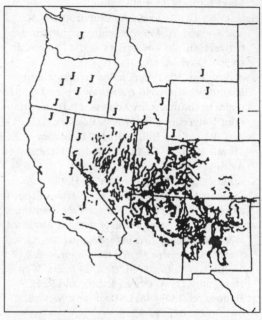

Figure 18.1. Geographic distribution of the juniper–piñon woodland vegetation type in the intermountain portion of the western United States (from Küchler 1970). J indicates pure stands of *Juniperus occidentalis* in the Pacific Northwest and *J. osteosperma* in the northern Rocky Mountains and Great Plains.

exception is *Juniperus scopulorum*, which occurs well within the montane zone in the north. Glades occur largely on topographic positions, where there is potential for occasional water saturation of the soil profile (Van Pelt 1978). The dominant conifers from the surrounding woodlands are easily killed by occasional soil anoxia on such sites.

From about 38° N latitude southward, piñons, junipers, and oaks (*Quercus* spp.)

Roger C. Anderson, James S. Fralish, and Jerry M. Baskin, eds. *Savannas, Barrens, and Rock Outcrop Plant Communities of North America*. Copyright © 1999 Cambridge University Press. All rights reserved.

Table 18.1. *Distribution of principal tree species in juniper–piñon savannas in various sections of western North America. Nomenclature follows Flora of North America Editorial Committee 1993.*

Area	Pines	Junipers	Others
British Columbia and Alberta		*Juniperus scopulorum*	
Interior Pacific Northwest (Oregon, Washington, Idaho)		*J. occidentalis*	
Northern Rocky Mountains and adjacent plains (Montana, Wyoming)		*J. scopulorum* *J. osteosperma*	
Eastern and Central Great Plains		*J. virginiana*	
Great Basin	*Pinus monophylla*	*J. osteosperma*	
Colorado Plateau	*P. edulis*	*J. osteosperma*	
Southern Great Plains and Edwards Plateau		*J. ashei* *J. pinchotii*	
Mogollon Rim (Arizona and New Mexico)	*P. edulis*	*J. monosperma* *J. deppeana*	*Cupressus arizonica*
Baja California Norte (Sierra Juarez)	*P. quadrifolia*	*J. californica*	
Sierra Madre Occidental	*P. cembroides*	*J. coahulensis*	*Quercus* spp.
Big Bend–Trans Pecos	*P. cembroides*	*J. deppeana* *J. flaccida*	*Quercus* spp.
Sierra Madre Oriental	*P. cembroides*	*J. coahulensis* *J. flaccida* *J. monosperma*	*Quercus* spp.
Serranias Meridionales del Altiplano Potosino	*P. ayacahuite* *P. cembroides* *P. joharinis*	*J. flaccida*	*Quercus* spp.
Sierra Madre del Sur	*P. teocote*		
Sierra Madre de Chiapas		*J. comitana* *J. gamboana* *J. monticola*	

become intermingled. Oaks are distinctly less tolerant of low temperatures than the piñons. Deciduous oaks are more cold tolerant than evergreen oaks. In areas where deciduous and evergreen oaks occur, the evergreens occupy more acidic and nutrient-poor soils than the deciduous oaks (Goldberg 1983). Evergreen oaks become increasingly dominant in these juniper–piñon–oak communities in Texas (Gehlbach 1967) and Mexico (Robert 1977, 1979; Rzedowski 1978; Brown 1982; Garcia Moya 1985; Passini 1985; Passini, Delgadillo and Salazar 1989; Segura and Snook 1992; Romero Manzanares et al. in press). On the southern part of the Mexican highlands, piñons and junipers are found mostly on north slopes or in ravines or barrancas (canyons) (Rzedowski 1978). Junipers are more common than piñons and oaks on rocky and/or shallow sites and are considered seral to a piñon–oak climax there (Rzedowski 1978).

Understory Plant Community Composition

Understories of juniper–piñon savannas and woodlands are floristically and structurally more variable than the overstory. Generally the understory is compositionally similar to that of adjacent grasslands, shrub steppes, chaparral, and forests (West, Rea and Tausch 1975). For instance, in the western juniper woodlands and savannas of the Pacific Northwest, the understory is mostly a mixture of sagebrushes (Section Tridentatae of *Artemisia*) and cool-season bunchgrasses. The relatively wet winters and dry summers in this region favor plants that either can complete their growth before midsummer, like the cool-season grasses, or can utilize deep soil moisture, as do the trees and shrubs (Flanagan, Ehleringer and Marshall 1992).

South and east of the Pacific Northwest, the proportion of warm-season bunch and sod grasses increases, and the abundance of shrubs declines as the fraction of total annual precipitation received during the summer increases. Juniper and piñon stands of New Mexico, Texas, and northern Mexico thus have more half-shrubs (suffrutescents), such as *Senecio longilobus*, *Gutierrezia* spp., *Brickellia* spp., *Haplopappus* spp., and *Salvia* spp., and succulents, such as various cacti and monocots (e.g., *Agave* spp., *Nolina* spp., *Yucca* spp., *Dasylirion* spp.) than true shrubs. Warm-season, C4 grasses that dominate are from the nearby semidesert grasslands or from the southern mixed- and short-grass prairies, including species of *Aristida*, *Digitaria*, *Eragrostis*, *Bouteloua*, *Hilaria*, *Sporobolus*, *Muhlenbergia*, *Schizachyrium*, *Botriochloa*, *Lycurus*, *Piptochaetium*, and *Leptochloa*, where grazing is not excessive (Moir 1979; Pieper 1992).

Forbs associated with juniper-piñon savanna or woodlands have distinctive geographic distributions. Understory forbs in juniper stands of the Pacific Northwest and Great Basin are derivatives of the tree-dominated Arcto-Tertiary Geoflora (D. I. Axelrod 1976). Principal genera are *Lupinus*, *Penstemon*, *Castelleja*, *Balsamorhiza*, *Allium*, and others. On the Colorado Plateau and south and east of that region, forbs associated with junipers and piñons are derived mostly from the Madro-Tertiary Geoflora (D. Axelrod 1958), a heat-tolerant group of plants. Example genera are *Croton*, *Euphorbia*, *Ipomoea*, *Polygala*, and herbaceous *Salvia* (Pieper 1992; Romero Manzanares et al. in press). Abundance of annuals varies greatly from year to year (Treshow and Allan 1979), making them of little value as indicators of other than near term climatic influences.

Vegetation Dynamics

The composition and structure of the vegetation associated with junipers and piñons is very dynamic and has changed greatly since the Pleistocene. Juniper–piñon savannas and woodlands occurred only in the lowlands of the present-day Mohave, Sonoran, and Chihuahuan Desert regions at the peak of the Wisconsinan glacial advances (Van Devender and Spaulding 1977; Betancourt, Van Devender and Marten 1990). Junipers migrated northward much more rapidly than piñons during post-Pleistocene warming (Wells 1983). During the xerothermic interval of about 7,000 to 4,000 yr BP, there was a rapid migration of these species, especially junipers, northward and to higher elevations (Betancourt 1987; Nielson 1987). Because some montane conifers were eliminated from the tops of moderate-sized mountains in the Great Basin during the hot, dry, xerothermic interval, juniper–piñon woodland belts are especially broad in that region now (West et al. 1978; Tueller et al. 1979).

Climates have become cooler and wetter since about 4,000 yr BP, especially during the Little Ice Age of about the 1400s to mid-1800s. Whether this climatic shift resulted in fewer trees because of an increase in fine grassy fuels to carry fire, or encouraged tree growth because of favorable climatic conditions for tree establishment, is debatable (Betancourt 1987; Gottfried et al. 1995). Nevertheless, Miller and Wigand (1994) found greater net establishment of western juniper during dry rather than wet periods. Charcoal abundance in sediments is not uniform, indicating variable fire regimes in the past. Aboriginal peoples were an additional source of ignition during the entire post-Pleistocene period, making this debate more than just a climatic issue.

Prior to the arrival of Europeans in North America, piñons and junipers probably were found primarily on steep and/or rocky sites where moisture easily infiltrates (Sauerwein 1981), but soils do not develop well enough to produce a continuous understory of fine fuels (Bunting 1994; Gottfried et al. 1995). The formerly more restricted distribution of junipers and piñons is known from examination of size–age–vigor class analysis of existing stands of trees, reexamination of original land survey records, and repeat photography. Records in woodrat middens are not helpful in addressing this issue, because long-persisting middens occur only on rocky parts of the landscape.

By examination of tree size–form–growth classes, Blackburn and Tueller (1970), A. H. Johnson and Smathers (1976), Burkhardt and Tisdale (1976), and Young and Evans (1981) found considerable invasion of these trees into former shrub steppe or grassland at selected sites in the Pacific Northwest and in the Great Basin. A systematic assessment of 486 stands of *Juniperus osteosperma* and of *Pinus monophylla* or *P. edulis* on 66 mountain ranges across the Great Basin showed that slightly over 50% of the stands had no trees over about 120 years in age (Tausch, West and Nabi 1981). Maximum age of trees was inversely related to steepness and rockiness of the sites. Young stands predominated on level to gently sloping sites, especially on the upper bajadas.

Sparks, West and Allen (1990) and Creque (1996) revisited some witness trees used in the original land surveys (done in the 1870s) in western Utah. Sparks, West and Allen (1990) found considerable increase in juniper density since 1871 on two townships in Tooele County, Utah, despite considerable evidence of wood harvest, chaining, and fire. To document vegetation changes, Creque

compared original land survey records and data obtained by repeating the survey, and old photographs. He found that although the area occupied by trees diminished from 1874 to about 1940, it has increased since then to presently occupy about as much acreage as in 1874, albeit now in smaller, more numerous patches. Tintic Valley, Creque's (1996) study area, was very heavily affected by mining-related activities from about 1880 to 1920.

Comparing 19th-century photographic images with those taken more recently demonstrates increased tree density and average size within the original area of tree concentration. Trees had also invaded adjacent grasslands, shrub steppe, or mountain browse community types (G. G. Rogers 1982; Bahre 1991; Miller and Rose 1995; Creque 1996). Miller and Wigand (1994) summarized all lines of evidence in the northern intermountain region and concluded that more junipers exist today than at any time in the past for which records are available. More recently, Betancourt et al. (1993) noted the extensive loss of trees in the Southwest during the intense 1950s drought. However, the generally very wet conditions from 1976 to 1994 favored tree growth (Gottfried et al. 1995). Nevertheless, dry conditions and widespread fires of 1995–96 in the Southwest may reverse some of the earlier tree expansion.

Plant Community Structure and Its Interactions with Dynamics

An overview of systems with juniper and piñon shows at least a two-phase structure (Figure 18.2): a tree-centered phase where microclimates and soils are controlled by the trees; and a non–tree-dominated open interspace where some mixture of shrubs, grasses, and forbs generally provides the major cover. The term *savanna* can be applied to vegetation where the nontree phase dominates the matrix (major fraction of the landscape) and trees are widely scattered. Where trees are denser, the proper descriptor is *woodland*.

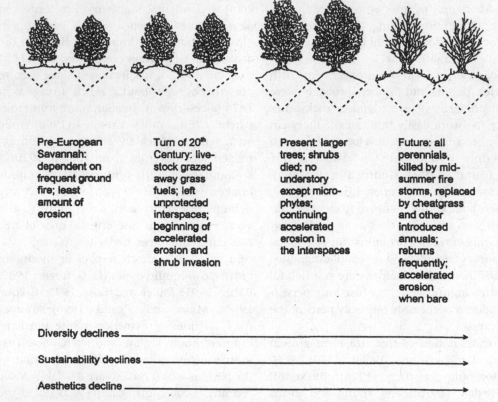

Pre-European Savannah: dependent on frequent ground fire; least amount of erosion

Turn of 20th Century: livestock grazed away grass fuels; left unprotected interspaces; beginning of accelerated erosion and shrub invasion

Present: larger trees; shrubs died; no understory except microphytes; continuing accelerated erosion in the interspaces

Future: all perennials, killed by mid-summer fire storms, replaced by cheatgrass and other introduced annuals; reburns frequently; accelerated erosion when bare

Diversity declines ————————————————————————————→

Sustainability declines ————————————————————————→

Aesthetics decline ———————————————————————————→

Figure 18.2. Depiction of how juniper–piñon woodland structure changes through time (earlier to left, later to right). Broken lines are outer limits of tree roots.

Forest occurs where tree coverage increases. Boundaries between savanna/woodland/forest are ultimately arbitrary human constructs. All these vegetation types are found over the area covered. Deciding which structure is to be expected because of the interactions of physical environmental causes and human influences is a major challenge to our interpretive abilities. Vegetational structure influences many other attributes of land, such as native animal communities, water quantity and quality, risk of wildfire, and so forth.

Most studies in the northern, and a few in the southern, portion of the juniper–piñon region have shown that the proportion of landscape dominated by trees has increased in recent decades, whereas the proportion in interspace generally has declined. Many studies from the southern border states and southward into Mexico show net declines in

piñon- and juniper-dominated areas over recent decades. Whether these trends are natural and sustainable, or human-accelerated and degradational (and thus should be acted upon by land managers), has generated an enormous amount of acrimony (e.g., Bahre 1991; Belsky 1996).

A steady reduction of livestock-carrying capacity has been observed at several locations; for example, see Figure 18.3. Although increase in the number of trees was a major factor in this decline in value for livestock grazing, changing species composition in the herbaceous components, and soil erosion may be involved (E. Huston, personal correspondence, 13 February 1997). Associated with this reduction of lower-statured vegetation has been a concomitant decline in surface water yields and aquifer recharge, resulting from increased interception of pre-

cipitation by trees, evapotranspiration, and a decline in the
infiltration, storage, and down-
stream release of water
(Thurow and Carlson 1994;
Thurow and Taylor 1995).

Mounds commonly develop
under trees, and rilled concav-
ities form in the interspaces
(Figure 18.2). Although the
cause of this differential
microrelief is disputed, most
workers (e.g., Price 1993)
attribute mounding to
increased soil erosion in the
less-protected interspace following excessive grazing and fire control.
Davenport, Wilcox and Breshears (1996),
however, point out that the tree-dominated
mound has soil characteristics indicating that
the trees catch aeolian sediments and slow
soil horizon maturation. Thus, the mound–
hollow microrelief apparently represents the
net effect of aggradational and degradational
forces at work on the different microsites.
The area studied by Davenport, Wilcox and
Breshears (1996) was an ungrazed relict, a
rare circumstance in today's world. In northwest Colorado, Carrara and Carroll (1979)
used the date of the exposure of tree roots
well beyond the tree canopies to demonstrate
more than a 400% increase in soil erosion
during the past century compared to the previous three centuries.

Numerous forces have independent influences on tree- or interspace-dominated
phases of juniper–piñon savannas or woodlands (Table 18.2). Causes of vegetational
change are rarely singular or simple, however. Thus, a rationale is provided for syner-
gism, and further interactions are considered.

Fire was, by far, the most important and
extensive factor in maintaining juniper-
piñon savannas before the introduction of
livestock (Wright, Neuenschwander and
Dritton 1979). Although most ecologists and
land managers agree that fire was once

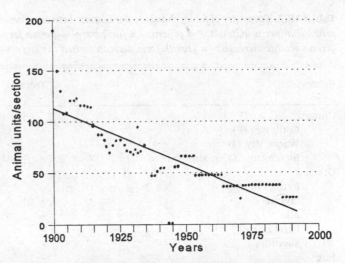

Figure 18.3. Decline in stocking rates for juniper
savanna at the Sonora Station, Edwards Plateau,
Texas, 1900–1992 (from Taylor and Smeins, 1994,
with permission of E. Huston). Data points prior to
purchase of the Sonora Station by Texas A&M in
1914 are intermittent, not annual, estimates based
on interviews with, and diaries of, previous owners or managers. Values from 1915 to 1949 are
subjective estimates. Experimental data were first
collected in 1949 (personal communication from
E. Huston, 13 February 1997).

important and should probably be reintroduced, there are few data to indicate means
and ranges of natural variability in fire return
intervals. piñons tend to burn either completely or not all, and thus they do not form
basal fire scars ("cat-faces"). Although
junipers occasionally have fire scars at the
stem base (Despain and Mosely 1990), they
rarely have been examined because frequent
occurrence of multiple stems and false and
missing annual stem growth rings make dating unreliable. Ground fires easily kill non-
sprouting juniper seedlings and saplings less
than about 3 m tall (Johnsen 1962), effectively excluding them from grasslands that
are not excessively grazed. Larger junipers,
however, commonly survive ground fires to
form savanna. If further fires are excluded,
junipers and shrubs will serve as nurse plants
to the faster growing piñons, which eventually dominate the tree stratum (Tausch and

Table 18.2. *Summary of forces changing the balance between trees and perennial grasses in piñon–juniper woodlands. P = piñons; J = junipers; + = growth form increases when the given variable increases; - = growth form decreases when the given variable increases.*

Forces	Trees	Grasses
Climate		
Cool, wet (P)	+	-
Warm, dry (J)	-	+
Increasing CO_2 in atmosphere	+	-
Grazing		
Extinct browsers	-	+
Livestock	+	-
Elk	+	
Feral horses	+	
Saw flies	-	+
Fire	-	+
Tree harvest	-	
Animals		
Jays and nutcrackers (P)	+	-
Chipmunks and ground squirrels (P)	+	-
Thrushes (J)	+	-
Rabbits and hares (J)	+	-
Livestock	+	-
Parasites	-	+
Pathogens	-	+

West 1988, 1995). The root-sprouting junipers of the southern Great Plains and Arizona and New Mexico (e.g., *J. deppeana, J. ashei, J. pinchotii*) have additional means of recovery.

Over the past several centuries in New Mexico, ground fires covering at least 10 ha occurred every 15–20 years on the most mesic sites on moderately rugged terrain (Gottfried et al. 1995). Fires probably were more frequent and covered larger areas on gentler, drier landscapes, where the vegetational matrix was grassland or savanna. On the most dissected topography, sites could escape frequent fires, and eventually the trees could become large and dense enough to allow stand-replacing crown fires at intervals of several hundred years (Erdman 1969; Barney and Frischknecht 1974; Gottfried et al. 1995). Milne et al. (1996) reported that when trees dominate about 60% of a landscape, the grassland proportion quickly decreases. Thereafter, tree growth accelerates and the chances of crown fire greatly increase. Minnich (1991) notes that in the Mexican province of Baja California Norte, where fire suppression policies have not existed, juniper–piñon woodlands are more open and patchy than in adjacent southern California. The chance of large, devastating wildfires is much greater now in California where fires have been controlled. This is because contiguous and highly flammable fuels have accumulated.

Interrelationships with Animals

Although juniper–piñon savannas and woodlands are important habitat for a variety of wild animals, little is known about the roles of most animals in the dynamics of these ecosystems. However, a few animals are known to influence the growth and survival of the trees or major understory plants, thereby affecting the whole system.

One of the most obvious ways in which the tree/interspace balance is influenced is via direct herbivory. Since the demise of Pleistocene large browsers, such as the shrub ox (*Euceratherium*), saiga (*Saiga*), elk-moose (*Cervalces*), ground sloths (*Megalonyx, Nothrotheriops,* and *Glossotherium*), camels (*Camelops*), and mastodons (*Mammut*), about 11,000 to 10,000 yr BP (Grayson 1994), trees have been subject to little vertebrate herbivory. Extant deer and elk do not eat much conifer foliage except during the most severe winters. Of the introduced vertebrates, only goats consume substantial amounts of conifer foliage and thus prune off the lower branches, allowing more penetration of light from the side (Figure 18.4). Porcupines (*Erethizon dorsatum*) girdle the cambium of piñons (Spencer 1964). Although the diet of Stephen's woodrat (*Neotoma stephensii*) (Vaughn 1982) concentrates on young junipers, its distribution is limited to the Mogollon Rim of Arizona and New Mexico.

Most current interactions between native vertebrates and juniper and piñon are commensal (Table 18.2). For instance, hares, rabbits, and thrushes consume and distribute juniper seeds (V. R. Johnson 1994). The large, nutritious seeds of piñons are cached by chipmunks, golden-mantled ground squirrels, and corvid birds (Van der Wall 1990), particularly the piñon jay (Ligon 1978). Deer mice (*Peromyscus* spp.) take advantage of these caches (Martinez-Delgado et al. 1996). Apparently, the only mammals exclusively associated with juniper or piñon trees are the piñon mouse, cliff chipmunk, and the Boyle's deer mouse (Parmenter et al. 1995). Small mammal density and richness generally is greater in savannas or in areas cleared of trees than in the tree-dominated terrain (Austin and Urness 1976).

At least 73 species of birds breed in juniper–piñon savannas or woodlands (Balda and Masters 1980, Balda 1987). However, the piñon jay, montezuma quail, and bush-tit are the only birds strictly associated with these

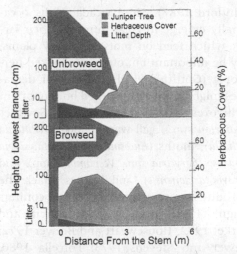

Figure 18.4. Vertical and horizontal patterns of Ashe Juniper, grass, and litter from ungrazed and goat-utilized pastures at the Sonora Station, Edwards Plateau, Texas (from Smeins, Owens and Fuhlendorf 1994). Note that goats create a browse line and allow understory to move toward the base of tree.

ecosystems (Parmenter et al. 1995). Ground-dwelling larks, buntings, doves, and quail prefer the more open terrain of savannas. Canopy-gleaning fringillids favor denser woodlands. Of these, the black-capped vireo and golden-cheeked warbler are endangered. The gray fly catcher and black-throated gray warbler are on the sensitive list (Gottfried et al. 1995). Where these species remain, there are restrictions on habitat alteration (Rollins and Armstrong 1994) that reduce ranchers' efforts to favor deer, turkey, quail, and livestock by partial removal of trees (Rollins et al. 1988).

Although many reptiles are found in the juniper–piñon savanna/woodlands, the only one restricted to this vegetation type is the mountain king snake (Parmenter et al. 1995). Larger vertebrate predators range over wide areas and include juniper-piñon stands as only a portion of their home range.

Parmenter et al. (1995) lists 115 species of surface-dwelling arthropods that prefer juniper- and piñon-dominated habitat. Many of these are important in nutrient cycling

(Whitford 1987). Since cicada adults occasionally emerge in great numbers, their larvae, which feed on roots of woody plants, may be important but often overlooked herbivores (Gottfried et al. 1995). Soil nematodes might also be little noted but important herbivores (Riffle 1967, 1972). Sawflies (*Zadiprion* spp.), gall wasps, midges (*Pinonia edulicola*), moths (*Eucosma babana, Halysidota imagens, Dioryctria* spp., *Vespamima* spp., and *Petrovia arizonensis*), and beetles attack the reproductive structures of these conifers (Gagne 1970; Brewer 1971; Barcia and Merkel 1972; Houseweart and Brewer 1972; Brewer and Stevens 1973; Forcella 1980; Gillet and Brewer 1983; Jenkins 1984; Whitham and Mopper 1985; Cibrian-Tovar et al. 1986; Mopper and Whitham 1992; T. J. Rogers 1993).

Very few invertebrate consumers utilize the vast food reserves in juniper and piñon foliage and wood. This may be because it is well-protected both chemically and mechanically. *Ips* bark beetles attack the cambium of piñons, especially during drought periods (Wilson and Tkacz 1992). Twig beetles (*Pityophythorus* spp. and *Pityogenes* spp.) are frequent pests of piñons that are in deep shade or have been damaged by storms (e.g., ice and snow breakage). The piñon needle miner (*Coleotechnites edubicola*) and piñon needle scale (*Matsucoccus acalyptus*) most commonly are found during droughts (T. J. Rogers 1993). Junipers are attacked by the buprestid western cedar borer (*Trachykele blondelii*), cerambycid juniper twig pruner (*Styloxus bicolor*) (T. J. Rogers 1993), and twig beetles (*Phloeosinus* sp.) (Gottfried et al. 1995).

Mutualists, Pathogens, and Parasites

Piñons are ectomycorrhizal, whereas junipers have vesicular arbuscular mycorrhizae (Klopatek 1987). Fresquez (1990) found many more kinds and densities of fungi under juniper and piñon trees than in the interspaces. Ferrera and Saenz (1987) reported similar differences in Mexico. Few parasitic fungi attack aboveground portions of junipers or piñons, although junipers have a few rusts (*Gymnosporangium* spp.) (Peterson 1967, 1972), and blister rust outbreaks on singleleaf piñon have been recently noted in western Nevada (R. J. Tausch, personal communication, 1996). Standing dead and fallen piñons, however, are more susceptible to stem rots, and they decompose much more rapidly than junipers. Very little research has been done on root diseases of junipers or piñons. Only recently has *Armillaria* sp. been observed on piñons in northern New Mexico (T. R. Rogers 1993). From their results on comparisons of trees on cinder fields and adjacent sandy loams of northern Arizona, Gehring and Whitham (1995) speculated that global warming would increase the likelihood that these trees will experience additional herbivore and parasite attack. This, in turn, would reduce the abundance of mycorrhizae, thus making it more difficult for the trees to maintain themselves on the marginal sites they occupy.

Junipers and piñons are hosts for mistletoes (*Arceuthobuim* spp. on piñons and *Phoradendron* spp. on junipers). Hreha and Weber (1979) concluded that in recent decades the degree of infestation of these parasites has increased with increasing tree density. *Pedicularis* is known to parasitize the roots of both junipers and piñons (Riffle 1968).

Interrelationships with Humans

Human impacts on these savannas and woodlands date back to early hunters and gatherers (Archaic period, before 100 B.C.), who may have been at least partially responsible for the demise of large browsers at the end of the Pleistocene. Their careless (or purposive?) use of fire may have favored savanna over woodland on gentle slopes and deep, rock-free soils (Taylor and Smeins 1994; Creque 1996).

During the Basket Maker Period (100 B.C. to A.D. 700), firewood was used for warming homes and firing pottery, as well as for clearing juniper and piñon to make fields and erect pit houses. During the Pueblo Period (A.D. 900–1600), there was accelerated human population growth and an increased intensity of land use, particularly after A.D. 1200 in the Four Corners Region (Kohler 1988, 1992; Floyd and Kohler 1989; Cartledge and Propper 1993). In fact, overutilization of the land may be one of the reasons for the demise of the Pueblo civilization in the north (Scurlock and Johnson 1993). For instance, Samuels and Betancourt's (1982) model indicates that fuel needs in Chaco Canyon exceeded the replenishment capacity of the woodlands near Chaco Canyon at about the time of its abandonment. Tree loss also contributed to flooding and sedimentation on the canyon floors.

The first European impact on indigenous peoples was the introduction of pandemic diseases new to the continent. These may have been introduced after Columbus's first voyage and may have spread via trade well before further European explorations on land. Direct European impacts on juniper–piñon systems occurred first in central Mexico. They quickly radiated northward with the establishment of Catholic missions in the area that now includes Texas, New Mexico, and Arizona. Among the direct influences of Europeans, introduction of livestock was the most important and extensive. Horses, cattle, sheep, and goats were raised to provide transport, fiber, and food to miners and agriculturists. Once native peoples obtained the horse, their whole culture took a decidedly different turn. The Navajos also acquired sheep and goats and became a pastoral people.

Livestock grazing reduced the chance of fire and competition between grasses and trees for moisture and nutrients. Livestock also trampled vegetation and exposed interspace soils. Tree seedlings thus had more available habitat (Jameson 1970b). Junipers and piñons have tap roots and also fine fibrous roots that occur near the surface (Foxx and Tierney 1987). Furthermore, the root systems extend 2–5 times the width of the canopy (West and Young in press); thus, interspaces are primary places of water and nutrient absorption for the trees (Breshears 1993; Davenport et al. 1998). As tree density increased, the herbaceous understory plants received decreasing amounts of light, precipitation, and nutrients, and there was an increasing amount of litter (Jameson 1966). Litter accumulation under trees concentrates nutrients (Charley and West 1975; Barth 1980; Everett, Sharrow and Thran 1986; Tiedemann and Klemmedson 1995), increases hydrophobicity of the soil surface (Scholl 1972), reduces N fixation (Klopatek, DeBano and Klopatek 1988), increases N mineralization (Charley and West 1977; Klopatek 1987), and inhibits seedling establishment of some species (Jameson 1961, 1970a; Everett 1987). Since conifer litter is acidic, whereas grassland soils are neutral to slightly alkaline, recent shifts in dominance can be detected by determining the spatial patterns of soil pH and carbonate content (Davenport, Wilcox and Breshears 1996).

A disproportionately high fraction of soil carbon, nitrogen, and phosphorus is concentrated in the surface horizons, thus loss of a few centimeters of the soil surface results in a large portion of the nutrient pool being lost for long periods (West 1984; Tiedemann 1987). With decreased vegetation and litter, interspaces are less protected than areas under the trees, and thus microclimatic extremes and freeze/thaw cycles disrupt more seedlings. With increased amounts of bare soil, surface runoff increases, and less precipitation infiltrates into the soil. This leads to a reduction in soil water available to shallow-rooted herbs. This sequence of positive feedbacks leading toward desertification ends in a dominance of physical over biological processes. Albedo increases when the relatively impermeable subsoil (B horizon), which is rich in clay and calcium but low in organic matter, is exposed (Gottfried et al. 1995). Juniper–piñon rocklands (Moir 1979) are the expected result of this site degradation. From the human perspective, these changes essentially have led to permanent reductions in site productivity (McDaniel and Graham 1992).

The increase in trees and associated changes in understory have decreased the value of these lands for supporting equids and most ungulates (Figure 18.3). An interim increase in shrubs resulted in increases of mule deer for about a decade during the mid-20th century, but shrubs and deer now are declining (Peek and Krausman 1996). Even with livestock removed, understory recovers only very slowly (Yorks, West and Capels 1994). Wild herbivores, including feral horses and burros, elk, deer, and rabbits, can respond quickly to any increases in grasses and can keep them suppressed.

The response of savannas and woodlands to increased atmospheric CO_2 and global warming is not clear. Gottfried et al. (1995) concluded that evidence for global warming already is available in juniper–piñon systems of the Middle Rio Grande Basin, New Mexico. Gosz (1992), Keeley and Mooney (1993), and Nielson and Marks (1994) project further increases in woody dominance, as well as in total standing crops of phytomass in these woodlands and savannas across the West. Although the current stocks of carbon, nitrogen, and other essential elements have been increasing as the phytomass of juniper and piñon trees has increased (Lajtha and Barnes 1991), their availability to most herbivores and detritivores has diminished. Most tissues of the trees are protected chemically and physically against consump-

tion and decomposition. The increased C/N ratio of the phytomass and its detritus inhibits biological activity of most consumers (Scholes 1996). Fire is the only natural mechanism left to reverse the process.

Only at particular times and places has wood production in these communities been of value to humans. For instance, the Anasazi cleared trees from fields to raise corn, beans, and squash. They used the trees for home fires, pottery firing, and roof beams. Paiutes and Goshutes used juniper posts for wing traps to corral pronghorn (Fowler 1986). Early miners used the trees for railroad ties, mine props, and charcoal to smelt ores (Young and Budy 1979; Creque 1996). Ranchers cut juniper for fenceposts. Piñon is still widely used for Christmas trees and firewood. The latter use has been exceeding the growth increment of these woodlands of northern New Mexico for at least several decades (Gray, Fowler and Bray 1982). There are probably places in Mexico (Cavazos-Doria and Medina 1996) where extensive use of juniper and piñon for charcoal making is still common.

During the late 1940s through 1960s, vast acreages of juniper–piñon woodland in the United States were chained or cabled and planted to introduced grasses. This increased livestock forage to grow meat animals that the public desired during the post–World War II recovery period (West 1984). The downed trees usually were burned either shortly after treatment or in subsequent maintenance operations. Such treatments recently have decreased, largely for economic reasons on private lands and for protection of wildlife and archeological values on public lands. Maintaining the objectives of greater forage and water production, yet diminished soil erosion, proved elusive (Belsky 1996). The necessary follow-up burning and grazing management rarely was done. Because wood in the chainings and cablings usually was not harvested, most people viewed the process as "wasteful." Recently, western juniper has begun to be harvested for chipboard manufacture in northeastern California (Swan 1995).

Juniper–piñon woodlands eventually become dense enough for fire storms to occur (Figure 18.2). The apparent thresholds for crown fires are about 1,100 trees per ha, relative humidity below 30%, and winds exceeding 55 km per hour (Wright and Bailey 1982). Crown fires instantaneously release considerable stores of aboveground carbon and other nutrients in living tissues, standing dead material, and litter. Because seed rain (Everett and Sharrow 1983) and soil seed banks of understory plants usually have been depleted (Koniak and Everett 1982), only ruderals invade after such hot fires (Koniak 1985; Barber and Josephson 1987). Immediate seeding of native perennial species could be done to obtain dense herbaceous cover (Koniak 1983); it is rarely accomplished due to financial costs and bureaucratic inertia (West 1984).

The Future

The juniper–piñon systems first encountered by Europeans in North America were not wilderness in equilibrium with environments uninfluenced by humans. Rather, these systems had been manipulated toward aboriginal management objectives. Lewis (1993) has argued that these original Americans had to have a sophisticated understanding of their surroundings in order to survive. Besides the well-known hunting, gathering, and burning, these early people coppiced, weeded, tilled, planted, and irrigated a wide variety of food and ceremonial plants (Fowler 1986).

These human-directed perturbations (or management?), particularly fire, probably kept large patches of these landscapes in the stage of "exploitation" (Stage #1, Figure 18.5) of Holling's et al. (1995) adaptive cycle. Relatively small amounts of energy and nutrients accumulated in trees on landscapes

that could support fine, continuous fuels near the soil surface. The more open environment, occupied by herbaceous plants, provided a greater variety of foods (game animals, grass seeds, forb roots) for early people and was more predictable in its resource availability than an unmanaged, tree-dominated system.

Whereas aboriginals actively used fire to shape their environment, Europeans regarded it as evil and did their best to stop burning (Pyne 1982; Bailey 1996). Furthermore, grazing by European-introduced livestock reduced the fine, continuous fuel provided by the once-abundant herbaceous understory and grassland matrix. These direct and indirect lengthenings of fire-free intervals first led to dominance by shrubs, and then the much-longer-lived trees were favored in the "conservation" stage (#2, Figure 18.5). As total aboveground phytomass increased, most plants of the herbaceous understory were outcompeted by trees for light, moisture, and nutrients. Tree canopies and litter reduced throughfall, microbial activity, understory establishment, and infiltration of water into the soil. Accelerated soil erosion occurred in the interspaces, and the mound-and-hollow microrelief became accentuated (Figure 18.2).

Europeans regarded prescribed burning by aboriginals as wasteful, but the "disclimax" produced by their actions as desirable and stable. This was an illusion, however, as the predictable patterns of early succession were replaced by a slowly increasing tree phytomass. As tree cover increased, there was an "accident waiting to happen" (Goodloe 1993). Indeed, fire storms have become more frequent, creating the "release" stage (#3, Figure 18.5). Since these fires can occur only during periods of extreme drought and wind, fire temperatures are high, and nearly all organisms on a site are destroyed.

Without human-assisted restoration efforts during the reorganization stage (#4, Figure 18.5), only vegetatively reproducing perennials such as some sod grasses and half-shrubs

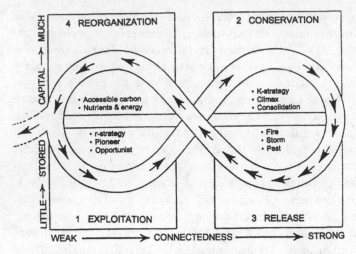

Figure 18.5. Holling's et al. (1995) adaptive cycle. Arrows close to each other indicate rapidly changing state; those far apart, a slowly changing state. Amount of accumulated capital (nutrients, stored energy) is represented on Y axis, degree of connectedness on the X axis, and chance of permanent change in reorganization stage is outgoing arrow on left.

are established. In the Great Basin, self-sown, introduced winter annuals such as cheatgrass (*Bromus tectorum*) and mustards (*Descurania* spp., *Sisymbrium* spp., etc.) prevail (Figure 18.2) (West and Young in press). Since these are green only a small portion of the year, they produce forage that provides adequate supplies of nutrients for graminvores for only a short time each spring; only granivores can survive year-long. Water, soils, and nutrients are not retained efficiently on these burned areas. If the annuals are allowed to flourish, the chance of reburning is increased several orders of magnitude during the exploitative and conservation stage systems. Accelerating this "reorganization" stage in more desirable directions (#4, Figure 18.5) requires the expeditious attention of the land manager.

Unfortunately, in most juniper–piñon woodlands, opportunities to use light ground fires during spring or fall (Bruner and Klebenow 1979) to gradually reduce tree dominance have already been lost. Presently, most land management policies regard Stage

#2 as both desirable and sustainable. For instance, the U.S. National Park Service at Bandelier National Monument in northern New Mexico has resisted killing trees and, until recently, feral burros (Craig Allen, personal communication, 1996). This resistance occurred even though accelerated soil erosion has been linked to increased tree dominance, and burro grazing has destroyed archaeological treasures that the monument was created to protect (Traylor et al. 1990). At Bandelier in 1994, tree canopy cover was 12%–45%, understory cover 0.4%–9% (basal), and bare soil covers 38%–75% (Gottfried et al. 1995).

The gradual, linear, deterministic, and reversible Clementsian model of succession toward a steady state (the "climax") remains the predominant paradigm in the land management agencies. This view of succession persists even though scientists (Sauerwein 1984; Jameson 1987; West and Van Pelt 1987; Tausch, Wigand and Burkhardt 1993) have pointed out for some time that this model is no longer reliable for choosing desired future conditions or dealing with the issue of sustainability. Rapid reversion to herbaceous dominance has been already triggered by midsummer fire storms over an increasing area of western juniper and juniper–piñon woodlands in the intermountain West. However, a parallel change in thinking about the role of fire needed by land management agencies and the public has not occurred (Gunderson, Holling and Light 1995; Holling and Meffe 1996). Accelerated soil erosion has diminished the potential of many areas to recover the kinds and amounts of vegetation, animals, and watershed protection they once had (West 1984; McDaniel and Graham 1992; Wilcox et al. 1996).

Although massive chaining of trees and

seedings of exotics in the 1950s and 1960s did not lead to sustainable grasslands (Belsky 1996), it should be recognized that custodial-style management also has its consequences. Management must become proactive and begin to reverse the overdominance of trees in the present system if soils and biodiversity are to be maintained. A more holistic view that includes sophisticated combinations of prescribed burning, mechanical removal of selected individual trees, and time-controlled livestock grazing applied consistently over decades seems to be the only sustainable management solution (Goodloe 1995; Marston 1996).

Junipers and piñon are long lived (to ca. 500 years for piñons and to ca. 2,000 years for junipers), and soil erosion and desertification are accelerating on most watersheds occupied by these trees. If proactive management does not occur, crown fires eventually will destroy the trees, standing dead, and litter, and weedy herbs will replace woodlands. Dominance by these short-lived herbaceous species will lead to quick release of nutrients and burning at too short intervals for shrubs or trees to reestablish, for soils and nutrient reserves to recover, or for the diverse native cover of watersheds to return. The major invaders are noxious species of little value to any subset of society. When these areas are reburnt early in the growing season, the land remains unprotected and winds and rains associated with summertime convectional storms can redistribute fine soil particles. The system will not restabilize until a ground surface of gravel, an indurated horizon, or bedrock is exposed. By then, sites will have substantially diminished potential for most values society desires from these lands. To avoid the fate of once-similar ecosystems in the Middle East, this trend must be reversed through proactive management.

Acknowledgments

Dr. Edmundo García Moya, Colegio de Postgraduados, Chapingo, Mexico, kindly gave suggestions of recent, relevant literature and a review of an early draft so as to make inclusion of Mexican conditions more complete. Jeff Creque provided a helpful review of an early draft. David Breshears sent reprints and reports of relevant work in and near Los Alamos National Laboratory. Craig Allen alerted the author to useful information about Bandelier National Monument.

REFERENCES

Austin, D. D., and Urness, P. J. 1976. Small mammal densities related to understory cover in a Colorado Plateau piñon–juniper forest. *Transactions of the Utah Academy of Sciences, Arts, and Letters* 53:5–12.

Axelrod, D. 1958. Evolution of the Madro-tertiary Geoflora. *Botanical Review* 24:433–509.

Axelrod, D. I. 1976. *History of the Coniferous Forests, California and Nevada*. University of California Publications. in Botany 70. Berkeley, Calif.: University of California Press.

Bahre, C. J. 1991. *A Legacy of Change: Historic Human Impact on Vegetation in the Arizona Borderlands*. Tucson, Ariz.: University of Arizona Press.

Bailey, A. W. 1996. Future role of fire in rangeland vegetation dynamics. In *Proceedings of the Fifth International Rangeland Congress*, vol. II, ed. N. E. West, pp. 160–163. Denver, Colo.: Society for Range Management.

Balda, P. P. 1987. Avian impacts on piñon–juniper woodlands. In *Proceedings – Piñon–Juniper Conference*, compiled by R. L. Everett, pp. 525–533. USDA Forest Service, Gen. Tech. Rep. INT-215.

Balda, R. P., and Masters, N. 1980. Avian communities in the piñon–juniper woodland: A descriptive analysis. In *Management of Western Forests and Grasslands for Nongame Birds*, Tech. Coord. R. M. DeGraff, pp. 146–169. USDA Forest Service, Gen. Tech. Rep. INT-86.

Barber, M. J., and Josephson, W. R. 1987. Wildfire patterns and vegetation response in east-central Nevada. In *Proceedings – Piñon–Juniper Conference*, compiled by R. L. Everett, pp. 158–160. USDA Forest Service, Gen. Tech. Rep. INT-215.

Barcia, D. R., and Merkel, E. P. 1972. *Bibliography on Insects Destructive to Flowers, Cones, and Seeds of*

North American Conifers. USDA Forest Service, Res. Pap. SE-92.

Barney, M. C., and Frischknecht, N. C. 1974. Vegetation changes following fire in the piñon–juniper type of west-central Utah. *Journal of Range Management* 27:91–96.

Barth, R. C. 1980. Influence of piñon pine trees on soil chemical and physical properties. *Soil Science Society of America Journal* 44:112–114.

Belsky, A. J. 1996. Western juniper expansion: Is it a threat to arid northwestern ecosystems? *Journal of Range Management* 49:53–59.

Betancourt, J. L. 1987. Paleoecology of piñon–juniper woodlands: Summary. In *Proceedings – Piñon–Juniper Conference*, compiled by J. L. Everett, pp. 129–139. USDA Forest Service, Gen. Tech. Rep. INT-215.

Betancourt, J. L., Pierson, E. A., Rylander, K. A., Fairchild-Parks, J. A., and Dean, J. S. 1993. Influence of history and climate on New Mexico piñon–juniper Woodlands. In *Managing Piñon–Juniper Ecosystems for Sustainability and Social Needs*, tech. coords. E. F. Aldon and D. W. Shaw, pp. 42–62. USDA Forest Service, Gen. Tech. Rep. RM-236.

Betancourt, J. L., Van Devender, T. R. and Marten, P. S., eds. 1990. *Packrat Middens: The Last 40,000 Years of Biotic Change.* Tucson, Ariz.: University of Arizona Press.

Blackburn, W., and Tueller, P. T. 1970. Pinyon and juniper invasion in black sagebrush communities of east-central Nevada. *Ecology* 51:841–848.

Breshears, D. D. 1993. Spatial partitioning of water use by herbaceous and woody life-forms in semiarid woodlands. Ph.D. dissertation, Colorado State University, Fort Collins, Colo.

Brewer, J. W. 1971. Biology of the piñon stunt needle midge. *Annals of the Entomological Society of America* 64:1099–1102.

Brewer, J. W., and Stevens, R. E. 1973. The piñon Pitch Nodule Moth in Colorado. *Annals of the Entomological Society of America* 66:789–792.

Brown, D. E., ed. 1982. Biotic communities of the American Southwest-United States and Mexico. *Desert Plants* Vol. 4, No. 1–4. Special Issue.

Bruner, A. D., and Klebenow, D. A. 1979. *Predicting Success of Prescribed Fires in Piñon–Juniper Woodlands in Nevada.* USDA Forest Service, Res. Paper INT-219.

Bunting, S. C. 1994. Effects of fire on juniper woodland ecosystems in the Great Basin. In *Proceedings – Ecology and Management of Annual Rangelands,* ed. S. B. Monsen and S. G. Kitchen, pp. 53–55. USDA Forest Service, INT-GTR-313.

Burkhardt, J. W., and Tisdale, E. W. 1976. Causes of juniper invasion in southwestern Idaho. *Ecology* 76:472–484.

Carrara, P. E., and Carroll, T. R. 1979. The determination of erosion rates from exposed tree roots in the Piceance Basin, Colorado. *Earth Surface Processes* 4:307–317.

Cartledge, T. R., and Propper, J. G. 1993. Piñon–juniper ecosystems through time: Information and insights from the past. In *Managing Piñon–Juniper Ecosystems for Sustainability and Social Needs,* tech. coords. E. F. Aldon and D. W. Shaw, pp. 63–71. USDA Forest Service, Gen. Tech. Rep. RM-236.

Cavazos-Doria, J. R., and Medina, A. L. 1996. Multiple use of rangelands in Northern Mexico In *Proceedings of the Fifth International Rangeland Congress,* vol. II, ed. N. E. West, pp. 46–51. Denver, Colo.: Society for Range Management

Charley, J. L., and West, N. E. 1975. Plant-induced soil chemical patterns in some shrub-dominated semi-desert ecosystems in Utah. *Journal of Ecology* 63:945–963.

Charley, J. L., and West, N. E. 1977. Micro-patterns of nitrogen mineralization activity in soils of some shrub dominated semi-desert ecosystems of Utah. *Soil Biology and Biochemistry* 9:357–365.

Cibrian-Tovar, D. B., Ebel, H. D. and Yates, J. T. III. 1986. *Cone and Seed Insects of the Mexican Conifers.* USDA Forest Service, Gen. Tech. Rep. SE-40.

Creque, J. A. 1996. An ecological history of Tintic Valley, Juab County, Utah. Ph.D. dissertation, Utah State University, Logan, Utah.

Davenport, D. W., Breshears, D. D., Wilcox, B. P., and Allen, C. D., 1998. Viewpoint: Sustainability of piñon–juniper ecosystems – a unifying perspective of soil erosion thresholds. *Journal of Range Management* 51:231–240.

Davenport, D. W., Wilcox, B. P., and Breshears, D. D. 1996. Soil morphology of canopy and inter-canopy sites in a piñon–juniper woodland. *Soil Science Society of America Journal.* 60:1881–1887.

DeLucia, E. H., and Schlesinger, W. H. 1991. Resource use efficiency and drought tolerance in adjacent Great Basin and Sierran plants. *Ecology* 72:51–58.

Despain, D. W., and Mosley, J. C. 1990. *Fire His-*

tory and Stand Structure of a Piñon–Juniper Woodland at Walnut Canyon National Monument, Arizona. Tucson, Ariz.: USDI National Park Service, Coop. Park Service Studies Unit, Univ. Tech. Rep. 34.

Erdman, J. A. 1969. piñon–juniper succession after fires on residual soils of Mesa Verde, Colorado. Brigham Young University Science Bulletin, Biological Series 11:1–24.

Everett, R. L. 1987. Allelopathic effects of piñon and juniper litter on emergence and growth of herbaceous species. In Proceedings: Symposium on Seed and Seedbed Ecology of Rangeland Plants, ed. G. W. Frasier and R. A. Evans, pp. 62–67. USDA, Agricultural Research Service.

Everett, R. L., and Sharrow, S. H. 1983. Understory seed rain on tree-harvested and unharvested piñon–juniper sites. Journal of Environmental Management 17:349–358.

Everett, R. L., Sharrow, S., and Thran, D. 1986. Soil nutrient distribution under and adjacent to singleleaf piñon crowns. Soil Science Society of America Journal 50:788–792.

Ferrera, C. R., and Saenz, G. J. 1987. Association simbiotica entre Pisolithus tinctorius y do especies de pinos pinoneros. In Memorias del II Simposium Nacional sabre Pinos Pinoneros Centre de Etudes Mexicaines et Centramericaines, compiled by M. F. Passini, D. C. Tovar, and T. E. Piedra, pp. 93–99. Chapingo, D. F.: Divion de Ciencias Forestales, Universidad Autonoma de Chapingo, Mexico.

Flanagan, L. B., Ehleringer, J. R., and Marshall, J. D. 1992. Differential uptake of summer precipitation among co-occurring trees and shrubs in a pinyon-juniper woodland. Plant, Cell and Environment 15:831–836.

Flora of North America Editorial Committee. 1993. Flora of North America, vol. 2. Pteridophytes and Gymnosperms. New York: Oxford University Press.

Floyd, M. L., and Kohler, T. A. 1989. Current productivity and prehistoric use of piñon (Pinus edulis) in the Dolores Archeological Project Area, southwestern Colorado. Economic Botany 44:141–156.

Forcella, F. 1980. Cone predation by piñon cone beetle (Conophthorus edulis: Scolytidae): Dependence on frequency and magnitude of cone production. The American Naturalist 116:594–598.

Fowler, C. S. 1986. Subsistence. In Handbook of North American Indians. vol. II. The Great Basin,

ed. W. L. D'Azevedo, pp. 64–97. Washington, D.C.: Smithsonian Institution.

Foxx, T. S., and Tierney, G. D. 1987. Rooting patterns in the piñon–juniper woodland. In Proceedings – Piñon–Juniper Conference, compiled by R. L. Everett, pp. 69–79. USDA Forest Service, Intermountain Research Station, Gen. Tech. Rep. INT-215.

Fresquez, P. R. 1990. Fungi associated with soils collected beneath and between piñon and juniper canopies in New Mexico. Great Basin Naturalist 50:167–172.

Gagne, R. J. 1970. A new genus and new species of Cecidomyiidae on piñon pine (Diptera). Entomology News (Philadelphia) 81:153–156.

Garcia Moya, E. 1985. Estado actual conocimento des los Pineños. In Memorias ler Simposium Nacional Sobre Pinos Piñoneros, pp. 1–18. Universidad Autonoma de Nuevo Leon Facultad de Silvicultura y Manejo de Recursos Renovables.

Gehlbach, F. R. 1967. Vegetation of the Guadalupe Escarpment, New Mexico, Texas. Ecology 48:404–418.

Gehring, C., and Whitham, T. 1995. Environmental stress influences aboveground pest attack and mycorrhizal mutualism in piñon–juniper woodlands: Implications for management in the event of global warming. In Desired Future Conditions for Piñon–Juniper Ecosystems, tech. coords. D. W. Shaw et al., pp. 30–37. USDA Forest Service, Gen. Tech. Rep. RM-258.

Gillet, C., and Brewer, J. W. 1983. Biology of Janetiella coloradensis Felt., a needle gall midge on Pinus edulis Engelm. Zeitschrift für Angewandte Entomologie 95:326–335.

Goldberg, D. E. 1983. The distribution of evergreen and deciduous trees relative to soil type: An example from the Sierra Madre, Mexico and a general model. Ecology 63:942–951.

Goodloe, S. 1993. The piñon–juniper invasion: An inevitable disaster. In Managing Piñon–Juniper Ecosystems for Sustainability and Social Needs, tech. coords. E. F. Aldon and D. W. Shaw, pp. 153–154. USDA Forest Service, Gen. Tech. Rep. RM-236.

Goodloe, S. 1995. Watershed restoration through integrated resource management on public and private rangelands. In Desired Future Conditions for Piñon–Juniper Ecosystems, tech. coords. D. W. Shaw, E. F. Aldon, and C. LoSapio,. pp. 136–140. USDA Forest Service, Gen. Tech. Rep. RM-258.

Gosz, J. R. 1992. Gradient analysis of ecological change in time and space: Implications for forest management. *Ecological Applications* 2:248–261.

Gottfried, G. F., Swetnam, T. W., Allen, C. D., Betancourt, J. L., and Chung-MacCoubrey, A. L. 1995. Piñon–juniper woodlands. In *Ecology, Diversity, and Sustainability of the Middle Rio Grande Basin*, tech. eds. D. M. Finch and J. A. Tainter, pp. 95–132. USDA Forest Service, Gen. Tech. Rep. RM-GTR-268.

Gray, J. R., Fowler, J. F., and Bray, M. A. 1982. Free-use fuelwood in New Mexico: Inventory, exhaustion, and energy equations. *Journal of Forestry* 80:23–26.

Grayson, D. K. 1994. The extinct Late Pleistocene mammals of the Great Basin. In *Natural History of the Colorado Plateau and Great Basin*, ed. K. T. Harper, L. L. St. Clair, K. H. Thorne, and W. M. Hess, pp. 55–85. Niwot, Colo.: University Press of Colorado.

Gunderson, L. H., Holling, C. S., and Light, S. S., eds. 1995. *Barriers and Bridges to the Renewal of Ecosystems and Institutions*. New York: Columbia University Press.

Holling, C. S., and Meffe, G. K. 1996. Command and control and the pathology of natural resource management. *Conservation Biology* 10:328–337.

Holling, C. S., Schindler, D. W., Walker, B. W., and Roughgarden, J. 1995. Biodiversity in the functioning of ecosystems: An ecological synthesis. In *Biodiversity Loss: Economies and Ecological Issues*, ed. C. Perrings, K. G. Maler, C. Folke, C. S. Holling, and B. D. Jansson, pp. 44–83. New York: Cambridge University Press.

Houseweart, M. W., and Brewer, J. W. 1972. Biology of a piñon spindle gall midge (Diptera: Cecidomyiidae). *Annals of the Entomological Society of America* 65:331–336.

Hreha, A. M. and Weber, D. J. 1979. Distribution of *Arceuthobium divaricatum* and *Phoradendron juniperum* (Viscaceae) on the South Rim of Grand Canyon, Arizona. *The Southwestern Naturalist* 24:625–636.

Jameson, D. A. 1961. *Growth Inhibitors in Native Plants of Northern Arizona*. USDA Forest Service, Res. Note RM-61.

Jameson, D. A. 1966. Piñon–juniper litter reduces growth of blue grama. *Journal of Range Management* 20:247–249.

Jameson, D. A. 1970a. Degradation and accumulation of inhibitory substances from *Juniperus*

osteosperma (Torr.) Little. *Plant and Soil* 33:213–224.

Jameson, D.A. 1970b. Juniper root competition reduces basal area of blue grama. *Journal of Range Management* 23:217–218.

Jameson, D. A. 1987. Climax vs. alternative steady states in woodland ecology. In *Proceedings – Pinyon-Juniper Conference*, compiled by R. L. Everett, pp. 9–13. USDA Forest Service, Gen. Tech. Rep. INT-215.

Jenkins, M. J. 1984. Seed and cone insects associated with *Pinus monophylla* in the Raft River Mountains, Utah. *Great Basin Naturalist* 44:310–312.

Johnsen, T. N. 1962. One seed juniper invasion of northern Arizona grasslands. *Ecological Monographs* 32:187–207.

Johnson, A. H., and Smathers, G. A. 1976. Fire history and ecology, Lava Beds National Monument. *Proceedings – Tall Timbers Fire Ecology Conference* 15:103–115.

Johnson, V. R. 1994. *California Forests and Woodlands*. Berkeley, Calif.: University of California Press.

Keeley, S. C., and Mooney, H. A. 1993. Vegetation in western North America, past and future. In *Earth System Responses to Global Change: Contrasts between North and South America*, ed. H. A. Mooney and E. R. Fuentos, pp. 209–237. New York: Academic Press.

Klopatek, C. C. 1987. Nitrogen mineralization and nitrification in mineral soils of piñon–juniper ecosystems. *Soil Science Society of America Journal* 51:453–457.

Klopatek, C. C., De Bano, L. F., and Klopatek, J. M. 1988. *Impact of Fire on the Microbial Processes in Piñon–Juniper Woodlands: Management Implications*. USDA Forest Service, Gen. Tech. Rep. RM-191.

Kohler, T. A. 1988. Long-term Anasazi land use and forest reduction: A case study from southwest Colorado. *American Antiquity* 53:537–564.

Kohler, T. A. 1992. Prehistoric human impact on the environment in the upland North American Southwest. *Population and Environment* 13:255–268.

Koniak, S. 1983. *Broadcast Seeding Success in Eight Piñon–Juniper Stands after Wildfire*. USDA Forest Service, Res. Note INT-334.

Koniak, S. 1985. Succession in piñon–juniper woodlands following wildfire in the Great Basin. *Great Basin Naturalist* 45:556–566.

Koniak, S., and Everett, R. L. 1982. Seed reserves

in soils of successional stages of piñon wood-lands. *The American Midland Naturalist* 108:295–303.

Küchler, A. W. 1970. Potential natural vegetation (map at scale 1:7,000,000). In *The National Atlas of the U.S.A.*, pp. 90–91. Washington, D.C.: Government Printing Office.

Lajtha, K., and Barnes, F. J. 1991. Carbon gain and water use in piñon–pine juniper woodlands of northern New Mexico. *Tree Physiology* 9:59–67.

Lajtha, K., and Getz, J. 1993. Photosynthesis and water use efficiency in piñon–juniper commu-nities along an elevational gradient in northern New Mexico. *Oecologia* 94:95–101.

Lewis, H.T. 1993. In retrospect. In *Before the Wilderness: Environmental Management by Native Californians*, ed. T. C. Blackburn and K. Anderson, pp. 389–400. Menlo Park, Calif.: Ballena Press.

Ligon, J. D. 1978. Reproductive interdependence of piñon jays and piñon pines. *Ecological Monographs* 48:111–126.

Malusa, J. 1992. Phylogeny and biogeography of the piñon pines (*Pinus* subsection Cembroides). *Systematic Botany* 17:42–66.

Marston, E. 1996. Raising a ranch from the dead. *High Country News* 28(7):1, 10–14.

Martinez-Delgado, E., Mellink, E., Aguirre-Rivera, J. R., and Garcia-Moya, E. 1996. Removal of piñon seeds by birds and rodents in San Luis Potosi. *The Southwestern Naturalist* 41:270–274.

McDaniel, P. A., and Graham, R. C. 1992. Organic carbon distributions in shallow soils of piñon–juniper woodlands. *Soil Science Society of America Journal* 56:499–504.

Miller, R. F., and Rose, J. A. 1995. Historic expan-sion of *Juniperus occidentalis* (western juniper) in southeastern Oregon. *Great Basin Naturalist* 55:37–45.

Miller, R. F., and Wigand, P. E. 1994. Holocene changes in semi-arid piñon–juniper woodlands. *BioScience* 44:465–474.

Milne, B. T., Johnson, A. R., Keitt, T. H., Hatfield, C. A., David, J., and Hraber, P. T. 1996. Detection of critical densities associated with piñon–juniper woodland ecotones. *Ecology* 77:805–821.

Minnich, R. A. 1991. Conifer forest fire dynamics and distribution in the mountains of southern California (Part Two). *Crossoma* 17:1–10.

Moir, W. H. 1979. Soil–vegetation patterns in the central Peloncillo Mountains, New Mexico. *The American Midland Naturalist* 102:317–331.

Mopper, S., and Whitham, T. G. 1992. The plant stress paradox: Effects on piñon sawfly sex ratios and fecundity. *Ecology* 73:515–525.

Nielson, R. P. 1987. On the interface between current ecological studies and the paleobotany of piñon–juniper woodlands. In *Proceedings – Piñon–Juniper Conference*, ed. R. Everett, pp. 93–98. USDA Forest Service, Gen. Tech. Rep. INT-215.

Nielson, R. P., and Marks, D. 1994. A global per-spective of regional vegetation and hydrologic sensitivities from climatic change. *Journal of Vegetation Science* 5:715–730.

Padien, D. J., and Lajtha, K. 1992. Plant spatial pattern and nutrient distribution in piñon–juniper woodlands along an elevational gradient in northern New Mexico. *International Journal of Plant Science* 153:425–433.

Parmenter, R. R., Brantley, S. L., Brown, J. H., Crawford, C. S., Lightfoot, D. C., and Yates, T. L. 1995. Diversity of animal communities on southwestern rangelands: Species patterns, habitat relationships, and land management. In *Biodiversity on Rangelands. Natural Resources and Environmental Issues*, vol. IV, ed. N. E. West, pp. 50–71. Logan: Utah State University.

Passini, M. F. 1985. Les forêts de *Pinus cembroides* Zucc. de la Sierra de Urica, Réserve de la Biosphère "La Michilla" État de Durango, Mexique. *Bulletin of Ecology* 16:161–168.

Passini, M. F., Delgadillo, J., and Salazar, M. 1989. L'écosystème forestier de Basse-California: Composition foristique, variables écologues principales dynamiques. *Acta Oecologia/Oecologia Plantarum* 10:275–293.

Peek, J. M., and Krausman, P. R. 1996. Grazing and mule deer. In *Rangeland Wildlife*, ed. P. R. Krausman, pp. 183–192. Denver, Colo.: Society for Range Management.

Peterson, R. S. 1967. Studies of juniper rusts in the west. *Madroño* 19:79–91.

Peterson, R. S. 1972. On *Coleosporium crowellii* (Uredinales). *Plant Dis. Report* 56:474–475.

Pieper, R. D. 1992. Species composition of wood-land communities in the Southwest. In *Proceedings – Symposium on Ecology and Management of Oak and Associated Woodlands: Perspectives in the Southwestern U.S. and Northern Mexico*, tech. coords. P. F. Ffolliott, G. Gottfried, D. Bennett, V. Hernandez, A. Ortega-Rubio, and R. Hamre, pp. 119–124. USDA Forest Service, Gen. Tech. Rep. RM-218.

Price, K. P. 1993. Detection of soil erosion within piñon–juniper woodlands using Thematic Mapper (TM) data. *Remote Sensing of Environment* 45:233–248.

Pyne, S. 1982. *Fire in America: A Cultural History of Wildland and Rural Fire.* Princeton, N.J.: Princeton University Press.

Riffle, J. W. 1967. Effect of an *Aphelenchoides* species on the growth of a mycorrhizal and a pseudomycorrhizal fungus. *Phytopathology* 57:541–544.

Riffle, J. W. 1968. *Pedicularis centrantheca,* a parasite of three southwestern tree species. *The Southwestern Naturalist* 13:99–100.

Riffle, J. W. 1972. Effect of certain nematodes on the growth of *Pinus edulis* and *Juniperus monosperma* seedlings. *Journal of Nematology* 4:91–94.

Robert, M. F. 1977. Notás sobré el estudio ecologico y fitogeographico de los bosques de *Pinus cembroides* Zucc. en Mexico. *Ciencia Forestal* 2:49–58.

Robert, M. F. 1979. Ensayo sobre la evolucion de los bosques de coniferas de la Sierra Madre Occidental. *Ciencia Forestal* 4:3–16.

Rogers, G. G. 1982. *Then and Now: A Photographic History of Vegetation Change in the Central Great Basin Desert.* Salt Lake City, Utah: University of Utah Press.

Rogers, T. J. 1993. Insect and disease associates of the piñon–juniper woodlands. In *Managing Piñon–Juniper Woodlands for Sustainability and Social Needs,* tech. coords. E. F. Aldon and D. W. Shaw, pp. 124–125. USDA Forest Service, Gen. Tech. Rep. RM-236.

Rollins, D., and Armstrong, B. 1994. Cedar through the eyes of wildlife. In *Juniper Symposium,* ed. C. A. Taylor, Jr., pp. 53–60. Texas A&M University Research Station, Sonora, Tex., Tech. Rep. 94-2.

Rollins, D., Bryant, F. C., Ward, D. W., and Bradley, L. C. 1988. Deer response to different intensities of brush removal in central Texas. *Wildlife Society Bulletin* 16:277–284.

Romero Manzanares, A., Garcia Moya, E., Oyama Nakagawa, K., and Passini, M. F. In press. Los inventarios floristicos y la Inercia-resiliencia en los piñonares meridionales de San Luis Potosi. *Boletin Sociedad Botanica de Mexico.*

Rzedowski, J. 1978. *Vegetacion de Mexico.* Mexico City, D. F.: Editorial Limusa.

Samuels, M. L., and Betancourt, J. L. 1982. Modeling the long-term effects of fuel wood harvests on piñon–juniper woodlands. *Environmental Management* 6:505–515.

Sauerwein, W. J. 1981. *Piñon–Juniper Management.* Portland, Ore.: USDA Soil Conservation Service, West Technology Service Center, Tech. Notes, Woodland No. 13.

Sauerwein, W. J. 1984. Too many trees? *Journal of Soil and Water Conservation* 39:348.

Scholes, R. J. 1996. Global change and global rangelands: Some mechanisms of interaction. In *Proceedings of the Fifth International Rangeland Congress,* ed. N. E. West, pp. 135–138. Denver, Colo.: Society for Range Management.

Scholl, D. G. 1972. Soil wetability in Utah juniper stands. *Soil Science Society of America Proceedings* 35:344–345.

Scurlock, D., and Johnson, A. R. 1993. Piñon–juniper in Southwest history: An overview of eco-cultural use. In *Human Ecology: Crossing Boundaries,* ed. S. D. Wright, pp. 272–286. Ft. Collins, Colo.: Colorado State University.

Segura, G., and Snook, L. C. 1992. Stand dynamics and regeneration patterns of a piñon pine forest in east central Mexico. *Forest Ecology and Management* 47:175–194.

Smeins, F. E., Owens, M. K., and Fuhlendorf, S. D. 1994. Biology and ecology of Ashe (Blueberry) Juniper. In *Juniper Symposium,* ed. C. A. Taylor, Jr., pp. 9–24. Sonora, Tex.: Texas A&M University Research Station, Tech. Rep. 94-2.

Sparks, S. R., West, N. E., and Allen, E. B. 1990. Changes in vegetation and land use at two townships in Skull Valley, western Utah. In *Proceedings – Symposium on Cheatgrass Invasion, Shrub Die-off, and Other Aspects of Shrub Biology and Management,* compilers E. D. McArthur, E. M. Romney, S. D. Smith, and P. T. Tueller, pp. 26–36. USDA Forest Service, Gen. Tech. Rep. INT-276.

Spencer, D.A. 1964. Porcupine population fluctuations in past centuries revealed by dendrochronology. *Journal of Applied Ecology* 1:127–150.

Swan, L. 1995. Western juniper: An evolving case study in commercialization, ecosystem management, and community development. In *Desired Future Conditions for Piñon–Juniper Ecosystems,* tech. coords. D. W. Shaw, E. F. Aldon, and C. LoSapio, pp. 179–183. USDA Forest Service, Gen. Tech. Rep. RM-258.

Tausch, R. J., and West, N. E. 1988. Differential establishment of pinyon and juniper following fire. *The American Midland Naturalist* 119:174–184.

Tausch, R. J., and West, N. E. 1995. *Plant Species Composition Patterns with Difference in Tree Dominance on a Southwestern Utah Piñon–Juniper Site*, tech. coords. D.W. Shaw, E. F. Aldon, and C. LoSapio, pp. 16–23. USDA Forest Service, Gen. Tech. Rep. RM-258.

Tausch, R. J., West, N. E., and Nabi, A. A. 1981. Tree age and dominance patterns in Great Basin piñon–juniper woodlands. *Journal of Range Management* 34:259–264.

Tausch, R. J., Wigand, P. E., and Burkhardt, J. W. 1993. Plant community thresholds, multiple steady states, and multiple successional pathways: Legacy of the Quaternary? *Journal of Range Management* 46:431–438.

Taylor, C. A., Jr., and Smeins, F. E 1994. A history of land use of the Edwards Plateau and its effect on the native vegetation. In *Juniper Symposium*, ed. C. A. Taylor, Jr., pp. 1–8. Sonora, Tex.: Texas A&M University Research Station, Tech. Rep. 94-2.

Thurow, T. L., and Carlson, D. H. 1994. Juniper effects on rangeland watersheds. In *Juniper Symposium*, ed. C. A. Taylor, Jr., pp. 31–43. Sonora, Tex.: Texas A&M University Research Station, Tech. Rep. 94-2.

Thurow, T. L., and Taylor, C. A., Jr. 1995. Juniper effects on the water yield of central Texas rangeland. In *Proceedings of the 24th Water for Texas Conference*, ed. R. Jensen, pp. 657–665. Austin, Tex.: Texas Water Research Institute.

Tiedemann, A. R. 1987. Nutrient accumulations in piñon–juniper ecosystems: managing for future site productivity. In *Proceedings – Piñon–Juniper Conference*, compiler R. L. Everett, pp. 352–359. USDA Forest Service, Gen. Tech. Rep. INT-215.

Tiedemann, A. R., and Klemmedson, J. O. 1995. The availability of Western Juniper development on soil nutrient availability. *Northwest Science* 69:1–8.

Traylor, D., Hubbel, L., Wood, N., and Fielder, B. 1990. *The 1977 La Mesa Fire: An Investigation of Fire and Fire Suppression Impact on Cultural Resources in Bandelier National Monument*. USDI National Park Service, Southwest Cultural Resources Center, Division of Anthropology Professionals Paper 28.

Treshow, M., and Allan, J. 1979. Annual variations in the dynamics of a woodland plant community. *Environmental Conservation* 6:231–236.

Tueller, P. T., Beeson, C. D., Tausch, R. J., West, N. E., and Rea, K. H. 1979. *Piñon–Juniper Woodlands of the Great Basin: Distribution, Flora, Vegetal Cover*. Chicago, Ill.:USDA Forest Service, Res. Pap. INT-229.

Van der Wall, S. B. 1990. *Hoarding in Animals*. Chicago, Ill.: University of Chicago Press.

Van Devender, T. R., and Spaulding, W. G. 1977. Development of vegetation and climate in the southwestern United States. *Science* 204:701–710.

Van Pelt, N. S. 1978. Woodland parks of southeastern Utah. M.S. thesis, University of Utah, Salt Lake City, Utah.

Vaughn, T. A. 1982. Stephen's woodrat, a dietary specialist. *Journal of Mammalogy* 63:53–62.

Wells, P. V. 1983. Paleobiogeography of montane islands in the Great Basin since the last glaciopluvial. *Ecological Monographs* 53:341–382.

West, N. E. 1984. Successional patterns and productivity potentials of piñon–juniper ecosystems. In *Developing Strategies for Range Management*. pp. 1301–1332. Boulder, Colo.: Westview Press.

West, N. E., Rea, K. H., and Tausch, R. J. 1975. Basic synecological relationships in piñon–juniper woodlands. In *The piñon–juniper ecosystem: A symposium*, ed. G. F. Gifford and F. E. Busby, pp. 41–58. Logan, Utah: Utah State University Agriculture Experiment Station.

West, N. E., Tausch, R. J., Rea, K. H., and Tueller, P. T. 1978. Phytogeographical variation within juniper–pinyon woodlands of the Great Basin. *Great Basin Naturalist Memoirs* 2:119–136.

West, N. E., and Van Pelt, N. A. 1987. Successional patterns in piñon–juniper woodlands. In *Proceedings – Piñon–Juniper Conference*, compiler R. L. Everett, pp. 43–52. USDA Forest Service, Gen. Tech. Rep. INT-215.

West, N. E., and Young, J. A. In press. Vegetation of intermountain valleys and lower mountain slopes. In *Terrestrial Vegetation of North America*, 2nd ed., ed. M. G. Barbour and W. D. Billings. New York: Cambridge University Press.

Whitford, W. G. 1987. Soil fauna as regulators of decomposition and nutrient cycling in piñon–juniper ecosystems. In *Proceedings – Piñon–Juniper Conference*, Compiler R. L. Everett, pp. 365–368. USDA Forest Service, Gen. Tech. Rep. INT-215.

Whitham, T. G., and Mopper, S. 1985. Chronic

herbivory: Impacts and architecture and sex expression of piñon pine. *Science* 228:1089–1091.

Wilcox, B. P., Pitlick, J., Allen, C. D., and Davenport, D. B. 1996. Runoff and erosion from a rapidly eroding piñon–juniper hillslope. In *Advances in Hillslope Processes*, vol. 1, ed. M. G. Anderson and S. M. Brooks, pp. 62–77. London: John Wiley & Sons.

Wilson, J .L., and Tkacz, B. M. 1992. Piñon *Ips* outbreak in piñon–juniper woodlands in northern Arizona: A case study. In *Proceedings – Symposium on Ecology and Management of Oak and Associated Woodlands: Perspectives in the Southwestern U.S. and Northern Mexico*, tech. coords. P. F. Ffolliott, G. Gottfried, D. Bennett, V. Hernandez, A. Ortega-Rubio, and R. Hamre, pp. 187–190. USDA Forest Service, Gen. Tech. Rep. RM-218.

Wright, H. A., and Bailey, A. W. 1982. *Fire Ecology: United States and Southern Canada*. New York: John Wiley & Sons.

Wright, H. A., Neuenschwander, L. F., and Britton, C. M. 1979. *The Role and Use of Fire in Sagebrush–Grass and Piñon–Juniper Plant Communities. A State-of-the-Art Review*. USDA Forest Service, Gen. Tech. Rep. INT-58.

Yorks, T. P., West, N. E., and Capels, K. M. 1994. Changes in piñon–juniper woodlands of western Utah's Pine Valley between 1933–1989. *Journal of Range Management* 47:359–364.

Young, J. A., and Budy, J. D. 1979. Historical use of Nevada's piñon–juniper woodlands. *Journal of Forest History* 23:113–121.

Young, J. A., and Evans, R. A. 1981. Demography and fire history of a western juniper stand. *Journal of Range Management* 34:501–506.

19 Serpentine Barrens of Western North America

A. R. KRUCKEBERG

Introduction

Serpentine habitats in western North America have been recognized for their special geological, floristic, and ecological significance since the mid-20th century. Their intimate link with plate tectonics was established in the revolution that remade geological science. The occurrence of serpentines is not uncommon in the three Pacific Coast states (California, Oregon, and Washington) and in adjacent British Columbia, Canada. The uniqueness of biota on serpentines has commanded much attention. Their rich endemic floras and remarkable physiognomy, which contrasts with typical vegetation on adjacent soils, coupled with specialized ecophysiological attributes, have interested a variety of plant biologists from physiologists and taxonomists to population and evolutionary biologists. Comprehensive reviews of serpentine biology can be found in Proctor and Woodell (1975), Kruckeberg (1985), Brooks (1987), Baker, Proctor and Reeves (1992), Roberts and Proctor (1992).

This review focuses on an aspect of Pacific Coast serpentines that is one of the major themes of the present volume, namely barrens. The word *barren* and its companion word *serpentine* have taken on diverse meanings. *Serpentine* has been used adjectivally (and even as a noun) by botanists in a loose generic sense to stand for rock, minerals, soils, flora, vegetation, habitats, and even landscapes. For the geologist, the word embraces a family of ferromagnesian miner-als, whereas *serpentinite* is used for a class of metamorphic rocks that contain serpentine minerals. The geological concept of serpentine will be considered later. However, the loose, all-inclusive meaning of serpentine is retained in this chapter.

The word *barren* with its several meanings has been intensively, though inconclusively, reviewed in a recent symposium on eastern North American barrens (Homoya 1994; Tyndall 1994; White 1994). In eastern North America, the barren is at least one thing – an opening in deciduous forest. But in the west, barrens are defined in a more literal sense: tracts of land almost devoid of vegetation, very nearly "moonscapes." Here, the word *barren* is largely restricted in use to serpentine outcrops with scarcely any plant cover, in marked contrast to adjacent non-serpentine habitats with more diverse plant cover. Further, the serpentine barren has no western counterpart on other substrates, as do the eastern North American shale or limestone barrens. I define serpentine barrens as those serpentine habitats that are nearly devoid of vegetation; they include the sparsely vegetated serpentine grasslands of the Coast Ranges of California. Nevertheless, not all Pacific Coast serpentine habitats qualify as barrens, as will be discussed later.

Two major themes will be amplified: (1) descriptive accounts of geology, soils, flora, and fauna, and (2) a focus on biological processes in the realm of ecophysiology and community dynamics.

Roger C. Anderson, James S. Fralish, and Jerry M. Baskin, eds. *Savannas, Barrens, and Rock Outcrop Plant Communities of North America*. Copyright © 1999 Cambridge University Press. All rights reserved.

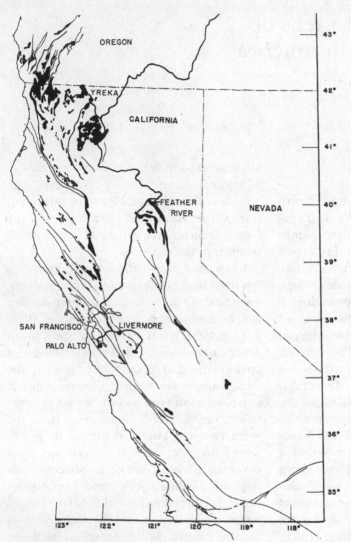

Figure 19.1. Distribution of serpentine and other ultramafic rock outcrops in southwestern Oregon and California. Note the strong north–south alignment that is related to plate movements, and the discontinuous pattern of exposures. (Map courtesy of R. Colman.)

Distribution

The connection between vulcanism of the Pacific Rim of Fire and ultramafics (serpentines and related lithology) is more than coincidental. Where there are vulcanism and other evidence of plate tectonics, there is likely to be the emergence of ultramafic man-

tle material at the sutures between continental and oceanic plates. Serpentine occurrences in California extend from Santa Barbara County, in the south, northward in two south–north-oriented bands (Figure 19.1). In western California, serpentine outcrops occur along the inner and outer Coast Ranges to Oregon; a narrower ribbon of serpentine outcrops east of the Great Valley extends north from Tulare County to Plumas County in the western Sierra Nevada. Several California counties south of the Bay Area have exposed serpentine, some abundantly so, like Santa Clara, San Luis Obispo, San Benito, and Monterey. Serpentine is abundant in the Bay Area counties as well: Contra Costa, San Mateo, and San Francisco. North of the Bay in the outer and inner Coast Ranges, serpentine is common in Sonoma, Napa, and Lake counties. Northward to the Oregon border, serpentines are increasingly frequent in all the counties of northwestern California (Figures 19.1, 19.2). In the Sierra Nevada of California, serpentine occurs frequently in a narrow band from the southern Sierra Nevada foothills to the Feather River country in Plumas County.

Oregon has two remotely separated serpentine areas. The main occurrence is in the geologically complex terrain of the Klamath–Siskiyou Mountains in the southwestern counties of the state (Curry, Josephine, and Jackson). These extensive outcrops are actually a northward continuation of those occurring in northwestern California. In this western sector of Oregon and adjacent

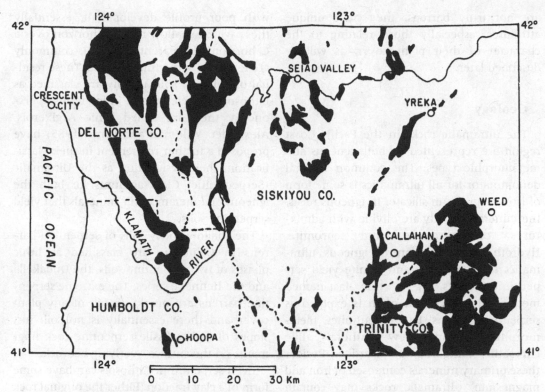

Figure 19.2. Distribution of serpentine outcrops is commonly insular, promoting genetic isolation. Weed Quadrangle, northwestern California. (Redrawn from Geologic Map of California, California Division of Mines and Geology, 1958–67, by F. Bardsley.)

California, the outcrops are in the coastal conifer biome. In central Oregon, serpentine is concentrated in the John Day country (the Strawberry and Aldrich mountains). Here serpentine ranges from low-elevation sagebrush and yellow pine (*Pinus ponderosa*) habitats to timberline on Baldy Mountain.

Further north in Washington State, serpentine occurrences are fewer and are smaller in size; the main outcrops are in the Wenatchee Mountains (Kittitas and Chelan counties) and in the North Cascades (Twin Sisters Mountain and the Stillaguamish River drainages). All outcrops are associated with conifer ecosystems.

British Columbia serpentines are scattered widely throughout the province; notable outcrops can be found in the Tulameen River district, in the Okanogan Highlands; scattered outcrops occur in the upper Fraser River country, along the Pinchi fault, and in the Cassiar Mountains of northern British Columbia.

For details of the geographic distribution of Pacific Coast serpentine (Figure 19.1), consult maps and descriptions in Kruckeberg (1969, 1985), and for British Columbia, consult Holland (1961). Serpentine barrens, as previously defined, are most common in the xeric Mediterranean climate of California, although barrens also can be found within the conifer biomes of Oregon and Washington. However, many of the serpentine occurrences in the region are not barrens. Rather, the truly barren landscapes, which are sparsely vegetated or nearly devoid of plant cover, woody or herbaceous, appear intermittently, mostly on unstable steeper slopes of raw serpentine rock outcrops. Although the majority of serpentine habitats are not truly barrens, they have unique attributes, especially those relating to the character of their plant cover, as will be described later.

Geology

The ultramafic rocks in the Pacific Coast region are represented by both igneous and metamorphic types. The common mineral denominator for all ultramafics is some form of iron magnesium silicate. In igneous rocks, the minerals usually are olivine with admixtures of pyroxene and chromite. Hydrothermal alteration of igneous ultramafics like peridotite and dunite yield serpentinite rocks. It is this last-named metamorphic rock that often is exposed as serpentine barrens. For serpentinites, metamorphism yields new minerals like chrysotile, antigorite, or lizardite. Besides these primary minerals composed of iron and magnesium, ultramafic rocks may contain several "impurities," notably minerals with nickel, chromium, and cobalt. This exceptional mineral composition of ultramafic rocks, so radically different from most other rock types, is largely responsible for the striking biological responses to serpentines.

The north–south-oriented belts of ultramafics along the Pacific Rim are associated with major fault zones and vulcanism; the rocks were derived from mantle magma and emplaced tectonically from oceanic lithosphere at the continental margin. Further discussions of ultramafic geology are in Kruckeberg (1985) and Coleman and Jove (1992). A fascinating account of the significance of plate tectonics for the region is given in John McPhee's dialogues with the geologist Eldridge Moores in *Assembling California* (McPhee 1993).

Soils

The weathered products of ultramafic rocks form a variety of soil types. Those soils with poor profile development, essentially those with a shallow rocky A horizon over a C horizon (the parent material), commonly are found on serpentine barrens. These residual or colluvial soils usually are classified as Lithosols or skeletal soils; in modern soil taxonomy they are called lithic Argixerols. Alexander, Wildman and Lynn (1985) have proposed a further refinement in their classification; their designation as the Ultramafic (Serpentinitic) Class accounts for both the igneous and metamorphic minerals that yield serpentine soils.

The shallow rocky soils of serpentine barrens in California are classified as lithic phases of two serpentine soils, the Dubakella and the Henneke series. The extreme serpentine barrens are nearly devoid of any plant cover and there essentially is no soil: just slopes of highly fissile serpentine rock fragments on the surface overlying bedrock.

Even serpentinitic Lithosols can have some form of a clay fraction. Either the original rock minerals become the clay component, or new minerals are chemically transformed into soil clay minerals like montmorillonite and smectite. In either form, the clay fractions have a cation exchange capacity, ranging from 5 to 50 milliequivalents per 100 g of soil. Exchangeable magnesium far exceeds calcium as the major divalent cation. The combination of high Mg and low Ca apparently is the major cause of infertility of serpentine soils on barrens and the more vegetated serpentines (Walker 1954; Kruckeberg 1985). The Mg:Ca ratio of more than 1.0 (Table 19.1) leads to low levels of other available soil nutrients, especially nitrogen and phosphorus. This deviation from normal soil nutrient composition is thought to be the primary cause of what Jenny (1980) calls the "serpentine syndrome," a nutritional imbalance coupled with mineral deficiencies, leading to sparse plant cover. Furthermore, the syndrome includes adaptive plant responses to the soil, such as xeromorphic plant form,

Table 19.1. *Characteristics of ultramafic soils in western North America. Cation exchange capacity (CEC) and nutrient values are expressed as milliequivalents per 100 g of oven-dry soil. Data are from Kruckeberg (1969, 1985). For comparison, sample 150 is a non-ultramafic soil.*

Sample no.	Locality	CEC	Ca	Mg	Ca + Mg	Mg/Ca	pH
87	Olivine Mt., British Columbia	20.1	6.5	13.9	20.4	2.1	6.6
3	Wenatchee Mts.	18.4	4.7	11.1	15.8	2.4	7.0
9	Wenatchee Mts., Washington	27.8	4.9	16.0	20.9	3.3	6.4
63	Twin Sisters Mt., Washington	16.1	1.5	16.5	18.0	11.0	7.1
51	Baldy Mt., Oregon	25.0	2.3	15.2	17.5	6.6	6.6
138	San Benito County, California	15.0	2.4	11.4	13.8	4.8	7.2
135	Lake County, California	16.0	2.8	11.8	14.6	4.2	7.0
152	Del Norte County, California	14.0	2.8	7.1	9.9	2.5	7.1
150	San Luis Obispo County, California	34.0	52.5	2.7	55.2	0.1	7.5

endemism, and overall tolerance to the serpentine habitat. Detailed accounts of serpentine soils can be found in Proctor and Woodell (1975), Jenny (1980), Kruckeberg (1985), and Brooks (1987).

However, there is a major gap in our knowledge of the soil–plant link for the serpentine syndrome. What part do soil algae, fungi, and small invertebrates play in the nutritional status of serpentine barrens, soils, and associated plants? The scanty literature on this subject was reviewed by Kruckeberg (1992). Two studies suggest that serpentine soils are highly selective for microorganisms, and that on the most extreme barrens the few tolerant plants are likely to be mycorrhizal. In the Wenatchee Mountains, Washington, Pegtel (1980) found that a nitrogen-fixing *Rhizobium* associated with a serpentine-tolerant legume (*Lupinus laxiflorus*) was a serpentine-tolerant strain. Hopkins (1987) found that 25 of the 27 serpentine grassland species in central California had vesicular–arbuscular mycorrhizae. Further study of plant microorganisms and their influences on plant growth on serpentine soils will provide much-needed information on the serpentine soil–plant linkage.

Vegetation

Because ultramafic outcrops occur from British Columbia to southern California, the nature of the serpentine vegetation reflects the regional climate and its associated bioregional flora with which the serpentine habitats are linked. In British Columbia, Washington, Oregon, and northwestern California, conifer forest is the predominant ecosystem type. Vegetational diversity is greater in central and southern California. Serpentines can be associated with mixed conifer–hardwood forest in the Coast Ranges, often with an admixture of chaparral at lower elevations. Further south, serpentines occur with hard (inland) to soft (coastal) chaparral, chaparral woodland, or grassland. Often, serpentine community is a depauperate version of the surrounding nonserpentine associations. For example, conifer woodland on nearby nonserpentine soils is transformed into sparsely stocked, open conifer stands composed of either (1) a single tree species or (2) a unique mix of the regional tree species and either species endemic to serpentine (e.g., *Cupressus sargentii, C. macnabiana*) or "extralimital" species in their occurrence on serpentine (*Calocedrus decurrens, Pinus jeffreyi,* and *P. sabiniana*) (Whittaker 1960; Kruckeberg 1969, 1985). Similarly, chaparral of nonserpentine sites is replaced by serpentine chaparral, often with species not found in nonserpentine chaparral (Hanes 1977; Kruckeberg 1985).

Figure 19.3. Serpentine barren in conifer biome, Wenatchee Mountains, Kittitas County, Washing- ton. Sparse herbaceous cover on the barren consists mostly of serpentine endemics. (Photo by author.)

Barrens may occur in any of the aforementioned vegetation types. Thus, in the coniferous serpentine landscape in Washington, Oregon, or California, barrens will be openings in the midst of conifer stands on the same serpentine outcrops; these clearings, often many hectares in size, are devoid of any woody species and may support a low-density herbaceous flora. On these barrens, endemic taxa are well represented in the sparse cover. Typical serpentine barrens are shown in Figures 19.3 and 19.4.

Plant community types have been recognized for some Pacific Coast serpentine habitats; most are tree or shrub communities. Serpentine chaparral, serpentine grassland, and serpentine woodland are recognized for

California (Barbour and Major 1977). For the Siskiyou Mountains of Oregon and adjacent northwestern California, Franklin and Dyrness (1988) describe a *Pinus jeffreyi*–grass–woodland on xeric serpentine sites. Without specifying community types, Whittaker (1960), in his classic study of Siskiyou vegetation on contrasting ultramafic and diorite soils, singled out several plant associations that occur along moisture gradients, with serpentine barrens at the most xeric end of a gradient. No community types have been formally named for West Coast serpentine barrens, although most have consistent floristic assemblages of herbaceous taxa. Plant associations have been identified for the several Research Natural Areas (RNA) on serpentine

Figure 19.4. Serpentine barren in the New Idria area of San Benito County, California. The barren is thinly populated with endemic herbs such as *Layia discoidea* and *Camissonia benitensis*, with *Pinus* *sabiniana* and *P. jeffreyi* along the border. This Clear Creek barren is the most extensive and spectacular barren in California. (Photo courtesy J. Griffin.)

in California. For the Frenzel Creek RNA (Keeler-Wolf 1990, p. 89), one of the three serpentine associations is called "Serpentine Barrens." It has an extremely sparse herbaceous vegetation, including nine serpentine endemic or indicator species.

Apparently, no quantitative data on the vegetation of western serpentine barrens per se have been published. Several authors have given cover and abundance data for those serpentine habitats that support woody vegetation (trees and shrubs) along with herbaceous plants. Where these authors have recorded vegetation data along a moisture gradient, one can infer that the xeric end of the gradient approximates the barren habitat. For the serpentine areas of the Siskiyou Mountains of southwestern Oregon, Whittaker (1960, Tables 9–11) includes quantitative data for communities of mesic to

xeric conditions, where the latter habitat is likely to be a barren. A similar inference can be drawn from vegetation analysis of serpentines in the Wenatchee Mountains of Washington (del Moral 1982). The xeric end of the gradient is a steep serpentine scree that fits the character of a barren.

Serpentine grassland communities are treated here along with serpentine barrens. Serpentine grasslands are found mainly in the Bay Area counties of central California (Marin, Sonoma, San Francisco, San Mateo, and Santa Clara). They were first recognized by Howell (1970) in his flora of Marin County. They differ markedly from nonserpentine grasslands. Density and abundance of herbs (grasses and forbs) are substantially lower on serpentines than on nonserpentines (McNaughton 1968). Further, alien annual grasses are much less common, or even absent, in the serpentine grasslands. The presence of several serpentine endemic herbs in these serpentine grasslands is noteworthy.

Attributes of Serpentine Floras

Serpentine soils can support a variety of vegetation types, from forested landscapes and chaparral to barrens. With this diversity of vegetation types, the distinctive feature of species composition stands out as uniquely "serpentinitic." Even given the paucity of plant cover and species diversity on barrens, three rather distinct categories of higher plant taxa can be recognized as major floristic components of most serpentine habitats: (1) endemic taxa, both species and infraspecific variants, which are narrowly restricted to serpentine, either locally or regionally; (2) indicator species, which are nonendemic taxa that locally or regionally are conspicuous on serpentine, but not restricted to it; and (3) indifferent species, which are widespread and common native species occurring on a wide range of parent materials/soils. These last are also called *bodenvag* (soil-wandering) or ubiquist taxa, and their populations usually include serpentine-tolerant and serpentine-intolerant races (Kruckeberg 1951, 1967, 1995). Representatives from these floristic types can be found on barrens; proportions of each type will differ, depending on habitat; for example, forest, chaparral, grassland, or barren.

Some species of adjacent nonserpentine sites are excluded from serpentine (Kruckeberg 1985). This absence on serpentine of certain species in the nearby nonserpentine flora is a major cause of the unique floristic composition and physiognomy of serpentine vegetation; for example, blue oak savanna is replaced by serpentine chaparral or the barren habitat in the Sierra Nevada of California.

For serpentine barrens, narrow endemism is a keynote feature. In fact, barrens may harbor only endemics. Thus, the barren openings

in serpentine coniferous woodlands in the Wenatchee Mountains of Washington support sparse populations of such narrow endemics as *Poa curtifolia, Lomatium cuspidatum,* and *Cryptantha thompsonii* (Table 19.2). This preponderance of endemics in the floras of serpentine barrens also occurs in California. The several serpentine endemics in the brassicaceous genus *Streptanthus* are confined to barrens, and species like *S. brachiatus, S. barbiger,* and *S. morrisonii* exist only on the most extreme barrens. In fact, these species may be the only plant life on some barrens of the central Coast Range serpentines (Kruckeberg and Morrison 1983).

Why should this high frequency of endemism be the hallmark of serpentine barrens? Two attributes of serpentine barrens may be prime selective forces. First, the barrens habitat is the severest of serpentine environments because almost nonexistent soil and soil formation lead to nutritional sterility coupled with high drought stress. The second factor is the discontinuous occurrence of barrens. Whereas serpentine areas may be extensive in a region, barrens tend to occur sporadically and discontinuously within a serpentine landscape.

Therefore, isolation, coupled with the challenge of a stressful habitat, can promote speciation, the genesis of endemics. A model for speciation on serpentine barrens has been proposed by Kruckeberg (1986). A possible sequence of stages in speciation begins with (1) preadaptation to serpentine by individuals of an ubiquist species, (2) acquisition of genetically fixed tolerance to serpentine by a small population, and (3) further divergence in character states of the adapted population induced by spatial (ecogeographic) isolation. If this model is correct, it is certainly the serpentine barren that best serves as the milieu on which it is played out.

Since barrens of all kinds are usually discontinuously distributed, they are ideal examples of edaphic "islands." Their insularity on continental mainlands can be tested to

Table 19.2. *Selected endemic (E) and indicator (I) species of Pacific Coast serpentine barrens. In California, only endemics mainly on barrens are listed; there are about 215 serpentine endemics on both barren and nonbarren habitats (Kruckeberg 1985).*

Washington

Arenaria rubella	I	*Cryptantha thompsonii*	E
Chaenactis ramosa	I	*Douglasia nivalis*	I
Chaenactis thompsonii	E	*Lomatium cuspidatum*	E
Claytonia megarhiza var. *nivalis*	E	*Poa curtifolia*	E
Cerastium arvense	I		

Oregon and northwestern California

Antennaria suffrutescens	E	*Haplopappus racemosus congestus*	E
Arenaria howellii	I	*Hieracium bolanderi*	E
Arabis aculeolata	E	*Horkelia sericata*	E?
Arabis koehleri stipitata	E	*Iris innominata*	E?
Arabis serpentinicola	E	*Juniperus communis jackii*	E
Astragalus whitneyi siskiyouense	E?	*Lomatium howellii*	E?
Balsamorhiza platylepis	E?	*Lomatium engelmannii*	E?
Brodiaea greenei	E?	*Lomatium tracyi*	E?
Brodiaea hendersonii	E?	*Phacelia capitata*	E
Calochortus greenei	E?	*Phacelia dasyphylla ophitidis*	E
Calochortus howellii	E?	*Sanicula peckiana*	E?
Calochortus tolmiei	I	*Sanicula tracyi*	E?
Castilleja elata	E?	*Sedum laxum laxum*	E
Castilleja brevibracteata	E?	*Sedum laxum heckneri*	E
Ceanothus pumilus	E	*Sedum moranii*	E
Dicentra formosa oregana	E?	*Senecio fastigiatus*	E
Epilobium rigidum	E?	*Streptanthus howellii*	E
Eriogonum ternatum	E?	*Tauschia howellii*	E?
Erythronium howellii	E?	*Thlaspi montanum siskiyouense*	E
Fritillaria atropurpurea	E?	*Viola lobata psychodes*	E?

California (monocots)

Allium falcifolium	*Calochortus raichei*
Allium fimbriatum diabolense	*Chlorogalum grandiflorum*
Allium hoffmannii	*Erythronium,* 4–5 spp.
Allium howellii sanbenitensis	*Festuca microstachys complex*
Allium sanbornii	*Fritillaria,* ca. 7 spp.
Calochortus obispoensis	*Stipa lemmonii pubescens*
Calochortus tiburonensis	

California (dicots)

Acanthomintha spp.	*Cordylanthus,* 4 spp.
Arabis macdonaldiana	*Collinsia greenei*
Arabis constancei	*Eriogonum alpinum*
Arenaria rosei	*Eriogonum kelloggii*
Arenaria howellii	*Eriogonum libertini*
Astragalus, 3 spp.	*Eriophyllum jepsonii*
Benitoa occidentalis	*Galium ambiguum siskiyouense*
Castilleja neglecta	*Hesperolinon,* several spp.
Clarkia franciscana	*Hemizonia halliana*
Calystegia stebbinsii	*Layia discoidea*

(cont.)

Table 19.2. (cont.)

Linanthus spp.	Streptanthus brachiatus
Lomatium, 6 spp.	Streptanthus breweri
Madia hallii	Streptanthus drepanoides
Mimulus nudatus	Streptanthus hesperidis
Navarretia spp.	Streptanthus insignis
Phacelia corymbosa	Streptanthus morrisonii
Senecio lewisrosei	Streptanthus niger
Streptanthus barbatus	Streptanthus polygaloides
Streptanthus barbiger	Streptanthus tortuosus optatus
Streptanthus batrachopus	Veronica copelandii

British Columbia to California (ferns)

Aspidotis densa	I	Polystichum lemmonii (mesic barrens)	E

determine the extent to which they fit the classic island biogeography theory of MacArthur and Wilson (1967). Pacific Coast serpentine barrens epitomize this insularity (Figures 19.1, 19.2). Barrens, even more than serpentine chaparral or serpentine woodlands, are highly discontinuous in their spatial array. This kind of insularity fosters genetic isolation, a key promoter of evolutionary divergence and speciation.

I proposed (Kruckeberg 1991) that the spatial isolation of serpentine barrens only partially fits the MacArthur–Wilson model. Size of barren "island" and distance between barrens does fit the theory; that is, the smaller the island, the fewer the number of species, and species richness may decrease with distance between barrens. For example, a small serpentine outcrop in Santa Barbara County, California, has only one endemic (Streptanthus amplexicaulis var. barbarae), whereas the extensive barren at New Idria (San Benito County), about 130 airline miles to the north, has several serpentine endemics as well as indicator species. But the recruitment onto mainland islands differs. Potential immigrants are often on nearby normal substrates, unlike on oceanic islands where recruits must migrate over inhospitable sea. Further, serpentine barrens are not only insular, they are extremely stressful habitats quite unlike most oceanic islands.

In addition to tolerating the stressful physical and nutritionally demanding environments of the serpentine barren, the plants must also tolerate the presence of heavy metals, especially nickel. Most serpentine plants, especially endemics, exclude nickel. They grow on soils with high levels of nickel, but do not accumulate major amounts toxic to most plants. A less common way of surviving on soils with high nickel levels is by hyperaccumulation. A moderate number of hyperaccumulators have been identified worldwide. Yet, for Pacific Coast serpentines, only three of the several taxa tested took up nickel in excess of 1,000 ppm, the defining attribute of hyperaccumulator species (Reeves, Brooks and MacFarlane 1981; Brooks 1987; Kruckeberg, Peterson and Samiullah 1993; Kruckeberg and Reeves 1995). The most impressive of these hyperaccumulators is Streptanthus polygaloides (Brassicaceae), which occurs on barrens and in serpentine chaparral of the California Sierra Nevada foothill serpentines. Every population sampled throughout the extensive north–south range of S. polygaloides accumulated nickel, often far above the 1,000-ppm level (Kruckeberg and Reeves 1995). The other two Pacific Coast hyperaccumulators are Thlaspi montanum ssp. siskiyouense (Reeves, MacFarlane and Brooks 1983), endemic to the Siskiyou Mountains of northwestern California and adjacent Oregon, and *Arenaria rubella*, an indicator species on the Twin Sisters Mountain ultramafic rock dunite in Washington (Kruckeberg, Peterson and Samiullah 1993). The two substantially different strategies for surviving high levels of heavy metals (exclusion versus hyperaccumulation) pose intriguing questions for ecophysiologists and evolutionary biologists. Both strategies allow plants to tolerate high soil nickel levels and thus avoid their toxic effects. To exclude nickel, the plant must utilize a particular cellular mechanism, whereas hyperaccumulation requires a different mechanism. Accumulation requires that this element be moved into cells and then sequestered in a nontoxic form or "safe" site. There remains a strong need for cellular and molecular biologists to seek causal mechanisms for these fascinating phenomena. Furthermore, additional sampling for nickel levels of serpentine barren plants is needed, particularly in Oregon and California. It is likely that more hyperaccumulators will be found. Yet, I predict that the higher proportion of excluders to hyperaccumulators will remain the same as it is at present.

Fauna

Although some vertebrates (rodents, snakes, and lizards) inhabit barrens, none is distinctly a "serpentine" animal. It is the arthropod fauna that is most likely to have some intimate links to serpentine soils and flora. Certain butterflies have serpentine affinities. The long-term studies by Paul Ehrlich and colleagues of the population biology of the checkerspot butterfly (*Euphydryas editha*) indicate that certain races are monophagous on serpentine grassland plants (Singer et al. 1995). Shapiro (1981) reported a remarkable instance of mimicry involving species of the serpentinicolous genus *Streptanthus* (Brassicaceae) and butterflies. Whitish callosities on the teeth of leaf margins resemble insect eggs, these callosities dissuade adult butterflies from laying eggs on the plants. Shapiro calls this novel syndrome "egg mimicry." Shapiro (1995) also has found the ecotypically variable *Papilio zelicaon* to be univoltine (one brood per season) on serpentine barrens in central California; they feed on species of *Lomatium* (Apiaceae) native on serpentine soils. Harrison and Shapiro (1988) describe features of population biology of butterflies on serpentine in northern California.

There are both need and opportunity for further work on the faunal links to serpentine barrens as well as other kinds of serpentine habitats. Do herbivores have to cope with the nutritional peculiarities of serpentine soils? Are there endemic and/or indicator animals on serpentines?

Summary

Barrens on outcrops of serpentine and other ultramafic (ferromagnesian) rocks occur frequently in the Pacific Coast states and in British Columbia, but they are most abundant in southwestern Oregon and western California. The skeletal soils of the barrens are high in magnesium, low in calcium and other plant nutrients, and may contain toxic levels of nickel. These stressful chemical and associated physical properties are responsible for a depauperate as well as unique flora and vegetation on the barrens. Yet, evolutionary accommodation to serpentines often has resulted in ecotypic (racial) tolerance and narrow endemism. The greatest number of serpentine endemics occur in southwestern Oregon and cismontane California. The discontinuous, insular occurrence of serpentine barrens and their highly stressful environment are probable causes of their remarkable floristic composition. As defined here, the vegetation of the barrens is mainly herbaceous, consisting primarily of widely scattered annuals and perennials, in contrast to those serpentine habitats (nonbarren in nature) with woody as well as herbaceous species. Species richness of the barrens is low and often consists mostly of narrow endemics.

REFERENCES

Alexander, E. B., Wildman, W. E., and Lynn, W. C. 1985. Ultramafic (Serpentinitic) mineralogy class. In *Mineral Classification of Soils*, Special Publication No 16, pp. 135–146. Madison, Wis.: Soil Science Society of America.

Baker, A. J. M., Proctor, J., and Reeves R. D., eds. 1992. *The Vegetation of Ultramafic (Serpentine) Soils*. Andover, U.K.: Intercept Ltd.

Barbour, M. G., and Major, J. 1977. *Terrestrial Vegetation of California*. New York: John Wiley & Sons.

Brooks, R. R. 1987. *Serpentine and Its Vegetation: A Multidisciplinary Approach*. Portland, Ore.: Dioscorides Press.

Coleman, R. G., and Jove, C. 1992. Geological origin of serpentinites. In *The Vegetation of Ultramafic (Serpentine) Soils*, ed. A. J. M. Baker, J. Proctor, and R. Reeves. Andover, U.K.: Intercept Ltd.

del Moral, R. 1982. Control of vegetation on contrasting substrates: Herb patterns on serpentine and sandstone. *American Journal of Botany* 69:227–238.

Franklin, J. F., and Dyrness, C. T. 1988. *Vegetation of Oregon and Washington*. Corvallis, Ore.: Oregon State University Press.

Hanes, T. L. 1977. California chaparral. In *Terrestrial Vegetation of California*, ed. M. G. Barbour and J. Major, pp. 417–469. New York: John Wiley & Sons.

Harrison, S., and Shapiro, A. M. 1988. Butterflies of northern California serpentines. *Fremontia* 15:17–20.

Holland, S. S. 1961. Jade in British Columbia. In *Placer* (Annual Report Minister of Mines and Petroleum Resources), pp. 118–126. Province of British Columbia.

Homoya, M. A. 1994. Barrens as an ecological term: an overview of usage in the scientific literature. In *Proceedings of the North American Conference on Barrens and Savannas*, ed. J. S. Fralish, R. C. Anderson, J. E. Ebinger, and R. Szafoni, pp. 295–303. Chicago, Ill.: U.S. Environmental Protection Agency, Great Lakes National Program Office.

Hopkins, N. A. 1987. Mycorrhizae in a California serpentine grassland. *Canadian Journal of Botany* 65:484–487.

Howell, J. T. 1970. *Marin Flora*. 2d ed. Berkeley, Calif.: University of California Press.

Jenny, H. 1980. *The Soil Resource: Origin and Behavior*. New York: Springer Verlag.

Keeler-Wolf, T. 1990. *Ecological Surveys of Forest Service Research Natural Areas in California*. Berkeley, Calif.: USDA Forest Service, Pacific Southwest Research Station.,Gen. Tech. Rep. PSW-125.

Kruckeberg, A. R. 1951. Intraspecific variability in response of certain native plant species to serpentine soil. *American Journal of Botany* 38:408–419.

Kruckeberg, A. R. 1967. Ecotypic response to ultramafic soils by some plant species of northwestern United States. *Brittonia* 19:133–151.

Kruckeberg, A. R. 1969. Plant life on serpentinite and other ferromagnesian rocks in northwestern North America. *Syesis* 2:15–114.

Kruckeberg, A. R. 1985. California serpentines: flora, vegetation, geology, soils and management problems. *University of California Publications in Botany* 78:1–180.

Kruckeberg, A. R. 1986. An essay: The stimulus of unusual geologies for plant speciation. *Systematic Botany* 11:455–463.

Kruckeberg, A. R. 1991. An essay: Geoedaphics and island biogeography for vascular plants. *Aliso* 13:225–238.

Kruckeberg, A. R. 1992. Serpentine biota of western North America. In *The Vegetation of Ultramafic (Serpentine) Soils*, ed. A. J. M. Baker, J. Proctor, and R. D. Reeves, pp. 19–33. Andover, U.K.: Intercept Limited.

Kruckeberg, A. R. 1995. Ecotypic variation in response to serpentine soils. In *Genecology and Ecogeographic Races*, ed. A. R. Kruckeberg, R. B. Walker, and A. Leviton, pp. 57–66. San Francisco: American Association for the Advancement of Science (Pacific Division).

Kruckeberg, A. R., and Morrison J. L. 1983. New *Streptanthus* taxa (Cruciferae) from California. *Madrono* 10:230–244.

Kruckeberg, A. R., Peterson, P. J., and Samiullah, Y. 1993. Hyperaccumulation of nickel by *Arenaria rubella* (Caryophyllaceae) from Washington State. *Madrono* 40:25–30.

Kruckeberg, A. R., and Reeves, R. D. 1995. Nickel accumulation by serpentine species of *Streptanthus* (Brassicaceae): Field and greenhouse studies. *Madrono* 42:458–469.

MacArthur, R. H., and Wilson, E. O. 1967. *The Theory of Island Biogeography*. Princeton, N.J.: Princeton University Press.

McNaughton, S. J. 1968. Structure and function in California grasslands. *Ecology* 49:962–972.

McPhee, J. 1993. *Assembling California.* New York: Farrar, Strauss and Giroux.

Pegtel, D. M. 1980. Evidence for ecotypic differentiation in *Lupinus*-associated *Rhizobium. Acta Botanica Neerlandica* 29:429–441.

Proctor, J., and Woodell, S. J. R. 1975. The ecology of serpentine soils. *Advances in Ecological Research* 9:255–365.

Reeves, R. D., Brooks, R. R., and MacFarlane, R. M. 1981. Nickel uptake by Californian *Streptanthus* and *Caulanthus* with particular reference to the hyperaccumulator, *S. polygaloides* Gray (Brassicaceae). *American Journal of Botany* 68:708–712.

Reeves, R., MacFarlane, R. M., and Brooks, R. R. 1983. Accumulation of nickel and zinc by western North American genera containing serpentine-tolerant species. *American Journal of Botany* 70:1297–1303.

Roberts, B., and Proctor, J., eds. 1992. *The Ecology of Areas with Serpentinized Rocks – A World View.* Dordrecht: Kluwer Academic Publishers.

Shapiro, A. M. 1981. Egg-mimics of *Streptanthus* (Cruciferae) deter oviposition by *Pieris sisymbrii* (Lepidoptera: Pieridae). *Oecologia* 48:142–143.

Shapiro, A. M. 1995. From the mountains to the prairies to the oceans white with foam: *Papilio zelicaon* itself at home. In *Genecology and Ecogeographic Races*, ed. A. R. Kruckeberg, R. B. Walker, and A. Leviton, pp. 67–100. San Francisco: American Association for the Advancement of Science (Pacific Division).

Singer, M. C., White, R. R., Vasco, D. A., Thomas, C. D., and Broughton, D. R. 1995. Multi-character ecotypic variation in Edith's checkerspot butterfly. In *Genecology and Ecogeographic Races*, ed. A. R. Kruckeberg, R. B. Walker, and A. Leviton, pp. 67–100. San Francisco: American Assoc. for the Advancement of Science (Pacific Division).

Tyndall, R. 1994. Contributions of the Barrens Symposium, 1993 (Preface). *Castanea* 59:182–183.

Walker, R. B. 1954. Factors affecting plant growth on serpentine soils. *Ecology* 35:258–266.

White, J. 1994. How the terms savanna, barrens and oak openings were used in early Illinois. In *Proceedings of the North American Conference on Barrens and Savannas*, ed. J. S. Fralish, R. C. Anderson, J. E. Ebinger, and R. Szafoni, pp. 25–63. Chicago, Ill.: U. S. Environmental Protection Agency, Great Lakes National Program Office.

Whittaker, R. H. 1960. Vegetation of the Siskiyou Mountains, Oregon and California. *Ecological Monographs* 30:279–338.

20 California Oak Savanna

BARBARA ALLEN-DIAZ, JAMES W. BARTOLOME,
AND MITCHEL P. McCLARAN

Introduction

California oak savannas are dominated by blue oak (*Quercus douglasii*), valley oak (*Q. lobata*), interior live oak (*Q. wislizenii*), coast live oak (*Q. agrifolia*), and Engelmann oak (*Q. engelmannii*), occurring in mixed or monospecific stands occupying less than 30% cover. Nine tree species of oaks occur in the state (Pavlik et al. 1991), but four are limited to forests or woodlands. Annual grassland is the major understory type, although some savannas may have a shrub understory component. Nomenclature follows Hickman (1993).

The oak savannas of California generally occur at elevations ranging from 60 to 700 m between annual grasslands at lower elevations and mixed conifer or ponderosa pine (*Pinus ponderosa*) forests at the higher margin. Precipitation occurs primarily during winter to early spring, generally late October through April. Thus, oak savannas persist through a severe, annual drought period lasting 2–11 months.

Ninety percent of the oak savannas in California are privately owned (Ewing et al. 1988). Over 300 species of vertebrates live in oak savannas and woodlands of California (Jensen, Tom and Harte 1990). Livestock are the primary users and the primary economic products of the savannas; firewood, wildlife, and water are secondary products (McClaran and Bartolome 1985; Standiford and Tinnin 1996).

Past and Present Geographic Distribution

Oak distributions have been in flux throughout the geologic record because of changes in climatic patterns, which led to migration and juxtaposition of communities, with each individual savanna species responding according to its own tolerances (Evett 1994). Axelrod (1977) described a Miocene (25 M yr BP) Oak Woodland–Savanna vegetation type in interior southern California dominated by live oaks and sclerophyllous shrubs that extended into areas now desert. At lower elevations during this period, the dense oak woodlands gave way to open oak savannas with grassland understory. By the Pliocene (10 M yr BP), live oak woodland–savanna with scattered juniper occupied Great Basin lowlands and moist areas of the Central Valley.

Most of the dominant genera of the oak savanna occupied their current distribution by 10 million yr BP in California. Thus, adaptations to the emerging summer–dry Mediterranean climate, which was well established by 5 million yr BP, probably were preexisting physiological modifications within the dominant savanna oaks. The trend throughout the Tertiary was toward less total rainfall, with a distinct summer drought and milder winters and higher summer temperatures (Rundel 1987). Based on pollen evidence, Byrne, Edlund and Mensing (1991) suggested that savanna oaks moved to higher elevations in the Sierra Nevada between 10,000 and 5,000 years ago. By the middle 1800s, oaks had retreated to lower elevations in the Coast Ranges (Byrne, Edlund and Mensing 1991).

The degree to which dominant oaks in the savanna were adapted to a regime of summer drought and higher mean temperatures served to differentiate their current distributions (Evett 1994). In addition, the physiography of the California floristic province, which is characterized by a long north–south gradient, large elevational differences, complex soil parent material, and steep gradients of rainfall, temperature, and nutrients, directly influences the current distribution of oak savannas (Pavlik et al. 1991).

In the 1800s, oak savannas and grasslands of California were estimated to occupy about 13 million ha (Burcham 1957). Valley oak occupied the vast Central Valley of California, which extended 560 km, north to south. The foothill woodland, dominated by coast live oak, blue oak, interior live oak, and Engelmann oak in southern California, occupied approximately 4 million ha that were discontinuous and widely distributed along valley borders and lower slopes of the Coast Ranges and Sierra Nevada (Burcham 1957).

By 1980, the oak savannas/grasslands of California had been reduced to approximately 6.5 million ha, of which approximately 3.8 million ha were occupied by oak species (Ewing et al. 1988). Presently, oak savanna types occur in the Coast Ranges and in a ring around the Central Valley in the foothills of the Sierra Nevada and Coast Ranges (Figure 20.1) (Allen et al. 1989).

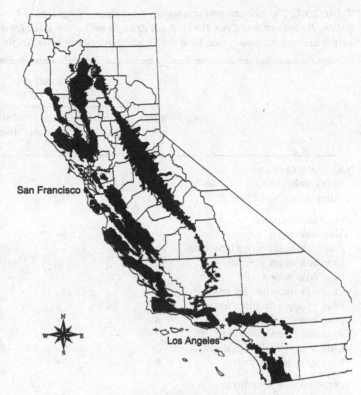

Figure 20.1. Distribution of California oak savannas and woodlands. Savanna types include communities dominated by valley oak, blue oak, coast live oak, interior live oak, and mixtures of these species. Engelmann oak communities have a very narrow range of distribution, occurring only in southern California. See text for details of specific savanna community distributions.

Environment

Climate

California oaks occur in a Mediterranean climate characterized by cool, wet winters and hot, dry summers. Nearly all of the annual precipitation comes during winter and early spring. The summer drought period can vary from 2 to 11 months, but averages 6 months (Pavlik et al. 1991). Considerable variation in the seasonal distribution and pattern of rainfall is typical (George et al. 1985). Oaks generally are found in regions where rainfall is at least 200 mm annually and mean annual temperatures are between 13 and 15 °C (Evett 1994). However, oaks toler-

Table 20.1. *The oak savanna community types from descriptions of hardwood rangelands in California (Allen, Holzmann and Evett 1991). Basal areas in m²/ha are given for dominant oak tree species only. Grass values are in percentage cover. Data for Engelmann oak are from Griffin (1977).*

| | Basal area (m² ha⁻¹) | | | | | Cover (%) |
	Blue oak	Valley oak	Coast live oak	Interior live oak	Engelmann oak	Grass
Valley oak savannas						
Valley oak–grass	--	17	--	--	--	90
Valley oak–coast live oak	--	15	11	--	--	97
Blue oak savannas						
Blue oak–grass	9	--	--	--	--	90
Blue oak–understory blue oak	11	--	--	--	--	85
Blue oak–coast live oak	9	--	11	--	--	83
Blue oak–valley oak	7	4	--	--	--	99
Blue oak–interior live oak	5	--	--	5	--	76
Blue valley–coast live oaks	11	14	5	--	--	95
Blue oak–foothill pine	8	--	--	--	--	92
Blue oak–wedgeleaf ceanothus	5	--	--	--	--	50
Blue oak–whiteleaf manzanita	7	--	--	--	--	52
Blue oak–goldenbush	6	--	--	--	--	67
Evergreen oak savannas						
Coast live oak–grass	--	--	17	--	--	83
Coast live–valley oaks–poison oak	--	10	21	--	--	76
Interior live oak–blue oak–pine	7	--	--	9	--	61
Mixed oak savannas						
MO–interior live oak–pine	--	--	--	4	--	78
MO–California buckeye	6	--	--	--	--	96
MO–grass	6	3	9	--	--	68
MO–pine	10	--	9	--	--	82
Engelmann oak savannas	--	--	--	--	5–14	79

ate temperatures from below freezing in winter to >40 °C in summer.

Evett (1994) described mean annual precipitation (MAP) and mean annual temperature (MAT) in °C for four common savanna oaks across their geographic ranges. Ranges of MAP and MAT are in parentheses.

Coast live oak: MAP = 551 (203–1524)
                  MAT = 14.5 (12.2–17.1)
Interior live oak: MAP = 771 (274–1524)
                  MAT = 14.8 (10.5–17.1)

Valley oak: MAP = 544 (275–1439)
            MAT = 14.5 (11.6–16.7)
Blue oak: MAP = 557 (203–1439)
            MAT = 15.0 (11.7–17.1)

Geology and Soils

In California's complex geologic mosaic, parent materials of a wide variety of origins, chemistries, and physical properties underlie oak savannas. Blue, valley, and coast live oaks occupy sites composed primarily of hard and soft sedimentary parent materials, sandstones, and shales. On the north and east side of the Central Valley these species occupy granitic parent materials (Allen et al. 1989). Sites occupied by interior live oak, with or without foothill pine (*Pinus sabiniana*), manzanita (*Arctostaphylos* spp.), and wedgeleaf ceanothus (*Ceanothus cuneatus*), tend to occur on mafic, metamorphic, and deep igneous parent materials (Allen et al. 1989). Oak savanna types occur on a wide variety of substrates, from rich loams in the valleys, to thin, rocky soils on upland slopes.

Vegetation

A total of 18 tree and shrub oak species, commonly divided into 30 distinct varieties, occupy the valleys and foothills of California (Pavlik et al. 1991). The 5 primary savanna species include 3 deciduous oaks (the stately valley oak, and the smaller-statured blue and Engelmann oaks) and two evergreen oaks (coast live and interior live oak). Understory species composition generally is dominated by introduced Mediterranean annual grasses and forbs (Allen, Holzmann and Evett 1991; M. E. Borchert et al. 1993), but occasionally poison oak (*Toxicodendron diversiloba*), wedgeleaf ceanothus, and/or manzanita may dominate the understory (Allen, Holzmann and Evett 1991).

Oak savannas have been described in various classification systems devised to meet specific needs (e.g., Cheatham and Haller 1975; Küchler 1977; R. F. Holland 1986; Mayer and Laudenslayer 1988). Most often, oak savanna and oak woodland have not been distinguished. For example, Munz and Keck (1968) described oak communities for their flora of California in geographic terms: northern, southern, and foothill woodland. Griffin (1977) expanded the Munz and Keck scheme to discuss ecological relationships within geographic zones and divided the foothill woodland into four phases: valley oak, blue oak, interior live oak, and north slope. Southern oak woodland was expanded to include Engelmann oak and coast live oak

phases, and the northern oak woodland was divided into the foothill woodland transition and bald hills phases.

Each of the communities just described is classified by geographic regions and/or phases within regions and is characterized by specific oak species and their associates. Allen et al. (1989) and Allen, Holzmann and Evett (1991) analyzed and classified data collected from 2,300 one-fifth-acre plots in the 1920s, to develop a comprehensive, data-driven classification of California oak communities. Of the 57 oak-dominated community types described, 19 are common savanna types with tree canopy cover less than 30% (Table 20.1). Blue oak community types, or blue oak with mixtures of the other savanna oak species, are by far the most common types in the state.

Valley Oak Savanna

Valley oak savannas were once more widespread, but they now cover only about 2.7%, or 109,000 ha, of the state (Bolsinger 1988; Jensen, Tom and Harte 1990). This savanna type generally occurs on the rich loam soils of the valleys and foothills below 730 m. Valley oak savanna is bounded by the presence of black oak (*Q. kelloggii*) and pines at higher elevations and by open grassland at lower elevations. Valley oaks are endemic to California and among the oldest and largest oak species in North America (Evett 1994), with maximum recorded measurements >4 m dbh (diameter at breast height = 1.37 m) and crowns over 45 m tall and wide (Griffin 1973). Tree age can exceed 500 years. Mean basal area of these tall, stately trees is relatively high, from 15 to 17 m^2 ha^{-1} in the savanna types (Allen, Holzmann and Evett 1991). Common associates include coast live oak, blue oak, black oak, walnut (*Juglans hindsii*), and western sycamore (*Platanus racemosa*) (Allen, Holzmann and Evett 1991).

Parklike valley oak savannas (Figure 20.2) have been subject to massive clearing because of the high potential of valley oak

Figure 20.2. Valley oak savannas generally occupied deep, fertile soils of the Central Valley and foothills, and thus were some of the first savannas subject to conversion to crop agriculture.

sites for agriculture (Evett 1994; Jensen, Tom and Harte 1990).

Blue Oak Savannas

Blue oak savanna community types occupy about 1.2 million ha in the state (Bolsinger 1988), at elevations ranging from 300 to 760 m. Blue oak is the most widespread savanna species; it is endemic to the state and does not reproduce well in most areas (Muick and Bartolome 1987; Jensen, Tom and Harte 1990). Mean basal area of blue oak in the savanna type varies from 5 to 11 m² ha⁻¹ (Allen, Holzmann and Evett 1991).

Stand structure generally is characterized by a single tree layer dominated either by widely spaced, large-diameter trees or by dense, smaller-diameter trees, almost always with a grassy understory (Figure 20.3). The rounded tree canopy is characterized by distinctive blue-green leaves. Trees are usually less than 70 cm dbh, and vary between 7 and 20 m in height (Jepson 1910). Blue oak savanna most often is found on dry, shallow, loamy or gravelly, rocky soils (Allen et al. 1989; McDonald 1990), and apparently it does not compete well on deeper, fertile soils (McDonald 1990). Blue oaks have many adaptations to drought, from waxy, summer-deciduous leaves to a fast-growing, extensive root system, that allow them to survive on shallow, rocky soils in summer (Pavlik et al. 1991).

Some blue oak savannas are characterized by the presence of shrubs, specifically wedgeleaf ceanothus, whiteleaf manzanita (*Arctostaphylos viscida*), or narrowleaf goldenbush (*Haplopappus linearifolius*). Other common woody components are *Aesculus californica, Heteromeles arbutifolia, Juniperus californica* (in the southwest), valley oak, interior live oak (north slopes and higher elevations), and poison oak; saplings of blue oak are rare

Figure 20.3. Oak-dominated landscapes include deciduous blue oak savannas associated with live oak riparian corridors.

(Allen-Diaz and Holzmann 1991). *Ceanothus* shrub cover averages 18% in the blue oak–wedgeleaf ceanothus–grass type; manzanita and narrowleaf goldenbush average 21% and 22% cover, respectively, on their respective types (Allen et al. 1989). Blue oak density averages 14–16 trees ha^{-1} and is similar for the three shrub understory types. Mean basal area of blue oak in the savanna type varies from an average of 5 m^2 ha^{-1} to 11 m^2 ha^{-1} (Allen, Holzmann and Evett 1991).

Engelmann Oak Savannas

Engelmann oak communities are described by Griffin (1977) as a semi-evergreen southern California version of blue oak savanna. Englemann oak savannas have a very limited range, covering 15,700 ha in southern California and Baja (Jensen, Tom and Harte 1990). Zuill (1967) reported that Englemann oak savanna had a tree density of 27 ha^{-1}.

However, calculations by Griffin (1977) using 16 Vegetation Type Map plots, which were selected from the thousands of one-fifth-acre (0.08-ha) California vegetation survey plots collected in the 1920s (Wieslander 1934), showed that Engelmann oak communities averaged 147 trees ha^{-1}.

Evergreen Oak Savannas

Coast live oak savannas extend in a patchy distribution from Sonoma County in the north-middle part of the state to San Diego County in the south (Figure 20.4). It is the most coastally restricted of the California oaks, seldom extending more than 96 km inland from the Pacific Ocean (Griffin and Critchfield 1972); thus, it generally occurs within the coastal fog belt of California (Evett 1994). The savanna is dominated by low, broad trees, generally about 10 m high with a crown spread of up to 40 m across (Jepson

Figure 20.4. Coast live oaks dominate the woodland chaparral of southern California. They are the most coastal of the California oaks.

1910). Age of coast live oaks commonly exceeds 250 years (Pavlik et al. 1991). California prairie forms the ground layer where the canopy is open. Other woody components include *Heteromeles arbutifolia*, *Juglans californica* (southern California), and Engelmann oak (Los Angeles to San Diego counties). Other common coast live oak savanna associates include valley oak, blue oak, and sometimes interior live oak (Allen et al. 1989).

Coast live oak communities in general have the highest oak basal area of any of the hardwood rangeland types (Allen, Holzmann and Evett 1991), but the savanna types, by definition, are more open, averaging between 17 and 21 m² ha⁻¹ (Table 20.1). The coast live oak–grass type occurs on a variety of aspects and slope angles, whereas the coast live oak–valley oak–poison oak community tends to occupy mesic sites (Allen, Holzmann and Evett 1991).

Interior live oak, although similar in appearance to coast live oak, rarely is found near the coast, and usually not in monospecific stands. Common associates of interior live oak include blue oak, foothill pine, wedgeleaf ceanothus, and California buckeye (Allen et al. 1989).

Of the interior live oak communities described by Allen, Holzmann and Evett (1991), only interior live oak–blue oak–foothill pine type is considered a savanna type, and then only the more open stands. It is found from Butte to Kern to San Benito counties in the Sierra Nevada and Coast Ranges (Allen, Holzmann and Evett 1991). Mean tree basal area averages 23 m² ha⁻¹ (Table 20.1), and cover of grass understory averages 61%.

The Mixed Oak Savannas

The mixed oak savannas of California are composed of communities that generally contain three or more species of oaks (Allen, Holzmann and Evett 1991). Of the four mixed oak savanna types, only the mixed oak–interior live oak–foothill pine community is found in the Sierra Nevada at elevations averaging 580 m (Figure 20.3). The other three mixed oak savanna types (Table 20.1) occur in the Coast Ranges: the mixed oak–grass and mixed oak–California buckeye–grass at lower elevations (305 m), and the mixed oak–foothill pine–grass community at higher elevations (457 m).

Spring and Riparian Communities within the Oak Savanna

Spring communities within the oak savanna are small in extent but are important sources of water and habitat in the oak savanna (Figure 20.3). These systems range in size from a few meters to 25 m in diameter. Springs may flow year round or intermittently.

Of the 68 springs located in Contra Costa and Alameda counties near Oakland, California, all had a sharp ecotone between the spring and upland areas (Allen-Diaz 1995). A total of 200 plant species were found in the springs, including 16 trees and 4 shrubs. Four oak species – coast live oak, valley oak, blue oak, and interior live oak – were found at the springs. In addition, willow (Salix sp.) and bay (Umbellularia californica), and sometimes alder (Alnus rhombifolia), buckeye (Aesculus californica), and/or maple (Acer macrophyllum), were common on springs that had a tree overstory. Spring herbaceous species were diverse (for example, 6 species of rush were identified) and were important in identifying spring types.

Rare Animal Species

Currently, 313 species of vertebrates have been identified as using oak habitats for food, cover, and/or breeding (Johnson 1995). Six federally listed endangered animal species inhabiting oak savannas in California are San Joaquin kit fox (Vulpes macrotis), California condor (Gymnogyps californianus), Peregrine falcon (Falco peregrinus anatum), blunt-nosed leopard lizard (Gambelia silus), San Francisco garter snake (Thamnophis sirtalis tetrataenia), and the long-toed salamander (Ambystoma macrodactylum). One bird, the bald eagle (Haliaeetus leucocephalus), is listed as federally threatened. Human development, poisoning, grazing by livestock, hunting, and major public works projects appear to be the major threats to species survival.

Ecological Relationships

Canopy Effects on Understory Species Composition

Studies in the North Coast Ranges and the Sacramento Valley foothills consistently have shown differences in understory species composition between blue oak canopy and open grassland (Murphy 1980; Jansen 1987; and Jackson et al. 1990). For example, under sheep-grazed blue oak canopy there were more forbs, fewer legumes, and more perennials compared to grazed open grassland (Murphy (1980). Similarly, Jackson et al. (1990) found species composition markedly different between ungrazed understory and ungrazed open grassland. Shared species between the two habitats were Bromus mollis and Lolium multiflorum. Whereas the open grassland supported Hordeum hystrix, Avena barbata, Bromus madritensis, Cynosurus echinatus, Anagalis arvensis, Daucus pusillus, Geranium molle, Madia spp., and Trifolium spp., the oak understory was primarily Brachypodium distachyon, Bromus diandrus, L. multiflorum, B. mollis, Hordeum leporinum, D. pusillus, Geranium molle, Silene sp., and Silybium marianum.

Differences in species composition between understory and open grasslands also occurs across relatively broad environmental gradients. For example, along a precipitation gradient from 40 to 90 cm yr^{-1}, the grazed understory of blue oaks, with about 50% canopy, was consistently different in composition from the adjacent grazed open grassland (McClaran and Bartolome 1989b). *Bromus mollis, Erodium cicutarium, A. barbata,* and *Lupinus bicolor* dominated the grassland, whereas *Festuca megalura* and *B. diandrus* were the most common understory species.

There are also differences in understory composition under individual trees. In ungrazed woodlands and savannas in the central coast, some blue oak trees facilitated understory, whereas others suppressed it (Callaway, Nadkarni and Mahal 1991). Trees enhancing understory had relatively more *H. leporinum* compared to open grasslands, whereas savanna trees suppressing understory development had relatively more *Nasella pulchra*. In the San Joaquin Valley and Central Coast, understory species composition was different under living blue oak canopies than it was under old (>25 years) dead trees. Under living canopies, *B. mollis, A. fatua,* and *B. diandrus* were the dominant species, and under dead trees the dominant species were *B. diandrus, A. fatua, Erodium botrys, and H. leporinum* (V. L. Holland 1980).

Canopy Effects on Herbage Productivity

Expression of tree canopy dominance has been variously ascribed to shade (Maranon and Bartolome 1994), allelopathy (Parker and Muller 1982), water relations (Calloway, Nadkarni and Mahal 1991), and nutrient dynamics (Jackson et al. 1990). The seasonal pattern of growth and reproduction in California annual grasslands consists of rapid fall growth following germination, slow winter growth, rapid spring growth, and summer seed set (Jackson et al. 1990). This basic pattern of production is altered by presence of an oak overstory, with the amount and quality of forage differing according to oak species and canopy cover.

In areas receiving more than 50 cm of precipitation annually, the canopy suppresses understory biomass throughout the growing season. The degree of suppression depends on canopy density and site characteristics, and it is greater with evergreen species than with deciduous species on similar sites (Pitt and Heady 1978). For example, at the Hopland Research and Extension Center near the coast in Mendocino County, forage yield averaged 2,300 kg/ha in open grassland and 1,300 kg/ha under a mostly blue oak canopy (Murphy 1980). Bartolome and McClaran (1992) also found that there was less productivity in the understory at Hopland, and more variation among years in fall–winter production under canopy than under open grassland. Productivity under the canopy was 50% of open in fall–winter and 33% of open in spring.

Comparisons of cut and uncut areas provide most of the information about canopy effects in the northern part of the state. At the Sierra Foothill Range Field Station (SFRFS) near Marysville, California, clear-cut areas always produced more total herbage than areas with trees or even grassland, and grassland produced more herbage than areas with trees (Jansen 1987). However, there was considerable variation in total biomass production between years regardless of any manipulation or understory/open association. Similarly, Kay and Leonard (1980) found more grass under killed blue oaks than in a dense woodland. Compared to productivity under intact canopies, forage productivity averaged 66% more with root kill, and only 45% more if tops were killed. Kay (1987) reported that 21 years of productivity measurements after tree removal showed enhanced herbage production only for the first 15 years. The improved forage yield on cleared sites was greatest in dry years, and there was greater yearly variation under the canopy than in open grassland. However, Jansen (1987) did not find a canopy effect on variability on a nearby site. On another SFRFS site, there were no differences in herbaceous biomass between canopy and open areas (Jackson et al. 1990).

In regions with less precipitation, such as the San Joaquin Valley, the effect of canopy on understory biomass becomes neutral or it enhances understory productivity. Production data from northern California are derived from experimental removal of canopy, whereas most of the information about the south is derived from comparisons of areas under canopy and existing openings. The effect of canopy on production appears to be highly variable by season and year for a given site.

At the San Joaquin Experimental Range, there was enhanced early season production under canopy in two drought years (Frost and McDougald 1989). Compared to herbage on open grassland, the peak herbage under blue oak canopy was about 1,000 kg/ha greater, and peak herbage production under interior live oak was 700 and 1,000 kg/ha greater for the two years. Frost, McDougald and Demment (1991) also reported more herbaceous biomass under blue oaks than in adjacent openings. Forage production was higher under scattered blue oak (6% cover) than under foothill pine and interior live oak (12% cover) (Frost and Edinger 1991).

Blue oak removal in San Luis Obispo County caused an increase in cover from 24.3% to 32.6% but there was no increase in herbage production (Bartolome, Allen-Diaz and Tietje 1994). These results are congruent with those of McClaran and Bartolome (1989b). They reported that understory peak forage production was not enhanced by canopy removal at sites with 50% canopy cover of blue oak and less than 50 cm mean annual precipitation. Other work based on comparison of canopy and open suggests that tree-occupied sites may not have the same productive potential as nearby open grassland.

In the driest savannas, with sparse canopy cover, forage production under individual trees may be higher than in adjacent grassland (Callaway, Nadkarni and Mahal 1991). Forage production under blue oaks was twice that in open grassland and in an ungrazed area Holland (1980). Similarly, Duncan and Clawson (1980) reported more forage under blue oak in winter and spring than in the open grassland. However, once trees die, the level of productivity gradually declines to grassland levels.

It appears that the generalizations about overstory–understory patterns do not have a simple suite of causal explanations. Likewise, management recommendations need to be developed conservatively, with canopy removal for forage enhancement limited to dense stands exceeding 50% canopy and with more than 50 cm mean annual precipitation.

Animal–Plant Interactions

Grazing in the Understory. Few studies have investigated grazing effects on oak understories. In contrast, there are many studies of grazing in open grassland. Hatch et al. (1992) found that recovery of native species from grazing was highly erratic over a 25-year period in exclosures at the Hopland Field Station, 160 km north of San Francisco. In an ungrazed grassland, perennials increased from 1.8 plants m^{-2} in 1958 to 2.6 m^{-2} in 1991 after a decline to 1.1 m^{-2} in 1979. In adjacent ungrazed woodland, perennials increased from 0.3 plants m^{-2} in 1958 to 2 m^{-2} in 1991. Change in the understory was mainly due to a steady increase in Elymus glaucus. The successional pathways differed strikingly in woodland and grassland.

Most observers have commented subjectively on the differences in grazing pressure in open grassland and the oak understory. For example, Duncan and Clawson (1980) reported that cattle prefer forage under blue oak canopy over that in open grasslands. Nevertheless, utilization data collected by Bartolome and McClaran (1992) showed no important seasonal or overall differences in forage utilization between canopy and open at Hopland. These authors also found few within-year effects of seasonal grazing by sheep on composition or productivity, but many between-year effects, which apparently were tied closely to weather patterns and the level of residual dry matter.

Forage quality and management of understory forage. Understory forage consistently is higher in important nutrients than open grassland. Frost, McDougald and Demment (1991) reported more protein under canopies; ADF and lignin often were lower under a canopy than in open. Herbs under blue oak had significantly more N, protein, P, K, and biomass than adjacent grassland (Holland and Norton 1980). Kay (1987) observed significantly more N in tree understory forage and also more phosphates and sulfates.

Better forage quality resulted from tree killing at SFRFS, mainly due to an increase in *B. mollis* (Kay and Leonard 1980). This response also was noted by Menke (1989). In contrast, McClaran and Bartolome (1989b), Murphy and Crampton (1964), and Pitt and Heady (1978) observed negligible differences in forage quality due to changes in species composition, suggesting that overall quality due to species differences balances out. Jansen (1987) claimed that there are more undesirable forage plants in open than in understory.

Reports of forage species composition differences in relation to forage quality are inconsistent, but differences in chemical composition are not. Forage quality varies so much seasonally within sites that small differences in composition and quality between understory and open usually are unimportant as management considerations. An exception could be early season enhancement under canopy in the San Joaquin Valley. Frost, McDougald and Nelson (1990) suggest that moving cattle to areas with scattered blue oaks helps deal with drought because of better early season forage under trees.

Clearing of oaks frequently was recommended for range improvement in the 1950s and 1960s. Where stands are dense and dominated by live oaks, understory forage production can be considerably enhanced by tree removal. In areas with <50 cm annual precipitation and with deciduous canopies of <50% cover, the understory will not respond reliably to tree removal with increased productivity. However, there is also considerable evidence indicating that scattered trees enhance overall forage productivity.

Forage quality under oak canopies is related to chemical composition of herbage and to species composition. The oak understory consistently is higher in nitrogen and other minerals, which should result in better nutrient quality and increased palatability compared to plants in adjacent grassland. However, understory species composition includes several unpalatable species. Although seasonal livestock preference for understory forage frequently has been noted, species composition includes desirable and undesirable species in open grassland and oak understory. Understory forage quality should not be a major consideration in forage management.

Regeneration Issues

McClaran and Bartolome (1989a) found that most blue oak recruitment since the 1700s was associated with fire, and they suggested that the lack of recruitment since about 1940 was related to fire suppression. They reported that past blue oak recruitment was through seedling establishment and resprouting. Allen-Diaz and Holzmann (1991) resampled 21 plots established in blue oak savanna in 1932 and found little recruitment, but they found an increase in tree size after 60 years and some mortality of large trees. Two plots with valley oaks showed no recruitment.

Regeneration varied depending on species in two extensive surveys summarized by Bolsinger (1988). He characterized recruitment as being sparse to absent in coast live oak, blue oak, and valley oak, although valley oak seedlings were fairly common within non–valley oak woodland types. There was an apparent decrease in the extent of oak woodlands, primarily due to range improvement prior to 1970, and urban and suburban development since then. In their statewide survey of oak regeneration, Muick and Bartolome (1987) discovered considerable site-specific variation. Based on the ratio of saplings to trees, they concluded that interior live oaks were regenerating well, blue oaks were regenerating better in the Sierra Nevada than in the Coast Ranges, and coast live oaks and blue oaks were regenerating poorly in the Coast Ranges.

Based on size distributions, Lathrop et al. (1991) reported that Engelmann oak stands in Southern California exhibited inadequate recruitment. In contrast, Lawson (1993) characterized recruitment in coast live oaks and Engelmann oaks in coastal Southern California as adequate, although she concluded that the relative abundance of coast live oaks increased on frequently burned sites.

Predation by birds and rodents, and herbivory on seedlings by invertebrates and vertebrates apparently are important factors in explaining apparent lack of oak regeneration (Griffin 1971). Most recent work suggests that recruitment is not limited by reproduction but by establishment and survival of saplings. Various studies have shown that grazing by vertebrates reduces growth and survival of young oaks (e.g., M. I. Borchert et al. 1989; Adams et al. 1991) and that protection of small plants is important for successful regeneration (Allen-Diaz and Bartolome 1992; Hall et al. 1992).

Other studies have focused on physical factors and plant competition as they relate to germination, establishment, and survival of recruits. These studies identified several important site- and species-specific factors, including weather patterns (Griffin 1971), water stress (Gordon, Rice and Welker 1991), competition for water and nutrients between oak seedlings and exotic grasses (Davis et al. 1991), and shading (Muick 1991). Despite this work and the development of successful

techniques for artificial regeneration (Standiford and Tinnin 1996), it still is not clear how oaks successfully regenerated in the past, which makes management of natural regeneration problematic for the future.

Role of Humans

Over the last 12,000 years, Amerindians, Spanish, Mexican, and United States people have occupied and used the natural resources in the oak savanna. The savanna has provided food, livestock grazing and wood resources, and shelter. Early records indicate that at 15 people km^{-2}, the oak savanna and woodland supported the second most dense Amerindian population in California (Cook 1976).

Spanish settlement began in 1769, and by 1823 there were 21 missions near the West Coast (Burcham 1981). Many of these settlements were in oak savanna and woodland. The Mexican presence, starting with independence from Spain in 1821 and ending in 1848 with annexation by the United States, was only slightly different from that of the Spanish. Settlement was limited to the coastal valleys and mountains until small land grants were made in the interior valley in the 1840s, secularization of the missions in 1833 transferred title to the State, and Anglo settlers were lured by land grants in exchange for settlement (Gates 1991).

United States annexation in 1848, statehood in 1850, and the discovery of gold in 1848 brought burgeoning populations that spread to the interior of the savanna and a conversion to privately held titles to land and resources. From a population of 93,000 in 1850, California grew to 380,000 in 1860, 1.4 million in 1900, 28 million in 1988, and is projected to be 33 million in 2020 (Ewing et al. 1988). Land and resource titles were converted, somewhat tumultuously, from usufructuary to private within the first decade of statehood (Gates 1991). Currently, over 80% of the savanna is privately owned (Ewing et al. 1988). Average land holding

Table 20.2. *General effect of oak canopies on understory biomass and species composition in the oak savannas of California.* Under *refers to under the canopy of the oaks;* open *refers to the surrounding grassland not directly under an oak canopy.* Quag = *valley oak;* Qudo = *blue oak;* Quem = Quercus emory *(Emory oak); an Arizona species is shown for comparison purposes.*

Precipitation (cm)	Species	Vegetation type	Biomass under/open	Species composition between under and open	Source
68	Quag	woodland	less/more	different	Maranon and Bartolome 1994
<50	Quag	savanna	less/more	different	Parker and Muller 1982
<50	Qudo	woodland	no difference	different	McClaran and Bartolome 1989b
>50	Qudo	woodland	less/more	different	M. I. Borchert et al. 1989
>50	Qudo	woodland	less/more	different	McClaran and Bartolome 1989b
95	Qudo	savanna	less/more	different	Bartolome and McClaran 1992
48	Qudo	woodland	more/less	different	V. L. Holland 1980
37	Quem	savanna	less/more	no difference	Haworth and McPherson 1994

size has declined since statehood. The decline has escalated since the 1950s and is likely to continue (Huntsinger and Fortmann 1990). Since about 1950, there has been a decoupling of occupation from the provision of food and shelter. Thus, the burgeoning human population relies less and less on the savanna for livelihood.

Human Use

Acorns were a major source of energy and nutrients for Amerindians in the savanna. The richness of this resource, and the variety and sophistication of techniques used by these people to prepare the acorns, was a major determinant for the high human population density in the oak savanna and woodland (Gifford 1936; Cook 1976). However, there is no evidence that oaks were cultivated (McCarthy 1993).

Fuelwood has been a longstanding use of oak from the savanna and, until recently, it was the sole source of fuel for human occupants. Charcoal production was a vital con-

cern in the oak savannas and woodlands until the 1920s (May 1976). The rise in energy prices during the 1970s started a fuelwood boom, and currently about 160,000 cords are harvested annually from oak savannas and woodlands (Ewing et al. 1988).

Human-caused fires in the savanna have occurred since Amerindian occupation, but there have been changes in its frequency. Lewis (1973) suggests that Amerindians frequently used fire to drive game and to stimulate favored plants, but Sampson (1944) was unable to substantiate a significant use of fire by Amerindians. McClaran and Bartolome (1989a) used fire scars in blue oak to describe a pattern of burning of about one fire per 25 years from 1681 to 1852. Following European occupation of the Sierra foothills, fire frequency increased to about one per 7 years, until fire suppression efforts beginning in 1948 greatly reduced human burning of the savanna.

Grazing by domestic cattle, and to a lesser extent sheep, was the basis for the Spanish,

Mexican, and early United States economies in California. Only the grasslands supported more livestock than the oak savanna and woodland (Burcham 1981; Ewing et al. 1988). Beginning with about 200 Andalusian cattle in 1769, numbers rose to 150,000–400,000 in the 1830s and supported an international trade in hides and tallow (Burcham 1981). By 1860, cattle numbers reached one million head in California in an effort to feed a growing human population.

Recurring droughts caused declines in livestock numbers, but the average cattle population in the state was about 800,000 between 1850 and 1950. Burcham (1981) suggests that the carrying capacity for cattle in the grasslands and savanna declined by 50% during this period because of overstocking and recurring droughts. Now, approximately 80% of privately owned savanna is grazed (Huntsinger and Fortmann 1990) at 5–10 ha/cow/year, which accounts for about 30% of all grazing in the state (Ewing et al. 1988). Cattle production remains one of the top five industries in most counties that have a substantial amount of oak savanna (Ewing et al. 1988).

Human Threats to the Savanna

The greatest threat to oak savanna is conversion to croplands, housing areas, or more open grasslands. Between 1950 and 1980, about 240,000 ha were converted from savanna to other land uses, and about 75% of that was for cropland (Ewing et al. 1988). The conversion of savannas is likely to continue because of increasing population pressures, low economic returns for cattle grazing, and the large proportion of oak savanna in private ownership (Ewing et al. 1988).

However, there have been some sociological and political developments in the past 20 years that may slow the rate of conversion. The perception of the oak savanna as a public resource (Huntsinger and Hopkinson 1996) has been growing as a result of a settlement of the savanna by small land owners who appear to be more receptive to conserva-

tion values and public regulation of private property than cattle ranchers (Huntsinger and Fortmann 1990). Also, growing activism by an open space–starved and oak-sensitized urban public, and negative repercussions with wildlife populations and urban watersheds, are focusing political efforts to conserve large areas of oak savanna (Pavlik et al. 1991). The State Board of Forestry established the Integrated Hardwood Range Management Program, which combines University of California expertise in extension, research, and teaching with state natural resources agencies to distribute information about land use alternatives and to monitor land use conversions (Bartolome and Standiford 1992).

Summary and Future Research Needs

The oak savannas of California are divided into five main types dominated individually or in combination by savanna species of valley, blue, Engelmann, coast live, and interior live oaks. Annual grassland is the dominant understory cover, although some savannas have woody understory species.

Savanna trees have a significant effect on understory composition and productivity depending on oak species and annual precipitation (Table 20.2). In general, in areas receiving more than 50 cm precipitation annually, the canopy suppresses understory biomass throughout the growing season. The degree of suppression is greater with evergreen species than with deciduous species on similar sites.

Management of savanna understory species composition and productivity has been assumed to fit the residual dry-matter model, which has been successfully applied to grasslands (George et al. 1985). This model assumes that if land managers leave recommended levels of mulch at the end of the dry season, the best combination of positive effects to the grassland ecosystem as a whole will result, such as optimum forage production, biodiversity, protection from soil erosion, and wildlife habitat. However, there is little credible research to back up this assumption. Because of the highly variable nature of canopy effects on understory composition and productivity, many more site-specific studies are needed to fully examine ecosystem response and management options.

Much of the work in the California oak savanna has focused on reproduction of oaks, incorrectly assuming that it is synonymous with recruitment. Future research will have to focus on recruitment success, which appears to vary with species, past stand structure, and tree mortality, and seems to vary at multiple spatial and temporal scales (Bartolome, Muick and McClaran 1987; Standiford and Tinnin 1996).

Finally, more research focused on whole-ecosystem interactions, including development of viable economic alternatives for land use, will be most productive in ensuring the long-term preservation of California oak savannas.

Acknowledgments

The authors would like to thank members of Environmental. Science, Policy and Management class, ESPM 268, Fall 1995, for their thoughtful inputs into this chapter. Class members include Allisa Neves, Robin Liffmann, Keith Labnow, Mark Homrighausen, Kate Rassbach, Deborah Waller, Clay Taylor, and Dr. Guy McPherson.

REFERENCES

Adams, T. E., Jr., Sands, P. B., Weitkamp, W. H., and McDougald, N. 1991. Blue and valley oak seedling establishment on California's hardwood rangelands. In *Proceedings of Symposium on Oak Woodlands and Hardwood Rangeland Management*, tech. coord. R. B. Standford, pp. 41–47. USDA Forest Service, Gen. Tech. Rep. PSW-126.

Allen, B. H., Evett, R. R., Holzmann, B. A., and Martin, A. 1989. *Rangeland Cover Type Descriptions for California Hardwood Rangelands*. A report for California Department of Forestry and Fire Protection. Sacramento, Calif.

Allen, B. H., Holzmann, B. A., and Evett, R. R. 1991. A classification system for California's hardwood rangelands. *Hilgardia* 59(2):1–45.

Allen-Diaz, B. H. 1995. *Classification and Description of East Bay Regional Park District and East Bay Municipal District Springs*. Report to East Bay Municipal District, Contra Costa, Calif.

Allen-Diaz, B. H., and Bartolome, J. W. 1992. Survival of *Quercus douglasii* (Fagaceae) seedlings under the influence of fire and grazing. *Madrono* 39:47–53.

Allen-Diaz, B. H. and Holzmann, B. 1991. Blue oak communities in California. *Madrono* 38:80–95.

Axelrod, D. I. 1977. Outline history of California vegetation. In *Terrestrial Vegetation of California*, ed. M. G. Barbour and J. Major, pp. 139–193. New York: John Wiley and Sons.

Bartolome, J. W., Muick, P. C., and McClaran, M. P. 1987. Natural regeneration of California hardwoods. In *Proceedings of Symposium on Multiple-Use Management of California's Hardwood Resources*, tech. coords., T. R. Plumb and N. H. Pillsbury, pp. 29–31. Forest Service, Gen. Tech. Rep. PSW-100.

Bartolome, J. W., Allen-Diaz, B. H., and Tietje, W. D. 1994. The effect of *Quercus douglasii* removal on understory yield and composition. *Journal of Range Management* 47:151–154.

Bartolome, J. W., and McClaran, M. P. 1992. Composition and production of California oak savanna seasonally grazed by sheep. *Journal of Range Management* 45:103–107.

Bartolome, J.W., and Standiford, R.B. 1992. Ecology and management of Californian oak woodlands. In *Proceedings Symposium on Ecology and Management of Oak and Associated Woodlands: Perspectives in the Southwestern United States and Northern Mexico*. tech. coords., P. Ffolliott, G. Gottfried, D. Bennett, V. Hernandez, A. Ortega-Rubio, and R. Hamre, pp. 115–118. USDA Forest Service, Gen. Tech. Rep. RM-218.

Bolsinger, C. L. 1988. *The Hardwoods of California's Timberlands, Woodlands, and Savannas*. USDA Forest Service Resource Bulletin PNW-RB-148.

Borchert, M. E., Cunha N. D., Krosse, P. C., and Lawrence, M. 1993. *Blue Oak Plant Communities of Southern San Luis Obispo and Northern Santa Barbara Counties*. USDA Forest Service, Gen. Tech. Rep. PSW-GTR-139.

Borchert, M. I., Davis F. W., Michaelsen, J., and Oyler, L. D. 1989. Interactions of factors affecting seedling recruitment of blue oak in California. *Ecology* 70:389–404.

Burcham, L. T. 1957. *California Range Land: An Historical Study of the Range Resource of California.* Sacramento, Calif.: California Dept. of Natural Resources.

Burcham, L. T. 1981. *California Rangeland.* Davis, Calif.: University of California,.Center for Archeological Research at Davis.

Byrne, R., Edlund, E., and Mensing, S. 1991. Holocene changes in the distribution and abundance of oaks in California. In *Proceedings of Symposium on Oak Woodlands and Hardwood Rangeland Management,* tech. coord. R. B Standford, pp. 182–188. USDA Forest Service, Gen. Tech. Rep. PSW-126.

Callaway, R. M., Nadkarni, N. M, and Mahal, B. E. 1991. Facilitation and interference of *Quercus douglasii* on understory productivity in central California. *Ecology* 72:1484–1499.

Cheatham, N. H., and Haller, J. R. 1975. An annotated list of California habitat types. Berkeley, Calif.: University of California Natural Land and Water Reserves System. Unpubl.

Cook, S. F. 1976. *The Population of California Indians 1769–1970.* Berkeley, Calif.: University of California Press.

Davis, F. W., Borchert, M., Harvey, L. E., and Michaelsen, J. C. 1991. Factors affecting seedling survivorship of blue oak (*Quercus douglasii*) in Central California. In *Proceedings of Symposium on Oak Woodlands and Hardwood Rangeland Management,* tech. coord. R. B. Standford, pp. 81–86. USDA Forest Service, Gen. Tech. Rep. PSW-126.

Duncan, D. A., and Clawson, W. J. 1980. Livestock utilization of California's oak woodlands. In *Proceedings of Symposium on Ecology, Management, and Utilization of California Oaks,* tech. coord., T. R. Plumb, pp. 306–319. USDA Forest Service Gen. Tech. Rep. PSW-44.

Evett, R. R. 1994. Determining environmental realized niches for six oak species in California through direct gradient analysis and ecological response surface modeling. Ph.D. dissertation, University of California, Berkeley, Calif.

Ewing, R. A. Tosta, N., Tuazon, R., Huntsinger, L., Morose, R., Nielson, K., Motroni, R., and Turan, S. 1988. *California's Forests and Rangelands: Growing Conflict over Changing Uses.*

Sacramento, Calif.: California Dept. Forestry and Fire Protection, Forest and Rangeland Resources Assessment Program.

Frost, W. E., and Edinger, S. B. 1991. Effects of tree canopies on soil characteristics of annual rangeland. *Journal of Range Management* 44:286–288.

Frost, W. E., and McDougald, N. K. 1989. Tree canopy effects on herbaceous production of annual rangeland during drought. *Journal of Range Management* 42:281–283.

Frost, W. E., McDougald, N. K., and Demment, M. W. 1991. Blue oak canopy effect on seasonal forage production and quality. In *Proceedings of Symposium on Oak Woodlands and Hardwood Rangeland Management,* tech. coord. R. B. Standiford, pp. 307–311. USDA Forest, Service Gen. Tech. Rep. PSW-126.

Frost, W. E., McDougald, N. K., and Nelson, A. O. 1990. *Coping with Drought: Range and Livestock Management Recommendations.* San Joaquin Experimental Range Research Bulletin, January, Calif.

Gates, P. 1991. *Land and Law in California – Essays on Land Policies.* Ames, Iowa: Iowa State University Press.

George, M. R., Clawson, W. J., Menke, J. W., and Bartolome, J. W. 1985. Annual grassland forage productivity. *Rangelands* 7:17–19.

Gifford, P. W. 1936. California balanophagy. In *Essays in Anthropology.* Presented to A. L. Kroeber on his sixtieth birthday, ed. Anonymous, pp. 87–98. Berkeley, Calif.: University of California Press.

Gordon, D. R., Rice, K. J., and Welker, J. M. 1991. Soil water effects on blue oak seedling establishment. In *Proceedings of Symposium on Oak Woodlands and Hardwood Rangeland Management,* tech. coord. R. B. Standiford, pp. 54–58. USDA Forest Service, Gen. Tech. Rep. PSW-126.

Griffin, J. R. 1971. Oak regeneration in the upper Carmel Valley, California. *Ecology* 52:862–868.

Griffin, J. R. 1973. Xylem sap tension in three woodland oaks of central California. *Ecology* 54:152–159.

Griffin, J. R. 1977. Oak woodland. In *Terrestrial Vegetation of California,* ed. M. G. Barbour and J. Major, pp. 383–416. New York: John Wiley and Sons.

Griffin, J. R., and Critchfield, W. B. 1972. *The Distribution of Forest Trees in California.* USDA Forest Service Research Paper PSW-82.

Hall, L. M., George, M. R., McCreary, D. D., and Adams, T. E. 1992. Effects of cattle grazing on blue oak seedling damage and survival. *Journal of Range Management* 45:503–506.

Hatch, D. A., Bartolome, J. W., Heady, H. F., and McClaran, M. P. 1992. Long-term changes in perennial grass populations in oak savanna. *Bulletin Ecological Society of America* 73(supplement):202.

Haworth, K., and McPherson, G. R. 1994. Effects of *Quercus emoryi* on herbaceous vegetation in a semi-arid savanna. *Vegetatio* 112:153–159.

Hickman, J. C. 1993. *The Jepson Manual.* Berkeley, Calif.: University of California Press.

Holland, R. F. 1986. *Preliminary Description of the Terrestrial Natural Communities of California.* Sacramento, Calif.: California Resource Agency, Dept. of Fish and Game.

Holland, V. L. 1980. Effect of blue oak on rangeland forage production in central California. In *Ecology, Management and Utilization of California Oaks,* tech. coord. T. R. Plumb, pp. 323–328. USDA Forest Service, Gen. Tech. Rep. PSW-44.

Holland, V. L., and Norton, J. 1980. Effect of blue oak on nutritional quality of rangeland forage in central California. In *Ecology, Management and Utilization of California Oaks,* tech. coord. T. R. Plumb, pp. 319–322. USDA Forest Service, Gen. Tech. Rep. PSW-44.

Huntsinger, L., and Fortmann, L. P. 1990. California's privately owned oak woodlands: owners, use, and management. *Journal of Range Management* 42:147–152.

Huntsinger, L., and Hopkinson, P. 1996. Sustaining rangeland landscapes: a social and ecological process. *Journal of Range Management* 49:167–173.

Jackson, L. E., Strauss, R. B., Firestone, M. K., and Bartolome, J. W. 1990. Influence of tree canopies on grassland productivity and nitrogen dynamics in deciduous oak savanna. *Agriculture, Ecosystems and Environment* 32:89–105.

Jansen, H. C. 1987. The effect of blue oak removal on herbaceous production on a foothill site in the Northern Sierra Nevada. In *Proceedings of Symposium on Multiple-Use Management of California's Hardwood Resources,* tech. coord. T. R. Plumb and N. H. Pillsbury, pp. 343–350. USDA Forest Service, Gen. Tech. Rep. PSW-100.

Jensen, D. B., Tom, M., and Harte, J. 1990. *In Our Own Hands: A Strategy for Conserving Biological Diversity in California.* California Policy Seminar Research Report. Berkeley, Calif.: University of California Press.

Jepson, W. L. 1910. *The Silva of California.* Memoirs Univ. of Calif. (2). Berkeley, Calif.: University of California Press.

Johnson, S. G. 1995. *Wildlife Among the Oaks: A Management Guide for Landowners.* Berkeley, Calif.: University of California, Integrated Hardwood Range Management Program.

Kay, B. L. 1987. Long-term effects of blue oak removal on forage production, forage quality, soil, and oak regeneration. In *Proceedings of Symposium on Multiple-Use Management of California's Hardwood Resources.* tech. coords., T. R. Plumb and N. H. Pillsbury, pp. 351–357. USDA Forest Service, Gen. Tech. Rep. PSW-100.

Kay, B. L., and Leonard, O. A. 1980. Effect of blue oak on rangeland forage production in central California. In *Ecology, Management, and Utilization of California Oaks.* tech. coord. T. R. Plumb, pp. 323–328. USDA Forest Service, Gen. Tech. Rep. PSW-44.

Küchler, A. W. 1977. Natural vegetation of California (map). In *Terrestrial Vegetation of California,* ed. M. C. Barbour and J. Major, pp. 909–938. New York: Wiley-Interscience.

Lathrop, E. W., Osborne, C. , Rochester, A., Yeung, K., Sorest, S., and Hopper, R. 1991. Size class distribution in southern oak woodland on the Santa Rosa plateau, Riverside county, California. In *Proceedings of Symposium on Oak Woodlands and Hardwood Rangeland Management,* tech. coord. R. B. Standiford, pp. 371–376. USDA Forest Service, Gen. Tech. Rep. PSW-126.

Lawson, D. M. 1993. The effects of fire on stand structure of mixed *Quercus agrifolia* and *Quercus englemannii* woodlands. M.A. thesis, San Diego State University, San Diego, Calif.

Lewis, H. 1973. The origin of diploid neospecies in *Clarkia. American Naturalist* 107:161–170.

Maranon, T., and Bartolome, J. W. 1994. Coast live oak (*Quercus agrifolia*) effects on grassland biomass and diversity. *Madrono* 41:39–52.

May, R. H. 1976. *Wood Charcoal in California.* Berkeley, Calif.: USDA Forest Service, Forest Survey Release.

Mayer, K. E., and Laudenslayer, W. F., Jr., eds. 1988. *A Guide to Wildlife Habitats of California.* Sacramento, Calif.: California Dept. of Forestry and Fire Protection.

McCarthy, H. 1993. Managing oaks and the acorn crop. In *Before the Wilderness: Environmental Management by Native Californians,* ed. T. C. Blackburn and K. Anderson, pp. 213–228. Menlo Park, Calif.: Ballena Press.

McClaran, M.P., and Bartolome, J.W. 1985. The importance of oak to ranchers in the California foothill woodland. *Rangelands* 7(4):158–161.

McClaran, M. P., and Bartolome, J. W. 1989a. Fire-related recruitment in stagnant *Quercus douglasii* populations. *Canadian Journal of Forest Research* 19:580–585.

McClaran, M. P., and Bartolome, J. W. 1989b. Effect of *Quercus douglasii* (Fagaceae) on herbaceous understory along a rainfall gradient. *Madrono* 36:141–153.

McDonald, P. M. 1990. *Quercus douglasii* – blue oak. In *Silvics of North America: 2. Hardwoods,* tech. coords. R. M. Burns and B. H. Honkala, pp. 631–639. Agriculture Handbook 654. Washington, D.C.: USDA Forest Service.

Menke, J. W. 1989. Management controls on productivity. In *Grassland Structure and Function: California Annual Grassland,* ed. L. F. Huenneke and H. A. Mooney, pp. 173–200. Dordrecht: Kluwer Academic.

Muick, P. C. 1991. Effects of shade on blue oak and coast live oak regeneration in California annual grasslands. In *Proceedings of Symposium on Oak Woodlands and Hardwood Rangeland Management,* tech. coord. R. B. Standiford, pp. 21–24. USDA Forest Service, Gen. Tech. Rep. PSW-126.

Muick, P. C., and Bartolome, J. W. 1986. Oak regeneration on California hardwood rangelands. *Transactions, Western Section of the Wildlife Society* 22:121–125.

Muick, P. C., and Bartolome, J. W. 1987. *An Assessment of Natural Regeneration of Oaks in California.* Unpublished report to California Department of Forestry and Fire Protection, Sacramento, Calif.

Munz, P. A., and Keck, D. D. 1968. *A California Flora.* Berkeley, Calif.: University of California Press.

Murphy, A. H. 1980. Oak trees and livestock-management options. In *Proceedings of Symposium on Ecology, Management, and Utilization of California Oaks,* tech. coord. T. R. Plumb, pp. 329–332. USDA Forest Service, Gen. Tech. Rep. PSW-44.

Murphy, A. H., and Crampton, B. 1964. Quality and yield of forage as affected by chemical removal of blue oak (*Quercus douglasii*). *Journal Range Management* 17:142–144.

Parker, V. T., and Muller, C. H. 1982. Vegetational and environmental changes beneath isolated live oak trees (*Quercus agrifolia*) in a California annual grassland. *The American Midland Naturalist* 107:69–81.

Pavlik, B. M., Muick, P. C., Johnson, S., and Popper, M. 1991. *Oaks of California.* Los Olmos, Calif.: Cachuma Press.

Pitt, M. D., and Heady, H. F. 1978. Response of annual vegetation to temperature and rainfall patterns in Northern California. *Ecology* 59:335–350.

Rundel, P. W. 1987. Origins and adaptations of California hardwoods. In *Proceedings of Symposium on Multiple-Use Management of California's Hardwood Resources.* tech. coords. T. R. Plumb and N. H. Pillsbury, pp. 11–17. USDA Forest Service, Gen. Tech. Rep. PSW-100.

Sampson, A. W. 1944. *Plant Succession on Burned Chaparral Lands in Northern California.* University of California Agriculture Experiment Station Bulletin 685.

Standiford, R. B., and Tinnin, P. 1996. *Guidelines for Managing California's Hardwood Rangelands.* University of California, Division of Agriculture and Natural Resources Publication 3368.

Wieslander, A. E. 1934. *Vegetation Types of California, Romona Quadrangle* (181) (map). Berkeley, Calif.: U.S. Forest Service.

Zuill, H. J. 1967. Structure of two cover types of southern oak woodland in California. M.S. thesis, Loma Linda University, Loma Linda, Calif.

NORTHERN REGION

21 Jack Pine Barrens of the Northern Great Lakes Region

KURT S. PREGITZER AND SARI C. SAUNDERS

> The grassy and sweet fern barrens bear no mark of present utilization, but are desolate open tracts where only an occasional stump, a cluster of jack pines, or a scrub oak bush breaks the monotonous sweep of the rolling, thinly clad ground surface.
>
> Description of northwest Wisconsin pine barrens, 1931
> (cited by Curtis 1959)

Introduction

Surveyor records and other historical data indicate that at the time of European settlement, there were approximately 920,000 ha of pine barrens landscape in Wisconsin. Approximately 3,500 ha remain (Chequamegon National Forest 1993). Pre-European jack pine (*Pinus banksiana*) barrens are estimated to have covered 20,000 km² in northern Minnesota, Wisconsin, and Michigan (Vora 1993; Figure 21.1). Barrens in the western Upper Peninsula of Michigan, such as the Baraga and Yellow Dog plains, were also dominated by jack pine. The Baraga Plains support serviceberry (*Amelanchier*) and cherry (*Prunus*) species, as well as scattered red oak (*Quercus rubra*), red maple (*Acer rubrum*), big-tooth aspen (*Populus grandidentata*), and black spruce (*Picea mariana*). The Yellow Dog Plains are less xeric than the Baraga Plains and thus support a larger component of white spruce (*P. glauca*), and some red (*P. resinosa*) and white pine (*P. strobus*) (Bourdo 1954).

Barrens communities similar to those found in Michigan, Wisconsin, and Minnesota (the Great Lakes states) also are found in New York, Manitoba, and Saskatchewan, and, historically, in eastern Ontario. The New Jersey pine barrens are similar in structure to barrens in the Great Lakes region. However, the predominant woody species in New Jersey is pitch pine (*P. rigida*) instead of jack pine, and the shrub stratum is also compositionally different (Curtis 1959; also see Chapter 3).

Historically, pine barrens of the Great Lakes region were used for agriculture, logged in the late 1800s, and replanted to jack and red pine during the 1930s (Curtis 1959; Vogl 1970). Increased understanding of the role that fire played in maintaining this ecosystem prior to European settlement and of the impact of continued human-induced disturbance suggests that this system always has been dynamic (Figure 21.2a). Results of initial restoration efforts are encouraging and suggest that much of the floral and faunal diversity characteristic of the barrens can be maintained through landscape-level planning and management (Figure 21.2b).

Paleobotanical History

Jack pine reached the northern Great Lakes states between 8,000 and 10,000 years ago (Brubaker 1975; Davis 1983). Apparently, it

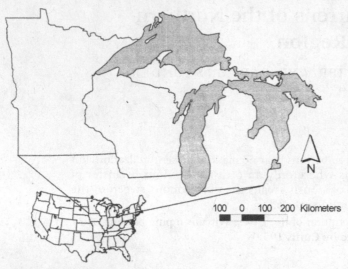

Figure 21.1. The Great Lakes region of North America. Jack pine barrens typically occur on sandy soils of northern Michigan, Wisconsin, and Minnesota.

survived the last glacial period (Wisconsin) at low elevations in the Appalachian Mountains and in the western Ozarks (Rudolph and Laidly 1990). Based on pollen analysis, jack and red pine migrated into Minnesota and became more common in Michigan and Wisconsin during the rapid warming and glacial retreat of the Midwestern region between 11,000 and 9,500 yr BP (Brubaker 1975). Over the next 3,000 years, the abundance of jack and red pine fluctuated within Minnesota and Wisconsin as the prairie–forest border expanded and contracted.

By 9,000 yr BP, areas of northern lower Michigan and Wisconsin had a pine-dominated forest wedged between the deciduous forests to the south and the boreal systems to the north (Webb, Cushing and Wright 1983). Macrofossil evidence suggests that jack pine reached Upper Michigan before red pine and was the dominant species of the early postglacial pine forests within the central and southern parts of the Great Lakes region (Critchfield 1985).

Studies of pollen profiles from the Yellow Dog Plains in Baraga and Marquette counties

of Upper Michigan provide specific examples of the Holocene history of vegetation on outwash plains. Aspen (*Populus tremuloides* and *P. grandidentata*) and juniper (*Juniperus* spp.), which, like jack pine, are shade intolerant, were important early colonizers of these geologic features. Spruce also reached its maximum abundance about 9,500 yr BP (Brubaker 1975). Brubaker's (1975) pollen data suggest that species in the genera *Artemisia* and *Hudsonia* (currently common on jack pine barrens) covered large areas of the Yellow Dog Plains from 8,500 to 9,000 yr BP. Based on pollen percentages, jack pine apparently increased in abundance around 9,000 yr BP and grasses increased approximately 8,400 yr BP. Pollen deposits indicate dominance of jack pine throughout the postglacial period (Brubaker 1975). Scattered spruce was present throughout the mid-postglacial period. Minimal amounts of oak (*Quercus* spp.), maple (*Acer* spp.), and hemlock (*Tsuga canadensis*) appeared within the last 3,000 years. Brubaker's study demonstrates that the jack pine of the Yellow Dog Plains has been the most stable community in this region over the last 9,000 years, despite the fact that jack pine is an intolerant species.

Environmental Relationships

Climate and Site Factors

Factors important in the development and maintenance of the pine barrens include climate, edaphic features, and fire. Collectively, these factors reduce the growth of woody plants and favor a plant community dominated by grasses and forbs (Heikens and Robertson 1994). No single abiotic factor maintains the pine barrens community. Rather, the system is preserved through the

Figure 21.2. (a) Rolling pine barrens, June 18, 1888, Kalkaska County, Michigan. (Photograph courtesy of Michigan State University Archives and Historical Collections, Michigan State University Department of Forestry Collection – Beal Exped-ition.) (b) Diverse structure of the current pine barrens of Moquah Wildlife area, Chequamegon National Forest, Wisconsin. (Photograph S. C. Saunders.)

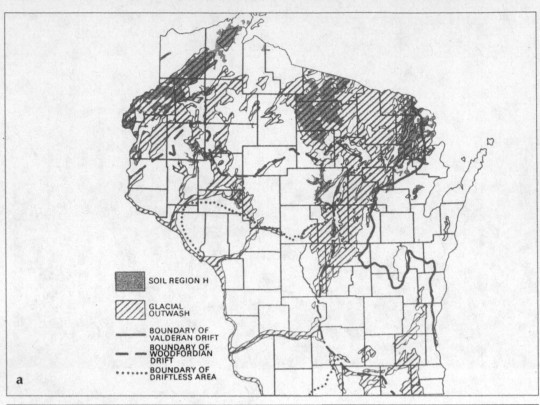

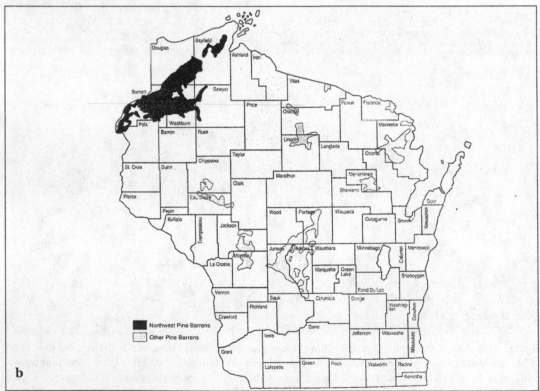

Figure 21.3. (a) The geographic relationship of Soil Region H to bodies of glacial outwash and major glacial boundaries (from Hole 1976). (b) The historical distribution of pine barrens in Wisconsin (from Borgerding, Bartlet and McCown 1995).

interaction of the sandy, outwash substrate (and associated topography) and a dynamic fire regime. Sandy soils favor the establishment of jack pine, and the community is maintained by the flammability of dominant taxa and by reduced competition from understory species.

In general, jack pine barrens establish in areas with a continental climate. Summers are typically short and are warm to cool (average July, 13–22 °C); winters are cold (average January, -29 to -4 °C). Annual precipitation averages 250–1,400 mm (usually 380–890 mm). The growing season averages 50–173 frost-free days (commonly 80–120 days); this number increases from the northwest to the southeast part of the species' range (Rudolph and Laidly 1990). Historically, the region of the Wisconsin barrens had a mean annual temperature of 4.3–5.3 °C and a mean annual precipitation of 72–80 cm (Murphy 1931; see also Curtis 1959). Vogl (1970) suggested that seasonal extremes of temperature in these ecosystems maintained open areas.

Physiography delimits jack pine barrens through its influence on fire frequency and spread, moisture availability, and the interaction of these factors. Fire tends to readily move upslope and across level uplands; however, in ravines or drainage bottoms, fire spread and frequency are reduced (e.g., Romme and Knight 1981; Grimm 1984). Jack pine generally is found in almost pure stands on the most xeric, level, or upland sites, whereas red pine is more abundant in depressions or downslope areas where fire frequency is lower and the accumulation of soil organic matter increases water-holding capacity (Whitney 1986).

Jack pine barrens often occur on pitted outwash plains that have coarse-textured, rapidly drained, acidic soils (Stern and Buell 1951; Curtis 1959), although jack pine can be found on loamy, clay, or organic soils (Schoenike 1976). For example, the soils of Michigan's Yellow Dog glacial outwash plains are sorted, well-drained, medium sands

(more than 95% sand by weight; Pregitzer and Barnes 1984). These soils have thin A and B horizons and an unweathered C horizons less than 60 cm below the surface. In Michigan and Wisconsin, the depth of these outwash sands can be >200 m. Soil region H of Wisconsin, which encompasses the region of pitted outwash topography, corresponds well to the distribution of jack pine barrens in the state (Figures 21.3a, b). This region is characterized by extreme seasonal temperatures, a short spring season, and droughty conditions because of the rapid hydraulic conductivity of the soil (Hole 1976).

It appears that reduced tree growth in barrens communities results from high soil acidity and scarcity of Ca, Mg, or N, as well as moisture deficiency in the upper soil horizons (Sweet 1880; Heikens and Robertson 1994). Poor nutrient status is due to limited nutrient exchange capacity and recurrent loss of nutrients following fire (Hole 1976). Bockheim and Leide (1991) recorded low levels of extractable macronutrients, such as Ca, K, P, and Mg, in pitted outwash soils of Wisconsin. Rudolph and Laidly (1990) reported that additional fine sand and silt or clay in the upper soil layer increased available water capacity (AWC) and cation exchange capacity (CEC). Increased AWC and CEC resulted in higher site indices for the jack pine stands in Minnesota and Wisconsin. Foliar carbon-to-nitrogen ratios in jack pine systems suggest nitrogen often limits tree growth (M. B. David et al. 1988).

Almendinger (1990) determined that, in Minnesota soil, organic matter was inversely related to the length of time that jack pine had dominated the site. According to radiocarbon dating of stratified samples of lake sediment, sites occupied by jack pine longer than 1,300 years have <2.2% organic matter in the surface horizon of the mineral soil. The decrease in soil organic matter with time probably is related to the frequency and intensity of wildfire, with the jack pine forest burning more intensely than other vegetation types. Available water capacity also is inversely related to number of years after jack pine establishment (Almendinger 1990), because of the positive correlation between AWC and organic matter in the soil.

Nutrient Allocation in the Ecosystem

Studies of fuel loads in jack pine and aspen stands of northeastern Minnesota showed that the undecomposed litter layer contained 65% foliage, 31% wood, 2% herbaceous vegetation, and 2% unidentified material by weight (Loomis 1977). The relative importance of herbaceous material in litter probably is significantly greater for an open barrens ecosystem than forested stands because of the order of magnitude fewer trees present (approximately 20 ha^{-1} versus 680 ha^{-1}; Loomis 1977) and the high coverage of shrubs and herbs in an open barrens landscape.

MacLean and Wein (1978) examined biomass distribution in jack pine stands in New Brunswick, with climatic regimes and understory species similar to those that might be expected in pine barrens of the Great Lakes area. Their study indicated that understory biomass peaked at about 500 g m^{-2} 10–20 years after a fire disturbance. Understory biomass appeared to increase faster in pine stands relative to deciduous stands until age 50 years. Although in many forest systems the relative percentage of biomass in the understory declines following crown closure, the barrens ecosystem retains a high proportion of its productivity in herbaceous and shrub layers. Factors contributing to the dominance of lower and midcanopy strata versus overstory also may include the ability of common species of the ground flora in the barrens to grow relatively quickly after fire and the slow accumulation of biomass in the pine overstory. Thus, the percentage of the nutrient pool in the understory layer is expected to be relatively high in the barrens. Even in well-stocked, 57-year-old jack pine forests, up to 30% of some macronutrients were found in the understory (MacLean and Wein 1978).

Litter decomposition rates are relatively slow in jack pine ecosystems because litter from coniferous species contains organic compounds that resist decay. The C:N ratios also indicate that net immobilization of N is occurring in the forest floor, further suggesting that nutrient availability may be limiting growth (Alban 1982; M. B. David et al. 1988). Jack pine stands in northeastern Minnesota had foliar N concentrations of 1.04%–1.13% and foliar P concentrations of 0.08%–0.11%. The upper values of both ranges are below critical values (Green and Grigal 1980). Fertilization of young jack pine stands increased the weight and N content of needles, the length of shoots, and the production of male cones (McCullough and Kulman 1991). Thus, availability of N may limit both vegetative and reproductive growth.

Although these studies were done in forests with higher densities of jack pine than is typical of barrens, similar nutrient limitations occur in the barrens systems. Organic material is lost periodically due to burning. However, jack pine conserves nutrients in these limited systems by incorporating relatively low amounts of nutrients into biomass. Retranslocation and subsequent mobilization of nutrients within the tree may be an important method of nutrient conservation. Cycling of macronutrients also appears to be slow in jack pine ecosystems (Bockheim, Jepsen and Heisey 1991; Bockheim and Leide 1991). Further, S, P, and Ca are immobilized in decomposing needles of jack pine, and this accentuates limited nutrient availability. Fire mobilizes nitrogen and other nutrients that otherwise are unavailable. Thus, fire plays a critical role in regulating nutrient availability in these nutrient-deficient ecosystems.

Role of Disturbance

Fire

Fire is the predominant disturbance in the jack pine community, and prior to the arrival of European settlers, it influenced the dynamics and structure of this ecosystem in Wisconsin, northern Minnesota (Heinselman 1973), and Michigan (Simard and Blank 1982). Different compositions of floral and faunal communities are expected in the absence of fire (Stern and Buell 1951). Pine communities generally burn more readily than hardwood-dominated ecosystems because of the low moisture content of leaves and the high content of flammable, organic compounds (Whitney 1986). Although a severe crown fire will kill most trees, seedling establishment is promoted by the reduction of competition from other herbaceous species and by seedbed preparation. Fire also releases seeds from serotinous cones by melting the resin on cone scales (Cayford 1970; Vogl 1970; Benzie 1977). However, fires with a return time of less than 20–40 years can reduce establishment and recruitment of jack pine and increase the vigor of understory shrub species (Vogl 1970). Areas with a fire return time of less than 50 years will be dominated by jack pine, whereas a 50–200-year burning frequency may favor a system dominated by red pine (Givnish 1995). Thus, variation in return time and in intensity of fire has important impacts on the species composition and structure of the landscape. It seems likely that the jack pine barrens burned frequently and that fire was a key factor in maintaining the low density of trees in these communities.

Historically, the average return time for fire in jack pine forests of northern lower Michigan was about 80 years (Whitney 1986). Similar return times (50–100 years) have been estimated by Heinselman (1981) in Minnesota. According to studies of jack pine forests in northern lower Michigan, less severe, surface fires may have had a return interval of about 25 years in presettlement times (Whitney 1986). Heinselman (1981) also suggested that, in the northern Great Lakes area, light to moderate surface fires may have burned jack pine forests on drier

sites every 20–40 years. Though few large-scale studies have been done, research from the Boundary Waters Canoe Area (BWCA) of Minnesota and the Mack Lake area of northern Lower Michigan suggests similar trends in fire frequencies from presettlement to recent times. At Mack Lake, the mean interval for a fire <4,000 ha in size was 27 years during the presettlement period (1830–49), although this estimate may be low because of limited data. During the settlement period (1850–1909), the fire return interval decreased to 10 years. It then increased to 18 years during the early years of fire suppression (1910–29), and eventually to 30 years during the most recent study period (1930–69) (Simard and Blank 1982). A similar trend in fire return time was associated with settlement and fire suppression in the BWCA (Heinselman 1973).

Examination of historical fire data for xeric areas dominated by jack pine in the eastern Upper Peninsula of Michigan indicates considerable variation in the size of burned areas (Table 21.1). General Land Office survey data (1840–56; Comer et al. 1995) indicate that some barrens landscapes were disturbed by both wind and fire. This interaction between windthrow and fire may have played a role in structuring landscape heterogeneity.

Managers from the U.S. Forest Service in Wisconsin plan to implement a fire regime that approximates the periodicity, intensity, scale, and diversity of fire in this ecosystem during pre-European times. It appears that restoration of areas such as the Moquah barrens of northwest Wisconsin may require frequent burns. Biology of jack pine in this area and recent experience suggest that catastrophic crown fires may have occurred as frequently as every 20–40 years (Linda Parker, Forest Ecologist, Chequamegon National Forest, Park Falls, Wisconsin, personal communication, 1996).

Fire in the jack pine barrens probably maintains a mixture of grasses and sedges (e.g., *Carex pensylvanica*) by allowing herba-

Table 21.1. Patch sizes of natural disturbances for Land Type Associations (LTAs) dominated by jack pine in the eastern Upper Peninsula of Michigan. Data are taken from the Michigan Natural Features Inventory database (Comer et al. 1995).

LTA	LTA area (ha)	No. samples		Mean patch size (ha)		Max. patch size (ha)		Min. patch size (ha)	
		Fire	Wind	Fire	Wind	Fire	Wind	Fire	Wind
Mint Farm	19,603	1	1	3,436	29				
Raco Plains	26,502	7	2	220		318	393	15	74
Wetmore Outwash Plains	11,037	1	1	34	128				
Two-Hearted Sands	36,628	7	3	265	539	725	1,011	5	5

ceous plants to exploit resources such as light, space, and nutrients released after burning, and by increasing soil temperature (Abrams and Dickmann 1984). Sites recently disturbed by either logging or fire have a higher coverage of vascular plants than older jack pine forests (Abrams, Sprugel and Dickmann 1985). Thus, periodic burning appears to be important for maintenance of annual and biennial species.

Biotic Factors

Jack pine is susceptible to diseases such as stem rusts and *Armillaria* root rot, and insects such as the jack pine budworm (*Choristonuera pinus pinus*). Moisture deficiency increases the susceptibility of trees to attack by pine budworm (Benzie 1977). The budworm defoliates trees stressed by moisture deficiency and causes reductions in the production of both vegetative and reproductive biomass. Budworm outbreaks typically occur every decade and last 2–4 years (Weber 1995). Although the effect on mature trees is devastating, these disturbances probably naturally interacted with fire, with the subsequent periodic regeneration of large areas (Weber 1995).

Community Characteristics

Composition

Curtis (1959) described the Wisconsin barrens as regions that were historically dominated by grasses, forbs, and shrubs, with scattered stands of trees. The most common tree was jack pine, but red pine was common and dominant in some areas (Table 21.2). Trembling and bigtooth aspen, red and scrub (Hill's, northern pin) oak (*Q. ellipsoidalis*), white pine, willow (*Salix* spp.), and hazel (American, *Corylus americana*, and beaked, *C. cornuta*) also were found in the understory. When present with jack pine, red pine dominated the larger diameter classes in the period prior to European settlement; modal diameter class was 30–40 cm for red pine ver-

sus 10–20 cm for jack pine (Whitney 1986). Sweet (1880), as quoted in Curtis (1959), reported that the average diameter of jack pine or scrub oak on the barrens of northern Wisconsin was only 7–9 cm and that the trees often reached a height of only 4–5 m.

The lower vegetation layers of the pine barrens of Wisconsin were dominated by shrubs (41% of total species and 33% of common ground flora; Table 21.3; Curtis 1959). Currently, over half of the species of ground flora are from five families: Asteraceae, Gramineae, Rosaceae, Liliaceae, and Ericaceae. Common species include blueberries (*Vaccinium* spp.), hazel, cherry, spreading dogbane (*Apocynum androsaemifolium*), and sweet fern (*Comptonia peregrina*) (Curtis 1959; Chequamegon National Forest 1993). Bracken fern (*Pteridium aquilinum*) also is common. Of these species, the blueberries and sweet fern often are considered the most common component of the barrens vegetation. Although the soils are extremely sandy, only a few species, such as sand cherry (*Prunus pumila*), sweet fern, and sand jointweed (*Polygonella articulata*), are obligate sand plants. Sedges (*Carex* spp.) and grasses, such as *Danthonia*, *Andropogon*, and *Oryzopsis*, dominate the ground layer after burning.

Structure

Historical descriptions of the pine barrens focus on the rolling landscape and diverse structure. Open pine barrens typically contained <20 trees ha^{-1} prior to European settlement; Curtis (1959) and Brown (1950) documented as few as 5–20 trees ha^{-1}. Vogl (1970) also described the barrens as landscapes with localized, dense stands of even-aged jack pine and scattered, towering red pines. Drake and Faber-Langendoen (1994) used multivariate techniques to reanalyze vegetation data from stands originally studied by Curtis (1959) and from 40 additional stands. Using overstory composition and the degree of canopy cover, they distinguished five subtypes of the pine barrens landscape in

Table 21.2. *Tree composition of pine barrens (modified from Curtis 1959).*

Species	Average importance value[1]	Constancy (%)[2]
Pinus banksiana	187.3	100
Quercus ellipsoidalis	57.7	86
Quercus macrocarpa	28.0	21
Populus grandidentata	10.2	21
Pinus resinosa	8.4	14
Populus tremuloides	2.7	7
Quercus rubra	1.2	7

[1] IV_{300}; sum of relative density, relative frequency, and relative dominance (basal area); values range from 0 to 300.
[2] Percentage of occurrences within samples of 80 trees per stand from each of 26 stands.

comparison to one type described by Curtis (1959) and two described by Vogl (1970): jack pine–oak sparse woodland, jack pine–oak woodland, jack pine forest, Hill's oak forest, and aspen forest. Jack pine, white pine, bigtooth aspen, and Hill's oak occurred in all five communities, with jack pine dominant in three of them. Basal area and canopy cover ranged from 14.1 m² ha⁻¹ (512 stems ha⁻¹) and 34%, respectively, in the jack pine–oak sparse woodland, to 25.5 m² ha⁻¹ (730 stems ha⁻¹) and 54%, respectively, in the jack pine forest.

During their studies of avian communities, Mossman and Sample (1995) differentiated five types of pine barrens in northern and central Wisconsin: open barrens, brush prairie, conifer dominated, barrens/woods, and diverse. Open barrens had substantial herbaceous and heath cover (up to 35%) but <5% cover of scattered pine or hardwoods. The brush prairie was characterized by 30%–60% cover of heath and shrub. Jack pine was the dominant species in all height classes of the conifer-dominated landscapes, although with historical fire regimes, red pine is thought to have been more prevalent. The barrens/woods landscape is mostly forested and dominated by aspen, oak, and jack pine. The diverse barrens described by Mossman and Sample (1995) covered areas greater than 80 ha and, because of the diverse topography and soil, exhibited

structural features of at least two of the other types of pine barrens (see also Mossman, Epstein and Hoffman 1991).

Succession

Periodic fires are necessary to maintain the barrens communities of the northern Great Lakes region, and species composition and structure of these communities often change if fire is suppressed. In Wisconsin, fire suppression after European settlement allowed an increase in tree density from <20 to >200 trees ha⁻¹ within a 100-year period (Brown 1950). Although there was no measurable change in the composition and dominance of tree species during this period, fire suppression often permits development of a community with greater portions of red and white pine than would otherwise occur on these sites. Frequent fires generally favor jack pine, whereas infrequent fires or fire-free periods favor the establishment of red pine and hardwoods (Vogl 1970). Shrubby species, which represent up to 30% of the species present in northern dry communities, also become more prevalent in the absence of fire (Chequamegon National Forest 1993). Dunn and Stearns (1980) determined that forests with jack pine, aspen, chokecherry (*Prunus virginiana*), and pin cherry (*P. pensylvanica*) would grow if fire were not frequent enough

Table 21.3. *Common species of ground flora in the pine barrens (modified from Curtis 1959). Percentages are occurrence in 26 stands.*

Species	Presence (%)	Species	Presence (%)
Vaccinium angustifolium	85	Chimaphila umbellata	50
Corylus americana	77	Gaultheria procumbens	50
Euphorbia corollata	77	Lysimachia quadrifolia	50
Fragaria virginiana	73	Monarda fistulosa	50
Apocynum androsaemifolium	69	Aster macrophyllus	46
Rosa sp.	65	Danthonia spicata	46
Rubus pubescens	65	Gaylussacia baccata	42
Maianthemum canadense	62	Salix discolor	42
Pteridium aquilinum	58	Helianthemum canadense	39
Smilacina stellata	58	Lithospernum canescens	39
Ceanothus ovatus	54	Lupinus perennis	39
Comptonia peregrina	54	Leptoloma cognatum	35
Smilacina racemosa	54	Coreopsis palmata	34

to maintain open barrens conditions. Over time, black cherry (*P. serotina*), maples, and paper birch (*Betula papyrifera*) also would become more common. Mossman and Sample (1995) proposed a successional sequence for Wisconsin from open barrens to brush prairie and eventually to barrens/woods in absence of fire.

In Minnesota, jack pine usually is succeeded by red pine, followed by eastern white pine, and finally a mixed hardwood forest composed of sugar maple (*Acer saccharum*), basswood (*Tilia americana*), and northern red oak. The red and white pine stages occasionally are absent, and speckled alder (*Alnus rugosa*), American and beaked hazel, paper birch, and quaking aspen succeed jack pine; the final stage is mixed hardwood forest (Rudolph and Laidly 1990).

Fauna

Avian Species

Diversity in plant composition and forest structure supports an interesting and diverse avian community. According to Mossman, Epstein and Hoffman (1991) and the Chequamegon National Forest (1986), about 100 species of breeding birds occur in the

pine barrens of Wisconsin. In addition to those restricted to barrens landscapes, this avifauna is composed of species associated with xeric pine and hardwood forest, boreal forest, and prairie ecosystems. At least 27 of these species are common breeding birds in pine barrens systems (Table 21.4) (Mossman, Epstein and Hoffman 1991), and 13 are threatened or of special concern within the state of Wisconsin (Table 21.5) (Wisconsin Department of Natural Resources 1995).

Successional transitional stages and natural disturbances play an important role in determining the relative abundance of bird species. Frequent burning helps to maintain prairie-associated species (Table 21.4) and species, including Brewer's blackbird (*Euphagus cyanocephalus*), that require fine-scale habitat features such as dead and down material (Mossman, Epstein and Hoffman 1991). As the vegetation is replaced by young, dense pine or mixed forest, the avian community also tends to include more species characteristic of northern dry forests (Hoffman and Mossman 1990). Other multiscale features, such as litter depth and high coverage of herbs and blueberry, affect the presence of characteristic barrens species such as the rufous-sided towhee (*Pipilo erythroph-*

Table 21.4. *Common breeding birds of Wisconsin pine barrens, found in at least 50% of sites surveyed by Mossman, Epstein and Hoffman (1991).*

Species	Barrens affiliation[1]	Species	Barrens affiliation[1]
American crow	D	Field sparrow	BP, BW
American goldfinch	O, BP, BW, D	Gray catbird	BP
American robin	CD, BW	Indigo bunting	BW, D
Black-capped chickadee	CD	Northern flicker	D
Blue jay	CD, D	Ovenbird	BW
Brown-headed cowbird	CD, BW, D	Red-eyed vireo	BW
Brewer's blackbird	O	Rose-breasted grosbeak	BW, D
Brown thrasher	C	Rufous-sided towhee	CD, BP, BW
Cedar waxwing	BW, D	Song sparrow	BP, D
Chestnut-sided warbler	BW, D	Tree swallow	D
Chipping sparrow	CD, BW	Upland sandpiper	OD
Common yellowthroat	D	Vesper sparrow	BP, D
Eastern bluebird	D	Yellow warbler	BP
Eastern kingbird	D		

[1] Type of barrens in which species is common: O = open barrens; BP = brush prairie; CD = conifer dominated; BW = barrens/woods, D = diverse. (See text for explanation of vegetation structure and composition.)

thalmus) (Morimoto and Wasserman 1991).

In northern Lower Michigan, the federally endangered Kirtland's warbler (*Dendroica kirtlandii*) breeds only in relatively dense, but patchy, jack pine barrens of a restricted age (5–23 years) and height (1.4–5.0 m) (Zou, Theiss and Barnes 1992; Probst and Weinrich 1993). The warbler is highly dependent on the periodic regeneration of this patch type within the landscape through either prescribed or natural burning, or through harvest and artificial regeneration. The warbler's decline appears to be associated primarily with habitat loss and parasitism by the brown-headed cowbird (*Molothrus ater*). Frequency and size of wildfires have decreased because of (a) habitat fragmentation by land development for human use and (b) fire suppression. Thus, the combined effects of changes in land use and altered disturbance regimes have reduced the successional stage of pine barrens required by the Kirtland's warbler (Bocetti 1994). Probst and Weinrich (1993) indicate that specific habitat

features such as tree height and percentage cover control habitat suitability for this species. For example, areas of dense regeneration after wildfire are preferred for foraging. Populations of Kirtland's warbler increased substantially after the Mack Lake fire of 1980, which burned approximately 10,000 ha and created areas of ideal habitat for this species over the next 15 years (Zou, Theiss and Barnes 1992; Probst and Weinrich 1993). Bocetti (1994) suggested that management of sites for features characteristic of post-wildfire areas, including increased density of jack pine and increased cover of ground species like blueberry, is important for warbler conservation.

The sharp-tailed grouse (*Tympanunchus phasianellus*), another species of special concern, also requires large tracts of open barrens habitat. It is sensitive to both the size and connectivity of regions of brushy barrens. Management agencies are cooperating to reintroduce and maintain a viable population of sharp-tailed grouse in areas such as

Table 21.5. *Species of birds that are associated with pine barrens ecosystems described by Mossman, Epstein and Hoffman (1991) and that are threatened or of special concern within the state of Wisconsin (adapted from* Natural Heritage Working List, *Wisconsin Department of Natural Resources, 1995).*

Species	Barrens affiliation[1]					Status in Wisconsin[2]
	O	BP	CD	BW	D	
Upland sandpiper	C[3]				C	SC
Greater prairie-chicken	R	R				Thr
Sharp-tailed grouse	U	U			U	SC
Connecticut warbler					R	SC
Tennessee warbler					U	SC
Kirtland's warbler[4]						SC
Dickcissel					R	SC
Field sparrow	FC	C	A	R	C	SC
Vesper sparrow	A	C	A	R	C	SC
Lark sparrow			R			SC
Grasshopper sparrow	FC				R	SC
Bobolink	U	U				SC
Western meadowlark	U				U	SC

[1] O = open barrens; BP = brush prairie; CD = conifer dominated; BW = barrens/woods; D = diverse. (See text for further explanation of vegetation structure and composition.)
[2] SC = special concern; Thr = threatened.
[3] A = abundant; C = common; FC = fairly common; U = uncommon; R = rare.
[4] Habitat not classified by Mossman, Epstein and Hoffman (1991).

the Moquah pine barrens in northern Wisconsin (P. David 1995). The reduction in numbers of this species and the extirpation of the prairie chicken (*Tympanuchus cupido*) in the Wisconsin pine barrens appear to be caused by the change from savanna vegetation to denser jack pine forest. This replacement is the result of burn control measures instituted in the 1920s (Brown 1950) and the reforestation of barrens with dense jack and red pine plantations by the Civilian Conservation Corps in the 1930s (Vora 1993). The red-shouldered hawk (*Buteo lineatus*) and northern harrier (*Circus cyaneus*) also are species of federal concern that sometimes occur in the pine barrens ecosystem.

Mammals

Wydeven (1995) noted that approximately 30 species of mammals occur in the pine barrens of Wisconsin, including shrews (*Sorex cinereus*, *Microsorex hoyi*, and *Blarina brevicauda*), bats (*Lasiurus borealis*, *L. cinereus*, *Myotis lucifugus*, *Eptesicus fuscus*, and *Lasionycteris noctivagans*), mice (*Peromyscus* spp.), voles (*Microtus* spp.), ground squirrels (*Spermophilus* spp.), woodchuck (*Marmota monax*), pocket gopher (*Geomys bursarius*), coyote (*Canis latrans*), badger (*Taxidea taxus*), white-tailed deer (*Odocoileus virginiana*), red fox (*Vulpes vulpes*), bobcat (*Lynx rufus*), black bear (*Ursus americanus*), and wolf (*Canis lupus*). There is little information on the habitat preferences for most of the small or nongame mammal species. As new fire management regimes are undertaken, efforts should be made to assess the suitability for mammalian species of burned sites compared to clear-cut and undisturbed sites of different ages. Although many sites initially appear similar in habitat structure and composition, some important features may differ substantially between the disturbance types. Initial efforts to address these questions are being undertaken by researchers at the Great Lakes Indian Fish and Wildlife Commission (Peter David, Wildlife Biologist, Great Lakes Indian Fish and Wildlife Commission, Odanah, Wisconsin, personal communication, 1996).

Plans for preharvest silvicultural activities and integrated resource management should incorporate recommendations on specific, fine-scale habitat features, forage species, proportions of vegetation cover, and landscape structure. For example, jack pine and associated understory shrubs and herbaceous species are considered a medium-preference food for white-tailed deer, although populations of this species are not limited in the Great Lakes region. Eastern cottontail rabbits (*Sylvilagus floridanus*) and snowshoe hares (*Lepus americanus*) are found in barrens landscapes, and they prefer the young, herbaceous plants that sprout after burning of the ground cover (Niemi and Probst 1990). Black bears forage extensively on blueberries. Wolves primarily use this type of habitat for hunting and as travel corridors (Wydeven 1995).

Herptiles

Twenty-five species of reptiles and amphibians are associated with the sandy soils of pine barrens in Wisconsin. Fifteen of these are at the edges of their geographic ranges in this region (Hay 1995) and, thus, they may be of management concern in the event of climate change or human-related losses of habitat. The genetic distinctness of populations of these species within isolated areas of barrens is not well understood.

Five species are strong "indicators" of barrens habitat: bullsnake (*Pituophis melanoleucus sayi*), eastern hognose snake (*Heterodon platyrhinos*), Blanding's turtle (*Emydoidea blandingi*), five-lined skink (*Eumeces fasciatus*), and northern prairie skink (*E. septentrionalis septentrionalis*). Both the Blanding's turtle and the bullsnake are of special concern and are monitored by the Wisconsin Natural Heritage Program (Wisconsin Department of Natural Resources 1995). Microhabitats in areas of sparse canopy and low vegetative cover of the barrens are important for thermal regulation, foraging, and nesting of these five vertebrates. Periodic fire also may be an important factor for maintaining landscape structure, community composition, and associated invertebrate fauna required by these herptiles (Hay 1995).

Invertebrates

Up to 43 species of rare invertebrates are associated to some degree with barrens and savanna habitat in Wisconsin (Wisconsin Department of Natural Resources 1995). The Karner blue butterfly (*Lycaeides melissa samuelis*), a federally endangered species, and the phlox flower moth (*Schinia indiana*), under consideration for federal protection, are strongly associated with barrens habitat. Although the metapopulation dynamics of the Karner blue butterfly is poorly understood, this species requires an abundance of its larval host plant, wild lupine (*Lupinus perennis*), and a fine-scale mosaic of sun and shade (Leach 1995). The frosted elfin butterfly (*Incisalia irus*), threatened within Wisconsin, also is specific to barrens landscapes.

Influence of Fire

Fire affects the habitat structure of each of the taxa just listed. The intensity of burn influences the amount of standing dead wood available for cavity-nesting birds, perches for avian species, and cover used by small mammals. The size, orientation, and patchiness of a burn influence the habitat encountered by larger species such as black bear and deer (Niemi and Probst 1990). Following a fire, the release of nutrients and reduction of competition for water and light often result in an increase in growth of small young plants, which are preferred forage for these large mammals. Invertebrates that require a mosaic of sun and shade provided by periodic burning will be killed if fire is too expansive or recurrent (Leach 1995). Prescribed burning may not mimic a wildfire in extent or patchiness or in type or degree of vegetation burned (e.g., crown versus ground-cover fires). Therefore, this technique may not provide ideal habitat for certain species. More information is needed on the linkages between prescribed fire and wildlife population dynamics.

Timber Harvesting

Logging can affect species composition in the pine barrens. For example, *Carex* meadows often form after clear-cutting. In Michigan, jack pine forests that were both clear-cut and burned had fewer species and a sizable reduction in *Carex* cover (Abrams 1977; Abrams and Dickmann 1984). These results highlight the importance of considering interactions among disturbances in any management regime.

Historically, logging, which began in about 1860 in Wisconsin barrens, facilitated the growth of jack pine relative to other tree species. Cutting of white and red pine left large amounts of slash, which kindled frequent fires during the early industrial period and promoted growth of jack pine and Hill's oak over other less fire-tolerant species (Murphy 1931). Prior to the initiation of large-scale logging by European immigrants, both aboriginal and natural fires probably played roles in creating and maintaining barrens ecosystems. Although some authors have concluded that certain "savanna" systems are relatively stable communities maintained by climatic and edaphic factors (Whitford and Whitford 1971), it seems unlikely that the jack pine barrens could exist in the absence of fire.

Management Implications

Maintenance of the pine barrens ecosystem has ecological, economic, and recre-ational implications. Species such as the Kirtland's warbler, phlox moth, and Karner blue butterfly are dependent on barrens systems (Vora 1993; Temple 1995). Currently, large numbers of people enjoy picking edible fruits (blueberries, raspberries, blackberries, *Rubus* spp.) as well as hiking and camping in these areas. The barrens also provide habitat for large game species such as black bear and white-tailed deer. Finally, the aesthetic appeal of the barrens is dependent on the persistence of the mosaic of patchy vegetation dominated by expanses of open rolling plains.

It is important to plan and manage at a landscape scale to maintain the floral and faunal diversity of the barrens ecosystem. Patch size and connectivity may influence the persistence of species such as the sharp-tailed grouse. Currently, managers of the Chequamegon National Forest, Wisconsin, are undertaking measures to maintain open areas up to 200 ha with high-frequency, high-intensity burns. Although the U.S. Forest Service is aware that this technique will not always mimic natural wildfire, it provides habitat for the reintroduction of sharp-tailed grouse and is beneficial to a number of other vertebrate and invertebrate species (Vora 1993). Unfortunately, there is limited knowledge of the habitat requirements of most nongame species. This presents a challenge to managers who hope to conserve or restore pine barrens ecosystems.

A focal point of management in the pine barrens is knowledge of how prescribed fire can closely mimic the natural return time and spatial scale of this disturbance. Do barrens ecosystems occupy the same position in the landscape through time, or is this ecosystem simply a patch type that follows certain fire regimes? Obviously, jack pine barrens occur on sandy, glacial–fluvial landforms in the Lake States. However, it is not clear if there are any prerequisite topographic or edaphic conditions for persistence of this ecosystem. It may be that the frequency and intensity of wildfire were the main factors responsible for the extensive areas of jack pine barrens that were once a dominant feature in the Lake States. Clearly, a better understanding of fire management and its interaction with topography, soil, and vegetation is important to the restoration of jack pine barrens.

Summary

Jack pine barrens have been reduced to a fraction of the area they covered prior to pre-European settlement. Habitat fragmentation associated with human land use and fire suppression have contributed to the loss of this ecosystem in the northern Great Lakes states. Jack pine barrens characteristically occur on sandy, glacial outwash soils that have limited moisture and nutrient reserves. However, nutrient status and edaphic factors do not appear to maintain barrens in the absence of fire. Fire also affects species composition and habitat structure. Most importantly, frequent burning promotes growth of jack pine relative to other tree species, and creates expansive, open areas with high coverage of herbaceous vegetation. Jack pine barrens can be subdivided based on the dominant vegetation in different strata and successional status. Numerous species of birds, herptiles, and invertebrates require a mosaic of fine and coarse-scale habitat features associated with one or more of these barrens types. Specific management regimes currently are aimed at conserving populations that are federally endangered or locally threatened because of the decline in quantity and quality of barrens habitat. Recent initiatives in the northern Great Lakes states suggest that prescribed burning can be used successfully to restore the diversity of composition and structure historically found in this landscape. Further information is required on the life history attributes and habitat requirements of both jack pine flora and fauna to expedite these efforts.

REFERENCES

Abrams, M. D. 1977. Post-fire revegetation of jack pine sites in Michigan: an example of successional complexities. *Proceedings of the Annual Tall Timbers Fire Ecology Conference* 17:197–209.

Abrams, M. D., and Dickmann, D. I. 1984. Floristic composition before and after prescribed fire on a jack pine clear-cut site in northern lower Michigan. *Canadian Journal of Forest Research* 14:746–749.

Abrams, M. D., Sprugel, D. G., and Dickmann, D. I. 1985. Multiple successional pathways on recently disturbed jack pine sites in Michigan. *Forest Ecology and Management* 10:31–48.

Alban, D. H. 1982. Effects of nutrient accumulation by aspen, spruce, and pine on soil properties. *Soil Science Society of America Journal* 46:853–861.

Almendinger, J. C. 1990. The decline of soil organic matter, total-N, and available water capacity following the late-Holocene establishment of jack pine on sandy mollisols, north-central Minnesota. *Soil Science* 150:680–694.

Benzie, J. W. 1977. *Managers Handbook for Jack Pine in the North Central States*. USDA Forest Service, St. Paul, Minn., Gen. Tech. Rep. NC-32.

Bocetti, C. I. 1994. Density, demography, and mating success of Kirtland's Warblers in managed and natural habitats. Ph.D. dissertation, Ohio State University, Columbus, Ohio.

Bockheim, J. G., Jepsen, E. A., and Heisey, D. M. 1991. Nutrient dynamics in decomposing leaf litter of four tree species on a sandy soil in northern Wisconsin. *Canadian Journal of Forest Research* 21:803–812.

Bockheim, J. G., and Leide, J. E. 1991. Foliar nutrient dynamics and nutrient-use efficiency of oak and pine on a low-fertility soil in Wisconsin. *Canadian Journal of Forest Research* 21:925–934.

Borgerding, E. A., Bartelt, G. A., and McCown, W. M., eds. 1995. *The Future of Pine Barrens in Northwest Wisconsin: A Workshop Summary*, PUBL-RF-913-94. Madison, Wis.: Wisconsin Department of Natural Resources.

Bourdo, E. A., Jr. 1954. A validation of methods used in analyzing original forest cover. Ph.D. dissertation, University of Michigan, Ann Arbor, Mich.

Brown, R. T. 1950. Forests of the central Wisconsin sand plains. *Bulletin of the Ecological Society of America* 22:217–234.

Brubaker, L. B. 1975. Postglacial forest patterns associated with glacial till and outwash in northcentral Upper Michigan. *Quaternary Research* 5:499–527.

Cayford, J. H. 1970. The role of fire in the ecology and silviculture of jack pine. *Proceedings of the Annual Tall Timbers Fire Ecology Conference* 10:221–244.

Chequamegon National Forest. 1986. *Birds of the Chequamegon.* Park Falls, Wis.: USDA Forest Service.

Chequamegon National Forest. 1993. *Landscape Level Analysis, Desired Future Vegetative Condition.* Park Falls, Wis.: USDA Forest Service.

Comer, P. J., Albert, D. A., Wells, H. A, Hart, B. L., Raab, J. B., Price, D. L., Kashian, D. M., Corner, R. A., and Schuen, D. W. 1995. *Michigan's Presettlement Vegetation, as Interpreted from the General Land Office Surveys 1816–1856.* Lansing, Mich.: Michigan Natural Features Inventory.

Critchfield, W. B. 1985. The late Quaternary history of lodgepole and jack pines. *Canadian Journal of Forest Research* 15:749–772.

Curtis, J. T. 1959. *Vegetation of Wisconsin: An Ordination of Plant Communities.* Madison, Wis.: University of Wisconsin Press.

David, M. B., Grigal, D. F., Ohmann, L. F., and Gertner, G. Z. 1988. Sulfur, carbon, and nitrogen relationships in forest soils across the northern Great Lakes States as affected by atmospheric deposition and vegetation. *Canadian Journal of Forest Research* 18:1386–1391.

David, P. 1995. The Great Lakes and Wisconsin's pine barrens. In *The Future of Pine Barrens in Northwest Wisconsin: A Workshop Summary*, ed. E. A. Borgerding, G. A. Bartelt, and W. M. McCown, PUBL-RF-913-94. p. 48. Madison, Wis.: Wisconsin Department of Natural Resources.

Davis, M. B. 1983. Holocene vegetational history of the eastern United States. In *Late Quaternary Environments of the United States: The Holocene*, ed. H. E. Wright, pp. 166–181. Minneapolis, Minn.: University of Minnesota Press.

Drake, J., and Faber-Langendoen, D. 1994. A reanalysis of the pine barrens of Wisconsin. Barrens as an ecological term: an overview of usage in the scientific literature. In *Proceedings of the North American Conference on Savannas and Barrens*, ed. J. S. Fralish, R. C. Anderson, J. E. Ebinger, and R. Szafoni, pp. 349–354. Chicago,

Ill.: U.S. Environmental Protection Agency, Great Lakes National Program Office.

Dunn, C. and Stearns, F. 1980. Vegetation of the Moquah Barrens Research Natural Area. Unpublished report for the USDA Forest Service, Chequamegon National Forest, Park Falls, Wis.

Givnish, T. 1995. A national perspective of the pine barrens community. In *The Future of Pine Barrens in Northwest Wisconsin: A Workshop Summary*, ed. E. A. Borgerding, G. A. Bartelt, and W. M. McCown, p. 3. Madison, Wis.: Wisconsin Department of Natural Resources, PUBL-RF-913-94.

Green, D. C., and Grigal, D. F. 1980. Nutrient accumulations in jack pine stands on deep and shallow soils over bedrock. *Forest Science* 26:325–333.

Grimm, E. C . 1984. Fire and other factors controlling the big woods vegetation of Minnesota in the mid-nineteenth century. *Ecological Monographs* 54:291–311.

Hay, B. 1995. Herptiles of Wisconsin's pine barrens. In *The Future of Pine Barrens in Northwest Wisconsin: A Workshop Summary*, ed. E. A. Borgerding, G. A. Bartelt, and W. M. McCown, pp. 13–16. Madison, Wis.: Wisconsin Department of Natural Resources, PUBL-RF-913-94.

Heikens, A. L., and Robertson, P. A. 1994. Barrens of the Midwest: a review of the literature. *Castanea* 59:184–194.

Heinselman, M. L. 1973. Fire in the virgin forests of the Boundary Waters Canoe Area, Minnesota. *Journal of Quaternary Research* 3:329–382.

Heinselman, M. L. 1981. Fire and succession in the conifer forests of northern North America. In *Forest Succession: Concepts and Applications*, ed. D. C. West, H. H. Shugart, and D. B. Botkin, pp. 374–405. New York: Springer-Verlag.

Hoffman, R. M., and Mossman, M. J. 1990. Birds of northern Wisconsin pine forests. *Passenger Pigeon* 52:339–355.

Hole, F. D. 1976. *Soils of Wisconsin.* Madison, Wis.: University of Wisconsin Press.

Leach, M. K. 1995. The Karner blue butterfly and the phlox moth. In *The Future of Pine Barrens in Northwest Wisconsin: A Workshop Summary*, ed. E. A. Borgerding, G. A. Bartelt, and W. M. McCown, p. 12. Madison, Wis.: Wisconsin Department of Natural Resources, PUBL-RF-913-94.

Loomis, R. M. 1977. *Jack Pine and Aspen Forest Floors in Northeastern Minnesota*. St. Paul, Minn.: USDA Forest Service, Research Note NC-222.

MacLean, D. A., and Wein, R. W. 1978. Weight loss and nutrient changes in decomposing litter and forest floor material in New Brunswick forest stands. *Canadian Journal of Botany* 56:2730–2749.

McCullough, D. G., and Kulman. H. M. 1991. Effects of nitrogen fertilization on young jack pine (*Pinus banksiana*) and on its suitability as a host for jack pine budworm (*Choristonuera pinus pinus*) (Lepidoptera: Tortricidae). *Canadian Journal of Forest Research* 21:1447–1458.

Morimoto, D. C., and Wasserman, F. E. 1991. Dispersion patterns and habitat associations of rufous-sided towhees, common yellowthroats, and prairie warblers in the southeastern Massachusetts Pine Barrens. *Auk* 108:264–276.

Mossman, M. J., Epstein, E., and Hoffman, R. M. 1991. Birds of Wisconsin pine and oak barrens. *Passenger Pigeon* 53:137–163.

Mossman, M. J., and Sample, D. 1995. Birds of the pine barrens. In *The Future of Pine Barrens in Northwest Wisconsin: A Workshop Summary*, ed. E. A. Borgerding, G. A. Bartelt, and W. M. McCown, pp. 20–22. Madison, Wis.: Wisconsin Department of Natural Resources, PUBL-RF-913-94.

Murphy, R. E. 1931. The geography of the northwestern pine barrens of Wisconsin. *Wisconsin Academy of Sciences, Arts and Letters* 26:69–120.

Niemi, G. J., and Probst, J. R. 1990. Wildlife and fire in the upper Midwest. In *Management of Dynamic Ecosystems*, ed. J. M. Sweeney, pp. 34–49. West Lafayette, Ind.: The Wildlife Society.

Pregitzer, K. S., and Barnes, B. V. 1984. Classification and comparison of upland hardwood and conifer ecosystems of the Cyrus H. McCormick Experimental Forest, Upper Michigan. *Canadian Journal of Forest Research* 14:362–375.

Probst, J. R., and Weinrich J. 1993. Relating Kirtland's Warbler population to changing landscape composition and structure. *Landscape Ecology* 8:257–271.

Romme, W. H., and Knight, D. H. 1981. Fire frequency and subalpine forest succession along a topographic gradient in Wyoming. *Ecology* 62:319–326.

Rudolph, T. D., and Laidly, P. R. 1990. *Pinus banksiana* Lamb., Jack Pine. In *Silvics of North America*, vol. 1, *Conifers*, tech. coords. R. M. Burns and B. H. Honkala, pp. 280–292. Agricultural Handbook 654. Washington, D.C.: USDA Forest Service.

Schoenike, R. E. 1976. *Geographic Variation in Jack Pine* (Pinus banksiana). Technical Bulletin. St. Paul, Minn.: University of Minnesota Agricultural Experiment Station.

Simard, A. J., and Blank, R. W. 1982. Fire history of a Michigan jack pine forest. *Michigan Academician* 15:59–71.

Stern, W. L., and Buell, M. F. 1951. Life-form spectra of New Jersey pine barrens forest and Minnesota jack pine forest. *Bulletin of the Torrey Botanical Club* 78:61–65.

Sweet, E. T. 1880. Geology of the western Lake Superior district; climate, soils, and timber. *Geology of Wisconsin* 3:323–329.

Temple, S. A. 1995. Biodiversity, landscape-scale management and the ecological importance of the pine barrens community. In *The Future of Pine Barrens in Northwest Wisconsin: A Workshop Summary*, ed. E. A. Borgerding, G. A. Bartelt, and W. M. McCown, p. 2. , Madison, Wis.: Wisconsin Department of Natural Resources, PUBL-RF-913-94.

Vogl, R. J. 1970. Fire and the Northern Wisconsin pine barrens. *Proceedings of the Tall Timbers Fire Ecology Conference* 10:175–209.

Vora, R. S. 1993. Moquah Barrens: pine barrens restoration experiment initiated in Chequamegon National Forest. *Restoration and Management Notes* 11:39–44.

Webb, T., III, Cushing, E. J., and Wright, H. E., Jr. 1983. Holocene changes in the vegetation of the Midwest. In *Late Quaternary Environments of the United States: The Holocene*, vol. 2, ed. H. E. Wright, Jr., pp. 143–165. Minneapolis, Minn.: University of Minnesota Press

Weber, S. D. 1995. Jack pine budworm and the Wisconsin pine barrens. In *The Future of Pine Barrens in Northwest Wisconsin: A Workshop Summary*, ed. E. A. Borgerding, G. A. Bartelt, and W. M. McCown, p. 8. Madison, Wis.: Wisconsin Department of Natural Resources, PUBL-RF-913-94.

Whitford, P. B., and Whitford, K. 1971. Savanna in central Wisconsin, U.S.A. *Vegetatio* 23:77–88.

Whitney, G. G. 1986. Relation of Michigan's presettlement pine forests to substrate and disturbance history. *Ecology* 67:1548–1559.

Wisconsin Department of Natural Resources. 1995. *Natural Heritage Working List*. Madison, Wis.: Bureau of Endangered Resources.

Wydeven, A. P. 1995. Wolf and other mammals of the pine barrens. In *The Future of Pine Barrens in Northwest Wisconsin: A Workshop Summary*, ed. E. A. Borgerding, G. A. Bartelt, and W. M. McCown, pp. 24–26. Madison, Wis.: Wisconsin

Department of Natural Resources, PUBL-RF-913-94.

Zou, X., Theiss, C., and Barnes, B. V. 1992. Pattern of Kirtland's warbler occurrence in relation to the landscape structure of its summer habitat in northern Lower Michigan. *Landscape Ecology* 6:221–231.

22 The Cliff Ecosystem of the Niagara Escarpment

D. W. LARSON, U. MATTHES-SEARS,
AND P. E. KELLY

Introduction

The Niagara Escárpment is a 50–100-m high cuesta following the rim of a 450-million-year-old saucer-shaped geological structure known as the Michigan Basin. It extends in a roughly circular shape from Niagara Falls north to Manitoulin Island, across to the junction of Lakes Michigan and Huron at Sault Ste. Marie, then south along the western shore of Lake Michigan. Generally, the Silurian-aged rim is buried by glacial till, but 150 km of exposed near-vertical cliff face is present between Niagara Falls and Manitoulin Island, a straight-line distance of approximately 360 km.

The escarpment lies within the Great Lakes–St. Lawrence Forest (Rowe 1972). The climate of the area is temperate with a strong local influence of the Great Lakes. Mean January air temperature ranges from -5 °C in the south to -7.5 °C in the north; mean July temperature varies between 21 °C in the south to 18 °C in the north. Total annual precipitation of 900 mm is evenly distributed throughout the year. Lumbering activities were intense from 1800 to 1920, and the area was almost totally deforested by European settlers. Fire consumed almost all of the forested horizontal landscape of the Bruce Peninsula in 1908 (Gillard and Tooke 1975). However, no evidence of fire was found in cores or cross-sections obtained from trees growing on the exposed vertical cliffs (Larson and Kelly 1991).

Natural History

Cliffs of the Niagara Escarpment support a stable, ancient forest ecosystem whose structure probably has changed little since the melting of the Laurentide ice sheet. At the peak of Wisconsin glaciation, northern white cedar (white cedar, *Thuja occidentalis*) and its associated cliff community probably was restricted to marginal habitats such as exposed limestone cliffs in Tennessee, Kentucky, and the Carolinas (Walker 1987; Young 1996). In such locations, a large collection of fugitive plant and animal species probably was assembled. From this viewpoint, cliff habitats represent a glacial refugium, a conclusion also suggested by Jackson and Sheldon (1949), Davis (1951), Oppenheimer (1956), Brunton and Lafontaine (1974), and Holmquist (1962). This refugial status of cliff ecosystems suggests that considerably greater biodiversity may be present on cliffs than is currently recognized.

Literature on the northern postglacial migration of white cedar suggests that cliffs and adjacent swamps were colonized by 10,500 yr BP (Bell 1972; Warner 1982; Warner, Hebda and Hann 1984; Walker 1987). Samples of Holocene-aged trees collected in growth position from cliff edges submerged during the refilling of glacial Lake Hough (now Georgian Bay, Lake Huron) have been recently identified as white cedar (Larson and Melville 1996). The growth rate and morphology of these subfossil trees indicate that the forest of the immediate post-glacial environment had the same tree species composition, growth rate, and stem orientation as that presently found on the cliff. Radiocarbon dating carried out on snags of cliff-face northern white cedar found lying on talus slopes of the Niagara Escarpment show that tree density, growth rates, and stem morphology have changed little during the past 3,400 years (Kelly, Cook and Larson 1992, 1994). In addition, growth and photosynthesis in white cedar (Matthes-Sears and Larson 1990; Kelly, Cook and Larson 1994) are negatively influenced by temperatures exceeding 30 °C in the summer. This information, combined with knowledge of the distribution patterns of herbaceous species (Morton and Venn 1987), suggests that the forest ecosystem on the exposed cliffs has an arctic/alpine affinity and that composition has been relatively stable since deglaciation. The occurrence of bones of an extinct species of pika (genus *Ochotona*) supports the idea that, unlike the rest of southern Ontario, the exposed cliffs and talus slopes of the Niagara Escarpment had an ecological structure similar to alpine or arctic tundra (Morisset 1971; Riley 1993).

Community Relationships

The first studies on the ecology of the Niagara Escarpment were initiated in 1984. Morton and Venn (1984, 1987) presented qualitative habitat descriptions and species lists for a variety of islands in the northern part of the Niagara Escarpment; several of these islands included large limestone/dolostone cliffs. Quantitative studies were first

Figure 22.1. (a) Oblique aerial photograph of a typical section of cliff face along the Niagara Escarpment near Wiarton, Ontario. (b) Oblique photograph of the Niagara Escarpment at Milton, Ontario. (Photos taken from Kelly, Cook, and Larson 1992, courtesy University of Chicago Press)

started on the exposed cliffs in 1985 (Larson et al. 1989). Two features of the escarpment determine the extent and type of vegetational work. First, it is topographically abrupt (Figure 22.1a, b). Second, the escarpment is essentially ribbonlike across 740 km of landscape. Defining the structural and functional characteristics of this unusual forest presents some difficult problems as described in detail in Larson et al. (1989).

The change in species composition across the topographic gradient of the Niagara Escarpment is as rapid as the change in the

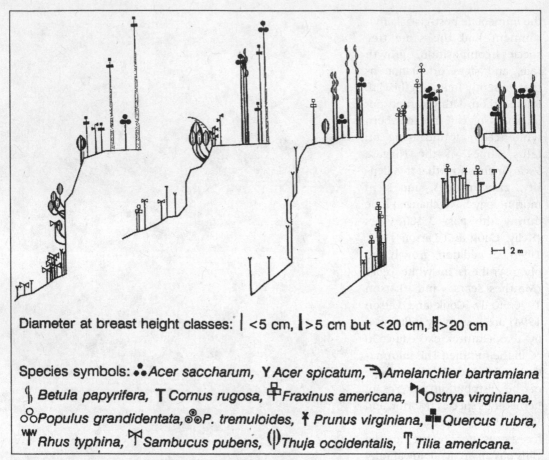

Diameter at breast height classes: | <5 cm, ▮ >5 cm but <20 cm, ▯ >20 cm

Species symbols: •• *Acer saccharum*, Y *Acer spicatum*, ⌐ *Amelanchier bartramiana*
\| *Betula papyrifera*, T *Cornus rugosa*, ✚ *Fraxinus americana*, ⌐ *Ostrya virginiana*,
○○ *Populus grandidentata*, ◉◉ *P. tremuloides*, ❀ *Prunus virginiana*, ◼ *Quercus rubra*,
ⱲⱲ *Rhus typhina*, ⋈ *Sambucus pubens*, (I) *Thuja occidentalis*, T *Tilia americana*.

Figure 22.2. Diagrammatic profile diagrams of the cliffs at the five sites sampled along the Niagara Escarpment. (Taken from Larson et al. 1989.)

physical environment (Larson et al. 1989). None of the taxa forming the canopy in the level-ground woodland at the top of the cliff occurred on the cliff face or on the talus slopes at the base of the cliff (Figure 22.2). The level-ground woodland is dominated by highly competitive species such as sugar maple (*Acer saccharum*) and northern red oak (*Quercus rubra*); the cliff face supports only white cedar in the canopy, and the talus slope is dominated by disturbance-tolerant shrubs such as roundleaf dogwood (*Cornus rugosa*) and red elderberry (*Sambucus pubens*). An analysis using understory vegetation, soil depth, and the level of photosynthetically active radiation (PAR) indicated that the

plateau to cliff face to talus slope environment was divided into three distinct biotic zones. Plateaus on the cliff tops have deep soil (ca. 40 cm), little ground-level irradiance, and a wide diversity of highly competitive vascular taxa (Figure 22.3a–c). Cliff faces have little or no soil but high irradiance and a depauperate vascular flora; talus slopes have highly variable amounts of soil because of the irregular face of the rocky substrate, moderate surface-level irradiance, and high levels of taxonomic diversity. An analysis of variance of these physical and biotic variables showed that "position" along transects, which were positioned across the gradient from plateau to cliff face to talus slope,

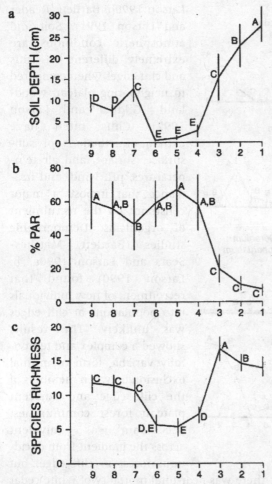

Figure 22.3. Relationship between plot position and (a) soil depth, (b) percentage of photosynthetically active radiation, and (c) local species richness. Plot 1 is located in plateau deciduous forest; plot 9, on the talus slope; intermediate numbers are for plots on the cliff face and transition areas. Identical letters indicate mean values that are not significantly different as determined by a multiple F-test. (Taken from Larson et al. 1989.)

accounted for most of the variance in soil depth (65%), available PAR (52%), species richness (77%), and species diversity (92%) as measured by the Shannon–Weiner index (p < 0.001). Neither transect nor geographical location of sites explained much of the variance (2%–18%) (Larson et al. 1989).

Common species have a specific, characteristic distribution pattern. Sugar maple

and wild lily-of-the-valley (*Maianthemum canadense*) are abundant in the plateau woodland and absent from face and talus habitats (Figure 22.4a, b), whereas white cedar and rock polypody (*Polypodium virginianum*) peak in abundance at the cliff edge (Figure 22.4c, d). Recent field sampling indicated a density of 1,013 stems per hectare of white cedar on the cliff face although many stems were stunted, giving the appearance that actual density was much lower. Smooth cliff-brake (*Pellaea glabella*) was found only on faces but declined in abundance toward the bottom of the face (Figure 22.4e), whereas Robert's geranium (*Geranium robertianum*) peaked in abundance in the area of talus closest to the cliff face (Figure 22.4f). *Brachythecium salesbrosum* and roundleaf dogwood were moderately abundant in both talus and plateau habitats (Figure 22.4g, h).

These results indicate that a predictable assemblage of physical and biological characteristics exist on the exposed cliffs of the Niagara Escarpment. However, there is no published evidence indicating that the Niagara Escarpment is geobotanically or ecologically unique. Work by Morisset (1971, 1979), Morisset, Bedard and Lefebvre (1983), and Walker (1987) showed that Silurian- and Ordovician-aged limestone outcrops in Quebec, New York, Ohio, Tennessee, and Virginia support communities similar to those found on exposed cliffs of the Niagara Escarpment. The exposed limestone cliffs along the New River in Giles County, Virginia, have many structural and functional characteristics similar to those of Niagara Escarpment cliffs (Larson 1997).

Autecological Relationships

White cedar is abundant on cliff edges and faces, despite an exceptionally low seedling recruitment rate (Spring 1988). Three hypotheses for this apparent contradiction have been suggested. The first is that this species was capable of pulsed recruitment as

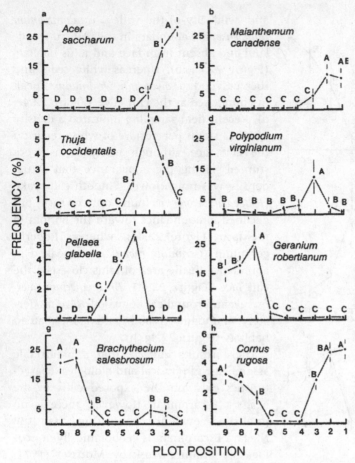

Figure 22.4. Relationship between plot position and species frequency for eight important taxa. Plot 1 is located in deciduous forest; plot 9, on the talus; intermediate numbers are for plots on the cliff face and intermediate areas. Identical letters indicate mean values that are not significantly different as determined by a multiple F-test. (Taken from Larson et al. 1989.)

Barden (1988) has shown for table-mountain pine (*Pinus pungens*). The second is that the restriction of white cedar to cliff edges and faces may be due to a single pulse of recruitment subsequent to forest clearing at the end of the last century. Third is that recreational hiking and climbing may eliminate seedlings before they become established.

Conditions at cliff edges of the Niagara Escarpment are stressful to all plants, including white cedar (Bartlett, Matthes-Sears and

Larson 1990b; Bartlett, Reader and Larson 1991). Soil and atmospheric conditions are extremely different on cliffs and cliff edges when compared to neighboring plateau woodland (Gildner and Larson 1992). Cliffs often have extreme fluctuations of subsurface, surface, and air temperatures, pH, wind, and irradiance that impose a major constraint on the recruitment of organisms. Demographic studies (Bartlett, Matthes-Sears and Larson 1990a, b; Larson 1990) found that recruitment of new individuals into populations at cliff edges was unlikely. The results showed a complex and temporally variable form of mutual exclusion between members of the cliff edge and adjacent plateau forest communities. Seed rain was equivalent across the gradient from deciduous forest to cliff edge, but there was near-total mortality of white cedar beneath deciduous forest litter (Bartlett and Larson 1990; Bartlett, Matthes-Sears and Larson 1990a, b; Bartlett, Reader and Larson 1991). In contrast, maple mortality was low in the understory away from the cliff edge, and increased to high levels only near the cliff edge when herbivory and drought became intense. These results also suggest a temporally varying pattern of physiological activity but, more importantly, it appears that there is no specialization of either taxon to the environment in which the species are normally dominant (Matthes-Sears and Larson 1990).

White cedar growing on cliffs were not nutrient deficient (Matthes-Sears and Larson 1991) even though their growth rates were exceptionally low. One 25-year-old stem was only 10 cm high and 0.7 cm in diameter.

Nevertheless, the slow growth of cliff-face white cedar apparently was not strongly controlled by either water supply or nutrient availability (Matthes-Sears and Larson 1995; Matthes-Sears, Nash and Larson 1995). Roots of all trees were found to be equally and highly (80%) colonized by vesicular– arbuscular mycorrhizae (Matthes-Sears, Neeser and Larson 1992).

Populations of white cedar located at the cliff edge were uneven aged, and living individuals were as old as 700 years. White cedar also had a pulse recruitment pattern, and populations were extremely sensitive to intense hiking disturbance (Larson 1990). Larson and Kelly (1991) and Kelly, Cook and Larson (1992) have demonstrated the extended longevity of white cedar at all sites along the escarpment. During 1993, individuals were found with ages up to 1,653 years, four times those quoted in Burns and Honkala (1990).

Near and Functional Endemism

No strict endemics are known to occur along the Niagara Escarpment. Nevertheless, certain taxa (e.g., Hart's tongue, *Phyllitis scolopendrium*; rue spleenwort, *Asplenium rutamuraria*) are found exclusively or mainly on limestone cliffs in eastern North America (Cody and Britton 1989). Other more common taxa (Table 22.1), such as walking fern (*Camptosorus rhizophyllus*), Steller cliff-brake (*Cryptogramma stelleri*), green spleenwort (*Asplenium viride*), and a sedge (*Carex eburnea*), also occur on limestone cliffs or associated talus slopes in Quebec and the southern United States (Morisset 1979; Walker 1987; Young 1996). In addition to these species, Morton and Venn (1987) reported two grasses (*Poa glauca, Poa canbyi*), spikemoss (*Selaginella selaginoides*), and butterwort (*Pinguicula vulgaris*) on cliff faces of the escarpment and concluded that the presence of such species indicated an arctic origin of at least part of the flora. Recent discovery of a cryptoendolithic community within the

limestone cliffs of the escarpment (Gerrath, Gerrath and Larson 1995) indicates that determining the range of biodiversity in these habitats requires further study.

Despite the absence of endemics that have evolved within the habitat, plants of the escarpment may show features that occur nowhere else. For example, white cedar has a distinctive morphology, development, longevity, and recruitment pattern on the cliff face (Figure 22.5) that differs from that in the plateau woodland (Larson 1990; Larson and Kelly 1991; Kelly, Cook and Larson 1992). These differences are not of genetic origin (Matthes-Sears and Larson 1991; Matthes-Sears, Stewart and Larson 1991) and are entirely induced by the environmental conditions on the cliffs. The existence of ecological forms of taxa that function in a distinctive fashion may be important to conservation and preservation of these habitats.

Fauna

The fauna of escarpment cliffs has been studied less than the vegetation. Sinclair and Marshall (1986) reported that madicolous insects, including caddice flies (Trichoptera), flies and mosquitoes (Diptera), and solitary midges (Thaumaleidae), are found on escarpment cliffs in areas where there is a continuous seepage of water. A large assemblage of arachnids occurs on limestone cliff faces and talus slopes in Forillon National Park (Koponen 1990), and recent collections from cliffs of the Niagara Escarpment (Buddle and Larson unpublished) showed that of 222 trapped spiders (Arachnida), 35% were members of Linyphidae. Spiders in this group were less than 8% of the total trapped on the talus slope and level-ground woodland on the cliff plateau. Thus, spatial patterns in the distribution of invertebrates clearly are present. Similarly distinctive distribution patterns of amphibians occur along the escarpment. Genotype frequencies of salamanders (*Ambystoma*) suggest that these salamanders

Table 22.1. *List of species found on level-ground plateau woodland (cliff top), cliff face,and talus slope habitats of the Niagara Escarpment in southern Ontario, Canada. Species lists are taken from Spring (1988) and Cox and Larson (1993).*

Species primarily of plateau woodland on cliff tops

Trees

Acer saccharum	Ostrya virginiana	Prunus serotina
Betula lutea	Picea glauca	Quercus rubra
Betula papyrifera	Populus deltoides	Tsuga canadensis
Fagus grandifolia	Populus grandidentata	Ulmus americana

Shrubs and vines

Amelanchier bartramiana	Lonicera canadensis	Shepherdia canadense
Diervilla lonicera	Lonicera involucrata	Sorbus decora
Euonymus obovatus	Lonicera tartarica	Viburnum edule

Herbaceous vascular plants

Achillea millefolium	Chrysanthemum leucanthemum	Polypodium virginianum
Actaea rubra	Circaea quadrisulcata	Prenanthes alba
Antennaria neglecta	Epipactis helleborine	Pteridium aquilinum
Aralia nudicaulis	Festuca ovina	Silene cucubalus
Arenaria stricta	Fragaria virginiana	Smilacina racemosa
Aster cordifolius	Hystrix patula	Solidago flexicaulis
Aster macrophyllus	Maianthemum canadense	Solidago puberula
Botrychium virginianum	Poa annua	Trifolium repens
Bulbostylis capillaris	Poa palustris	Trillium grandiflorum
Carex plantaginea	Poa trivialis	Viola pallens
Chenopodium hybridum	Polygonatum biflorum	

Cryptogams

Dicranum scoparium	Hypnum pallescens

Species primarily of cliff faces

Trees and vines

Thuja occidentalis	Solanum dulcamara (vine)

Herbaceous vascular plants

Agropyron trachycaulum	Blephilia ciliata	Cystopteris fragilis
Asplenium trichomanes	Cerastium vulgatum	Pellaea glabella
Asplenlium ruta-muraria	Cryptogramma stelleri	Poa pratense
Asplenium viride	Cystopteris bulbifera	Ranunculus acris

Cryptogams

Anomodon attenuatus	Homomalium adnatum	Tortella fragilis
Anomodon rostratus		

Species primarily of talus slopes

Trees

Juglans cinerea	*Thuja occidentalis*	*Tsuga canadensis*
Picea rubens	*Tilia americana*	*Ulmus americana*
Prunus virginiana		

Shrubs and vines

Acer spicatum	*Rhus typhina*	*Shepherdia canadensis*
Cornus rugosa	*Ribes glandulosum*	*Solanum dulcamara*
Juniperus communis	*Rubus allegheniensis*	*Taxus canadensis*
Parthenocissus quinquefolia	*Rubus idaeus*	*Vitis riparia*
Rhus radicans	*Sambucus pubens*	

Herbaceous vascular plants

Arisaema atrorubens	*Dryopteris marginalis*	*Nepeta cataria*
Camptosorus rhizophyllus	*Erysimum cheiranthoides*	*Phyllitis scolopendrium*
Carex eburnea	*Geranium robertianum*	*Tiarella cordifolia*
Dryopteris filix-mas	*Impatiens pallida*	*Verbascum thapsus*
Dryopteris goldiana	*Mitella diphylla*	*Viola selkirkii*
Dryopteris intermedia		

Cryptogams

Brachythecium salesbrosum	*Isopterygium pulchellum*	*Radula complanata*
Dicranum flagellare	*Mnium cuspidatum*	*Rhytidiadelphus triquetrus*
Dicranum scoparium	*Plagiothecium cavifolium*	*Tortella tortuosa*
Grimmia alpicola	*Ptilidium pulcherrimum*	*Thuidium abietinum*

have used the cliffs as a postglacial migrational corridor (Bogart and Cook 1991). These conclusions agree with those of other researchers (Bateman 1961; Churcher and Dods 1979; McAndrews and Jackson 1988; Riley 1993) who examined a variety of extinct species that used the escarpment during seasonal migrations. The conclusions also agree with other work suggesting that extreme cliff topography can act as a refugium (Guilday 1971; Ward and Anderson 1988) from both natural and anthropogenic disturbance. Jeffries and Lawton (1984) describe such habitats as "enemy free space" in which plants and animals can escape intense competition and predation that occurs elsewhere in the landscape.

Matheson (1995) reported that bird and small mammal communities were locally structured by the habitat heterogeneity from plateau woodland to cliff faces. Birds such as the American goldfinch (*Carduelis tristis*), Nashville warbler (*Vermivora ruficapilla*), and cliff swallow (*Petrochelidon pyrrhonota*) tended to roost on cliff faces but not in plateau forest. Conversely, some forest interior birds, including the red-eyed vireo (*Vireo olivaceus*) and wood thrush (*Hylocichla mustelina*), were found in plateau forest surrounding the cliffs, but not on the cliff. Small mammals such as the white-footed mouse (*Peromyscus leucopus*), deermouse (*P. maniculatus*), and raccoon (*Procyon lotor*) had extremely high capture rates on the cliff face; the two species of mouse had low capture rates in the plateau forest. When combined with anecdotal reports of bobcat (*Lynx rufus*) dens on the escarpment cliffs, these results suggest that the cliff environment strongly organizes behavior of vertebrates as well as distribution of plants.

Figure 22.5. Cliff edge and cliff-face forests of the Niagara Escarpment are dominated by slow-growing and deformed trees such as this individ- ual growing in the inverted position. (Photograph taken from Larson and Kelly 1991.)

Palaeohumans reportedly sought out the cliffs of the Niagara Escarpment as shelter, as a source of weapons, and as burial sites (Fox 1990).

Disturbance and Threats

Inaccessibility appears to have been responsible for the limited exposure of the Niagara Escarpment cliffs to fire, logging, and other intense disturbances during the European colonization of Ontario. As population size, standard of living, and industrial activity expand, however, presettlement forests are exposed to increasing pressure from recreational activity. Along the length of the Niagara Escarpment, there is a well-publicized hiking trail known as the "Bruce Trail." Access provided by this trail has exposed the cliff edge to disturbance from hikers, all-terrain vehicles, horses, and bicycles. McLean (1989) and Taylor, Reader and Larson (1993) reported that community composition markedly changes with moderate

(500 passes per year) levels of hiking disturbance, but neither species richness nor individual species frequency showed such effects at this level of use. However, when the intensity of hiking exceeded 25,000 passes per year, there was a major change in species frequency (Figure 22.6a, b). The sensitivity of cliff-edge forests to hiking disturbance also has been documented for the northern sections of the escarpment in the Bruce Peninsula (Parikesit, Larson and Matthes-Sears 1995). The combination of shallow soils and slow-growing species suggests that the cliff-edge vegetation may have little buffering or rapid regrowth capacity when exposed to chronic disturbance.

Recently, recreational climbing also has increased in popularity, and enthusiasts are expanding their search for alternate and little-used cliffs. Although a responsible use of natural environments is advocated by the leaders of recreational hiking and climbing organizations (Bracken, Barnes and Oates 1991), there is little that they can do to stop habitat degradation. The effects of these increased pressures have not been formally studied, but they cannot help in the maintenance of the cliff ecosystem of the escarpment.

Unlike the chronic hiking and climbing disturbances that slowly degrade the cliff-edge habitat, real estate development in the Niagara, Grimsby, Hamilton, Dundas, and Owen Sound areas of the Niagara Escarpment completely removes the cliff-plateau deciduous forest ecosystem and, in many cases, the cliff-edge and cliff-face forest as well. In the Hamilton and Niagara regions, sections of cliff edge were scaled with dynamite during the period 1880–1930 to protect against future rockfall injury. Other sections of escarpment have been wholly transformed for ski resort construction. It seems reasonably safe to conclude that private ownership of cliff ecosystems that permits destruction is worse than public ownership that involves multiple nonconsumptive uses.

Another potentially destructive influence on the Niagara Escarpment cliff ecosystem is uncontrolled quarry operations. The dolomitic limestone of the Silurian deposits of the Niagara Escarpment is extremely valuable as a building and manufacturing material (Gorrie and Dutka 1993). Since European colonization began in the 1800s, over 2,000 quarry sites have been operated and then abandoned in southern Ontario. Research on the form and rate of natural regeneration of vegetation on limestone cliffs of anthropogenic origin indicates that natural recolonization in abandoned quarries follows a predictable sequence (Ursic 1994). For the first 50 years, a wide diversity of alien ruderals col-

Figure 22.6. Effects of trampling on cliff edges for (a) cliff-edge undisturbed plateau forest receiving fewer than 100 visitors per year and (b) cliff-edge plateau forest exposed to chronic trampling by 40,000 visitors per year. (Taken from Taylor, Reader and Larson 1993.)

onize the cliffs. Starting at about 60 years, a small array of species, including white cedar, smooth cliff-brake, bulblet fern (*Cystopteris bulbifera*), and a variety of lichens, dominates the face. This colonization is similar to that for sites in western Europe (Usher 1989).

Importance of the Niagara Escarpment

Unfortunately, few places in eastern North America retain presettlement ecosystems.

These systems are useful for research, monitoring, conservation, and education (Hunter 1989) to determine how humans have altered the natural landscape. A recent model predicts that presettlement forests in the central United States are likely to be found in locations with low commercial value and high topographic relief (Stahle and Chaney 1994). If these predictions prove correct, there may be additional undocumented presettlement forest ecosystems in the landscape.

In the meantime, the structure, function, genesis, and maintenance of known old-growth forest such as that along the Niagara Escarpment must be studied as intensively as resources permit, as this forest is the only remaining system in the landscape that operates under essentially natural controls. Our ability to recognize and manage global environmental change generally is dependent on the knowledge gained from systems that have been operating without human intervention for a long period of time.

Summary

Exposed dolomitic cliffs of the 50–100-m-high Niagara Escarpment, Ontario, Canada, support a presettlement-forest ecosystem that was unknown prior to 1989. Structurally and functionally, the cliff ecosystem is distinct from the adjacent plateau forest. On cliff faces, the dominant canopy tree species is white cedar, with white birch (*Betula papyrifera*) and mountain maple (*Acer spicatum*) as subdominant woody plants. Cliff-face trees are slow growing, deformed, and extremely old (1,653 years for white cedar). A consistent assemblage of herbaceous perennial vascular plants, pteridophytes, bryophytes, lichens, and endolithic organisms also occurs on the cliffs. Although there are no strict endemics, many species, such as rue spleenwort, bulblet fern, and rock polypody, occur only on exposed rock. In addition, the functional ecology of plants in the forest is relatively distinctive because of extreme microclimatic conditions and a high degree of stress imposed by gravity. The cliffs of the Niagara Escarpment also represent a habitat rich with refugia for vertebrates. The lack of anthropogenic disturbance since deglaciation allows the Niagara Escarpment to be studied as an example of a system operating under primarily natural control.

Acknowledgments

This research has been supported by the Natural Sciences and Engineering Research Council of Canada, Parks Canada, the Ontario Heritage Foundation, the Ontario Ministry of the Environment and Energy, the Ontario Ministry of Natural Resources, and the World Wildlife Fund, Canada.

REFERENCES

Barden, L. S. 1988. Drought and survival in self-perpetuating a *Pinus pungens* population: equilibrium or non-equilibrium? *American Midland Naturalist* 119:253–257.

Bartlett, R. M., and Larson, D. W. 1990. The physiological basis for the contrasting distribution patterns of *Acer saccharum* and *Thuja occidentalis* at cliff edges. *Journal of Ecology* 78:1063–1078.

Bartlett, R. M., Matthes-Sears, U., and Larson, D. W. 1990a. Organization of the Niagara Escarpment cliff community. II: Characterization of the physical environment. *Canadian Journal of Botany* 68:1931–1941.

Bartlett, R. M., Matthes-Sears, U., and Larson, D. W. 1990b. Microsite- and age-specific processes controlling natural populations of *Acer saccharum* at cliff edges. *Canadian Journal of Botany* 69:552–559.

Bartlett, R. M., Reader, R. J., and Larson, D. W. 1991. Multiple controls of cliff-edge distribution patterns of *Thuja occidentalis* and *Acer saccharum* at the stage of seedling recruitment. *Journal of Ecology* 79:183–197.

Bateman, R. M. 1961. Mammal occurrences in escarpment caves. *Ontario Field Biologist* 15:16–18.

Bell, L. 1972. *Fathom Five Field Survey and Recommendations Report*. Internal report. Toronto: Ontario Ministry of Natural Resources.

Bogart, J. P., and Cook, W. C. 1991. *Ambystoma Survey on the Niagara Escarpment*. Final report. Toronto: Ontario Heritage Foundation.

Bracken, M., Barnes, J., and Oates, C. 1991. *The Escarpment, a Climbers Guide*. Toronto: Borealis Press.

Brunton, D. F., and Lafontaine, D. 1974. An unusual escarpment flora in western Quebec. *Canadian Field Naturalist* 88:337–344.

Burns, R. M., and Honkala, B. H. 1990. *Silvics of North America: Conifers*, vol. 1. USDA Forest Service Handbook 654. Washington D.C.: Superintendent of Documents.

Churcher, C. S., and Dods, R. R. 1979. *Ochotona* and other vertebrates of possible Illinoisan age from Kelso Cave, Halton County, Ontario. *Canadian Journal of Earth Science* 16:1613–1620.

Cody, W. J., and Britton, D. M. 1989. *Ferns and Fern Allies of Canada*. Ottawa: Research Branch, Agriculture Canada, Publication 1829/E.

Cox, J. E., and Larson, D. W. 1993. Spatial heterogeneity of vegetation and environmental factors on talus slopes of the Niagara Escarpment. *Canadian Journal of Botany* 71:323–332.

Davis, P. H. 1951. Cliff vegetation in the eastern Mediterranean. *Journal of Ecology* 39:63–93.

Fox, W. A. 1990. The Odawa. In *The Archeology of Southern Ontario to A.D. 1650*, ed. C. J. Ellis and N. Ferris, pp. 457–474. Occasional Publication of the London Chapter, No. 5. London, Ontario: Ontario Archeological Society.

Gerrath, J. F., Gerrath, J., and Larson, D. W. 1995. A preliminary account of endolithic algae of limestone cliffs of the Niagara Escarpment. *Canadian Journal of Botany* 73:788–793.

Gildner, B. S., and Larson, D. W. 1992. Photosynthetic response to sunflecks in the desiccation-tolerant fern *Polypodium virginianum*. *Oecologia* 89:390–396.

Gillard, W. H., and Tooke, T. R. 1975. *The Niagara Escarpment*. Toronto: University of Toronto Press.

Gorrie, P., and Dutka, D. 1993. Quandary at the quarry. *Canadian Geographic* 113:76–84.

Guilday, J. E. 1971. Vertebrates. In *The Distributional History of the Biota of the Southern Appalachian*, ed. P. C. Holt., R. A. Patterson, and J. P. Hubbard, pp. 233–261, Monograph 4. Blacksburg, Va.: Virginia Polytechnic Institute and State University.

Holmquist, C. 1962. The relict concept – is it merely a zoological conception? *Oikos* 13:262–292.

Hunter, M. L. 1989. What constitutes an old-growth stand? *Journal of Forestry* 87:33–35.

Jackson, G., and Sheldon, J. 1949. The vegetation of Magnesian Limestone Cliffs at Markland Grips near Sheffield. *Journal of Ecology* 37:38–50.

Jeffries, M. J., and Lawton, J. H. 1984. Enemy free space and the structure of ecological communities. *Biological Journal of the Linnean Society* 23:269–286.

Kelly, P. E., Cook, E. R., and Larson, D. W. 1992. Constrained growth, cambial mortality, and dendrochronology of ancient *Thuja occidentalis* on cliffs of the Niagara Escarpment: an eastern version of Bristlecone pine? *International Journal of Plant Sciences* 153:117–127.

Kelly, P. E., Cook, E. R., and Larson, D. W. 1994. A 1397-year tree-ring chronology of *Thuja occidentalis* from cliff faces of the Niagara Escarpment, southern Ontario, Canada. *Canadian Journal of Forest Research* 24:1049–1057.

Koponen, S. 1990. Spiders (Araneae) on the cliffs of the Forillon National Park, Quebec. *Naturaliste Canadien* 117:161–165.

Larson, D. W. 1990. Effects of disturbance on old-growth *Thuja occidentalis* at cliff edges. *Canadian Journal of Botany* 68:1147–1155.

Larson, D. W. 1997. Dendroecological potential of *Juniperus virginiana* L. growing on cliffs in western Virginia. *Banisteria* 10:13–18.

Larson, D. W., and Kelly, P. E. 1991. The extent of old-growth *Thuja occidentalis* on cliffs of the Niagara Escarpment. *Canadian Journal of Botany* 69:1628–1636.

Larson, D. W., and Melville, L. 1996. Stability of wood anatomy of living and Holocene *Thuja occidentalis* L. derived from exposed and submerged portions of the Niagara Escarpment. *Quaternary Research* 45:210–215.

Larson, D. W., Spring, S. H., Matthes-Sears, U., and Bartlett, R. M. 1989. Organization of the Niagara Escarpment cliff community. *Canadian Journal of Botany* 67:2731–2742.

Matheson, J. 1995. Habitat structure of birds and small mammals on the Niagara Escarpment. M.S. thesis, University of Guelph, Guelph, Ontario.

Matthes-Sears, U., and Larson, D. W. 1990. Environmental controls of carbon uptake in two woody species with contrasting distributions at the edge of cliffs. *Canadian Journal of Botany* 68:2371–2380.

Matthes-Sears, U., and Larson, D. W. 1991. Growth and physiology of *Thuja occidentalis* L.

from cliffs and swamps: is variation habitat or site specific? *Botanical Gazette* 152:500–508.

Matthes-Sears, U., and Larson, D. W. 1995. Rooting characteristics of trees in rock: a study of *Thuja occidentalis* on cliff faces. *International Journal of Plant Science* 156:679–686.

Matthes-Sears, U., Nash, C. H., and Larson, D. W. 1995. Constrained growth of trees in a hostile environment: the role of water and nutrient availability for *Thuja occidentalis* on cliff faces. *International Journal of Plant Science* 156:311–319.

Matthes-Sears, U., Neeser, C., and Larson, D. W. 1992. Mycorrhizal colonization and macronutrient status of cliff-edge *Thuja occidentalis* and *Acer saccharum*. *Ecography* 15:262–266.

Matthes-Sears, U., Stewart, S. C., and Larson, D. W. 1991. Sources of allozymic variation in *Thuja occidentalis* in southern Ontario, Canada. *Silvae Genetica* 40:100–105.

McAndrews, J. H., and Jackson, L. J. 1988. Age and environment of late Pleistocene mastodont and mammoth in Southern Ontario. In *Late Pleistocene and Early Holocene Paleontology and Archeology of the Eastern Great Lakes Region*, ed. R. S. Laub, N. G. Miller, and D. W. Steadman. *Bulletin of the Buffalo Society of Natural Sciences* 33:161–170.

McLean, K. C. 1989. Effects of trampling disturbance on the richness of forest vegetation: a model and its conservation implications. M.Sc. thesis, University of Guelph, Guelph, Ontario.

Morisset, P. 1971. Endemism in the vascular plants of the Gulf of St. Lawrence. *Naturaliste Canadien* 98:167–177.

Morisset, P. 1979. *Localisation et Abondance des Plantes Vasculaires Arctiques-Alpines et Rares des Falaises du Parc National Forillon*. Quebec: Parcs Canada.

Morisset, P., Bedard, J., and Lefebvre, G. 1983. *The Rare Plants of Forillon National Park*. Quebec: Parks Canada.

Morton, J. K., and Venn, J. M. 1984. *The Flora of Manitoulin Island*. 2d ed. Biology Series No. 28. Waterloo, Ontario: University of Waterloo.

Morton, J. K., and Venn, J. M. 1987. *The Flora of the Tobermory Islands, Bruce Peninsula National Park*. Biology Series No. 31. Waterloo, Ontario: University of Waterloo.

Oppenheimer, H. R. 1956. Pénétration active des racines de buissons Méditerranéens dans les roches calcaires. *Bulletin of the Research Council of Israel* 5d:219–222.

Parikesit, P., Larson, D. W., and Matthes-Sears, U.

1995. Impacts of trails on cliff-edge forest structure. *Canadian Journal of Botany* 73:943–953.

Riley, J. 1993 . There are bones down there. *Cuesta* 1993:11–13.

Rowe, J. S. 1972. *Forest Regions of Canada*. Ottawa: Department of Fisheries and Environment, Canadian Forestry Service, Publication No. 1300.

Sinclair, B. J., and Marshall, S. A. 1986. The madicolous fauna in southern Ontario. *Proceedings of the Entomological Society of Ontario* 117:9–14

Spring, S. H. 1988. Organization of the Niagara Escarpment cliff community. M.S. thesis, University of Guelph, Guelph, Ontario.

Stahle, D. W., and Chaney, P. L. 1994. A predictive model for the location of ancient forests. *Natural Areas Journal* 14:151–158.

Taylor, K. C., Reader, R. J., and Larson, D. W. 1993. Scale-dependent inconsistencies in the effects of trampling on a forest understory community. *Environmental Management* 17:239–248.

Ursic, K. 1994. Comparative vegetation structure on limestone escarpments of anthropogenic origin in southern Ontario. M.S. thesis, University of Guelph, Guelph, Ontario.

Usher, M. B. 1989. Scientific aspects of nature conservation in the United Kingdom. *Journal of Applied Ecology* 26:813–824.

Walker, G. L. 1987. Ecology and population biology of *Thuja occidentalis* L. in its southern disjunct range. Ph.D. dissertation, University of Tennessee, Knoxville, Tenn.

Ward, J. P., and Anderson, S. H. 1988. Influences of cliffs on wildlife communities in south central Wyoming. *Journal of Wildlife Management* 52:673–678.

Warner, B. G. 1982. Late glacial fossil leaves of *Thuja occidentalis* from Manitoulin Island, Ontario. *Canadian Journal of Botany* 60:1352–1356.

Warner, B. G., Hebda, R. J., and Hann, B. J. 1984. Postglacial paleoecological history of a cedar swamp, Manitoulin Island, Ontario, Canada. *Palaeogeography, Palaeoclimatology, and Palaeoecology* 45:301–345.

Young, J. 1996. The cliff ecology and genetic structure of northern white cedar (*Thuja occidentalis* L.) in its southern disjunct range. M.S. thesis, University of Tennessee, Knoxville, Tenn.

23 Alvars of the Great Lakes Region

PAUL M. CATLING AND VIVIAN R. BROWNELL

Introduction

"Alvars are naturally open areas of thin soil over essentially flat limestone or marble rock with trees absent or at least not forming a continuous canopy" (Catling et al. 1975; Catling and Brownell 1995). Workers familiar with similar habitats in Sweden, Denmark, and Estonia first applied the term *alvar* in North America to areas near Kingston, Ontario. The term is still widely used in Europe (e.g., Krahulec, Rosen and van der Maarel 1986). Although alvars of the Great Lakes region may be structurally similar to sites in northern Europe and subject to similar ecological processes, they are not necessarily more closely related to the European alvars than to habitats in adjacent regions of North America that are classified by other names. To the west, alvars grade into dry prairies over limestone or calcareous gravel (Curtis 1959; Erickson, Breener and Wraight 1942). Northward, similar habitats exist within the boreal forest region, where they are referred to as "limestone barren." The "cedar glades" of Kentucky and Tennessee (e.g., Freeman 1933) are similar, but differ in having more endemic species (Baskin and Baskin 1986, 1988; Catling and Brownell 1995) and a different floristic composition (Catling and Brownell 1995). Nevertheless, the term *alvar* has been widely adopted, and over the past several years alvars have become a major focus of conservation efforts.

A mosaic of plant associations is a characteristic feature of alvars. These associations are related to periodic flooding and severe drought, which are mediated by soil depth, water table, and runoff patterns (Stephenson 1983; Stephenson and Herendeen 1986). Drought, inundation, and fire prevent invasion of woody species into alvar openings and promote high species diversity. However, periodic catastrophic drought effects (e.g., Stephenson and Herendeen 1986), rather than seasonal droughts, may be the most important factor excluding woody species. Alvars are restricted in occurrence and distinctive in biological composition and history. Not surprisingly, they are recognized as globally imperiled by The Nature Conservancy. The recent discovery of gnarled and stunted *Thuja occidentalis* trees up to 524 years old on open alvar pavement (Schaefer 1996a, b) supports earlier contentions that many dry alvar openings are natural occurrences rather than the result of postsettlement disturbance. The presence of old trees also indicates that not all portions of all alvars were subjected to fires. Charcoal on many alvars, however, indicates that periodic fires have contributed to the openness of the habitat.

Considerable attention has been focused on alvars in the Great Lakes region since the review by Catling and Brownell (1995). Since then, new ideas have been formed relating to the range of sites to be included and the importance of surrounding woodland habitats and degraded sites. There also is new knowledge relating to history, fauna, and regional occurrences.

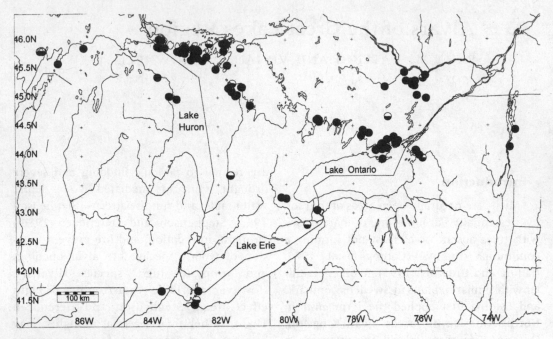

Figure 23.1. Distribution of alvars in the Great Lakes region.

Distribution

Alvars occur primarily in the Great Lakes region along the contact line of the granitic Canadian Shield upland with the Ordovician limestones and dolomites (Figure 23.1). This area extends from the Upper Peninsula of Michigan to the islands of northern Lake Huron, such as Drummond Island (Stephenson 1983), east across Manitoulin Island, southeast to the Bruce Peninsula and the Georgian Bay region of Ontario, discontinuously east to the Napanee limestone plain at the eastern end of Lake Ontario (Reschke 1990). Following a gap due to the Frontenac Axis of granite rocks, alvar landscapes appear again in New York State at the east end of Lake Ontario and on the Smith Falls limestone plain, and along and near to the Ottawa River in the Ottawa Valley (Huggett 1993; Catling and Brownell 1995; Cayouette et al. 1997). This distribution includes alvars on limestone plains and plateaus as well as shoreline sites. Isolated sites occur on the Flamborough limestone plain at the western

end of Lake Ontario, in the western Lake Erie region, and in the Lake Champlain regions. Shoreline alvars occur along the Lake Huron shorelines of the Bruce Peninsula and Manitoulin Island. In river valleys, alvar sites are found along the Ausable (Brownell 1984) and Maitland (Oldham et al. 1993) rivers draining the Ontario shore of Lake Huron; the upper Salmon River drainage; and the Ottawa River in eastern Ontario. The Escanaba River in Michigan (Chapman 1986) and the Black River in upper New York State are also good examples of riverine alvar.

Approximately 90% of Great Lakes region alvars are in Ontario. Not surprisingly, the alvars of this province contain more endemic, restricted, and disjunct species than in the surrounding region.

Environment

Alvars are frequently wet in spring and fall but very dry in summer. The overlying soil is usually alkaline and calcareous (pH ca. 8.0) due to the influence of, or derivation from, the underlying limestone. However, if the soil is relatively deep sand, as on some parts of the Carden Plain, a neutral or even acid (pH ca. 5.0) soil reaction may develop. On these sites the vegetation may include blueberries (*Vaccinium angustifolium* and *Vaccinium myrtilloides*) and other plants characteristic of acid substrate.

Limestone and related rocks may vary widely in hardness and in the extent to which they weather to form parent material. The prevalence of limestone savanna in the Trent River valley (Catling and Catling 1993) and absence of open alvar probably result from the tendency of soft limestone to form soil that can support trees. The dolomite cap of the Niagara Escarpment on Manitoulin Island may contribute to the floristic differences between the alvars on the north and south shores. The Blackwater limestones in the North Channel are softer and weather more readily than dolomite.

Climate varies substantially throughout the alvar region. This variation is most evident in the occurrence of southern floristic elements in the western Lake Erie sites and boreal floristic elements in the northern Lake Huron sites. However, climatic effects also may operate over short distances. Floristic differences between alvars on the north and south shores of Manitoulin may be influenced by climate, with the south shore being a little more moderate and cooler. The unifying climatic feature of alvars is the periodic occurrence of hot and dry conditions in general, which helps explain the presence of species of the western plains.

Community Types

Two kinds of alvars have been recognized (Catling and Brownell 1995). Shoreline alvars occur as bands along river and lake shores and are subject to predictable seasonal flooding and erosion. The characteristically drier plateau alvars occur on limestone plateaus, which do not experience the scour-

ing action of spring floods. This alvar type often includes extensive areas of open ground, savanna, and open woodland, separated by denser woodlands developed on deeper soils.

Shoreline Alvar

Open areas on shallow soil (usually 0–5 cm) over limestone or marble along the shores of rivers and lakes are a consequence of periodic flooding. Turbulent water and ice scouring limit development of woody cover. Erosion of soil-building materials is characteristic, and there are fewer plant associations than on plateau alvars. The effects of high, turbulent waters may be evident for only a few days each year, but they create an open vegetation zone that extends as far back as 50 m or more from lake and river shores. High winds, especially on lakeshores, also contribute to removal of soil-building materials and result in damage to woody vegetation. Morton (1996) described the effect of erosion due to rising water levels on Half Moon Island, where the grassland was "rolled up like a carpet" and "everything over several acres was destroyed." Levels of Lake Huron are subject to erratic cyclical changes every few years. By summer of the following year, Morton (1996) found that most of the area that was destroyed was dominated by *Deschampsia cespitosa* and *Poa palustris*. The strong root systems of perennial prairie grasses, including *Andropogon gerardii*, *Schizachyrium scoparium*, and *Sporobolus heterolepis*, are able to resist some erosion and thus maintain soil, and these species are characteristic dominants of alvars damaged by high water. Vegetation, including both woody and herbaceous species, may be nearly confined to cracks that accumulate debris.

The relatively low position of shoreline alvars on the landscape and periodic flooding result in moister conditions than on plateau alvars, as described later. A relatively high water table results in frequent surface seepage. Nevertheless, periodic severe droughts also occur. Lack of tree cover on shoreline alvars may be largely a result of flooding and erosion, whereas fire may be the most important factor in limiting woody cover on plateau alvars. Portions of alvars of the Lake Huron shoreline of the Bruce Peninsula do not burn and thus differ from the inland sites (Schaefer 1996a, b).

Because of erosion, and relatively less severe and shorter periods of drought, the vegetation of shoreline alvars is different from that of plateau alvars. The shoreline alvars have many mesic prairie or prairie/fen species, including the dominant grasses. Some of these, such as *Andropgon gerardii*, generally are not found on plateau alvars (except on a few western Lake Erie plateau sites, where the soil is relatively deep). Other species more or less restricted to shoreline alvars include *Cacalia plantaginea, Carex meadii, Helenium autumnale, Hypericum kalmianum, Lobelia kalmii, Physocarpus opulifolius, Selaginella apoda, Solidago ohioensis, Valeriana edulis* ssp. *ciliata*, and *Zizia aptera*. Pavement communities, including moss and lichen cushions, which are so characteristic of plateau alvars, are less well developed on shoreline alvars, probably largely due to erosion. Even in areas where lower-elevation shoreline alvars merge into higher plateaus, such as on the Bruce Peninsula and on Manitoulin Island, significant differences in algae and moss cover between plateau and shoreline alvars were demonstrated (Schaefer 1996a). Communities of groundwater seepage areas and lower shoreline areas may include an assemblage of fen and wetland species, including *Carex garberi, Eleocharis elliptica, Juncus alpinoarticulatus, Juncus balticus* var. *littoralis*, and *Tofieldia glutinosa*.

Although differences between shoreline and plateau alvars are of interest, there are also substantial similarities. Among the numerous dominant or conspicuous species that occur in both alvar types are *Calamintha arkansana, Carex crawei, Carex scirpoidea, Hedyotis longifolia, Juniperus horizontalis, Schizachyrium scoparium,* *Senecio pauperculus, Solidago nemoralis, Sporobolus heterolepis, Viola nephrophylla,* and *Zigadenus glaucus*. In the upper Great Lakes, shoreline alvars grade into plateau alvars and are much less distinctive than some alvar sites that occur along rivershores. The rivershore sites vary regionally. *Andropogon gerardii*, for example, is absent from many of the upper Great Lakes limestone shorelines, and the upper Great Lakes sites on the shores of Lake Huron appear to have the highest plant species diversity.

Plateau Alvar

Limestone or marble tablelands inland from shorelines are subject to seasonal flooding. Tablelands have lower water tables, and more severe and prolonged summer droughts than shorelines. Flooding of alvar grasslands and pavements is mostly the result of precipitation runoff, unlike the shoreline alvar situations, where some of the water may originate as seepage groundwater with higher calcium and magnesium levels. Groundwater seepage may be periodic, but it is likely to be less influenced by drought.

Parallel openings and bands of woodland are characteristic of plateau alvar landscapes. The woody cover develops on ridges with cracks or low escarpment edges, or it may be associated with sand and gravel deposited over the top of otherwise exposed limestone pavement. Six major open or semi-open habitat associations may be recognized, including alvar pavement, alvar grassland, alvar shrubland, alvar pavement savanna, alvar grassland savanna, and alvar shrubland savanna (Table 23.1). The most open areas of plateau alvar may be divided into alvar grassland (also called "alvar meadows," e.g., Belcher, Keddy and Catling 1992), alvar shrubland, and alvar pavement (also called "rock flats"). A semi-closed habitat association is also recognized that includes alvar pavement woodland, alvar grassy woodland, and alvar shrubby woodland. Communities larger than 1.2 acres (0.5 ha) have been recognized and mapped in

Table 23.1. *Alvar habitat matrix based on a modification of earlier classifications including Catling and Brownell (1995).*

Substrate character	Open (<1 tree/acre or 10% tree cover)	Semi-open (>1 tree/acre or 10%–50% tree cover)	Semi-closed (50%-80% tree cover)
Much exposed rock (>50%), soil <2 cm deep	Alvar pavement (annual and some perennial herbs)	Alvar pavement savanna (annual and perennial herbs with scattered trees)	Alvar pavement woodland (annual and perennial herbs prominent in understory)
Little exposed rock (<50%) , soil >2 cm deep	Alvar grassland (mostly perennial graminoids)	Alvar grassland (mostly perennial graminoids)	Alvar grassy woodland (perennial graminoids prominent in understory)
Little exposed rock (<50%), soil >2 cm deep	Alvar shrubland (shrubs prominent, >25% cover)	Alvar shrubland savanna (trees and shrubs prominent)	Alvar shrubby woodland (shrubs prominent in understory, >25% cover)

recent studies, but much smaller areas can provide significant habitat for invertebrates. Transition zones are often abrupt or nonexistent, as reported for some Bruce Peninsula sites (Schaefer 1996a). A combination of soil moisture and soil depth determines limits of each alvar type. Each is subject to regional variation in plant associations (Belcher, Keddy and Catling 1992; Catling and Brownell 1995), and any of a number of associated species may assume dominance as a result of differences in disturbance regimes.

Alvar Pavement. Pavement edge vegetation zones are characterized by soil less than 2 cm deep and are dominated by mosses, lichens, and annual vascular plants. Phytosociological differences in annual pavement and perennial grassland communities appear to be most pronounced on the warmer and drier alvars, such as those of the Napanee Plain in eastern Ontario. Three major groupings within this category were

reported on the Burnt Lands Alvar complex in eastern Ontario (Belcher, Keddy and Catling 1992). Periodically inundated sites and those accumulating runoff are characterized by vascular plants such as *Agrostis scabra*, *Epilobium coloratum*, *Gratiola neglecta*, and *Portulaca oleracea*, with the moss *Physcomitrium pyriforme*, the liverwort *Riccia sorocarpa*, and the cyanobacterium or blue-green algae *Nostoc commune*. Many species, including some with a relatively broad moisture tolerance, such as *Arabis hirsuta*, *Myosurus minimus*, and *Trichostema brachiatum*, occupy sites of intermediate moisture availability. Drier pavements have *Minuartia michauxii, Hedeoma hispida, Saxifraga virginiensis*, and *Selaginella rupestris* in moss cushions composed of *Tortella tortuosa, Tortula ruralis*, or *Ceratodon purpureus*. These moss cushions are very important to the community as a whole, since they accumulate water like sponges following a rain. Winter annuals such as *Draba reptans, Gratiola neglecta, Myosotis verna,* and *Veronica peregrina* flower and fruit in the spring wet period, whereas summer annuals such as *Trichostema brachiatum, Panicum flexile,* and *Panicum philadelphicum* become apparent later in the season. Some species are very irregular in presence and abundance from year to year on alvars. Notable examples include *Euphorbia commutata, Gratiola neglecta,* and *Juncus secundus.* These species may be locally dominant some years, but subsequently are absent (as a green leafy plant) for one to several years. Moisture appears to be a controlling factor in these irregular occurrences. Recent studies have demonstrated root competition in the open, annual-dominated associations (Belcher 1992).

On a per unit area basis, pavements often have the highest species richness of alvar communities, and single dominants often are lacking. Dominant species include annual grasses and *Trichostema brachiatum,* which is widespread on alvars, or *Calamintha arkansana,* which is confined to the Lake Huron sites. *Cerastium nutans, Draba nemorosa, D. reptans, Geranium bicknellii, G. carolinianum, Lepidium* spp., *Myosotis verna, Potentilla norvegica,* and *Scutellaria parvula* have a frequent, but often patchy, occurrence. Areas of shallow soil dominated by the annual grasses *Panicum flexile, Panicum philadelphicum,* or *Sporobolus vaginiflorus* probably are treated best as pavement, since the dominants are annuals that frequently occur on very shallow substrates; however, they have been included in alvar grasslands by some authors (e.g., Belcher 1992). The annual-dominated pavement associations sometimes grade into the grassland associations dominated by perennials such as *Poa compressa,* but a line of demarcation a few decimeters wide often is quite clear.

Ponding and periodic inundation with runoff probably are important factors accounting for local variation in pavement communities. On a broader scale, drought and heat probably play a major role in defining the community, but the effects of frost heaving of soil also may affect vegetation development. The nature of the vegetation probably also influences the direction of successional changes. For example, creeping shrubs such as *Juniperus horizontalis* and *Prunus pumila* var. *depressa* on the upper Lake Huron alvars trap debris on pavements, thus promoting colonization by other species. Sites where these shrub species are prevalent, however, should be referred to as alvar shrubland.

Small cracks in pavements that may provide little more than a space for roots sometimes enable perennials with storage capabilities to be abundant on pavements; these perennials include grasses such as *Schizachyrium scoparium* or *Sporobolus heterolepis* and other species such as *Allium schoenoprasum, Hymenoxys herbacea, Liatris cylindracea,* and *Penstemon hirsutus.* Such situations may be considered as perennial grassland extending into a pavement matrix due to an abundance of small cracks.

Alvar Grassland. The alvar grassland category includes sites dominated by perennial graminoid species. In the Great Lakes region, one of several perennial graminoids, such as *Carex scirpoidea, Danthonia spicata, Poa compressa, Schizachyrium scoparium, Sporobolus heterolepis,* or *Bouteloua curtipendula,* may dominate dry, open areas, where soil exceeds 2 cm in depth. Sites where soil moisture is higher or periodically inundated may be dominated by *Deschampsia cespitosa. Scirpus cespitosus* may be dominant in certain fenlike shoreline sites along Lake Huron. In the wettest sites, various combinations of *Carex sartwellii, Carex lanuginosa,* and *Glyceria striata* dominate the associations. Although these perennial graminoids generally occur on soils deeper than 2 cm, they may extend into areas with shallower soils where rainwater or meltwater accumulates.

Poa compressa is an abundant alvar grassland dominant with a broad tolerance range for soil moisture, and it may be more characteristic of disturbed than of undisturbed sites.

Eleocharis compressa may occur with *P. compressa* or in pure stands. *E. compressa* is also sometimes characteristic of grazed sites, where its avoidance by cattle confers an advantage. Grazed sites with more permanent moisture may have associations dominated by *Eleocharis erythropoda* and various common grasses, including *Poa* spp. and *Agrostis* spp. There are at least 12 alvar grassland associations that may intergrade and vary a great deal in their diversity and abundance of nongraminoid species. For example, *Geum triflorum* may be a codominant in a *Poa compressa* grassland, and many other species, such as *Anemone cylindrica, Calystegia spithamea, Cerastium arvense, Comandra umbellata, Fragaria virginiana, Polygala senega, Potentilla arguta, Prunella vulgaris, Ranunculus fascicularis, Scutellaria parvula, Senecio pauperculus, Solidago juncea, Solidago nemoralis, Virgulus ericoides,* and *Zigadenus glaucus,* may be present, or the grassland may be composed entirely of *Poa compressa.*

Where grazing is intense on alvar grasslands, the grasses may be reduced, and thus the species avoided by cattle, such as *Eleocharis compressa,* may increase in abundance. Rosette-forming species such as *Aster ciliolatus* and *Solidago* species also may increase. *Ranunculus fascicularis* is much more frequent on some alvars subject to grazing than on adjacent nongrazed sites. Nevertheless, even light grazing tends to result in elimination of certain species, such as the disjunct *Orobanche fasciculata* (Catling and Brownell 1995). Perennial and annual introduced and ruderal species, such as *Hieracium* spp., *Chrysanthemum leucanthemum, Hypericum perforatum, Trifolium* spp., *Phleum pratense,* and others, replace the dominant native grassland species under continuous intensive grazing, as has been shown on alvars of the Baltic islands (Titlyanova, Rusch and van der Maarel 1988).

Alvar Shrubland. This community usually develops on soils deeper than 2 cm, except where cracks are frequent. This category was established largely to accommodate extensive areas dominated by *Juniperus communis,* sometimes with *J. horizontalis, Potentilla fruticosa, Prunus pumila* var. *depressa,* and *Amelanchier alnifolia* var. *compacta.* Of these shrubs, *Juniperus communis* is the most frequent and widespread dominant.

Based on charred wood and observations of long-time residents, it can be concluded that fire may be an important factor in developing and maintaining alvar shrublands on Manitoulin Island (J. Jones, personal communication). Since the shrublands are a successional stage, they contain species of both open and wooded habitat, and diversity may be relatively high. Some uncommon animal species, such as dark crescent butterfly, *Phyciodes batesii,* are associated with *Aster ciliolatus* and *Aster cordifolius* (Catling 1997), which are abundant in some alvar shrublands.

Alvar Pavement Savanna (Pavement Ridge). Pavement ridges usually are elevated areas with deep, moist cracks where trees and shrubs can become established. Unlike grassland and shrubland savannas, pavement ridges have shallow soils and much exposed rock. The trees and shrubs may be species of dry or wet sites. *Picea glauca, Pinus banksiana,* and/or *Pinus strobus* may be dominant, but many other species often are present, including *Betula papyrifera, Juniperus virginiana, Populus tremuloides,* and *Thuja occidentalis. Cornus stolonifera, Cornus racemosa, Fraxinus* spp., *Spiraea alba,* and *Tilia americana* are characteristic of mesic microsites. Woody plants characteristic of alvar savanna and alvar edges establish in drier microsites. Ferns, such as *Dryopteris marginalis, Asplenium trichomanes,* and *Cystopteris fragilis,* occur in the cracks. The rare *Pellaea atropurpurea* and *Gymnocarpium robertianum* occur on flat rock or on crevice edges. *Saxifraga virginiana* and other species of dry pavement often are prevalent in the open places. *Euphorbia commutata* is highly sporadic in its appearance. In some years, it is abundant on very shallow soils of the Napanee Plain of Ontario, and in other years, it is absent.

Alvar Grassland Savanna. Open alvar may grade into deciduous, mixed, or evergreen woodland, or there may be a sharp line of separation between open alvar and woodland. Possibly, drought and/or fire, and/or excessive browsing by deer favors dominance of herbs and grasses in areas that would otherwise be more shrubby, that is, tree/shrubland. Nevertheless, savanna should not be regarded as simply a brief successional stage between open and wooded conditions. It may be long persisting and is a primary habitat for certain species, such as *Astragalus neglectus*, *Erigeron pulchellus*, *Helianthus* spp., and many others.

Major variations in alvar savanna have a regional basis and are usually recognized by the tree dominants. *Quercus macrocarpa* often is dominant on alvar savanna, but not in other habitats. *Quercus muehlenbergii* and *Ulmus thomasii* also are characteristic of alvar savanna and limestone slopes and crests. However, none of these species are confined to alvar savanna.

The most widespread type of savanna is characterized by dominance of *Quercus macrocarpa* with or without other species including *Q. alba*, *Q. borealis*, *Q. muehlenbergii*, and *Carya ovata*. Characteristic herbs include *Astragalus neglectus*, *Bromus pubescens*, *Carex cephalophora*, *Carex pensylvanica*, *Carex siccata*, *Galium boreale*, *Erigeron pulchellus*, *Helianthus divaricatus*, *Hystrix patula*, *Schizachne purpurascens*, and *Taenidia integerrima* (Catling and Catling 1993). Shrubs include *Ceanothus americana*, *Cornus racemosa*, *Juniperus communis*, *Rhus aromatica*, *Symphoricarpos albus*, and *Viburnum rafinesquianum*.

Pinus banksiana savannas occur on the upper Lake Huron region. The prominent herbs in these savannas are *Sporobolus heterolepis*, *Deschampsia caespitosa*, and *Eleocharis compressa*. Among the interesting understory plants here are *Coreopsis lanceolata*, *Cypripedium arietinum*, *Hymenoxys herbacea*, and *Iris lacustris*.

Another major type of alvar savanna is the *Juniperus virginiana* type. This varies in species composition depending on past disturbance. It is also a frequent seral stage in old pastures, but in these situations native plant diversity is low, older trees are absent, and *Poa compressa* is a frequent herb dominant with introduced grasses.

An unusual type of savanna in the Great Lakes region is the *Quercus muehlenbergii–Fraxinus quadrangulata* savanna on the Stone Road Alvar on Pelee Island. The openings are prairielike, with *Andropogon gerardii*. Thickets of *Zanthoxylum americanum* harbor the provincially rare *Blephilia ciliata*, which appears to increase following reduction in shrub density due to drought.

Alvar Shrubland Savanna. This habitat is most readily visualized as an alvar shrubland with scattered trees. Grasses including *Danthonia spicata* and/or sedges such as *Carex pensylvanica* and *Carex siccata* may be an important part of the community. At least three subgroups that may intergrade may be recognized based on the dominant trees: (1) *Carya ovata*, *Quercus alba*, *Quercus macrocapa*, *Quercus muehlenbergii*; (2) *Juniperus virginiana*; (3) *Betula papyrifera*, *Picea glauca*, *Pinus banksiana*, *Pinus strobus*, *Populus* spp, and *Thuja occidentalis*. These groups may intergrade. The shrubs include *Cornus racemosa*, *Juniperus communis*, *Juniperus horizontalis*, *Prunus pumila* var. *depressa*, *Prunus virginiana*, *Rhus aromatica*, *Symphoricarpos albus*, and *Viburnum rafinesquianum*.

The most frequent type is probably the one where *Juniperus communis* is prevalent. Characteristic herbs of natural sites include *Aster cordifolius*, *Carex backii*, *Carex siccata*, *Carex pensylvanica*, *Comandra umbellata*, *Fragaria virginiana*, *Monarda fistulosa*, and *Poa compressa*; the moss *Thuidium abietinum* also is often common. This community and the more or less open woodlands surrounding are the habitats of the rare *Carex juniperorum* (Catling, Reznicek and Crins 1993). Shrubs include *Rhus* spp. and *Juniperus communis*.

Alvar Pavement Woodland. Elevated areas with numerous cracks in the pavement support this type of community, which has a more complete tree cover than the alvar pavement savanna, but is otherwise similar. Large cracks (grykes) with ferns may be a prominent feature of this type. Many species characteristic of shadier sites can occur, including *Carex oligocarpa, Claytonia virginica,* and *Dicentra* spp. in the more southern deciduous pavement woodlands. Small "islands" of vegetation may be present under the tree cover, but generally the open rock supports few herbs. In the north, where *Pinus banksiana* is a dominant in pavement woodland, *Aster macrophyllus, Carex eburnea, Cornus canadensis, Iris lacustris, Linnaea borealis, Oryzopsis asperifolia,* and *Pteridium aquilinum* may be frequent.

Alvar Grassy Woodland. The increased tree cover in grassy woodlands reduces extreme temperature and drought. *Carex pensylvanica* and other sedges often are dominant. In the southern deciduous pavement woodlands, many species characteristic of forests occur, such as *Allium tricoccum* and other spring ephemerals. Alvar grassy deciduous woodlands are quite distinctive, however, and yet they have not been distinguished clearly in vegetation classifications and literature. In southern Ontario, these communities often are affected by timber cutting and past grazing, which may give the impression that their openness is artificial. Substrates vary from slopes with limestone rubble or boulders to relatively flat pavement with a thin layer of glacial drift. Periodic dry conditions exist. Some of the characteristic trees in alvar grassy deciduous woodlands are *Acer nigrum, Quercus muehlenbergiii, Carya* spp., and *Juglans cinerea*. Species such as *Phlox divaricata, Oryzopsis racemosa,* and *Millium effusum* often are characteristic of this type. Other notable species include *Galium boreale, Lathyrus ochroleucus,* and *Zizia aurea*. *Carex juniperorum* is largely restricted to this habitat in alvars. Rare species of this habitat in the Carolinian zone of southern Ontario include *Carex davisii, Carex jamesii, Carex oligocarpa, Corydalis flavula, Myosotis macrosperma,* and *Phacelia purshii*.

In the north, alvar grassland savanna may grade into alvar grassy woodland, often with *Carex eburnea,* depending on the topography and amount of surficial drift. *Pinus banksiana* often is the most common tree. Some areas of this type have tree cover that may exceed 80%, but they are interspersed with small grassy openings.

Alvar Shrubby Woodland. This habitat is identified most easily in the southern portions of the Great Lakes region, where deciduous trees such as *Carya ovata, Quercus alba, Q. macrocarpa,* and *Q. muehlenbergii* are prevalent and form an apparently stable association. Where *Abies balsamea, Picea glauca, Pinus banksiana,* and *P. strobus* are the prevalent trees, shrubs often are less well developed in the semi-closed woodland. This may be a successional stage following fire. The distinctive features are the prevalence of shrubs and a combination of plants characteristic of either open or wooded sites. Shrubs may form dense, continuous thickets, or they may occur in patches. Some herbs attain maximum frequency where shrub density is low in this habitat, including *Aster* spp. and *Helianthus* spp. Many species of herbs, including *Carex siccata* and *Carex pensylvanica,* are the same as those of alvar grassland savanna. A particularly prevalent shrub in the more southern sites is *Cornus racemosa*. Others include *Ceanothus americanus, Rhus aromatica, Prunus virginiana, Symphoricarpos albus,* and *Viburnum rafinesquianum* (Catling and Catling 1993).

Flora

Although alvars are known primarily as a distinct entity because of the rare and restricted species found on them, the characteristic flora includes many species that frequently occur in a variety of habitats off alvars. During their study of alvars in the eastern Great Lakes area, Catling and Brownell

Table 23.2. *Characteristic species of alvars in the Great Lakes region based on occurrence in five or more of seven alvar regions defined by Catling and Brownell (1995). Authorities and synonymy for scientific names are available from Morton and Venn (1990).*

Abies balsamea	*Lilium philadelphicum*
Agalinus tenuifolia	*Lonicera dioica*
Agrostis scabra	*Minuartia michauxii*
Ambrosia artemsiifolia	*Muhlenbergia glomerata*
Amelanchier alnifolia	*Oryzopsis asperifolia*
Antennaria neglecta	*Panicum flexile*
Aquilegia canadensis	*Panicum implicatum*
Arabis hirsuta	*Panicum philadelphicum*
Arctostaphylos uva-ursi	*Penstemon hirsutus*
Arenaria serpyllifolia	*Picea glauca*
Asclepias syriaca	*Pinus strobus*
Bromus kalmii	*Poa compressa*
Campanula rotundifolia	*Poa pratensis*
Carex aurea	*Polygala senega*
Carex castanea	*Populus tremuloides*
Carex crawei	*Potentilla arguta*
Carex eburnea	*Potentilla norvegica*
Carex flava	*Prunella vulgaris*
Carex granularis	*Prunus virginiana*
Carex lanuginosa	*Quercus macrocarpa*
Carex molesta	*Ranunculus fascicularis*
Carex pensylvanica	*Rhus aromatica*
Carex richardsonii	*Rhus typhina*
Carex umbellata	*Rosa blanda*
Castilleja coccinea	*Saxifraga virginiensis*
Celastrus scandeus	*Schizachyrium scoparium*
Cerastium velutinum	*Scutellaria parvula*
Chenopodium vulgare	*Senecio pauperculus*
Comandra umbellata	*Shepherdia cauadensis*
Cornus racemosa	*Silene antirrhina*
Cornus stolonifera	*Sisyrinchium montanum*
Cystopteris fragilis	*Smilacina stellata*
Danthonia spicata	*Solidago hispida*
Deschampsia cespitosa	*Solidago juncea*
Dryopteris marginalis	*Solidago nemoralis*
Eleocharis compressa	*Solidago ptarmicoides*
Elymus trachycaulus	*Sporobolus heterolepis*
Epilobium coloratum	*Sporobolus vaginiflorus*
Fragaria virginiana	*Symphoricarpos albus*
Geranium bicknellii	*Thuja occidentalis*
Geranium carolinianum	*Toxicodendron radicans radicans*
Glyceria striata	*Trichostema brachiatum*
Hedeoma hispida	*Veronica peregrina*
Hedeoma pulegioides	*Viburnum rafinesquianum*
Hedyotis longifolia	*Viola nephrophylla*
Juncus dudleyi	*Zigadenus glaucus*
Juniperus communis	

(1995) compiled a list of 93 species characteristic of alvars, based on their occurrence in five or more of the seven regions surveyed (Table 23.2). The introduced species range from highly localized (e.g., *Polycnemum arvense* on the Smiths Falls Plain) to widespread and characteristic of all disturbed sites (e.g., *Chrysanthemum leucanthemum, Hieracium piloselloides, Hypericum perforatum,* and *Medicago lupulina*).

With regard to ecological restriction, 70% of the locations of each of 13 species are on alvars: *Astragalus neglectus, Carex crawei, Coreopsis lanceolata, Eleocharis compressa, Euphorbia commutata, Geum triflorum, Hymenoxys herbacea, Panicum philadelphicum, Piperia unalascensis, Scutellaria parvula, Solidago ptarmicoides, Sporobolus heterolepis,* and *Trichostema brachiatum* (Catling 1995). In addition to the vascular plants, some mosses (e.g., *Scorpidium turgescens*) and liverworts (e.g., *Riccia sorocarpa*) are narrowly confined to alvars. The more restricted species have helped to identify higher-quality alvar sites and to define alvar vegetation.

Alvars contain many rare plant species. Forty-three are noted for Ontario (Catling and Brownell 1995), including southern species (e.g., *Allium cernuum, Blephilia ciliata, Carex juniperorum, Celtis tenuifolia*), western species (e.g., *Bouteloua curtipendula, Sporobolus heterolepis, Myosurus minimus*), and endemics (Table 23.3). The majority of endemics found on alvars (*Hymenoxys herbacea, Iris lacustris, Cirsium hillii, Hypericum kalmianum, Solidago houghtonii, Solidago ohioensis*) are largely restricted to the alvars near the shores of the upper Great Lakes (Bruce Peninsula and Manitoulin Island; Figure 23.1). The daisy (*Hymenoxys herbacea*) occurs also on the Lakeside Plains of the Marblehead Peninsula, western Lake Erie (DeMauro 1990). On alvars, juniper sedge (*Carex juniperorum*) occurs only on the Napanee Plain at the eastern end of Lake Ontario.

The recent discovery of 46 species of algae on Bruce Peninsula alvars (Schaefer 1996a, b)

is noteworthy. Many species of lichens also are present, and three previously unknown in Ontario were collected on the Bruce Peninsula by Schaefer (1996a, b).

Fauna

As with plants, many widespread animal species are common in dry, open habitats of alvars; others are somewhat confined to alvars. Here again, the latter category includes various western and boreal elements, as well as endemics. Catling and Brownell (1995) provide a more extensive discussion of the alvar fauna. A bird, the loggerhead shrike (*Lanius ludovicianus*), probably is the best-known animal associated with alvars. This rapidly declining species is found in a few other habitats, and it breeds on some alvars that are degraded from a botanical perspective (Chabot 1994).

Inventory of invertebrates and documentation of invertebrate bioindicators has been initiated only recently in the Great Lakes region, and discussions of the invertebrate fauna (e.g., Catling and Brownell 1995) are incomplete. The most extensive works on alvar animals currently available are those of Bouchard (1997, 1998) and Hamilton (1994) on leafhoppers, which provide support for the sidewalk and xerothermic models of the derivation of the alvar flora. The dark crescent butterfly, *Phyciodes batesii* ssp. *batesii*, a declining species in the northeast, is associated with the semi-open alvar savanna and semi-closed alvar woodlands, where the populations of asters used as larval food plants probably are promoted by periodic fires (Catling 1997). A recent interesting faunal discovery on alvars is the diverse assemblage of snails, including subarctic and prairie disjuncts and new taxa, some of which may be endemic (Grimm 1995).

Biogeography and Origin

In the Great Lakes region, eastern floristic elements have poor representation on alvars

Table 23.3. *Species found on alvars that are rare and/or restricted in the Great Lakes region. + = rare in Ontario; X = not known from Ontario; E = endemic to/or including the Great Lakes region; A = with 81%–100% of occurrences in Great Lakes region on alvars; a = with 50%–80% of sites in the Great Lakes region on alvars. Authorities for scientific names and synonymy may be found in Morton and Venn (1990).*

+A	*Agalinus gattingeri*	+	*Chaerophyllum procumbens* var. *shortii*
+a	*Allium cernuum*		*Cirsium discolor*
	Allium schoenoprasum var. *sibiricum*	+E	*Cirsium hillii*
	Anemone multifida	a	*Coreopsis lanceolata*
	Arabis lyrata	+	*Coreopsis tripteris*
+	*Arabis shortii*	+	*Cornus drummondii*
	Arctostaphylos uva-ursi		*Corydalis flavula*
+	*Asclepias purpurascens*	+a	*Crataegus dilatata*
+	*Asclepias quadrifolia*	+	*Cypripedium arietinum*
	Asclepias tuberosa		*Deschampsia cespitosa*
+	*Asclepias verticillata*		*Desmodium paniculatum*
+	*Asclepias viridiflora*	+	*Draba reptans*
	Asplenium trichomanes	A	*Eleocharis compressa*
+	*Aster dumosus*		*Epilobium hornemanii*
	Aster oolentangiensis	+a	*Eupatorium altissimum*
+	*Aster shortii*	+A	*Euphorbia commutata*
+	*Astragalus neglectus*	+	*Euphorbia spathulata*
+a	*Blephilia ciliata*	+	*Fraxinus quadrangulata*
X	*Boltonia asteroides*		*Geranium bicknellii*
+a	*Bouteloua curtipendula*	a	*Geranium carolinianum*
	Bromus kalmii	a	*Geum triflorum*
+	*Cacalia plantaginea*	+	*Geum vernum*
	Calystegia spithamea	+	*Gymnocarpium robertianum*
+	*Camassia scilloides*	X	*Hedyotis nigricans*
+	*Carex annectens*	+	*Heuchera americana*
+	*Carex bicknellii*	+AE	*Hymenoxys herbacea*
+	*Carex conoidea*	E	*Hypericum kalmianum*
A	*Carex crawei*	+	*Hypoxis hirsuta*
+	*Carex davisii*	+E	*Iris lacustris*
+	*Carex formosa*	+a	*Juncus secundus*
+	*Carex gracilescens*	+	*Lactuca floridana*
+	*Carex jamesii*		*Lespedeza intermedia*
+AE	*Carex juniperorum*	+a	*Leucospora multifida*
+	*Carex leavenworthii*	+a	*Liatris cylindracea*
+	*Carex meadii*	+	*Liatris spicata*
+	*Carex oligocarpa*		*Lilium philadelphicum*
	Carex richardsonii	+	*Linum sulcatum*
	Carex sartwellii	+	*Muhlenbergia sobolifera*
	Carex siccata	+a	*Myosurus minimus*
	Castilleja coccinea	+	*Myosotis macrosperma*
	Ceanothus americanus	+a	*Myosotis verna*
	Ceanothus herbaceus	+A	*Orobanche fasciculata*
+	*Celtis tenuifolia*		*Panicum lindheimeri*
+a	*Cerastium brachypodum*		*Panicum oligosanthes*
+	*Cerastium velutinum*	+a	*Pellaea atropurpurea*
+	*Cercis canadensis*	+	*Phacelia purshii*
+	*Chaerophyllum procumbens* var. *procumbens*	a	*Piperia unalascensis*

	Poa alpina	aE	*Solidago ohioensis*
	Poa canbyi		*Solidago ptarmicoides*
	Poa glaucantha	+	*Solidago rigida*
	Polygonum douglasii		*Solidago spathulata* ssp. *randii*
+	*Polygonum erectum*	+	*Solidago ulmifolia* var. *ulmifolia*
	Potentilla arguta	+	*Spiranthes magnicamporum*
	Potentilla fruticosa	+	*Spiranthes ovalis* var. *erostellata*
	Prenanthes racemosa	+	*Sporobolus asper*
	Primula mistassinica	+a	*Sporobolus heterolepis*
a	*Prunus pumila* var. *depressa*	+	*Sporobolus ozarkanus*
+	*Ptelea trifoliata*	+	*Stipa spartea* (sandy ridges)
	Pycnanthemum flexuosum		*Thalictrum confine*
	Quercus muehlenbergii	+	*Thaspium trifoliatum* var. *aureum*
a	*Ranunculus fascicularis*		*Tofieldia glutinosa*
+	*Ratibida pinnata*	X	*Tomanthera auriculata*
	Rhus aromatica		*Trichostema brachiatum*
+	*Rhus copallina*		*Triodanis perfoliata*
+	*Rosa setigera*	+A	*Triosteum angustifolium*
	Scirpus caespitosus		*Trisetum spicatum*
a	*Scutellaria parvula*	+	*Valerianella umbilicata*
+	*Senecio obovatus*		*Verbena simplex*
+	*Silphium terebinthaceum*	+	*Vernonia altissima*
+	*Sisyrinchium albidum*	+a	*Viola rafinesquii*
	Sisyrinchium angustifolium	+	*Woodsia oregana* ssp. *cathcartiana*
	Sisyrinchium mucrontatum		*Zigandenus glaucus*
+E	*Solidago houghtonii*		*Zizia aurea*

compared to western elements. Either the latter are relics of the *Picea* parkland corridor that extended as a narrow band from the north and west, eastward along the former Wisconsin ice front, or they have more recently invaded during the Xerothermic period (Hypsithermal Interval). Examples include characteristic plants such as *Bouteloua curtipendula*, *Geum triflorum*, and *Sporobolus heterolepis*, and animals such as garita skipper butterfly (*Oarisma garita*) (Catling 1977), various ground beetles, flightless leafhoppers (Hamilton 1994), sharp-tailed grouse (*Tympanuchus phasianellus campestris*), and various molluscs (Grimm 1995).

The recent classification of Great Lakes region alvars into three groups (Catling and Brownell 1995) is related to geography and affinity of species: (1) the eastern Lake Erie alvars with a proportionally high component of species occurring to the south, but with a relatively small proportion occurring to the north; (2) the alvars of the Bruce Peninsula and Manitoulin Island, with a high proportion of northern and endemic species, and few southern species; and (3) the alvars of central Ontario, eastern Ontario, and northern New York, with a major representation of southern species and moderate representation of northern species. The high component of southern and midwestern elements of the first group suggests a relatively recent origin of the flora, probably involving migration of southern and western species northward and eastward during warmer and drier conditions that occurred at different times in different places during the Xerothermic Interval (ca. 8,000–4,000 yr BP).

The upper Great Lakes alvar flora, the second group, probably is a relict of the *Picea* parkland community that existed near the continental ice front that covered much of

Canada and the northern United States more than 11,000 years ago (Wright 1987). The possibility that alvars are relicts of this community is suggested by the western elements and endemics that probably evolved along the ice front or along the periglacial sidewalk following its fragmentation to Great Lakes shorelines and Gaspésian slopes. Also present are boreal and cordilleran disjuncts, which are plants of open, disturbed habitats. Both the "periglacial sidewalk" and "xerothermic" theories are supported by distribution patterns of insects (e.g., Hamilton 1994; Catling and Brownell 1995) and plants (Catling and Brownell 1995).

Species Richness

Within alvar sites, species richness increases with soil depth and biomass, but beyond a certain point, where shoot and root competition become significant factors, it declines with dominance by one or a few species (Belcher, Keddy and Catling 1992). The pavement edge associations on substrates 1–3 cm deep are particularly rich, with up to 20 species per m².

Alvars of Manitoulin Island and the Napanee Plain appear to have the highest species richness (Catling and Brownell 1995). A few openings of less than 2 ha in these areas have over 140 species of native vascular plants, which is exceptional. Preliminary censuses of other groups of organisms, such as molluscs (Grimm 1995), suggest a similar pattern of high species richness.

Species richness appears to increase with low levels of disturbance, such as grazing, but to decline with replacement by alien species under conditions of heavy grazing. There is a very slow return to diverse native cover, if this happens at all. Death of woody cover due to drought, fire, or cutting substantially increases species richness. One of the areas of highest species richness on the Burnt Lands alvar near Ottawa is the one where woody vegetation has been removed around radio towers. This suggests that recent fire prevention may have contributed to a substantial decline in species richness per surface area at this site, although extirpation of a species has not been documented. Species such as *Aster ciliolatus, Astragalus neglectus, Cirsium discolor, Cypripedium arietinum*, and *C. calceolus*, were more common on this cutover area than elsewhere.

Protection and Protection Needs

Alvars of the Great Lakes region have been described as important for protection because they are (1) sources of genes of wild crop relatives for crop improvement; (2) areas of concentration of rare, endangered, and disjunct species of plants and animals; (3) areas containing imperiled plant communities; (4) areas where natural history recreation and education can be very successful; and (5) particularly valuable for biological research and monitoring of climatic change. Outdoor activities such as birdwatching are among the most rapidly growing pastimes in North America. One alvar region, the Carden Plain, east of Orillia, Ontario, is one of Ontario's top 10 birdwatchers' locations, partially because birds of naturally open habitats, such as loggerhead shrikes, bobolinks, rufous-sided towhees, and brown thrashers, still can be observed at this site (Reid 1996). Catling and Brownell (1995) provide a detailed account of the values of alvars.

The most serious current threats to alvars involve quarrying, and housing and cottage development (Reid 1996). Alvars are prime sites for limestone extraction for road materials, concrete, and armour stone. Sites such as the Carden Plain in Ontario and Marblehead Peninsula in Ohio are threatened by both current and proposed quarry operations. Cottage development is a major concern on Manitoulin Island, Ontario, and residential development is increasing in the Napanee and Smith Falls plains of Ontario.

Although 90% of alvars in the Great Lakes

region occur in Ontario, this jurisdiction has yet to develop an adequate protection strategy for this globally imperiled ecosystem. In a recent summary of the threats to, and current protection of, alvars in Ontario, Reid (1996) noted that U.S. protection efforts have been far more active, sometimes even where only a few sites of reasonably good quality exist. In 1995, the Great Lakes Protection Fund and The Nature Conservancy's Great Lakes Program provided funds for an International Alvar Conservation Initiative. Its goal was to provide a unified, consistent approach to understanding the Great Lakes alvar ecosystem and developing basinwide conservation strategies to ensure its protection and stewardship. This three-year project has involved community, rare plant, and invertebrate inventories; research on hydrological processes and soil moisture regimes; and data collection on the effects of invasive exotics, fire, and browsing and grazing. The final goal is to prioritize sites for conservation strategies through comparative evaluations and conservation planning.

To identify future protection needs for alvars, a comprehensive comparison is required. Ecological criteria, such as community representation, community and species diversity, species of conservation concern, site condition, surrounding landscape attributes, and site fragmentation, should be utilized in the initial evaluation process. In addition, information on current levels of protection and management would enable the development of a final conservation strategy. Catling and Brownell (1995) provide a detailed account of protected sites and a discussion of site condition, representation, and rare species in southern Ontario alvars (V. R. Brownell unpublished). Although the recent focus on alvars has not resulted in substantial additional protection or flourish of scientific publications, it has indicated that the knowledge base for decisions has been inadequate and that there is more to protect than was realized only a few years ago. For

example, botanically degraded alvar sites may be important for certain fauna, such as the loggerhead shrike (*Lanius ludovicianus*), which extensively uses botanically degraded alvar in some areas (Chabot 1994).

Research Needs

Unlike research on the similar cedar glades further to the south, information and research relating to Great Lakes region alvars has been limited, and most of it is recent. Since the first paper published by Catling et al. (1975), work has been done by Stephenson (1983), Stephenson and Herendeen (1986), Catling (1995), Catling and Brownell (1995), and others (see Belcher, Keddy and Catling 1992; Gilman 1995; Goodban 1995; Reschke 1995a, b; Schaefer 1996a). Much of the research is descriptive. Now, with the beginning of an essential descriptive base, research to better understand the effects and interactions of ecological processes is possible. These processes include drought, hydrology, grazing and browsing, fire, and invasion by alien plants and animals.

Alvars provide opportunities to study species' strategies for survival under extreme environmental conditions and to test hypotheses relating to diversity and competition where some of the apparent controlling factors are measured relatively easily. Alvars also are sensitive indicators of recent and distant climatic change, as a consequence of sensitivity of the flora and of the record in tree rings dating back over 500 years. Studies of genetic variation in widespread species with respect to occurrence on alvars are needed, especially of crop relatives (e.g., *Amelanchier, Prunus,* and *Fragaria*), to determine the extent of need for *in situ* protection of germplasm resources. The interactions of various factors, such as moisture retention, frost heaving, and root competition, in controlling vegetation composition of alvar pavements also requires more study.

Acknowledgments

W. Grimm, J. Jones, M. J. Oldham, J. L. Riley, and C. Schaefer were particularly helpful in providing information gathered recently. Dr. R. Anderson and Dr. J. Baskin provided helpful reviews of the manuscript.

REFERENCES

Baskin, J. M., and Baskin, C. C. 1986. Distribution and geographical/evolutionary relationships of cedar glade endemics in southeastern United States. *Association of Southeastern Biologists Bulletin* 33:138–154.

Baskin, J. M., and Baskin, C. C. 1988. Endemism in rock outcrop communities of unglaciated eastern United States: evaluation of the roles of the edaphic, genetic and light factors. *Journal of Biogeography* 15:829–840.

Belcher, J. W. 1992. The ecology of alvar vegetation in Canada: description, patterns, competition. M.S. thesis, University of Ottawa, Ontario.

Belcher, J. W., Keddy, P. A., and Catling, P. M. 1992. Alvar vegetation in Canada: a multivariate description at two scales. *Canadian Journal of Botany* 70:1279–1291.

Bouchard, P. 1997. Insect diversity of four alvar sites on Manitoulin Island, Ontario. M.S. thesis, McGill University, Montreal, Quebec.

Bouchard, P. 1998. *Insect Diversity in Alvars of Southern Ontario.* Toronto, Ontario: Federation of Ontario Naturalists.

Brownell, V. R. 1984. *A Life Science Inventory and Evaluation of the Ausable River Valley.* Prepared for the Ontario Ministry of Natural Resources, Southwestern Region, Aylmer District, Ontario.

Catling, P. M. 1977. On the occurrence of *Oarisma garita* (Reakirt) (Lepidoptera: Hesperiidae) in Manitoulin District, Ontario. *Great Lakes Entomologist* 10: 59–63.

Catling, P. M. 1995. The extent of confinement of vascular plants to alvars in southern Ontario. *Canadian Field-Naturalist* 109:172–181.

Catling, P. M. 1997. Notes on the ecology of *Phyciodes batesii* ssp. *batesii. Holarctic Lepidoptera* 4:35–36.

Catling, P. M., and Brownell V. R. 1995. A review of the alvars of the Great Lakes Region: distribution, floristic composition, biogeography and protection. *The Canadian Field-Naturalist* 109:143–171.

Catling, P. M., and Catling, V. R. 1993. Floristic composition, phytogeography and relationships of prairies, savannas and sand barrens along the Trent River, eastern Ontario. *The Canadian Field-Naturalist* 107:24–45.

Catling, P. M., Cruise, J. E., McIntosh, K. L., and McKay, S. M. 1975. Alvar vegetation in southern Ontario. *Ontario Field Biologist* 29:1–25.

Catling, P. M., Reznicek, A. A., and Crins, W. J. 1993. *Carex juniperorum* (Cyperaceae), a new species from northeastern North America, with a key to *Carex* sect. *Phyllostachys. Systematic Botany* 18:496–501.

Cayouette, J., Sabourin, A., Paquette, D., and Fillion, N. 1997. *Preliminary List of Plants Surveyed in the Alvars along the Outaouais, Western Québec (Ottawa River Valley).* Ottawa, Ontario: Eastern Cereal and Oilseed Research Centre, Agriculture and Agri-food Canada, Central Experimental Farm.

Chabot, A. A. 1994. *Habitat Selection and Reproductive Biology of the Loggerhead Shrike in Eastern Ontario and Québec.* Montreal, Québec: Faculty of Science, Macdonald College, McGill.

Chapman, K. A. 1986. Alpine Hedysarum (*Hedysarum alpinum*) discovered in Michigan. *Michigan Botanist* 25:45–46.

Curtis, J. T. 1959. *The Vegetation of Wisconsin.* Madison, Wis.: The University of Wisconsin Press.

DeMauro, M. M. 1990. *Recovery Plan for the Lakeside Daisy* (Hymenoxys acaulis var. glabra). Twin Cities, Minn.: U.S. Fish and Wildlife Service.

Erickson, R. O., Brenner, L. G., and Wraight, J. 1942. Dolomitic glades of east-central Missouri. *Annals of the Missouri Botanical Garden* 29:89–101.

Freeman, C. P. 1933. Ecology of the cedar glade vegetation near Nashville, Tennessee. *Journal of the Tennessee Academy of Science* 8:143–228.

Gilman, B. A. 1995. Vegetation of Limerick Cedars: pattern and process in alvar communities. Ph.D. dissertation, State University of New York, Syracuse, N.Y.

Goodban, A. G. 1995. *Alvar Vegetation on the Flamborough Plain: Ecological Features, Planning Issues and Conservation Recommendations.* North York, Ontario: Faculty of Environmental Studies, York University.

Grimm, F. W. 1995. *Molluscs of the Alvar Arc and the Niagara Cuesta Uplands and Barren Zones.* Toronto, Ontario: Ministry of Environment and

Energy Ecosystem Planning Series, pp. 1–17.

Hamilton, K. G. A. 1994 Leafhopper evidence for origins of northeastern relict prairies (Insecta: Homoptera: Cicadellidae). In *Spirit of the Land, Our Prairie Legacy, Proceedings of the 13th North American Prairie Conference*, ed. R. G. Wickett, P. D. Lewis, A. Woodliffe, and P. Pratt, pp. 61–70. Windsor, Ontario: Department of Parks and Recreation.

Huggett, I. 1993. The discovery of alvars at Alymer. *Trail and Landscape* 27:55–57.

Krahulec, F., Rosen, E, and van der Maarel, E. 1986. Preliminary classification and ecology of dry grassland communities on Olands Stora Alvar (Sweden). *Nordic Journal of Botany* 6:797–809.

Morton, J. K. 1996. Grassland communities on Manitoulin Island. *Wildflower* 12(3):16–18.

Morton, J. K., and Venn, J. M. 1990. *A Checklist of the Flora of Ontario Vascular Plants*. University of Waterloo Biology Series 34.

Oldham, M. J., Lobb, T. J., Reznicek, A. A., Bowles, J. M., and Kilgour, D. 1993. *Field Trip Report, Maitland River, Huron County, Ontario*. Peterborough, Ontario: Natural Heritage Information Centre.

Reid, R. 1996. Habitat for the hardy. *Seasons* 36:14–22.

Reschke, C. 1990. *Ecological Communities of New York State*. Latham, N.Y.: New York Natural Heritage Program.

Reschke, C. 1995a. *Development of Research Method-*

ologies for the International Alvar Conservation Initiative. Latham, N.Y.: New York Natural Heritage Program.

Reschke, C. 1995b. *Biological and Hydrological Monitoring at the Chaumont Barrens Preserve*. Latham, N.Y.: New York Natural Heritage Program.

Schaefer, C. 1996a. Alvars in the Bruce Peninsula, Ontario. M.S. thesis, University of Guelph, Ontario.

Schaefer, C. 1996b. The survivors. *Wildflower* 12:19–21.

Stephenson, S. N. 1983. Maxton Plains, prairie refugia of Drummond Island, Chippewa County, Michigan. In *Proceedings of the Eighth North American Prairie Conference*. ed. R. Brewer, pp. 56–60. Kalamazoo, Mich.: Western Michigan University.

Stephenson, S. N., and Herendeen, P. S. 1986. Short-term drought effects on the alvar communities of Drummond Island, Michigan. *Michigan Botanist* 25:16–27.

Titlyanova, A., Rusch, G., and van der Maarel, E. 1988. Biomass structure of limestone grasslands on Öland in relation to grazing intensity. *Acta phytogeographica suecica* 76:125–134.

Wright, H. E., Jr. 1987. Synthesis; the land in front of the ice sheets. In *The Geology of North America*, vol. K-3, *North America and Adjacent Oceans during the Last Deglaciation*, ed. W. F. Ruddiman and H. E. Wright, pp. 479–488. Boulder, Colo.: Geological Society of America.

24 The Flora and Ecology of Southern Ontario Granite Barrens

PAUL M. CATLING AND VIVIAN R. BROWNELL

Naturally open habitats consisting of exposed granite rock with grassy areas, shrubs, and scattered trees often are referred to as granite barrens. These areas are never entirely barren, but nevertheless, openness and bare rock or rock with only lichen and moss cover is a characteristic feature (Figure 24.1). In Ontario, granite barrens are a restricted and special habitat with unique environmental factors and specialized plants and animals.

The terrain is composed of a ridge and trench system of extensively folded granite rockland. A faulting network (differential displacement of bedrock blocks resulting in long, steep-sided depressions) often exists perpendicular to the folds. The trenches often contain oligotrophic ponds created by beaver, or they may contain bog mats. One of the most striking features is the contrast between adjacent wet and dry land, a consequence of the fact that the granite rock is impervious and holds water in depressions. Lakes, ponds, and other wetlands are often as characteristic of granite rock barren landscapes as are the dry rock exposures. Due to the retention of water in small or shallow depressions, extreme wetness may be followed by extreme drought. In fact, granite rock barrens are characterized by a wide variation in soil depth and in water-holding capacity over a short distance. These variations are responsible for the mosaic patterns of vegetation. The rock is acidic and generally contributes to the formation of acidic soils. The wetlands are generally bogs or acidic lakes. However, within the region, barrens also develop on basic metasediments (e.g., marble, amphibolite), but the barrens developed on such substrates differ, often substantially, in their floristic composition from those on granite.

When discussing "barrens" it is often unclear whether the reference is to the patches of bare rock or to the entire landscape. The term most often is used in a landscape sense and indeed it is often difficult to draw a line separating components. Is a vernal pool surrounded by bare rock a feature of the barrens, or is it a wetland that should be considered in a separate context? All assemblages are closely related and determined by scale. Naturally fluctuating water levels and beavers substantially contribute to the maintenance of open areas by reducing tree cover and creating continuous disturbances necessary for the maintenance of open-habitat species. If parts of the assemblage are to be discussed, it seems most appropriate to discuss aquatic, wet, mesic, or dry portions. Here the concern is primarily with the dry zone, but recognizing interdependence, there will be an occasional reference to the entire landscape.

Distribution

Rock barrens exist to a greater or lesser extent throughout the Canadian Shield region, a horseshoe-shaped area that extends around Hudson Bay and includes much of eastern and central Canada. The southern portion of this area is within the Great

Figure 24.1. Granite rock barren near Kaladar, Ontario, with lichen mat in the foreground, open rock, grassland dominated by *Deschampsia flexuosa*, and scattered trees and groves of *Pinus banksiana*.

Lakes–St. Lawrence Forest region (Rowe 1977), and the rest, within the boreal forest. The granite barrens are a particularly distinctive habitat type within the southern, temperate, and largely forested portions of the shield. Although small open areas with exposed granite rocks occur throughout the southern Canadian Shield region, extensive areas of barrens are more restricted. Within southern Ontario, extensive areas are present within a discontinuous band extending east of Georgian Bay to the Kaladar area of eastern Ontario. This region of granite barrens was mapped by Chapman and Putnam (1984) as "bare rock ridges and shallow till" (Figure 24.2) and identified by them as the "Georgian Bay fringe" physiographic region.

Smaller and more isolated areas exist elsewhere in the Canadian Shield, particularly in the Frontenac Axis region of the upper St. Lawrence River in both New York and Ontario (Kloet 1973; Seischab and Bernard 1991; Bernard and Seischab 1995). In this area, barrens with pitch pine (*Pinus rigida*) have developed on both granite and Potsdam sandstone.

Habitat Ecology

Lack of Tree Cover

Tree cover is limited by lack of soil development, periodic drought, and interaction of these two factors. The barrens areas along the edges of the shield apparently had much of

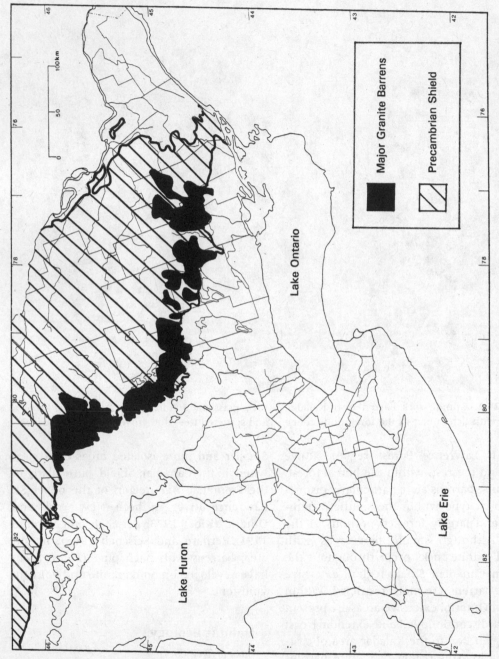

Figure 24.2. Granite barrens region (shaded) of southern Ontario. (After the "bare rock ridges and shallow till" of Chapman and Putnam 1984.)

Lake Ontario

Lake Huron

Lake Erie

Major Granite Barrens

Precambrian Shield

100 km

the overlying glacial till removed by glacial meltwater and possibly also by wave action. The hard rock itself does not readily form parent material. Where the rocks are rounded rather than cracked there is little opportunity for the establishment of trees, and many of the barrens areas are characterized by a relatively smooth, undulating, rather than blocky, surface. In spring and fall, the granite barrens may be very wet due to water held on the surface by the impervious rock, but as evaporation increases over summer they are sometimes extremely hot and dry for extended periods. Both the extremes and severe heat and dryness limit establishment of tree growth. Burned stumps and fire scars throughout the barrens region suggest that natural and anthropogenic fire also has been a major factor in determining vegetation cover. Several severe fires burned on the eastern Kaladar barrens in the late 1880s, and the most recent severe and extensive fire occurred between 1931 and 1936. In the 1950s, fire was used briefly to enhance blueberry production. No major fires have occurred in the past 40 years.

However, severe heat and drought and fire are not the only factors limiting tree establishment and growth. In 1981, the introduced gypsy moth (*Lymantria dispar*) arrived, and the abundance of red oak in the barrens region provided favorable conditions for population expansion. Defoliation over much of the area was severe for several consecutive years, and by the mid-1980s a large proportion of the trees were dead. The population of gypsy moths crashed in 1986. A spray program was initiated with the bacterium *Bacillus thuringensis* on government-owned lands. This spraying was continued, but the moth population increased again and then crashed in 1992.

During the most recent peak in the gypsy moth population, there also was an outbreak of forest tent caterpillar (*Malacosoma disstria*), which feeds mainly on maple and ash.

However, either species may defoliate a variety of trees as the populations expand. Trees may recover from forest tent caterpillar, which often has a single year of major defoliation followed by near absence for a relatively long period of 10–20 years. Gypsy moth is more likely to cause death of trees than are tent caterpillars, because gypsy moth causes heavy defoliation for many consecutive years with less time between population explosions. Trees weakened by defoliation have an increased susceptibility to drought and, undoubtedly, recent drought has played a role in the extensive reduction of tree cover.

The many wetlands characteristic of granite rock barrens provide habitat for beavers, and these animals also contribute to the lack of tree cover in some areas, particularly reducing fast-growing species such as poplars and birches. Beaver activity is largely confined to areas near shorelines. Shorelines of larger lakes and rivers also are important to the barrens ecosystem, because they tend to be more continuously open due to the disturbance of fluctuating water levels and the action of strong desiccating or abrading winds on adjacent barrens. When succession results in closure of sites distant from shore, or a lapse in catastrophic events leads to a loss of certain seral stages, the continuously disturbed shoreline acts as a refuge for rare species. Furthermore, shorelines are very significant portions of the barrens ecosystem. For example, in the Salmon River drainage, most of the occurrences of bear oak (*Quercus ilicifolia*) are along shores. Big bluestem (*Andropogon gerardii*) is often dominant at high water level and up to 20 m upslope; however, it does not occur away from major river and lake shores. Although not community dominants, other species (e.g., *Allium canadense, Aristida dichotoma*) have a similar distribution. When water levels are stabilized with dams, there can be major impacts to the barrens ecosystem as well as to the aquatic ecosystem (P. A. Keddy 1985; P. A. Keddy and Wisheu 1989), in addition to pollution effects due to reduced flushing. The effect of fluctuating water levels has been associated with high diversity of shoreline flora, but its significance to the dry barrens vegetation on shores has not been investigated.

Edaphic Characteristics

The soil is frequently extremely shallow and low in nutrients, and it may be organic, sandy, or gravelly. In general, the soils belong to the Brunisol group. Since the massive rock is impervious, the interaction between depth of substrate and slope of underlying and surrounding bedrock is a major factor in the presence or absence of particular species of vascular plants. The extremes of wetness and dryness are greater than those on nearby alvars, where the rock is porous and substrate depth alone is a predominant factor.

Granite rock barrens generally have an acid soil reaction, but the extent of acidity depends upon the nature of the underlying rock and the presence or absence of overlying calcareous till. Many other types of rocks are present in the barrens regions (e.g., Harding 1944; Hewitt 1964; Wolff 1982; Bright 1986), and these weather to varying degrees to parent material for soil. The metasediments have barrens floras different from those of the granite rocks, and distinctive floras exist in some places where marbles and biotite- and hornblende-rich metasediments accompany the granite.

Climatic Aspects

In general, the region where extensive granite barrens occur is one with warm summers and mild to cool winters, a long growing season, and more or less reliable, year-round precipitation. The barrens shown in Figure 24.2 are distributed in three climatic regions: Those barrens near Georgian Bay are in the Muskoka region; those further east, in the Haliburton Slopes region; and those in the easternmost areas, in the Eastern Counties region (Brown, McKay and Chapman 1980).

The mean annual growing season ranges from 190 to 200 days in the Muskoka region and the southern portion of the Eastern Counties region, whereas it is 180–190 days in the Haliburton Slopes region and the northern portion of the Eastern Counties (Brown, McKay and Chapman 1980). Some of the major differences in the floras of Ontario granite barrens are associated with these different climatic regions, but regional differences have yet to be fully quantified. The Western Muskoka region is the mildest (mean daily maximum temperature in July is 24.4 °C and in January it is 3.3 °C) and has proportionately more precipitation (96–102 cm of precipitation annually compared to 81–86 cm). The central Haliburton Slopes region is the coolest and has the shortest season. The easternmost barrens are the warmest (mean daily maximum temperature in July is 27 °C and in January it is 3.3 °C) and contain a number of southern species.

Plant Characteristics

A number of the plants present are able to grow rapidly during the spring, when water is held on top of the rock. *Saxifraga virginiensis* sets seed by June, but growth of rosettes begins again during the autumn wet period in September and October. The biennial *Corydalis sempervirens* similarly completes flowering and fruiting early, then plants appear again in the autumn. Many species flower in May, with fruit production dependent upon adequate moisture extending into at least early summer. This is true of the juneberries (*Amelanchier* spp.), cherries (*Prunus* spp.), and blueberries (*Vaccinium* spp.). Other species, such as *Aster ciliolatus* and *Solidago nemoralis*, are characterized by growth of rosettes in spring followed by a slower growth rate that increases as the environment becomes cool and moist in autumn. Flowering extends from September to past the first frosts of October.

Species that flower during the summer, such as *Spiranthes lacera* and *Juncus secundus*, appear irregularly, apparently depending on moisture of current and previous years.

With rather severe summer droughts, many species tend to grow primarily during the spring or fall moist periods. Some annuals, such as *Bulbostylis capillaris*, may initiate growth during summer if moisture is adequate but may not appear at all in some years.

Flora

On granite barrens there are 70 characteristic vascular plant species, of which 36 are considered dominant or codominant (Table 24.1). Apart from scattered trees and groves, the prominent vascular plant species are the grasses *Danthonia spicata* and *Deschampsia flexuosa*, the sedge *Carex pensylvanica*, the shrubs *Comptonia peregrina*, *Diervilla lonicera*, *Rhus typhina*, *Vaccinium angustifolium* and the fern *Pteridium aquilinum*. A few alien species, such as *Hieracium piloselloides* and *Hypericum perforatum*, have spread into natural granite barrens that have not been substantially influenced by human activity. *Trifolium aureum*, *Melilotus alba*, and *Medicago lupulina* are characteristic of areas previously grazed by cattle. Heavy grazing leads to invasion of mostly introduced grasses, such as *Poa compressa* and *Agropyron repens*, generally not present or dominant in rock barrens, as well as invasion of alien species such as *Bromus inermis*, *Phleum pratense*, and *Dactylis glomerata*.

Distinctive zones of vegetation exist, but often on a small scale and less well defined than on nearby alvars, because the slopes are both more variable and usually steeper. The granite barrens frequently undulate in all directions. Thin substrates up to 1 cm deep (shallow soil) have a lichen (*Cladina* spp., *Cladonia* spp.) and moss (e.g., *Grimmia* spp., *Polytrichum* spp.) cover that often grades into a combination of bryophytes and herbs (Wong and Brodo 1973; Crowder, Greer and Fongern 1979). The major herbs on moss mats and shallow soil are mostly perennial (*Saxifraga virginiensis*, *Selaginella rupestris*).

Corydalis sempervirens is an annual or biennial. *Agrostis scabra* is the only prominent annual on shallow soil (which often accumulates rainwater temporarily); others, such as *Polygonum douglasii* and *Bulbostylis capillaris*, are rarely common or dominant and often are inconspicuous in dry years. The lack of annuals is a distinct contrast to alvars (Catling and Brownell 1995) and, evidently, to the granite barrens in the granite outcrops of the Piedmont Plateau of the southeastern United States, where annuals such as *Diamorpha smallii* and *Arenaria uniflora* are characteristic of open shallow soil (Burbanck and Platt 1964; Shure and Ragsdale 1977). Perennial grasses (*Danthonia spicata* and *Deschampsia flexuosa*) or shrubs (e.g., *Comptonia peregrina*, *Vaccinium angustifolium*, and *Rhus typhina*) occur as soil depth increases. Different environmental tolerances and competitive abilities probably characterize some of the species along the soil depth gradient, as has been shown for the flat-rock areas of the southeastern United States (see Chapter 6). In general, the heaths and/or *Carex pensylvanica* appear to occupy the mesic sites, whereas the perennial grasses occupy drier sites of equivalent soil depth.

Where substrates are a few decimeters deep or where there are deeper cracks, trees or patches of woodland have established. Oaks (*Quercus rubra*, *Q. alba*) occur in dry, relatively deep substrates, and red maple (*Acer rubrum*) occurs in more moist sites. Jack pine (*Pinus banksiana*) occupies the driest sites and those prone to the greatest extremes of wetness and drought. *Deschampsia flexuosa*, *Gaylusaccia baccata*, and *Vaccinium angustifolium* dominate the understory of jack pine groves. In oak groves, *Aralia nudicaulis*, *Carex pensylvanica*, *Gaultheria procumbens*, and *Pteridium aquilinum* often dominate the understory. *Helianthus divaricatus* is also a conspicuous species. Shrubs are few except in areas where metasediments and/or marble is present, in which case dense thickets of *Cornus* spp. may develop.

Table 24.1. *Characteristic vascular plant species of the dry portions of granite rock barrens in the southern Canadian Shield region (Frontenac Axis) of eastern Ontario. Species found at most eastern Ontario sites and often dominant or codominant in some of the plant associations are preceded by an asterisk (*).*

* *Acer rubrum* L.
  Agropyron trachycaulum (Link) Malte ex H.F. Lewis
* *Agrostis scabra* Willd.
  Antennaria neglecta Greene (incl. *A. neodioica* & *A. canadensis*)
* *Apocynum androsaemifolium* L.
* *Aralia hispida* Vent.
  Arctostaphylos uva-ursi (L.) Spreng.
* *Aronia prunifolia* (Marsh.) Rehd. (incl. *A. melanocarpa*)
* *Aster ciliolatus* Lindl.
* *Aster macrophyllus* L.
  Aster umbellatus Mill.
  Betula papyrifera Marsh.
  Bromus kalmii A. Gray
  Campanula rotundifolia L.
* *Carex pensylvanica* Lam.
* *Carex rugosperma* Mack.
  Carex tonsa (Fern.) Bickn.
* *Comandra umbellata* (L.) Nutt.
* *Comptonia peregrina* (L.) Coult.
  Corydalis sempervirens (L.) Pers.
* *Danthonia spicata* (L.) R.&S.
* *Deschampsia flexuosa* (L.) Beauv.
* *Diervilla lonicera* Mill.
* *Fragaria virginiana* Dcne.
* *Gaultheria procumbens* L.
* *Gaylussacia baccata* (Wang.) K. Koch
  Helianthus divaricatus L.
  Juniperus communis L. var. *depressa* Pursh
* *Lechea intermedia* Legg.
  Lilium philadelphicum L.
* *Maianthemum canadense* Desf.
  Malaxis unifolia Michx.
* *Melampyrum lineare* Desr.
* *Oryzopsis asperifolia* Michx.
* *Oryzopsis pungens* (Spreng.) Hitchc.
  Panicum depauperatum Muhl. (*Dichanthelium depauperatum* (Muhl.) Gould)

  Panicum lanuginosum Ell. var. *implicatum* (Scribn.) Fern.
* *Panicum latifolium* L. (*Dichanthelium latifolium* (L.) Gould & Clark)
* *Panicum linearifolium* Britt. (*Dichanthelium linearifolium* (Scribn.) Gould)
  Pinus banksiana Lam.
* *Pinus strobus* L.
  Poa compressa L.
  Populus tremuloides Michx.
  Potentilla simplex Michx. (incl. *P. canadensis*)
* *Prunus pensylvanica* L. f.
  Prunus susquehanae Willd.
  Prunus virginiana L.
* *Pteridium aquilinum* (L.) Kuhn var. *latiusculum* (Desv.) Underw.
* *Quercus alba* L.
* *Quercus rubra* L.
* *Rhus typhina* L.
* *Rubus allegheniensis* Porter
  Rubus flagellaris L. (incl. *R. arundelanus* Blanch.)
  Rubus hispidus L.
  Rubus strigosus Michx.
* *Rumex acetosella* L.
  Salix humilis Marsh.
  Saxifraga virginiensis Michx.
  Selaginella rupestris (L.) Spring
  Solidago bicolor L.
  Solidago graminifolia (L.) Salisb. (*Euthamia graminifolia* (L.) Nutt.)
* *Solidago juncea* Ait.
* *Solidago nemoralis* Ait.
* *Spiraea alba* DuRoi
  Spiranthes casei Catling and Cruise
  Spiranthes lacera (Raf.) Raf. var. *lacera*
* *Vaccinium angustifolium* Ait.
  Vaccinium angustifolium Ait. var. *nigrum* (Wood) Dole
  Viburnum rafinesquianum Schultes
  Viola fimbriatula Sm.

The extent to which bryophyte mats expand over smooth, bare rock surfaces is unknown, but some expansion of bryophyte mats appears likely on wet and mesic sites (including temporary pools). Soil depth probably increases extremely slowly, especially on drier sites, and a slow successional change related to soil depth is likely, as in granite barrens further south (Burbanck and Platt 1964; Shure and Ragsdale 1977; also see Chapter 6). A sward of *Deschampsia flexuosa*, for example, might be regarded as stable vegetation on very dry, open sites, due to the slowness of change.

A relatively rapid succession is noticeable in groves of trees where there is a strong development of herbs following death of woody cover (due to fire, defoliation by insects, or drought). Within one to three years low shrubs become prevalent, probably due to resprouting. Occasionally, shrubs that were nearly dead appear to have an uncanny ability to persist in large numbers (e.g., *Amelanchier* spp., *Ceanothus ovatus*). Taller shrubs and trees replace these in turn, as light becomes increasingly limited. Catastrophic events, such as cutting by beaver, may interrupt succession at any time and return the system to an earlier stage. As with the composition of vegetation in the pitch pine communities in northeastern New York state (Bernard and Seischab 1995), the composition of vegetation in southern Ontario granite barrens probably depends not only on simple succession, but also on the degree, nature, and timing of past disturbances. Although an extensive literature exists on succession in the granite barrens of the southeastern United States (e.g., Shure and Ragsdale 1977; Quarterman, Burbanck and Shure 1993; also see Chapter 6), similar literature for Ontario sites is not available.

Wetlands ranging from small temporary pools to extensive and permanently wet bogs or lakes occupy the characteristic depressions between dry open ridges. Even on the tops of the highest and driest ridges, an isolated depression 1 m deep may contain characteristic bog plants such as *Sphagnum* mosses and Virginia chain fern (*Woodwardia virginica*).

Biogeography and Origin

Species that are dominant and frequent in the granite rock barrens of the southern Canadian Shield region generally have widespread distributions. The western element is not as conspicuous as in the prairies and alvars of southern Ontario, but *Carex siccata* is a notable example. A few of the species, such as *Pinus banksiana* around Kaladar in eastern Ontario and *Potentilla tridentata* in southern Muskoka, are boreal in their affinity and near to the southern limits of their distribution. The second largest group is the southern species that reach their northern limit on the eastern Ontario rock barrens. Included in this group are *Asclepias exaltata*, *Desmodium paniculatum*, *Cirsium discolor*, *Juniperus virginiana*, *Panicum oligosanthes*, *Pinus rigida*, *Quercus ilicifolia*, *Rhus copallina*, *Solidago arguta*, *Solidago puberula*, *Spiranthes ochroleuca*, and *Woodsia obtusa*.

Unlike the flat-rocks of the southern Appalachians, where more than one-third of the 44 characteristic species are endemic (McVaugh 1943; also see Chapter 6), no endemic vascular plants have been reported from southern Ontario granite barrens. Flat-rock endemism decreases rapidly north of North Carolina (Harvill 1976), and no granite barren endemics in the Appalachians occur north of Virginia (Murdy 1968). The endemism in the south is associated with a long-term habitation by plants as compared to the short-term habitation in the north due to recent Wisconsin glaciation.

Although there are no strict endemics in Ontario, several species have most of their Ontario occurrences on the granite barrens. *Juncus secundus* is a good example, but it also occurs on thin soil over limestone. *Aster ciliolatus*, *Prunus susquehanae*, *Quercus ilicifolia*, *Rhus copallina*, and *Spiranthes casei* var. *casei* are concentrated in granite rock barren ecosystems in Ontario, but also occur on sand barrens. *Bulbostylis capillaris* is mainly a species of the granite barrens, but also occurs on gravelly roadsides and sandy, receding shorelines.

Granite barrens in Ontario are also quite different from the flat-rock areas of the southeastern United States in their general floristic composition. Only a few of the prominent species of the southeastern sites, such as *Polytrichum commune* and *Bulbostylis capillaris*, are also present in the north. Characteristic species of flat-rock, such as

Andropogon virginicus, Agrostis elliotiana, Arenaria brevifolia, Crotonopsis elliptica, Diamorpha smallii, Hypericum gentianoides, Linaria canadensis, and *Virguiera porteri,* are absent from the granite barrens in southern Ontario. However, the southern Ontario granite barrens are similar in many respects to the basalt glades of Wisconsin and Minnesota (Glenn-Lewin and Ver Hoef 1988a), which are approximately the same distance away, but near to the same latitude. In common is the abundance of lichens and bryophytes and dominants such as *Carex pensylvanica, C. siccata, Danthonia spicata, Quercus* spp., *Juniperus communis, Rhus* spp., *Rubus* spp., and *Vaccinium angustifolium.* The Wisconsin and Minnesota sites differ in having characteristic prairie species among the dominants, including *Amorpha canescens, Andropogon gerardii, Lespedeza capitata,* and *Schizachyrium scoparium. Sorgastrum nutans* is the only tall grass generally distributed on the Ontario barrens sites. The fewer midwestern species make the southern Ontario barrens less likely to be regarded as forest–grassland transition as the Wisconsin basalt barrens are (Glenn-Lewin and Ver Hoef 1988a).

Species Richness Trends

The greatest species richness of vascular plants is in areas near to metasediments, where oaks are the dominant trees. Fire, drought, and infestations of gypsy moth and forest tent caterpillar, which killed many of the woody plants, have enlarged the open areas. Areas where pine is dominant have fewer total species, but not necessarily fewer rare species. A 2-ha granite barren (not including the wetlands) may have up to 100 native species whereas sites with lower richness may have 30–40 species (Brownell 1994; Brownell 1997a, b). The richest associations are found in the open or semishaded depressions with 5–10 cm deep soil. Here, up to 15 species may be present in a 1-m^2 quadrat. The moderate to relatively high species richness in the dry granite barrens may be attributed to the patch diversity, which is a consequence of local variations in soil depth over rock and in water-holding capacity, as reported by Glenn-Lewin and Ver Hoef (1988b) for Wisconsin basalt balds.

Summary of Importance
Berry Production, Germplasm

Blueberries have been harvested in the Kaladar region for many decades, and blueberries once were a significant source of income to the local inhabitants. Presently, the local harvest (*Vaccinium angustifolium* var. *angustifolium* and *V. angustifolium* var. *nigrum*) is sold at roadside stands along with the larger, imported blueberries (*Vaccinium corymbosum*) from the Maritime Provinces, New England, Pennsylvania, or New Jersey. The generally preferred local crop sometimes sells for twice the price of the imported blueberry. Local blueberries are in peak production in mid to late July. Prior to the 1950s, blueberry harvesters in the area burned the barrens to improve production of blueberries. When the Kaladar area came under the Fire Control Act in 1952, residents were no longer allowed to burn the blueberry areas, so residents of Kaladar, Sheffield, and Kennebec townships requested a prescribed burning program from the Department of Lands and Forests. The request was granted, and in 1957 a prescribed burning management program for blueberry production was initiated, mostly on 200-acre tracts of government-owned land. Interest in local blueberries declined after 1957 and the management program was discontinued.

In addition to blueberries, germplasm of other native berry crops is potentially valuable, including three species of cherries (*Prunus*), of which *Prunus susquehannae* is restricted in Ontario. Seven native species of raspberries and blackberries (*Rubus*) are present, as well as five species of saskatoons (*Amelanchier*) and two species of strawberries (*Fragaria*). Some of these species probably are represented by ecotypes adapted to extreme periodic drought.

Significant Plant Species

Although no plants in the dry granite barrens of Ontario are considered endemic, 16 provincially rare species are present along with 29 regionally rare plants (Table 24.2). Most of these are southern species at their northern range limit. This number of rare species is less than in other ecosystems, such as prairie or deciduous forest. Some of the provincially rare species are confined within Ontario largely or entirely to the dry granite barrens of the Frontenac Axis (e.g., *Opuntia fragilis, Pinus rigida, Quercus ilicifolia, Rhus copallina, Solidago puberula,* and *Woodsia obtusa*) (Kloet 1968, 1973; Brownell, Blaney and Catling 1996). Some of the regionally rare species have a Canadian Shield distribution to a greater or lesser extent, but they are more geographically widespread and occur on sandplains as well as on granite rock (e.g., *Aster ciliolatus, Comptonia peregrina, Prunus susquehannae,* and *Selaginella rupestris*). In addition to the rare species of dry sites, the barrens landscapes contain rare coastal plain elements (P. A. Keddy and Reznicek 1982; C. J. Keddy and Sharp 1994; Reznicek 1994), most of which are wetland plants, including *Isoëtes engelmannii, Panicum spretum, Polygonum careyi, Potamogeton bicupulatus, Rhexia virginica, Xyris difformis,* and *Bartonia paniculata.* These are prominent in the Georgian Bay region. Outside of this region, there are few detailed reports on the rare species of barrens landscapes, the one exception being for the St. Lawrence Islands (Brownell 1984).

Significant Animals

The granite barrens support a diverse fauna with at least 30 mammal, 136 breeding bird, 10 reptile, 9 amphibian, and 46 butterfly species (e.g., Brownell 1997; White 1994). Probably more than 80% of the occurrences of a rather restricted lizard, five-lined skink

(*Eumeces fasciatus*), are on granite barrens, and the distribution of this rare (in Ontario) lizard is closely correlated with the major areas of barrens along the southern Canadian Shield between Georgian Bay and Kingston (Seburn 1990). Black rat snake (*Elaphe obsoleta obsoleta*), rare in Ontario, also has most of its Ontario localities within the granite rock barrens region of the Frontenac Axis, although it is not necessarily confined to the open areas.

The olympia marblewing butterfly (*Euchloe olympia*) occurs on granite and limestone rock barrens, and the only Ontario locations outside of rock barrens are the dunes at Grand Bend, where a different morphotype is present (Wagner 1977). It appears that the olympia was either scarce or absent from the Ontario rock barrens prior to the 1970s (D. Lafontaine, personal communication). Another butterfly, chryxus arctic (*Oeneis chryxus*), is also largely confined in Ontario to rock barrens, although it also occurs on a few limestone and sand barren areas. Gray hairstreak (*Strymon melinus*) is the only butterfly regarded as rare in Ontario that has a rock barrens habitat, but it is less confined to the barrens than are the preceding two species. Although not rare in Ontario, dark crescent, *Phyciodes batesii* ssp. *batesii,* occurs in some rock barrens sites and is extirpated over much of its range in the United States. There are almost certainly many other insects that are more or less restricted in Ontario to dry granite rock barrens, but distributions of many are poorly documented.

More than 80% of Ontario populations of prairie warblers (*Dendroica discolor*) are in granite rock barrens areas (Lambert and Smith 1984a, b). These birds do not occur on the limestone rock barrens to the same extent, and are scarce in southern Ontario outside of the prairie scrub areas of Norfolk County and the sand dunes at Grand Bend. Although not as strictly confined to granite rock barrens as prairie warbler, yellow-billed cuckoo (*Coccyzus americanus*) is prominent in the eastern Ontario barrens, but scarce over much of the remainder of Ontario. Its recent abundance in

Table 24.2. *Regionally and provincially (P) rare species found on dry granite rock barrens of the Frontenac Axis region. Provincial rarity is based on Argus et al. (1982–1987).*

Allium canadense L.	*Lysimachia quadrifolia* L.
Andropogon gerardii Vitman	*Minuartia michauxii* (Fenzl) Farw.
Asclepias exaltata L. (semi-shaded)	P *Opuntia fragilis* (Nutt.) Haw.
Asclepias tuberosa L.	*Panicum oligosanthes* Schultes (*Dichanthelium*
Asplenium platyneuron (L.) Oakes ex D.C. Eat.	*oligosanthes* (Schultes) Gould)
P *Aristida dichotoma* Michx.	*Polygala polygama* Walter
P *Bartonia virginica* (L.) B.S.P.	*Polygonum douglassii* Greene
P *Bulbostylis capillaris* (L.) C.B. Clarke	*Potentilla arguta* Pursh
P *Carex artitecta* MacKenzie	*Pinus banksiana* Lambert
Carex backii Boott	P *Pinus rigida* Miller
Carex cumulata (Bailey) Fern.	*Prunus susquehanae* Willd.
Carex siccata Dewey	P *Quercus ilicifolia* Wang.
Carex lucorum Willd. ex Link	P *Rhus copallina* L.
Carex tonsa (Fern.) Bickn.	P *Solidago arguta* Ait. (semi-shaded)
Cirsium discolor (Muhl.) Spreng.	*Solidago ptarmicoides* (Nees) B. Boivin
Desmodium paniculatum (L.) DC. (semi-shaded)	P *Solidago puberula* Nutt.
Hedeoma hispida Pursh	*Solidago squarrosa* Muhl.
Hieracium canadense Michx.	*Sorghastrum nutans* (L.) Nash
Hieracium scabrum Michx.	*Spiranthes casei* Catling & Cruise
P *Hieracium venosum* L. var. *nudicaule* (Michx.)	P *Spiranthes ochroleuca* (Rydb. ex Britt.) Rydb.
Farw. (semi-shaded)	P *Vaccinium stamineum* L. (semi-shaded)
P *Juncus greenei* Oakes & Tuckerman	*Viola fimbriatula* Smith
P *Juncus secundus* Beauv. ex Poir.	P *Woodsia obtusa* (Spreng.) Torr.
Lonicera hirsuta Eat.	

the area may be related to the outbreaks of gypsy moth and forest tent caterpillar.

A High-Priority Community

Granite rock barrens have a limited occurrence in southern Ontario and deserve a relatively high priority for protection. Although the dry barrens themselves may not have an unusual level of biodiversity within southern Ontario, they are part of a landscape that does have a high level of biodiversity. As many as 500–600 species of vascular plants may be present within an area of several square miles.

Recreation

Although they have a limited occurrence in southern Ontario, the barrens areas are extensive where they occur and have a great potential for development of excellent hiking trails. The terrain is open and interesting but, except for blueberry picking and limited

hunting, they are little used. Summer cottages, however, surround many of the larger lakes and rivers. As discussed by Chapman and Putnam (1984), the westernmost barrens region of Georgian Bay and Muskoka include the major recreational area of Ontario with summer cottages, which number over 20,000. It includes the rugged landscape of bare rocks with pines reflected in clear water for which Ontario is so well known.

Protection and Protection Needs

The flora and fauna of barrens developed on acid rocks has been far less affected than those developed on acid sand, in both Ontario (Carbyn and Catling 1995) and New York (Bernard and Seischab 1995). Among the important biological influences on southern Ontario granite barrens are flooding through dams (which destroyed some of the outstanding Precambrian marble barrens in southern Ontario), mineral extraction, cessation of fire due to fire control and stand isolation, and recreational development. Increasing recreational development in the form of cottages on shorelines is probably the principal current influence. As noted previously, shorelines and their associated naturally fluctuating water levels are a very significant portion of barrens ecosystems and require special consideration for protection.

Researchers have largely neglected the flora and fauna of dry granite rock barrens in Ontario, but the protection of representative examples has been considered in various reports (e.g., White 1993). A few sites are protected, and one of these, the Kaladar Jack Pine Barrens, was selected for protection based on representative features (Brownell 1994).

Research and Research Needs

Unlike the flat-rock areas in the southern Appalachians, where the foundation for research on rock barrens was established many decades ago (e.g., Harper 1939; Oosting and Anderson 1939; McVaugh 1943) and has been followed by more recent comprehensive ecological studies, the foundation for research on rock barrens in Ontario is still incomplete. Until recently, no attempt has been made to classify dry granite rock barrens in Ontario or to characterize their major associations in a comprehensive manner. Characterization of the barrens flora and fauna currently is limited to several consultant reports designed to assist in establishing a network of protected sites (e.g., Brownell 1984, 1994, 1997a, b; Macdonald 1986; Varga 1988; Bergsma 1994; White 1994, 1995a, b). Ecological studies are essentially lacking, with the exception of those concerning the barrens near Sudbury that resulted from smelter emissions (Freedman 1989, pp. 30–39). An understanding of the significance of granite rock barrens was so incomplete that there was no research aimed at determining the effect of spraying *Bacillus thuringensis* on the community as a whole, either prior to or during the spray program to control gypsy moth. This is a lamentable situation considering current world commitments to the protection of biodiversity. Research is essential with regard to potential influences such as the construction of dams and the limitations on natural water level fluctuations, as well as spraying to control insects. Apart from the research essential to effectively manage barrens landscapes, there are some very promising opportunities involving evolution of specialized ecotypes, drought tolerance, environmental monitoring, and comparison with better-studied nonglaciated sites further south.

Acknowledgments

The Ontario Ministry of Natural Resources provided funds to conduct biological inventories and evaluations of several granite barrens. T. Norris supervised the studies. T. Scar of the Ontario Ministry of Natural Resources and D. Lafontaine of Agriculture Canada provided information on gypsy moth and forest tent caterpillar. D. Mumford, formerly of the Ministry of Natural Resources, provided information on the management of the barrens for blueberry production. S. Porebski of Agriculture Canada prepared the distribution map.

REFERENCES

Argus, G. W., Pryer, K. M., White, D. J., and Keddy, C. J. eds. 1982–1987. *Atlas of the Rare Vascular Plants of Ontario*, Parts 1–4. Ottawa: Botany Division, National Museum of Natural Sciences.

Bergsma, B. M. 1994. *Torrance Barrens: Vegetation and Significant Features Inventory and Evaluation*. Gartner Lee Ltd.; prepared for the Ontario Ministry of Natural Resources, Parry Sound District, Bracebridge.

Bernard, J. M. and Seischab, F. K. 1995. Pitch Pine (*Pinus rigida* Mill.) communities in north eastern New York State. *The American Midland Naturalist* 134:294–306.

Bright, E. G. 1986. *Geology of the Mellon Lake Area.* Ontario Geological Survey, Open File Report 5598.

Brown, D. M., McKay, G. A., and Chapman, L. J. 1980. The climate of southern Ontario. *Environment Canada Climatological Studies* 5.

Brownell, V. R. 1984. *A Management Plan for Rare Vascular Plants in St. Lawrence Islands National Park.* Cornwall: Parks Canada, Ontario Region.

Brownell, V. R. 1994. *A Biological Inventory and Evaluation of the Kaladar Jack Pine Barrens Area of Natural and Scientific Interest.* Ontario Ministry of Natural Resources Open File Ecological Report 9402, Tweed District, Tweed, Ontario.

Brownell, V. R. 1997a. *A Biological Inventory and Evaluation of the Puzzle Lake Area of Natural and Scientific Interest.* Ontario Ministry of Natural Resources Open File Ecological Report, Peterborough District, Kingston, Ontario.

Brownell, V. R. 1997b. *A Biological Inventory and Evaluation of the Mellon Lake Area of Natural and Scientific Interest.* Ontario Ministry of Natural Resources Open File Ecological Report, Peterborough District, Kingston, Ontario.

Brownell, V. R., Blaney, C. S., and Catling, P. M. 1996. Recent discoveries of southern vascular plants at their northern limits in the granite barrens area of Lennox and Addington County, Ontario. *Canadian Field-Naturalist* 110:255–259.

Burbanck, M. P., and Platt, R. B. 1964. Granite outcrop communities of the Piedmont Plateau in Georgia. *Ecology* 45:292–306.

Carbyn, S. E., and Catling, P. M. 1995. Vascular flora of sand barrens in the middle Ottawa valley. *Canadian Field-Naturalist* 109:242–250.

Catling, P. M., and Brownell, V. R. 1995. A review of the alvars of the Great Lakes Region: distribution, floristic composition, biogeography and protection. *Canadian Field-Naturalist* 109:143–171.

Chapman, L. J., and Putnam, D. F. 1984. Physiography of southern Ontario. *Ontario Geological Survey* Special Vol. 2: Map P.2715.

Crowder, A., Greer, D., and Fongern, N. 1979. *The Ecology of Terricolous Mosses in St. Lawrence Islands National Park.* Cornwall: Parks Canada, Ontario Region.

Freedman, B. 1989. *Environmental Ecology, the Impacts of Pollution and Other Stresses on Ecosystem Structure and Function.* New York: Academic Press.

Glenn-Lewin, D. C., and Ver Hoef, J. M. 1988a. Prairies and grasslands of the St. Croix National Scenic Riverway, Wisconsin and Minnesota. *Prairie Naturalist* 20:65–80.

Glenn-Lewin, D. C., and Ver Hoef, J. M. 1988b. Scale, pattern analysis, and species diversity in grasslands. In *Diversity and Pattern in Plant Communities*, ed. H. J. During, M. J. A. Werger, and H. J. Willems, pp. 115–129. The Hague: Academic Publishing.

Harding, W. D. 1944. Geology of Kaladar and Kennebec townships. *Ontario Department of Mines, Annual Report* 51(4):57–74.

Harper, R. M. 1939. Granite outcrop vegetation in Alabama. *Torreya* 39:153–159.

Harvill, A. M., Jr. 1976. Flat-rock endemics in Gray's Manual range. *Rhodora* 78:145–147.

Hewitt, D. F. 1964. *Geology Map 2053 – Madoc Area.* Ontario Department of Mines 6715.

Keddy, C. J., and Sharp, M. J. 1994. A protocol to identify and prioritize significant coastal plain plant assemblages for protection. *Biological Conservation* 68:269–274.

Keddy, P. A. 1985. Lakeshores in the Tusket River valley, Nova Scotia: distribution and status of some rare species, including *Coreopsis rosea* Nutt. and *Sabatia kennedyana* Fern. *Rhodora* 87:309–320.

Keddy, P. A., and Reznicek, A. A. 1982. The role of seed banks in the persistence of Ontario's coastal plain flora. *American Journal of Botany* 69:13–22.

Keddy, P. A., and Wisheu, I. C. 1989. Ecology, biogeography, and conservation of coastal plain plants: some general principles from the study of Nova Scotian wetlands. *Rhodora* 91:72–94.

Kloet, L. vander. 1968. Occurrence of *Rhus copallina* in Leeds County, Ontario. *Canadian Field-Naturalist* 82:291–293.

Kloet, L. vander. 1973. The biological status of Pitch Pine, *Pinus rigida* Miller, in Ontario and adjacent New York. *Canadian Field-Naturalist* 87:249–253.

Lambert, A. B., and Smith, R. B. H. 1984a. The status and distribution of the Prairie Warbler in Ontario. *Ontario Birds* 2:99–115.

Lambert, A. B., and Smith, R. B. H. 1984b. *The Status of the Prairie Warbler* (Dendroica discolor). Toronto: Nongame Program, Wildlife Branch, Ontario Ministry of Natural Resources.

Macdonald, I. D. 1986. *A Preliminary Ecological Inventory of Killarney Provincial Park, Ontario.* Environmental Planning Series IV(12), Environmental Section, Ontario Division of Parks.

McVaugh, R. 1943. The vegetation of the granitic flat-rocks of the southeastern United States. *Ecological Monographs* 13:119–165.

Murdy, W. H. 1968. Plant speciation associated with granite outcrop communities of the southeastern Piedmont. *Rhodora* 70:394–407.

Oosting, H. J., and Anderson, L. E. 1939. Plant succession on granite rock in eastern North Carolina. *Botanical Gazette* 100:750–768.

Quarterman, E., Burbanck, M. P., and Shure, D. J. 1993. Rock outcrop communities: limestone, sandstone and granite. In *Biodiversity of the South-eastern United States: Upland Terrestrial Communities*, ed. W. H. Martin, S. G. Boyce, and A. E. Echternacht, pp. 35–86. New York: John Wiley & Sons, Inc.

Reznicek, A. A. 1994. The disjunct coastal plain flora in the Great Lakes region. *Biological Conservation* 68:203–215.

Rowe, J. S. 1977. *Forest Regions of Canada.* Ottawa, Ontario: Canadian Forestry Service, Department of Fisheries and the Environment.

Seburn, C. N. L. 1990. *A Status Report for the Five-lined Skink* (Eumeces fasciatus) *in Ontario*. Draft. Toronto: Nongame Program, Wildlife Branch, Ontario Ministry of Natural Resources.

Seischab, F. K., and Bernard, J. M.. 1991. Pitch Pine (*Pinus rigida* Mill.) communities in central and western New York. *Bulletin of the Torrey Botanical Club* 118:412–423.

Shure, D. J., and Ragsdale, H. L. 1977. Patterns of primary succession on granite outcrop surfaces. *Ecology* 58:993–1006.

Varga, S. 1988. *A Biological Inventory and Evaluation of the Big Chute Rocklands Area of Natural and Scientific Interest.* Ontario Ministry of Natural Resources, Central Region, Richmond Hill.

Wagner, W. H., Jr. 1977. A distinctive dune form of the Marbled White butterfly, *Euchloe olympia* (Lepidoptera: Pieridae) in the Great Lakes area. *Great Lakes Entomologist* 10:107–112.

White, D. J. 1993. *Life Science Areas of Natural and Scientific Interest in Site District 6-10: A Review and Assessment of Significant Natural Areas.* Draft report. Ontario Ministry of Natural Resources, Kemptville & Tweed Districts.

White, D. J. 1994. *Life Science Survey and Evaluation of the Blue Mountain ANSI and Charleston Lake Crown Islands.* Ontario Ministry of Natural Resources, Open File Report 9401, Kemptville District, Kemptville.

White, D. J. 1995a. *Life Science Survey and Evaluation of the Elzevir Peatlands and Barrens ANSI.* Ontario Ministry of Natural Resources, Tweed District, Tweed.

White, D. J. 1995b. *Life Science Survey and Evaluation of the Hungry Lake Barrens ANSI.* Ontario Ministry of Natural Resources, Tweed District, Tweed.

Wolff, J. M. 1982. *Geology of the Kaladar Area, Lennox and Addington Counties.* Ontario Geological Survey Report 215.

Wong, P. Y., and Brodo, I. M.. 1973. Rock-inhabiting lichens of the Frontenac Axis, Ontario. *Canadian Field-Naturalist* 87:255–259.

25 The Aspen Parkland of Canada

O. W. ARCHIBOLD

Introduction

The aspen parkland is a mosaic of grassland and woodland plant communities forming a transitional zone between boreal forests and prairies of the interior plains of central Canada. The main tract of aspen parkland forms a crescent stretching from northern Montana northeastward along the foothills of the Rocky Mountains in southern Alberta, then curving through central Saskatchewan into southwestern Manitoba and adjacent areas of northwestern Minnesota (Figure 25.1). The largest area of parkland is in Saskatchewan, where it occupies about 72,000 km² in a belt varying in width from 40 to 160 km. Trembling aspen (*Populus tremuloides*) is the dominant tree species throughout the parkland and typically grows in pure stands or groves (Figure 25.2). The southern boundary is fragmented, and the widely scattered groves of aspen are replaced by mixed-grass and shortgrass prairie. Grove size increases northwards, and eventually the distinctive, isolated character of the groves is lost as the trees merge with the mixedwood (conifer–hardwood) stands of the southern boreal forest. Extirpation of the bison and suppression of fire following settlement in the 1890s resulted in the spread of aspen throughout the parkland region (Archibold and Wilson 1980). Some authors (Sauer 1950; Stewart 1956) have suggested that aspen forest is the climax vegetation of the northern Great Plains; however, Looman (1979) considered the forest to be a subclimax formed by the southward extension of this pioneer boreal species.

Environment

Climate

The Rocky Mountains effectively block the flow of warm, moist air from the Pacific, and consequently, the interior plains are drier and experience more extreme temperature conditions in summer and winter than other parts of the continent. For example, in Saskatchewan, mean monthly temperature drops below freezing in October (Figure 25.1). January is the coldest month, with average temperature typically falling to -20 °C. Mean air temperature rises above freezing in April, but frosts can occur as late as July. July is the warmest month, with mean air temperatures of 18–20 °C. Climatic conditions become drier toward the west. The dryness of the climate traditionally has been emphasized as a factor in the origin of the grasslands, but other factors such as fire and grazing also have been considered important (Pyne 1982; Axelrod 1985; Knapp and Seastedt 1986; Hildebrand and Scott 1987). The Canadian prairies are prone to drought (Maybank et al. 1995). These periods of lower-than-normal precipitation are associated with higher temperatures (Borchert 1950). Such conditions favor the spread of fire; thus, the southward expansion of aspen during the last century probably is related to fire suppression.

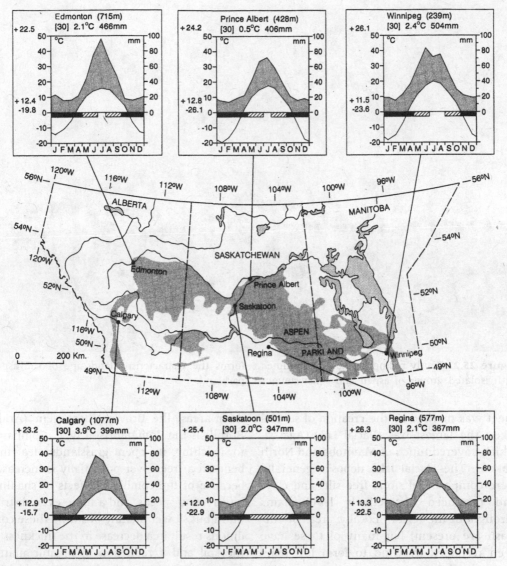

Figure 25.1. Location of the aspen parkland, and representative climate diagrams. (For a key to the climate diagrams, see Walter 1973.)

Topography and Soil

The aspen parkland region consists of areas of low rolling topography interspersed with occasional uplands and featureless plains. The surficial landforms were produced during the Wisconsin glaciation, the ice sheets withdrawing from the area about 12,000– 15,000 yr BP. The greatest relief, generally between 10 and 20 m, is found in areas of hummocky moraine that character-

istically covers preexisting uplands. In some areas, melting of remnant ice blocks within these deposits created extensive pits and hollows that now are permanently or intermittently filled with water. These prairie sloughs provide ideal habitat for waterfowl that annually migrate through the parklands. Similar depressions occur in areas of ground moraine and provide a local relief of 5–6 m in otherwise rather flat terrain. The retreat of the ice

Figure 25.2. Gently rolling farmland interrupted with isolated groves of aspen and small sloughs forms the characteristic landscape of the aspen parkland.

sheet was marked by the creation of several lakes; the largest was Glacial Lake Agassiz, which covered much of Manitoba and North Dakota. The glacial lake deposits generally were composed of stone-free silts and clays and produced distinctively level plains throughout the region. Locally, areas of outwash are present, but many of these have been reshaped by wind to form low-relief dune complexes.

The major climatic regimes are reflected in the zonal character of soils of the Canadian prairies, with darker soils associated with the cool, moist areas in the northern and eastern parkland. The soil of the aspen parkland typically is classed as Dark Brown and Black Chernozems (Typic and Udic Borolls) developed on moderately to well drained, unsorted loamy till and clayey lacustrine sediments. The profile is characterized by a dark surface horizon and brownish B horizon underlain by light-colored parent materials usually containing calcium carbonate (C horizon). In drier areas, the Brown Chernozem (Aridic Boroll) is the dominant soil type. Compared to leaching in open grasslands, leaching beneath a tree cover potentially is increased because of the combined effects of shading, lower wind speeds, and a more even distribution but slower melting of snow. These conditions result in a decrease in the thickness of the soil and a reduction in organic matter content of the surface horizons. Organic acids in forest litter promote leaching of clay and constituent nutrients such as N, S, and P, with a corresponding increase in soil acidity (Anderson 1987). The boreal forest–parkland transition to the north is characterized by Eluviated Chernozemic soils (Dark Gray Boralfic Borolls) that intergrade with the Gray Luvisols (Boralfs) and Brunisols (Eutrochrepts) of the mixedwood forests (Canadian Soil Survey Committee 1978).

Changes in soil properties along transects from grassland into woodland in east-central Saskatchewan support the idea that the aspen range is expanding (Fuller and Anderson 1993). The amount of grass opal present in the soil under woodland indicates that the area previously supported prairie vegetation. High opal content does not decline at the edge of the woodland, which suggests that the present boundary was established quite recently.

Parkland Vegetation

The grassland associations of the Great Plains extend into the southern parkland. In southeastern Manitoba, elements of true prairie are present with mixed prairie species represented in Saskatchewan and Alberta. The remnant tallgrass region is restricted to a low-lying area of heavy clay soils in the extreme southeastern part of the parkland (Bird 1961). Wet soils support cordgrass (*Spartina pectinata*) and big bluestem (*Andropogon gerardii*). Blue grass (*Poa palustris*) dominates mesic sites, and western wheatgrass (*Agropyron smithii*), June grass (*Koeleria cristata*), and speargrass (*Stipa comata*) become increasingly important on dry sites. Plant nomenclature follows Looman and Best (1979).

The character of the grassland changes westward in response to increasing aridity, and distinctive associations generally related to major soil zones are recognized. The Black Chernozem soil zone supports fescue (*Festuca altaica*) grasslands, which extend across west-central Saskatchewan into central Alberta. This is the most northern of the grassland associations and is associated with aspen groves throughout most of its range. Fescue is the dominant grass species; sub-dominant species include wheatgrass (*Agropyron subsecundum*, *A. trachycaulum*) and June grass. Various sedges (*Carex* spp.) add to the diversity, and together with shortgrasses, such as mat muhly (*Muhlenbergia richardsonis*), form a discontinuous understory. A third layer composed of a sparse cover of low-growing, densely matted club-moss (*Selaginella densa*) may also be present.

Grasses and sedges account for over 90% of the basal cover, with the remainder composed of forbs such as pasture sage (*Artemisia frigida*) and prairie crocus (*Anemone patens*).

Speargrass and blue grama (*Bouteloua gracilis*) occur extensively throughout southern Saskatchewan, but the rolling nature of the terrain creates subtle changes in the composition of prairie cover. On moist sites within the Dark Brown Chernozem soil zone, grasslands are dominated by medium-height speargrass (*Stipa comata*, *S. spartea*) and wheatgrass (*Agropyron dasystachyum*, *A. smithii*). Associated herbs include hairy golden aster (*Chrysopsis villosa*), many-flowered aster (*Aster ericoides*), and long-headed coneflower (*Ratibida columnifera*). Patches of western snowberry (*Symphoricarpos occidentalis*) and wolfwillow (*Eleagnus commutata*) add to the diversity of the plant cover, which, together with rose (*Rosa acicularis*, *R. woodsii*), is common around the periphery of aspen groves.

In the comparatively dry Brown Chernozem soil zone, speargrass and blue grama (*Bouteloua gracilis*) are dominant species, with the latter increasing in abundance on dry knolls or where coarse soils reduce moisture availability. However, the Brown Chernozem soil zone generally is beyond the limits of the parkland, and aspen is only occasionally present.

Over much of the region, trembling aspen grows in pure, isolated groves. Tree growth is enhanced northwards in response to an increasingly humid climate, and eventually the groves coalesce to form continuous woodland. Boreal species, including white spruce (*Picea glauca*) and white birch (*Betula papyrifera*), become increasingly conspicuous in the northern parklands, which give rise to transitional mixedwood forests. In wet–mesic sites, aspen may be found in association with Manitoba maple (*Acer negundo*), American elm (*Ulmus americana*), and green ash (*Fraxinus pennsylvanica*), but these communities typically are replaced by balsam poplar (*P. balsamifera*) where the soil is imperfectly drained. A transition from aspen to shrubby willows (*Salix* spp.) and red osier dogwood (*Cornus stolonifera*) characteristically is found around sloughs and other permanent water bodies. Bur oak (*Quercus macrocarpa*) increases in abundance in the southeastern part of the aspen parkland; basswood (*Tilia americana*) is a distinctive associate of bur oak in southeastern Manitoba.

Several shrub and herb species grow within the dense aspen stands found along the northern fringe of the parkland. Patches of saskatoon (*Amelanchier alnifolia*) are common on well-drained sites. On mesic sites, low-bush cranberry (*Viburnum edule*) is an important species in the tall shrub stratum. Boreal species present in the ground cover include wild sarsaparilla (*Aralia nudicaulis*), northern bedstraw (*Galium boreale*), two-leaved Solomon's-seal (*Maianthemum canadense*), palmate-leaved colt's-foot (*Petasites palmatus*), and bunchberry (*Cornus canadensis*). A gradual change in species composition occurs southwards, where the dry climate favors an open shrub layer. Species such as choke cherry (*Prunus virginiana*) and rose are conspicuous, and dense patches of western snowberry and wolfwillow often surround the groves. Ground cover within the groves varies with soil texture and moisture availability. Maini (1960) recorded 183 plant species within the ground flora of selected aspen stands across Saskatchewan; the most common are listed in Table 25.1. A study of a 130-ha tract of undisturbed prairie in central Saskatchewan (Pylypec 1986) recorded a total of 165 species of plants, of which 91% were native to the region. Introduced, weedy species had invaded the site from the surrounding agricultural land; within the aspen groves, the dominant species in the ground cover was smooth brome (*Bromus inermis*).

Numerous water-filled depressions throughout the parkland provide a unique habitat for plants and wildlife. These potholes vary in size and depth; some are filled permanently with water, whereas others are temporarily flooded

for only a brief period in the spring. This region is referred to by Zoltai and Pollett (1983) as the "Aspen Parkland Prairie Wetland District" and is distinguished from the adjacent "Grassland Prairie Wetland District" by willow (*Salix bebbiana, S. discolor, S. petiolaris*) and aspen bordering the waterbodies. Because of climatic conditions, these wetlands are subject to annual and seasonal fluctuations in water depth; this fluctuation, together with local differences in water chemistry, greatly influences the nature of the vegetation (B. H. Walker and Coupland 1970). Fresh water sloughs generally are surrounded by a zone of willow that may be partly submerged during periods of high water. Aspen grows next to willow on higher ground but is killed by prolonged flooding on low-lying sites (Bird 1961). Aspen and willow are absent from alkaline sloughs, which typically are separated from the prairie by a zone of salt-tolerant species such as seaside arrow grass (*Triglochin maritima*), alkali grass (*Distichlis stricta*), and Nuttall's alkali grass (*Puccinellia nuttalliana*). In extremely saline sites, red goosefoot (*Chenopodium rubrum*), western sea blite (*Suaeda depressa*), and red samphire (*Salicornia rubra*) may be present.

In shallow marshy areas, slough grass (*Beckmannia syzigachne*) and spangletop (*Scholochloa festucacea*) commonly are found in association with sedge, such as *Carex aquatilis* and *C. lanuginosa*, and Baltic rush (*Juncus balticus*). However, the exact composition of the vegetation depends upon the water regime, water chemistry, and land-use practices. Thus, seasonally flooded grazed sites of low salinity may be characterized by tall manna grass (*Glyceria grandis*), bur-reed (*Sparganium eurycarpum*), and water parsnip (*Sium suave*). Sites used for hay may support awned sedge (*Carex atherodes*) and reed canary grass (*Phalaris arundinacea*), with water foxtail (*Alopecurus aequalis*) and swamp persicaria (*Polygonum coccineum*) present in cultivated slough margins (Kantrud, Millar and van der Valk 1989). Deep sloughs are fringed with cat-

Table 25.1. *Average percentage frequency for common species in aspen communities on four soil types in Saskatchewan (after Maini 1960). Frequency is based on 10 quadrats of 0.5 m² on each soil type.*

Species	Soil type			
	Grey Luvisol	Black Chernozem	Dark Brown Chernozem	Brown Chernozem
Herbs				
Achillea lanulosa	5.0	11.6	6.3	2.9
Actaea rubra	1.7			
Anemone canadensis		0.6	18.7	6.4
Aralia nudicaulis	49.1	9.7	5.6	
Artemisia ludoviciana		11.0	4.5	7.9
Aster cordifolius	36.7	25.4	13.3	
Aster spp.	15.0	0.6	6.8	22.9
Campanula rotundifolia		8.7	1.7	2.1
Cerastium arvense	1.7	6.9	1.2	
Cornus canadensis	73.3	6.8	2.5	
Epilobium angustifolium	37.5	4.2	4.5	
Fragaria glauca	33.3	31.4	27.3	1.4
Galium boreale	35.8	59.0	43.7	14.3
Glycyrriza lepidota				28.6
Hedysarum alpinum		17.7	10.6	12.9
Lathyrus ochroleucus	50.0	23.1	3.7	
Maianthemum canadense	52.5	11.1	3.9	
Mertensia pilosa	29.2			
Mitella nuda	18.3			
Petasites palmatus	47.5			
Pyrola asarifolia	50.8	11.1		
Sanicula marilandica	18.3	1.2		
Senecio spp.		7.8		
Smilacina stellata	18.3	16.0	7.7	15.7
Solidago canadensis	2.5	3.1	8.5	1.9
Solidago spp.	2.3		4.3	
Thalictrum spp.	19.2	3.2	13.9	
Trientalis borealis	20.0			
Vicia americana	22.5			
Vicia spp.		1.2	7.2	
Viola rugulosa	25.8	7.5	2.2	
Zizia aptera		2.3	8.3	5.0
Grasses and sedges				
Agropyron dasystachyum	5.0	17.5	2.4	20.7
Agropyron smithii			8.0	
Agropyron trachycaulum	30.0	21.1	3.5	1.4
Agropyron subsecundum	32.5	29.0	29.7	0.7
Calamagrostis canadensis	10.8	5.5	3.6	
Calamagrostis spp.	21.7	1.2		
Carex spp.	34.2	80.6	66.8	82.9
Elymus spp.			3.0	
Festuca altaica		24.8	12.4	
Juncus ater			11.6	
Juncus spp.				20.0
Koeleria cristata	1.7		2.7	22.9
Muhlenbergia cuspidata			2.5	7.1

(cont.)

Table 25.1 (cont.)

Species	Grey Luvisol	Black Chernozem	Dark Brown Chernozem	Brown Chernozem
			Soil type	
Shrubs				
Amelanchier alnifolia	19.1	21.7	4.7	
Arctostaphylos uva-ursi	9.1	3.9	6.1	2.1
Eleagnus commutata		9.8	6.4	
Juniperus horizontalis		1.2	13.3	19.4
Ledum groenlandicum	4.2			
Linnaea borealis	42.5			
Lonicera dioica		7.0	1.8	
Prunus pensylvanica	1.7	12.5	0.8	3.5
Prunus virginiana		9.0	4.0	9.3
Rosa acicularis	0.8	10.1	17.4	2.9
Rosa woodsii	62.5	49.9	27.6	34.3
Rubus pubescens	60.8		1.7	
Spiraea alba			22.6	
Symphoricarpos albus		10.0	14.8	
Symphoricarpos occidentalis	8.3	52.4	46.2	34.3
Viburnum edule	32.5	12.5	2.2	

tail (*Typha latifolia*) and bulrush (*Scirpus validus*). These emergent plants grow in water up to one meter deep, but are replaced by submersed or floating aquatic plants where open water persists throughout the year. Pondweed (*Potamogeton richardsonii*), water crowfoot (*Ranunculus subrigidus*), and water milfoil (*Myriophyllum exalbescens*) typically are present as part of the rooted aquatic flora of freshwater and brackish sloughs, along with floating duckweed (*Lemna minor*) and bladderwort (*Utricularia vulgaris*). Increasing salinity favors macrophytic species, such as horned-pondweed (*Zannichellia palustris*), sago pondweed (*Potamogeton pectinatus*), and, less commonly, ditch grass (*Ruppia maritima*).

Aspen Groves

Physiognomic Characteristics

Aspen is a small, fast-growing tree, rarely exceeding 60 years of age in the parkland region (Bird 1961). In the northern transitional forest areas, the average height of dominant aspen trees is about 23 m, with a basal diameter of 42 cm (Maini 1960), but tree size decreases southward. Thus, in the Black Chernozem soil zone, aspen grows to a height of 8–11 m and a basal diameter of 11–17 cm. The controlling effect of topography and soil texture on moisture availability becomes increasingly important in the growth and distribution of aspen in the driest parts of the parkland. On medium-textured soils in the Dark Brown Chernozem soil zone, aspen grows mainly in depressions and around the margin of sloughs. Heights of 12 m are reported for trees growing on favorable sites, compared to only 4–5 m on dry slopes (Maini 1960). In the Brown Chernozem soil zone, representative heights range from 9 to 12 m, with basal diameter ranging from 15 to 18 cm.

There is a strong correlation between the height and age of aspen trees in the parkland region (Maini 1960). Thus, the general form of a grove provides some evidence of its past history. Flat-topped aspen groves are even-aged and originated vegetatively following the destruction of previous stands. In undisturbed grassland, a distinct age gradient develops from the center of a grove to the periphery as young trees continue to expand the grove into the surrounding land. As groves expand, they develop a distinctive dome shape as trees become progressively shorter toward the edge. Aspen stands do not stagnate from overstocking because periods of accelerated mortality result in natural thinning (Peterson and Peterson 1992). In the parkland region, this mortality typically results in tree death near the center of the grove. Analysis of spatial patterns in old, established aspen groves suggests that three zones can be distinguished (Archibold and Wilson 1978). The periphery of the grove represents a zone of active vegetative regeneration characterized by an abundance of young trees. Toward the center, the number of young trees steadily declines, and the number of dead trees and stumps increases. At the center core area, a few old trees persist, but typically this is an area where younger stems are developing.

Regeneration

Although aspen is a prolific seed producer, the seeds quickly lose viability (Moss 1938) and few seedlings become established. Consequently, regeneration generally is by root suckering. Thus, most groves are clonal in origin, as shown by morphological, phenological, and physiological characteristics such as leaf size and shape, time of leaf flush and leaf fall, and susceptibility to decay (Peterson and Peterson 1992).

In groves, aspen trees typically develop extensive lateral root systems radiating as much as 12–15 m from the trunk. Tap roots penetrate to depths of nearly 2 m (Maini 1960). Suckers normally originate along shallow horizontal lateral roots 0.5–2.5 cm in diameter. Trees growing outside the groves characteristically develop only lateral roots, some of which can be traced back to older trees within the grove. In the absence of a major disturbance, only a limited number of suckers are formed at the periphery of an aspen grove. Prolific suckering occurs during the first growing season following a major disturbance such as fire, probably because increased insolation raises soil temperature following the death of the mature trees whose presence also would normally inhibit new shoot development (Maini and Horton 1966). Suckering also is promoted by the return to favorable conditions following a period of drought. However, natural thinning of suckers occurs and their density rapidly declines in the first few years.

Young trees at the grove margins indicate a continuing expansion into adjacent grasslands. A comparison of vegetation between 1907 and 1966 in south-central Alberta indicated that aspen cover increased at an annual rate of 0.05% (Bailey and Wroe 1974). On fertile soils in Saskatchewan, the rate of invasion ranges from 42 to 76 cm yr^{-1}, depending on such factors as aspect, topographic gradient, and soil texture (Maini 1960). In rolling terrain, aspen groves typically are restricted to moist depressions and tend to encircle water bodies, with little encroachment onto drier, upslope sites. Expansion toward sloughs is limited by excessive moisture, and in wet years some trees may be killed. On these sites, the groves respond to climatic cycles, and their boundaries may move upslope or downslope depending on available soil moisture. The widespread dieback of aspen groves that occurred in the mid-1960s in the parkland of Saskatchewan was linked to severe drought and the prevalence of the endemic canker caused by *Cytospora chrysosperma* (Dix and Steen 1968).

Agricultural Impact

The development and morphology of aspen groves are altered by agricultural practices. Where groves adjoin arable land, repeated cultivation tends to suppress suckers, resulting in an abrupt transition to older trees. The invasion of pasture land by aspen also is a concern to farmers, and control methods have been tested to prevent grove redevelopment. Usually, the groves are cleared by pushing the trees into windrows and then the land is broken with discs or plows. In trials carried out in central Alberta, regrowth of aspen suckers was reduced by 58% in areas where the land was subsequently seeded to forage species. Regrowth of native shrubs was reduced by 22%–85% depending on the species (Bailey 1972). Further control of aspen was achieved by applications of the herbicide 2,4-D. Repeated burning in the spring followed by 2,4-D also was an effective way of reducing the density of aspen suckers on rangeland (Bailey and Anderson 1979). However, when prescribed spring burning or herbicide application alone is used to kill the mature aspen trees, suckers quickly reinvade.

The invasion of grassland by aspen is a problem for ranchers and farmers because it may reduce available herbage by as much as 90% (Bailey and Wroe 1974). However, aspen suckers are palatable to cattle and may provide supplemental forage as long as leaves remain accessible. Regeneration studies following spring burning indicate that aspen suckers defoliated by cattle late in the growing season, but prior to leaf fall, are nearly eliminated (Bailey, Irving and Fitzgerald 1990). Conversely, when grazing occurs early in the growing season, suckers continue to regenerate, and 6–8 years may elapse before food reserves are totally exhausted. In undisturbed groves, the exposed marginal trees typically have branches along most of the trunks, whereas in heavily grazed pasture, many lower branches are removed. The loss of branches, together with the absence of young suckers, results in an open grove appearance. However, unpalatable species such as snowberry, which also vegetatively regenerate after fire, usually increase in density with grazing and may alter grove appearance.

Parkland Fauna

Each of the dominant plant communities within the aspen parkland provides different habitats for wildlife, although human activities have substantially altered the range and abundance of native animal species. Aspen parkland originally supported large populations of herbivores including bison, elk, and deer. Prior to settlement, the dominant species was bison, which, through grazing and trampling, maintained the area as open grassland (Roe 1970). Other grassland ungulates, such as pronghorn antelope (*Antilocapra americana*) and white-tailed deer (*Odocoileus virginianus*), were able to extend their ranges well into the parkland. Elk (*Cervus canadensis*) also browsed and grazed the aspen groves. The white-tailed jack rabbit (*Lepus townsendii*) similarly is found throughout the parkland. It feeds on grasses and herbs in the summer and on twigs, buds, and bark in the winter; the last-named activity often is damaging to shrubs and trees because of girdling. The thirteen-lined ground squirrel (*Citellus tridecemlineatus*) is common among the grasses with other small mammals, including the deer mouse (*Peromyscus maniculatus*) and meadow vole (*Microtus pennsylvanicus*), that inhabit the thatch and litter layers. Mounds of earth excavated from underground burrows indicate the presence of colonies of the pocket gopher (*Thomomys talpoides*), which is the principal food of the badger (*Taxidea taxus*). Other predatory species include the coyote (*Canis latrans*), red fox (*Vulpes vulpes*), and weasel (*Mustela* spp.)

The only resident bird species native to the grassland communities is the sharp-tailed grouse (*Pediocetes phasianellus*), although it tends to be more abundant near woodland, where it seeks shelter in winter. Migratory grassland species include the horned lark (*Eremophila alpestris*), western meadowlark (*Sturnella neglecta*), and chestnut-collared longspur (*Calcarius ornatus*). The horned lark is one of the first species to return to the grasslands, typically arriving in late February, compared to late March or early April for the western meadowlark, and late April for the chestnut-collared longspur (Maher 1973). These species feed on grasshoppers, cut-worms, and other insects during the summer, then change to a diet of seeds in the fall and remain in the parkland until October. Density of these species is highest on grazed pasture, which again suggests the important role that bison once played in the ecosystem. The cow-bird (*Molothrus ater*) followed the bison herds and fed on the attendant insects. However, the cowbird lays its eggs in nests of other species, and this parasitic habit potentially is an important source of mortality in other grassland birds. Species such as the vesper sparrow (*Pooecetes gramineus*) and clay-colored sparrow (*Spizella pallida*) are more abundant on prairie adjacent to woodland than in the woodland. During migration, flocks of Lapland longspur (*Calcarius lapponicus*) are observed in the grasslands, and in winter, snow buntings (*Plectrophenax nivalis*) feed on the seeds of grasses protruding above the snow (Bird 1961).

Common birds of prey include the red-tailed hawk (*Buteo jamaicensis*) and Swainson's hawk (*Buteo swainsoni*). These species typically nest in tall trees but hunt over the grasslands, where they feed primarily on small rodents. The marsh hawk (*Circus cyaneus hudsonius*) nests on the ground in patches of tall grass or snowberry. Tall grass also is used by the ground-nesting short-eared owl *(Asio flammeus)*. Both these species feed principally on mice. All of these birds are summer residents. In winter, the snowy owl (*Nyctea scandiaca*) is occasionally seen.

Breeding bird surveys indicate that because of the ecotonal nature of the region, species diversity and density are greater in the aspen parkland than in most other habi-tats in North America (Johns 1993). Several species appear to have minimum grove size requirements; for example, the clay-colored

and vesper sparrows are present in groves of all sizes; the mourning dove (*Zenaida macroura*) and woodpeckers (*Picoides pubes-cens, P. villosus*) occur in groves exceeding 1 ha; and the hermit thrush (*Catharus guttatus*) is found only in groves larger than 24 ha. Analysis of niche requirements suggests that small groves are dominated by short-dis-tance migrating omnivores, with larger groves used mainly by long-distance migra-tory insectivores.

The aspen groves provide a microclimatic refugium for a number of wildlife species on the Canadian prairies. Mule deer (*Odocoileus hemionus*) and elk once were common in the aspen groves, but these species largely have been replaced by white-tailed deer. The solitary porcupine (*Erethizon dorsatum*) still is common in the woodland, where it remains active throughout the year. In spring and summer, it consumes tree flowers, leaves, and young plants, but in fall and winter, it peels off tree bark and feeds on the sugar and starch in the sap beneath. The striped skunk (*Mephitis mephi-tis*) is common around the grove margins; it is a predatory animal that feeds mostly on bird eggs and young nestlings. Scattered groves also are the preferred habitat for Franklin's ground squirrel (*Citellus franklinii*). They feed mainly on green leaves and stems in the spring and summer, and later on seeds, but also will consume insects, earthworms, eggs, and occa-sionally smaller rodents and birds. The ground squirrels are true hibernators and are well adapted to survive winter in deep burrows.

The dominant mammal in the sloughs and marshes is the muskrat (*Ondatra zibethicus*). It eats the roots and bulbs of many plants, but cattail and bulrushes are preferred; occasion-ally it feeds on frogs, clams, and fish (Fritzell 1989). The tough leaves and stems of both species are used to construct houses and feeding platforms, which often are used as basking sites by ducks and as nest sites by the black tern (*Chlidonias niger*). Beaver (*Castor canadensis*) is found throughout the parkland, but generally is associated with streams and rivers, where it feeds on aquatic plants and on the bark of deciduous trees and shrubs, especially aspen, cottonwood, and willow. Muskrat tunnels and abandoned lodges provide den sites for other mammals, such as meadow voles, raccoons (*Procyon lotor*), and mink (*Mustela vison*). The omnivorous raccoon feeds on leafy plant materials in spring, but is increasingly dependent on aquatic insects, snails, and especially waterfowl and their eggs during summer. Mink are well adapted to the wetland habitat; they are the most common carnivore in the wetlands and usually prey on small mammals and marsh birds at night.

Wetland areas support large, diverse bird populations. One of the commonest species is the red-winged blackbird (*Agelaius phoeniceus arctolegus*), which builds its nests in cattails and willows. Also common is the yellow-headed blackbird (*Xanthocephalus xanthocephalus*), which nests in the bulrushes and reed grass that grow around open water bodies. The killdeer (*Charadrius vociferous*) is a common shore bird inhabiting exposed mud-flats around sloughs (Bird 1961).

The prairie wetlands also provide ideal habitat for a variety of ducks. Dabbling ducks such as mallard (*Anas platyrhynchos platyrhynchos*), northern pintail (*A. acuta*), and blue-winged teal (*A. discors*) usually are found in shallow sloughs and seasonally flooded wetlands where patches of open water are interspersed with emergent vegetation. Mallards prefer to nest in tall, dense cover, especially patches of western snowberry and rose. Blue-winged teal usually nest in grassy vegetation. Dead thatch in patches of native grasses such as green needlegrass (*Stipa viridula*) and western wheatgrass provide ideal nest sites. However, introduced grasses such as Kentucky bluegrass (*Poa pratensis*) and smooth bromegrass frequently are selected (Swanson and Duebbert 1989). Diving ducks such as canvasback (*Aythya valisineria*), redhead (*Aythya americana*), and lesser scaup (*Aythya affinis*) usually are found in permanent deep-water areas. Nest sites for diving ducks usually are found among cattails, bulrushes, or other tall emergent plants, especially where the vegetation cover is patchy; occasionally, upland sites are used. Nests built over water have a low chance of predation, although mink and raccoons often take breeding females and eggs. Predatory losses vary with the seasonal life cycle of the ducks, with losses increasing during incubation and molt; ducklings also are vulnerable and many are lost to marsh hawks.

Human Impact on the Parkland

Records of human habitation in the parkland extend back 7,000–8,000 years, and include campsites, burial sites, bison (*Bison bison*) kills, and processing sites. The Plains Indians culture was based on the bison, which not only provided food, clothing, shelter, and tools, but also was an important element in social, economic, and political patterns. Hunting of bison varied from single encounters to communal events that resulted in mass kills. The bison had several traits that made it susceptible to mass hunting. They congregated in large herds, had poor eyesight, and followed lead animals. Controlled stampeding of bison herds was so successful that quantities of surplus meat were conserved and traded by Native Americans to the early European trappers and explorers (Linnamae 1988). Bison were especially abundant in the parkland during fall and winter, and here, the preferred method of capture was in a pound. A pound consisted of a 1.5–2.5-m-high containment fence made of spaced trees or posts interwoven with brush and branches. The enclosed space typically was 60–100 m wide and equipped with a 3–6-m-wide ramp. Often it was flanked with fencing that narrowed from the drive lane toward the pound entrance. The technique was well established by about A.D. 1000 and, as reported in early journals, continued to be used into recent historic times.

Archaeological dating of bone beds at Wanuskewin Heritage Park near Saskatoon indicates that intensive human use of bison began in the central aspen parkland about 1,500 years ago (E. G. Walker 1988). In the early 1800s, as many as 70 million bison roamed the Great Plains, usually in large herds. However, by the late 1880s, over-hunting had reduced the bison population almost to extinction and ended the unique life of the Plains Indian. The demise of the bison has been linked to the expansion of the aspen groves in historic times (Campbell et al. 1994). Analysis of fossil pollen from sites in Saskatchewan and Alberta indicates that the spread of aspen preceded European settlement by at least a decade (McAndrews 1988). The activities of the bison, such as trampling and toppling mature trees, would have restricted the expansion of aspen groves. Thus, extirpation of the bison, the accompanying exodus of the Plains Indian, and the changing fire regime may have been important factors influencing the present distribution of aspen groves in the prairie.

Prior to A.D. 1700, human activities had little effect on the parkland. Apart from bison hunts, there is evidence that the indigenous populations used fire in a precautionary manner; grassland often was burned while snow remained in adjacent woodland in order to reduce timber losses later in the year. Natural fire starts were invariably caused by lightning. In southern Saskatchewan, the combination of suitable fuel and "dry storm" weather occurs perhaps once every 6 years (Rowe 1969), possibly often enough to favor grassland. Historically, extensive areas probably were affected. However, fire frequency substantially increased in the early period of settlement because of homesteader activities and railway development. As a result, with the enactment of fire ordinances in the late 1890s, plowed fireguards began to be constructed. By 1913, 6,500 km of guards were in place in the Prairie provinces, mostly in the drier grasslands. Fireguards, breaking land for arable crops, and eventually the construction of grid roads produced the familiar checkerboard landscape in the prairie and substantially reduced the spread of fire. Lower fire frequency, combined with the loss of the bison, resulted in an expansion of aspen in the early 1900s, and since the 1930s, appearance of the wooded landscape has been enhanced by planted shelterbelts.

Problems and Threats

Most of the parkland now is intensively used for arable crops, principally wheat and barley; the remaining grassland is seeded to improved grass species and is strongly affected by grazing. Grassland species are classified according to their response to grazing (Dyksterhuis 1949). Palatable species, such as wheatgrass and fescue, decrease in abundance and vigor with repeated defoliation. This decrease favors the spread of increasers such as blue grama, pasture sage, and many shrubs that are less desirable to livestock. Overgrazing favors the spread of weedy invaders, such as crested wheatgrass (*Agropyron cristatum*) and smooth brome. Improper grazing practices that result in bare patches of soil allow annual weeds to become established. The source of the weed seeds ultimately can be traced to cultivated or disturbed land infested with a wide range of species, including lamb's-quarters (*Chenopodium album*), stinkweed (*Thlaspi arvense*), and thistle (*Cirsium arvense*). Noxious species such as leafy spurge (*Euphorbia esula*) also are present. Weeds of cultivated areas also have invaded the aspen groves, thereby altering the composition of the ground cover. In addition, European buckthorn (*Rhamnus cathartica*), a tall shrub introduced into the Saskatoon district in the 1930s as a potential shelterbelt species, has invaded local aspen groves, shrub communities, and riparian woodland (Archibold et al. 1997).

Changes in animal populations as a result

of human alteration of the parkland habitat are equally apparent. The demise of the bison led to the loss of food and extirpation of the wolf and grizzly bear. Species such as the elk were driven from the parkland, and the thirteen-lined ground squirrel and Franklin's ground squirrel became less abundant as their native habitat was destroyed. Similarly, duck populations have been affected by wetland drainage, increased tillage of wetlands, and the impact of agrochemicals on aquatic invertebrates. Conversely, some species have increased in abundance as a result of human settlement. For example, the herring gull (*Larus argentatus*) and Franklin's gull (*L. pipixcan*) often are seen around garbage dumps, and the house sparrow (*Passer domesticus*) now is ubiquitous.

Although the demand for agricultural land continues to increase, there is a growing awareness that adverse environmental changes have occurred in the parkland. Consequently, various efforts are being undertaken to minimize impacts or reverse deleterious trends. Native grassland not only provides a source of forage, but it also serves as a refuge for rare and endangered species. Grassland restoration projects using locally harvested native seeds are becoming more widespread, and wetland preservation and restoration of wetland are extensive. Reintroduction of birds such as the burrowing owl (*Athene cunicularia*) and development of nest sites to encourage species such as ferruginous hawks are occurring throughout the region. These activities increase the biodiversity of the parkland and help restore some of the natural elements to this greatly altered ecosystem.

Summary

In its natural state, aspen parkland consists of two major plant communities, woodland and grassland, intermingled in an irregular patchwork and interspersed with numerous wetlands. Grasslands predominate in the southern part of the parkland; groves of trembling aspen trees increase in size and abundance northward until they are eventually replaced by the mixedwood communities of the southern boreal forest. Today, much of the parkland is used for intensive grain production. Consequently, native grassland and aspen groves now are found only where the land is not suitable for cultivation. Significant changes in wildlife occurred during the period of agricultural settlement, and habitat loss continues to threaten species in this diverse ecotonal region.

REFERENCES

Anderson, D. W. 1987. Pedogenesis in the grassland and adjacent forests of the Great Plains. *Advances in Soil Science* 7:53–93.

Archibold, O. W., Brooks, D., and Delano, L. 1997. An investigation of the invasive shrub European buckthorn, *Rhamnus cathartica* L., near Saskatoon, Saskatchewan. *Canadian Field Naturalist* 111:617–621.

Archibold, O. W., and Wilson, M. R. 1978. Spatial patterns and population dynamics of *Populus tremuloides* in a Saskatchewan aspen grove. *Canadian Field Naturalist* 92:369–374.

Archibold, O. W., and Wilson, M. R. 1980. The natural vegetation of Saskatchewan prior to agricultural settlement. *Canadian Journal of Botany* 58:2031–2042.

Axelrod, D. I. 1985. Rise of the grassland biome, central North America. *Botanical Review* 51:163–201.

Bailey, A. W. 1972. Forage and woody sprout establishment on cleared, unbroken land in central Alberta. *Journal of Range Management* 25:119–122.

Bailey, A. W., and Anderson, H. G. 1979. Brush control on sandy rangelands in central Alberta. *Journal of Range Management* 32:29–32.

Bailey, A. W., Irving, B. D., and Fitzgerald, R. D. 1990. Regeneration of woody species following burning and grazing in aspen parkland. *Journal of Range Management* 43:212–215.

Bailey, A. W., and Wroe, R. A. 1974. Aspen invasion in a portion of the Alberta parklands. *Journal of Range Management* 27:263–266.

Bird, R. D. 1961. *Ecology of the Aspen Parkland of*

Western Canada in Relation to Land Use. Publication 1066. Ottawa: Research Branch, Agriculture Canada,

Borchert, J. R. 1950. The climate of the central North American grassland. *Annals of the Association of American Geographers* 61:1–39.

Campbell, C., Campbell, I. D., Blyth, C. B., and McAndrews, J. H. 1994. Bison extirpation may have caused aspen expansion in western Canada. *Ecography* 17:360–362.

Canadian Soil Survey Committee. 1978. *The Canadian System of Soil Classification.* Publication No. 1646. Ottawa: Research Branch, Agriculture Canada.

Dix, R. L., and Steen, O. A. 1968. Aspen dieback in the parklands of Saskatchewan. *Bulletin of the Ecological Society of America.* Late Spring: 81.

Dyksterhuis, E. J. 1949. Condition and management of range land based on quantitative ecology. *Journal of Range Management* 2:104–115.

Fritzell, E. K. 1989. Mammals in prairie wetlands. In *Northern Prairie Wetlands,* ed. A. van der Valk, pp. 268–301. Ames, Iowa: Iowa State University Press.

Fuller, L. G., and Anderson, D. W. 1993. Changes in soil properties following forest invasion of black soils of the aspen parkland. *Canadian Journal of Soil Science* 73:613–627.

Hildebrand, D. V., and Scott, G. A. J. 1987. Relationships between moisture deficiency and amount of tree cover on the pre-agricultural Canadian Prairies. *Prairie Forum* 12:203–216.

Johns, B. W. 1993. The influence of grove size on bird species richness in aspen parklands. *Wilson Bulletin* 105:256–264.

Kantrud, H. A., Millar, J. B., and van der Valk, A. G. 1989. Vegetation of wetlands of the prairie pothole region. In *Northern Prairie Wetlands,* ed. A. van der Valk, pp. 132–187. Ames, Iowa: Iowa State University Press.

Knapp, A. K., and Seastedt, T. R. 1986. Detritus accumulation limits productivity of tallgrass prairie. *BioScience* 36:662–668.

Linnamae, U. 1988. The Tschetter site: a prehistoric bison pound in the parklands. In *Out of the Past: Sites, Digs and Artifacts in the Saskatoon Area,* ed. U. Linnamae and T. E. H. Jones, pp. 91–115. Saskatoon: Saskatoon Archaeological Society.

Looman, J. 1979. The vegetation of the Canadian Prairie Provinces: 1. An overview. *Phytocoenologia* 5:347–366.

Looman, J., and Best, K. F. 1979. *Budd's Flora of the Canadian Prairies.* Publication No. 1662. Ottawa: Research Branch, Agriculture Canada, Ontario.

Maher, W. J. 1973. *Birds: I. Population Dynamics.* Technical Report No. 34, Matador Project. Saskatoon: University of Saskatchewan.

Maini, J. S. 1960. Invasion of grassland by *Populus tremuloides* in the northern Great Plains. Ph.D. dissertation, University of Saskatchewan, Saskatoon.

Maini, J. S., and Horton, K. W. 1966. Vegetative propagation of *Populus* spp. I. Influence of temperature on formation and initial growth of aspen suckers. *Canadian Journal of Botany* 44:1183–1189.

Maybank, J., Bonsal, B., Jones, K., Lawford, R, O'Brien, E. G., Ripley, E. A., and Wheaton, E. 1995. Drought as a natural disaster. *Atmosphere–Ocean* 33:195–222.

McAndrews, J. H. 1988. Human disturbance of North American forests and grasslands: the fossil pollen record. In *Vegetation History,* ed. R. Huntley and T. Webb, pp. 673–697. Dordrecht: Kluwer Academic Publishers.

Moss, E. H. 1938. Longevity of seed and seedlings in species of *Populus. Botanical Gazette* 99:529–542.

Peterson, E. B., and Peterson, N. M. 1992. *Ecology, Management and Use of Aspen and Balsam Poplar in the Prairie Provinces of Canada.* Special Report 1. Edmonton: Northern Forestry Centre, Forestry Canada.

Pylypec, B. 1986. The Kernen Prairie – a relict fescue grassland near Saskatoon, Saskatchewan. *The Blue Jay* 44:222–231.

Pyne, S .J. 1982. *Fire in America: A Cultural History of Wildlife and Rural Fire.* Princeton, N.J.: Princeton University Press.

Roe, F. G. 1970. *The North American Buffalo: A Critical Study of the Species in Its Wild State.* Toronto: University of Toronto Press.

Rowe, J. S. 1969. Lightning fires in Saskatchewan grassland. *Canadian Field-Naturalist* 83:317–324.

Sauer, C. O. 1950. Grassland climax, fire and man. *Journal of Range Management* 3:16–21.

Stewart, O. C. 1956. Fire as the first great force employed by man. In *Man's Role in Changing the Face of the Earth,* ed. W. L. Thomas, pp. 115–133. Chicago, Ill.: University of Chicago Press.

Swanson, G. A., and Duebbert, H. F. 1989.

Wetland habitats of waterfowl in the prairie pothole region. In *Northern Prairie Wetlands*, ed. A. van der Valk, pp. 228–267. Ames, Iowa: Iowa State University Press.

Walker, E. G. 1988. The archaeological resources of the Wanuskewin Heritage Park. In *Out of the Past: Sites, Digs and Artifacts in the Saskatoon Area*, ed. U. Linnamae and T. E. H. Jones, pp. 75–89. Saskatoon: Saskatoon Archaeological Society.

Walker, B. H., and Coupland, R. T. 1970. Herbaceous wetland vegetation in the aspen grove and grassland regions of Saskatchewan. *Canadian Journal of Botany* 48:1861–1878.

Walter, H. 1973. *Vegetation of the Earth in Relation to Climate and the Eco-physiological Conditions*. New York: Springer-Verlag.

Zoltai, S. C., and Pollett, F. C. 1983. Wetlands in Canada: their classification, distribution, and use. In *Mires: Swamp, Bog, Fen and Moor*, ed. A. J. P. Gore, pp. 245–268. Amsterdam: Elsevier.

26   Subarctic Lichen Woodlands

E. A. JOHNSON AND K. MIYANISHI

Introduction

Between the tundra and the closed-canopied boreal forest is a vast region of largely undisturbed, open-canopied conifers with a ground cover of lichens (Figure 26.1). This region has been called various names in the North American literature: subarctic forest, lichen woodland, hemiarctic, and spruce woodland. Refer to Löve (1970) and Blüthgen (1970) for the semantics of these terms.

The subarctic woodlands are primarily climatic in origin, occupying the transitional areas between the summer location of Arctic airstreams and more southern airstreams. As in all transition zones, small variation in the primary forcing variable changes the vegetation composition and structure. This change in airstream position interacts with the available plants, landforms, elevations, substrates, and fire regime to create a vegetation mosaic reflecting past and present interactions.

The North American subarctic woodland is one of the last remaining extensive and continuous ecosystems (Figure 26.2a). It covers approximately 2 million square kilometers, most of which is without roads or permanent settlements. Hunting, trapping, and some mineral extraction are the primary land uses. The subarctic has a population of indigenous people whose way of life is still strongly related to the land. Furthermore, during part of the year it is the home of caribou (*Rangifer tarandus*), the last large migratory ungulate herd in North America.

Distribution

The subarctic woodland in North America (Figure 26.2a) occurs from the northern interior of Alaska, cutting through the Richardson Mountains and north of the Mackenzie Mountains, to the south end of Hudson Bay. East of Hudson Bay, the woodland extends north toward the coast of Labrador and Quebec.

The general position of the subarctic woodland coincides with the mean position of the Arctic front in summer (Figure 26.2b). The Arctic front separates maritime and continental Arctic airstreams from Pacific, Atlantic, and tropical airstreams. This baroclinic structure is a characteristic of the mean westerly air flow and influences the large-scale dynamics of climate, particularly precipitation, and temperature patterns. Thus, the subarctic is positioned between the colder, drier climate of the tundra and the warmer, wetter climate of the boreal forest (Bryson 1966; Barry 1967; Krebs and Barry 1970; Barry and Hare 1974).

Although large-scale climate patterns determine the general distribution pattern of the subarctic woodland, regional features also have an effect (Figure 26.2a). The woodland does not reach the west coast in Alaska, which is covered instead by cool, wet tundra due to the influence of the cold, extensively ice-covered Bering Sea. Similarly, the subarctic zone narrows and is forced southward around Hudson Bay because of the influence

Roger C. Anderson, James S. Fralish, and Jerry M. Baskin, eds. *Savannas, Barrens, and Rock Outcrop Plant Communities of North America.* Copyright © 1999 Cambridge University Press. All rights reserved.

Figure 26.1. Subarctic lichen woodland in the Northwest Territories, Canada. The trees are black spruce (*Picea mariana*) and white birch (*Betula papyrifera*). Note the layering around the base of the black spruce and basal sprouts around the white birch. The white ground cover consists of lichens, primarily *Stereocaulon paschale*, *Cetraria* spp., and *Cladonia* spp. The darker ground cover is mostly ericaceous shrubs, particularly *Ledum groenlandicum* and *Empetrum nigrum*.

of this large, cold mass of water. East of Hudson Bay, the woodland extends northward in Quebec and Labrador due to the diminishing effect of the cold water and due to the maritime influence of the north Atlantic. The influence of mountain ranges is indicated by the bend in the subarctic zone around the Northern Cordillera in the Northwest Territories.

In delineating the subarctic woodland, an important consideration is its transitional nature from forest to tundra. For example, Timoney et al. (1992) used the ratio of trees to upland tundra cover to define the northern section of the subarctic. Most "tree line" maps generally are based on this approach. Hustich (1966, 1979), for example, defined transition from forest to tundra in terms of the economic forest line (limit of sexually reproduc-ing trees), the physiognomic forest line (limit of vegetatively reproducing trees), the tree line (limit of tree growth form), the tree species line (limit of all species that normally grow as trees), and the historical tree line. The physiognomic forest line, tree line, and tree species line are observable on air photos and are persistent because of tree longevity, whereas the economic forest line changes from year to year. The historical tree line is determined from paleoecological evidence such as macrofossils and microfossils (Payette and Gagnon 1979; Ritchie 1984, 1987).

For convenience, the divisions of the subarctic recognized by the Ecoregions Working Group (1989) are used here. This classification (Figure 26.2a) is a compromise between a large number of divisions needed to reflect the transitional nature of the region and a

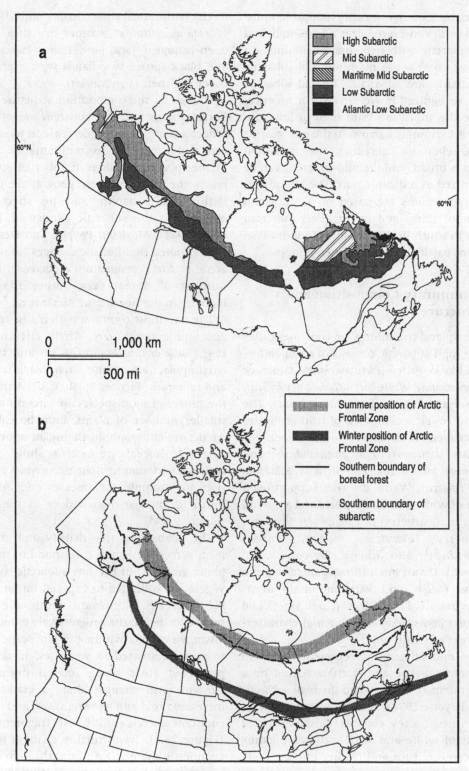

Figure 26.2. (a) The range of the subarctic lichen woodland in Canada. (Redrawn from Ecoregions Working Group 1989.) (b) The coincidence of the subarctic lichen woodlands with the summer position of the Arctic front. (Redrawn from Barry 1967.)

reasonable number of easily mappable units. The classification recognizes a high, mid, and low subarctic with two smaller subunits on the east coast – the maritime mid-subarctic and Atlantic low subarctic. The mid-subarctic is not recognized in western North America, where the transition from high to low subarctic is extremely narrow. In the east (northern Quebec to Labrador), the transition region is broad, and the mid-subarctic can be recognized as a distinct unit. The boundaries of these regions are based on density and height of trees, and on the decrease from north to south in the amount of tundra vegetation, particularly on exposed hilltops.

Community Composition and Structure

The upland community on cool, mesic sites of the high subarctic consists of an open forest of black spruce (*Picea mariana*), tamarack (*Larix laricina*), white birch (*Betula papyrifera*), and some white spruce (*Picea glauca*). The ground cover is composed of feather mosses (*Pleurozium*, *Hylocomium*, and *Dicranum*), ericaceous shrubs (*Ledum groenlandicum* and *Vaccinium vitis-idaea*), and lichens (*Cladonia* and *Peltigera*). Warm, dry sites support open forest of white and black spruce and a ground cover of bearberry (*Arctostaphylos uva-ursi*), crowberry (*Empetrum nigrum*), mosses (*Polytrichum*), and lichens (*Stereocaulon* and *Cladonia*). Dwarf birch (*Betula glandulosa*) and willow (*Salix* spp.) with mountain avens (*Dryas integrifolia*), sedges (*Carex* spp.), and lichens (*Stereocaulon* and *Cladonia*) characterize areas of shrub tundra.

The mid-subarctic is represented best in northern Quebec and is characterized by a higher density of trees than the high subarctic (e.g., Payette 1976). Within the low subarctic, cool, mesic sites support closed-canopied stands of white and black spruce containing ericaceous shrubs and feather mosses. Cool, dry–mesic sites are covered by open-canopied stands of black spruce and white

birch with ericaceous shrubs and lichen (*Cladonia*), whereas warmer dry sites have open-canopied jack pine (*Pinus banksiana*) and black spruce woodlands with a ground cover of lichen (*Stereocaulon*).

Regionally, the composition, structure, and floristic range of the subarctic vegetation reflect a south-to-north decrease in temperature and a maritime-to-continental decrease in precipitation. A large number of species reach the limit of their geographic range within the subarctic and no species is endemic to the subarctic (Rousseau 1968; Cody 1971). Morisset, Payette and Deshaye (1983) noted that floristic changes from subarctic to Arctic result from a decrease in the number of boreal taxa, rather than an increase in the number of Arctic taxa.

The two most common patterns in species geographic ranges are Arctic circumpolar (e.g., *Salix arctica*; Figure 26.3a) and boreal circumpolar (e.g., white birch, black spruce, and tamarack; Figures 26.3b, c, d). Although the ranges of most species are circumpolar, a smaller number of plants, both boreal and Arctic, are either amphi-Beringian or amphi-Atlantic. Boreal trees also show some regional variation in their northward extension; for example, tamarack extends further north in Quebec than it does in the west (Figure 26.3d).

The changes in tree density and growth form across the region correspond to the climatic gradient. In the low subarctic, tundra vegetation is limited to exposed hilltops and high plateaus. Northward within the high subarctic, the tundra progressively extends to lower elevations within the landscape until trees are restricted to wet sites, usually in protected areas along incised drainages. Growth form changes along a gradient of increasing cold and blowing snow, and a concomitant increase in forest fragmentation (Figure 26.4). Tree structure changes from a symmetric, open growth form through a progressive series of flag, verticil, skirted, fruticoide, and mat growth forms (Payette 1974).

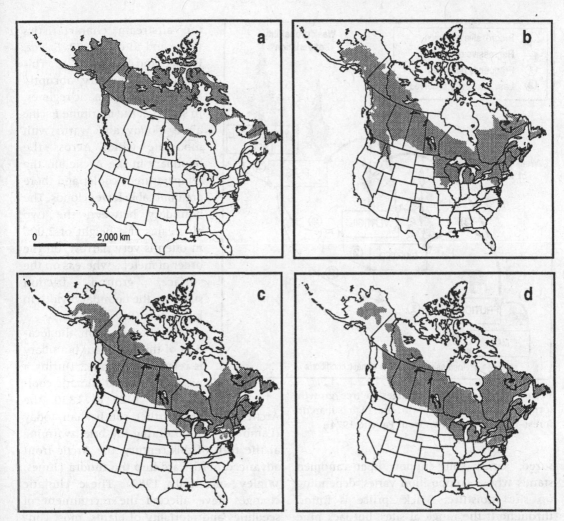

Figure 26.3. (a) Arctic willow (*Salix arctica*), showing an Arctic circumpolar range; (b) white birch (*Betula papyrifera*); (c) black spruce (*Picea mariana*); and (d) tamarack (*Larix laricina*), showing boreal circumpolar ranges. (Redrawn from Hultén 1968 and Elias 1987.)

Locally, the principal topographically related factors affecting vegetation composition, canopy structure, and species diversity are moisture, nutrients, and heat (Maikawa and Kershaw 1976; Payette 1976; Payette and Gagnon 1979; E. A. Johnson 1981; Ritchie 1984; Ecoregions Working Group 1989; Timoney et al. 1992). At the moist, high-nutrient, low-heat end of the gradient are found closed black and white spruce stands with a ground cover of sphagnum, feather mosses, and lichen (*Cladonia rangiferina*). These stands often are relatively rich in species, containing both tundra and boreal representatives because of the high availability of nutrients. The sites tend to be located at the base of north-facing slopes that occasionally contain permafrost. On slightly drier but still cold sites, closed black spruce and white birch stands occur with a ground cover of feather mosses and ericaceous shrubs. Stands are found on sites that generally are north-facing on bottom to mid slopes, have fine- to moderate-textured soils, and lack permafrost. Warm, low-nutrient sites found on mid to top slopes, or on level plains with course-tex-

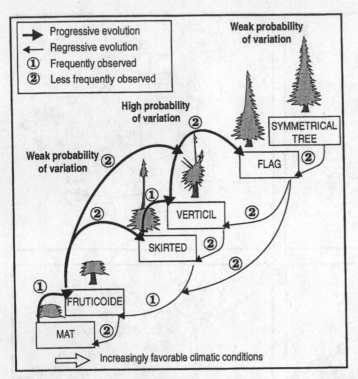

Figure 26.4. Observed changes in tree growth form as climate changes in the high subarctic forest–tundra. (Redrawn from Payette 1974.)

tured, sandy soil, support open-canopied stands whose composition varies, depending on site moisture. Black spruce is found throughout the range of sites, but jack pine occurs only on dry, warm, low-nutrient sites. The dominant lichen composition also changes along a moisture gradient from mesic to dry as follows: *Cladonia rangiferina–Cladonia alpestris* to *Cladonia mitis* to *Cetraria* spp.–*Stereocaulon pascale*.

Ecosystem Dynamics

Climate

The subarctic is related to the summer contact zone between the cool, stable, Arctic airstream and warm, unstable, maritime Pacific or tropical airstreams (Bryson 1966; Barry 1967; Krebs and Barry 1970; Barry and Hare 1974). In turn, these airstreams pass over, and interact with, surfaces that change

the airstream characteristics (Hare and Ritchie 1972; Szeicz, Petzold and Wilson 1979). This airstream boundary abruptly delineates two climatic regimes. In summer, the maritime Pacific air is cloudy and warm with moderate wind. Across the boundary in the Arctic air, the temperature is cooler and there are noticeably fewer clouds. The boundary between the two airstreams at a height of 2,000 m often is very narrow, on the order of meters, whereas on the surface, ground friction stretches the boundary width to kilometers (Bryson 1966).

Over the long term, the location of this airstream boundary is relatively dynamic. During a period of general climatic cooling from 1425 to 1850, the Arctic front was further south than today (Lamb 1977), and during the brief warming at the turn of this century, the Arctic front advanced northward into the tundra (Jones, Wigley and Kelly 1982). These climatic changes have affected the recruitment of seedlings and mortality of plants, most conspicuously trees. Having established during warmer periods in the past, black and white spruce survive in what is currently tundra, although they are not producing seed (Payette and Lajeunesse 1980; Payette and Filion 1984; Payette and Gagnon 1985). On the other hand, tundra plants such *as Diapensia lapponica* can be found over 50 km south of present tree line on eskers and ridges, where they appear to have survived in the subarctic from the 1500s, when this forested area was tundra.

Fire

The other main force in the dynamics of subarctic vegetation is forest fire. Generally, these fires are quite large, measuring in thousands of hectares, with intensities greater than 4,000 kW m⁻¹ and relatively rapid rates of spread (tens of meters per minute). This fire behavior results from the open forest canopy, which leads to relatively high winds at the surface (Szeicz, Petzold and Wilson 1979) and rapid drying of surface fuels, relatively low precipitation, and the long day length during the summer fire season. The low crown base of the open grown trees (Figure 26.5) leads to passive crown fires. Lightning is the major cause of fire, accounting for approximately 85% of fires and 99% of the total area burned (E. A. Johnson and Rowe 1975).

Fire is less frequent in the high subarctic than in the low subarctic (E. A. Johnson and Rowe 1975; E. A. Johnson 1979; Payette et al. 1989). E. A. Johnson (1979) showed a decrease in the expected fire cycle (time required to burn an area equivalent to the study area) from 100 years at 100 km from the black spruce tree line to 50 years at 400 km from the tree line. Payette et al. (1989) found a decrease in number of fires as well as a 100-fold decrease in fire size from the low to the high subarctic.

In general, subarctic forests are open forests. Closed canopies are present on certain sites with higher nutrients and moisture, as noted earlier. Of the tree species, black spruce, white birch, trembling aspen (*Populus tremuloides*), and balsam poplar (*Populus balsamifera*) have the capacity for vegetative regeneration, whereas jack pine, white spruce and tamarack do not. All of these species are capable of living several hundred years, either as trees, krummholz, or vegetative sprouts.

The age distributions of these upland subarctic tree populations are remarkably similar (Figure 26.6). In all cases, the stands have one or more age classes of trees separated by age classes with either no or few individuals. The challenge is to understand how this pattern developed as a result of the forces that operate on the recruitment and mortality of

Figure 26.5. Passive crown fire in black spruce (*Picea mariana*) lichen woodland. The fire burns along the ground and, because of the low branches, burns into the canopy of each tree.

the individual tree species populations. The regeneration cohort of a local population is a group of individuals recruited at approximately the same time and generally experiencing a similar mortality schedule. A local population is made up of one or more regeneration cohorts. These cohorts form a population that contributes seed (along with long-distance dispersed seed) to produce new cohorts. Consequently, the dynamics of a local population are based on the mortality rate of these cohorts and the original recruitment number. Unfortunately, age distributions do not indicate the number of individuals recruited in a cohort, the rate of cohort mortality, or if cohorts have become extinct (E. A. Johnson, Miyanishi and Kleb 1994). Despite these limitations, ecologists have gained a reasonable understanding of

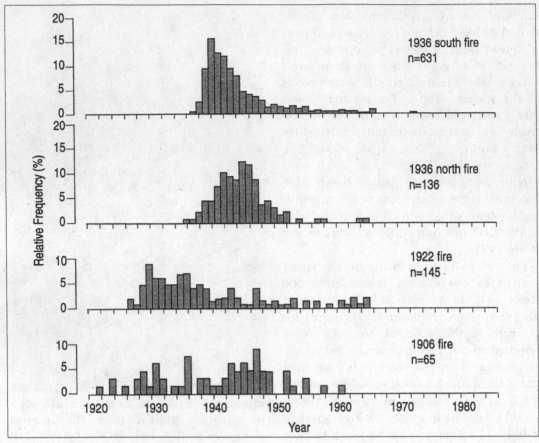

Figure 26.6. Age structures of living black spruce (*Picea mariana*) recruited after fire in 1936, 1922, and 1906. (Redrawn from Sirois and Payette 1989.)

the tree dynamics in the subarctic.

Generally, large numbers of seedlings establish after a disturbance (usually fire) that removes the herbaceous and lichen cover and exposes large areas of mineral soil (Cowles 1982; Thomas and Wein 1985). Effective germination and early seedling survival tend not to occur on intact forest floors, probably because of low moisture supply caused by drying of litter, duff, and lichen layers (Kershaw 1977). Low moisture and heat, as well as a limited seed supply, lead to low seedling density, even when there is a mineral soil surface. Many seeds germinate in the burned-out bases of trees (C. E. Van Wagner, unpublished manuscript). The number of

seedlings recruited after a fire depends to a large degree on the regional climate; cooler episodes lead to few or no recruits, whereas warmer periods have a greater number (e.g., Payette and Filion 1984). Cohort survival varies depending on the specific historical sequence through which the cohorts live. Trees recruited in the cool period of the 1400s were subjected to a different mortality schedule than those recruited in the warm period of the early 1900s. However, mortality of established trees generally seems to be low. Persistence of a cohort also will be influenced by fire frequency and species maximum life span. Black spruce appears to live well over 400 years, whereas white birch rarely lives more than 150 years and jack pine rarely more than 200 years. Exceptions, of course, do occur. Individual trees of black spruce persist by layering around the base. This also allows them to increase their density, at least locally, and to modify their microclimate.

Climate–Fire Interaction

Explanations of the low tree density and open canopy of the subarctic vegetation have primarily centered on the interaction of fire and climate. Maikawa and Kershaw (1976) and Kershaw (1977) took a traditional successional view that tree regeneration after fire was a slow process, requiring a long period for development of a self-reproducing closed-canopy forest with a feather moss ground cover. Closed-canopy forest was thought to be rare, because the harsh surface conditions of high temperature and low soil moisture immediately after fire (Rouse and Kershaw 1971; Rouse 1976) lead to poor tree establishment but abundant regeneration of mosses and lichens. However, the few trees that are established ameliorate the high summer temperatures and low soil moisture, thereby allowing increased vegetation cover. Canopy closure was thought to be retarded by the inhibitory effect of lichens on tree seedling establishment and on the recurrence of fire.

However, this traditional view of succession is not well supported by detailed studies of forest dynamics. Studies of tree regeneration patterns in open- and closed-canopy boreal forest indicate that most regeneration occurs in the few years after a fire (Black and Bliss 1980; E. A. Johnson 1981; Cogbill 1985; Carleton and Wannamaker 1987; Bergeron and Dubuc 1989; Morneau and Payette 1989; Sirois and Payette 1989; E. A. Johnson 1992). In the subarctic, the stressful environment and low seed rain following a fire limit seedling recruitment (Sirois and Payette 1989, 1991). Furthermore, after this limited postfire seedling establishment, regeneration

is reduced because the lichen mat prevents seeds from making contact with mineral soil. Any germination that occurs is associated with "cracks" in the lichen mat that allow seeds to penetrate to mineral soil (Cowles 1982). These few seedlings also are subject to high mortality. In fact, most regeneration occurring 30 or more years after fire is by vegetative means. Black spruce reproduces by layering, and white birch by basal sprouts. However, vegetative reproduction does not progressively create a closed canopy but instead results in a balance between production of new vegetative stems and mortality of older ones.

Recurrence of fire determines the frequency of periods during which the seedling recruitment "window" is open, and climate determines the seed bed condition. Furthermore, since there is no soil seed bank (E. A. Johnson 1975), tree seeds must come from prefire trees. Serotinous cones of jack pine and semiserotinous cones of black spruce release viable seeds for several years after the tree has been killed by fire and heat from the fire has opened the cones (E. A. Johnson and Gutsell 1993; J. S. Johnson and Johnson 1994). The number of seeds from these aerial seed banks depends on the density and fecundity of the prefire trees. White spruce and tamarack have no serotinous cones and, consequently, the trees must either survive the fire or their seeds must disperse into the burned area from unburned edges.

In the high subarctic, fires are infrequent, and tree seed crops are small and highly variable. Furthermore, conditions for tree establishment are generally poor. Consequently, when a fire occurs there is a high probability that tree density will decrease and forest fragmentation will increase. The persistence of old trees in the high subarctic is probably a result of the long interval between fires. On the other hand, in the low subarctic – with its more frequent fires, larger seed rains, and

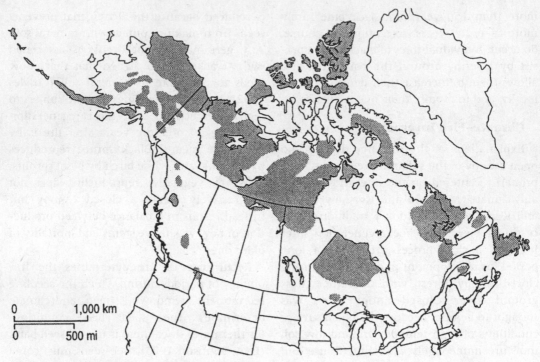

0 1,000 km

0 500 mi

Figure 26.7. Location of some of the major caribou herds in North America. (Redrawn from Bergerud 1978.)

generally better tree establishment conditions – there is a greater probability of maintaining existing areas of forest at their current density (e.g., Sirois and Payette 1991). If the Arctic front were to move south and produce a cooler period in the mid and low subarctic, fire could have a greater impact. Increased forest fragmentation resulting from reduced postfire tree regeneration during this cooler period likely would persist for some time.

Caribou

The subarctic forest is occupied during part of the year by large migratory herds of barren-ground caribou (*Rangifer tarandus* var. *groenlandicus*) and woodland caribou (*Rangifer tarandus* var. *caribou*) (Figure 26.7; Bergerud 1978). These are the last migratory herds of free-ranging ungulates in North America.

Caribou spend the summer on the tundra.

Females, yearlings, and calves form large herds for insect abatement and protection from predators while males are in smaller, bachelor herds. In August, as insect pests decline and calves become larger, the herds break into smaller groups. By September, they have moved south into the subarctic forest; the males move first, followed by the females and young. The rut occurs in October and November. The first heavy winter snow results in movement of the caribou deeper into the forested region, where food is easier to obtain.

Winter is spent close to large lakes, which provide escape from wolves, and in upland areas, where lichen, herbaceous, and shrub browse are available and accessible due to the reduced snow cover. Caribou dig small holes with their hooves to reach browse. Food apparently is detected below the snow by odors diffusing up along the branches of shrubs extending above the snow. Thus, conditions of food availability and snow depth largely dictate the location and dispersion of herds.

In April, the caribou migrate northward toward tundra. Traditional routes are used because they are generally snow free early. Pregnant females lead the migration to reach their calving grounds by late May to early June. Caribou reproduction rate is low, with one calf per female being the rule. Female pregnancy rates vary by age, but for mature females (3–4 years old) the rate is around 80% (Bergerud 1980). Calf death rate is high, with weather and predators accounting for most of the mortality (F. L. Miller 1974; F. L. Miller and Broughton 1974). Calves are particularly vulnerable in their first few weeks, with 50%–90% mortality in areas where wolf density is high. Mortality of adults is 4%–15% annually, depending on predator density. Caribou are not particularly effective in escaping predators. Their general strategy is to run from the forest to hard snow and ice on large lakes, where they can see the predator and try to keep a safe distance. However, wolves are successful in more than 50% of the chases (Haber 1977).

Until the last few decades, the estimated number of caribou has been based on speculation. For the three large herds in Ungava, Northwest Territories, and Alaska, estimates of population size vary considerably but are thought to be between one and two million (Bergerud 1974). Herd size probably has decreased in the last 100 years or so. The extent to which the herds have declined is a politically volatile question. Several explanations for the decline have been proposed: (1) overhunting, (2) predators, particularly wolves, and (3) decrease in carrying capacity of the winter range due to destruction of the lichens by an increased frequency of forest fire.

Despite the low number of indigenous hunters, overhunting probably has been an important factor in the decline of caribou. Most hunting occurs during winter, when caribou are in the forest and native hunters are on their trap lines. The introduction of the high-powered repeating rifle has made caribou herds particularly vulnerable. Because the normal response of a herd to danger is to run onto a frozen lake and mill around looking for the predator, shooting from a distance is easy. Caribou, like bison, seem to have no adaptive behavior to respond to a hunter who can kill from a distance. These facts suggest that hunting probably was important in reducing the caribou population during this century.

Predators seem to be particularly effective in harvesting caribou. With the low population growth rate of caribou, it appears that in the past a delicate balance existed between recruitment and mortality from predators. Moreover, a case can be made that multiequilibria exist, depending on hunting pressure and predator density (Haber 1977; Haber and Walters 1980). An understanding of caribou population dynamics and impacts of predation and hunting requires not only an estimate of birth and death rates but also metapopulation estimates of dispersal between herds. If calf mortality due to predators is consistently high (assuming predator densities normal or high), then an above-normal adult mortality (e.g., by hunting) will result in a decrease in population. However, a decrease in a local population could be prevented by immigration of animals from another herd.

The suggestion that the carrying capacity of caribou winter range in the subarctic forest has been reduced significantly (Scotter 1967) is based on several premises, none of which seem to have much support (Bergerud 1974; E. A. Johnson and Rowe 1975; D. R. Miller 1980). The first premise is that caribou are climax forest animals that depend on lichens (*Cladonia rangiferina* and *Cladonia alpestris*) found only in climax forest. However, compared to other ungulates, caribou have highly variable diets, of which lichens are only one part (Murie 1935; Skoog 1968; Bergerud 1974; D. R. Miller 1980). It has been reported that caribou feeding only on lichens lose weight (Bergerud 1974). The second premise is that climax (old-growth) forest stands of the subarctic have been replaced largely by younger, successional stands due to fires set by Europeans (Scotter 1967, 1971). Evidence from subarctic forest fires does not support the suggestion that human-caused fires account for much area burned (E. A. Johnson and Rowe 1975; Payette et al. 1989). For the last 300 years, fire frequencies have changed due to climatic shifts associated with the Little Ice Age and not humans (Payette et al. 1989; E. A. Johnson 1992). Also, there is no documentation that reduced range quality has resulted in changes in birth or death rates in caribou (Bergerud 1974). The only possible effect of fire is local and short term, causing caribou to seek older, unburned forests, which contain more lichens rather than younger, recently burned stands (Klein 1982).

Native People

Indigenous hunters have always occupied the subarctic seasonally, usually during winter when caribou migrate into the woodland. Bands of hunters migrated south into the boreal forest or north onto the tundra during the summer, and into the subarctic in the winter, after the caribou arrived. The prehistoric era (prior to 1680) was, in the context of the time, a period of plenty for the subarctic people (Yerbury 1986). Most took advantage of the summer abundance of caribou, fish, and waterfowl on the tundra. Winters were spent in the subarctic lichen woodlands, fishing and hunting caribou, moose, and smaller game (Jarvenpa 1980). Food generally was reliable and plentiful (Helm 1965; Smith 1975; Yerbury 1986).

The beginning of changes in migration pattern caused by the evolution to a more trading–trapping economy occurred during the protohistoric era (1680–1769) (Bishop and Ray 1976; Yerbury 1986). During this period, little direct contact occurred with European traders, but goods were passed along through native middlemen. Although the mark-up by middlemen on European goods was high because of their technological superiority, the trade was not initially exploitive of the fur resource for two reasons: the difficulty of transport by the middlemen, and cultural wealth accumulation not being part of the native culture (Ray 1974; Ray and Freeman 1978). However, to keep fur supplies at satisfactory levels, the European traders resorted to even higher mark-ups, which then were passed along to native trappers. Furthermore, hunting of food animals, including caribou, increasingly occurred to provision the fur traders and not simply to provide for the subsistence of the local people.

Further changes in the human population and in their location took place during the historic era (1770–1890) due to the increasing importance of the fur trade and due to secondary effects such as disease (Bishop and Ray 1976; Yerbury 1986). Many tribes were forced into the low subarctic, either from further south or from the tundra. The low subarctic is not as good a subsistence habitat as the high subarctic and tundra. This forced relocation of people, together with the increasing fur trade economy, meant that hunting/trapping pressure increased on the populations of caribou and fur-bearing animals. Particularly during the period between 1763 (Treaty of Paris) and 1821 (merger of the Northwest and Hudson Bay companies), fur and game animals over increasing areas of the boreal forest were exploited with little concern for conservation. Fur animals were exhausted commercially in many areas (Ray 1974). The beaver, which was the mainstay of the early fur trade, was completely eliminated from large areas. Since beaver feed on birch, aspen, and balsam poplar, and build dams that modify the water table, their elimination probably had a major effect on the local vegetation and hydrology. As exploitation progressed, trapping activity moved west and north, allowing the "trapped out" parts of the southern and eastern subarctic time to recover.

In both prehistoric and historic times, caribou had a cultural and survival (food, shelter, clothing) significance to native hunters. However, caribou have been severely hunted, particularly in recent times. The indigenous people seemed to take the existence of caribou for granted. When caribou were abundant, large numbers were killed and only partially used. Periodic shortages were attributed to changes in migration routes due to snow conditions or to the necessity of trapping in areas where there were no caribou. It appears that in traditional resource utilization, indigenous people under-utilized available resources due to the unimportance of accumulating wealth and the reciprocity of exchange within kinship groups.

Summary

Despite the occurrence of fire and the presence of large grazing herds of caribou in the subarctic, the major factor determining the open-canopy nature of the subarctic spruce–lichen woodland is climate. Thus, unlike other transitional open-canopy vegetation such as pine barrens and oak openings, the lichen woodlands would not be converted into closed-canopy forest in the absence of disturbance by fire or grazing. The location of the subarctic woodlands is closely related to that of the Arctic front in summer, which is determined by large-scale atmospheric (climatic) processes. Changes in the average location of the Arctic front in summer occur over long time scales due to long-term climatic shifts. Also, fires in this region are primarily influenced by large-scale atmospheric processes. Furthermore, because of the extremely low human population density, the subarctic woodlands have not suffered the fragmentation that has occurred in other transition vegetation, such as temperate oak savanna. As a consequence, natural large-scale processes continue to be the dominant influence on the extent, location, composition, and structure of the lichen woodland.

Acknowledgments

We gratefully acknowledge the financial support of a Natural Sciences and Engineering Research Council of Canada (NSERC) operating grant, as well as the NSERC Networks of Centres of Excellence in Sustainable Forest Management. We also thank Marie Puddister for the figures, and Don Freeman, Sheri Gutsell, and Kara Webster for their helpful comments on earlier drafts of this manuscript. Finally, we acknowledge Serge Payette, Louise Filion, and their co-workers for their enormous and sustained contribution to our understanding of the subarctic region.

REFERENCES

Barry, R. G. 1967. Seasonal location of the Arctic front over North America. *Geographical Bulletin* 9:79–95.

Barry, R. G., and Hare, F. K. 1974. Arctic climate. In *Arctic and Alpine Environments*, ed. J. D. Ives and R. G. Barry, pp. 17–54. London: Methuen.

Bergeron, Y., and Dubuc, M. 1989. Succession in the southern part of the Canadian boreal forest. *Vegetatio* 79:51–63.

Bergerud, A. T. 1974. Decline of caribou in North America following settlement. *Journal of Wildlife Management* 38:757–770.

Bergerud, A. T. 1978. Caribou. In *Big Game of North America, Ecology and Management*, ed. J. L. Schmidt and D. L. Gilbert, pp. 83–101. Harrisburg, Pa.: Stackpole.

Bergerud, A. T. 1980. A review of the population dynamics of caribou and wild reindeer in North America. In *Proceedings of the Second International Reindeer/Caribou Symposium, Røros, Norway*, ed. E. Reimers, E. Gaare, and S. Skjenneberg, pp. 556–581. Trondheim, Norway: Direktoratet for vilt og ferskvannsfisk.

Bishop, C. A., and Ray, A. J. 1976. Ethnohistorical research in the central subarctic: some conceptual and methodological problems. *The Western Canadian Journal of Anthropology* 4:116–144.

Black, R. A., and Bliss, L. C. 1980. Reproductive ecology of *Picea mariana* (Mill.) BSP. at tree line near Inuvik, Northwest Territories, Canada. *Ecological Monographs* 50:331–354.

Blüthgen, J. 1970. Problems of definition and geographical differentiation of the subarctic with special regard to northern Europe. In *Ecology of the Subarctic Regions, Proceedings of the Helsinki Symposium*, pp. 11–33. Paris: UNESCO.

Bryson, R. A. 1966. Airmasses, streamlines and the boreal forest. *Geographical Bulletin* 8:228–269.

Carleton, T. J., and Wannamaker, B. A. 1987. Mortality and self-thinning in postfire black spruces. *Annals of Botany* 59:621–628.

Cody, W. J. 1971. A phytogeographic study of the floras of the continental Northwest Territories and Yukon. *Le Naturaliste Canadien* 98:145–158.

Cogbill, C. V. 1985. Dynamics of the boreal forests of the Laurentian Highlands, Canada. *Canadian Journal of Forest Research* 15:252–261.

Cowles, S. 1982. Preliminary results investigating the effects of lichen ground cover on the growth of black spruce. *Le Naturaliste Canadien* 109:573–581.

Ecoregions Working Group. 1989. *Ecoclimatic Regions of Canada, First Approximation*. Ottawa: Ecoregions Working Group of the Canada Committee on Ecological Land Classification, Sustainable Development Branch, Canadian Wildlife Service, Conservation and Protection, Environment Canada. Ecological Land Classification Series, No. 23.

Elias, T. S. 1987. *The Complete Trees of North America*. New York: Gramercy Publishing Company.

Haber, G. C. 1977. Socio-ecological dynamics of wolves and prey in a subarctic ecosystem. Ph.D. thesis, University of British Columbia, Vancouver.

Haber, G. C., and Walters, C. J. 1980. Dynamics of the Alaska Yukon caribou herds, and management implications. In *Proceedings Second International Reindeer Caribou Symposium, Røros, Norway*, ed. E. Reimers, E. Gaare, and S. Skjenneberg, pp. 645–663. Trondheim, Norway: Direktoratet for vilt og ferskvannsfisk.

Hare, I. K., and Ritchie, J. C. 1972. The boreal bioclimates. *Geographical Review* 62:333–365.

Helm, J. 1965. Patterns of allocation among the Arctic Drainage Dene. In *Essays in Economic Anthropology*, ed. J. Helm, P. Bohannan, and M. D. Sahlin, pp. 33–45. Proceedings of the 1965 Annual Spring Meeting of the American Ethnological Society.

Hultén, E. 1968. *Flora of Alaska and Neighbouring Territories*. Stanford, Calif.: Stanford University Press.

Hustich, I. 1966. On the forest tundra and the northern tree lines. *Annals University of Turku* AII 36:7–47 (Republic of Kevo Subarctic Research Station 3).

Hustich, I. 1979. Ecological concepts and biogeographical zonation in the North: the need for generally accepted terminology. *Holarctic Ecology* 2:208–217.

Jarvenpa, R. 1980. *The Trappers of Patuanak: Toward a Spatial Ecology of Modern Hunters*. Ottawa: National Museums of Canada. National Museum of Man Mercury Series, Canadian Ethnology Service Paper No. 67.

Johnson, E. A. 1975. Buried seed populations in the subarctic forest east of Great Slave Lake, Northwest Territories. *Canadian Journal of Botany* 53:2933–2941.

Johnson, E. A. 1979. Fire recurrence in the subarctic and its implications for vegetational composition. *Canadian Journal of Botany* 57:1374–1379.

Johnson, E. A. 1981. Vegetational organization and dynamics of lichen woodland communities in the Northwest Territories, Canada. *Ecology* 62:200–215.

Johnson, E. A. 1992. *Fire and Vegetation Dynamics, Studies from the North American Boreal Forest*. Cambridge: Cambridge University Press.

Johnson, E. A., and Gutsell, S. L. 1993. Heat budget and fire behavior associated with the opening of serotinous cones in two *Pinus* species. *Journal of Vegetation Science* 4:745–750.

Johnson, E. A., Miyanishi, K., and Kleb, H. 1994. The hazards of interpretation of static age structures as shown by stand reconstructions in a *Pinus contorta–Picea engelmannii* forest. *Journal of Ecology* 82:923–931.

Johnson, E. A., and Rowe, J. S. 1975. Fire in the subarctic wintering ground of the Beverly Caribou Herd. *The American Midland Naturalist* 94:1–15.

Johnson, J. S., and Johnson, E. A. 1994. Opening of semi-serotinous cones of *Picea mariana* by fire and ambient heating. *Supplement to Bulletin of the Ecological Society of America* 75:123.

Jones, P. D., Wigley, T. M. L., and Kelly, P. M. 1982. Variations in surface air temperatures: Part 1. North Hemisphere, 1881–1980. *Monthly Weather Review* 110:59–70.

Kershaw, K. A. 1977. Studies on lichen dominated systems XX: An examination of some
aspects of the northern boreal lichen woodlands of Canada. *Canadian Journal of Botany*
55:393–410.

Klein, D. R. 1982. Fire, lichens and caribou.
Journal of Range Management 35:390–395.

Krebs, J. S., and Barry, R. G. 1970. The Arctic
front and the tundra–taiga boundary in
Eurasia. *Geographical Review* 60:548–554.

Lamb, H. H. 1977. *Climatic History and the Future.*
London: Methuen.

Löve, D. 1970. Subarctic and subalpine: where
and what? *Arctic and Alpine Research* 2:63–73.

Maikawa, E., and Kershaw, K. A. 1976. Studies
on lichen-dominated systems XIX: The postfire
recovery sequence of black spruce–lichen
woodland in the Abitau Lake Region, NWT.
Canadian Journal of Botany 54:2679–2687.

Miller, D. R. 1980. Wildfire effects on barren-
ground caribou wintering on the Taiga of
northcentral Canada: a reassessment. In
*Proceeding Second International Reindeer Caribou
Symposium, Røros, Norway*, ed. E. Reimers, E.
Gaare, and S. Skjenneberg, pp. 84–98.
Trondheim, Norway: Direktoratet for vilt og
ferskvannsfisk.

Miller, F .L. 1974. *Biology of the Kaminuriak
Population of Barren-Ground Caribou. Part 2.
Dentition as an Indicator of Age and Sex:
Composition and Socialization of the Population.*
Ottawa: Canadian Wildlife Service. Department
Series 31.

Miller, F. L., and Broughton, E. 1974. *Calf Mortal-
ity on the Calving Ground of Kaminuriak Caribou.*
Ottawa: Canadian Wildlife Service. Department
Series 26.

Morisset, P., Payette, S., and Deshaye, J. 1983.
The vascular flora of the northern
Quebec–Labrador peninsula: phytogeographical
structure with respect to the tree-line. In *Tree-
line Ecology: Proceedings of the Northern Québec
Tree-line Conference*, ed. P. Morisset and S.
Payette, pp. 141–151. Québec: Centre d'Études
Nordiques de l'Université Laval.

Morneau, C., and Payette, S. 1989. Postfire
lichen–spruce woodland recovery at the limit
of the boreal forest in northern Quebec.
Canadian Journal of Botany 67:2770–2782.

Murie, O. 1935. *Alaska-Yukon Caribou.*
Washington D.C.: United States Department of
Agriculture, North American Fauna. No. 54.

Payette, S. 1974. Classification écologique des
formes de croissance de *Picea glauca* (Moench)

Voss et de *Picea mariana* (Mill.) BSP. en milieux
subarctiques et subalpins. *Le Naturaliste
Canadien* 101:893–903.

Payette, S. 1976. Les limites écologiques de la
zone hémi-arctique entre la mer d'Hudson et la
baie d'Ungava, Nouveau-Québec. *Cahiers de
Géographie de Québec* 20:347–364.

Payette, S., and Filion, L. 1984. White spruce
expansion at the treeline and recent climatic
change. *Canadian Journal of Forest Research*
15:241–251.

Payette, S., and Gagnon, R. 1979. Treeline
dynamics in Ungava Peninsula, Northern
Quebec. *Holarctic Ecology* 2:239–248.

Payette, S., and Gagnon, R. 1985. Late Holocene
deforestation and tree regeneration in the forest-tundra of Quebec. *Nature* 313:570–572.

Payette, S., and Lajeunesse, R. 1980. Les combes
à neige de la rivière aux feuilles (Nouveau-
Québec): indicateurs paléoclimatiques
Holocènes. *Géographie Physique et Quaternaire*
34:209–220.

Payette, S., Morneau, C., Sirois, L., and Desponts,
M. 1989. Recent fire history of the northern
Quebec biomes. *Ecology* 70:656–673.

Ray, A. J. 1974. *Indians in the Fur Trade: Their Role
as Trappers, Hunters, and Middlemen in the Lands
Southwest of Hudson Bay, 1660–1870*. Toronto:
University of Toronto Press.

Ray, A. J., and Freeman, D. B. 1978. *'Give us good
measure': An Economic Analysis of Relations
between the Indians and the Hudson's Bay Company
before 1763*. Toronto: University of Toronto
Press.

Ritchie, J. C. 1984. *Past and Present Vegetation of the
Far Northwest of Canada*. Toronto: University of
Toronto Press.

Ritchie, J. C. 1987. *Postglacial Vegetation of Canada*.
Cambridge: Cambridge University Press.

Rouse, W. R. 1976. Microclimatic changes accompanying burning in subarctic lichen woodland.
Arctic and Alpine Research 8:357–376.

Rouse, W. R., and Kershaw, K. A. 1971. The
effects of burning on the heat and water relations of lichen-dominated subarctic surfaces.
Arctic and Alpine Research 3:291–304.

Rousseau, J. 1968. The vegetation of the
Quebec–Labrador Peninsula between 55° and
60° N. *Le Naturaliste Canadien* 95:469–563.

Scotter, G. W. 1967. *Effects of Forest Fires on the
Winter Range of Barren-Ground Caribou in
Northern Saskatchewan*. Ottawa: Canadian
Wildlife Service. Wildlife Management Bulletin
Series 1, 18.

436   E. A. JOHNSON AND K. MIYANISHI

Scotter, G. W. 1971. Fire, vegetation, soil and barren ground caribou relations in northern Canada. In *Fire in the Northern Environment*, pp. 209–226. Portland, Ore.: USDA Forest Service, Pacific Northwest Forest and Range Experimental Station.

Sirois, L., and Payette, S. 1989. Postfire black spruce establishment in subarctic and boreal Quebec. *Canadian Journal of Forest Research* 19:1571–1580.

Sirois, L., and Payette, S. 1991. Reduced postfire tree regeneration along a boreal forest–forest–tundra transect in Northern Quebec. *Ecology* 72:619–627.

Skoog, R. O. 1968. Ecology of the caribou (*Rangifer tarandus granti*) in Alaska. Ph.D. dissertation, University of California, Berkeley, Calif.

Smith, J. G. E. 1975. The ecological basis of Chipewyan socio-territorial organization. In *Proceedings of the Northern Athapascan*

Conference, 1971, Vol. 2, ed. A. M. Clark, pp. 390–461. Ottawa: National Museums of Canada. National Museum of Man Mercury Series, Canadian Ethnology Service Paper No. 27.

Szeicz, G., Petzold, D. E., and Wilson, R. G. 1979. Wind in the subarctic forest. *Journal of Applied Meteorology* 18:1268–1274.

Thomas, P. A., and Wein, R. W. 1985. Delayed emergence of four conifer species on postfire seedbeds in eastern Canada. *Canadian Journal of Forest Research* 15:727–729.

Timoney, K. P., La Roi, G. H., Zoltai, S. C., and Robinson, A. L. 1992. The high subarctic forest-tundra of northwestern Canada: position, width, and vegetation gradients in relation to climate. *Arctic* 45:1–9.

Yerbury, J. C. 1986. *The Subarctic Indians and the Fur Trade, 1680–1860*. Vancouver: University of British Columbia Press.

Indexes

There are three indexes: plant, animal, and topic. Plants and animals are indexed by their scientific names or in some cases by generic or family names. Pages cited in the index may, however, include text in which an organism is referred to by its common name only.

Index of Plants

Abies balsamea 383, 384
Abies fraseri 123–125
Abies sp. 60
Abronia fragrans (sweet sand verbena) 266
Acacia (acacia) 33
Acacia pinetorum (Florida acacia) 34
Acalypha gracilens 215
Acanthomintha 317
Acer macrophyllum (big-leaf maple) 328
Acer negundo (boxelder, Manitoba maple) 249, 409
Acer nigrum 383
Acer rubrum (red maple) 32, 88, 143, 195, 198, 199, 343, 397, 398
Acer saccharum (sugar maple) 159, 177, 178, 226, 237, 353, 364–366
Acer spicatum (mountain maple) 364, 369, 372
Achillea lanulosa (yarrow) 197, 368, 411
Actaea rubra (red baneberry) 368, 411
Adiantum aleuticum 75–77
Adiantum pedatum ssp. calderi 75
Adiantum pedatum var. aleuticum 75–77
Adiantum viridimontanum 75
Aesculus californica (California buckeye) 324, 326, 328
Agalinis acuta (sandplain gerardia) 74
Agalinus fasciculata 197
Agalinus gattingeri 386
Agalinus tenuifolia 101, 197, 384
Agave spp. 290
Agropyron cristatum (crested wheatgrass) 417
Agropyron dasystachyum (northern wheatgrass) 409, 411
Agropyron repens 397
Agropyron smithii (western wheatgrass) 409, 411
Agropyron spicatum (bluebunch wheatgrass) 250, 256
Agropyron subsecundum (awned wheatgrass) 409, 411
Agropyron trachycaulum (slender wheatgrass) 368, 398, 409, 411
Agrostis elliotiana 108, 400
Agrostis mertensii 127
Agrostis perennans 123, 125, 176

Agrostis scabra 379, 384, 397, 398
Alepocurus aequalis (water foxtail) 410
Aleurodiscus sp. (fungus) 144
Allium canadense 395, 402
Allium cernuum 385, 386
Allium falcifolium (sickle-leaf onion) 317
Allium fimbriatum diabolense 317
Allium hoffmanii 317
Allium howellii sanbenitensis 317
Allium oxyphilum 91
Allium sanbornii 317
Allium schoenoprasum var. sibiricum 380, 386
Allium spp. (onions) 290
Allium tricoccum 383
Allium vineale 197
Alnus maritima (seaside alder) 238
Alnus rhombifolia (white alder) 328
Alnus rugosa (speckled alder) 353
Alnus viridis 125
Amaranthus retroflexus (redroot pigweed) 266
Ambrosia artemisiifolia 164, 197, 215, 384
Ambrosia bidentata (ragweed) 197, 224
Ambrosia psilostachya (western ragweed) 266
Amelanchier alnifolia (Saskatoon) 381, 384, 410, 412
Amelanchier arborea (serviceberry) 90, 177, 179, 182, 185
Amelanchier bartramiana (oblong fruited June-berry) 364
Amelanchier spp. (Saskatoon, serviceberry) 343, 400
Amelanchier utahensis (serviceberry) 250
Amorpha canescens (lead-plant) 136, 139, 161, 162, 400
Amphianthus pusillus 101, 103, 104, 115
Amphicarpaea bracteata (hog peanut) 139, 161
Anagallis arvensis (scarlet pimpernel) 329
Andrachne phyllanthoides 225
Andropogon (bluestem) 24, 31, 351
Andropogon cabanisii (firegrass) 31
Andropogon cirratus (Texas bluestem) 278
Andropogon gerardii (big bluestem) 30, 76, 139, 143, 161, 162, 196, 199, 223, 237, 267, 377, 378, 382, 395, 400, 402, 409

439

Andropogon gyrans 195, 198, 199
Andropogon hallii (sand bluestem) 267
Andropogon ischaemum (King Ranch bluestem)
  240
Andropogon ternarius 39, 198
Andropogon virginicus (broom sedge) 104, 110,
  400
Anemone canadensis (Canada anemone) 411
Anemone cylindrica 161, 381
Anemone multifida 384
Anemone patens (prairie croccus) 409
Angadenia sagraei 34
Angelica triquinata 124, 125
Annona glabra (pond apple) 34
Anomodom atenuatus 368
Anomodon rostratus 368
Antennaria neglecta (field cat's foot) 162, 368,
  384, 398
Antennaria plantaginifolia (pussytoes) 162, 164,
  177, 180, 182
Antennaria spp. (pussytoes) 139
Antennaria suffrutescens 317
Antennaria virginica 91
Apocynum androsaemifolium (spreading dog-
  bane) 161, 351, 353, 398
Apocynum cannabinum 197
Aquilegia canadensis 384
Arabis aculeolata 317
Arabis constancei 317
Arabis hirsuta 379, 384
Arabis koehleri stipitata 317
Arabis laevigata 92
Arabis lyrata 71, 72, 74, 75, 386
Arabis macdonaldiana 317
Arabis serotina (shale barren rock cress) 91–93,
  96
Arabis serpentinicola 317
Arabis shortii 386
Aralia hispida 398
Aralia nudicaulis (wild sarsaparilla) 161, 368,
  397, 410, 411
Arceuthobium spp. (dwarf mistletoe) 296
Arceuthobium vaginatum (dwarf mistletoe) 256
Arctostaphylos spp. (manzanita) 325
Arctostaphylos uva-ursi (bearberry) 55, 384, 386,
  398, 412, 424
Arctostaphylos viscida (whiteleaf manzanita) 324,
  326
Ardisia (marbleberry) 33
Ardisia escallonioides (marbleberry) 34
Arenaria brevifolia 400
Arenaria howellii 317
Arenaria macrophylla 76
Arenaria marcescens 77

Arenaria patula 210, 213–216
Arenaria rosei 317
Arenaria rubella 317, 319
Arenaria serpyllifolia 384
Arenaria stricta (rock sandwort) 368
Arenaria uniflora (sandwort) 101, 104, 106, 108,
  397
Argythamnia blodgettii 34
Arisaema atrorubens (jack in the pulpit) 369
Aristida (wiregrass, three-awn) 31, 36, 267,
  278, 290
Aristida beyrichiana (southern wiregrass) 11, 30,
  31, 39, 41
Aristida dichotoma 71, 72, 74, 75, 395, 402
Aristida longespica 75
Aristida longiseta (red three-awn) 265
Aristida purpurascens (three-awn) 71–74, 237
Aristida stricta (northern wiregrass) 31
Aristida wrightii (Wright's three-awn) 267
Armillaria spp. (white root rot fungus) 296
Aronia arbutifolia 125
Aronia prunifolia 398
Artemisia arbuscula (low sagebrush) 250
Artemisia cana (silver sagebrush) 250
Artemisia caudata (threadleaf sagewort) 266
Artemisia filifolia (sand sagebrush) 265
Artemisia frigida (pasture sage) 409
Artemisia ludoviciana 411
Artemisia spp. (sagebrush) 290
Artemisia tridentata (big sagebrush) 249, 250,
  257
Arundinaria gigantea 32
Asclepias amplexicaulis 197
Asclepias arenaria (sand milkweed) 266
Asclepias exaltata 399, 402
Asclepias meadii (Mead's milkweed) 182, 183
Asclepias pumila (plains milkweed) 266
Asclepias purpurascens 386
Asclepias quadrifolia 386
Asclepias spp. (milkweed) 139
Asclepias syriaca 384
Asclepias tuberosa, 197, 386, 402
Asclepias verticillata 71, 72, 162, 210, 386
Asclepias viridiflora 386
Asclepias viridis 214, 215
Aspidotis densa (rock brake fern) 317
Asplenium montanum 121
Asplenium platyneuron (ebony spleenwort) 180,
  402
Asplenium ruta-muraria (wall rue) 367
Asplenium septentrionale 92
Asplenium trichomanes (spleenwort) 368, 382,
  386
Asplenium viride (green spleenwort) 367

Aster 210
Aster acuminatus 125, 129
Aster azureus 161
Aster ciliolatus 381, 382, 388, 396, 398, 399, 401
Aster concinnus 75
Aster cordifolius (rice button aster) 92, 382, 368, 411
Aster depauperatus 71–74
Aster divaricatus 126
Aster dumosus 195, 197, 199, 216, 386
Aster ericoides (many flowered aster, heath aster) 161, 162, 409
Aster laevis 75, 92, 161
Aster linariifolius (stiff aster) 56, 139, 141
Aster macrophyllus (big-leafed aster) 353, 368, 383, 398
Aster oblongifolius 216
Aster oolentangiensis (prairie heart-leaved aster) 139, 386
Aster parviceps 72
Aster patens 177, 180
Aster pilosus 197
Aster pilosus var. *pringlei* 72
Aster schistosus 92
Aster sericeus 161
Aster shortii 386
Aster solidagineus 195, 199
Aster umbellatus 398
Astragalus 317
Astragalus bibullatus 211
Astragalus distortus (bent milkvetch) 91, 93
Astragalus mollissimus (woolly loco) 266
Astragalus neglectus 382, 385, 386, 388
Astragalus tennesseensis 211, 212
Astragalus whitneyi siskiyouense 317
Aulacomnium palustre 104, 109
Aureolaria flava (false foxglove) 182
Avena barbata (slender wild oats) 329, 330
Avena fatua (wild oats) 330
Ayenia euphrasifolia 34

Baccharis halmifolia (eastern baccharis) 34
Baccharis texana (prairie baccharis) 267
Bacillus thuringensis (bacterium) 395, 403
Balsamorhiza platylepis (balsamroot) 317
Balsamorhiza spp. (balsam root) 290
Baptisia leucantha 162
Bartonia paniculata 401
Bartonia virginica 402
Beckmannia syzigachne (slough grass) 410
Befaria racemosa (tarflower) 11
Denitoa occidentalis 317
Berchemia scandens 214
Besseya bullii (kittentail) 149

Betula glandulosa (dwarf birch) 424
Betula lutea (yellow birch) 368
Betula papyrifera (white birch) 76, 353, 364, 368, 373, 382, 398, 409, 422, 424, 425
Blephilia ciliata (downy blephilia) 368, 382, 385, 386
Boltonia asteroides 386
Boltonia diffusa 195, 199
Bonamia grandiflora (scrub morning-glory) 11
Bothriochloa barbinodis (cane bluestem) 267
Bothriochloa saccharoides (silver bluestem) 267
Botrychium virginianum (Virginia grape-fern) 368
Bouteloua curtipendula (side-oats grama) 161, 224, 237, 249, 267, 278, 281, 380, 385–387
Bouteloua eriopoda (black grama) 267
Bouteloua gracilis (blue grama) 249, 267, 278, 409
Bouteloua hirsuta (hairy grama) 237, 267, 278
Bouteloua radicosa (purple grama) 278
Bouteloua spp. (grama grass) 290
Brachiaria ciliatissima (fringed signalgrass) 267
Brachypodium distachyon 329
Brachythecium salesbrosum 365, 366, 369
Brickellia spp. 290
Brodiaea greenei 317
Brodiaea hendersonii 317
Bromus diandrus (ripgut brome) 329, 330
Bromus inermis (smooth bromegrass) 397, 410
Bromus kalmii 384, 386, 398
Bromus madritensis (foxtail chess) 329
Bromus mollis (soft chess) 329, 330, 332
Bromus pubescens 163, 164, 382
Bromus tectorum (cheatgrass) 300
Buchloë dactyloides (buffalo grass) 237, 267
Bulbostylis capillaris 106, 108, 113, 368, 397, 399, 400, 402
Bumelia lanuginosa (chittamwood) 111, 237
Bumelia lycioides 214
Bumelia reclinata 34
Bumelia salicifolia 34
Bupleurum rotundifolium 215
Bursera simaruba (gumbo–limbo) 31, 34
Byrsonima 33
Byrsonima lucida 34, 40

Cacalia plantaginea 378, 386
Cacalia rugelia 124
Caesalpinia jamesii (James' rushpea) 267
Calamagrostis cainii 124–126, 128, 129
Calamagrostis canadensis (marsh reed grass) 411
Calamintha arkansana 378, 380
Calamintha ashei (Ashe's savory) 10, 12

Calamovilfa gigantea (giant sandreed) 267
Callicarpa americana 32–34, 214
Callirhoe scabriuscula (Texas poppy-mallow) 265
Calocedrus decurrens (incense cedar) 313, 317
Calochortus greenei (mariposa lily) 317
Calochortus howellii (mariposa lily) 317
Calochortus obispoensis (mariposa lily) 317
Calochortus raichei (mariposa lily) 317
Calochortus tiburonensis (mariposa lily) 317
Calochortus tolmiei (mariposa lily) 317
Caloplaca flavorubescens (crustose lichen) 144
Calylophus serrulatus (yellow evening primrose) 266
Calystegia spithamaea (low bindweed) 381, 386
Calystegia stebbinsii 317
Camassia scilloides (wild hyacinth) 156, 164, 165, 386
Camissonia benitensis 315
Campanula rotundifolia (harebell) 75, 124, 127, 384, 398, 411
Camptosorus rhizophyllus (walking fern) 367, 369
Campylopux (moss) 104
Candelaria concolor (foliose lichen) 144
Candelariella xanthostigma (crustose lichen) 144
Cardamine parviflora (small-flowered bittercress) 179
Carex 210
Carex annectens 386
Carex aquatilis (water sedge) 410
Carex artitecta 402
Carex atherodes (awned sedge) 410
Carex aurea 384
Carex backii 162, 382, 386, 402
Carex biltmoreana 123, 124
Carex blanda 163, 164
Carex brunnescens 125
Carex castanea 382
Carex cephalophora 163, 164, 382
Carex communis 111
Carex complanata 198
Carex conoidea 386
Carex crawei 378, 384–386
Carex cumulata 402
Carex davisii 383, 386
Carex debilis 123
Carex eburnea (bristle-leaved sedge) 367, 369, 383, 384
Carex festucacea 176
Carex flava 384
Carex foenea 399, 400
Carex formosa 386
Carex garberi 378

Carex gracilescens 386
Carex granularis 384
Carex hirsutella 164, 176
Carex hystericina 75
Carex jamesii 163, 383, 386
Carex juniperorum 382, 383, 385, 386
Carex lanuginosa (woolly sedge) 380, 384, 410
Carex leavenworthii 386
Carex lucorum 402
Carex meadii 378, 386
Carex misera 121, 123–125, 128
Carex molesta 384
Carex oligocarpa 383, 386
Carex pensylvanica (Pennsylvania sedge) 136, 139, 161, 164, 176, 382–384, 397, 398, 400
Carex plantaginea (plaintain-leaved sedge) 368
Carex richardsonii 75, 384
Carex rugosperma 398
Carex ruthii 124
Carex sartwellii 380, 386
Carex scirpoidea 378, 380
Carex siccata 382, 383, 386, 402
Carex sparganoides 164
Carex spp. (sedge) 139, 177, 180, 182, 184, 351, 409, 411
Carex tonsa 398, 402
Carex umbellata 71, 72, 75, 123, 125, 384
Carya alba see *Carya tomentosa* (mockernut hickory)
Carya cordiformis (bitternut hickory) 237
Carya floridana (scrub hickory) 11
Carya glabra (pignut hickory) 90, 111, 237
Carya glabra-ovalis (pignut or red hickory) 173, 174, 176, 179, 182, 185
Carya illinoensis (pecan) 237
Carya ovata (shagbark hickory) 138, 148, 173, 174, 176, 185, 382, 383
Carya pallida (sand hickory) 32
Carya spp. (hickory) 201
Carya texana (Texas or black hickory) 174, 182, 226, 234, 236
Carya tomentosa (mockernut hickory) 31, 32, 173, 174, 176, 185
Cassia deeringiana (cassia) 31, 34
Cassia fasciculata rostrata (partridge pea) 266
Castanea dentata (American chestnut) 88
Castilleja spp. (Indian paintbrush) 290
Castilleja brevibracteata (Indian paintbrush) 317
Castilleja coccinea (Indian paintbrush) 384, 386
Castilleja elata (Indian paintbrush) 317
Castilleja neglecta (Tiburon paintbrush) 317
Ceanothus americanus (New Jersey tea) 75, 139, 143, 162, 382, 383, 386

Ceanothus cuneatus (wedgeleaf ceanothus) 324, 325, 327, 328

Ceanothus herbaceus 386

Ceanothus ovatus (New Jersey tea) 353, 399

Ceanothus pumilus 317

Celastrus scandens 384

Celtis laevigata 213

Celtis occidentalis 90

Celtis reticulata (western hackberry) 267

Celtis spp. 210, 214

Celtis tenuifolia 385, 387

Cenchrus incertus (common sandbur) 267

Cerastium arvense (chickweed, field chickweed) 317, 381, 411

Cerastium arvense var. *villosissimum* 71, 73–75

Cerastium arvense var. *villosum* 71, 72

Cerastium brachypodum 386

Cerastium nutans 380

Cerastium terrae-novae 77

Cerastium velutinum 384, 386

Cerastium vulgatum (large mouse-eared chick-weed) 368

Ceratiola ericoides (Florida rosemary) 9, 12, 18

Ceratocystis fagacearum (oak wilt) 240

Ceratodon purpureus 379

Cercis canadensis (eastern redbud) 90, 177, 182, 214, 237, 386

Cercocarpus ledifolius (curlleaf mountain-mahogany) 250

Cercocarpus montanus (mountain-mahogany) 249

Cetraria islandica 93

Cetraria spp. 422

Chaenactis ramosa 317

Chaenactis thompsonii 317

Chaerophyllum procumbens var. *procumbens* 386

Chaerophyllum tainturieri 215

Chamaecrista fasciculata 197

Chamaecyparis thyoides (Atlantic white cedar) 52

Cheilanthes castanea 92

Cheilanthes siliquosa 76

Chelone lyonii 123, 124

Chelone obliqua 124, 126

Chenopodium album (lamb's quarters) 417

Chenopodium hybridum (maple-leaved goose-foot) 368

Chenopodium leptophyllum (slimleaf goosefoot) 266

Chenopodium rubrum (red goosefoot) 410

Chenopodium vulgare 384

Chimaphila umbellata (pipsissewa) 353

Chiococca parvifolia 34

Chloris cucullata (hooded windmill grass) 267

Chloris verticillata (tumble windmill grass) 267

Chlorogalum grandiflorum (soap plant) 317

Chrysanthemum leucanthemum (ox-eye daisy) 197, 368, 381, 385

Chrysobalanus icaco (cocoplum) 34

Chrysoma pauciflosculosa (woody goldenrod) 10, 11

Chrysophyllum oliviforme (satinleaf) 34

Chrysopsis floridana (Florida golden-aster) 13

Chrysopsis villosa (hairy golden aster) 409

Chrysothamnus pulchellus (Southwest rabbit-brush) 267

Cinna arundinacea (reed grass) 173, 176

Circaea quadrisculata 368

Cirsium arvense (Canada thistle) 417

Cirsium discolor 197, 386, 388, 399, 402

Cirsium hillii 385, 386

Citharexylum fruiticosum (fiddlewood) 34

Cladina rangiferina also see *Cladonia rangiferina* (reindeer lichen) 184

Cladina spp. (reindeer lichen) 178, 179, 224

Cladium jamaicense (sawgrass) 30, 33

Cladonia 93, 422, 424

Cladonia alpestris 425, 431

Cladonia caroliniana 104

Cladonia leporina 102, 104, 107, 108, 114

Cladonia mitis 425

Cladonia perforata (perforate cladonia) 15

Cladonia rangiferina see also *Cladina rangiferina* (reindeer moss) 425, 431

Cladonia strepsilis (lichen) 56

Clarkia franciscana 317

Claytonia megarhiza var. *nivalis* 317

Claytonia virginica 164, 383

Clematis albicoma 91

Clematis coactilis 91, 92, 95

Clematis viticaulis 91

Clethra alnifolia 32

Clintonia borealis 124

Coccoloba diversifolia (doveplum) 34

Coccothrinax (silver palm) 33

Coccothrinax argentata (silver palm) 34

Collinsia greenei 317

Colubrina arborescens (coffee colubrina) 34

Comandra pallida (western comandra) 266

Comandra umbellata also see *C. richardsiana* (bastard toad-flax) 139, 143, 381, 382, 384, 398

Comandra richardsiana also see *C. umbellata* 161, 162

Commelina erecta (erect dayflower) 266

Comptonia peregrina (sweet fern) 55, 56, 351, 353, 397, 401

Conocarpus erectus (buttonbush) 34

Conyza canadensis (horsetail conyza) 266

Corchorus siliquosus 34

Cordylanthus 317

Corema conradii (broom crowberry, corema) 56, 58, 61, 62

Coreopsis grandiflora (large-flowered coreopsis) 226

Coreopsis lanceolata (tickseed coreopsis) 224, 382, 385, 386

Coreopsis major 124, 125, 197

Coreopsis palmata (finger-tickseed, prairie coreopsis) 139, 161, 196, 353, 386

Cornus canadensis (bunchberry) 383, 410, 411

Cornus drummondii (Drummond's dogwood, roughleaf dogwood) 163, 237, 386

Cornus florida (flowering dogwood) 32, 177, 182, 185, 198

Cornus obliqua (pale dogwood) 163

Cornus racemosa (gray dogwood) 139, 161, 163, 382–384

Cornus rugosa (roundleaf dogwood) 364–366, 369

Cornus spp. (dogwood) 143

Cornus stolonifera (red oiser dogwood) 163, 382, 384, 410

Corydalis flavula 383, 386

Corydalis sempervirens 396–398

Corylus americana (American hazel) 75, 139, 160–163, 165, 351, 353

Corylus cornuta (beaked hazel) 140, 351

Coryphantha macromeris (long mamma) 266

Crataegus dilatata 386

Crataegus uniflora (hawthorn) 32

Cronartium ribicola (white pine blister rust) 256–258

Crossopetalum ilicifolium 34

Crossopetalum rhacoma 34

Crotalaria avonensis (Avon Park harebell) 15

Crotalaria pumila (rattlebox) 34

Croton (croton) 33

Croton capitatus 210, 213–215

Croton dioicus (rubaldo) 266

Croton lindheimerianus (three-seed croton) 266

Croton linearis (croton) 34

Croton monanthogynus 210, 213–216

Croton pottsii (leather weed croton) 266

Croton spp. 290

Croton texensis (Texas croton) 266

Crotonopsis elliptica (crotonopsis, rushfoil) 104, 106, 108, 179, 180, 182, 224, 226, 400

Cryphonectria parasitica (chestnut blight) 240

Cryptantha jamesii (James' cryptantha) 266

Cryptantha minima (small cryptantha) 266

Cryptogramma stelleri (steller cliff-brake) 367

Ctenium (toothache grass) 31

Cunila origanoides (dittany) 177, 178, 180, 182

Cuppressus arizonica (Arizona cypress) 289

Cupressus macnabiana (Macnab cypress) 313

Cupressus sargentii (Sargent cypress) 313

Cuscuta campestris 212

Cycloloma atriplicifolium (tumble ringwing) 267

Cynodon dactylon (Bermuda grass) 240

Cynosurus echinatus (dog tail) 329

Cyperus aristatus 210

Cyperus granitophilus 101, 102, 114

Cyperus spp. (sedge) 267

Cyperus uniflorus (oneflower flatsedge) 267

Cypripedium arietinum 382, 386, 388

Cypripedium calceolus 388

Cyrilla racemiflora (swamp titi) 32

Cystopteris bulbifera (bulblet fern) 371

Cystopteris fragilis 382, 384

Cytospora chrysosperma (canker) 413

Dactylis glomerata 397

Dalea foliosa 211, 212

Dalea formosa (feather plume) 267

Dalea gattingeri 210–216

Dalea lanata (woolly dalea) 266

Dalea nana (dwarf dalea) 266

Danthonia 351, 353

Danthonia compressa 120

Danthonia intermedia 77

Danthonia spicata (poverty grass or curly grass) 123, 164, 176–180, 182, 216, 237, 380, 382, 384, 397, 398, 400

Dasylirion spp. (desert spoon) 290

Daucus carota 197

Daucus pusillus 329

Delphinium alabamicum 211

Delphinium carolinianum ssp. *calciphilum* 211, 214

Dentaria laciniata (toothwort) 182

Dermatocarpon hepaticum 213

Deschampsia caespitosa 124, 377, 380, 382, 384, 386

Deschampsia flexuosa 393, 397, 398

Descurania spp. (tansy mustard) 300

Desmanthus cooleyi 267

Desmodium canadense 162, 197

Desmodium ciliare 197

Desmodium glutinosum (naked tick-trefoil) 149

Desmodium illinoense 162

Desmodium nudiflorum (cluster-leaf tick-trefoil) 149

Desmodium paniculatum 197, 386, 399, 402

Desmodium rigidum 75

Desmodium sessilifolium 197

Diamorpha cymosa also see *D. smallii* and *Sedum*

smallii (stonecrop) 101, 102, 104–106, 108, 110
Diamorpha smallii also see D. cymosa and Sedum smallii (stonecrop) 101, 104, 397, 400
Diapensia lapponica 426
Dicentra formosa oregana (yellow bleedingheart) 317
Dicerandra christmanii (Garrett's mint) 15
Dicerandra frutescens (Lake Placid scrub mint) 11, 13, 15
Dichanthelium acuminatum 123, 125, 164, 176
Dichanthelium spp. 210, 215
Dicranum (moss) 424
Dicranum flagellare (moss) 369
Dicranum scoparium (moss) 184, 368, 369
Diervilla lonicera (bush honeysuckle) 368, 397, 398
Diervilla sessilifolia 124
Digitaria sanguinalis (hairy crabgrass) 267
Digitaria spp. (crab grass, finger grass) 33, 290
Diodia teres (bottonweed, rough buttonweed) 179, 195, 197, 199, 210, 224
Diospyros virginiana (persimmon) 31, 32, 195, 198, 199, 214, 239
Distichlis stricta (alkali grass) 410
Dithyrea wislizenii (spectaclepod) 266
Dodecatheon meadia (shooting star) 156, 162, 164, 165
Dodonaea 33
Dodonea viscosa 34
Douglasia nivalis 317
Draba nemorosa 380
Draba reptans 379, 380, 386
Drosera (sundew) 28
Dryas integrifolia (mountain avens) 424
Dryopteris felix-mas (male fern) 369
Dryopteris goldiana (Goldie's fern) 369
Dryopteris intermedia (American shield fern) 369
Dryopteris marginalis (wood-fern) 369, 382, 384

Echinacea paradoxa var. paradoxa 225
Echinacea tennesseensis 209, 211, 213
Eleagnus commutata (wolfwillow) 409, 412
Eleocharis compressa 210, 381, 382, 384, 386
Eleocharis elliptica 378
Eleocharis erythropoda 381
Eleocharis montevidensis (sand spikesedge) 266
Eleocharis nitida 77
Eleocharis verrucosa 176
Elymus glaucus (blue wildrye) 331
Elymus hystrix also see Hystrix patula (bottlebrush grass) 163
Elymus spp. (wild rye) 411
Elymus trachycaulus 384

Elymus villosus 164
Elymus virginicus 198
Empetrum nigrum (crowberry) 422, 424
Ephedra antisyphilitica (vine ephedra) 267
Epigaea repens 56
Epilobium angustifolium (fireweed) 411
Epilobium coloratum 379, 384
Epilobium hornemanii 386
Epilobium rigidum (willowherb) 317
Epipactis helleborine (helleborine orchid) 368
Eragrostis curtipedicellata (gummy lovegrass) 267
Eragrostis curvula (weeping lovegrass) 267
Eragrostis intermedia (plains lovegrass) 278
Eragrostis lehmanniana (Lehmann lovegrass) 267
Eragrostis mexicana (Mexican lovegrass) 278
Eragrostis oxylepis (red lovegrass) 267
Eragrostis sessilispica (tumble lovegrass) 267
Eragrostis spectabilis 198
Eragrostis spp. 290
Eragrostis trichodes (sand lovegrass) 267
Erianthus alopecuroides 198
Erigeron modestus (plains fleabane) 266
Erigeron pulchellus 382
Erigeron strigosus 162, 164, 197, 210, 213, 216
Eriogonum allenii (yellow buckwheat) 88, 91, 93–95
Eriogonum alpinum 317
Eriogonum annuum (annual wild buckwheat) 266
Eriogonum kelloggii (wild buckwheat) 317
Eriogonum libertini (wild buckwheat) 317
Eriogonum ternatum (wild buckwheat) 317
Erioneuron pulchellum (fluffgrass) 267
Eriophyllum jepsonii 317
Erodium botrys (filaree) 330
Erodium cicutarium (redstem filaree) 330
Eryngium cuneifolium (wedge-leaved button snakeroot) 11, 12, 14
Erysimum capitatum 93
Erysimum cheiranthoides (worm seed) 369
Erythrobalanus (red/black oak subgenus) 135
Erythronium (fawn lily) 317
Erythronium albidum 164
Erythronium howellii (fawn lily) 317
Eugenia axillaris (white stopper) 34
Eugenia foetida 4
Euonymous obovatus (running strawberry bush) 368
Eupatorium altissimum 196, 197, 386
Eupatorium coelestinum 34
Eupatorium hyssopifolium 195, 197, 199
Eupatorium mikanoides 34
Eupatorium rotundifolium 195, 197, 199

Euphorbia commutata 380–381, 385
Euphorbia corollata (flowering spurge) 139, 161, 162, 196, 197 180, 353
Euphorbia dentata 214, 215
Euphorbia esula (leafy spurge) 417
Euphorbia fendleri (Fendler's euphorb) 266
Euphorbia missurica (prairie euphobia) 266
Euphorbia spathulata 214, 215, 386
Euphorbia spp. 290
Eurytaenia texana (Texas spread-wing) 266
Euthamia graminifolia 197
Evolvulus nuttallianus (hairy evolvulus) 266
Evolvulus pilosus 225
Exothea paniculata (butterbough) 34

Fagus grandifolia (American beech) 159, 177, 179, 368
Ferns 182
Ferocactus wislizenii (southwestern barrel cactus) 266
Festuca altaica (northern rough fescue) 77, 409, 411
Festuca arizonica (Arizona fescue) 249
Festuca kingii (king spikefescue) 256
Festuca megalura (annual fescue) 330
Festuca microstachys (annual fescue) 317
Festuca obtusa 164
Festuca ovina (sheep's fescue grass) 368
Ficus aurea (strangler fig) 34
Ficus citrifolia 34
Filipendula rubra 126
Forestiera ligustrina 210, 213, 214
Forestiera segregata (Florida privet) 34
Fragaria glauca (smooth wild strawberry) 411
Fragaria spp. 400
Fragaria virginiana (wild strawberry) 139, 162, 197, 216, 353, 368, 381, 382, 384, 398
Fraxinus americana (American ash, white ash) 90, 214, 224, 364
Fraxinus pennsylvanica (green ash) 238, 249, 409
Fraxinus quadrangulata 382, 386
Fritillaria 317
Fritillaria atropurpurea (checker lily) 317
Froelichia floridana (Florida snakecotton) 266
Froelichia gracilis (snakecotton) 266

Gaillardia pulchella (Indian blanket) 266
Galium ambiguum siskiyouense (bedstraw) 317
Galium boreale (northern bedstraw) 75, 139, 161, 164, 180, 197, 382, 383, 410, 411
Galium circaezans 164
Galium pilosum 180, 197
Galium spp. (bedstraw) 180

Galium virgatum 214, 215
Gaultheria procumbens (wintergreen) 56, 353, 398
Gaura biennis 197
Gaura coccinea (scarlet gaura) 266
Gaura villosa (hairy gaura) 266
Gaylussacia (huckleberry) 28, 33
Gaylussacia baccata (black huckleberry, huckleberry) 55, 56, 74, 75, 90, 138, 139, 184, 353, 397, 398
Gaylussacia dumosa 32
Gaylussacia frondosa 32, 56
Gentiana andrewsii 75
Gentiana linearis 124, 125, 129
Gentianopsis crinita (fringed gentian) 74
Geranium bicknellii 380, 384, 386
Geranium carolinianum 384, 386
Geranium maculatum 161
Geranium molle 329
Geranium robertianum (Robert's geranium) 365, 366, 369
Gerardia flava (false foxglove) 180
Gerardia tenuifolia 195, 199
Geum canadense 164
Geum radiatum 123, 124, 126, 128, 129
Geum triflorum 381, 385–387
Geum vernum 386
Glyceria grandis (tall manna grass) 410
Glyceria nubigena 124
Glyceria striata 380, 384
Glycyrrhiza lepidota (wild licorice) 411
Gnaphalium obtusifolium 197
Gratiola neglecta 379, 380
Grimmia alpicola 369
Grimmia laevigata 102, 114
Grindelia lanceolata 213, 215
Guapira discolor 34
Guettarda (velvetseed) 33
Guettarda elliptica (Everglades velvetseed) 34
Guettarda scabra (roughleaf velvetseed) 34
Gutierrezia sarothrae (broom snakeweed) 266
Gymnocarpium robertianum 382, 386
Gymnoderma lineare 127
Gymnosporangium spp. (rust) 296

Haplopappus linearifolius (narrowleaf goldenbush) 324, 326
Haplopappus racemosus congestus 317
Haplopappus spp. 290
Hedeoma hispida 379, 384, 402
Hedeoma pulegioides 384
Hedyotis caerulea 74
Hedyotis crassifolia 215
Hedyotis humifusa (mat bluet) 266

Hedyotis longifolia 378, 384
Hedyotis nigricans 213, 386
Hedyotis nuttalliana 215
Hedyotis purpurea 197, 215
Hedyotis pusilla (bluet) 179
Hedyotis spp. 210
Hedysarum alpinum (American hedysarum) 411
Helenium autumnale 196, 379
Helianthemum canadense (frostweed) 139, 353
Helianthus angustifolius 195, 199
Helianthus annuus (common sunflower) 266
Helianthus divaricatus (woodland sunflower) 176, 177, 180, 182, 382, 397, 398
Helianthus grosseserratus 162
Helianthus hirsutus 196, 197
Helianthus laetiflorus 16, 162
Helianthus laevigatus (smooth sunflower) 92
Helianthus mollis 197
Helianthus occidentalis 197
Helianthus petiolaris (narrowleaf sunflower, plains sunflower) 266, 278
Helianthus porteri also see *Viguiera porteri* 101, 104, 106, 108–111, 113
Helianthus strumosus 161
Helictotrichon hookeri (oat grass)
Heliotropium convolvulaceum (bindweed heliotrope) 266
Heliotropium tenellum 210, 213
Hemizonia halliana (tarweed) 317
Herteromeles arbutifolia (toyon) 326, 328
Hesperolinon 317
Heterotheca latifolia (camphorweed) 266
Heterotheca villosa (golden aster) 266
Heuchera americana 386
Heuchera richardsonii (prairie alum-root) 139, 162
Hexastylis arifolia var. *ruthii* 76
Hieracium bolanderi (hawkweed) 317
Hieracium canadense 402
Hieracium gronovii (hairy hawkweed) 180, 182
Hieracium kalmii (Canada hawkweed) 140
Hieracium longipilum (long-haired hawkweed) 140
Hieracium piloselloides 385
Hieracium scabrum 402
Hieracium traillii 91, 92
Hieracium venosum 402
Hoffmanseggia glauca (Indian rushpea) 266
Hoffmanseggia jamesii (James' rushpea) 266
Homomalium adnatum 368
Hordeum hystrix (Mediterranean barley) 329
Hordeum leporinum (foxtail) 329, 330
Horkelia sericata 317

Houstonia purpurea 124–126
Houstonia serpyllifolia 126
Hudsonia ericoides (false heather) 56
Huperzia appalachiana 124, 127
Huperzia porophila 124, 126
Hylocomium 424
Hymenopappus flavescens (yellow woollywhite) 266
Hymenoxys herbacea 380, 382, 385–386
Hymenoxys scaposa (stemmed bitterweed) 266
Hypericum 210
Hypericum brachyphyllum 34
Hypericum buckleyi 124, 125
Hypericum cumulicola (scrub hypericum) 12, 14
Hypericum dolabriforme 210, 216
Hypericum drummondii 197
Hypericum frondosum 214
Hypericum gentianoides (pinweed) 104, 106, 108, 182, 197, 400
Hypericum graveolens 124
Hypericum hypericoides 34
Hypericum kalmianum 378, 385, 386
Hypericum mitchellianum 124
Hypericum perforatum 381, 397
Hypericum piloselloides 397
Hypericum punctatum 197
Hypericum reductum (sand hypericum) 11
Hypericum sphaerocarpum 210, 213–216
Hypnum pallescens 368
Hypoxis hirsuta 162, 386
Hystrix patula also see *Elymus hystrix* (bottle brush grass) 368, 382

Ilex (holly) 33
Ilex cassine (dahoon holly) 34
Ilex coriacea 32, 40
Ilex glabra (gallberry, inkberry) 32, 39, 56
Ilex krugiana (tawnyberry holly) 34
Ilex opaca (American holly) 32
Ilex vomitoria (yaupon) 32
Impatiens pallida (pale touch-me-not) 369
Imperata cylindrica (cogongrass) 16
Ipomoea spp. 290
Ipomopsis longiflora (whiteflower ipomopsis) 266
Iris innominata (Oregon iris) 317
Iris lacustris also 382, 383, 385, 386
Isanthus brachiatus also see *Trichostema brachiatum* 213
Isoetes butleri 210, 214, 215
Isoetes engelmannii 401
Isoetes melanospora (quillwort) 101, 103, 113
Isoetes tegetiformans (quillwort) 101, 103, 115
Isopterygium pulchellum 369

Iva microcephala 34

Jacquemontia curtisii 35
Janquinia keyensis (joewood) 35
Juglans californica (California black walnut) 328
Juglans cinerea (butternut) 369, 383
Juglans hindsii (northern California walnut) 325
Juglans nigra (black walnut) 238
Juncus alpinoarticulatus 378
Juncus balticus (Baltic rush) 410
Juncus balticus var. *littoralis* 378
Juncus dudleyi 384
Juncus georgianus 101
Juncus greenei 402
Juncus scirpoides (needlepod rush) 267
Juncus secundus 380, 386, 396, 399, 401
Juncus trifidus 123, 124, 127, 128
Juniperus ashei (Ashe's juniper) 238, 289
Juniperus californica (California juniper) 289, 326
Juniperus coahulensis 289
Juniperus comitana 289
Juniperus communis (common juniper) 76, 256, 369, 381, 382, 384, 398, 400
Juniperus communis jackii (common juniper) 317
Juniperus deppeana (alligator bark juniper) 278, 289
Juniperus flaccida 289
Juniperus gamboana 289
Juniperus horizontalis (creeping juniper) 412, 378, 380, 381, 382
Juniperus monosperma (one-seeded juniper, single-seeded juniper) 278, 289
Juniperus monticola 289
Juniperus occidentalis (western juniper) 288, 289
Juniperus osteosperma (Utah juniper) 249, 251, 278, 288, 289
Juniperus pinchotii (Pinchot's juniper) 289
Juniperus scopulorum (Rocky Mountain juniper) 249, 250, 257, 278, 289
Juniperus spp. (juniper) 344
Juniperus virginiana (eastern juniper, eastern redcedar, redcedar) 67, 71, 73–75, 90, 95, 111, 137, 148, 171, 174, 178, 179, 185, 210, 213, 214, 216, 222, 223, 226, 234, 236, 267, 289, 382, 399

Kalmia latifolia (mountain laurel) 32, 56, 76, 90, 125
Koeleria cristata (June grass) 409, 411
Kosteletzkya virginica 35
Krameria lanceolata (trailing ratany) 266
Krigia biflora (orange dwarf-dandelion) 139, 140

Krigia dandelion (false dandelion) 179, 182
Krigia montana 123–125, 128, 129
Krigia spp. 180
Krigia virginica 215
Kuhnia eupatorioides (false boneset) 196, 266
Kummerowia stipulacea 197

Lactuca canadensis 75
Lactuca floridana 386
Lantana (lantana) 33
Lantana depressa 35
Lantana involucrata 35
Larix laricina (tamarack) 424, 425
Larrea tridentata (creosote bush) 267
Lathyrus ochroleucus (cream-colored vetchling) 383, 411
Layia discoidea (tarweed) 315, 317
Leavenworthia alabamica 214
Leavenworthia alabamica var. *alabamica* 211
Leavenworthia alabamica var. *brachystyla* 211
Leavenworthia crassa var. *elongata* 211
Leavenworthia crassa var. *crassa* 211
Leavenworthia exigua var. *exigua* 211
Leavenworthia exigua var. *laciniata* 211
Leavenworthia exigua var. *lutea* 211
Leavenworthia spp. 210, 216
Leavenworthia stylosa 211, 214, 215
Leavenworthia uniflora 214
Lechea intermedia 398
Lechea mucronata (pinweed) 141
Lechea pulchella (pinweed) 141
Ledum groenlandicum (Labrador tea) 412, 422, 424
Leiophyllum buxifolium 56
Lemna minor (duckweed) 412
Lepidobalanus (white oak subgenus) 135
Leptochloa dubia (green sprangletop) 278
Leptochloa spp. 290
Leptoloma cognatum (fall witchgrass) 267, 353
Lespedeza (lespedeza) 180, 184
Lespedeza capitata (bush-clover) 139, 162, 400
Lespedeza intermedia 386
Lespedeza procumbens 197
Lespedeza virginica 195, 197, 199
Lesquerella gracilis 225
Lesquerella lyrata 211
Lesquerella spp. (bladderpod) 266
Leucobryum glaucum (cushion moss) 184
Leucospora multifida 386
Leucothoe recurva (fetterbush) 90
Liatris aspera 196
Liatris cylindracea 380, 386
Liatris helleri (blazing star) 115, 124, 126, 128
Liatris punctata (dotted gayfeather) 266

Liatris scariosa var. *nieuwlandii* 165
Liatris spicata 386
Licania michauxii 32, 35
Lichen 179, 182
Lilium philadelphicum 384, 386, 398
Linanthus 318
Linaria canadensis 400
Linaria texana (Texas toadflax) 266
Linnaea borealis (twinflower) 383, 412
Linum rigidum (stiff-stem flax) 266
Linum striatum 197
Linum sulcatum 73, 74, 386
Liquidambar styraciflua (sweetgum) 32, 179
Lithospermum canescens (hoary puccoon) 139, 162, 353
Lithospermum incisum (narrowleaf gromwell) 266
Lobelia appendiculata var. *gattingeri* 211, 215
Lobelia kalmii 378
Lobelia puberula 197
Lobelia spicata 162, 196
Lolium multiflorum (annual ryegrass) 329
Lomatium (biscuitroot, desert parsley) 318, 319
Lomatium cuspidatum (biscuitroot) 318
Lomatium engelmannii 318
Lomatium howellii (biscuitroot) 318
Lomatium tracyi (biscuitroot) 318
Lonicera canadensis (American fly-honeysuckle) 368
Lonicera dioica (twining honeysuckle) 384, 412
Lonicera hirsuta 402
Lonicera involucrata (involucred fly-honey-suckle) 368
Lonicera japonica 198
Lonicera tartarica (Tartarian bush honeysuckle) 368
Lupinus bicolor (miniature lupine) 330
Lupinus laxiflorus (lupine) 318
Lupinus perennis (wild blue lupine) 139, 140, 144, 147, 353, 356
Lupinus spp. (bluebonnet) 290
Luzula multifolia (wood rush) 182
Lycium berlandieri (wolfberry) 267
Lycopodium selago 76
Lycurus phleoides (wolftail) 267
Lycurus spp. (wolfgrass) 290
Lyonia ligustrina (lyonia) 32
Lyonia lucida 39
Lyonia mariana (lyonia) 32
Lyonia spp. (fetterbush) 7
Lysiloma latisiliquum (wild tamarind) 31, 35
Lysimachia quadrifolia (prairie loosestrife) 353, 402

Machaeranthera australis (spiny haplopappus) 266
Machaeranthera pinnatifida (spiny haplopappus) 266
Machaeranthera tanacetifolia (Tahoka daisy) 266
Madia hallii (tarweed) 318
Madia spp. (tarweed) 329
Magnolia virginiana (sweetbay) 32
Mahonia repens (Oregon grape) 256
Maianthemum canadense (two-leaved Solomon's-seal, wild lily-of-the-valley) 353, 365, 366, 368, 398, 410, 411
Malaxis unifolia 398
Malus coronaria 163
Malus ioensis (wild crab apple) 163
Malvastrum hispidum 214
Manfreda virginica (American agave) 179, 212, 214, 216
Maurandya wislizenii 266
Medicago lupulina 215, 385, 397, 398
Melampodium leucanthum (plains blackfoot daisy) 266
Melampyrum lineare 56
Melilotus alba 164, 397
Mentzelia nuda (bractless mentzelia) 266
Menziesia pilosa (minnie-bush) 90, 124, 125, 411
Metopium toxiferum (poisonwood) 35
Milium effusum 383
Mimosa biuncifera (cat-claw mimosa) 267
Mimulus nudatus (monkey flower) 318
Minuartia biflora 77
Minuartia brevifolia 114
Minuartia groenlandica 124, 125, 127
Minuartia marcescens 77
Minuartia michauxii 379, 384, 402
Mirabilis linearis (linearleaf four o'clock) 266
Mitella diphylla (mitre-wort) 369
Mitella nuda (bishop's-cap) 411
Mnium cuspidatum 369
Mollugo verticillata (Indian chickweed) 266
Monarda bradburiana (wild bergamot) 182
Monarda fistulosa (wild bergamot) 139, 161, 162, 353, 382
Monarda pectinata (beebalm) 266
Monarda punctata (spotted beebalm) 266
Morinda 33
Morinda royoc 35
Morus rubra (red mulberry) 176
Mosiera longipes 35
Moss 179, 180, 182
Muhlenbergia (muhly grass) 33
Muhlenbergia capillaris (muhly grass) 30, 33

Muhlenbergia cuspidata (prairie muhly) 411
Muhlenbergia emersleyi (bullgrass) 278
Muhlenbergia glomerata 124, 383
Muhlenbergia porteri (bush muhly) 267
Muhlenbergia richardsonis (mat muhly) 409
Muhlenbergia sobolifera 386
Muhlenbergia virescens (screwleaf muhly) 249
Munroa squarrosa (false buffalograss) 267
Myosotis macrosperma 383, 386
Myosotis verna 380, 386
Myosurus minimus 379, 385, 386
Myrica cerifera (wax myrtle) 28, 33, 35, 37
Myrica heterophylla 32
Myriophyllum exalbescens (water milfoil) 412
Myrsine (myrsine) 33
Myrsine floridana (myrsine) 35

Nassella pulchra (purple needlegrass) 330
Navarretia 318
Nepeta cataria (catnip) 369
Nerisyrenia camporum (mesa greggia) 266
Nolina spp. (beargrass) 290
Nostoc commune 206, 213, 379
Nothoscordum bivalve (false garlic) 179, 210,
    214, 215
Nyssa biflora (swamp tupelo) 32
Nyssa sylvatica (black gum) 31, 32, 90, 174, 198

Oenothera argillicola (shale barren evening prim-
    rose) 91, 94, 95
Oenothera fruticosa (sundrops) 71, 72, 101
Oenothera linifolia 101
Oenothera missouriensis var. *missouriensis* 225
Oenothera rhombipetala (fourpoint evening
    primrose) 266
Onosmodium molle 211
Onosmodium subsetosum 225
Ophioglossum engelmannii 210, 214, 215
Opuntia fragilis 401, 402
Opuntia humifusa (opuntia cactus) 179, 210,
    212, 214
Opuntia leptocaulis (tasajillo) 266
Opuntia phaeacantha (brownspine pricklypear)
    266
Opuntia spp. (prickly pear cactus) 237, 266
Orobanche fasciculata 381, 386
Oryzopsis 351
Oryzopsis asperifolia 383, 384, 398
Oryzopsis racemosa 383
Osmanthus americanus (devilwood) 32
Ostrya virginiana (hop hornbeam, ironwood)
    177, 182, 364
Oxalis priceae subsp. *priceae* 210
Oxalis spp. (wood sorrel) 164, 179

Oxalis stricta 198, 215
Oxalis violacea 162, 164, 215
Oxydendrum arboreum (sourwood) 32, 174, 184
Oxypolis rigidior 126

Palafoxia feayi (palafoxia) 11
Palafoxia sphacelata (rayed palafoxia) 266
Panicium boscii 180
Panicum 210
Panicum abscissum (cutthroat) 14
Panicum anceps 198
Panicum capillare (witchgrass) 210, 213, 267
Panicum commutatum 195, 199
Panicum depauperatum 71, 72, 74, 398
Panicum dichotomum 177, 178, 180
Panicum flexile 73, 75, 210, 213, 216, 380, 384
Panicum havardii 267
Panicum implicatum 162, 383
Panicum lanuginosum 104, 195, 199, 398
Panicum latifolium 398
Panicum laxiflorum 181
Panicum leiburgii 161
Panicum lindheimeri 386
Panicum linearifolium 398
Panicum lithophilum 101
Panicum obtusum (vine-mesquite) 267
Panicum oligosanthes (panic-grass, small panic
    grass) 75, 76, 139, 237, 386, 399, 402
Panicum philadelphicum 75, 380, 384, 385
Panicum ramisetum (bristle panicum) 267
Panicum sphaerocarpon 71, 72, 74, 75
Panicum spp. (panic grass) 178, 180, 184
Panicum spp. 111, 215
Panicum spretum 401
Panicum villosissimum (panic-grass) 139
Panicum virgatum (switchgrass) 267
Parnassia asarifolia 124
Paronychia argyrocoma 123
Paronychia jamesii (James'nailwort) 266
Paronychia montana 91
Paronychia virginica 93
Parthenium integrifolium 162
Parthenocissus quinquefolia (Virginia creeper)
    176, 177, 180, 369
Paspalum (paspalum) 33
Paspalum setaceum (thin paspalum) 265
Pectis angustifolia 266
Pedicularis spp. (elephant trunk) 296
Pediomelum subacaule 210–215
Pellaea atropurpurea 216, 382, 386
Pellaea glabella (smooth cliff-brake) 365, 366,
    368
Peltigera 424
Penstemon buckleyi (penstemon) 266

Penstemon grandiflorus (large beard-tongue) 145
Penstemon hirsutus 380, 384
Penstemon spp. 290
Persea borbonia (redbay) 35
Persea palustris (redbay) 32
Petalostemum purpureum (purple prairie clover) 161, 266
Petalostemum villosum (silky prairie clover) 266
Petasites palmatus (palmate-leaved colt's-foot) 410, 411
Phacelia capitata 317
Phacelia corymbosa 318
Phacelia dasyphylla ophitidis 317
Phacelia dubia 101, 102, 114
Phacelia dubia var. *interior* 211
Phacelia integrifolia (phacelia) 266
Phacelia maculata 101, 102, 114
Phacelia purshii 383, 386
Phaeophyscia rubropulchra (foliose lichen) 144
Phalaris arundinacea (reed canary grass) 410
Phleum pratense 125, 381, 397
Phlox buckleyi 91, 92
Phlox divaricata 164, 383
Phlox pilosa (prairie phlox) 139, 225
Phoradendron spp. (mistletoe) 296
Phyllanthus abnormis 266
Phyllitis scolopendrium (Harts tongue fern) 367, 369
Physalis heterophylla 215
Physalis spp. (ground-cherry) 139, 143
Physalis viscosa (groundcherry) 266
Physcia millegrana (foliose lichen) 144
Physciella chloantha (foliose lichen) 144
Physcomitrium pyriforme 379
Physocarpus opulifolius 378
Picea glauca (white spruce) 343, 368, 382–384, 409, 424
Picea mariana (black spruce) 343, 422–428, 430
Picea rubens (red spruce) 125, 369
Pinguicula (butterwort) 28
Pinguicula vulgaris (butterwort) 368
Pinus aristata (Colorado bristlecone pine) 256
Pinus ayacahuite 289
Pinus banksiana (jack pine) 137, 138, 147, 158, 343, 352, 382, 383, 393, 397, 398, 399, 402, 424
Pinus cembroides (cembroid pinon, pinyon pine) 278, 289
Pinus clausa (sand pine) 7, 11, 15
Pinus echinata (shortleaf pine) 29, 32, 76, 223
Pinus edulis (pinyon pine, true or Colorado pinon) 249, 251, 289
Pinus elliottii densa (south Florida slash pine) 9, 24, 29, 31, 35, 36, 39, 40, 42

Pinus elliottii elliottii (north Florida slash pine) 29
Pinus flexilis (limber pine) 249–251, 253, 256–258
Pinus jeffreyi (Jeffrey pine) 249–251, 313, 315
Pinus joharinus 289
Pinus longaeva (intermountain bristlecone pine) 256
Pinus monophylla (single-leaf pinyon, singleneedle pinon) 249, 289
Pinus palustris (longleaf pine) 11, 24, 29–33, 36, 37, 39, 40, 42
Pinus ponderosa (ponderosa pine, yellow pine) 249–251, 253, 256, 258, 311, 322
Pinus pungens (table-mountain pine) 90, 366
Pinus quadrifolia (peninsular pinon) 289
Pinus resinosa (red pine) 343, 352
Pinus rigida (pitch pine) 75, 76, 343, 393, 399, 401, 402
Pinus sabiniana (foothill pine, gray pine) 313, 315, 325, 328
Pinus serotina (pond pine) 29, 32
Pinus strobus (white pine) 95, 343, 383, 384, 398
Pinus taeda (loblolly pine) 29, 32, 111
Pinus teocate 289
Pinus virginiana (Virginia pine) 67, 71, 74, 75, 88, 90
Piperia unalascensis 383, 386
Piptochaetium fimbriatum (pinyon ricegrass) 278
Piptochaetium spp. 290
Pityopsis graminifolia (golden-leafed aster) 41
Plagiothecium cavifolium 369
Plantago patigonica (woolly plantain) 266
Plantago rugelii 164
Plantago virginica 214, 215
Platanthera leucophaea (western prairie fringed orchid) 241
Platanthera praeclara (eastern prairie fringed orchid) 241
Platanus racemosa (western sycamore) 325
Pleurochaete squarrosa 206, 213, 216
Pleurozium 424
Poa alpina 387
Poa canbyi (Canby blue-grass) 367, 387
Poa compressa (Canada bluegrass) 125, 380–382, 384, 397, 398
Poa curtifolia 317
Poa fendleriana (muttongrass) 249
Poa glauca (glaucous spear-grass) 367
Poa glaucantha 387
Poa palustris (fowl bluegrass, fowl meadowgrass) 368, 377, 409
Poa pratense (blue-grass, Kentucky bluegrass) 164, 267, 384, 416

Poa spp. (bluegrass) 267
Poa trivialis (rough-stalked meadow-grass) 368
Polanisia jamesii (James' clammyweed) 266
Polycnemum arvense 385
Polygala alba (white milkwort) 266
Polygala polygama 402
Polygala sanguinea 195, 198, 199
Polygala senega 381, 384
Polygala spp. 290
Polygala verticillata 195, 199
Polygonatum biflorum (hairy Solomon's seal)
    366, 368
Polygonella articulata (sand jointweed) 351
Polygonella basiramia 12
Polygonum aviculare 125
Polygonum careyi 401
Polygonum coccineum (swamp persicaria) 410
Polygonum douglasii 387, 397, 402
Polygonum erectum 387
Polygonum tenue 71, 72
Polypodium appalachianum 123, 125
Polypodium polypodioides 211
Polypodium virginianum (rock polypody) 365
Polystichum lemmonii (serpentine hollyfern) 318
Polytrichum 424
Polytrichum appalachianum 128
Polytrichum commune 104, 109, 110, 113, 114, 399
Polytrichum ohioense 102, 113
Populus balsamifera (balsam poplar) 409, 427
Populus deltoides (cottonwood, eastern cotton-
    wood, plains cottonwood) 238, 249, 368
Populus grandidentata (big-toothed aspen, large-
    toothed aspen) 343, 352
Populus tremuloides (trembling aspen) 159, 256,
    364, 344, 352, 382, 384, 398, 406–409,
    413, 414, 427
Porteranthus stipulatus 176
Portulaca coronata (purslane) 278
Portulaca mundula (shaggy portulaca) 266
Portulaca oleracea (common purslane) 212, 266,
    379
Portulaca smallii 101, 102, 114
Potamogeton bicupulatus 401
Potamogeton pectinatus (sago pondweed) 412
Potamogeton richardsonii (Richardson's
    pondweed) 412
Potentilla arguta 381, 384, 387, 402
Potentilla canadensis 71, 72, 74
Potentilla fruticosa 381, 387
Potentilla norvegica 380, 384
Potentilla simplex 164, 198, 398
Potentilla tridentata 399
Prenanthes alba (rattlesnake root) 368
Prenanthes racemosa 387

Prenanthes roanensis 123, 124
Primula mistassinica 387
Prosopis glandulosa (mesquite) 265
Prosopis juliflora (mesquite) 237
Prunella vulgaris 198, 381, 384
Prunus americana (wild plum) 162, 163
Prunus angustifolia 198
Prunus depressa 380–382
Prunus gracilis (Oklahoma plum) 267
Prunus injucunda 111
Prunus mexicana (Mexican plum) 237
Prunus pensylvanica (pin cherry) 352, 398,
    412
Prunus pumila (sand cherry) 351
Prunus pumila var. *depressa* 387
Prunus serotina (black cherry) 31, 32, 111, 138,
    176, 198, 353, 368
Prunus spp. (cherry) 343
Prunus susquehanae 398–402
Prunus virginiana (choke cherry) 352, 364, 369,
    382, 384, 386, 398, 410, 412
Pseudotaenidia montana 92
Pseudotsuga menziesii (Douglas-fir) 256
Psychotria nervosa (wild coffee) 35
Ptelea trifoliata (hop-tree) 111, 387
Pteridium aquilinum (bracken fern) 138, 139,
    351, 353, 368, 383, 397, 398
Ptilidium pulcherrimum 369
Puccinellia nuttalliana (Nuttall's alkali grass) 410
Punctelia rudecta (foliose lichen) 144
Purshia tridentata (bitterbrush) 250
Pycnanthemum flexuosum 387
Pycnanthemum pilosum 198
Pycnanthemum tenuifolium 195, 196, 198, 199
Pycnanthemum torreyi 75
Pycnanthemum virginianum 162
Pyrola asarifolia (pink wintergreen) 411

Quercus (oak) 222, 223
Quercus ablocincta (cusi) 277
Quercus agrifolia (coast live oak) 322–325, 328,
    329, 332, 333, 335
Quercus alba (white oak) 78, 88, 135, 137, 138,
    141, 148, 173, 174, 176–178, 182, 185,
    201, 223, 382, 383, 397, 398
Quercus arizonica (Arizona white oak) 277, 280
Quercus borealis 382, 397
Quercus chapmanii (Chapman's oak) 9
Quercus chihuahensis (Chihuahua oak) 277
Quercus chuchuichupensis 277
Quercus coccinea (scarlet oak) 88, 140, 171, 174,
    179, 185
Quercus douglasii (blue oak) 322, 324, 325–328,
    330–335

Quercus ellipsoidalis (Hill's or northern pin oak, scrub oak) 137, 138, 147, 158, 351, 352

Quercus emoryi (Emory oak) 276–278, 280, 324

Quercus engelmannii (Engelmann oak) 322, 323, 324, 325, 327, 333, 335

Quercus falcata (southern red oak) 31, 32, 171, 174, 183, 185, 198

Quercus fusiformis (plateau live oak) 238

Quercus gambelii (Gambel oak) 249, 263

Quercus geminata (sand live oak) 9, 12, 32

Quercus georgiana (Georgia oak) 101, 111, 114

Quercus havardii (sand shinnery oak) 262, 263

Quercus hemisphaerica (laurel oak) 32

Quercus ilicifolia (scrub oak) 88, 90, 395, 399, 401, 402

Quercus imbricaria (shingle oak) 161, 163, 176

Quercus incana (bluejack oak) 31, 32

Quercus inopina (scrub oak) 9, 11

Quercus kelloggii (black oak) 325

Quercus laevis (turkey oak) 10, 31, 32

Quercus lobata (valley oak) 322, 324–328, 332, 335

Quercus macrocarpa (bur oak) 135, 137, 138, 157, 159, 223, 249, 382–384, 410

Quercus margaretta (sand-post oak) 31, 32

Quercus marilandica (blackjack oak) 31, 32, 71, 73, 74, 76, 78, 138, 141, 171, 174, 179, 182–183, 201, 226, 231, 233, 234, 240

Quercus minima (runner oak) 33, 39

Quercus mohriana (Mohr's oak) 263

Quercus muehlenbergii (chinquapin or yellow oak) 138, 148, 238, 382, 383, 387

Quercus myrtifolia (myrtle oak) 9, 11–12

Quercus nigra (water oak) 31, 32, 111

Quercus oblongifolia (Mexican blue oak) 277, 280

Quercus palustris (pin oak) 176, 223

Quercus prinoides acuminata (yellow chestnut oak) 224

Quercus prinus (chestnut oak, rock chestnut oak) 88, 90, 111, 171, 174, 179, 183–186

Quercus pumila (runner oak) 33

Quercus rubra (northern red oak, red oak) 88, 90, 138, 178, 179, 201, 343, 352, 364, 398

Quercus santaclarensis 277

Quercus shumardii (Shumard oak) 238

Quercus spp. (oak) 7, 9, 10, 249, 288, 289

Quercus stellata (post oak) 31, 33, 74, 75, 78, 171, 173, 174, 176–179, 182–185, 201, 224, 226, 231, 233, 234, 240, 263

Quercus undulata 263

Quercus velutina (black oak) 74, 78, 135, 137, 138, 141, 148, 162, 163, 173, 174, 176–179, 182, 183, 185, 201, 224, 234, 236

Quercus virginiana (live oak) 35

Quercus wislizenii (interior live oak) 322–328, 331, 335

Radula complanata 369

Randia aculeata 35

Ranunculus acris (tall buttercup) 368

Ranunculus fascicularis (early buttercup) 164, 165, 381, 384, 387

Ranunculus hispidus (bristly buttercup) 182

Ranunculus subrigidus (water crowfoot) 412

Ratibida columnaris (upright prairie-coneflower) 266

Ratibida columnifera (long-headed coneflower) 409

Ratibida pinnata (yellow cone flower) 161, 162, 387

Reverchonia arenaria (sand reverchonia) 266

Rhamnus caroliniana 214

Rhamnus cathartica (European buckthorn) 417

Rhamnus frangula (European buckthorn) 71

Rhexia virginica 401

Rhizobium (bacterium) 313

Rhododendron atlanticum (rhododendron) 33

Rhododendron catawbiense 123–125

Rhododendron minus 125

Rhododendron vaseyi 124

Rhus (sumac) 33

Rhus aromatica (skunkbrush) 210, 213, 214, 267, 382–384, 387

Rhus copallina (shining sumac, winged sumac) 33, 35, 111, 141, 195, 198, 199, 237, 387, 399, 401, 402,

Rhus glabra (smooth sumac) 75, 143, 237

Rhus radicans also see *Toxicodendron radicans* (poison ivy) 74, 369

Rhus spp. (sumac) 139

Rhus trilobata (skunkbush sumac) 249, 250

Rhus typhina (staghorn sumac) 141, 364, 369, 384, 397, 398

Rhynchelytrum repens (natal grass) 16

Rhynchospora saxicola 101, 102

Rhytidiadelphus triquestris 369

Ribes glandulosum (fetid currant) 369

Ribes spp. (currant) 257

Riccia sorocarpa 379, 385

Rinodina papillata (crustose lichen) 144

Rosa acicularis (prickly rose) 409, 412, 416

Rosa blanda 384

Rosa carolina 139, 162, 198, 213, 214

Rosa setigera 213, 387

Rosa sp. (rose) 161, 353

Rosa spp. 139, 143

Rosa woodsii (Wood's rose) 409, 412

Rubus alleghaniensis (blackberry) 369, 398

Rubus argutus 198
Rubus cuneifolius 74
Rubus flagellaris 71, 72, 198, 398
Rubus hispidus 398
Rubus idaeus (red raspberry) 124, 369
Rubus pubescens (dwarf raspberry) 353, 412
Rubus spp. 120
Rubus strigosus 398
Rudbeckia hirta 162, 164, 198
Rudbeckia missouriensis 225
Rudbeckia triloba 213
Ruellia humilis 210, 213–216
Rumex acetosella 125, 398
Ruppia maritima (ditch grass) 412

Sabal etonia (scrub palmetto) 7
Sabal palmetto (cabbage palm) 28–31, 33, 35
Sabatia angularis 198
Sagina caespitosa 77
Sagina saginoides 77
Salicornia rubra (red samphire) 410
Salix amygdaloides (peachleaf willow) 249
Salix arctica 77, 423, 424
Salix bebbiana (beaked willow) 410
Salix caroliniana (coastal plain willow) 35
Salix discolor (diamond willow, pussy willow) 353, 410
Salix goodingii (southwestern black willow) 267
Salix humilis (prairie willow) 162, 163, 398
Salix petiolaris (basket willow) 410
Salix spp. (willow) 326, 351, 410, 424
Salix tristis 75
Salsola iberica (Russian thistle) 266
Salvia spp. (sage) 290
Sambucus pubens (red elderberry) 364, 369
Sanguisorba canadensis 124
Sanicula gregaria 164
Sanicula marilandica (snakeroot) 411
Sanicula peckiana (sanicleroot) 317
Sanicula tracyi (sanicleroot) 317
Sapindus saponaria (wild chinaberry) 267
Sarcobatus vermiculatus (greasewood) 257
Sarracenia (pitcher plant) 28
Sassafras albidum (sassafras) 33, 74, 90, 176, 198
Saxifraga michauxii 121, 123–125, 127
Saxifraga virginiensis 71, 72, 379, 382, 384, 396–398
Schedonnardus paniculatus (tumblegrass) 267
Schinus terebinthifolius (Brazilian pepper) 16, 35
Schizachne purpurascens 382
Schizachyrium (bluestem) 31, 33

Schizachyrium rhizomatum (creeping bluestem) 30
Schizachyrium scoparium (little bluestem) 30, 31, 71–76, 124, 125, 139, 161, 162, 179, 180, 182, 183, 195, 196, 198, 199, 213, 216, 223, 225, 233, 237, 249, 265, 377, 378, 380, 384, 400
Schizachyrium stoloniferum (creeping bluestem) 39
Schoenolirion croceum (sunnybell) 104, 110, 114
Scholochloa festucacea (spangletop) 410
Schrankia occidentalis (eastern sensitive briar) 267
Schrankia uncinata (catclaw sensitive briar) 267
Scirpus acutus (Hardstem bulrush) 267
Scirpus cespitosus 123–125, 127, 128, 380, 387
Scirpus validus (bulrush) 412
Scleria oligantha (Nut rush) 111
Scleria pauciflora 72, 74, 75, 195, 199
Scleropogon brevifolius (burro grass) 267
Scorpidium turgescens 385
Scutellaria leonardii 210
Scutellaria ovata 91–93
Scutellaria parvula 210, 214, 215, 380, 381, 384, 385, 387
Sedum laxum heckneri 317
Sedum laxum laxum 317
Sedum moranii 317
Sedum pulchellum (rock-moss) 179, 210, 213–216
Sedum pusillum 101, 114
Sedum rosea 124
Sedum smallii also see Diamorpha cymosa and D. smallii (stone crop) 101, 104, 113
Selaginella apoda 378
Selaginella arenicola (sand spikemoss) 10
Selaginella densa (club-moss) 409
Selaginella rupestris (rock spikemoss) 102, 114, 224, 379, 397, 398, 401
Selaginella selaginoides (spikemoss) 367
Selaginella tortipila 121, 124–126
Senecio anonymus 71, 72, 74–76, 210
Senecio antennariifolius (pussytoes ragwort) 91
Senecio douglasii jamesii (threadleaf groundsel) 266
Senecio fastigiatus 317
Senecio lewisrosei 381
Senecio longlilobus 290
Senecio millefolium 124
Senecio multicapitatus (groundsel) 266
Senecio obovatus 387
Senecio pauperculus 378, 381, 384
Senecio plattensis 76
Senecio spartioides fremontii (broom groundsel) 266
Senecio spp. 411

Senecio tomentosus 104, 110, 114
Senna chapmanii 35
Serenoa repens (saw palmetto) 7, 31, 33, 35, 39–41
Sericocarpus linifolius 198
Setaria leucopila (plains bristlegrass) 267
Setaria macrostachya (plains bristlegrass) 267
Setaria parviflora 198
Shepherdia canadense (Canada buffalo-berry) 369, 384
Sibbaldiopsis tridentata 121, 124, 127, 128
Silene antirrhina 384
Silene cucubalus (bladder campion) 368
Silene spp. (campion) 329
Silphium integrifolium 162
Silphium terbinthinaceum 162, 387
Silybium marianum 329
Simarouba glauca (paradise tree) 35
Sisymbrium spp. (tumbling mustard) 300
Sisyrinchium albidum 164, 210, 215, 387
Sisyrinchium angustifolium 387
Sisyrinchium montanum 384
Sisyrinchium mucronatum 74, 162, 387
Sitanion hystrix (squirreltail) 278
Sium suave (water parsnip) 410
Smilacina racemosa (false Solomon's seal) 140, 353, 368
Smilacina stellata (star-flowered Solomon's-seal, starry false Solomon's-seal) 140, 161, 353, 384, 411
Smilax bona-nox (bullbrier) 179, 185, 214
Smilax glauca (catbrier, greenbrier) 111, 184, 195, 198, 199
Smilax rotundifolia 71–75
Solanum (nightshade) 33
Solanum donianum (nightshade) 35
Solanum dulcamara (nightshade) 369
Solanum elaeagnifolium (silverleaf nightshade) 266
Solidago altissima 196, 198
Solidago arguta 92, 399, 402
Solidago bicolor 398
Solidago canadensis (graceful goldenrod, tall goldenrod) 411
Solidago canadensis var. *scabra* 196
Solidago flexicaulis (zig-zag goldenrod) 368
Solidago glomerata 124, 125
Solidago graminifolia 398
Solidago harrisii (shale barren goldenrod) 91, 92
Solidago hispida 384
Solidago houghtonii 385, 387
Solidago juncea 162, 195, 199, 198, 381, 384, 398
Solidago nemoralis (gray goldenrod, oldfield goldenrod) 71, 72, 74, 140, 143, 164, 195,

198, 199, 210, 378, 381, 384, 396, 398
Solidago ohioensis 378, 385, 387
Solidago ptarmicoides (goldenrod) 140, 384, 385, 387, 402
Solidago puberula (downy goldenrod) 368, 399, 401, 402
Solidago rigida 162, 387
Solidago spathulata ssp. *randii* 387
Solidago spithamaea 123, 124, 127, 128
Solidago spp. (goldenrod) 180
Solidago squarrosa 402
Solidago ulmifolia 161, 164
Solidago ulmifolia var. *ulmifolia* 387
Sorbus americana 123, 125
Sorbus decora (mountain ash) 368
Sorghastrum (Indian grass) 31, 33, 71, 72, 162, 195, 196, 198, 199, 223, 237, 400, 402
Sorghastrum secundum 30, 31
Sorghum halepense (Johnson grass) 240
Sparganium eurycarpum (bur-reed) 410
Spartina pectinata (prairie cordgrass, cordgrass) 199, 223, 409
Spiraea alba (narrow-leaved meadowsweet) 382, 398, 412
Spiranthes casei 398, 399, 402
Spiranthes lacera 396, 398
Spiranthes magnicamporum 387
Spiranthes ochroleuca 399, 402
Spiranthes ovalis var. *erostellata* 387
Sporobolus (dropseeds) 31, 216
Sporobolus asper (tall dropseed) 237, 387
Sporobolus clandestinus (dropseed) 216, 223
Sporobolus contractus (spike dropseed) 267
Sporobolus cryptandrus (sand dropseed) 265
Sporobolus flexuosus (mesa dropseed) 267
Sporobolus giganteus (giant dropseed) 267
Sporobolus heterolepis (prairie dropseed) 71, 73, 75, 224, 377, 378, 380–382, 384, 385, 387
Sporobolus neglectus (sheathed dropseed) 216, 224
Sporobolus ozarkanus 387
Sporobolus spp. 216
Sporobolus vaginiflorus 210, 213–217, 380, 384
Stenosiphon linifolius 225
Stereocaulon (lichen) 424
Stereocaulon pascale (lichen) 422, 426
Stillingia sylvatica (queen's delight) 35, 266
Stipa avenacea 76
Stipa comata (needle-and-thread, speargrass) 267, 409
Stipa lemmonii pubescens (needlegrass) 317
Stipa spartea (porcupine grass) 161, 162, 387, 409
Stipa viridula (green needlegrass) 416
Streptanthus (jewel flower) 316, 317

Streptanthus amplexicaulis var. *barbarae* 318
Streptanthus barbatus 318
Streptanthus barbiger 317
Streptanthus batrachopus 318
Streptanthus brachiatus 316, 318
Streptanthus breweri 318
Streptanthus drepanoides 318
Streptanthus hesperidis 319
Streptanthus howellii (Howell's jewelflower) 317
Streptanthus insignis 318
Streptanthus morrisonii 316, 318
Streptanthus niger (Tiburon jewelflower) 318
Streptanthus polygaloides 318
Streptanthus tortuosus optatus 318
Strophostyles umbellata 198
Stylosanthes biflora 198
Suaeda depressa (western sea blite) 410
Symphoricarpos albus (snowberry) 382–384,
    412
Symphoricarpos occidentalis (western snowberry)
    409, 412, 416
Symphoricarpos orbiculatus (buckbrush) 213, 237
Symphoricarpos oreophilus (snowberry) 256
Symphoriocarpos spp. (snowberry) 140, 149
Symplocos tinctoria (sweetleaf) 33

Taenidia integerrima (yellow pimpernell) 92,
    164, 165, 382
Taenidia montana (Virginia mountain pimpernel)
    91
Talinum calcaricum 210, 211, 214
Talinum calycinum (rockpink flameflower) 266
Talinum mengesii 101, 113
Talinum rugospermum (sand fame-flower) 140
Talinum teretifolium 71–75, 101, 102, 104, 113,
    114
Tamarix spp. (salt cedar) 267
Taraxacum officinale 125, 164
Tauschia howellii 317
Taxodium ascendens (pond cypress) 29
Taxus canadensis (Canada yew) 369
Tephrosia virginiana (goat's rue) 140, 180
Tetrazigia bicolor (Florida tetrazygia) 35
Thalictrum confine 387
Thalictrum macrostylum 76
Thalictrum spp. 411
Thaspium trifoliatum var. *aureum* 387
Thelesperma megapotamicum (prairie green-
    thread) 266
Thelypteris simulata 76
Thlaspi arvense (stinkweed) 417
Thlaspi montanum var. *siskiyouense* (pennycress)
    317, 318

Thrinax (thatch palm) 33
Thuidium abietinum (moss) 369, 382
Thuja occidentalis (northern white cedar) 362,
    364–369, 371, 372, 375, 382, 384
Tiarella cordifolia (false mitrewort) 369
Tilia americana (basswood) 159, 353, 364, 369,
    382, 410
Tofieldia glutinosa 126, 378, 387
Tomanthera auriculata 387
Tortella fragilis (moss) 368
Tortella humilis (moss) 216
Tortella tortuosa (moss) 369, 379
Tortula ruralis (moss) 379
Townsendia exscapa (stemless townsendia) 266
Toxicodendron diversiloba (poison oak) 324–326,
    328
Toxicodendron pubescens 33
Toxicodendron radicans also see *Rhus radicans*
    (poison ivy) 139, 173, 176, 181
Toxicodendron radicans ssp. *radicans* 384
Toxicodendron vernix (poison sumac) 33
Trachypogon montufari (crinkleawn) 278
Tradescantia occidentalis (prairie spiderwort)
    266
Tradescantia ohiensis (smooth spiderwort, spider-
    wort) 104, 110, 140, 162
Tradescantia virginiana (short spiderwort, spider-
    wort) 164, 165, 182
Trema micrantha (Florida trema) 31, 35
Trichostema brachiatum also see *Isanthus brachia-
    tus* 210, 213–215, 379, 380, 384, 385, 387
Tridens flavus (purpletop) 198, 237
Trientalis borealis (northern starflower) 411
Trifolium aureum 397
Trifolium calcaricum 211
Trifolium pratense 125
Trifolium repens (white clover) 368
Trifolium spp. (clover) 329
Trifolium virginicum (Kate's Mountain clover) 92
Triglochin maritima (seaside arrow grass) 410
Trillium grandiflorum (trillium) 368
Trillium pusillum var. *ozarkanum* 225
Triodanis perfoliata 387
Triosteum angustifolium 387
Triosteum aurantiacum (horse-gentian) 149
Triosteum perfoliatum (horse-gentian) 149
Triplasis purpurea (purple sandgrass) 267
Tripsacum lanceolatum (Mexican gamagrass)
    278
Trisetum spicatum 124, 387
Tsuga canadensis (eastern hemlock) 344, 368,
    369
Typha latifolia (cattail) 412, 416

Ulmus alata (winged elm) 174, 179, 210, 213, 214, 234, 237

Ulmus americana (American elm) 222, 238, 368, 369, 409

Ulmus crassifolia (cedar elm) 238

Ulmus thomasii 382

Umbellularia californica (bay) 328

Utricularia (bladderwort) 28

Utricularia vulgaris (bladderwort) 412

Vaccinium angustifolium (blueberry, low sweet blueberry) 138–140, 353, 377, 397, 398, 400

Vaccinium arboreum (sparkleberry) 33, 111, 177–179, 182, 184–185

Vaccinium corymbosum 123, 125, 400

Vaccinium crassifolium 33

Vaccinium darrowii (dwarf blueberry) 11, 33

Vaccinium elliottii 33

Vaccinium erythrocarpum 124, 125

Vaccinium formosum 33

Vaccinium fuscatum 33

Vaccinium myrsinites (dwarf blueberry) 11, 33

Vaccinium myrtilloides 377

Vaccinium spp. (blueberry) 7, 11, 28, 33, 120, 123, 351

Vaccinium stamineum (deerberry) 33, 184, 402

Vaccinium tenellum 33

Vaccinium vacillans (sweet low blueberry) 74, 75, 177, 184–185

Vaccinium vitis-idaea (mountain cranberry) 424

Valeriana edulis ssp. *ciliata* 378

Valerianella umbilicata 387

Verbascum thapsus (mullein) 369

Verbena simplex 210, 213, 387

Verbesina enceloides (cowpen daisy) 266

Vernonia altissima 387

Vernonia gigantea 195, 199

Veronica copelandii 318

Veronica officinalis 215

Veronica peregrina 380, 384

Verrucaria nigrescens 216

Viburnum edule (low-bush cranberry) 368, 410, 412

Viburnum prunifolium (blackhaw) 176

Viburnum rafinesquianum 382–384, 398

Vicia americana (American vetch) 411

Viguiera porteri also see *Helianthus porteri* 106, 400

Viola egglestonii 215

Viola fimbriatula 398, 402

Viola lobata psychodes 317

Viola macloskeyi 124

Viola nephrophylla 378, 385

Viola pallens (northern white violet) 368

Viola papilionacea 162

Viola pedata (bird's-foot violet) 140

Viola pedatifida 162

Viola rafinesquii 387

Viola rugulosa (western Canada violet) 411

Viola sagittata 71, 72, 74

Viola selkirkii (great spurred violet) 369

Viola sororia 164

Viola spp. 164

Virgulus ericoides 381

Vitis aestivalis (summer grape) 184

Vitis riparia (frost-grape, wild grape) 140, 369

Vitis rotundifolia (muscadine grape) 184, 185

Vitis spp. (grape) 267

Vitis vulpina (fox grape) 237

Vulpia octoflora 181

Waltheria indica 35

Warea carteri (Carter's mustard) 11, 15

Woodsia obtusa 399, 401, 402

Woodsia oregana ssp. *cathcartiana* 387

Woodsia scopulina 93

Woodwardia virginica 399

Xanthisma texanum (sleepydaisy) 266

Xanthoparmelia conspersa 102

Xyris difformis 401

Yucca angustifolia (narrow-leaf yucca) 265, 267

Yucca campestris (plains yucca) 267

Yucca spp. 290

Zamia integrifolia (Florida coontie) 35

Zannichellia palustris (horned-pondweed) 412

Zanthoxylum americanum (prickly Ash) 163, 382

Zea mays (maize, corn) 227

Zigadenus glaucus 378, 381, 384, 387

Zigadenus leimanthoides 124

Zinnia grandiflora (plains zinnia) 266

Zizia aptera (heart-leaved alexander) 378, 411

Zizia aurea 383, 387

Ziziphus celata 15

Ziziphus obtusifolia (lotebush) 267

Index of Animals

Accipiter cooperi (Cooper's hawk) 268
Accipiter striatus (sharp-shinned hawk) 269
Adelges piceae (balsam wooly adelgid) 129
Agelaius phoeniceus (red-winged blackbird) 416
Agkistrodon contortrix phaeogaster (Osage copper-head) 227
Agrotis buchholze (Buchholz's dart moth) 59
Aimophila aestivalis (Bachman's sparrow) 36, 227
Ambystoma (salamander, mole salamander) 367
Ambystoma cingulatum (flatwoods salamander) 36
Ambystoma macrodactylum (long-toed salamander) 329
Ammodramus savannarum (grasshopper sparrow) 146, 335
Anas acuta (northern pintail) 416
Anas discors (blue-winged teal) 416
Anas platyrhynchos platyrhynchos (mallard) 416
Aneides aeneus (green salamander) 127
Antilocapra americana (pronghorn) 268, 414
Aphelocoma coerulescens (Florida scrub-jay) 10, 12–14, 18
Aphelocoma ultramarina (gray-breasted jay) 280
Aquila chrysaetos (golden eagle) 127, 268
Arizona elegans (glossy snake 269)
Asio flammeus (short-eared owl) 415
Athene cunicularia (burrowing owl) 268, 418
Atrytonopsis hianna (dusted skipper) 146
Aythya affinis (lesser scaup) 416
Aythya americana (redhead) 416
Aythya valisineria (canvasback) 416

Bartramia longicauda (upland sandpiper) 146, 354, 355
Bison bison (bison) 144, 227, 241, 416
Blarina brevicauda (shorttail shrew) 355
Bombycilla cedrorum (cedar waxwing) 354
Bonasa umbellus (ruffed grouse) 146, 148, 227
Bufo americanus (American toad) 146
Bufo quercicus (oak toad) 36
Bufo woodhousei (Fowler's toad) 146
Buteo jamaicensis (red-tailed hawk) 269, 415
Buteo lineatus (red-shouldered hawk) 355

Buteo regalis (ferruginous hawk) 268
Buteo swainsoni (Swainson's hawk) 268, 415

Calamospiza melanocorys (lark bunting) 268
Calcarius lapponicus (Lapland longspur) 415
Calcarius ornatus (chestnut-collared longspur) 414
Callipepla squamata (scaled quail) 268
Camelops spp. (camels) 295
Canis latrans (coyote) 146, 227, 268, 355, 414
Canis lupus (gray wolf) 227, 355
Cardinalis cardinalis (northern cardinal) 227
Carduelis tristis (American goldfinch) 146, 354, 369
Carphophis vermis (western worm snake) 227
Castor canadensis (beaver) 395, 415
Cathartes aura (turkey vulture) 269
Catharus guttatus (hermit thrush) 415
Catocala herodias gerhardi (pine barrens underwing moth) 59
Catocala jair (jair underwing moth) 59
Cecropia spp. 240
Centruroides vittatus (plains scorpion) 227
Cephaloziella obtusilobula 127
Cervalces spp. (fossil forms of elk-moose) 295
Cervus canadensis (elk) 227, 298, 414
Cervus elaphus (elk) 144
Cervus spp. (elk) 295
Charadrius vociferous (killdeer) 416
Chlidonias niger (black tern) 415
Chlosyne gorgone (gorgone checkerspot) 146
Chondestes grammacus (lark sparrow) 355
Chordeiles minor (common nighthawk) 60
Choristoneura lambertiana ponderosana (budworm) 256
Choristoneura pinus pinus (jack pine budworm) 351
Circus cyaneus (northern harrier) 269, 355, 415
Citellus franklinii (Franklin's ground squirrel) 146, 415
Citellus tridecemlineatus (thirteen-lined ground squirrel) 146, 415
Cnemidophorus sexlineatus (six-lined racerunner) 269

Cnemidophorus sexlineatus sexlineatus (six-lined racerunner) 227

Cnemidophorus tigris (western whiptail lizard) 269

Coccyzus americanus (yellow-billed cuckoo) 401

Coccyzus erythropthalmus (black-billed cuckoo) 146

Colaptes auratus (northern flicker) 354

Coleoptera (beetles) 269

Coleotechnites chillcotti (Louisiana longleaf needleminers) 37

Coleotechnites edubiola (pinon needleminer) 296

Colinus virginianus (bobwhite quail) 36, 227, 268

Coluber constrictor (blue racer, racer) 146

Contopus virens (eastern wood-pewee) 227

Conuropsis carolinensis (Carolina parakeet) 241

Corvidae 295

Corvus brachyrhynchos (American crow) 354

Corvus corax (common raven) 127

Corvus cryptoleucus (Chihuahuan white-necked raven) 268

Crotalus adamanteus (eastern diamondback rattlesnake) 28, 36

Crotalus atrox (western diamondback rattlesnake) 269

Crotalus horridus (timber rattlesnake) 59

Crotalus viridis (western rattlesnake) 269

Crotaphytus collaris collaris (eastern collared lizard) 227

Cyanocitta cristata (blue jay) 146, 354

Cyanocitta stelleri (Steller's jay) 256

Cydia ingens (longleaf pine seedworm) 37

Cynomys ludovicianus (prairie dog) 267

Dendroctonus frontalis (southern pine beetle) 37

Dendroctonus ponderosae (mountain pine beetle) 256

Dendroica discolor (prairie warbler) 58, 401

Dendroica kirtlandii (Kirtland's warbler) 354, 355

Dendroica pennsylvanica (chestnut-sided warbler) 354

Dendroica petechia (yellow warbler) 354

Diadophis punctatus arnyi (prairie ringneck snake) 227

Dioryctria amatella (southern pine coneworm) 37

Dioryctria spp. (moth) 296

Dipodomys ordii (Ord's kangaroo rat) 265

Dolichonyx oryzivorus (bobolink) 355, 388

Drymarchon corais (eastern indigo snake) 36

Dugesiella hentzi (Missouri tarantula) 227

Dumetella carolinensis (gray catbird) 146, 354

Ectopistes migratorius (passenger pigeon) 241

Elaphe guttata (corn snake) 59

Elaphe obsoleta obsoleta (black rat snake) 401

Empidonax virescens (Acadian flycatcher) 228

Emydoidea blandingi (Blanding's turtle) 356

Eptesicus fuscus (big brown bat) 355

Eremophila alpestris (horned lark) 414

Erethizon dorsatum (porcupine) 268, 295, 415

Erynnis martialis (mottled dusky wing) 146

Erynnis persius (Persius dusky wing) 147

Euceratherum spp. (shrub ox) 295

Euchloe olympia (Olympia marblewing) 96, 146, 401

Eucosma babana (pinon moth) 296

Eumeces egregius (mole skink) 36

Eumeces egregius lividus (bluetail mole skink) 13, 15

Eumeces fasciatus (five-lined skink) 146, 356, 401

Eumeces septentrionalis (northern prairie skink) 356

Euphagus cyanocephalus (Brewer's blackbird) 146, 353

Euphydryas edithae (checkerspot butterfly) 319

Falco femoralis (Aplomado falcon) 268

Falco mexicanus (prairie falcon) 269

Falco peregrinus (peregrine falcon) 127, 268, 329

Formicidae (ants) 269

Fringillids 295

Gambelia silus (blunt-nosed leopard lizard) 329

Gastrophryne carolinensis (eastern narrowmouth toad) 227

Geococcyx californianus (roadrunner) 227

Geomys bursarius (plains pocket gopher) 146, 268, 355

Geomys pinetis (southeastern pocket gopher) 36

Geothlypsis trichas (common yellowthroat) 354

Glossotherium spp. (ground sloth) 295

Goats 270

Gopherus polyphemus (gopher tortoise) 13, 17, 28, 36, 37

Gymnogyps californianus (California condor) 328, 329

Gymnorhinus cyanocephalus (pinyon jay) 256

Haliaeetus leucocephalus (bald eagle) 329

Halysidota imagens (moth) 296

Helmitheros vermivorus (worm-eating warbler) 228

Hesperia leonardus (Leonard's skipper) 146

Hesperia metea (cobweb skipper, metea skipper) 146

Hesperia ottoe (Ottoe skipper) 146

Heterocampa varia (notodontid moth) 59

Heterodon nasicus (western hognose snake) 269

Heterodon platirhinos (eastern hognose snake, hognose snake) 146, 336

Hirundo rustica (barn swallow) 268

Hyla andersoni (pine barrens treefrog) 59

Hyla chrysoscelis (Cope's gray treefrog) 146

Hylocichla mustelina (wood thrush) 228, 369

Hypochilus coylei (lampshade spider) 127

Hypochilus sheari (lampshade spider) 127

Hypsiglena torquata (night snake) 269

Icterus bullockii (northern Bullock's oriole) 269

Icterus galbula (Baltimore oriole) 146

Ictinia mississippiensis (Mississippi kite) 269

Incisalia irus (frosted elfin butterfly) 147, 356

Ips spp. (bark beetles) 296

Iridoprocne bicolor (tree swallow) 354

Itame sp. (an unnamed species of geometrid moth) 59

Lampropeltis calligaster calligaster (prairie kingsnake) 227

Lanius ludovicianus (loggerhead shrike) 268, 385, 389

Larus argentatus (herring gull) 418

Larus pipixcan (Franklin's gull) 418

Lasionycteris noctivagans (silver-haired bat) 355

Lasiurus borealis (red bat) 355

Latrodectus mactans (black widow spider) 227

Lepidoptera 59, 319

Leptoglossus corculus (leaf-footed pine seed bug) 37

Lepus americanus (snowshoe hare) 356

Lepus californicus (blacktailed jackrabbit) 268

Lepus townsendii (white-tailed jack rabbit) 414

Liomys (pocket mouse) 268

Lycaeides melissa samuelis (Karner blue butterfly) 147, 356

Lymantria dispar (gypsy moth) 395

Lynx rufus (bobcat) 268, 355, 369

Malacosoma disstria (forest tent caterpillar) 395

Mammut spp. (mastodons) 295

Marmota monax (woodchuck) 355

Masticophis (coachwhip) 269

Masticophis flagellum flagellum (eastern coachwhip) 227

Matsucoccus acalytus (pinon needle scale) 296

Megalonyx spp. (ground sloth) 295

Meleagris gallopava (turkey) 227, 295

Melospiza melodia (song sparrow) 146, 354

Mephites mephites (striped skunk) 146, 268, 415

Mexican ground squirrel 268

Microsorex hoyi 355

Microtus pennsylvanicus (meadow vole) 414

Microtus spp. (vole) 355

Mimus polyglottos (northern mockingbird) 269

Molothrus ater (brown-headed cowbird) 146, 228, 241, 354, 415

Mus musculus (house mouse) 268

Mustela nigripes (black-footed ferret) 241

Mustela sp. (weasel) 414

Mustela vison (mink) 416

Myiarchus crinitus (great crested flycatcher) 227

Myotis lucifugus (little brown myotis) 355

Neacoryphus bicrucis (ragwort seed bug, white crossedseed bug) 114

Neoseps reynoldsi (sand skink) 13, 15

Neotoma floridana (wood rat) 127

Neotoma lepida (southern plains woodrat) 268

Neotoma magister (wood rat) 127

Neotoma stephensii (Stephen's woodrat) 295

Nothrotheriops spp. (ground sloth) 295

Notophthalmus perstriatus (striped newt) 36

Nucifraga columbiana (Clark's nutcracker) 256, 257

Nyctea scandiaca (snowy owl) 415

Oarisma garita (garita skipper) 387

Ochotona (eastern giant pika, extinct) 363

Odocoileus hemionus (mule deer) 268, 295, 298, 415

Odocoileus spp. (deer) 295

Odocoileus virginiana (white-tailed deer) 58, 69, 96, 146, 227, 355, 414

Oeneis chryxus (chryxus arctic) 146, 401

Ondatra zibethicus (muskrat) 415

Onychomys leucogaster (northern grasshopper mouse) 268

Onychomys torridus (southern grasshopper mouse) 265

Opheodrys vernalis (green snake, smooth green snake) 146

Ophisaurus attenuatus (slender glass lizard) 146

Ophisaurus attenuatus attenuatus (western slender glass lizard) 227

Oporornis agilis (Connecticut warbler) 355

Otus asio and/or *O. kennicotti* (eastern and/or western screech owl) 269

Pandion haliaetus (osprey) 269

Papilio joanae (Missouri woodland swallowtail) 227

Papilio zelecaon 319

Parus atricapillus (black-capped chickadee) 354

Parus carolinensis (Carolina chickadee) 58

Passer domesticus (house sparrow) 418

Passerina cyanea (indigo bunting) 146, 227, 354

Pedioecetes phasianellus also see *Tympanuchus phasianelluus* (sharp-tailed grouse) 146, 149, 414

Peromyscus boylei (brush mouse) 227

Peromyscus leucopus (white-footed mouse) 58, 146, 268, 369

Peromyscus maniculatus (deermouse, prairie deer mouse) 146, 268, 369, 414

Peromyscus spp. (deer mice) 295, 355

Petrochelidon pyrrhonota (cliff swallow) 369

Petrovia arizonensis (moth) 296

Pheucticus ludovicianus (rose-breasted grosbeak) 354

Phloeosinus spp. (twig beetle) 296

Phrynosoma cornutum (Texas horned lizard) 269

Phyciodes batesii subsp. *batesii* (dark crescent butterfly) 382, 385, 401

Picoides borealis (red-cockaded woodpecker) 28, 36, 37

Picoides pubescens (downy woodpecker) 60, 415

Picoides villosus (hairy woodpecker) 415

Pinonia edulicola (pinon midge) 296

Pipilo erythrophthalmus (rufous-sided towhee) 36, 58, 146, 353, 354, 388

Piranga rubra (summer tanager) 227

Pituophis melanoleucus (bullsnake, pine snake) 146, 269

Pituophis melanoleucus mugitus (Florida pine snake) 36

Pituophis melanoleucus sayi (bullsnake) 227, 356

Pityophythorus spp. (twig beetle) 296

Plecotus townsendii virginianus (Virginia big-eared bat) 127

Plectrophenax nivalis (snow bunting) 415

Podomys floridanus (Florida mouse) 13

Polioptila caerulea (blue-gray gnatcatcher) 227

Pooecetes gramineus (vesper sparrow) 146, 354, 355, 415

Procyon lotor (raccoon) 146, 369, 416

Pyrgus wyandot (grizzled skipper) 96

Rana areolata (gopher frog, crawfish frog) 36

Rangifer tarandus (caribou) 421

   var. *groenlandicus* 430

   var. *caribou* 430

Rhadinaea flavilata (pine woods snake, yellow-lipped snake) 36

Rhinocheilus lecontei (longnosed snake) 269

Rhyacionia frustrana (Nantucket pine tip moth) 37

Riparia riparia (bank swallow) 146

Saiga spp. (Saiga) 295

Sceloporus undulatus (fence lizard) 36

Schinia indiana (phlox flower moth) 146, 356

Sciurus carolinensis (gray squirrel) 146

Sciurus niger (fox squirrel) 36, 146

Scleoporus woodi (Florida scrub lizard, scrub lizard) 13, 15

Seiurus aurocapillus (ovenbird) 228, 354

Sialia sialis (eastern bluebird) 146, 354

Sigmodon hispidus (hispid cotton rat) 268

Sistrurus catenatus (massasauga) 269

Sistrurus miliarius streckeri (pigmy rattlesnake) 227

Sitta pusilla (brown-headed nuthatch) 28, 36

Sorex cinereus (masked shrew) 355

Spermophilus franklinii see *Citellus franklinii*

Spermophilus spilosoma (spotted ground squirrel) 268

Spermophilus spp. (ground squirrel) 355

Spermophilus tridecemlineatus see *Citellus tridecem-lineatus*

Speyeria idalia (regal fritillary) 146, 147

Spharagemon saxatile (ledge locust) 127

Spiza americana (dicksissel) 355

Spizella pallida (clay-colored sparrow) 415

Spizella passerina (chipping sparrow) 354

Spizella pusilla (field sparrow) 146, 354, 355

Sterna antillarium (least tern) 241

Strymon edwardsii (Edward's hairstreak) 146

Strymon melinus (gray hairstreak) 401

Sturnella magna (eastern meadowlark) 146

Sturnella neglecta (western meadowlark) 355, 414

Styloxus bicolor (juniper twig pruner) 296

Sylvilagus audubonii (desert cottontail) 268

Sylvilagus floridanus (eastern cottontail rabbit) 356

Tamiasciurus hudsonicus (red squirrel) 256

Tantilla gracilis (flathead snake) 227

Tantilla oolitica (Miami black-headed snake, rim rock crowned snake) 36

Taxidea taxus (badger) 146, 268, 355, 414

Terrapene ornata (ornate box turtle, western box turtle) 146, 227, 269

Tetralopha robustella (pine webworm) 37

Thamnophis proximus (western ribbon snake) 146

Thamnophis sirtalis tetrataenia (San Francisco garter snake) 329

Thomomys talpoides (northern pocket gopher, pocket gopher) 414

Toxostoma rufum (brown thrasher) 146, 354, 388

Trachykele blondelii (western cedarborer) 296

Trimerotropis saxatilis (lichen grasshopper, winged locust) 127, 227

Turdus migratorius (American robin) 146, 354

Tympanuchus cupido (greater prairie-chicken) 355

Tympanuchus pallidicinctus (lesser prairie-chicken) 268

Tympanuchus phasianellus also see *Pedioecetes phasianellus* (sharp-tailed grouse) 354, 355, 387

Tyrannus forficata (scissor-tailed flycatcher) 269

Tyrannus tyrannus (eastern kingbird) 146, 354

Tyrannus verticalis (western kingbird)

Ursus americanus (black bear) 241, 256, 355

Ursus arctos horribilis (grizzly bear) 256

Uta stansburiana (side-blotched lizard) 269

Vermivora peregrina (Tennessee warbler) 355

Vermivora ruficapilla (Nashville warbler) 369

Vespamima spp. 296

Vireo atricapillus (black-capped vireo) 241, 295

Vireo olivaceus (red-eyed vireo) 228, 354, 369

Virginia striatula (rough earth snake) 227

Vulpes macrotis (San Joaquin kit fox) 328

Vulpes velox (swift fox) 268

Vulpes vulpes (red fox) 355, 414

Western kingbird 269

Wilsonia citrina (hooded warbler) 228

Xanthocephalus xanthocephalus (yellow-headed blackbird) 416

Zadipiron spp. (sawfly) 296

Zenaida asiatica (white-winged dove) 280

Zenaida macroura (mourning dove) 146, 269, 415

Zonotrichia leucophrys (white-crowned sparrow) 269

Topic Index

acclimate 87, 94
accumulation heavy metals 319
acorns 334
adaptive plant response 312
agriculture 222, 223, 228, 413, 414
air pollution 129
albedo 298
Alberta 406, 409–410
Aldrich Mountains, Oregon 311
Alfisol 27, 159, 235
algae 378, 379, 385
Allegheny Plateau 83
alpine treeline 249, 255, 256
alvar 375–390
    alien species 388, 389
    community 379–383, 388, 389
    grassland 378, 380, 382, 383
    grassland savanna 382, 383
    grassy woodland 378, 383
    pavement 375, 378, 381, 383
    pavement savanna 378, 379, 381, 382
    pavement woodland 378, 383
    protection 375, 388, 389
    rare species 381–385, 388, 389
    shrubby woodland 379, 380, 383
    shrubland savanna 379, 382
Amerindian 68, 69, 79, 333, 334
Anasazi 298
ancient woodland 362–374
animals of Great Lakes region dry oak savanna 144–148
annual grasses 315
Apache Indians 281
Appalachian Mountains, folded, 84
arachnids 367
Arbuckle Mountains, Oklahoma 238
archaeology 416, 417
Archaic Period 296, 297
Arcto-Tertiary Geoflora 290
argillic horizons 27, 279, 280
Argixerols 312
Arkansas River 234
aspect 177, 179, 182, 184
aspen grove dynamics 412, 413
aspen parkland

climate 406
    fauna 414–416
    geographic location 406, 407
    human impact 413, 414, 416–418
    physiography 407, 408, 410
    vegetation 409–412
atmospheric CO_2 283
avian communities 352–355

Baldy Mountains, Oregon 311
Bare Hills, Maryland 70–72
bark beetles and limber pine 256
    and ponderosa pine 254, 255
bark thickness 252, 253
barrens 67–72, 75–77, 79, 135, 136, 171, 184, 220–223, 225–228, 309, 316
barrens flora, Midwest, U.S.A. 163
barrens, pine 140, 141, 147, 148
basal area 175, 177, 182, 278, 324–326, 328
basalt glades 400
basket makers 297
Bay Area, California 310, 315
bedrock 69, 76, 77, 84, 177, 178, 184, 190, 191, 207, 220–223, 226, 392, 396
Big Barrens of Ky. and Tenn.
    boundaries and definition 190
    physiography 191, 192
biomass 106, 111, 112, 179, 182, 238
birds and seed dispersal, limber pine 256
Black Hills, South Dakota 249, 252, 253
blister rust 296
blue oak savannah 316
Blue Ridge shale barrens 86, 91, 95
Blue River, Oklahoma 238
blue-stain fungi and ponderosa pine 254, 255
bodenvag species 316
Boston Mountains, Arkansas 220, 221
Brallier shales 84, 86
Brazos River, Texas 234
British Columbia 311, 313
brood parasitism 228
brush prairie 160, 162, 165, 166
bryophytes 129, 171

C:N ratio 347, 348

C4 photosynthesis 279, 283
Caddo Canyons, Oklahoma 238
calcareous dry savanna 135, 148, 149
calcium 67, 77–79
    exchangeable 312
California 310, 313, 314
    northwestern 314
California oak savanna
    community classification 325
    development of 333, 335
    diameter at breast height (dbh) 325, 326
callosities 319
Canada 67, 75, 76
Canadian River, Oklahoma 234
Canadian Shield 392, 393, 398, 399, 401
canopy 87, 88
carbon cycle 298
caribou 430, 431
Cascade Mountains, Pacific Northwest, U.S.A.
    250
Cassiar Mountains, British Columbia 311
cation exchange capacity 312, 347
cattle 297
cavity nesters 356
cedar glades 375, 389
    definition 206
    flora 210
    geography 206, 207
    photosynthetic pathways in plants 212
    phytogeographical relationships 212
    plant communities 212–216
    plant life forms 210, 211
Central Lowland 234
Chaco Canyon, New Mexico 297
chaparral 13, 313
charcoal 298
Chelan County, Washington State 311
Cherry Hill, Maryland 68, 70, 72, 73
chert 221, 223
chipboard 299
chromite 62, 312
chromium 67, 77, 312
cicadas 296
Cimarron River, Oklahoma and Arkansas
    234
clear cut, jack pine barrens 357
cleistogamy 103
cliffs, Niagara Escarpment 362–374
climate 25, 26, 42, 43, 67, 68, 172, 192, 193,
    209, 396, 425, 426
Coast Ranges, California and Oregon 310, 323
coastal Florida scrub 10, 18
cobalt 312
Colorado Plateau 249, 252

communities of open woodlands of S. Ill.,
    western Ky., middle Tenn. 172, 174
community classification in California oak
    savanna 325
competition 106, 108, 109, 112, 366
conifer ecosystem 311
conifer forest 313, 314
connectivity and patch size of jack pine bar-
    rens 357
conservation 96, 115, 126, 128, 129
    strategies 17
continental margin 312
Contra Costa County, California 310
cool season grasses 290
Craig County, Virginia 90
Cretaceous 234
cryoturbation 77, 79
cryptogamic soil crust 144
cryptogams 93, 144, 225
Cumberland Plateau 92
Curry County, Oregon 310

deciduous oak
    California oak savanna, 325, 330
    southwestern oak savanna, 278
decomposition 348
deep-soil barrens 155–170, 190–205
deer 69, 96, 295, 298
defoliation 395, 399
demography of scrub plants 12
desert grassland 275, 276
desert shrublands 249
desertification 298
Detrended Correspondence Analysis 55, 56,
    123
Devil's Den State Park, Arkansas 225
Devonian 84, 86
disease of jack pine, 351
disjunct population 92
disjunct species 119, 121–123, 126, 127
disturbance 119–121, 125, 128, 129, 176, 183,
    353, 370, 371, 392, 399
disturbance interactions 344, 347–349, 351,
    357, 358
dolomite 221, 223
drainage 192
drought 70, 77, 79, 173, 182, 239, 240, 375,
    377, 380, 382, 383, 388, 389, 392, 393,
    395, 397, 399, 403
drought stress 9, 88, 316
dry sand savanna 135–148
Dubakella Series, 312
dunes 264
dunite 312

Eastern Highland Rim 200, 201
Eastern Sand Savanna 141
ecological processes 375, 389
ecophysiology 93, 94, 96
ecotypes 58, 59, 319
edaphic
    characteristics 396
    features 344, 347, 357
    islands 316
Edwards Plateau 238
egg mimicry 319
El Nino Southern Oscillation 277
elevation gradient 123–125
elk 298
encinal 275, 277, 278, 282, 283
endangered species 14, 15, 58, 329, 354, 356,
    388
endemism 15, 58, 83, 86–88, 90–96, 100–103,
    115, 225, 309, 313–316, 319, 367, 399,
    401
Entisols 27
equids 298
escarpment woodlands 249, 250, 252
escarpments 250
evergreen 325, 330
evergreen oak 275, 277, 278, 283
exclusion of heavy metal 319
exfoliation 103
exotic species 16, 96, 125, 166, 312

fault zones 312
fauna of SE pine savanna, 36–38
fauna of western serpentine, 319
felsic bedrock 121, 124, 125
fence law enactment 222
fire 7, 11, 14, 16, 53, 57–62, 67, 69, 141, 142,
    158–163, 165, 166, 171, 173, 176–178,
    182–186, 220, 222, 223, 225–228, 239,
    280, 282, 283, 290, 299, 334, 347, 375,
    378, 381–383, 388, 389, 395, 399, 400,
    403, 417, 426, 427
    and douglas-fir 252
    and habitat 354–356
    and limber pine 257
    and ponderosa pine 252–254 , 258
    and succession 352–353
    history in Devil's Tower National Park,
        Wyoming 253, 254
    management 16
    regime 136, 140–143, 147, 148
    return time 349
    suppression 352, 354, 355, 358
firewood 298
flat-rock 397, 399, 400, 103

flatwoods 172, 173, 176
flooding 375, 377, 378
flora of SE pine savanna, 29–33
Florida scrub types 9
fluctuating water 392, 395, 403
foothills 249
forage quality 332
forest encroachment 222, 225, 226, 228
Forest Service, United States 282–284
fossils 28, 29
fragipan 173
fragmentation 358
fragmentation and wildlife 354
Fraser River, British Colombia 311
Frontenac Axis 393, 398, 401
fuelwood 279, 281, 282, 322, 333, 334
fungi 254, 255, 296, 313

gall wasps 296
gap specialization 12
General Land Office Survey 136, 178, 199,
    222, 291
genetics, Florida scrub, 15, 18
geological history of SE pine savanna, 26
geology 53, 54, 121, 123–125, 128, 312
glade 220, 221
    sandstone 172
Goat Hill, Pennsylvannia and Maryland 73
goats 295, 297
Goshutes 298
Grandfather Mountain, North Carolina 121, 128
granite 392, 393, 395, 398, 402, 403
granite barrens 392, 393, 396, 401–403
    berry production in 400
    biogeography 399
    ecology 392, 393, 395
    flora 392, 395, 397–399, 402, 403
    importance 400, 401
    plant characteristics 396, 397
    protection 402, 403
    rare species 395, 398, 400, 401
    research needs 403
    significant animals 401
    significant plant species 401
    tree cover 392, 395
grassland 67, 71, 73, 74, 79, 313
grass–tree
    interactions 280, 283
    ratios 279–281, 283
grassy balds 119, 120, 125
gravel 183, 184
grazing 70, 79, 119, 120, 142, 144, 184, 222,
    225–228, 270, 271, 381, 383, 388, 389,
    417

Great Basin 249–251
Great Lakes 375, 376, 378, 381, 382, 385–389
Great Plains 249, 250, 252, 255, 406
Great Smoky Mountains 119–121
gypsy moth control 395, 403

habitat
    destruction 15, 16, 25, 95, 128
    restoration 42, 43, 46
    type, open woodland S. Ill., western Ky.,
        and middle Tenn. 173
hardwoods 29–33
Harper Shale formation 86
heath balds 119
heavy metals 318, 319
heliophytes 87, 94
Henneke Series, western serpentine soils, 312
herbicides 270
herbivores 319
    of longleaf pine 37, 38
herbivory 142, 144, 223, 240
herbs 171, 176–179, 182, 184
history of dry sand savanna 140
history of eastern serpentine barrens 67–71,
    74, 79
Holocene 158, 159, 235, 252
horses 297, 298
hunters and gatherers 296
hurricanes 26
hydrologic budgets 292, 293, 297
hydrology 27, 33, 42
hydrophobic soils 297
hydrothermal alteration of bedrock 312
hyperaccumulation of heavy metals 318, 319

igneous rock 221, 223, 312
imperiled plant communities 375, 388
Inceptisol 235
indicator species 83, 91, 316, 356
indifferent species of western serpentine 316
infertility 312
inorganic nutrient 86, 87, 94
insects 351, 356, 357
    and limber pine 256
    and ponderosa pines 254
insularity 126, 130, 311, 318, 319
integrated resource management 356
Interior Low Plateaus 172, 174
intermountain basins 252
introduced annual grasses or grassland 322,
    333, 335
invasive weeds 417
invasiveness of SE United States pines 42, 43
inverse soil texture effect 251
invertebrates 15, 145–147, 313, 356, 357

iron magnesium silicate 312
iron smelting 183
Ironto County, Virginia 86, 90
island biogeography 316–318
isolation 316

jack pine 354–357
    suppression 352, 354, 355, 358
jack pine barrens 343
    landscape 343, 357
    overstory 351, 352
    soil, available water capacity 347, 348
    species composition 343, 352, 353
    structure 351, 352, 356
    sub-types 351, 352
jack pine fire 347
    and succession 352, 353
    return time 349
Jackson County, Oregon 310
Jackson Purchase 183, 201, 202
Jeffrey pine 251
John Day, Oregon 311
Josephine County, Oregon 310

Karst Plain 190
key species of SE pine savanna 37
keystone species 15
Kittitas County, Washington State 311, 314
Klamath-Siskiyou Mountains, Oregon 310

Lake County, California 310
Lake Wales Ridge, Florida 8, 14–18
Land Between The Lakes of western Ky. and
    middle Tenn. 171, 182–186
landscape dynamics of Florida scrub 13,
    14, 17
leaf litter 269
lichens 13, 15, 73, 76, 102, 107, 115, 127, 129,
    179, 180, 378, 392, 397, 402
light compensation point, saturation point,
    response curves 94, 95
light intensity 87
lightning 9, 25, 38–42, 225
limber pine 249, 251, 253, 255–258
limestone 26, 28, 135, 148, 149, 177, 179,
    221, 223, 225, 226, 375–379, 382, 383,
    388
limestone barren 375
limiting factors 103, 104, 106, 108, 111–113, 115
lithology 84
Lithosol 312
Little Ice Age 290
livestock 20, 282, 283, 322, 329–335
livestock grazing 254, 292, 293
Llano Estacado 262

loess 173, 177, 178, 185, 222, 226, 228, 343, 357
logging impacts 119, 130

Madro-Tertiary geoflora 290
mafic bedrock 121, 123, 128
magnesium 67, 77–79, 312
mammals 58, 355, 356, 369, 414–416
management 96, 147, 148, 228
Manitoba 409, 410
mantle magma 312
marble 375, 377, 378, 392, 397, 403
Maryland serpentine barrens 67–79, 86, 87
Massanutten Mountain, Virginia 90
masting of pine 38
mechanical treatments 16
Mediterranean climate 311, 322, 335
mesophytes 177, 183
metamorphic rock 312
metamorphism 312
metapopulations 14, 18
Mg:Ca ratio 312
microbiotic soil crust 144
microenvironment 93, 100, 103, 105, 107, 109, 111
microorganisms 313
microsites 13
Mid-Appalachian shale barrens
    endemics 83, 86-88, 90-96
    geologic formations 84
    vegetation 83, 84, 86, 88, 90
midcanopy 171, 176, 182
Middle Appalachians 83, 84, 86, 93
mineral nutrition 67, 77, 79
mining 298
Miocene 235
Mississippi embayment 172, 174, 186
Mississippi River 93
mistletoe on ponderosa pine 255
mixed conifer–hardwood forest 313
mixed-grass prairie 250
Mogollon Plateau, Arizona 252
moisture gradient 183, 314
Mollisols 159
Monroe County, West Virginia 92
Monterey County, California 310
montmorillonite 312
moss 73, 76, 102, 109, 127–129, 378, 379, 382, 392, 397
mushrooms 255
mycorrhizae 54, 270, 296, 367, 313

Napa County, California 310
Native American, also see Amerindian, 69, 160, 223, 225, 227

native perennial grasses 331
natural divisions of Midwestern United States 172, 174
nematodes 296
New Idria, California 315, 318
Newfoundland 67, 76–79
Niagara Escarpment 362–374
nickel 77–79, 312, 318, 319
nitrogen 77, 313
nitrogen cycle 297, 298
nitrogen-fixing 313
nonserpentine grasslands 315
nonserpentine sites 316
North Cascades, Washington State 311
northern sand savanna 140, 141
nutrient
    availability 27, 28, 33, 42, 105, 108, 109, 111, 112, 366, 367
    dynamics 238
    status 347, 348
    stress 77, 79
nutritional sterility 316

oak 70, 71, 73–79
    forest 172, 177
    grubs 160–162, 165
    openings 156, 158–161, 163, 165, 166, 221, 228
    savanna 172
oak–chestnut forest 88
oak–hickory 223
oak–palmetto scrub 10
oceanic lithosphere 312
Octoraro Creek, Pennsylvania 75
Okanogan Highlands, British Columbia 311
olivine 312
Ontario 375, 376, 379, 382, 383, 385–389, 392, 393, 396, 398, 399, 402
Ordovician 84, 86
Osage Plains of Kansas, Oklahoma, and Texas 234
outcrop 121, 123, 311
outwash 347
overstory–understory relations of California oak savanna 329–331, 334
Ozark Hills, Illinois 172, 174, 176–178
Ozark Plateau 221
Ozarks Plateaus Province 92, 220, 221, 223, 225

Pacific Coast 314
Pacific Rim of Fire 310, 312
Paiutes 298
paleovegetation
    of Big Barrens of Ky. and Tenn. 193
    of the Prairie Peninsula 193

palynology 28
parkland corridor 387
patch dynamics, dry sand savanna 142, 147
patchiness in fire 12
peridotite 312
Permian 234
phosphorus 77, 78, 312
photosynthesis 94, 103, 113
photosynthetic pathways in plants 212
photosynthetically active radiation (PAR) 87, 178,
physiological adaptations 102, 103, 105, 106, 110, 113, 115
physiological ecology 67, 77, 78
Pilot Barrens, Maryland 68, 70, 71, 73
Pinchi Fault, British Columbia 311
Pinus jeffreyi-grassland-woodland 314
plant
    associations western serpentine 314
    life-forms 210, 211
    life stages, granite outcrops 105, 106, 110
plant-animal interactions 114
plate tectonics 309, 310, 312
plateau alvars 378
Pleistocene climate 158, 251
Plumas County, California 310
pollen profiles 344
pollination 96
ponderosa pine 252-258
population viability of Florida scrub 12-14
porcupines and ponderosa pine 254
postfire flowering 12
potassium 77, 78
potential natural vegetation of Big Barrens of Ky. and Tenn. 193, 194
prairie 171, 173, 183, 220-222, 226, 235, 236, 378, 382, 385
    peninsula 155, 190-194, 200, 202
prairie, brush 142, 145, 148
prairie, dry 143, 148
prairie-forest transition 155, 156
preadaptation to serpentine 316
precipitation 25-26, 88, 276-280, 283, 322-324, 329-335
prescribed fire in jack pine barrens 357
preservation of deep soil savanna 165
productivity 238
Puebloan people 297
pulse recruitment 366
pyroxene 312

quantum yield 95
quarries 371
Quartz Mountains, Oklahoma 238
Quebec 67, 76

radiocarbon dating 347, 363
rare species 119-121, 123, 125-128, 163, 221, 225, 240, 241
rarity, types of 15
reclamation 62
recreation 402
Red River, Arkansas 231
refugium 362
Research Natural Areas 314, 315
reserve design in Florida scrub 15
residential development 282-284, 333, 335
resins in SE pine 38, 41
resprouting 7, 11, 40
restoration and management 17, 128, 163, 164, 166, 226, 228, 418
Ridge and Valley 83, 88, 92
riparian woodlands 249
Robert E. Lee Park, Maryland 71, 73, 74
rock outcrops 249, 250
Rocky Mountain juniper 257
Rocky Mountains 249-251, 255, 256
rocky soils 250
rodents and ponderosa pine 254
Romney Shale 84
Rose Hill Shale 86
rosemary scrub 9, 14

Salem Plateau 221
San Benito County, California 310, 315, 318
San Francisco County, California 310
San Luis Obispo County, California 310
San Mateo County, California 310
sand pine scrub 9
sand savanna herbivory 142, 144
sandhill vegetation 11
sandstone 172, 178, 179, 221, 223, 225, 226
Santa Barbara County, California 310, 318
Santa Clara County, California 310
Saskatchewan 409
savanna 71, 73, 74, 76, 78, 79, 220-223, 225-228, 235, 236, 377, 378, 381-383, 387
    defined 1, 2, 23, 135, 155, 156, 171, 220, 221, 249
    dry sand, composition 140
scrubby flatwoods 9, 10
seed
    bank 11, 12, 105, 106, 108-111
    dispersal by birds in limber pine 256
    germination 106, 107, 109, 111
    in SE pine savanna 57
    predation 57, 60
    rain 299, 366
seedling 57, 58, 93
    recruitment 11, 332, 333, 336

seepage slopes 120, 126
serotiny 57–60
serpentine 309
    biology 309
    chaparral 313, 314
    endemics 314–316, 319
    floras 316
    grasslands 315, 316
    habitat 309
    Lithosol 312
    outcrops 309, 311
    Pacific Coast 309, 311, 312
    rocks 312
    scree 315
    soils, Dubakella and Henneke Series, 312
    syndrome 312
    tolerance 316
    vegetation 316
    western conifer ecosystem 311
    western conifer forest 313, 314
    western vertebrates 319
serpentine barrens 309, 312, 314, 315
    eastern postsettlement conditions 69–71, 73
    eastern presettlement conditions 67–71, 74
serpentinite 67, 71–79, 309, 312
shale 223
Shawnee Hills, Illinois 172, 174, 176, 178–182
sheep 297
shoreline alvar 377, 378
Sierra Nevada, California 250, 251, 310, 323
Silurian 84, 86
Siskiyou Mountains, Oregon 314, 315, 318
size class distribution see vegetation structure
skeletal soil 312, 319
smectite 312
soil 27, 33, 35, 67, 72, 74, 77–79, 171, 173,
        182–186, 192, 208, 209, 220–223, 225,
        226, 279, 280, 312, 347, 408
    algae 313
    chemistry 86, 87
    crusts 13, 144
    depth 104–113, 115, 375, 379, 388, 392,
        397, 398, 400
    development 102, 103, 105, 106
    disturbances 269–271
    erosion 222, 299
    fertility 251
    islands 103, 104
    moisture 103, 105–113, 115
    nutrients 9, 270, 312
    organic carbon 279
    pH 54, 86, 105, 109
    seedbanks 299
    taxonomy 312
    temperature 87, 88, 93

texture 249–251, 256, 257
    water holding capacity 171, 173, 179, 182,
        184, 185, 347, 348
Soldiers Delight, Maryland 69, 70, 71, 73, 74
Sonoma County, California 310
Southeastern Coastal Plain 29
southeastern pine savanna 23–43
    adaptation–modification fire model 41, 42
    biogeography 29
    endemism 25, 29, 42
    evolutionary models 40–42
    fire 53, 57–62
        adaptation 41–43
        ecology 25, 33, 36, 37, 40–43
        modification and facilitation 41–43
        regimes 38–42
        resistance 40, 42, 43
        suppression 24, 25, 38–42
    local distribution 33, 36
    old-growth pine savanna 25
    persistence 25, 39–42
    plant community composition 33–36
    plant life history characteristics 42
    postfire environments 41, 42
    pyrogenicity 25, 40–42
    regional distribution 29, 31
    resprouting 40
    second-growth pine savanna 25
    seepage zones 27
southern California 313
southwestern United States 251, 252
southwestern woodlands 249
Spanish and Mexican influence on California
        oak savanna 333–335
spatial patterns in shinnery oak 269
speciation 101, 102, 316
species diversity 106, 107, 110–113, 119, 179,
        182, 184, 185, 375, 378, 389
species richness 25, 35, 36, 125, 126, 179–181,
        265, 270, 271, 365, 380, 388, 400
Spodosols 27
Springfield Plateau 221
stable carbon isotopes 275, 279
stand dynamics in pines 39, 40
State Line Barrens, Maryland and
        Pennsylvania 76
Stillaguamish River, Washington State 311
Strawberry Mountains, Oregon 311
structure 137, 138, 141
subarctic woodland 421
    classification 422, 423
    climate 425, 426
    climate–fire interaction 429
    distribution 421, 422
    ecosystem dynamics 426

fire 426, 427
native people 432
species composition 424
succession 429
subtropical humid climate 233
suburban scrub-jays 14
succession 25, 38–40, 43,102–104, 107–109,
        111–113, 115, 116, 171, 182, 183, 226,
        353, 395, 398, 399, 429
    in limber pine 257
summer drought 322, 335
summer fires 160
Susquehannock Indians 69

temperature 25, 87, 88, 95, 100, 107, 109,
        113, 115, 171
temperature response curve 95
thermotolerance in Mid-Appalachian shale
        barrens plants 95
thunderstorms in SE Coastal Plain 38, 39
till plain 172, 176, 185
timber harvesting in open woodlands of S. Ill.,
        western Ky., and middle Tenn. 176, 178,
        182, 185
timber harvesting in ponderosa pine 253,
        258
tolerance in western serpentine plants 313,
        316, 319
topography 26, 27, 33, 35, 171, 173, 175–179,
        184, 185, 220–223
tree line of subarctic lichen woodland 422
Trinity River, Texas 234
truffles 255
Tulameen River, British Columbia 311
Tulare County, California 310
Twin Sisters Mountains, Washington State 311,
        319

ubiquist species 316
Ultisols 27
ultramafic outcrops 67, 76, 313
ultramafic rock 67, 79, 310, 312
understory 139, 142–144,
ungulates 298
univoltine 319

Upper Peninsula of Michigan 344, 349

vegetation
    continuum 156, 157, 163, 183
    dynamics 142–144
    history 120, 121, 123
    pre-European settlement 159, 160,
        171–173, 176–178, 183, 185, 220–225,
        228, 343, 344
    structure 176, 179, 184, 186, 278, 332, 333
    transition zone 136, 140,
vegetative reproduction 93
vertebrates 144–146, 319
Virginia Mountains 251
volcanism 310, 541

warm season grasses 290
Washita River, Oklahoma 234
water
    limitation 88
    potential 88
    relations 77
    table 375, 378
Wenatchee Mountains, Washington State 311,
        313–315
Western Highland Rim 172
wetlands 392, 395, 396, 399
white pine blister rust; on limber pine 256–258
Wichita Mountains, Oklahoma 238
wildlife 221, 227, 228
witness trees 177, 178, 291
woodland 235, 236, 375, 377–379, 382, 383
    defined 249
woodland, oak 137, 142
woodrat middens 291
woody understory 236, 237

xeric hammock 11
xeromorphic form 312
Xerothermic Interval 290, 387
xylem water potentials 78

Yellow Dog Plains, Michigan 344, 347
Yellowstone National Park, Wyoming 258